J. LYTTON.

List of Elements with their Symbols, Atomic Numbers and Atomic Masses
(based on $^{12}C = 12.00000$)

Element	*Symbol*	*Atomic Number*	*Atomic Mass*	*Element*	*Symbol*	*Atomic Number*	*Atomic Mass*
Actinium	Ac	89	(227)[a]	Manganese	Mn	25	54.9380
Aluminum	Al	13	26.98154	Mendelevium	Md	101	(256)[a]
Americium	Am	95	(243)[a]	Mercury	Hg	80	200.59
Antimony	Sb	51	121.75	Molybdenum	Mo	42	95.94
Argon	Ar	18	39.948	Neodymium	Nd	60	144.24
Arsenic	As	33	74.9216	Neon	Ne	10	20.179
Astatine	At	85	(210)[a]	Neptunium	Np	93	237.0482[c]
Barium	Ba	56	137.34	Nickel	Ni	28	58.71
Berkelium	Bk	97	(247)[a]	Niobium	Nb	41	92.9064
Beryllium	Be	4	9.01218	Nitrogen	N	7	14.0067
Bismuth	Bi	83	208.9804	Nobelium	No	102	(254)[a]
Boron	B	5	10.81	Osmium	Os	76	190.2
Bromine	Br	35	79.904	Oxygen	O	8	15.9994
Cadmium	Cd	48	112.40	Palladium	Pd	46	106.4
Calcium	Ca	20	40.08	Phosphorus	P	15	30.97376
Californium	Cf	98	(251)[a]	Platinum	Pt	78	195.09
Carbon	C	6	12.011	Plutonium	Pu	94	(242)[a]
Cerium	Ce	58	140.12	Polonium	Po	84	(210)[a]
Cesium	Cs	55	132.9054	Potassium	K	19	39.098
Chlorine	Cl	17	35.453	Praseodymium	Pr	59	140.9077
Chromium	Cr	24	51.996	Promethium	Pm	61	(147)[a]
Cobalt	Co	27	58.9332	Prolactinium	Pa	91	231.0359[c]
Copper	Cu	29	63.546	Radium	Ra	88	226.0254[c]
Curium	Cm	96	(247)[a]	Radon	Rn	86	(222)[a]
Dysprosium	Dy	66	162.50	Rhenium	Re	75	186.2
Einsteinium	Es	99	(254)[a]	Rhodium	Rh	45	102.9055
Erbium	Er	68	167.26	Rubidium	Rb	37	85.4678
Europium	Eu	63	151.96	Ruthenium	Ru	44	101.07
Fermium	Fm	100	(253)[a]	Samarium	Sm	62	150.4
Fluorine	F	9	18.99840	Scandium	Sc	21	44.9559
Francium	Fr	87	(223)[a]	Selenium	Se	34	78.96
Gadolinium	Gd	64	157.25	Silicon	Si	14	28.086
Gallium	Ga	31	69.72	Silver	Ag	47	107.868
Germanium	Ge	32	72.59	Sodium	Na	11	22.98977
Gold	Au	79	196.9665	Strontium	Sr	38	87.62
Hafnium	Hf	72	178.49	Sulfur	S	16	32.06
Hahnium[b]	Ha	105	(260)[a]	Tantalum	Ta	73	180.9479
Helium	He	2	4.00260	Technetium	Tc	43	98.9062[c]
Holmium	Ho	67	164.9304	Tellurium	Te	52	127.60
Hydrogen	H	1	1.0079	Terbium	Tb	65	158.9254
Indium	In	49	114.82	Thallium	Tl	81	204.37
Iodine	I	53	126.9045	Thorium	Th	90	232.0381[c]
Iridium	Ir	77	192.22	Thulium	Tm	69	168.9342
Iron	Fe	26	55.847	Tin	Sn	50	118.69
Krypton	Kr	36	83.80	Titanium	Ti	22	47.90
Kurchatavium[b]	Ku	104	(260)[a]	Tungsten	W	74	183.85
Lanthanum	La	57	138.9055	Uranium	U	92	238.029
Lawrencium	Lr	103	(257)[a]	Vanadium	V	23	50.9414
Lead	Pb	82	207.2	Xenon	Xe	54	131.30
Lithium	Li	3	6.941	Ytterbium	Yb	70	173.04
Lutetium	Lu	71	174.97	Yttrium	Y	39	88.9059
Magnesium	Mg	12	24.305	Zinc	Zn	30	65.38
				Zirconium	Zr	40	91.22

[a] *Mass number of most stable or best known isotope*
[b] *Tentative name*
[c] *Mass of most commonly available, long-lived isotope*

Atomic masses are based on $^{12}_{6}C = 12.00000$.
" Element 106, discovered in 1974, is as yet unnamed.

Biophysical Chemistry

Principles, Techniques, and Applications

Alan G. Marshall
University of British Columbia

JOHN WILEY & SONS

NEW YORK SANTA BARBARA CHICHESTER

BRISBANE TORONTO

Library of Congress Cataloging in Publication Data:

Marshall, Alan G 1944–
Biophysical chemistry.

Includes bibliographical references and index.
1. Biological chemistry. 2. Chemistry, Physical and theoretical. I. Title. [DNLM: 1. Biophysics. 2. Chemistry, Physical. QT34 M367b]
QH345.M325 541'.3'024574 77–19136
ISBN 0-471-02718-9

Printed in the United States of America

10 9 8 7 6 5 4 3 2 1

To Marilyn, Wendy, and Brian

PREFACE

The object of this book is to provide a *working knowledge* of basic physical chemistry to undergraduates whose primary academic interests lie in the biological sciences, including medicine and dentistry. Most existing texts aimed at a one- to two-semester undergraduate course in "physical chemistry for biologists" are of two types: (1) a mathematically and conceptually diluted abridgement of traditional physical chemistry texts, usually with strong emphasis on small molecules in the gas phase, with perhaps a single short chapter on "macromolecules"; or (2) a loosely connected series of chapters on particular biophysical methods, with strong emphasis on instrumental and practical details at the expense of principles. This textbook attempts to provide the minimal mathematical treatment of physical chemistry that will enable students to *use* their new knowledge to solve relevant biochemical and clinical problems.

The proposed goals call for a new organization of topics. Because most physical chemical techniques are understood in terms of a very few simple mathematical *models,* each of the six major sections of the book introduces a single mathematical model, such as the weight on a spring. The great advantage of this approach is that subsequent application of the *same* model to a variety of *different* biophysical techniques consists simply of changing the names of the mathematical variables. The generality of the approach provides for a natural entry into many modern topics not usually found in elementary texts, including: electron microscopy, "transient" chemical reaction kinetics and "relaxation" phenomena, electrical noise, Fourier methods in diffraction and spectroscopy, and a variety of "absorption" and "dispersion" phenomena. Since each of the six major sections is essentially self-contained (with extensive cross-referencing), the lecturer is free to choose the order of presentation of material according to his or her own taste.

All the material in this book has been class-tested by the author during his eight years of teaching physical chemistry to life science undergraduates at the University of British Columbia. Third-year undergraduates with a year of calculus and a year of introductory chemistry have demonstrated a high level of assimilation of Chapters 1 to 17 in a two-semester course, and more material could be covered by deleting some of the examples. Of course, topics could be deleted for those who would use the book in a one-term course. Problems at the end of each chapter are intended to challenge students, rather than have them "plug" numbers into memorized formulas. (A complete set of solved problems is available as a separate supplement to the text.)

Finally, the text attempts to be current in *modern* biophysical methods. Examples include: "disc" electrophoresis, affinity chromatography, X-ray scattering and diffraction techniques for macromolecules, ion-selective electrodes, circular dichroism, laser light scattering, fluorescence, ultrasonic imaging, and magnetic resonance "spin"-labels. In all cases, subject matter has been chosen for its direct biological impact, and student motivation is kept high with frequent modern biochemical and/or clinical applications of the physical results.

ACKNOWLEDGMENTS

The author particularly wishes to acknowledge the indulgence and support from his family during the writing process. The author is grateful for technical advice from many colleagues at the University of British Columbia and elsewhere, and takes special pleasure in thanking M. Comisarow, D. Clark, A. Addison, C. Meares, E. Burnell, L. Werbelow, D. C. Roe, J. Benbasat, and D. Coombe for their suggestions, corrections, and help.

Alan G. Marshall

CONTENTS

1

THERMODYNAMICS IN BIOLOGY

In general, a complete description of the behavior of matter would begin with *quantum mechanical* analysis of the motion and energy of *individual* atoms and molecules, followed by some *statistical mechanical* (averaging) procedures to provide a description of the behavior of *bulk* matter using *thermodynamics.* (The *rate of change* in properties of bulk matter is described by *kinetics.*) We will shortly encounter the mathematical elements of these various calculations: quantum mechanical mathematics may be developed from generalized *geometry* (Section 5); statistical averages are computed from the mathematics of *chance* (Section 2); and kinetic analysis is based on *calculus* (Section 3). Thermodynamics begins with the study of work and heat. Although the work or heat developed during a process depend on the particular *path* chosen, certain combinations of heat and work are independent of path. Thermodynamic descriptions of bulk matter are thus couched in the mathematics of *exact differentials,* which are simply mathematical objects whose evaluation does not depend on the particular path taken in proceeding from one place to another. The great *abstract* value of thermodynamics is that since the final result of going from one "state" of matter to another often does not depend on the path taken, we need know nothing about the *mechanism* involved along the way. The great *practical* use for thermodynamics comes from the fact that changes in such "state functions" as energy do not depend on the path from one state to another, and, therefore, we are able to equate the end results of proceeding from one state to another by two *different* paths. We thereby often obtain information about reactions that we could not study directly, such as reactions within living cells.

Specifically, thermodynamics is useful for predicting energy changes resulting from chemical reactions, often in cases where the particular reaction of interest is not easily studied directly. Thermodynamics predicts whether or not a given reaction can be expected to occur spontaneously, and how much energy must be supplied to make the reaction occur even if it is not spontaneous by itself—this information is crucial in determining the reaction routes involved in various metabolic sequences. The energies involved in breaking particular chemical bonds can be computed, and they provide a good measure of bond strength in various molecules. Thermodynamics explains the existence of equilibrium constants and their variation with temperature. Thermodynamics provides detailed information about the mechanism of enzyme-catalyzed reactions, in the sense of identifying the presence of certain intermediates and their associated energy stability. We will be able to account for the difference in pressure across a (semipermeable) membrane that separates a solution of neutral macromolecules from a solution with no macromolecules (osmotic pressure), as well as the difference in electrical potential (Donnan potential) across such a membrane when (charged) ions are present. We will explain the effects of dissolved salt on the solubility of macromolecules, a phenomenon of great practical importance in isolation and purification of proteins. Thermodynamics can be applied to electrochemical cells, so as to produce new types of electrodes that can be made sensitive to the presence of just one type of small molecule in a solution. The occurrence of distinct "phases" (solid, liquid, gas, and others) is readily treated by thermodynamics, and is of modern interest in accounting for the physiological functions of biological membranes at various temperatures, as well as for theories of "pre-biological" evolution of life. Finally, thermodynamics predicts the efficiency of certain types of engines (those that convert heat into work), and provides a suitable framework for discussion of trade-offs in power plant design and operation.

The full formal elegance of thermodynamics becomes evident from a careful orderly development of the first and second "laws" of thermodynamics, and is outside the scope of this discussion, largely because many of the necessary manipulations are most easily illustrated using the ideal gas, whose direct biological applications are limited. We will therefore begin with a brief review of the basic properties of thermodynamic "state functions," and then proceed at once to the biochemical applications. The reader is referred to any of several excellent monographs (see References) for more detailed justification of the basic thermodynamic relations and "laws," which are now presented in most elementary chemistry textbooks.

CHAPTER 1 Work, Heat, and Energy

The science of thermodynamics may be defined as the study of changes in energy of bulk matter and radiation. However, since it is conjectured that the *total* energy of the universe as a whole is constant ("first law" of thermodynamics), it is necessary to limit consideration to a special *isolated part* of the universe, a "system." The term "surroundings" thus denotes the rest of the universe. An "equilibrium" condition is said to apply when the *macroscopic* properties (volume, temperature, pressure, concentration, energy, etc.) of a system are constant with time. A "state" is defined as an *equilibrium* condition for a system. Thermodynamics provides us with a number of useful predictions of what must happen in going from one *state* to another *state,* as in a chemical reaction, phase change, translocation of solvent and/or ions across a semipermeable membrane, flow of electrical current, change in temperature, and the like. We are thus led in a natural way to consideration of "work" and "heat," which are forms of energy "in transit" during such changes in the state of a system.

Work may be expressed as the product of a generalized *force* ("intensive" factor) and a generalized *displacement* ("extensive factor"), as shown in Table 1-1. Because the generalized force can be taken as constant if the generalized displacement is made sufficiently small, it is most fruitful to define the work done over an infinitesimally small displacement, as in Table 1-1. Work is defined as positive when work is done *on* the system (at the expense of the surroundings), and thus is a negative number when the system does work on the surroundings. Finally, the magnitude of work done in going from one state of a system to another depends on the *path* (i.e., the course of the successive small displacements): for a special path for which the system is infinitesimally close to equilibrium throughout the process, the path is said to be "reversible," and the change in any parameter for the process is denoted by lower case "d" (e.g., dw for reversible work); for any other path, the ("irreversible") change will be denoted by a "δ" (e.g., δw for work done along an irreversible path).

Two bodies are said to be in "thermal contact" when energy can flow between them but matter cannot. *Heat* is that (energy) which is said to flow from a hotter to a colder body when the two are brought into thermal contact. Various types of heat are listed in Table 1-2, according to their origin. The heat evolved or applied in conducting a chemical reaction is often a good measure of the strengths of the chemical bonds that must be broken and re-formed in the process. The heat required to raise the temperature of a substance gives us a means of predicting the result of a process carried

Table 1-1 Types of Work (Reversible work, dw, is defined as the product of a generalized force and a generalized displacement)

dw =	(generalized force) ·	(generalized displacement)
$P\,dV$	P = pressure = force/area	V = volume
$\gamma\,dA$	γ = surface tension = force/distance	A = area
$E\,dq$	E = electromotive force	q = charge
$H\,d\mu$	H = magnetic field	μ = magnetic moment
$F\,dx$	F = force	x = distance

out at a temperature different from one at a given temperature—this is especially useful, since most thermodynamic properties of chemical reactions are tabulated for a particular temperature that may be far removed from the, say, physiological temperature of interest. The heat evolved during flow of current (Table 1-2) varies as the square of the current: therefore, since electrical power varies as $(V \cdot I)$, where V and I are voltage and current, it is clear why it is desirable to transmit electrical power at the highest possible voltage in order to minimize heat losses from the passage of current through power lines. Heat is conventionally positive when heat is absorbed *by* a system. A process is said to be "adiabatic" when no heat enters or leaves the system during the process. Steam engines, the major

Table 1-2 Sources of Heat

Process of Interest	δq = Heat Required or Generated During Process	
Raise the temperature of a substance	$\delta q = C \cdot m \cdot dT$,	where C = "heat capacity," m = mass of substance, and T = temperature.
Push electric current through a wire	$\delta q = I^2Rdt$,	where I = electric current, t = time, and R = electrical resistance of wire.
Change the (physical) phase of a substance (e.g., melt a solid; vaporize a liquid)	$\delta q = h\,dm$,	where dm is the mass of substance whose phase is changed, and h is the heat required per unit mass of substance ("heat of fusion;" "heat of vaporization")
Complete a chemical reaction	$\delta q = h\,dm$,	where dm is the mass of (say) reactant that is converted to product, and h is the heat required per unit mass of reactant ("heat of reaction").
Overcome friction	$\delta q = F \cdot dx$,	where F is the opposing frictional force and x is the distance over which the force acts.

present source of power today, act to convert *heat* into *work*. The amount of heat generated during a process also depends on the *path* from initial to final state, as we discuss next.

Although both heat and work for a given process depend on *path*, the first law of thermodynamics states that their *sum* ("energy" of the system) is *independent of path*. In our conventional notation, we would say that

$$dE = \delta q + \delta w \qquad \text{First Law of Thermodynamics} \tag{1-1}$$

in which E represents the energy of the system. *Physically*, Eq. 1-1 denies the existence of any device that generates energy with no other net change in the system: if net changes in energy could depend on path, then we could carry out a cyclic process that brought us back to the starting state at a higher energy (say) than when we started. Since no such processes have been found, it is conjectured that none are possible. *Physically*, Eq. 1-1 is based on the idea that

$$\oint dE = 0 \tag{1-2}$$

in which the $\oint$ denotes some (cyclic) path that ends up where it started. *Mathematically*, Eq. 1-2 defines an "exact differential," dE. Equivalently, any exact differential, dF, can be expressed in terms of its independent variables according to

$$\boxed{dF = \left(\frac{\partial F}{\partial x}\right)_{y,z} dx + \left(\frac{\partial F}{\partial y}\right)_{x,z} dy + \left(\frac{\partial F}{\partial z}\right)_{x,y} dz} \tag{1-3}$$

in which the various "partial" derivatives are evaluated as for ordinary derivatives, except that the subscripted variables are treated as constants. For example, for $F(x,y,z) = x^2y^3e^z$, $(\partial F/\partial x)_{y,z} = 2xy^3e^z$. Finally, since it is often particularly convenient to evaluate the heat and work developed for a *reversible* path, and since Eq. 1-1 is valid for *any* path, it is possible to compute the change in energy using

$$dE = dq + dw \qquad \text{(reversible path)} \tag{1-4}$$

and the resulting dE will be the same as for *any* chosen path.

Clausius in 1850 proposed a second "state function" (i.e., an exact differential). By breaking down an arbitrary cyclic path into infinitesimally small reversible steps, and working with an ideal gas to keep the calculations simple, it could be shown that

$$\oint (dq/T) = 0 \qquad \text{Second Law of Thermodynamics} \tag{1-5}$$

in which the dq notation indicates that any one infinitesimal step is reversible, and T represents (and may in fact be used to define) absolute temperature. It is thus possible to define a new state function, "entropy,"* S, which satisfies the definition of an exact differential:

$$dS = dq/T \tag{1-6}$$

Again, while changes in S are most easily *calculated* along the reversible path specified by Eq. 1-6, the final net change in S depends only on the specified initial and final states and *not* on the path between them.

We have presented the first and second laws of thermodynamics as a means for introducing two state functions, *Energy* (E) and *Entropy* (S),

* *Although* energy *is a quantity whose physical significance should be familiar to most readers from their earlier experience with classical and quantum mechanics, the intuitive meaning of* entropy *is probably less clear. Entropy may be thought of as a measure of the degree of "disorder" or "randomness" in a system, as illustrated by the example diagrammed in Fig. 1-1.*

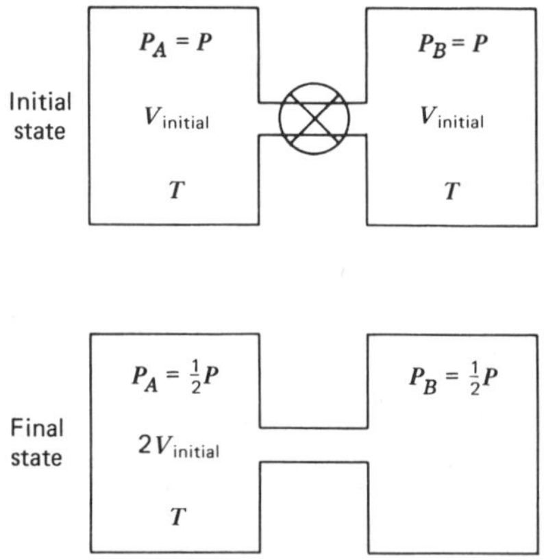

FIGURE 1-1. Entropy and disorder: entropy of mixing. This diagram shows a process in which two initially separated chambers containing different ideal gases, A and B, are connected and allowed to reach a new equilibrium state. Although the temperature (and energy—see text) of each gas does not change, the volume available to each gas doubles, and so the partial pressure of each gas halves as a result of the mixing process. Equations 1-7 to 1-12 show that there is an increase in entropy associated with this "disordering" process, even though the total volume, temperature, pressure, and energy remain unchanged.

Consider two gas-filled chambers (of equal volume and pressure, for simplicity), each initially containing two different ideal gases (e.g., helium in one chamber and neon in the other). The stopcock separating the two chambers is then opened and the two gases allowed to mix until an equilibrium condition is reached in which each type of gas is now spread out uniformly over both chambers. Since the total number of moles of gas, $n = n_A + n_B$, the total volume, V, and the total pressure, P, are unchanged after the mixing process, the ideal gas law,

$$PV = nRT, R = 8.314 \ Jmol^{-1} K^{-1} \tag{1-7}$$

indicates that the absolute temperature, *T, is also unchanged. Next, since the energy of an ideal gas depends only on temperature (kinetic energy = (3/2)RT for an ideal monoatomic gas; potential energy = zero, since there are no interatomic forces of attraction or repulsion between* whose changes are independent of the path taken between specified initial and final states. There are some other obvious state functions: *Temperature* (T), *Pressure* (P), and *Volume* (V). We have already stated that *heat* and *work* are *not* state functions, since their evaluation does depend on path—in fact, the reason that we can construct heat engines and refrigerators is precisely *because* there can be net work (or heat) changes in taking a working substance (steam or Freon in those cases) through a cyclic process. There are three additional useful state functions that may be formed by combinations of the previous ones: *Enthalpy* (H), *(Gibbs) Free Energy* (G), and *(Helmholtz) Free Energy* (A), according to

$$H = E + PV \tag{1-13}$$
$$G = H - TS \tag{1-14}$$
$$A = E - TS \tag{1-15}$$

---

ideal gas molecules), there is therefore no change in the energy *of the gases on mixing* ($dE = 0$). *From Equations 1-4 and 1-6, we can now calculate the entropy change for the mixing process:*

$$dS = dq/T = (dE - dw)/T = -dw/T \tag{1-8}$$

Since

$$dw = -P\,dV \qquad (\textit{ideal gas}) \tag{1-9}$$

Equation 1-8 may be rewritten in the form

$$dS = P\,dV/T = nR(dV/V) \tag{1-10}$$

We may thus treat the mixing process as a simple expansion, in which each gas expands from its initial volume, $V_{initial}$, *to a final volume,* $V_{final} = 2V_{initial}$. *The total entropy change may then be obtained as the sum of the individual entropy changes for the two gases, A and B:*

$$\Delta S = S_{final} - S_{initial} = \Delta S_A + \Delta S_B \tag{1-11}$$

$$= n_A R \int_{V_{initial}}^{V_{final}} dV/V + n_B R \int_{V_{initial}}^{V_{final}} dV/V$$

$$\begin{aligned} &= n_A R\,(\log_e(2V_{initial}) - \log_e(V_{initial})) \\ &+ \quad n_B R\,(\log_e(2V_{initial}) - \log_e(V_{initial})) \\ &= (n_A + n_B) R\,(\log_e(2V_{initial}/V_{initial})) = (n_A + n_B) R \log_e(2) \end{aligned} \tag{1-12}$$

Since the total pressure, volume, temperature, and energy for the two-chamber system are unchanged by the mixing process, but there is a well-defined entropy *change, we might associate that change in entropy with an increase in the* disorder *of the system, in the sense that the A and B gas molecules are now intermingled randomly rather than being separated into two groups. In statistical language, one would say that there is only one way of having the A and B molecules completely separated from each other, but there is an immense number of ways of assigning half the A molecules to each chamber; thus it should be overwhelmingly more likely to find the A and B molecules evenly distributed than to find them completely separated into the two chambers once the valve is opened. [We will encounter a similar argument in Section 5 in deciding how to distribute the total energy (rather than the number of molecules) of a system in various ways.] The relation between entropy and disorder is much more general than the present example, and is often invoked in accounting for the observed spontaneity of chemical reactions (Chapter 4) that are energetically unfavorable.*

Table 1-3 Thermodynamic State Functions and Their Basic Interrelations

State Function	Name	Basic Relation	Remarks
P	Pressure		
V	Volume	$dw = -P\,dV + dw'$	Since work done by system on surroundings is a negative number, and P and dV are both positive when the system expands, the minus sign is necessary. dw' indicates (reversible) nonPV work.
T	(Absolute) Temperature	$PV = nRT$	"Equation of State" for *ideal gas* only.
E	Energy	$dE = \delta q + \delta w$	General result (First Law of Thermodynamics).
		$dE = dq + dw$	Reversible process only.
		$dE = dq - PdV + dw'$	Reversible process only; dw' = nonPV work.
S	Entropy	$dS = dq/T$ (Second Law of Thermodynamics)	This equation defines changes in S along a *reversible* path. However, once the change in S has been computed, it will apply for *any* path between the same initial and final states (although dS will no longer be equal to $\delta q/T$).
		$dE = TdS - PdV + dw'$	Reversible process only.
H	Enthalpy	$H = E + PV$	Definition.
		$dH = dE + PdV + VdP$	General result.
		$dH = TdS + VdP + dw'$	Reversible process only.
		$dS = (dH/T)$	Constant P; $dw' = 0$ (PV work only).
G	Gibbs free energy	$G = H - TS$	Definition.
		$dG = dH - TdS - SdT$	General result.
		$dG = VdP - SdT + dw'$	Reversible process only.
A	Helmholtz free energy	$A = E - TS$	Definition.
		$dA = dE - TdS - SdT$	General result.
		$dA = dw = -PdV + dw'$	Reversible process only; constant temperature; dA represents the *maximum work* available when the system goes from specified initial to final state.

The basic useful relations between the various principal state functions (P,V,T,E,S,H,G,A) are listed in Table 1-3, and provide the framework for all the applications of thermodynamics considered in this section.

One of the major reasons for inventing some of the "state functions" in Table 1-3 is that they provide criteria for the existence of *equilibrium*. Equilibrium exists when *all* the macroscopic properties of a system are constant; however, if we specify *some* of the properties as constant in advance (say, T and V), then we often need examine only *one* other state function (in this case, A) to find out if equilibrium has been reached. Criteria for equilibrium for various choices of properties specified as constant in advance are listed in Table 1-4.

The contents of Tables 1-3 and 1-4 are helpful in explaining the need for the various thermodynamic state functions. P, V, and T are obvious choices but of little value by themselves. E and S arise directly from the first and second laws, but are again not especially useful as they stand. However, changes in H can be related directly to what goes on in a *calorimeter* and provide a measure of *bond strengths* when the net difference in bond enthalpies after a chemical reaction is manifested as heat produced in a calorimeter. (Gibbs) free energy (henceforth denoted simply as free energy) provides a simple criterion for *equilibrium* under the usual conditions of constant temperature and pressure, and also shows that equilibrium under those conditions is reached by a compromise between maximizing S and

Table 1-4. Criteria for Existence of Equilibrium in a System, Where Specified Properties Are Already Constant before the System is Examined

Previously Specified Conditions	Behavior of Specified State Function
Any reversible process	$dE = dq - PdV + dw'$
Heat-insulated bomb (constant V; adiabatic (no heat enters or leaves); $dw' = 0$ (PV work only))	↓ $dE = 0$ (E is a *minimum* at equilib.)
Any reversible process	$dS = (dq/T)$ ↓
Any reversible process	$dE = TdS - PdV + dw'$
Isolated system (constant E and V; $dw' = 0$)	↓ $dS = 0$ (S is a *maximum* at equilib.)
Any reversible process	$dH = dq + VdP + dw'$
Adiabatic calorimeter (constant P; $dq = 0$; $dw' = 0$)	↓ $dH = 0$ (H is a *minimum* at equilib.)
Any reversible process	$dG = VdP - SdT + dw'$
Typical conditions for chemical reaction (constant T and P; $dw' = 0$)	↓ $dG = 0$ (G is a *minimum* at equilib.)
Any reversible process	$dA = dE - TdS - SdT$
Temperature-controlled closed reaction vessel (constant T and V; $dw' = 0$)	↓ $dA = 0$ (A is a *minimum* at equilib.)

minimizing E at the same time. Helmholtz free energy provides a measure of the maximal *work* that may be derived from a given process—since most processes are not conducted under optimal conditions, the real work obtained in going from the same initial to final states is always less than dA in Table 1-3, but it is still useful to know the absolute maximal work possible for comparison.

THERMOCHEMISTRY: APPLICATIONS OF EXACT DIFFERENTIALS

In this section, we exploit the idea that changes in thermodynamic "state" functions (P,V,T,E,H,S,G,A) depend only on the *initial* and *final* states before and after the change, and not on the *path* between the initial and final states. It seems obvious that temperature, volume, and pressure possess this property, sometimes expressed as the "exact differential" equation (1-3). The observation that energy, E, possesses the same property constitutes the first law of thermodynamics. The second law of thermodynamics is equivalent to stating that entropy, S, also possesses this property. Since the remaining thermodynamic functions, H, G, and A are simply (linear) combinations of the preceding ones, they must also satisfy the exact differential condition. We shall now proceed with chemical examples in which the fact that, in particular, H is a "state function" (i.e., dH is an exact differential) makes it possible to construct a table of bond strengths using measurements of the heat produced or absorbed in related chemical reactions, and also to calculate the heat of reaction (containing bond strength information) for reactions that are not directly observable.

We first require an experimental means for determination of enthalpy changes, ΔH. From the definitions collected in Table 1-3, it quickly appears that ΔH may be measured as the heat evolved (or absorbed) when a process is conducted *reversibly* at *constant pressure* in the absence of electric or magnetic fields:

$$H = E + P\,V \tag{1-13}$$

or

$$dH = dE + P\,dV + V\,dP \tag{1-16}$$

But

$$dE = dq_{(\text{reversible})} + dw_{(\text{reversible})} + dw'_{(\text{reversible})} \tag{1-1}$$

so

$$dH = dq_{\text{rev}} - PdV + dw'_{\text{rev}} + PdV + VdP \tag{1-17}$$

Thus, at constant P, with $dw'_{\text{rev}} = 0$

$$dH = dq_{\text{rev}} \text{ at constant } P \tag{1-18a}$$

or,

$$\boxed{\Delta H = q_{\text{rev}} \text{ at constant } P} \tag{1-18b}$$

Because H is a state function, we know that once ΔH has been determined experimentally using a reversible constant-pressure path, ΔH would be the same for any other path, reversible or not. The measurement is experimentally simple. For example, for an "endothermic" (i.e., heat-absorbing) process, one might measure the electrical power (dissipated as heat in the reaction vessel) required to keep the temperature of the reaction constant; then by integrating (summing up) the electrical power over the course of the reaction time (power = energy/time), this "calorimetry" experiment yields the (heat) energy absorbed during the reaction, as equal to the total heat that was required to keep the temperature from falling.

Since the enthalpy change, ΔH, for a process will in general depend on the temperature, pressure, and physical state (e.g., liquid, solid, gas) of the reaction mixture, it is necessary to choose some *standard* condition(s) if we are to compare several different reactions on the same basis. The *standard state* of a pure chemical element is taken as the natural form (solid, liquid, gas) of that element at 25°C (about 298 K) at 1 atm (101,325 N m^{-2}). Although this definition is suitable for gases at low pressure, more elaborate definitions of standard state are useful for reactions in solution (see Chapter 3). A *standard enthalpy* (or free energy) *of reaction* is simply the enthalpy (or free energy) change when a specified number of moles of reactants are converted to products, when all reactants and products are in their standard states. A standard enthalpy (or free energy) change is usually designated with a superscript, "°", as for the simple chemical reaction shown below:

$$A + B \rightarrow C + D \qquad (A, B, C, \text{ and } D \text{ all in standard states})$$

$$\Delta H^\circ_{rx} = H^\circ_C + H^\circ_D - H^\circ_A - H^\circ_B \tag{1-19}$$

Finally, it is often convenient to designate a reference point, or *reference state,* for which H is defined to have zero-value. The standard enthalpy for a pure chemical element at 25°C and 1 atm pressure is usually zero as such a reference. These conventions are illustrated in the following examples.

EXAMPLE *Chemical Bond Strengths from Heats of Formation*

When one mole of a chemical compound is formed from its constituent elements, the corresponding change in enthalpy at standard conditions (25°C, 1 atm) is called the "standard heat of formation" of that compound, ΔH°_{298}. For example, for α-*d*-glucose,

$$6C_{(graphite)} + 3O_{2(gas)} + 6H_{2(gas)} \rightarrow C_6H_{12}O_{6(crystals)}$$

$$\begin{aligned}\Delta H^\circ_{298} \text{ for glucose} &= H^\circ_{298}(\text{products}) - H^\circ_{298}(\text{reactants})\\ &= H^\circ_{298}(\text{glucose}) - 6H^\circ_{298}(\text{graphite}) - 3H^\circ_{298}(O_2)_{gas}\\ &\quad - 6H^\circ_{298}(H_2)_{gas}\\ &= H^\circ_{298}(\text{glucose}), \text{ since the conventional reference state for}\\ &\qquad \text{each of the pure elements is } H^\circ_{298} = 0\\ &= -215.8 \text{ kcal/mole glucose} = -902.9 \text{ J/mole glucose}\end{aligned}$$

When ΔH°_{298} is a negative number (as in the glucose heat of formation), it means that heat is given off as a result of converting the reactants to products. Thus, the product(s) must have less enthalpy than the reactants, and the heat of formation is a measure of the stability of the chemical bonds that were formed during the reaction. Stated differently, $-\Delta H^\circ_{298}$ expresses the amount of heat (enthalpy) that must be put into the compound to break the bonds and produce the isolated elements.

For glucose, ΔH°_{298} represents a sum of the enthalpies of formation of *all* the chemical bonds in the molecule: C—H, C—O, C—C, and O—H bonds. In order to obtain the enthalpy of formation of a particular bond, we must turn to a simpler example, such as methane, CH_4, in order to find ΔH°_{298} for the C—H bond. Then, by comparing the enthalpies of formation of CH_4 and C_2H_6 (ethane), we can find ΔH°_{298} for a C—C single bond (see Problems at the end of the chapter). The O—H bond enthalpy can be found from the enthalpy of formation of water. In this way, it is possible to assemble a list of average bond strengths (as reflected by bond standard enthalpy of formation) for most chemical bonds. These bond strengths can then be used to compute the expected enthalpy of formation for reactions for which no data is available, and the calculations may also be compared with theoretical values from quantum mechanics.

For methane, CH_4, one measure of the average C—H bond strength is to find the enthalpy for forming all four C—H bonds at once, and then divide by 4 to obtain the standard enthalpy of formation of a single average C—H bond. In other words, we seek to determine ΔH°_{298} for the process,

Path No. 1 $C\cdot_{(gas)} + 4H\cdot_{(gas)} \rightarrow CH_{4(gas)}$

$$\begin{aligned}\Delta H_A &= \Delta H^\circ_{298}(CH_4)\\ &= 4\Delta H^\circ_{298}(\text{C—H average bond})\end{aligned}$$

For all practical purposes, there is no way to measure ΔH_A for this reaction directly. However, since H is a state function, we must obtain the same ΔH_A by following *any* reaction path leading from these reactants to the same product. Therefore, by assembling the data from the following three experimentally accessible reactions:

$$C_{(graphite)} \rightarrow C\cdot_{(gas)} \qquad \Delta H_B = \Delta H^\circ_{298} = 171.7 \text{ kcal/mole}$$
$$2H\cdot \rightarrow H_{2(gas)} \qquad \Delta H_C = \Delta H^\circ_{298} = -104.2 \text{ kcal/mole}$$

and

$$C_{(graphite)} + 2H_{2(gas)} \rightarrow CH_{4(gas)} \qquad \Delta H_D = \Delta H^\circ_{298} = -17.9 \text{ kcal/mole}$$

we may construct another path leading from the same reactants to the same product:

Path No. 2

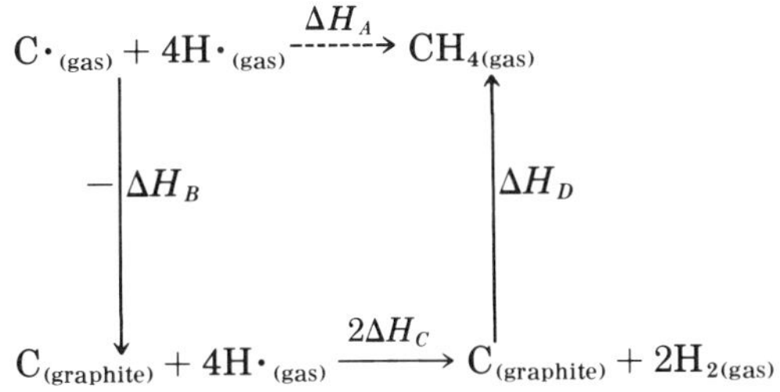

By the property that H is a state function, we then deduce immediately that

$$\Delta H_A = -\Delta H_B + 2\Delta H_C + \Delta H_D$$
$$= -171.7 + 2(-104.2) + (-17.9) = -398.0 \text{ kcal/mole } CH_4$$

so that the average C—H bond enthalpy in CH_4 becomes $-(398.0/4) = -99.5$ *kcal/mole.*

There are several noteworthy features of this result. First, we have obtained a number that is a direct measure of chemical bond strength, using experimentally measured enthalpy changes for some chemical reactions. Second, by exploiting the property that H is a state function, we were able to obtain the enthalpy change for the (unobservable) reaction of interest, by examining a different (lengthier) pathway involving reactions whose enthalpy changes are obtainable experimentally. Lastly, the final calculated bond enthalpy represents an *average* of the individual enthalpies corresponding to one-at-a-time removal of each of the four hydrogen atoms of methane:

$$CH_4 \rightarrow CH_3\cdot + H\cdot \qquad \Delta H_1$$
$$CH_3\cdot \rightarrow CH_2\cdot + H\cdot \qquad \Delta H_2$$
$$CH_2\cdot \rightarrow CH\cdot + H\cdot \qquad \Delta H_3$$
$$CH\cdot \rightarrow C\cdot + H\cdot \qquad \Delta H_4$$

(Current estimates for the above individual enthalpies suggest that $\Delta H_1 > \Delta H_3 > \Delta H_2 > \Delta H_4$.)

Bond *energies* may be determined spectroscopically, by finding the minimum *photon* energy that is just sufficient to completely dissociate the two bonded atoms. Alternatively, bond energies may be obtained from experiments in which a molecule is bombarded by high-energy electrons, based on the minimum *electron* energy sufficient to break the bond of interest, using a mass spectrometer to identify the product fragments from the process. Finally, infrared vibrational frequencies (see Section 5) give a measure of the "force-constant" of a chemical bond, where the chemical bond is regarded as a mechanical spring connecting the two atoms in question. Of all these measures of chemical bond strength, enthalpies of reaction obtained either from calorimetry (as detailed above) or from temperature-dependence of equilibrium constants (see Chapter 4) or emf (see Chapter 5) provided the first and most complete values available. Comparisons of the bond enthalpies for the same atom in different molecules (see Problems) have added much to our understanding of the nature of chemical bonds in molecules and complexes, and of the role of solvents in chemical reactions.

EXAMPLE *Heats of Reaction for Chemical Reactions Not Directly Observable*

Many of the most interesting chemical reactions, in particular those taking place within living cells, are often not amenable to direct observation. However, based on the method of the preceding example, we can obtain the change in any state function on proceeding from reactants to products, by resorting to another (usually lengthier, but more convenient) reaction pathway. When the heats of formation of all species from their elements are available (as for a very great number of biochemical molecular intermediates), we can thus obtain the heat of reaction (for example) by converting all the reactants back into their constituent elements, and then constructing all the products from their elements as illustrated in the ensuing numerical example.

Consider the formic hydrogenylase system from the bacterium, *E. coli*. In this system, an enzyme (see Section 3) catalyzes (increases the rate of) the reaction of hydrogen and bicarbonate to form formate ion and water. Now it can readily be shown (see Klotz reference) that the presence of any true catalyst cannot affect the *equilibrium* (i.e., the "state") properties of a system, even though the catalyst may greatly enhance the *rate* at which equilibrium is reached. Thus, we may ignore the enzyme and analyze the other participants in the reaction.

Path No. 1 $$H_{2(gas)} + HCO^-_{3(aqueous)} \xrightarrow[25°C,\ 1\ atm]{(enzyme)} HCOO^-_{(aqueous)} + H_2O_{(liquid)}$$

In order to find ΔH_{298} for the reaction written in Path No. 1, we begin by assembling all the available relevant thermodynamic enthalpy data:

$$C_{(graphite)} + H_{2(gas)} + O_{2(gas)} \rightarrow HCOOH_{(liq)} \qquad \Delta H_A = \Delta H^\circ_{298} = -99{,}750 \text{ cal/mole}$$
$$C_{(graphite)} + O_{2(gas)} \rightarrow CO_{2(gas)} \qquad \Delta H_B = \Delta H^\circ_{298} = -94{,}240 \text{ cal/mole}$$
$$H_{2(gas)} + (1/2)\ O_{2(gas)} \rightarrow H_2O_{(liq)} \qquad \Delta H_C = \Delta H^\circ_{298} = -68{,}310 \text{ cal/mole}$$
$$CO_{2(gas)} + \text{water} \rightarrow CO_{2(sat'd\ aq)} \qquad \Delta H_D = \Delta H_{298} = -4{,}844 \text{ cal/mole}$$
$$CO_{2(sat'd\ aq)} + H_2O_{(liq)} \rightarrow H_2CO_{3(aq)} \qquad \Delta H_E = \Delta H_{298} = 0 \text{ (same species)}$$
$$H_2CO_{3(aq)} \rightarrow H^+_{(aq)} + HCO^-_{3(aq)} \qquad \Delta H_F = \Delta H_{298} = +2{,}075 \text{ cal/mole}$$
$$HCOOH_{(liq)} + \text{water} \rightarrow HCOOH_{(aq)} \qquad \Delta H_G = \Delta H_{298} = -100 \text{ cal/mole}$$
$$HCOOH_{(aq)} \rightarrow H^+(aq) + HCOO_{(aq)} \qquad \Delta H_H = \Delta H_{298} = -13 \text{ cal/mole}$$

NOTE The definition of "standard" state for a solute (ionic or neutral) is somewhat more elaborate than for elements or pure compounds treated up to now. We discuss the issue further in Chapter 3; for now, the standard state of a solute may be approximated as 1 molar aqueous solution.

Next, we reconstruct the desired Path No. 1 by making the appropriate sequence of reactions whose heats of reaction are known:

Path No. 2

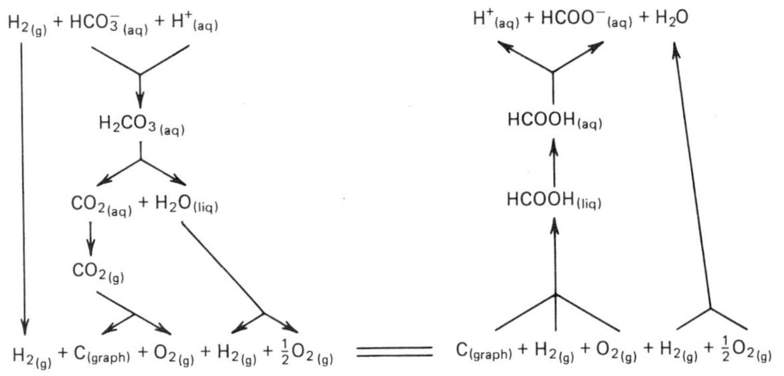

The reader will note that we have added an $H^+(aq)$ to each side of the original equation to balance the reaction. By keeping track of the signs of each of the ΔH_{298} as they appear in Path No. 2, we quickly calculate the desired heat of reaction for the quivalent Path No. 1 as,

$$\Delta H_{298} = -\Delta H_F - \Delta H_E - \Delta H_D - \Delta H_B - \Delta H_C + \Delta H_A + \Delta H_G + \Delta H_H + \Delta H_C$$
$$\Delta H_{298} = -2{,}854 \text{ cal/mole} \qquad \text{for the formic dehydrogenylase reaction.}$$

The reaction as written is thus *exo*thermic—in other words, heat is given off as the reaction proceeds from left to right under the stated concentrations, temperature, and pressure. We will later show how this sort of information can be used to deduce whether a given reaction is spontaneous or not, using changes in free energy, *G*, rather than enthalpy, *H*, as the state function of interest. We will also show how determination of ΔH_{298} makes it possible to predict how the equilibrium constant for this reaction will vary with temperature—qualitatively one would expect (correctly, as it turns out) that if the reaction as written absorbs heat, then the equilibrium should be shifted to the right as the temperature of the system is increased. The extent of the shift is related to the magnitude of ΔH, as we show in Chapter 4.

In concluding these thermochemical examples, we have shown how heats of formation can be used to provide a quantitative measure of chemical bond strengths and how to compute changes in a thermodynamic state function (in this case, enthalpy) for processes that we are not able to study directly.

We will later show how heats of reaction enable us to predict the variation of equilibrium constants with temperature.

PROBLEMS

1. Given the following heats of reaction at standard conditions, compute the average bond enthalpy for the carbon-carbon bond in ethane, in ethylene, and in acetylene (C_2H_6, C_2H_4, and C_2H_2). Assume that a C—H bond has the same bond enthalpy in each compound. What does this calculation show about the relative strengths of single, double, and triple carbon-carbon bonds?

$$2C_{(graphite)} + 3H_{2(gas)} \rightarrow C_2H_{6(gas)} \qquad \Delta H^\circ_{298} = -20{,}190 \text{ cal/mole}$$
$$2C_{(graphite)} + 2H_{2(gas)} \rightarrow C_2H_{4(gas)} \qquad \Delta H^\circ_{298} = 12{,}555 \text{ cal/mole}$$
$$2C_{(graphite)} + H_{2(gas)} \rightarrow C_2H_{2(gas)} \qquad \Delta H^\circ_{298} = 54{,}230 \text{ cal/mole}$$
$$H_{2(gas)} \rightarrow 2H\cdot_{(gas)} \qquad \Delta H^\circ_{298} = 104{,}200 \text{ cal/mole}$$
$$C_{(graphite)} \rightarrow C\cdot_{(gas)} \qquad \Delta H^\circ_{298} = 171{,}700 \text{ cal/mole}$$
$$C_{(graphite)} + 2H_{2(gas)} \rightarrow CH_{4(gas)} \qquad \Delta H^\circ_{298} = -17{,}865 \text{ cal/mole}$$

2. The enthalpies of formation of $H_2O_{(liq)}$ and $H_2O_{(gas)}$ are −68.32 kcal/mole and −57.80 kcal/mole, respectively. Compute the enthalpy for vaporization of water at 298 K,

$$H_2O_{(liq)} \xrightarrow{25^\circ C,\ 1 \text{ atm}} H_2O_{(gas)}$$

(This heat of vaporization corresponds to the energy required to break the intermolecular "bonds" that hold water molecules together in the liquid.)

A typical man produces about 2500 kcal of heat per day from metabolic activity. If 1 calorie of heat will increase the temperature of water by 1 K, calculate the increase in the man's temperature if all this heat were absorbed by his body (assume the man's body is mostly water, and will absorb the same amount of heat as the same amount of pure water). Assume the man weighs 70 kg.

In fact, the man is not a closed system, and a major mechanism for heat loss is evaporation of water (e.g., sweat). Calculate the amount of water that would have to be evaporated in order to release the heat from one day's metabolic activity.

REFERENCES

I. M. Klotz, *Energy Changes in Biochemical Reactions,* Academic Press, N. Y. (1967). Excellent short treatment.

J. Waser, *Basic Chemical Thermodynamics,* W. A. Benjamin, N. Y. (1966). Good coverage of basic material.

CHAPTER 2 Chemical Potential

Table 1-4 suggests that the existence and properties for any chemical equilibrium may best be understood from closer scrutiny of the *total* (Gibbs) free energy, G, in the vicinity of the equilibrium. However, since most equilibria of interest involve at least *two* chemical distinct constituents (*components*), we must first find out how to construct the total free energy from a suitable sum of the free energies of each of the component substances. Unfortunately, the total free energy of a mixture is *not* in general given by a simple sum of the free energies of the unmixed components—the mixing process itself can change the free energy of each component. This aspect may be grasped more directly by analogy to what happens to another *extensive* property (i.e., a property that depends on the *amount* of substance present, as opposed to an *intensive* property such as temperature or pressure that does *not* depend on the amount of substance present), namely the *volume of a mixture* of two components.

Figure 2-1 shows that when 1.00 liter of water and 1.00 liter of pure ethanol are mixed, the volume of the *mixture* is not 2.00 liter, but actually somewhat less. (In fact, when water and ethanol are mixed in *any* given proportion, the total volume of the solution is always less than the simple sum of the initial volumes of water and ethanol that were placed in the mixing vessel.) This result simply reflects the fact that the forces that attract a water molecule to an ethanol molecule are different from the attractive forces between two water molecules or two ethanol molecules. More generally, since the total volume (V) of the mixture depends on temperature (T), pressure (P), and the amount (expressed as number of moles, n) of each component of the mixture, and since V is a state function, we can express dV as an exact differential:

$$dV = \left(\frac{\partial V}{\partial T}\right)_{P,n_{\mathrm{EtOH}},n_{\mathrm{H_2O}}} dT + \left(\frac{\partial V}{\partial P}\right)_{T,n_{\mathrm{EtOH}},n_{\mathrm{H_2O}}} dP + \left(\frac{\partial V}{\partial n_{\mathrm{H_2O}}}\right)_{T,P,n_{\mathrm{EtOH}}} dn_{\mathrm{H_2O}} + \left(\frac{\partial V}{\partial n_{\mathrm{EtOH}}}\right)_{T,P,n_{\mathrm{H_2O}}} dn_{\mathrm{EtOH}} \tag{2-1}$$

or

$$dV = \bar{V}_{\mathrm{H_2O}} dn_{\mathrm{H_2O}} + \bar{V}_{\mathrm{EtOH}} dn_{\mathrm{EtOH}} \qquad \text{at constant } T \text{ and } P \tag{2-2}$$

where

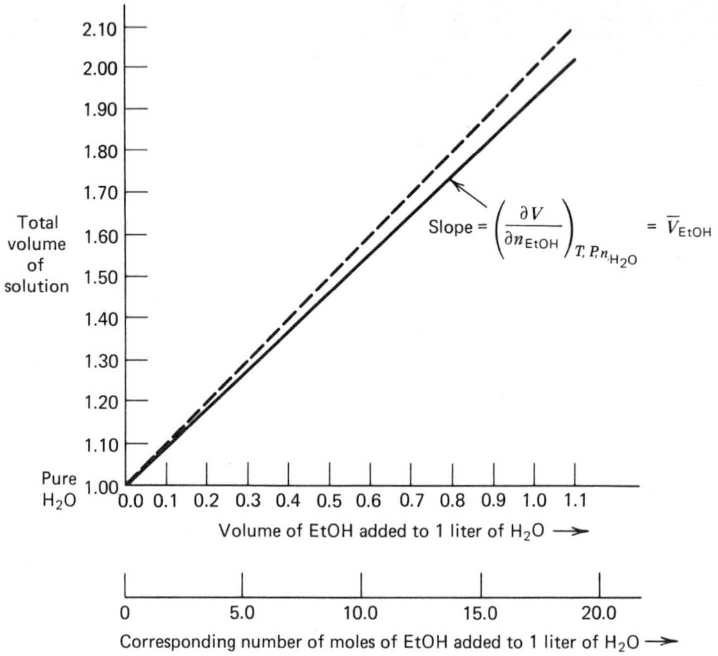

FIGURE 2-1. Total volume of a solution formed by mixing 1.00 liter of pure H_2O and a given volume of absolute ethanol. (Note that 1.0 liter of water plus 1.0 liter of ethanol gives less than 2.0 liter of solution.) The slope of the solid (experimental) curve is the "partial molal volume" of ethanol at this temperature (15.5°C), pressure (1 atm), and amount of water (n_{H_2O}).

$$\boxed{\bar{V}_i = \left(\frac{\partial V}{\partial n_i}\right)_{T,P,n_j} = \text{partial molal volume of the } i\text{th component}} \tag{2-3}$$

The meaning of $\bar{V}_i$ should become clear from Fig. 2-1. The main interest in Equations 2-1 to 2-3 and Fig. 2-1 is that they show how an extensive property (volume) of a mixture is not necessarily a simple sum of the extensive properties (volumes) of the two components before mixing. Experiments such as that in Fig. 2-1 are used to determine the partial molal volume of macromolecules in aqueous solution, and the partial molal volume may then in turn be used in determination of molecular weight, size, and shape from diffusion, sedimentation, and viscosity experiments (Section 2). [The particular example of Fig. 2-1 is even of some minor interest in calculating the "proof" of alcoholic beverages, since (U. S.) "proof" is defined as twice the volume concentration (volume of ethanol divided by total volume of mixture).]

The same analysis we have just applied to one extensive state function, V, can also be applied to other extensive state functions (e.g., E, S, H, G, and A in Table 1-3). In particular, we can write down the exact differential for (Gibbs) free energy, G, in terms of the independent variables, T, P, and n_i:

$$dG = (\partial G/\partial P)_{T,n_1,n_2 \ldots n_i} dP + (\partial G/\partial T)_{P,n_1,n_2 \ldots n_i} dT + \sum_i (\partial G/\partial n_i)_{T,P,\text{constant amounts of all but } i\text{th component}} dn_i \quad (2\text{-}4)$$

We can quickly identify the first two partial derivatives in Eq. 2-4 by some manipulation of the basic definitions of G, H, E, and S for a system of constant composition (i.e., all n_i are constant so that dn_i are zero in Eq. 2-4).

First,

$$G = H - TS \qquad \text{(definition)} \qquad (1\text{-}14)$$

so that

$$dG = dH - TdS - SdT \qquad (2\text{-}5)$$

But

$$H = E + PV \qquad \text{(definition)} \qquad (1\text{-}13)$$

so that

$$dH = dE + PdV + VdP \qquad (2\text{-}6)$$

and thus

$$dG = dE + PdV + VdP - TdS - SdT \qquad (2\text{-}7)$$

Now, for a reversible process,

$$dS = dq/T \qquad \text{(definition)} \qquad (1\text{-}6)$$

and

$$dE = dq - PdV + dw' \qquad (1\text{-}4)$$

so that

$$dG = dq - PdV + dw' + PdV + VdP - T(dq/T) - SdT$$

or just

$$dG = VdP - SdT + dw' \qquad (2\text{-}8)$$

Finally, for a system of constant composition, for a reversible process where only "PV" work is allowed (e.g., no electric or magnetic fields present), we have the useful result,

$$dG = VdP - SdT \qquad \text{constant composition, reversible process, } PV \text{ work only} \tag{2-9}$$

From Eq. 2-9, we can now immediately identify the desired partial derivatives for Eq. 2-4:

$$(\partial G/\partial P)_{T,n_1,n_2,\ldots,n_i} = V \tag{2-10}$$

and

$$(\partial G/\partial T)_{P,n_1,n_2,\ldots,n_i} = -S \tag{2-11}$$

Equation 2-4 may now be written in a more compact form, which we will use frequently:

$$dG = VdP - SdT + \sum_i \bar{G}_i \, dn_i \tag{2-12}$$

in which

$$\bar{G}_i = (\partial G/\partial n_i)_{T,P,\text{constant amounts of all but } i\text{th component}} \, dn_i = \text{Chemical Potential} \tag{2-13}$$

and the sum notation of Eq. 2-12 indicates a sum over all components in the mixture. Equations 2-10 and 2-11 are important, because they indicate how free energy varies with pressure or temperature. Since we will later show that free energy changes are related to equilibrium constants, we will thus be able to apply these equations in predicting the variation of equilibrium constants with pressure or temperature. For now, however, our principal interest is in the "partial molal free energy," $\bar{G}_i$, which is sometimes called the "chemical potential." *

The significance of chemical potential, $\bar{G}_i$, for *chemical* equilibrium is

* *Many texts denote $\bar{G}_i$ by the more general symbol, μ_i, to emphasize that the chemical potential may be defined (under suitable conditions) as the change in any of* several *forms of energy with amount of substance present:*

$$\mu_i = \left(\frac{\partial G}{\partial n_i}\right)_{T,P,n_j} = \left(\frac{\partial H}{\partial n_i}\right)_{S,P,n_j} = \left(\frac{\partial E}{\partial n_i}\right)_{S,V,n_j}.$$

For our purposes, $\bar{G}_i$ conveys all necessary meaning, since we will be more interested in applications of free energy (G) than of energy (E) or enthalpy (H).

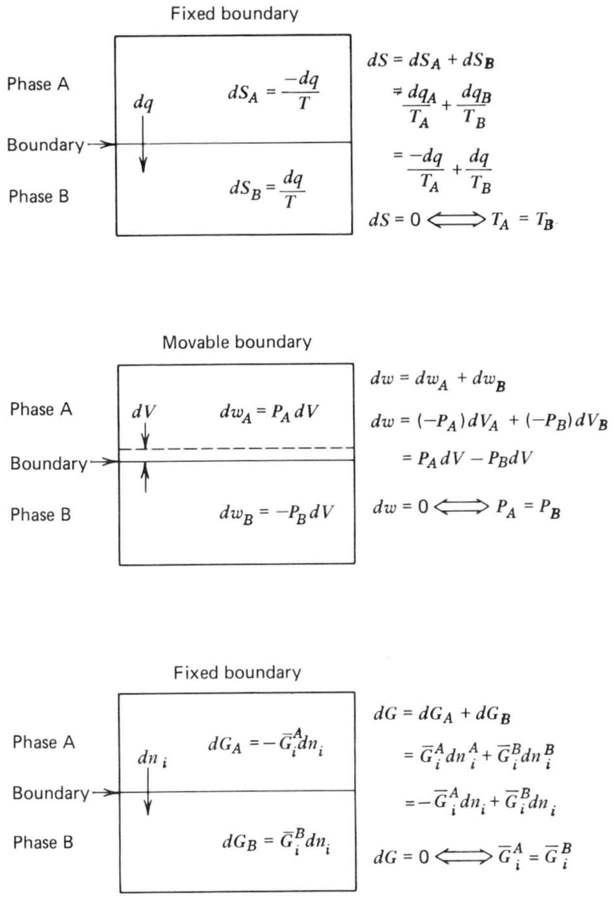

FIGURE 2-2. Behavior of a two-phase system near equilibrium. *Top:* Thermal equilibrium. *Middle:* Mechanical equilibrium. *Bottom:* Chemical equilibrium. In each case, a criterion for equilibrium is derived by considering the result of an infinitesimal displacement away from an existing equilibrium condition. Chemical potential is seen to play the same role in chemical equilibrium that temperature or pressure play in thermal or mechanical equilibria (see text and Table 2-1). The boundary between the two phases might be a liquid-vapor interface (top diagram), an impermeable membrane (middle diagram), or a semipermeable membrane (bottom diagram), as discussed later in this section.

perhaps best illustrated by analogy to criteria for *thermal* or *mechanical* equilibria, as shown in Fig. 2-2 and summarized in Table 2-1. Consider a system composed of two phases (Fig. 2-2), such as liquid-and-vapor, or two immiscible liquids, or liquid-and-solid. [A *phase* may be defined as a system that is *macroscopically* uniform throughout (thermodynamics is concerned with properties of bulk matter, not with individual atoms), both in chemical composition and in physical state.]

Thermal equilibrium between phases A and B is then said to apply when there is *no net flow of heat* from one phase to another, namely, when both phases are at the *same temperature.* Alternatively, by considering the (infinitesimal) change in entropy when an infinitesimal amount of heat, dq, is transferred from one phase to the other at thermal equilibrium (Fig. 2-2, top diagram), we can confirm that $dS = 0$ is a criterion for thermal equilibrium, as recorded in Table 2-1.

Similarly, we might define a *mechanical* equilibrium between two phases as a condition in which there is *no net "flow" of work* from one phase to the other. Figure 2-2 (middle diagram) shows that with this definition of equilibrium, the corresponding condition is that the *pressure* exerted by both phases be equal.

Finally, if we think of a *chemical* equilibrium as one in which there is *no net change in the amount of each chemical constituent* present, then the lower diagram of Fig. 2-2 shows that the appropriate condition for equilibrium (at constant temperature and pressure; PV work only), namely $dG = 0$, leads to the conclusion that the *chemical potential* of any component in one phase ($\bar{G}_i^A$) is equal to the chemical potential of that component in the other phase ($\bar{G}_i^B$). This conclusion is reached by considering the effect of moving an infinitesimal amount of the ith component, dn_i, from phase A to phase B in Fig. 2-2, just as we considered an infinitesimal change in volume or heat flow in the mechanical and thermal examples. The desired result is

$$\boxed{\bar{G}_i^A = \bar{G}_i^B \qquad \text{at chemical equilibrium (constant } T, P, dw' = 0)} \tag{2-14}$$

In the remarkable "Gibbs phase rule" of the following section, we will show that simply by using Eq. 2-14, we can predict how many phases are possible under a wide variety of conditions. The principal results of this introductory

Table 2-1 Chemical Potential as a Criterion for Chemical Equilibrium, Illustrated by Comparison to Corresponding Criteria for Thermal or Mechanical Equilibrium

Type of Equilibrium	Equilibrium Defined as No Net Flow of	Extensive[a] Property	Intensive[a] Property	Condition for Equilibrium Between Phases A and B
Thermal	heat	S	T	$dS = 0$ or $T_A = T_B$ (constant P, n_i)
Mechanical	work	V	P	$dV = 0$ or $P_A = P_B$ (constant T, n_i)
Chemical	substance	G or n_i	$\bar{G}_i$	$dG = 0$ or $\bar{G}_i^A = \bar{G}_i^B$ (constant T, P)

[a] *Extensive and intensive properties are those that do (or do not) depend on the total* amount *of substance present.*

section are summarized in Table 2-1. The reader should establish firm control over this material, because it forms the basis of much of this chapter.

2.A. PHASES AND PHASE TRANSITIONS

Phase transitions ("melting," "freezing," "boiling") are among the simpler "reactions," since often just one chemical component is considered. The first question to be answered is how many phases can be present under various conditions, and the answer forms the basis for the discussion in this section. Phase transitions are of increasing interest in biology, since the biological function of membranes can be correlated to the phase composition of the phospholipid bilayer (section 2.A.2.).

2.A.1. How Many Phases Are There? (Gibbs Phase Rule)

In the discussion leading up to Table 2-1, we argued that the state of a system at equilibrium may be specified by *temperature, pressure,* and the *concentration* (which determines the chemical potential) of each chemically distinct constituent (component).* For a system consisting of p phases and c components, the total number of specified variables is thus $pc + 2$, since there are c components (and thus c concentrations) per phase, while temperature and pressure are the same for all phases at equilibrium.

However, not all these variables are independent. For example, if we express the concentrations of various components as mole fractions, X_i,

$$X_i = \frac{n_i}{\sum_i n_i} = \text{mole fraction of } i\text{th component} \tag{2-15}$$

then in any one phase (say phase B), we know that the sum of the mole fractions of all the components in that phase must be unity:

$$X_1^B + X_2^B + X_3^B + \cdots + X_c^B = 1 \tag{2-16}$$

Thus, if we knew all but one of the component concentrations, we could compute the remaining concentration from Eq. 2-16. A similar argument can be made for each of the p phases. Therefore, the total number of variables needed to specify the state of the system may be reduced by p, to leave $(pc + 2 - p) = p(c - 1) + 2$.

Furthermore, we know from Eq. 2-14 that at equilibrium, the chemical potential of, say, the ith component in any one phase must be equal to the chemical potential of the ith component in any other phase:

* *The number of components, in this context, may be defined in practice as the total number of different chemical constituents (e.g., $CaCO_3$, CaO, and CO_2) minus the number of distinct chemical reactions that connect those constituents (e.g., the single reaction, $CaO + CO_2 \rightleftharpoons CaCO_3$, in this case, to give $3 - 1 = 2$ components).*

$$c \text{ Components} \begin{cases} \bar{G}_1^A = \bar{G}_1^B = \cdots = \bar{G}_1^p \\ \bar{G}_2^A = \bar{G}_2^B = \cdots = \bar{G}_2^p \\ \quad\vdots \qquad \vdots \qquad\quad \vdots \\ \bar{G}_c^A = \bar{G}_c^B = \cdots = \bar{G}_c^p \end{cases} \tag{2-17}$$

$p - 1$ Equations for each component

Inspection of Eq. 2-17 shows that these equations provide an additional $c(p-1)$ constraints, leaving the total number of *independently specifiable intensive variables* (sometimes called "degrees of freedom"), f, as

$$f = p(c-1) + 2 - c(p-1)$$

or just

$$\boxed{f = c - p + 2} \tag{2-18}$$

The significance of this "Gibbs phase rule," Eq. 2-18, is most readily understood by study of a simple one-component system, as shown in Fig. 2-3.

EXAMPLE *Phase Diagram of Water—A One-component System*

Equation 2-18 indicates that for a one-component system ($c = 1$), there will be $f = 2$, 1, or 0 degrees of freedom corresponding to $p = 1$, 2, or 3 phases as tabu-

Region of diagram	c	p	f
Area	1	1	2
Line	1	2	1
Triple point	1	3	0

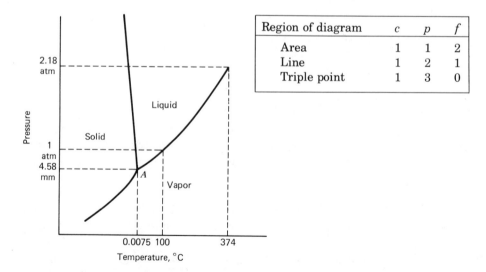

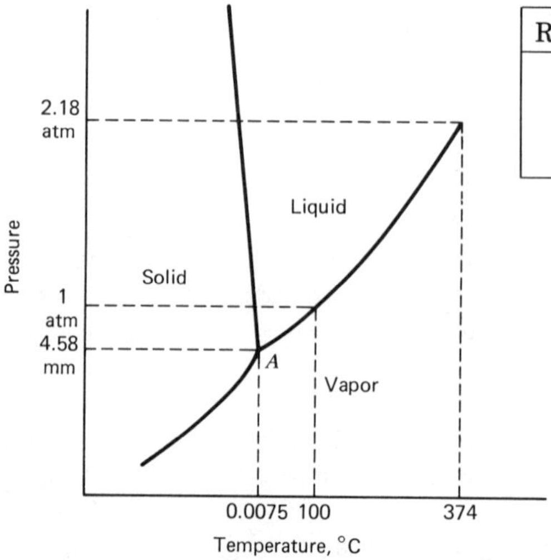

FIGURE 2-3. Schematic phase diagram (not to scale) for the one-component water system. See text for discussion.

lated at the upper right of Fig. 2-3. For example, in the temperature-pressure region where water is present as pure vapor, we may vary temperature and pressure independently anywhere in the region and still have just one phase (i.e., water vapor) present. However, along one of the "lines" in the phase diagram of Fig. 2-3, such as that between liquid and vapor, there are *two* phases present in equilibrium. Thus, if we specify a given pressure, the other degree of freedom (temperature) is constrained to the value given by the solid line. In other words, for a given pressure, water has a well-defined unique boiling temperature (boiling point). In particular, the boiling point of water is 100°C when the external pressure is 1 atm; however, the boiling point of water is much lower when the external pressure is reduced. Similarly, the melting point of ice is a function of pressure—ice skates slide easily, partly because the high pressure of the skate on the ice lowers the temperature at which the ice will melt, and the skater slides to some extent on a thin film of water. Finally, if we insist on having solid, liquid, and vapor phases all present simultaneously at equilibrium, then the degrees of freedom are zero, so that there is a unique temperature and pressure for this situation (point A in Fig. 2-3).

2.A.2. Phase Transitions in Biological Membranes

One of the most striking results in recent molecular biology has been the discovery that biological membranes are not static structures, but can have a degree of fluidity in their interiors. The degree of fluidity can be correlated with such biological functions as sugar transport across the membrane or with degree of anesthesia produced by certain drugs. The first important thing to realize is that biological membranes are multicomponent systems, in the language we have been using. Therefore, even with as few as two components ($c = 2$), Eq. 2-18 predicts that when two phases (say, solid and liquid) are present, there will still be two degrees of freedom. In other words, a two-component system will in general not have a "sharp" (i.e., single) melting temperature, since it is now possible to vary the temperature of the system and still have both phases present.* The biological value of this feature is now obvious: if biological membranes were made of just one kind of lipid, the membrane would have a sharp melting point, and the organism could respond to temperature variation only with an "all-or-none" change in membrane fluidity and thus in membrane function. However, with a multicomponent membrane composition, the membrane can gradually change its degree of fluidity over a wide temperature range, affording much finer "control" of cell function according to changes in temperature of the surroundings.

* *The Gibbs phase rule (Eq. 2-18) thus explains why a sharp melting point may be used as a criterion for* purity *of newly isolated or synthesized organic molecules: the "pure" compound is a one-component system, and should have a sharp melting point, while the presence of any contaminant changes the situation to a two-component system, for which we expect a range of temperature for melting.*

EXAMPLE *Phase Transitions in Synthetic Phospholipid Mixtures*

We will later observe (Section 4) that it is possible to introduce a paramagnetic (nitroxide) "spin-label" molecule into a phospholipid environment, and that the spin-label electron spin resonance absorption signal is sensitive to the rotational mobility of the spin-label molecule. This property has been used by McConnell and co-workers to detect phase changes in a phospholipid mixture as a function of temperature, because the ESR signal changes as soon as the surrounding phospholipid regions begin to "melt." Using ESR signal shape as the "indicator," it is then possible to construct a phase diagram for a mixture of two types of phospholipid molecules as shown in Fig. 2-4.

Because $c = 3$ (two types of phospholipid, plus water), we can specify that two phases (denoted as "liquid" and "solid") be present, $p = 2$, and we can vary $(c - p + 2) = 3$ intensive quantities and still retain a two-phase system. Thus, even when pressure is fixed (at 1 atm), we can still vary temperature and the relative proportion of the two phospholipids and remain in the "fluid" + "solid" region bounded by the solid curves in Fig. 2-4. For example, at a temperature of 30°C, and a composition of the total mixture of about 65 mole % DPL (point A in Fig. 2-4), there are two phases present: a "fluid" phase whose composition is about 45 mole % DPL (point B) and a "solid" phase whose composition is about 85 mole % DPL (point C). At a different temperature but with the same composition of the total mixture, we would still have "fluid" and "solid"

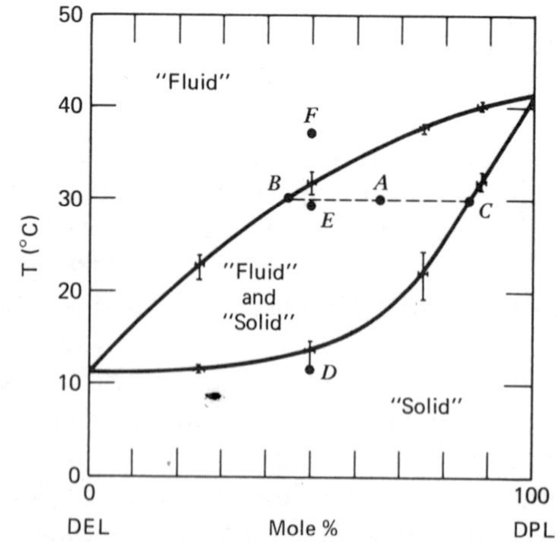

FIGURE 2-4. Phase diagram representing phase separation in aqueous mixtures of dielaidoylphosphatidylcholine (DEL) and dipalmitoylphosphatidylcholine (DPL), based on electron spin resonance data. The abscissa gives the mole per cent of DPL in the binary mixture. Conditions anywhere above the upper curve represent a totally "fluid" phase; conditions anywhere below the lower curve produce a totally "solid" phase. For points between the two curves (say, *A*), "fluid" and "solid" coexist, and their respective compositions are found from the intersection between a horizontal line through point *A* and the upper (point *B*) or lower (point *C*) curves (see text). Points *D, E,* and *F* represent mixtures that were rapidly frozen and then studied by electron microscopy (see Fig. 2-5). [From C. W. M. Grant, S. H. W. Wu, and H. M. McConnell, *Biochem. Biophys. Acta 363,* 151 (1974).]

present, but the composition of the "fluid" (or "solid") would be different from what it is at 30°C. Thus, not only does the multicomponent system "melt" over a wide temperature range, but the *composition* of the "fluid" or "solid" is different at different temperatures. This situation should be contrasted to the one-component system (say, water), in which liquid water has essentially the same composition no matter what temperature-pressure combination is used to condense it from vapor or melt it from solid.

Figure 2-5 shows direct electron micrographs, using the freeze-fracture technique (see Section 4) taken from phospholipid mixtures which were ini-

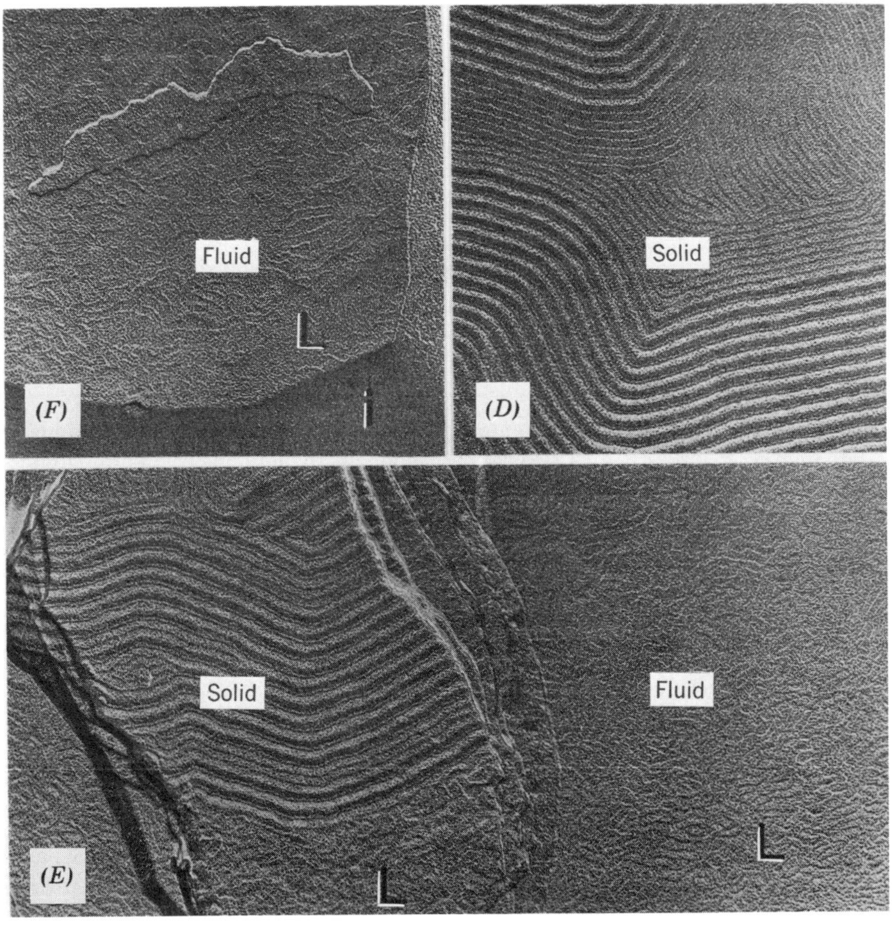

FIGURE 2-5. Electron microphotographs of platinum-carbon replicas of fracture faces for 50–50 mole per cent mixture of DEL and DPL, quick-frozen from about 36°C (*F* in diagram – see Fig. 2-4) 10°C (*D* in diagram – see Fig. 2-4) and 30°C (*D* in diagram – compare to Fig. 2-4). Note that the electron micrographs illustrate the predicted "fluid" (*F*), "solid" (*D*), and "fluid" + "solid" (*E*) patterns, as predicted from the phase diagram for the same system in Fig. 2-4. [From C. W. M. Grant, S. H. W. Wu, and H. M. McConnell, *Biochem. Biophys. Acta 363,* 151 (1974).]

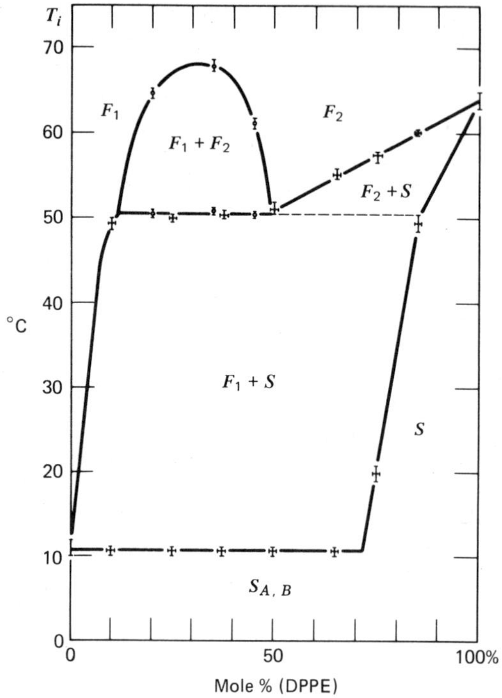

FIGURE 2-6. Phase diagram showing the existence of two different kinds of "fluid" phases, F_1 and F_2, for the system of aqueous dispersions of dielaidoylphosphatidylcholine (DEPC) and dipalmitoylphosphatidylethanolamine (DPPE). The diagram was constructed from the temperature-composition points at which characteristic changes appeared in the electron spin resonance spectrum of a paramagnetic "spin-label" tracer dissolved in the phospholipid phases. See text for interpretation of these results (Fig. 2-7). [From S. H. W. Wu and H. M. McConnell, *Biochemistry 14*, 847 (1974).]

tially at 36°C (upper left), 10°C (upper right), and 30°C (lower photograph), for the same DEL-DPL system at a composition of 50 mole % DPL. From the phase diagram of Fig. 2-4, we would expect that these images should show the "fluid," "solid," and "fluid" plus "solid" phases, and the micrographs confirm this view.

Figure 2-6 shows another interesting phenomenon regarding phase separations in multicomponent phospholipid systems. In this case, there are two distinct types of "fluid" phase, denoted as F_1 and F_2, one of which is richer in the first type of phospholipid and the other richer in the second type of phospholipid. Although it is possible that this phase separation is of the "lateral" type (Fig. 2-7b) that we have already observed in the "fluid" + "solid" example of Figs. 2-4 and 2-5, an intriguing possibility is that the separation might occur such as to enrich (say) the outer layer of the membrane in one phospholipid and the other (say, inner) layer in the other phospholipid (Fig. 2-7c). Then, because one would expect different phospholipid:water interactions for differ-

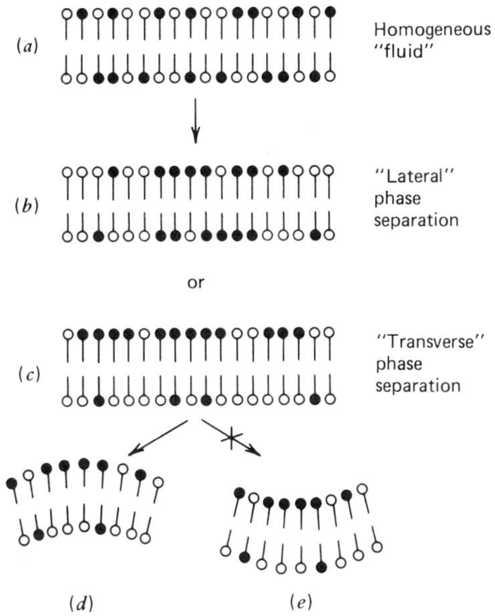

FIGURE 2-7. Schematic diagrams of possible types of "fluid" phases in phospholipid mixtures. (*a*) Homogeneous fluid solution of black and white lipids. (*b*) Partial fluid-fluid immiscibility arising from lateral phase separations into regions relatively rich in white lipids and regions relatively rich in black lipids—this is the sort of phase separation that does occur between "fluid" and "solid" phases of the type discussed in Figures 2-4 and 2-5. (*c*) Transverse phase separation such that one half of the bilayer is rich in black lipids and the other half is rich in white lipids. For situation (*c*), it would be expected that the membrane would be more stable when curved in one sense (*d*) than when curved in the other sense (*e*). [From S. H. W. Wu and H. M. McConnell, *Biochemistry 14*, 847 (1974).]

ent phospholipids, the effect of such a phase separation might be to induce *curvature* in the membrane.

EXAMPLE *Correlation Between Membrane Function and Membrane Phase State*

One biological function of cell membranes is the transport of various metabolites into or out of the cell. Figure 2-8a shows the temperature-dependence of a particular function of *E. coli* cells, namely the translocation of β-galactoside from one side of the cell membrane to the other. (A nitrophenyl group was attached to the β-galactoside as a chromophore in order that the process could be monitored by simple u.v.-visible absorption spectroscopy.) As we will soon explain (Chapter 4), a plot of log (transport rate) versus reciprocal of absolute temperature would be expected to give a straight line of negative slope, since most chemical reaction rate constants increase exponentially with temperature. The unusual feature of this particular transport rate is that there are two particular temperatures at which either the slope or the absolute trans-

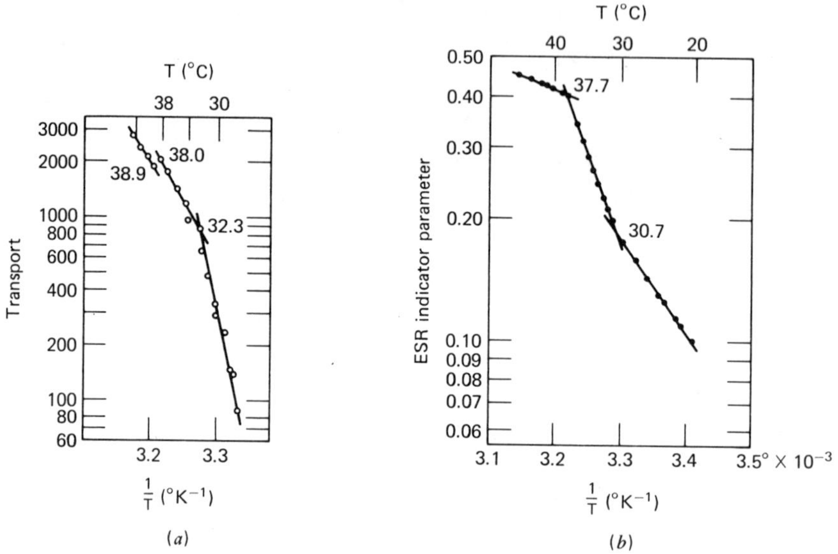

FIGURE 2-8. (*a*) Temperature-dependence (reciprocal scale) of the rate of transport of β-galactoside (log scale) by cells of *E. coli* (type βox^-) grown at 37°C in a medium supplemented with elaidic acid. The units for transport are nmol/20 min per 10^9 cells. (*b*) Electron spin resonance spectral parameter for a paramagnetic "spin-label" dissolved in the inner membranes of the same bacterium (log scale) as a function of temperature (reciprocal scale). Note that the temperatures at which phase changes begin and end as judged from the ESR data correspond closely to the temperatures at which there is a marked change in the membrane transport property (see text). [From C. Linden, K. Wright, H. M. McConnell, and C. F. Fox, *Proc. Natl. Acad. Sci. U.S.A. 70,* 2271 (1973).]

port rate change to different values ($T = 32.3$°C and $T = 38.0$°C). The explanation for this unusual behavior appears to result from the data in Fig. 2-8b, showing the variation of a parameter obtained from the ESR spectrum of a paramagnetic "spin-label" dissolved in the phospholipids of this same membrane system, as a function of reciprocal of absolute temperature. Figure 2-8b shows the two characteristic temperatures (30.7°C and 37.7°C in this case) at which this phospholipid system begins to melt and completes melting (compare with Fig. 2-5). It thus appears that the biological *function* of this bacterial membrane (sugar transport in this case) appears to correlate directly with the *phase state* of the membrane—the transport rate behaves differently for membrane that is "solid," "fluid," or "fluid" + "solid."

An especially striking example of this effect is provided by study of the phospholipid composition of a reindeer leg. With respect to adaptation to the environment, the immediate problem faced by the reindeer is that the temperature near one of its feet is much lower than where the leg joins the body. From what we now know, we would thus expect that the membranes of the cells in the reindeer leg should behave differently for cells near the foot compared to cells near the body, because the phase state of the membrane would vary from "solid" or "fluid" due to the temperature gradient going up the leg. However, the phase transition temperature for phospholipids varies according to the degree of "unsaturation" in the hydrocarbon chains of the phospholipid molecule (the more double bonds in the chain, the lower the "melting" temperature). Therefore, the reindeer's adaptive solution is that the phospholipids in the cells near its feet are richer in unsaturated phospholipid than the phospholipids in the leg region near the body. Thus, the membrane phase state (and thus membrane function) is relatively uniform throughout the leg, even though the temperature of the cells varies greatly over the same distance.

A schematic diagram summarizing much of what is believed about the phase state, structure, and function of biological membranes is shown in Fig. 2-9.

Finally it should be noted that "fluid"-"fluid" phase separations are not limited to phospholipids. Such separations have long been known to arise spontaneously when certain charged polysaccharides are mixed in suitable

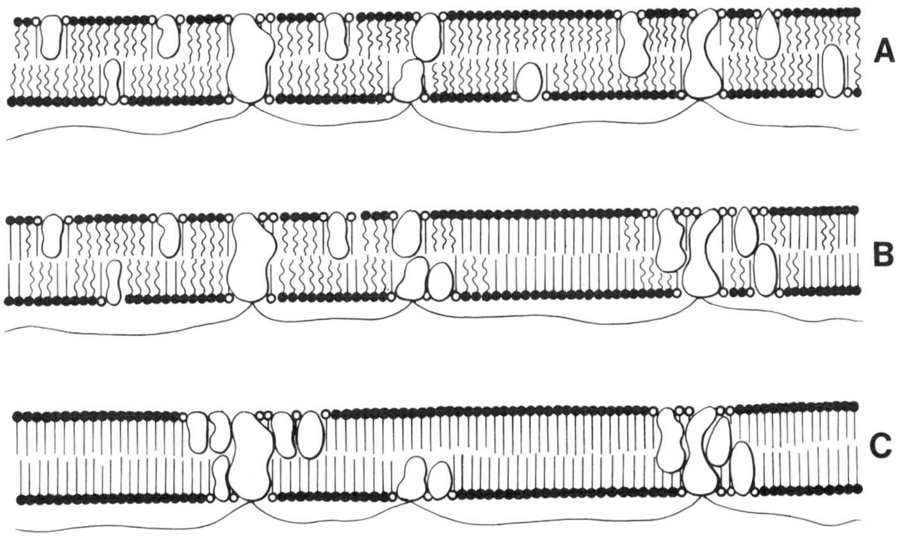

FIGURE 2-9. A diagrammatic representation of a cross section through a biological membrane containing proteins and lipids. The lipids are indicated by circles with tails, and the proteins as larger irregularly shaped bodies. Some proteins have been drawn as having points of attachment to a cytoskeletal structure. Lipids immediately adjacent to proteins (boundary lipids) are indicated by open rather than closed circles. This class of lipid may have properties unique from the bulk bilayer lipid (Section 4). (A) All the lipid in the membrane is in a fluid state, as indicated by the wavy tails. The proteins not restricted by cytoskeletal attachment are free to diffuse in the plane of the membrane. (B) The membrane is at a temperature between A and C. Some lipids have begun to freeze (straight tails), thus excluding the proteins into regions containing fluid lipid. As a consequence of their exclusion from regions containing frozen lipid, the proteins have begun to aggregate in patches. (C) The membrane is at a temperature at which all the membrane lipids are frozen, and the proteins have become maximally aggregated. [From C. D. Linden and C. F. Fox, *Accts. Chem. Res. 8*, 321 (1975).]

proportions. In some theories of the origin of life, it is conjectured that similar phase separations occurring in the primordial "soup" mixture of sugars, amino acids, and other molecules may have provided a kind of "cell" boundary within which chemical reactions could proceed differently from those in the rest of the solution. For deeper discussion of this and earlier topics, the reader is referred to the references at the end of this chapter.

2.B. SEMIPERMEABLE MEMBRANES AND NEUTRAL SPECIES IN SOLUTION: OSMOTIC PRESSURE

To understand how osmotic pressure comes about, it is first necessary to know that the vapor pressure of any solvent, P_A°, is *lowered* when an involatile solute (such as a protein) is dissolved in the solution, as shown in the top diagram of Fig. 2-10. Intuitively, we might explain this phenomenon as follows: since only a *fraction*, X_A (where X_A is the mole fraction of solvent molecules out of the total molecules in the solution), of the original solvent molecules now occupy the same volume, the vapor pressure of the solution, P_A, is reduced accordingly:

$$\boxed{P_A = P_A^\circ X_A} \qquad \text{Raoult's Law} \qquad (2\text{-}19)$$

If we place solutions of pure solvent (Fig. 2-10, top left) and solution (Fig. 2-10, top right) on either side of an impermeable membrane, there will thus

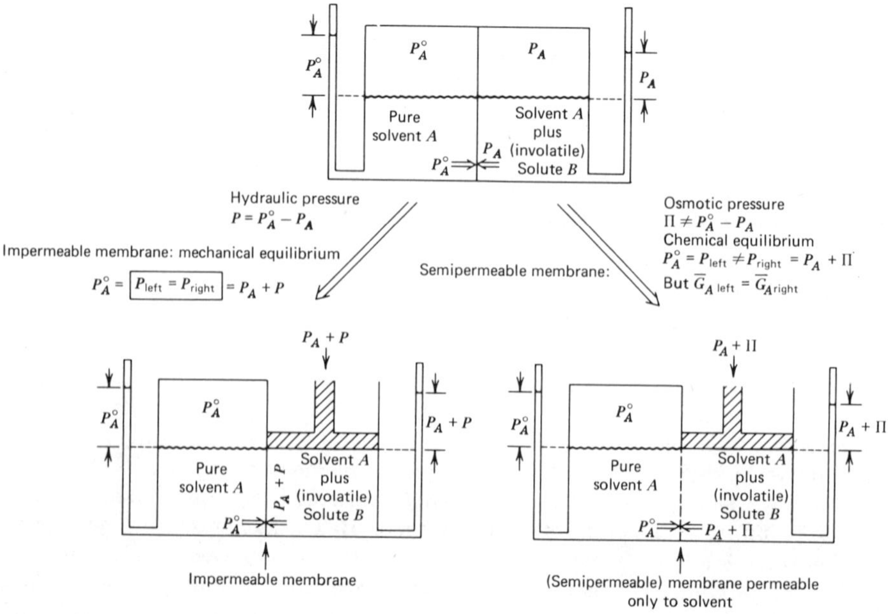

FIGURE 2-10. Schematic diagrams illustrating the distinction between *hydraulic* pressure and *osmotic* pressure. See text for discussion.

be a pressure difference, $P_A^\circ - P_A$, across the membrane. If the membrane is *impermeable,* then we may equalize the pressure across the membrane (i.e., bring the system to equilibrium) by applying an additional pressure, $(P_A^\circ - P_A)$ on the right-hand side, as shown at bottom left of Fig. 2-10. This equalizing pressure is not *osmotic* pressure, but rather ordinary *hydraulic* pressure.

However, if the membrane is permeable to solvent but not solute, then the system will try to relieve the nonequilibrium pressure difference across the membrane by driving solvent from left to right through the semipermeable membrane. We can prevent that solvent migration, and bring the system to equilibrium, by applying an external *osmotic* pressure, π, as shown at bottom right of Fig. 2-10. The important observation at this stage is that $\pi \neq (P_A^\circ - P_A)$, so that there is something different about this case than about the simpler hydraulic equilibrium just discussed. The difference is that we are now dealing with a "chemical" rather than "mechanical" equilibrium; the criterion for equilibrium (Table 2-1) is that the *chemical potential,* $\bar{G}_A$, of solvent should be the *same* on both sides of the membrane at equilibrium ($\bar{G}_{A_{\text{left}}} = \bar{G}_{A_{\text{right}}}$), rather than that the *pressure* be the *same* on both sides of the membrane ($P_{\text{left}} = P_{\text{right}}$ for hydraulic case).

In order to handle the osmotic pressure calculation, we first need to derive a relation between chemical potential and pressure. Beginning with a previously derived result for a two-component system (Eq. 2-12),

$$dG = VdP - SdT + \bar{G}_A dn_A + \bar{G}_B dn_B \tag{2-12}$$

we have already noted that

$$(\partial G/\partial P)_{T,n_A,n_B} = V \tag{2-10}$$

Now since

$$\left(\frac{\partial}{\partial n_A} V\right)_{T,P,n_B} \equiv \bar{V}_A = \text{partial molal volume of solvent } A \tag{2-3}$$

we can combine Equations 2-3 and 2-10 to give:

$$\left(\frac{\partial}{\partial n_A}\right)\left(\frac{\partial G}{\partial P}\right)_{T,n_A,n_B} = \bar{V}_A = \left(\frac{\partial}{\partial P}\right)\left(\frac{\partial G}{\partial n_A}\right)_{T,n_B,P} = \frac{\partial}{\partial P}\bar{G}_A, \tag{2-20}$$

where the second equality follows from the fact that the order of differentiation is immaterial. Finally, since we have already defined chemical potential as

$$(\partial G/\partial n_A)_{T,P,n_B} \equiv \bar{G}_A \tag{2-13}$$

we can rewrite Eq. 2-20 in the (desired) form:

$$\left(\frac{\partial \overline{G}_A}{\partial P}\right)_{T,n_A,n_B} = \overline{V}_A \tag{2-21}$$

We will use Eq. 2-21 in two forms. For *liquid* A, (constant temperature and composition)

$$\boxed{d\overline{G}_A = (\overline{V}_A)_{\text{liq}}\, dP} \quad \underset{\text{liquid}}{\uparrow} \tag{2-21a}$$

For *A vapor*, the ideal gas law predicts that $\overline{V}_A = RT/P_A$, so that Eq. 2-21 becomes

$$d\overline{G}_A = RT\frac{dP_A}{P_A}$$

$$\boxed{d\overline{G}_A = RT\, d\ln P_A} \quad \underset{\text{vapor}}{\uparrow} \tag{2-21b}$$

Although Equations 2-21a and 2-21b accurately describe *infinitesimally* small differences in chemical potential in terms of *infinitesimally* small differences in pressure for an ideal solution (one that obeys Eq. 2-19) and an ideal vapor (Eq. 1-7), we need to *sum* up many such small changes to describe a *macroscopic* change in chemical potential in terms of a macroscopic pressure change. Taking the liquid solvent first, we integrate Eq. 2-21a to obtain

$$\overline{G}_A = \overline{G}_{A\left(\substack{\text{standard}\\ \text{state}}\right)} + \int_{\substack{\text{standard}\\ \text{state}}}^{\substack{\text{experimental}\\ \text{state}}} d\overline{G}_A = \overline{G}_A^\circ + \int_{\substack{\text{standard}\\ \text{pressure}}}^{\substack{\text{experimental}\\ \text{pressure}}} \overline{V}_A\, dP \tag{2-22}$$

where $\overline{G}_A^\circ$ is the constant of integration and represents the chemical potential for the (arbitrarily) defined standard state of the solvent. For a *solvent,* the conventional choice of *standard state is pure solvent* in equilibrium with its vapor, so that the standard pressure becomes the vapor pressure, P_A°, of pure solvent. Moreover, if we suppose that the solvent is essentially incompressible (i.e., that $\overline{V}_A$ is essentially independent of pressure, and may thus be taken outside the integral of Eq. 2-22), we obtain

$$\overline{G}_A = \overline{G}_A^\circ + \overline{V}_A \int_{P_A^\circ}^{P} dP$$

or simply

$$\boxed{\bar{G}_A = \bar{G}_A^\circ + \bar{V}_A\,(P - P_A^\circ) \qquad \text{for incompressible liquid}} \qquad (2\text{-}23)$$

Similarly, for macroscopic changes in the chemical potential of the (ideal gas) vapor, we are already operating at constant temperature (so that T in Eq. 2-21b may be taken outside the following integral)

$$\bar{G}_A = \bar{G}_A^\circ + \int_{\substack{\text{standard}\\\text{pressure}}}^{\substack{\text{experimental}\\\text{pressure}}} RT\, d\,\ln(P) = \bar{G}_A^\circ + RT \int_{\substack{\text{standard}\\\text{pressure}}}^{\substack{\text{experimental}\\\text{pressure}}} d\,\ln(P) \qquad (2\text{-}24)$$

Finally, we define (according to the usual convention) the standard state of an ideal gas as 1 atm pressure, so that Eq. 2-23 may be written more compactly

$$\bar{G}_A = \bar{G}_A^\circ + RT \int_{1\text{ atm}}^{P\text{ atm}} d\,\ln(P) = \bar{G}_A^\circ + RT\,[\ln(P) - \ln(1)] = \bar{G}_A^\circ + RT\,\ln(P/1)$$

or

$$\boxed{\bar{G}_A = \bar{G}_A^\circ + RT\,\ln(P) \qquad \begin{array}{l}\text{constant temperature,}\\ \text{ideal gas, } P \text{ in atmospheres}^*\end{array}} \qquad (2\text{-}25)$$

Returning to the situation shown at the top of Fig. 2-10, it is clear that each solution is in equilibrium with its vapor. In other words,

$$(\bar{G}_A)_{\text{vapor, left}} = (\bar{G}_A)_{\text{liquid, left}} \qquad (2\text{-}26a)$$

and

$$(\bar{G}_A)_{\text{vapor, right}} = (\bar{G}_A)_{\text{liquid, right}} \qquad (2\text{-}26b)$$

so that

$$(\bar{G}_A)_{\text{liq, left}} - (\bar{G}_A)_{\text{liq, right}} = (\bar{G}_A)_{\text{vapor, left}} - (\bar{G}_A)_{\text{vapor, right}} \qquad (2\text{-}27)$$

Substituting Eq. 2-25 for $(\bar{G}_A)_{\text{vapor}}$ in Eq. 2-27 then gives the result

$$\begin{aligned}(\bar{G}_A)_{\text{liq, left}} - (\bar{G}_A)_{\text{liq, right}} &= \bar{G}_A^\circ + RT\,\ln(P_A^\circ) - [\bar{G}_A^\circ + RT\,\ln(P_A)]\\ &= RT\,\ln(P_A^\circ/P_A) = -RT\,\ln(P_A/P_A^\circ)\end{aligned} \qquad (2\text{-}28)$$

* *Note that although P must be expressed in atmospheres for Eq. 2-25 to be valid, the argument of the logarithm is nevertheless unitless because it represents a pressure* ratio—*we shall return to this issue in the next chapter.*

But for an ideal solution, Eq. 2-19 says that $(P_A/P_A^\circ) = X_A$, so Eq. 2-28 becomes

$$(\bar{G}_A)_{\text{liq, left}} - (\bar{G}_A)_{\text{liq, right}} = -RT \ln(X_A)_{\text{right}} \tag{2-29}$$

The importance of Eq. 2-29 is that since $(X_A)_{\text{right}}$ is less than unity, $\ln(X_A)_{\text{right}}$ is a negative number, and we conclude that the *chemical potential* of solvent A on the left-hand side of the membrane is *greater* than on the right-hand side. Thus, just as a rock rolls downhill to decrease its (*mechanical*) potential energy, solvent molecules have a tendency to migrate across a semipermeable membrane to decrease their *chemical* potential.

However, since the chemical potential of the liquid solvent can be *increased* by applying external pressure (Eq. 2-23), we can exactly nullify the tendency for solvent molecules to cross the semipermeable membrane by applying an external additional pressure, π, on the right-hand solution, so as to *increase* the chemical potential of the liquid on the right-hand side of the membrane by an amount exactly equal to the initial chemical potential difference.

$$\begin{aligned}\text{Increase in } (\bar{G}_A)_{\text{liq, right}} \quad &= (\bar{V}_A)_{\text{liq}}(P_A + \pi - P_A^\circ) + (\bar{V}_A)_{\text{liq}}(P_A - P_A^\circ) \\ \text{due to externally applied} &= \pi(\bar{V}_A)_{\text{liq}} \\ \text{additional pressure} &\end{aligned} \tag{2-30}$$

where the external pressure, π, has the property that

$$\boxed{\begin{aligned}\pi(\bar{V}_A)_{\text{liq}} &= -RT \ln(X_A)_{\text{right}} \\ &= \textit{osmotic pressure} \text{ for ideal solution}\end{aligned}} \tag{2-31}$$

We have developed an expression for osmotic pressure as the externally applied pressure necessary to *prevent* solvent migration across a semipermeable membrane—alternatively, one may consider osmotic pressure as a measure of the tendency for solvent migration to *occur* in the absence of external influences.

Since mole fraction of solvent (Eq. 2-31) is an inconvenient measure of concentration, it is useful to express Eq. 2-31 in terms of the concentration of (involatile) solute B. First, since

$$X_A + X_B = 1 \tag{2-32}$$

we may rewrite Eq. 2-31 as

$$\pi(\bar{V}_A)_{\text{liq}} = -RT \ln(1 - (X_B)_{\text{right}}) \tag{2-33}$$

Furthermore, for very dilute solutions

$$X_B << 1 \tag{2-34}$$

and we may approximate (see Appendix)

$$\ln(1 - X_B) \cong -X_B \tag{2-35}$$

to obtain

$$\pi(\bar{V}_A)_{\text{liq}} \cong RT\, X_B = RT \frac{n_B}{n_A + n_B} \cong RT \frac{n_B}{n_A} \tag{2-36}$$

where n_B and n_A are the respective number of moles of solute and solvent in the right-hand solution at lower right of Fig. 2-10. Next, we express the number of moles of solvent, n_A, in terms of the mass of solvent, w_A, and the solvent molecular weight, M_A:

$$n_A = w_A / M_A \tag{2-37}$$

so that Eq. 2-36 now becomes

$$\pi = RT \frac{n_B M_A}{w_A \bar{V}_A} \tag{2-38}$$

Finally, since $\bar{V}_A$ is the volume per *mole* of liquid A

$$\bar{V}_A = \bar{v}_A\, M_A \tag{2-39}$$

where $\bar{v}_A$ is called the "partial specific volume" per *gram* of A, we obtain

$$\pi = RT \frac{n_B}{w_A \bar{v}_A} = RT \frac{n_B}{V_A}$$

or more simply

$$\boxed{\pi = m\, RT \qquad \begin{pmatrix}\text{ideal} \\ \text{solution}\end{pmatrix}} \tag{2-40}$$

where we have recognized that the units of the gas constant (R) are l atm mol^{-1} K, so that the appropriate value of w_A for water is 1 kg to give a solvent volume, V_A, of 1 liter, and m is thus the "volume molal" concentration (i.e., moles solute per liter of solvent).

Equation 2-40 has formal similarity to the ideal gas law, $P = (n/V)RT$. However, Eq. 2-40 is valid only for dilute ideal solutions, for which the approximations of Equations 2-19, 2-35, and 2-36 apply. The principal biological significance of Eq. 2-40 is that it shows how very substantial pressure differences can develop across semipermeable (e.g., living cell) membranes when relatively small concentrations of large molecules are trapped on one side: a concentration difference of about 0.1 M leads to an osmotic pressure difference of about 2½ atm! Osmotic pressure provides much of the tremendous force required to burst open the protective coat of a wet germinating seed. Osmotic pressure contributes to the drawing of water from the soil into the roots and stems of plants. A convenient means for breaking open human blood (or other) cells is to immerse the cells in a solution of much lower osmotic pressure ("hypotonic" solution), so that the cells take up water until they burst. As a corollary, it is now clear why the solute concentration in medical injections should be adjusted to give the same osmotic pressure as the blood itself ("isotonic" solution) to avoid blood cell swelling or shrinking *in vivo.* We next proceed to some analytical uses for osmotic pressure in determining macromolecular size.

Molecular Weight from Osmotic Pressure Measurements

From Eq. 2-40, it is clear that a measurement of osmotic pressure, π, at known temperature suffices to determine the *molal* concentration of an involatile solute trapped on one side of a semipermeable membrane; thus, knowledge of the *weight* concentration of the same solute provides enough information to determine the *molecular weight* of the solute.

EXAMPLE There are 80 mg of an unknown protein dissolved in 1.00 ml of water. The osmotic pressure of this solution at 25°C is found to be 12 torr. What is the molecular weight of the unknown protein?

$$\pi = m\,RT \qquad (2\text{-}40)$$

$$\left(\frac{12}{760}\right)\text{atm} = \left(\frac{80\text{g}}{1}\right)\left(\frac{1}{M}\text{ mole/gram}\right)(0.08206\text{ l atm mole}^{-1}\text{ K}^{-1})(298\text{ K})$$

Solve for

$$M = 124{,}000$$

Because pressures much smaller than that of this example are difficult to measure accurately, it is clear that rather large amounts of solute are needed in order to conduct an osmotic pressure determination of molecular weight, particularly in comparison to the new electrophoretic techniques (see Section 2). Osmotic pressure determination of molecular weight is, however, historically important—it was the means by which the molecular weight of hemoglobin was shown to be 67,000, four times larger than the molecular weight per iron molecule (i.e., molecular weight per monomeric unit) in 1925. Osmotic pressure is also useful in characterizing the *polydispersity* of a mixture of macromolecules of different molecular weight, as discussed below.

Types of Average Molecular Weight in Polydisperse Systems

Suppose we were to measure the osmotic pressure of a mixture of three types of macromolecules: a_1 grams of molecular weight M_1, a_2 grams of molecular weight M_2, and a_3 grams of molecular weight M_3. The observed osmotic pressure would consist of a sum of the contributions from each of these species:

$$\pi = m_1RT + m_2RT + m_3RT = \left(\frac{(a_1/M_1)}{V} + \frac{(a_2/M_2)}{V} + \frac{(a_3/M_3)}{V}\right)RT \quad (2\text{-}41)$$

However, if we did not know that the mixture was heterogeneous, we would identify a single (average) molecular weight, M_n, for a total amount, $(a_1 + a_2 + a_3)$, of dissolved macromolecule. We will now relate this average molecular weight to the individual amounts and molecular weights of the components of the mixture:

$$\pi = \frac{(a_1 + a_2 + a_3)/M_n}{V}RT = \frac{(a_1/M_1) + (a_2/M_2) + (a_3/M_3)}{V}RT$$

$$M_n = \frac{a_1 + a_2 + a_3}{(a_1/M_1) + (a_2/M_2) + (a_3/M_3)}$$

$$= \frac{\left(\frac{a_1}{M_1}\right)M_1 + \left(\frac{a_2}{M_2}\right)M_2 + \left(\frac{a_3}{M_3}\right)M_3}{(a_1/M_1) + (a_2/M_2) + (a_3/M_3)}$$

$$= \frac{N_1M_1 + N_2M_2 + N_3M_3}{N_1 + N_2 + N_3}$$

$$M_n = \frac{\sum_i N_iM_i}{\sum_i N_i} = \text{Number-average molecular weight} \quad (2\text{-}42)$$

in which N_i is the number of moles of the ith component in the mixture.

As we will find in Section 2, it is important to note that the apparent (average) molecular weight determined by osmotic pressure measurement (Eq. 2-42) is not the same as the apparent (average) molecular weight that would be calculated for the *same mixture* based on diffusion (Eq. 2-43) or certain sedimentation experiments (Eq. 2-44).

$$M_w = \frac{\sum_i N_iM_i^2}{\sum_i N_iM_i} = \text{Weight-average molecular weight} \quad (2\text{-}43)$$

$$M_z = \frac{\sum_i N_iM_i^3}{\sum_i N_iM_i^2} = \text{“}z\text{”-average molecular weight} \quad (2\text{-}44)$$

Table 2-2 Types of (Average) Molecular Weight Obtained from Various Measurements

Measurement	Type of Average Molecular Weight (see Equations 2-42 to 2-44)
Osmotic pressure	M_n
Chemical analysis	M_n
Translational diffusion constant	M_w
Sedimentation velocity and diffusion	M_w
Light scattering	M_w
Sedimentation equilibrium	M_w, M_z
Viscosity	M_v

$M_n < M_v \leq M_w < M_z$

A brief summary of the type of average molecular weight obtained from each of several kinds of experiments is listed in Table 2-2.

EXAMPLE A mixture of macromolecules consists of 375 mg of molecular weight 48,000, 500 mg of molecular weight 67,000, and 220 mg of molecular weight 300,000. Compare the (average) molecular weights that would be deduced from osmotic pressure, diffusion, and sedimentation equilibrium experiments.

$$M_n = \frac{\left(\frac{375}{48{,}000}\right)(48{,}000) + \left(\frac{500}{67{,}000}\right)(67{,}000) + \left(\frac{220}{300{,}000}\right)(300{,}000)}{\left(\frac{375}{48{,}000}\right) + \left(\frac{500}{67{,}000}\right) + \left(\frac{220}{300{,}000}\right)} = 68{,}400$$

$$M_w = \frac{\left(\frac{375}{48{,}000}\right)(48{,}000)^2 + \left(\frac{500}{67{,}000}\right)(67{,}000)^2 + \left(\frac{220}{300{,}000}\right)(300{,}000)^2}{\left(\frac{375}{48{,}000}\right)(48{,}000) + \left(\frac{500}{67{,}000}\right)(67{,}000) + \left(\frac{220}{300{,}000}\right)(300{,}000)} = 107{,}000$$

$$M_z = \frac{\left(\frac{375}{48{,}000}\right)(48{,}000)^3 + \left(\frac{500}{67{,}000}\right)(67{,}000)^3 + \left(\frac{220}{300{,}000}\right)(300{,}000)^3}{\left(\frac{375}{48{,}000}\right)(48{,}000)^2 + \left(\frac{500}{67{,}000}\right)(67{,}000)^2 + \left(\frac{220}{300{,}000}\right)(300{,}000)^2} = 194{,}000$$

This example illustrates the general result that $M_z > M_w > M_n$. The three types of average molecular weight give the same value only when the mixture is monodisperse, that is, consists of a single molecular-weight component. All these average molecular weights are sensitive to small amounts of impurities whose molecular weight is substantially different from that of the solute of interest (see Problems).

Nonideal Solutions Just as real gases show *P-V* behavior that deviates from the ideal *gas* law, real solutions exhibit vapor pressure (and thus osmotic pressure) that deviates from the ideal *solution* equation (2-19). Furthermore,

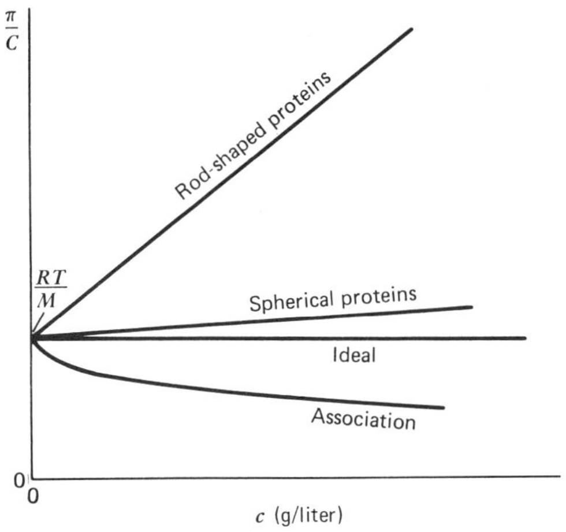

FIGURE 2-11. Deviations from ideality in osmotic pressure equation (Eq. 2-40), as shown by a plot of (π/c) versus c, in which c is the concentration (wt/vol) of macromolecular solute, and π is osmotic pressure. For examples of the various curves, see Problems. (Graph adapted from R. B. Martin, *Introduction to Biophysical Chemistry,* McGraw-Hill Book Co., New York, 1964, p. 111.)

just as any gas approaches ideal behavior in the limit of zero pressure, a solvent tends to become ideal in the limit of infinitely dilute solute, as shown by the convergence of the various curves in Fig. 2-11. Finally, just as deviations from the ideal gas law give information about the size and intermolecular forces between gas molecules, deviations from ideal osmotic pressure behavior give information about solute size and shape and degree of aggregation (see Fig. 2-11 and Problems).

2.C. SEMIPERMEABLE MEMBRANES AND CHARGED SPECIES IN SOLUTION: DONNAN EQUILIBRIUM

In the previous section, we compared the chemical potential of *solvent* on two sides of a membrane permeable to solvent (but not permeable to macromolecular solute), in order to show why solvent molecules tend to migrate from a region of higher solvent concentration (higher solvent chemical potential) to a region of lower solvent concentration (lower solvent chemical potential). In this section, we examine the behavior of small *charged solute* ions that can pass through the same semipermeable membrane in the presence of a charged macromolecular solute. (Small *neutral solute* molecules that can pass through the membrane will simply distribute themselves equally on the two sides of the membrane—see "equilibrium dialysis" example later in this chapter.)

Equation 2-29 (see Section 2.B) shows that the chemical potential, $\overline{G}_A$, of an ideal *solvent* increases with increasing solvent *concentration* according to:

$$d\bar{G}_A = RT\ d\ \ln(X_A) \qquad \text{for ideal solvent} \tag{2-29}$$

in which X_A is the mole fraction (a measure of concentration) of solvent. By similar reasoning based on the dependence of the vapor pressure of a dilute solution upon *solute* concentration, it is possible to describe the variation of *solute* chemical potential with solute *concentration* according to

$$d\bar{G}_B = RT\ d\ \ln(m_B) \qquad \text{for ideal solute} \tag{2-45}$$

in which m_B is the molal concentration (moles solute per kg solvent) of solute. Integrating Eq. 2-45 to describe a macroscopic change in $\bar{G}_B$, we find

$$\bar{G}_B = \bar{G}_{B\begin{pmatrix}\text{standard}\\ \text{state}\end{pmatrix}} + \int_{\substack{\text{standard}\\ \text{state}}}^{\substack{\text{experimental}\\ \text{state}}} d\bar{G}_B = \bar{G}_B^\circ + \int_{\substack{\text{standard solute}\\ \text{molality}}}^{\substack{\text{experimental}\\ \text{solute molality}}} RT\ d\ \ln(m_B)$$

$$= \bar{G}_B^\circ + RT\ \ln[(m_B)_{\text{expt'l}}/(m_B)_{\text{standard}}] \qquad \text{for constant temperature}$$

or

$$\boxed{\bar{G}_B = \bar{G}_B^\circ + RT\ \ln[(m_B)_{\text{experimental}}]} \tag{2-46}$$

for a standard state of $(m_B)_{\text{standard}} = 1$ molal*, constant temperature, ideal solute. Just as Eq. 2-29 formed the basis for our quantitative description of osmotic pressure, Eq. 2-46 leads directly to an explanation for the Donnan equilibrium.

Suppose that a small amount of the (sodium) salt of a (negatively) charged protein is dissolved in a solution on one side of a semipermeable membrane, and a small amount of sodium chloride is dissolved in a solution on the other side. Because there is initially no Cl^- on the left (see Fig. 2-12) and a finite amount of Cl^- on the right, chloride ions will have a tendency to move across the membrane from right to left (see Eq. 2-46). However, to preserve electrical neutrality, the *same* number of sodium ions must accompany the chloride ions in their migration. Therefore, equilibrium will

* *As for Eq. 2-25, we note that although* $(m_B)_{expt'l}$ *is expressed as molal concentration, the argument of the logarithm in Eq. 2-46 is unitless, because it represents a* ratio *of experimental to standard solute molality. The standard state for solute corresponding to Eq. 2-46 is a 1 molal solution, in which the environment of each solute molecule is that of an infinitely dilute solution—namely, that each solute molecule is surrounded only by solvent molecules and not by other solute molecules. Detailed justification of this convention would require several pages of discussion (which may be found in most detailed conventional treatments of thermodynamics, such as Walter J. Moore* Physical Chemistry), *and is omitted.* Use *of Eq. 2-46 does* not *depend on understanding of that standard state convention, as we will try to show in this and later chapters. For present purposes an ideal solute is one for which Eq. 2-46 is valid—nonideality is discussed in the next chapter.*

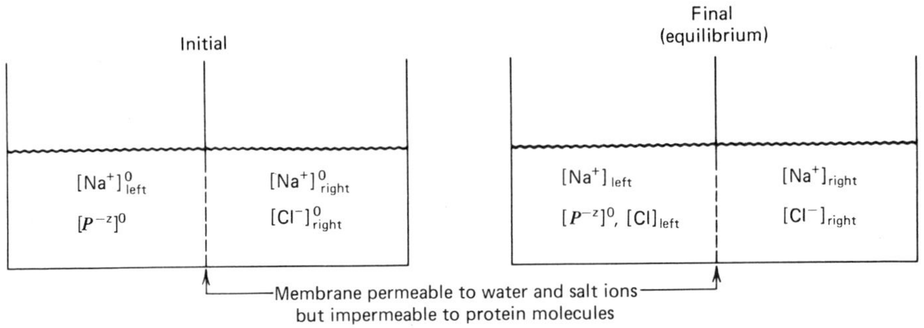

FIGURE 2-12. Donnan equilibrium distribution of ions (right) resulting from initial placement of two solutions of salt and protein on the two sides of a semipermeable membrane (left). At equilibrium, the sodium and chloride concentrations must satisfy Eq. 2-49. (Migration of solvent due to osmotic considerations has been neglected in this diagram.)

be reached when the chemical potential of (dissolved) NaCl is the same on both sides of the semipermeable membrane

$$\boxed{\bar{G}^{\text{left}}_{\text{NaCl}_{(\text{aq})}} = \bar{G}^{\text{right}}_{\text{NaCl}_{(\text{aq})}} \qquad \text{at equilibrium}} \tag{2-47}$$

Since $NaCl_{(aq)}$ is completely dissociated into Na^+ and Cl^- ions, it would be more appropriate to rewrite Eq. 2-47, recognizing that $\bar{G}_{NaCl_{(aq)}} = \bar{G}_{Na^+} + \bar{G}_{Cl^-}$

$$\bar{G}^{\text{left}}_{\text{Na}^+} + \bar{G}^{\text{left}}_{\text{Cl}^-} = \bar{G}^{\text{right}}_{\text{Na}^+} + \bar{G}^{\text{right}}_{\text{Cl}^-} \tag{2-48}$$

Substituting Eq. 2-46 for each of the terms in Eq. 2-48 gives

$$\bar{G}^\circ_{\text{Na}^+} + RT \ln m^{\text{left}}_{\text{Na}^+} + \bar{G}^\circ_{\text{Cl}^-} + RT \ln m^{\text{left}}_{\text{Cl}^-} = \bar{G}^\circ_{\text{Na}^+} + RT \ln m^{\text{right}}_{\text{Na}^+} + \bar{G}^\circ_{\text{Cl}^-} + RT \ln m^{\text{right}}_{\text{Cl}^-}$$

$$RT(\ln m^{\text{left}}_{\text{Na}^+} + \ln m^{\text{left}}_{\text{Cl}^-}) = RT(\ln m^{\text{right}}_{\text{Na}^+} + \ln m^{\text{right}}_{\text{Cl}^-})$$

$$\ln(m^{\text{left}}_{\text{Na}^+}\, m^{\text{left}}_{\text{Cl}^-}) = \ln(m^{\text{right}}_{\text{Na}^+}\, m^{\text{right}}_{\text{Cl}^-})$$

or just

$$m^{\text{left}}_{\text{Na}^+}\, m^{\text{left}}_{\text{Cl}^-} = m^{\text{right}}_{\text{Na}^+}\, m^{\text{right}}_{\text{Cl}^-}$$

usually written as

$$\boxed{\frac{[\text{Na}^+]_{\text{left}}}{[\text{Na}^+]_{\text{right}}} = \frac{[\text{Cl}^-]_{\text{right}}}{[\text{Cl}^-]_{\text{left}}} \qquad \text{at equilibrium}} \tag{2-49}$$

Equation 2-49 is a form of the Donnan equilibrium. When polyvalent ions are present, the same sort of derivation may be used to yield more general expressions

$$\left(\frac{[M^{+3}]_{\text{left}}}{[M^{+3}]_{\text{right}}}\right)^2 = \left(\frac{[X^{-2}]_{\text{right}}}{[X^{-2}]_{\text{left}}}\right)^3, \text{ and so on (see Problems)} \qquad (2\text{-}50)$$

EXAMPLE *Sample Calculation for Donnan Equilibrium Concentrations*

The initial concentration of $[Na_4P^{-4}]° = 10^{-3} M$ on the left side, and of [NaCl]° is 10^{-2} M on the right side of a semipermeable membrane. Find the final concentrations of all species.

The reader should first convince him(her)self that NaCl will move from right to left in this situation (Eq. 2-49). Let the volumes of solution be the same on both sides of the membrane; then if the NaCl that has moved from right to left is denoted as "x," the equilibrium ionic concentrations must satisfy

$$[Na^+]_{\text{left}} = 4 \times 10^{-3} + x$$
$$[Na^+]_{\text{right}} = 10^{-2} - x$$
$$[Cl^-]_{\text{left}} = x$$
$$[Cl^-]_{\text{right}} = 10^{-2} - x, \text{ to maintain electrical neutrality.}$$

Substituting these expressions into the Donnan equilibrium condition, Eq. 2-49,

$$\frac{4 \times 10^{-3} + x}{10^{-2} - x} = \frac{10^{-2} - x}{x}; \; x = 4.17 \times 10^{-3} M$$

to give

$$[Na^+]_{\text{left}} = 8.17 \times 10^{-3} M$$
$$[Na^+]_{\text{right}} = 5.83 \times 10^{-3} M$$
$$[Cl^-]_{\text{left}} = 4.17 \times 10^{-3} M$$
$$[Cl^-]_{\text{right}} = 5.83 \times 10^{-3} M$$

EXAMPLE *Donnan Effect on Osmotic Pressure Determination of Molecular Weight*

The alert reader will have noted in the preceding numerical example that because some NaCl moved across the membrane to the side where the protein was located, there will be an additional contribution to the osmotic pressure of the protein solution (compared to the salt solution on the other side of the membrane) according to this additional amount of salt. A direct (but tedious) algebraic treatment leads to the following expression for the observed osmotic pressure of the protein solution, measured against the salt solution on the other side of the membrane:

$$\pi = \frac{RT\,c_{\text{protein}}}{V_A^\circ\,M_{\text{protein}}}\left[1 + \frac{z_{\text{protein}}^2\,m_{\text{protein}}^2}{4\,\mu}\right] \qquad (2\text{-}51)$$

in which c_{protein} is the number of grams of protein per kg water, V_A° is the volume of solution containing 1 kg water, M_{protein} is the protein molecular weight, z_{protein} is the protein charge, m_{protein} is the protein molality (moles protein per kg water), and $\mu = 1/2\Sigma\, m_i z_i^2$ is the "ionic strength" of the solution (see Chapter 3.B.1). We need only observe that the second term of Eq. 2-51 will lead to a positive deviation in a plot of (π/c) versus c (see Fig. 2-11).

The net result is that for any finite protein concentration, the observed osmotic pressure will be *larger* than expected, since the (charged) protein effectively draws additional salt into its own compartment. The problem is that if we were unaware that the protein molecule was charged, we would suppose that the osmotic pressure was due only to protein molecules, and our calculated number-average (Eq. 2-42) molecular weight would be much *smaller* than the actual protein molecular weight. According to Eq. 2-51, this error increases as the square of protein charge and is inversely proportional to the concentration of added salt in the initial solution. Therefore, in order to minimize such errors, it is desirable to conduct experiments at relatively *high salt* concentration (see Problems) and as close as possible to the *isoelectric pH* (neutral charge) of the protein.

The same sorts of corrections appear in other kinds of solution measurements, such as sedimentation and electrophoresis, and can lead to apparent molecular weights that are more than an order of magnitude too low!

EXAMPLE *Dialysis and Its Uses (Fig. 2-13)*

A variety of experiments derive from the placement of a macromolecular solution inside a bag whose walls are semipermeable (i.e., impermeable to macromolecules but permeable to water and small neutral or charged species). The simplest experiment, "dialysis," consists of changing the composition (except for macromolecule) of the solution within the dialysis bag, by bathing the bag with a frequently (or even continuously) replaced external solution. The solution within the bag will eventually possess essentially the same (except for Donnan effects for charged species) composition as the outside solution, with the macromolecule still trapped within the bag. Since many

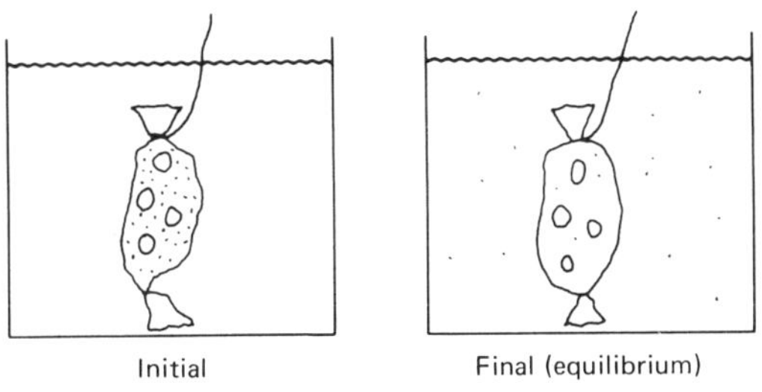

FIGURE 2-13. Schematic diagrams of simple dialysis experiment. A solution containing protein and concentrated salt is placed in a bag whose walls are impermeable to protein but permeable to solvent and small species (left). At equilibrium (right), the small molecules will have distributed essentially equally inside and outside the bag, while the protein will remain inside. The net effect is that most of the salt has been removed from the protein solution inside the bag. By using a large (and/or frequently changed) volume of external solution, it is possible to remove or introduce any desired species that will pass through the membrane. By monitoring the concentrations of small molecules or ions inside and outside the bag, it is possible to quantitate the binding of those species to the protein in the bag (see text).

protein preparations involve an isolation procedure using precipitation with some salt (such as ammonium sulfate), dialysis affords a convenient means for removing the salt from the re-suspended protein. Dialysis also provides a convenient means for changing pH, buffer, or salt concentration in a protein solution. Dialysis of essentially this form is used clinically for removal of metabolic wastes from the blood of kidney patients by passing the blood through an apparatus in which the blood is continuously circulated on one side of a semipermeable membrane, and an isotonic saline solution is continuously refreshed on the other side of the membrane. (One of the functions of a normal kidney is to allow for passive transport of metabolic waste molecules from blood into urine across the semipermeable membrane of the kidney tubules.)

Equilibrium dialysis: binding of small molecules or ions to macromolecules. Suppose that we are interested in measuring the binding constant for binding of, say, aspirin to serum albumin, and that radioactive aspirin is available. We could introduce the radioactive aspirin into a solution in which a bag containing albumin is immersed, as in Fig. 2-13, and let the system come to equilibrium. Ignoring for the moment any charge on the aspirin molecule, we can easily determine the total concentration of aspirin inside and outside the dialysis bag by radioactive counting measurements. Since the radioactivity counter does not distinguish between aspirin free in solution or bound to the albumin, the total aspirin concentration within the bag represents a sum of these two concentrations:

Measured concentrations

$$[\text{Aspirin}]_{\text{inside, total}} = [\text{Aspirin}]_{\text{inside, free}} + [\text{Aspirin:Albumin}]_{\text{inside}}$$
$$[\text{Aspirin}]_{\text{outside, total}} = [\text{Aspirin}]_{\text{outside, free}}$$

Ignoring for the moment any charge on the aspirin molecule, it is approximately correct to equate the *free* aspirin concentrations on both sides of the membrane:

$$[\text{Aspirin}]_{\text{inside, free}} = [\text{Aspirin}]_{\text{outside, free}}$$

since the aspirin molecules will be evenly distributed inside and outside the bag. Finally, we can solve these equations to obtain the free and bound aspirin concentrations inside the bag (i.e., where the protein is), and from some independent measure of total protein concentration (say, from optical absorption at 280 nm), we have,

$$[\text{Albumin}]_{\text{total}} = [\text{Albumin}]_{\text{free}} + [\text{Albumin:aspirin}]$$

But we already know the concentration of bound aspirin, so we now have determined the three concentrations necessary to establish an equilibrium constant for binding of aspirin to albumin:

$$K_{\text{binding}} = \frac{[\text{Albumin:aspirin}]}{[\text{Aspirin}]_{\text{free}}\ [\text{Albumin}]_{\text{free}}}$$

The equilibrium dialysis technique just described provides for determination of binding of any small neutral molecule to a macromolecule, provided that some suitable assay (such as radioactivity in this example) is available for the total concentration of small molecule within the dialysis bag. By taking advantage of the Donnan condition, the technique can also be used to study binding of ions (see Problems).

PROBLEMS

1. Distillation is a good example of the use of phase diagrams. For a two-component system of toluene and benzene (diagram (*a*) below), the phase diagram is similar to the phospholipid phase diagram of Fig. 2-4.
   (a) Compute the number of degrees of freedom for each region and line in the diagram.
   (b) Beginning with a liquid solution of composition shown at point *A* in diagram (*a*), suppose we collect the vapor and condense it by lowering the temperature of the vapor. Then, boil the new liquid, collect the *vapor*, and condense as before. Sketch these steps on the phase diagram and show that the net effect will be to produce pure benzene after many such steps. This process is called "fractional" distillation, and is the means by which many of the components of petroleum are separated.
   (c) Sketch the same procedure, beginning at point *B* of diagram (*b*), this time collecting the *liquid* at each stage, to show that in this case the final liquid will reach a constant composition that is neither pure acetone nor pure chloroform. This mixture is called an *azeotrope,* and its composition will remain unchanged by further distillation. Water and ethyl alcohol form an azeotrope whose composition is 95.6% ethanol—it is thus impossible to obtain ethanol more concentrated than 95.6% by repeated distillation of dilute ethanol-water mixtures.

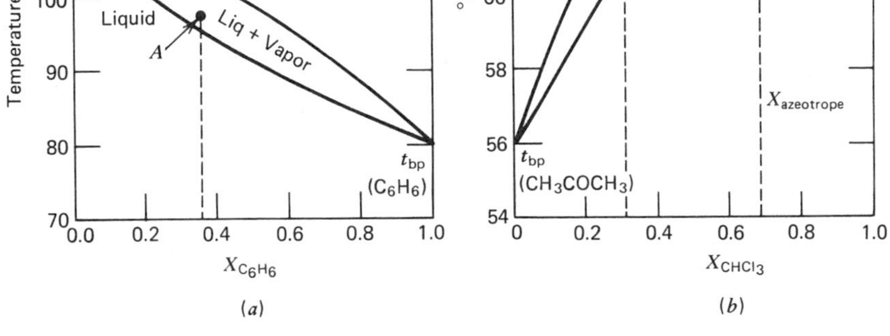

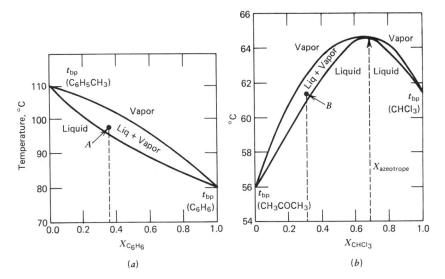

Temperature-composition phase diagrams for binary mixtures of benzene and toluene (*a*), and acetone and chloroform (*b*). (Data from *International Critical Tables,* McGraw-Hill Book Co., N. Y., 1926–1930.)

2. A certain mixture of proteins consists of 2.5 mg of a protein of molecular weight, 20,000; 1.5 mg of molecular weight 50,000; and 1.0 mg of molecular weight, 100,000. What would be the (average) molecular weight of this mixture, as determined from
   (a) osmotic pressure measurements
   (b) translational diffusion constant measurements
   (c) sedimentation equilibrium measurements?

3. Human blood has an osmotic pressure of about 7.7 atm at 40°C. Assuming ideal solution behavior,
   (a) What is the total concentration of solutes in the blood?
   (b) What would the osmotic pressure be at 4°C?

4. Just as the ideal gas law is based on gas molecules which behave as point masses with no intermolecular attractive forces, the formally analogous osmotic pressure equation 2-40, $\pi = mRT$, also applies to solutes that occupy negligible volume and for which solute-solute interactions may be neglected. Therefore, by analogy to the van der Waals equation for nonideal gases,

$$\left(P + \frac{a}{V^2}\right)(V - b) = RT \text{ for one mole of gas}$$

   in which b = excluded (nonaccessible) volume due to finite size of gas molecules, and a = constant related to attractive forces between gas molecules, develop an expression for nonideal osmotic pressure of the form,

$$\frac{\pi}{RTc} = \frac{1}{M} + \frac{Bc}{M^2}, \text{ where } B = \left(b - \frac{a}{RT}\right), c = \text{macromolecular concentration in } \frac{g}{l}$$

   Use this result to interpret the nonideal osmotic pressure behavior shown in Fig. 2-11.

5. Suppose that a polymer preparation consists of 40 grams of polymer of molecular weight, 100,000, contaminated by 0.04 grams of ethyl acrylic acid impurity (molecular weight, 100).
   (a) Calculate the average molecular weight of this mixture as determined by osmotic pressure measurement.
   (b) Considering that this preparation is composed of 99.9% polymer by weight, what does your result suggest about the importance of high purity in molecular weight determinations of macromolecules?

6. For the Donnan equilibrium (Eq. 2-49) example in the text, the salt present was NaCl.
   (a) Extend the Donnan equilibrium calculation to a case where the following ions are present: Na^+, K^+, Cl^-, and NO_3^-. *Hint:* consider the migration of each possible salt formed from these ions.
   (b) Suppose that polyvalent ions are present (e.g., $La_2(SO_4)_3$). Derive the Donnan equilibrium relation between the equilibrium concentrations

of cations and anions on both sides of a semipermeable membrane (Fig. 2-12, left), where the salt has the general formula, M_rX_s.

7. (a) At time zero, the two solutions shown below are placed on the two sides of a semipermeable membrane, where Prot$^-$ is a singly charged protein species. Neglecting any migration of solvent, calculate the equilibrium concentrations of Mg^{++} and Cl^- on both sides of the membrane.

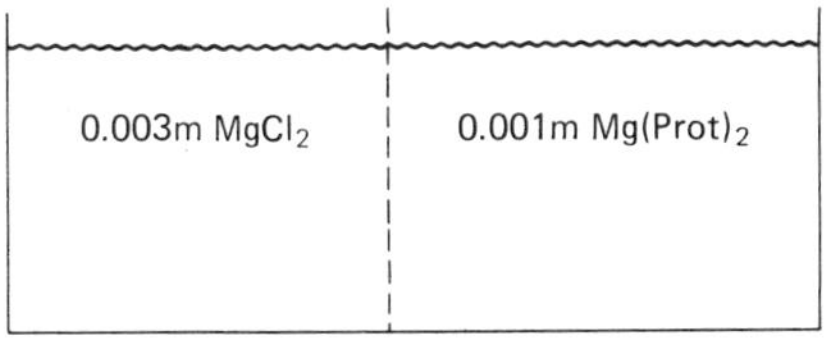

(b) Calculate the change (increase or decrease?) in osmotic pressure in the right-hand compartment, compared with the osmotic pressure that would have been observed for a neutral protein at the same concentration.

(c) Finally, if the true molecular weight of the protein is 70,000, what would be the apparent molecular weight determined from osmotic pressure in the right-hand compartment, if we were unaware of salt migration across the membrane?

8. In the equilibrium dialysis method for determining the binding constant for binding of a small molecule to a macromolecule as described in the text, it was assumed that the small molecule was neutral, so that its concentration inside or outside the semipermeable membrane is the same at equilibrium. Suppose now that we are interested in measuring the strength of binding of iodide ion to albumin:

$$\text{Albumin}^+ + \text{I}^- \rightleftharpoons \text{Albumin:I}; \qquad K_B = \frac{[\text{Albumin:I}]}{[\text{Albumin}^+][\text{I}^-]}$$

in which we have taken the binding stoichiometry as 1:1 for simplicity. Suppose that radioactive iodide is available and that we can measure the pH inside and outside the semipermeable membrane bag containing the albumin, and outline a step-by-step procedure for obtaining K_B from appropriate experimental measurements. *Hint:* Use the Donnan equilibrium condition.

REFERENCES

W. J. Moore, *Physical Chemistry,* any edition, Prentice-Hall, Englewood Cliffs, N. J. The most readable detailed treatment of chemical potential.

J. G. Morris, *A Biologist's Physical Chemistry,* 2nd ed., Edward Arnold, London (1974). Traditional treatment of thermodynamics of solutions.

K. E. van Holde, *Physical Biochemistry* Prentice-Hall, Englewood Cliffs, N. J. (1971). Good treatment of osmotic and Donnan effects.

R. B. Martin, *Introduction to Biophysical Chemistry,* McGraw-Hill, N. Y. (1964). Good section on osmotic pressure.

H. M. McConnell, "Molecular Motions in Biological Membranes," in *Spin Labeling,* edited by L. J. Berliner, Academic Press, N.Y. (1976), pp. 525–560. Good up-to-date review of phase transition phenomena in biological membranes.

CHAPTER 3 Chemical Reactions and Equilibrium Constants

In this chapter, we shall first apply our knowledge of chemical potential to show that equilibrium constants should exist. We then explain why "apparent" equilibrium constants (i.e., those constructed from molar or molal concentrations of reactants and products) seem to vary with concentration, and how that variation can be exploited in the isolation and purification of macromolecules using "salting-in" or "salting-out" procedures. Finally, we examine in some detail the particularly common and interesting equilibrium that describes the binding of one or more small molecules (neutral metabolites, ions, or protons) to macromolecules, and some useful ways (titration curves, Scatchard plots, etc.) for extracting the number of binding sites and the binding constants for binding at various sites.

3.A. EXISTENCE OF EQUILIBRIUM CONSTANTS

From our previous general expression for small changes in *total* free energy

$$dG = V\,dP - S\,dT + \sum_i \overline{G}_i\,dn_i \tag{2-12}$$

it is clear that for constant temperature and pressure,

$$dG = \sum_i \overline{G}_i\,dn_i, \text{ constant } T \text{ and } P \tag{3-1}$$

and thus

$$\boxed{G = \sum_i n_i\,\overline{G}_i} \quad \text{at constant } T \text{ and } P \tag{3-2}$$

Equation 3-2 simply states that the *total* free energy for a mixture is the number of moles of each component times the free energy *per mole* of that component (i.e., chemical potential) *at the stated concentration,* summed over all components of the mixture. We now need only know how $\overline{G}_i$ varies with concentration of the ith component:

$$\overline{G}_i = \overline{G}_i^\circ + RT\ln(X_i) \quad i = \text{(ideal) solvent} \tag{3-3a}$$
$$\overline{G}_i = \overline{G}_i^\circ + RT\ln(m_i) \quad i = \text{(ideal) solute (liquid sol'n)} \tag{3-3b}$$
$$\overline{G}_i = \overline{G}_i^\circ + RT\ln(P_i) \quad i = \text{(ideal) gas} \tag{3-3c}$$
$$\overline{G}_i = \overline{G}_i^\circ + RT\ln(X_i) \quad i = \text{(ideal) solute (solid sol'n)} \tag{3-3d}$$

in which X_i is mole fraction, m_i is molal concentration, and P_i is (partial) pressure. Equation 3-3a was obtained as Eq. 2-29 in the previous chapter, based on behavior of solvent in an "ideal" solution, namely, one for which Eq. 2-19 applies. We will shortly be able to deduce Equations 3-3b, 3-3c, and 3-3d as valid for infinitely dilute solutes or gases (see "activity" discussion 3.A.2.); in those limits, these equations are valid and define "ideal" behavior. $\overline{G}_i^\circ$ in each case represents the chemical potential that would be observed at unit concentration (X_i, m_i, P_i) for an ideal solvent, solute, or gas.

For a typical chemical reaction equilibrium situation

$$aA + bB \rightleftharpoons cC + dD \tag{3-4}$$

in which a moles of A react with b moles of B to form c moles of C and d moles of D, we can say that at equilibrium (constant temperature and pressure), the total free energy of the products must be the same as the total free energy of the reactants (Table 2-1 and Eq. 2-14), because otherwise the equilibrium would shift to one side or the other:

$$G_A + G_B = G_C + G_D \tag{3-5}$$

Substituting Eq. 3-2 into Eq. 3-5, we find that

$$a\,\overline{G}_A + b\,\overline{G}_B = c\,\overline{G}_C + d\,\overline{G}_D \tag{3-6}$$

and by further substitution of Eqs. 3-3 into Eq. 3-6, we obtain

$$a[\overline{G}_A^\circ + RT\ \ln(m_A)] + b[\overline{G}_B^\circ + RT\ \ln(m_B)]$$
$$= c[\overline{G}_C^\circ + RT\ \ln(m_C)] + d[\overline{G}_D^\circ + RT\ \ln(m_D)]$$

or

$$c\overline{G}_C^\circ + d\overline{G}_D^\circ - a\overline{G}_A^\circ - b\overline{G}_B^\circ = RT[-c\ \ln(m_C) - d\ \ln(m_D) + a\ \ln(m_A) + b\ \ln(m_B)]$$
$$(G_C^\circ + G_D^\circ) - (G_A^\circ + G_B^\circ) = -RT[\ln(m_C)^c + \ln(m_D)^d - \ln(m_A)^a - \ln(m_B)^b]$$

$$(G^\circ)_{\text{products}} - (G^\circ)_{\text{reactants}} = \Delta G_{rx}^\circ = -\text{RT}\ \ln\left[\frac{m_C^c m_D^d}{m_A^a m_B^b}\right] \tag{3-7}$$

where each of the reactants and products is treated as an ideal solute (Eq. 3-3b).

The principal significance of Eq. 3-7 is that because the left-hand side of the equation is a constant (i.e., independent of concentrations), the right-hand side must also be independent of the total concentration of all solutes. Thus, we may collect the terms in the bracket of Eq. 3-7 as a new "equilibrium" constant, K_{eq}:

$$K_{eq} = \frac{m_C^c m_D^d}{m_A^a m_B^b} \quad \begin{array}{l}\text{Constant } T \text{ and } P, \\ \text{ideal solutes}\end{array} \tag{3-8}$$

Furthermore, it should be clear that if one of the reactants were solvent, gas, or solid solute, we would simply express its respective concentration as mole fraction, partial pressure, or mole fraction in Eq. 3-8. In particular, since solvent is usually present in great excess, its mole fraction is usually very close to unity, and is thus often omitted from equilibrium constants. Similarly, the mole fraction of a solid solute is usually unity except for solid-solid mixtures such as amalgams.

Suppose we begin with the system of Eq. 3-4 at chemical equilibrium (constant T and P). Equation 3-8 then indicates that when an additional amount of one component, say, B is added to the equilibrium mixture, the reactions of Eq. 3-4 will proceed to the left or right (in this case, to the right) until the new equilibrium concentrations of all the components again satisfy Eq. 3-8. Elementary chemistry textbooks are replete with examples of this sort of calculation, and none are given here.

ΔG°_{rx} is the change in free energy that results when a moles of A combine with b moles of B to give c moles of C and d moles of D, where *each* of these components is present in its *standard state*, namely unit concentration (mole fraction, molality, or partial pressure) for ideal solvent, solute, or gas. Real solutions and gases can be appreciably nonideal, however, and so we next consider how to modify the previous treatment to account for nonideal behavior.

3.A.1. Activity: Thermodynamic "Concentration"

Although the equilibrium constant, K_{eq}, defined in Eq. 3-8 is adequate to describe equilibria in *dilute* solutions (particularly for *neutral* solutes), it is found that for more *concentrated* solutions (especially for *ionic* solutes), the right-hand side of Eq. 3-8 varies with the total concentration of the components of the equilibrium. In other words, the equilibrium "constant" is no longer constant! We will first try to account for this phenomenon, and then show how it may be exploited to aid in the separation and purification of macromolecular solutes.

Retracing the steps that led us to define an equilibrium constant by Eq. 3-8, we find that the mole fraction of solvent, X_A, was introduced as an approximation (Eqs. 2-19 and 2.29) to the vapor pressure ratio, (P_A/P_A°), of Eq. 2-28, where P_A is the vapor pressure of the solution, and P_A° is the vapor pressure of pure solvent. Although this approximation is valid in the limit of infinitely dilute solution, it breaks down for more concentrated solutions, where we should use (P_A/P_A°) directly: The mole fraction of solvent, X_A, may thus be thought of as an approximation to the true "thermodynamic concentration" ("activity", a_A), where we may obtain a_A from X_A by applying a suitable "fudge factor" known as the *activity coefficient*, γ_A:

$$a_A = P_A/P_A^\circ = \gamma_A X_A \qquad \text{for solvent } A \tag{3-9a}$$

where

$$\gamma_A \to 1 \text{ as } X_A \to 1 \text{ (infinitely dilute solution)} \tag{3-10a}$$

In other words, the equilibrium "constant" really *is* constant, provided that we use *activity* rather than *concentration* on the right-hand side of Eq. 3-8 (see below).

A similar argument leads to the notion of "activity" as the true thermodynamic measure of concentration of a solute (or gas or solid), a_B, related in a similar way to the ordinary molal concentration (or partial pressure or mole fraction) ratio by a suitable "activity coefficient" as indicated below.*

$$a_B = \gamma_B m_B, \text{ solute (liquid solution)} \tag{3-9b}$$
$$\gamma_B \to 1 \text{ as } m_B \to 0 \tag{3-10b}$$
$$a_B = \gamma_B P_B, \text{ gas} \tag{3-9c}$$
$$\gamma_B \to 1 \text{ as } P_B \to 0 \tag{3-10c}$$
$$a_B = \gamma_B X_B, \text{ solute (solid solution)} \tag{3-9d}$$
$$\gamma_B \to 1 \text{ as } X_B \to 1 \tag{3-10d}$$

Again, all the activities in Eq. 3-9 are unitless, since each represents a *ratio* of concentration relative to some standard state unit concentration (see Chapter 4 for examples). With the use of these new "activities," we can now rewrite Eqs. 3-7 and 3-8 in a form that is *independent* of total solute concentration:

$$\Delta G^\circ_{rx} = -RT \ln(K_a) \qquad \text{for constant } T \text{ and } P \tag{3-11}$$

where

$$K_a = \frac{a_C^c a_D^d}{a_A^a a_B^b} \qquad K_a \text{ is independent of total solute concentration} \tag{3-12}$$

It is important to recognize why the *approximate* definition of an equilibrium constant (Eq. 3-8) is often adequate for actual equilibrium concen-

** In this treatment, we have defined activity as the product of a particular type of concentration and a corresponding activity coefficient. Although our choices for the particular concentration measures used in Eq. 3-9 (namely, mole fraction for solvent or solid concentration, and molarity as solute concentration) are common choices, one could as easily define the activity of a solute, for example, as the product of molality or mole fraction of solute times the corresponding activity coefficient. In every case, unit activity corresponds to a (usually hypothetical) solution in which the actual concentration of solute (molality, molarity, mole fraction) is unity, but in an environment of an ideal solution ($\gamma = 1$). In practice, therefore, it would seem important to choose a consistent convention for defining activities of solute and solvent in constructing equilibrium constants. Fortunately, the problem can often be circumvented by conducting all experiments at constant ionic strength, or in dilute solution, as discussed above.*

tration calculations. First, if a solution is *dilute,* then all the activity coefficients (Eqs. 3-10) approach unity, and $K_{eq} \cong K_a$:

$$\lim_{\text{(infinite dilution)}} K_a = K_{eq} \tag{3-13}$$

Thus, Eq. 3-8 will usually suffice for dilute (0.01 M or less) solutions. Second, since (see 3.C.) activity coefficients vary primarily due to changes in solution ionic strength (at least for charged solutes), the various activity coefficients in Eq. 3-14 may be kept constant by conducting measurements at constant ionic strength, so that the K_{eq} concentration quotient also remains constant:

$$K_a = \frac{a_C^c a_D^d}{a_A^a a_B^b} = \frac{m_C^c m_D^d}{m_A^a m_B^b} \frac{\gamma_C^c \gamma_D^d}{\gamma_A^a \gamma_B^b} = K_{eq} \frac{\gamma_C^c \gamma_D^d}{\gamma_A^a \gamma_B^b} \tag{3-14}$$

Thus, since K_a is constant with concentration, K_{eq} will also be constant with concentration, provided that the various γ_i are kept constant by working at *constant ionic strength.*

Since activity and concentration may differ very substantially (up to a factor of 10 or more for some salt solutions—see Fig. 3-1), we are led at once to seek a method for determining the *activities* of each of the components of an equilibrium mixture in order to construct the true equilibrium constant, K_a. We can obtain the activity of the *solvent* immediately from measurement of the vapor pressure of the solution, according to Eq. 3-9a, as shown in Table 3-1. The activity of the *solute* is obtained from the Gibbs-Duhem equation, which we now derive.

For a system consisting of solvent A and solute B, we can express the total free energy, G, in terms of the free energies of the two components, G_A and G_B

$$G = G_A + G_B = n_A \overline{G}_A + n_B \overline{G}_B \text{ at constant } T,P \tag{3-2}$$

For an infinitesimal change in G, we thus find that

$$dG = n_A d\overline{G}_A + \overline{G}_A dn_A + n_B d\overline{G}_B + \overline{G}_B dn_B \tag{3-15}$$

But from Eq. 3-1, we know that

$$dG = \overline{G}_A dn_A + \overline{G}_B dn_B \tag{3-1}$$

so that by comparing Eqs. 3-15 and 3-1, we conclude that

$$\boxed{n_A d\overline{G}_A + n_B d\overline{G}_B = 0 \qquad \text{(Gibbs-Duhem equation)}} \tag{3-16}$$

Thus, since $\overline{G}_i = \overline{G}_i^\circ + RT \ln(a_i)$, as a generalization of Eqs. 3-3,

Table 3-1 Mole Fractions and Activities of Solvent and Solute in Sucrose Solutions at 50°C. Activity of Water Obtained from Vapor Pressure of the Solution (Equation 3-9a); Activity of Sucrose Then Obtained by Integration of the Gibbs-Duhem Equation (3-18).*

Mole Fraction of Water (solvent) X_A	Activity of solvent Water $a_A = P_A/P_A^\circ$	Mole Fraction of Sucrose solute X_B	Activity of Sucrose $a_B = \gamma_B X_B$
0.9940	0.9939	0.0060	0.0060
0.9864	0.9834	0.0136	0.0136
0.9826	0.9799	0.0174	0.0197
0.9762	0.9697	0.0238	0.0302
0.9665	0.9617	0.0335	0.0481
0.9559	0.9477	0.0441	0.0716
0.9439	0.9299	0.0561	0.1037
0.9323	0.9043	0.0677	0.1390
0.9098	0.8758	0.0902	0.2190
0.8911	0.8140	0.1089	0.3045

* *(See Problems.)*
* *Note that mole fraction (of either solvent or solute) is a good approximation to actual thermodynamic "concentration" (activity) at low solute concentrations, but that the activity and mole fraction become more and more different as the solution becomes more concentrated in solute. (From W. J. Moore,* Physical Chemistry, *Prentice-Hall, Englewood Cliffs, N. J., 3rd ed., 1963, p. 198.)*

$$d\overline{G}_i = RT\ d\ \ln(a_i) \tag{3-17}$$

in our new notation, we can rewrite Eq. 3-16 in the form

$$n_A\ d\ \ln a_A + n_B\ d\ \ln a_B = 0 \tag{3-18}$$

Finally, since we know the composition of a given solution (n_A and n_B), and since we can measure the activity of the *solvent* from Eq. 3-9a, we can use Eq. 3-18 to obtain the activity of the *solute,* as shown in Table 3-1. Solute activities obtained in this way can then be used to compute the *activity coefficient* (Eqs. 3-9b to 3-9d), which can be thought of as a measure of the ratio of the "thermodynamic concentration" (a_B) to the measured concentration (m_B), as shown in Fig. 3-1 for several solutes as a function of solute concentration.

The situation for ionic solutes is the same, but requires some additional notation. Since many ionic compounds are completely dissociated in aqueous solution,

$$MX \rightleftharpoons M^+ + X^- \tag{3-19}$$

the chemical potential of the aqueous (i.e., dissolved) salt, $\overline{G}_{MX(aq)}$ is equivalent to the sum of the chemical potentials of its component dissociated ions:

$$\bar{G}_{MX(aq)} = \bar{G}_{M^+_{(aq)}} + \bar{G}_{X^-_{(aq)}} \tag{3-20}$$

Substituting into Eq. 3-20 from the integrated form of Eq. 3-17, we obtain

$$\bar{G}^\circ_{MX_{(aq)}} + \mathrm{RT}\ \ln(a_{MX_{(aq)}}) = \bar{G}^\circ_{M^+_{(aq)}} + RT\ \ln(a_{M^+_{(aq)}}) + \bar{G}^\circ_{X^-_{(aq)}} + RT\ \ln(a_{X^-_{(aq)}}) \tag{3-21}$$

However, since Eq. 3-20 is true in particular for the (hypothetical) standard state of 1 molal aqueous MX in an ideal solution,

$$\bar{G}^\circ_{MX_{(aq)}} = \bar{G}^\circ_{M^+_{(aq)}} + \bar{G}^\circ_{X^-_{(aq)}} \tag{3-22}$$

combining Eqs. 3-21 and 3-22 gives the simple conclusion

$$a_{MX_{(aq)}} = (a_{M^+_{(aq)}})(a_{X^-_{(aq)}}) \tag{3-23}$$

However, because we cannot form a solution containing only positive or only negative ions, but must always have both present simultaneously, there is

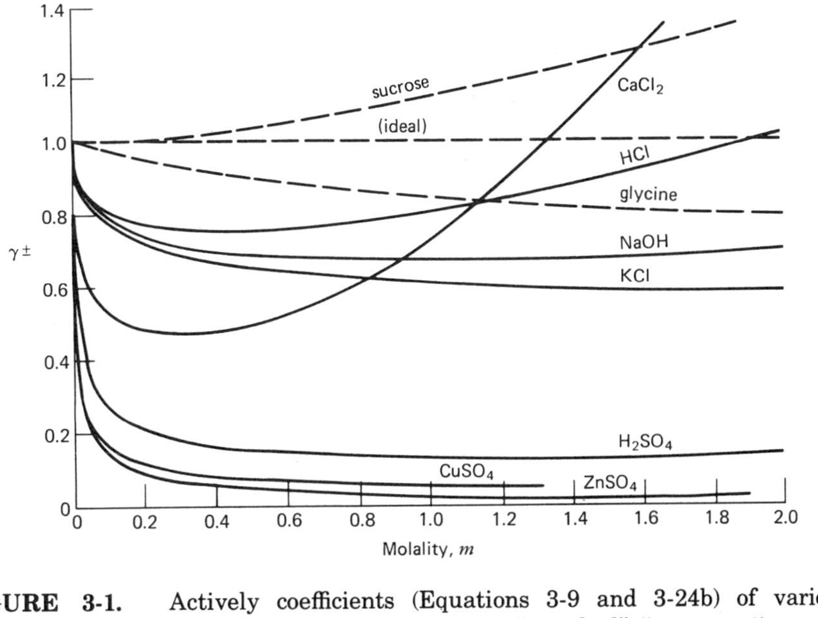

FIGURE 3-1. Actively coefficients (Equations 3-9 and 3-24b) of various solutes. Note that ionic solutes (solid lines) become "nonideal" (i.e., $\gamma_\pm \neq 1$) at relatively low concentrations, compared to neutral solutes (dotted lines). (All data in water at 25°C, taken from G. M. Barrow, *Physical Chemistry for the Life Sciences*, McGraw-Hill, N. Y., 1974, p. 238.)

no way of determining the activity for either $M^+_{(aq)}$ or $X^-_{(aq)}$ separately. The best we can do is to define a (geometric) *mean ionic activity*, $a_\pm$, according to

$$(a_\pm)^2 = a_{M^+} \, a_{X^-} \tag{3-24a}$$

along with

$$(a_\pm)^2 = (\gamma_\pm)^2 \, m_{M^+} \, m_{X^-} \tag{3-25b}$$

in which $\gamma_\pm$ is the *mean ionic activity coefficient;* examples are shown in Fig. 3-1. For more complex electrolytes,

$$M_rX_s \rightleftharpoons rM^{+s} + sX^{-r} \tag{3-25}$$

mean ionic activity and mean ionic activity coefficient may be defined (see Problems):

$$(a_\pm)^{(r+s)} = (a_M{}^{+s})^r (a_X{}^{-r})^s = (\gamma_\pm)^{(r+s)} (m_M{}^{+s})^r (m_X{}^{-r})^s \tag{3-26}$$

3.B. MACROMOLECULAR SOLUBILITY

3.B.1 "Salting-in" and "Salting-out"

The solubility of macromolecules in salt solutions is directly connected to the variation of $\gamma_\pm$ with salt concentration shown in Fig. 3-1. The basic idea is that since an ion of a given charge sign will be surrounded (on the average) by ions of the opposite charge sign,

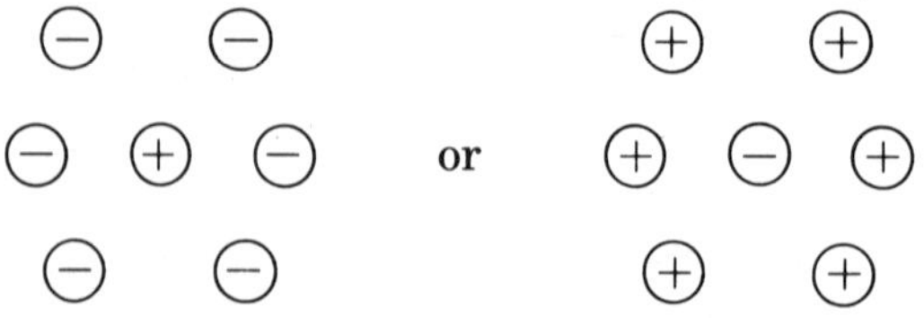

we expect that there will be a net attractive electrical potential energy between the ions. This electrical potential should therefore be added to the usual "ideal" chemical potential in expressing the free energy per mole of dissolved salt molecules:

$$\overline{G}_{MX_{(aq)}} = \left(\overline{G}_{MX_{(aq)}}\right)_{\text{chemical}} + \overline{G}_{\text{elec}}$$

$$\overline{G}_{MX_{(aq)}} = \overline{G}^\circ_{MX_{(aq)}} + RT \ln \left(m_{MX_{(aq)}}\right)^2 + \overline{G}_{\text{elec}} \tag{3-27}$$

Now for an "ideal" solution, $\gamma_{MX_{(aq)}} = 1$, so it is natural to associate any deviations from this condition as arising from the additional electrical ion-ion potential of Eq. 3-27:

$$\overline{G}_{\text{elec}} = RT \ln \gamma_{MX_{(aq)}} \tag{3-28}$$

Since the electrical potential is expected to increase with increased ionic charge and with increased salt concentration (for more concentrated salt solution, the average ion-ion distance will be closer, and the attractive potential varies inversely with ion-ion distance), it should seem reasonable that $\ln \gamma_\pm$ is of the form

$$\log_{10} \gamma_\pm = -A\,|z_+ z_-|\,I^{1/2} \qquad \text{(Debye-Hückel equation)} \tag{3-29}$$

where

$$I = \frac{1}{2} \sum_i m_i z_i^2 = \text{"ionic strength"} \tag{3-30}$$

and A is a constant, whose value is 0.509 for water at 25°C. Equation 3-29 follows from a lengthy, but straightforward calculation by Debye and Hückel of the potential for ion-ion interaction at an average ion-ion distance determined by the salt concentration. The terms z_+ and z_- represent the charges on the positive and negative ions. The "ionic strength," I, may be thought of as a weighted ion concentration average, in which the weight factor is the square of ionic charge. We are now ready to consider the solubility question.

Consider the equilibrium constant ("solubility product") for the dissolving of a (charged) protein, P^{+z}, in water:

$$PX_z \overset{K_a}{\rightleftharpoons} P^{+z} + zX^- \tag{3-31}$$

$$K_a = \frac{a_{P^+}\,(a_{X^-})^z}{a_{PX_z}} = a_{P^+}\,(a_{X^-})^z = K_{\text{S.P.}} \tag{3-32}$$

since the activity of solid PX_z is unity. Using our definition of activity coefficient for ionic solids in solution (Eq. 3-26), Eq. 3-32 becomes

$$K_{\text{S.P.}} = (m_{p^+})(m_{X^-})^z\,\gamma_\pm^{(z+1)} \tag{3-33}$$

If we now dissolve some *other* salt (say, M^+Y^-) in the *same* solution, Eq. 3-29 predicts that $\ln \gamma_\pm$ will become more negative, namely that $\gamma_\pm$ will *decrease*. However, $K_{\text{S.P.}}$ was defined in terms of activities, and is thus a true equilibrium constant. Therefore, if $K_{\text{S.P.}}$ is *constant*, and $\gamma_\pm$ *decreases* when salt is

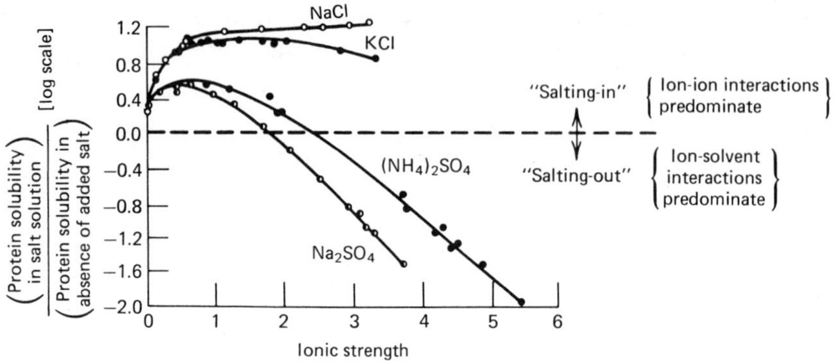

FIGURE 3-2. Relative solubility of carboxyhemoglobin as a function of ionic strength for various inorganic salts. The section of the curves at ionic strengths less than about 0.1 is linear and is described by Eq. 3-29. See text for qualitative explanation of "salting-in" and "salting-out" regions of the curves. (From E. J. Cohn, *Chem. Rev. 19,* 241 (1936).)

added to the solution, the concentrations m_{p^+} and m_{X^-} in Eq. 3-33 (i.e., the protein *solubility*) must *increase* proportionately. This increase in macromolecular solubility with increasing added salt is called "salting-in" (see Fig. 3-2). We have thus accounted for "salting-in," in terms of ion-ion attractive forces in a salt solution. Equations 3-29 and 3-30 result from a treatment valid only in the limit of infinitely small ionic strength—further analysis is required at higher salt concentrations, as we now discuss.

The results in Fig. 3-2 point up an additional effect of added salt at higher salt concentrations, namely an opposite effect resulting in *decreased* protein solubility with *increasing* salt ("salting-out"). The salting-out phenomenon may be explained as due to a "tying-up" of water molecules because of an electrical attraction between the charged *ion* and the electric *dipole moment* of the water molecule:

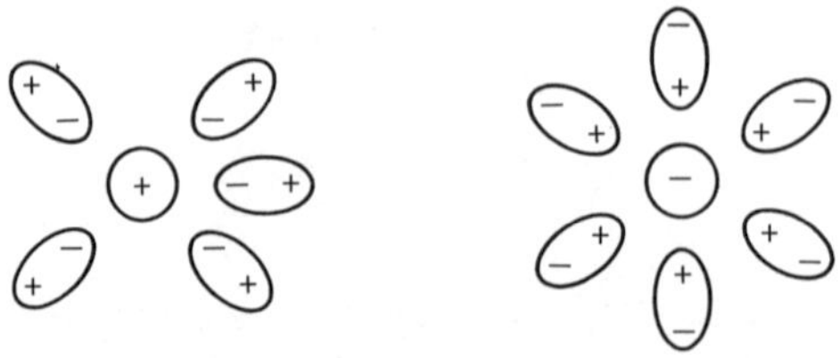

The net result is that water molecules are effectively "removed" from the solution, in terms of their availability for solvating the protein molecules. At high salt concentration, so much water is "tied-up" with salt ions that the effective protein concentration is increased, and the protein precipitates out of solution, as seen in the right-most portions of the curves in Fig. 3-2.

Although no real quantitative description of the salting-out effect is available from analysis of ion-dipole interactions, it is possible to make

Table 3-2 Ionic Radius and Relative Electric Field Strength at the Ionic Surface, for Various Cations and Anions.*

Ion	Ionic Radius	Relative Electric Field at Ion Surface
Li^+	0.60 Å	High
Na^+	0.95	↑
K^+	1.33	|
NH_4^+	1.48	|
Cs^+	1.69	Low
Mg^{++}	0.65	High
Zn^{++}	0.74	↑
Mn^{++}	0.80	|
Ca^{++}	0.99	Low
F^-	1.36	High
Cl^-	1.81	↑
Br^-	1.95	|
I^-	2.16	Low

* *The two parameters are inversely correlated, roughly speaking. For comparison, the water molecule has a mean diameter of about 2 Å.*

some qualitative statements about the results. Figure 3-2 shows that the choice of *anion* ($SO_4^=$ vs. Cl^-) is much more important than the choice of *cation* (Na^+ or K^+) in determining the magnitude of the salting-out effect, and this result is general. In comparing a series of anions or a series of cations, we would expect that the extent to which an ion can "tie-up" water molecules should depend on two (related) parameters: the ionic *size* (bigger ions should be able to accommodate more water molecules around them), and ionic *charge density* (a more highly charged surface should attract water molecules more strongly and thus draw more water molecules toward the ion). Trends in these two parameters are shown in Table 3-2 for various cations and anions. Since there is an inverse correlation between ionic radius and electric field at the ion surface, it is not obvious (in advance) which parameter will have a greater effect on ionic activity, and thus on protein solubility. Once we have the experimental result (e.g., Fig. 3-2), we can say that negative ions have a greater salting-out effect because negative ions are larger than positive ions (as a rule), and can thus accommodate more solvent molecules in the vicinity of the ion. Although after-the-fact explanations are much less useful than prior predictions, the results themselves are of interest. By comparing the relative solubility of a given protein in the presence of a large number of different added salts, it has proved possible to establish an empirical order of effectiveness of various cations and anions in salting-out of proteins (Eqs. 3-34a and 3-34b); this ordering is known variously as the "lyophobic," "lyophilic," or "Hofmeister" series

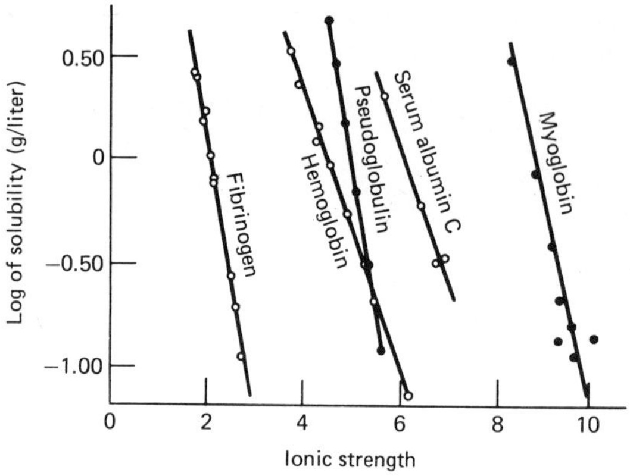

FIGURE 3-3. Selective precipitation of proteins by ammonium sulfate addition. Ionic strength is defined in Eq. 3-30. (After E. J. Cohn and J. T. Edsall, *Proteins, Amino Acids, and Peptides,* Reinhold, N. Y., 1943, p. 602, reprinted by permission of Hafner Publishing Co., N. Y., 1965.)

ANIONS $\text{citrate}^{-3} > \text{tartrate}^{-2} > SO_4^{-2} > \text{acetate}^- > Cl^- > NO_3^- > Br^- > I^- > CNS^-$ (3-34a)

CATIONS $Th^{+4} > Al^{+3} > H^+ > Ba^{+2} > Sr^{+2} > Ca^{+2} > K^+ > Na^+ > NH_4^+ > Li^+$ (3-34b)

From these sequences, we can choose the cation:anion combinations that should be most effective in precipitating proteins from solution.* The most commonly used salting-out compound is ammonium sulfate, because (Eq. 3-34) its effect is *selective* in precipitating *some* proteins from a $(NH_4)_2SO_4$ solution in which *other* proteins in the mixture are still soluble, as shown in Fig. 3-3. For example, Fig. 3-3 shows that an ammonium sulfate solution with an ionic strength of about 6 will precipitate hemoglobin but not myoglobin at the same protein concentration. Thus, ammonium sulfate addition provides a means for isolation of some proteins from a mixture, by selective precipitation, and is a widely used method for protein purification.

One reason for the practical value of the Hofmeister series (Eq. 3-34) is that even though we cannot really predict the ordering in advance, similar series are observed in a wide variety of phenomena other than solubility, including electrophoresis, viscosity, sedimentation, and the like. The reason can be traced to the effect of salt on ion *chemical potential* (which led to the effect of salt on protein solubility); we thus expect to find Hofmeister-type series for any phenomenon involving solutions containing dissolved salt. A variety of equilibrium constants and chemical reaction rate

* *In fact, at the extreme of the series, LiCNS has essentially no salting-*out *effect, and the salting-*in *effect will actually dissolve silk (i.e., protein) stockings in concentrated LiCNS solutions.*

constants (when plotted on a log scale) vary proportionately to the square root of ionic strength, as should seem reasonable from Equations 3-29 and 3-33 for the particular case of solubility equilibrium.

3.B.2. Hydrophobicity: Noncovalent Association between Nonpolar Molecules

With the exception of hydrogen bonds, most molecular associations of interest to chemists involve either covalent bonding or Coulomb attraction between charged species. In molecular biology, however, there are many instances in which essentially nonpolar molecules (i.e., neutral molecules with very small electric dipole moment) associate very strongly with one another when the molecules are in an aqueous environment. Examples include: protein-protein interactions (e.g., association of enzyme subunits to form a multimeric protein—see Chapter 11.B.2), protein folding to give a more functional shape; formation of lipid micelles and bilayer membranes, and the like. In order to understand the origin and nature of such "hydrophobic" ("water-fearing") interactions, it is first necessary to introduce some measure of the strength of the hydrophobic "bonds."

For ordinary covalent chemical bonds, the usual measure of bond strength is the change in *enthalpy* (ΔH) or *energy* (ΔE) on formation of the bond in question from unbonded atoms. The main reason for choosing these parameters is that they are readily measured experimentally from spectroscopy (ΔE), from calorimetry (ΔH), or from the temperature-dependence of equilibrium constants (ΔH). For the large molecules (often of limited volatility or solubility) encountered in biochemistry, heat changes are inconveniently small to measure in dilute solution, and spectra are generally too complicated for ready extraction of bond strength information. Hence, biochemical bond strengths are generally reported as changes in *free energy* (ΔG) or (equivalently—see Chapter 5) changes in electrochemical potential, because ΔG is readily obtained from measurement of equilibrium constant, as discussed in the preceding section.

Therefore, a logical measure of the strength of hydrophobic "bonds" might be the change in (standard) free energy (per mole), $\Delta\overline{G}°$, for transfer of solute from water (a solvent in which the solute molecules are uniformly dispersed in solution) to dioxane (a solvent in which hydrophobic interactions, such as solute:solute aggregation, are present):

$$\begin{matrix}\text{Hydrophobicity per mole} \\ \text{of solute molecules}\end{matrix} = \overline{G}^{\circ}_{\text{in dioxane}} - \overline{G}^{\circ}_{\text{in water}} \qquad (3\text{-}35)$$

Next, since the chemical potential of a given solute *at saturation* is the same in *any* solvent (i.e., as the saturating concentration is approached, the solute has the same tendency to precipitate, independent of the solvent chosen)

$$\bar{G}_{\text{in dioxane}} = \bar{G}_{\text{in water}}, \quad \begin{matrix}\text{saturated solution} \\ \text{in each solvent}\end{matrix}$$

$$\bar{G}^{\circ}_{\text{in dioxane}} + RT \ln[c]_{\substack{\text{at sat'n} \\ \text{in dioxane}}} + RT \ln \gamma_{\substack{\text{at sat'n} \\ \text{in dioxane}}} = \bar{G}^{\circ}_{\text{in water}} + RT \ln[c]_{\substack{\text{at sat'n} \\ \text{in water}}} + RT \ln \gamma_{\substack{\text{at sat'n} \\ \text{in water}}} \quad (3\text{-}36)$$

Since the concentrations in Eq. 3-36 are simply the solubilities of that solute in each of the two solvents, and since activity coefficients will be close to unity for the sparingly soluble (i.e., very dilute) solutes under consideration, we can thus combine Equations 3-35 and 3-36 to obtain

$$\boxed{\begin{matrix}\text{Hydrophobicity per mole} \\ \text{of solute molecules}\end{matrix} = RT \ln \frac{\text{(solubility in water)}}{\text{(solubility in dioxane)}}} \quad (3\text{-}37)$$

The hydrophobicity of Eq. 3-37 is thus readily determined experimentally from solubility measurements. We will now examine some applications of the relative hydrophobicities determined from Eq. 3-37.

EXAMPLE *Hydrophobicity of Amino Acid Side Chains in Proteins*

Equation 3-37 can be used to calculate the hydrophobicities of each of the amino acids that occur in native proteins and enzymes. Since glycine is the basic unit for peptide bonds that form the "backbone" of proteins, the folding of the polypeptide chain to form the particular configuration for a particular protein must be determined by the *side chains* of the remaining amino acids, since those amino acids (except for proline) differ from glycine simply in the

Table 3-3 Hydrophobicity Scale for Amino Acid Side Chains, Obtained by Subtracting the Hydrophobicity of Glycine from the Hydrophobicity of Each of the Listed Amino Acids*

Amino Acid Side Chain	Hydrophobicity
Tryptophan	−3400 cal/mole at 25°C
Norleucine	−2600
Phenylalanine	−2500
Tyrosine	−2300
Dihydroxyphenylalanine	−1800
Leucine	−1800
Valine	−1500
Methionine	−1300
Histidine	−500
Alanine	−500
Threonine	−400
Serine	+300

* *Most values are averages of the (usually similar) values obtained using either dioxane or ethanol as the relatively "nonpolar" solvent. [From Y. Nozaki and C. Tanford,* J. Biol. Chem. 246, *2211 (1971).]*

structure of the side chain branching out from the α-carbon. Thus, if we determine the hydrophobicity for each amino acid, and then subtract the hydrophobicity for glycine in each case, we expect to obtain a measure of the hydrophobicity of each of the individual *side chains* of the various amino acids, as tabulated in Table 3.3.

Table 3.3 shows that there can be a very large gain in free energy ($\Delta\overline{G}^\circ << 0$) on transfer of a nonpolar amino acid residue from an aqueous environment (protein surface) to the relatively nonpolar interior of the protein. The mechanism for this effect is subtle. It is *not* due to a "like-to-like" attraction between the hydrophobic (i.e., hydrocarbon, aromatic) groups themselves; neither is it due to any substantial repulsion between the hydrophobic group and the solvent. It is simply that there are very strong attractive forces between water molecules, which must be disrupted when a solute dissolves in water. For ionic solutes, the loss of these attractions is largely compensated by the replacement of water:water "bonds" by ion:water "bonds." However, for nonpolar (hydrophobic) solutes, no such compensation occurs, and the dissolving of solute is energetically unfavorable, compared to association of the solute molecules with each other so that the maximum number of water:water associations occur. The effect is more readily visualized using the correlation shown in Fig. 3-4, showing that hydrophobicity varies linearly with the accessible surface area of the amino acid side chain in question. The hydrophobicity of amino acid residues in proteins can thus be expressed as about −25 calorie per square angstrom of accessible surface area.

It is possible to use the relation between hydrophobicity and side chain surface area to compute the free energy gained by the association of two proteins or protein chains within a given macromolecule, in which the region of mutual

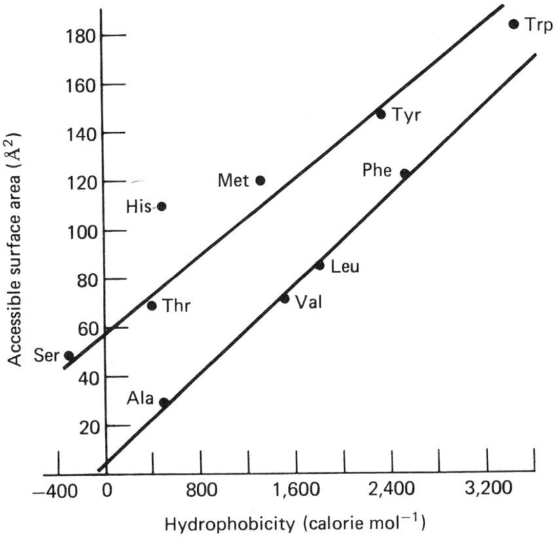

FIGURE 3-4. Accessible surface area (Å^2) of amino acid side chains versus hydrophobicity of that side chain. [From C. Chothia, *Nature 248,* 338 (1974).]

contact has known amino acid composition. These calculations show that only relatively small contact areas are sufficient to account for the observed stability of such aggregates. For example, the subunits of hemoglobin (see Section 3) are held together by contacts involving only about 12% of the accessible surface area of each monomer in the $\alpha_1\beta_1$ contact and only about 6% in the $\alpha_1\beta_2$ contact, in which α and β denote the two types of the four subunits in hemoglobin. Similar calculations can account for the stability of the insulin dimer with respect to its monomers, and for the combination of trypsin enzyme with the polypeptide pancreatic trypsin inhibitor. It thus appears that, to a large extent, it may be possible to account for the complicated forces that hold proteins together simply by considering the *relative areas* of the various side chains and the ways in which they may be juxtaposed.

EXAMPLE *Protein Folding: Theoretical Calculation of the Three-dimensional Structure of Bovine Pancreatic Trypsin Inhibitor (Figures 3-5 and 3-6)*

Perhaps the most spectacular of the successes of this type of reasoning is a recent theoretical calculation in which a particular protein, bovine pancreatic

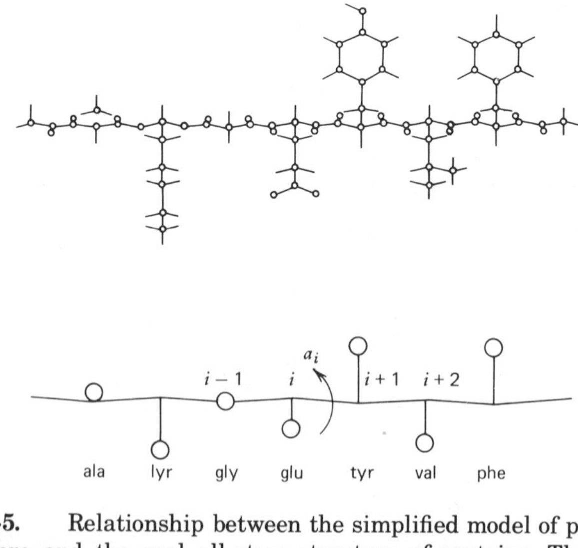

FIGURE 3-5. Relationship between the simplified model of protein structure introduced here and the real all-atom structure of proteins. The two reference points for each residue in the simplified model correspond to the centroid of the side chain and the C^α. Each residue is only allowed one degree of freedom: the torsion angle a between the 4 successive C^αs of residues $(i-1, i, i+1, i+2)$. All the side chains of a given type have the same simplified geometry. The bond lengths, bond angles, and torsion angles used to define the geometry of the simplified molecule were taken as the average values found in eight protein conformations, though they could just as well have been taken from amino-acid model compounds. [From M. Levitt and A. Warshel, *Nature 253,* 694 (1975).]

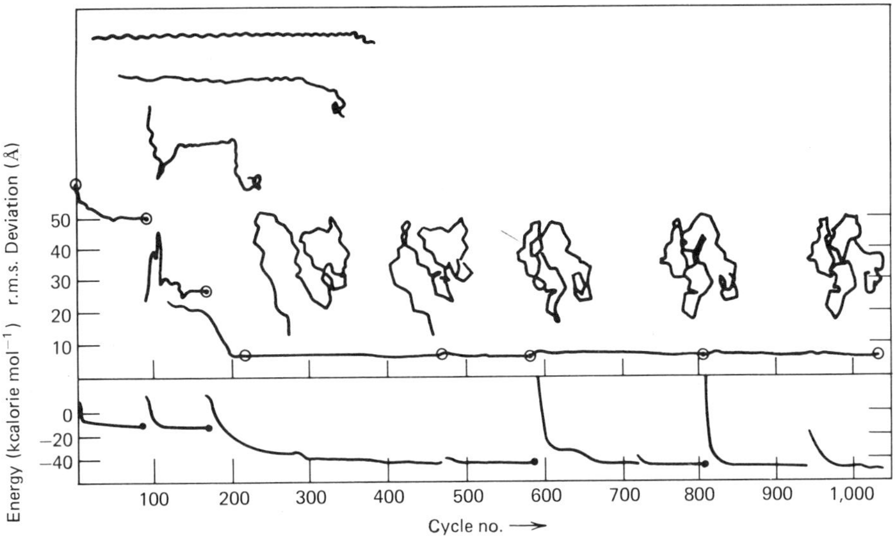

FIGURE 3-6. Calculated conformations of a simplified model (Fig. 3-5) of bovine pancreatic trypsin inhibitor, beginning with the macromolecule unfolded except for the terminal helix at one end (topmost diagram). In each calculational cycle, the computer varies the side chain positions and iterates to find a minimal total free energy. Then the backbone is allowed to move randomly by thermal motion and then frozen, and a new cycle begins. [From M. Levitt and A. Warshel, *Nature 253,* 694 (1975).]

trypsin inhibitor, was represented simply as a polypeptide backbone with side chains that are treated as if all the side chain mass were located at the center of mass of the side chain. (This is thus similar to the preceding example, in which a side chain is described only by its surface area; in other words, it is possible to describe a very complicated structure—750 atoms and 200 bond torsion angles for even a small protein with 50 amino acid residues—with a much simplified model.) Without going into the details of the calculation, which was based on determining the most favorable energy associated with the relative disposition of each possible pair of amino acids, the remarkable result was that by using only about 10 minutes of computer time, one could obtain the single conformation that is energetically most favorable. Moreover, this conformation placed the individual residues in positions that differed by a root-mean-square distance (see Section 2) of less than 3.4 angstrom from the positions occupied by those atoms in the crystalline protein (determined by X-ray crystallography—see Section 6)! Perhaps the most striking feature of this result is that it suggests that an actual protein might be expected to re-form spontaneously back into its functional configuration, even after the protein is first unfolded into a random shape. We give examples of this behavior in Section 2.

EXAMPLE *Micelles and Membranes*

A particular class of aggregates is formed by solute molecules that are "amphiphilic." These molecules have one end that is polar (often even electrically charged) and another end that is nonpolar (often a straight-chain hydrocarbon), and include soaps, detergents, and the phospholipids that give the basic structure to biological membranes. The form of the aggregate is shown below for low concentrations of solute, and is called a "micelle." In this case,

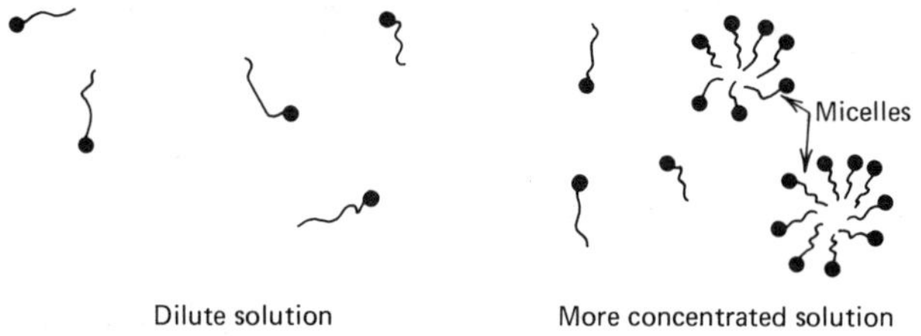

we can invoke hydrophobic effects due to the hydrocarbon nonpolar chains to explain the *formation* of the aggregate (see Fig. 3-7). (However, there also appears to be an *upper limit* to the *size* of the micelle, because there is Coulomb repulsion between the adjacent polar "head" groups in the micelle.) The overall process of micelle formation at increasing concentration of solute is

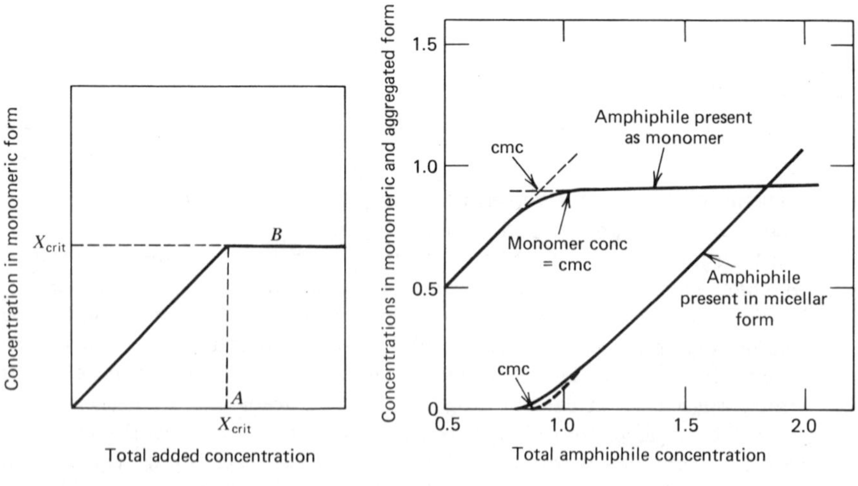

FIGURE 3-7. Relation between monomeric concentration in solution and total added concentration in (*a*) true phase separation and (*b*) micelle formation. The figure shows the concentration of amphiphile present in micellar form as well as the monomer concentration. The dashed lines show empirical procedures for determining the cmc: the point at which the monomer concentration is equal to the cmc is also indicated. Concentration units are arbitrary and could be expressed in grams of amphiphile or in moles based on monomer molecular weight. (From C. Tanford, *The Hydrophobic Effect,* John Wiley & Sons, 1973, p. 48.)

well-approximated as a phase transition (see Chapter 2.A.2), since there is a well-defined change in degree of aggregation of amphiphile (i.e., from pure monomer to monomer plus micelles) above a "critical micelle concentration."

The critical micelle concentration (cmc in Fig. 3-7) represents the minimum solute concentration at which micelles begin to separate out from the otherwise homogeneous solution, and is similar to the solubility (i.e., concentration at which solute begins to precipitate out of solution) that we used in the preceding example as a measure of hydrophobicity. It should thus seem reasonable to obtain a measure of the hydrophobicity of amphiphilic molecules based on critical micelle concentration, just as we defined hydrophobicity based on solubility for neutral species. An example of this new definition is given in Fig. 3-8, showing the variation of hydrophobicity with hydrocarbon chain length for a series of amphiphilic detergent molecules with two hydrocarbon chains per molecule:

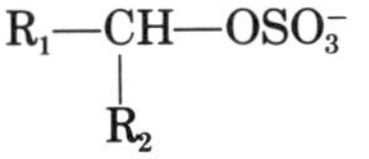

The results (Fig. 3-8) show that a second hydrocarbon tail, added to an amphiphilic molecule already possessing a longer tail, makes a smaller con-

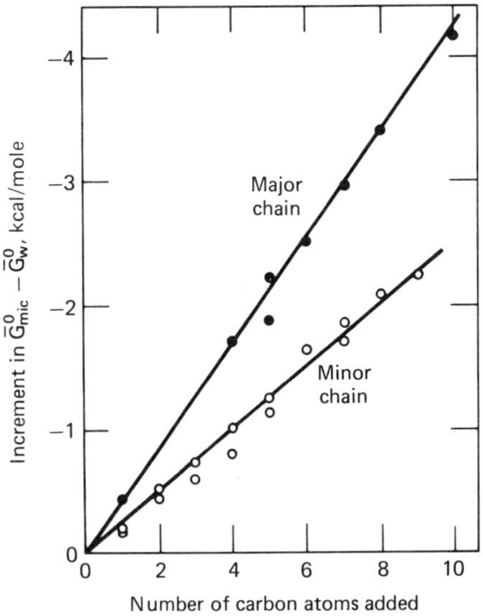

FIGURE 3-8. Incremental increase in hydrophobicity on lengthening the hydrocarbon chains of 1′(alkyl)-alkyl-1-sulfates. Major chain data are for single-chain alkyl sulfates with octyl sulfate as reference; for double-chain derivatives, 1-butyl-decyl sulfate and 1-pentyl-hexyl sulfate as reference. Minor chains data represent a variety of compounds, the parent major chain in each case being the reference compound. Hydrophobicity is determined from critical micelle concentrations in water. (From C. Tanford, *The Hydrophobic Effect,* John Wiley & Sons, N. Y., 1973, p. 53.)

tribution per unit chain length than does the longer chain. However, the hydrophobicity per carbon atom on the longer chain is just as large whether a second chain is present or not. In other words, the two chains on *one molecule* must associate *with each other*—this has very recently been confirmed from the correlated positions and motions of atoms in the two hydrocarbon chains of a biological phospholipid molecule in a membrane bilayer-type aggregate, using magnetic resonance techniques (see Section 4).

3.C. BINDING OF SMALL MOLECULES OR IONS TO MACROMOLECULES

Much of our existing knowledge about the *mechanisms* of chemical and biochemical processes is based on measurement of *equilibrium constants* for *binding* of one molecule to another. In the next section, we will use such data to account for the action of enzymes, drugs, poisons, and hormones. In most of those applications, the basic process is the binding of one or more small molecules, I, to equivalent sites on a given macromolecule, E:

$$E + nI \rightleftharpoons EI_n \tag{3-38}$$

where the equilibrium concentrations satisfy the relation*

$$K_I = \frac{[E][I]}{[EI]} \tag{3-39}$$

For the simplest possible case that $n = 1$ (i.e., just one binding site per macromolecule), there are several graphical means for extracting the desired equilibrium (dissociation) constant, K_I, from experimental data. [*Graphical* data reduction is always preferred to *tabulated* data reduction, since it is easy to identify random or systematic trends in plotted data (particularly when the data plot is designed to give a theoretical straight line), although such trends may not be nearly so evident from tabulated results.] We will begin by reviewing the one-site binding case in some detail, to see how to extend the treatment to many-site binding. Finally, the reason for presenting *several* different graphical means for extracting the same equilibrium constant is that certain measurements lend themselves most directly to certain graphs—for example, a pH meter provides a direct measure of the *logarithm* of H^+ concentration, so the appropriate data reduction ("titration" plot—see below) is plotted as a function of $\log[H^+]$.

** In this section, we use the approximate representation of equilibrium constants in terms of concentrations rather than activities (Eq. 3-8 rather than 3-12) for simplicity. The conditions under which this approximation applies (see discussion following Eq. 3-12) will often be valid for applications of biochemical interest.*

Data Reduction for One-Site Binding

We begin with the one-step, one-site binding situation

$$E + I \rightleftharpoons EI \tag{3-40}$$

approximately characterized by an equilibrium (dissociation) constant, K_I:

$$\boxed{K_I = \frac{[E][I]}{[EI]}} \tag{3-41}$$

It will be convenient to consider the *fraction*, ν, of total binding sites occupied:

$$\boxed{\nu = \frac{[EI]}{[EI] + [E]}} \tag{3-42}$$

The various means for extracting K_I from experimental data are based on various ways of combining Eqs. 3-41 and 3-42; ν itself may often be determined directly from experiments. For example, once the ultraviolet or visible absorbance (Section 4) for a solution of macromolecules exclusively in the pure E (and in the pure EI) form is known, the fraction in the EI form for an unknown solution is readily computed by interpolation between the absorbances for the pure E and pure EI forms. We may also obtain ν from equilibrium dialysis measurements, as discussed in Chapter 2.C.

"Direct" Plot If we solve Eq. 3-41 for $[EI]$ and substitute for $[EI]$ in Eq. 3-42, we obtain

$$\boxed{\nu = \frac{[I]}{K_I + [I]}} \tag{3-43}$$

Equation 3-43 suggests that one means for obtaining K_I is simply a "direct" plot of ν versus $[I]$ at constant $[E]_{total}$. The plot will be a hyperbola, and Eq. 3-43 shows that when exactly *half* the binding sites are occupied (i.e., $\nu = 1/2$), $[I] = K_I$. The direct plot thus provides a means for obtaining K_I from a set of experimental ν-values measured at various choices for $[I]$, as shown in Fig. 3-9*a*. The difficulty with the "direct" plot is that it is nonlinear, and it is thus necessary to obtain data over a wide range in $[I]$ concentration in order to establish the midpoint of the curve from which K_I is obtained. We will return to this point shortly.

"Bjerrum" ("Titration") Plot For the particular case in which $I = H^+$, the most convenient experimental measure of $[H^+]$ is from a pH meter:

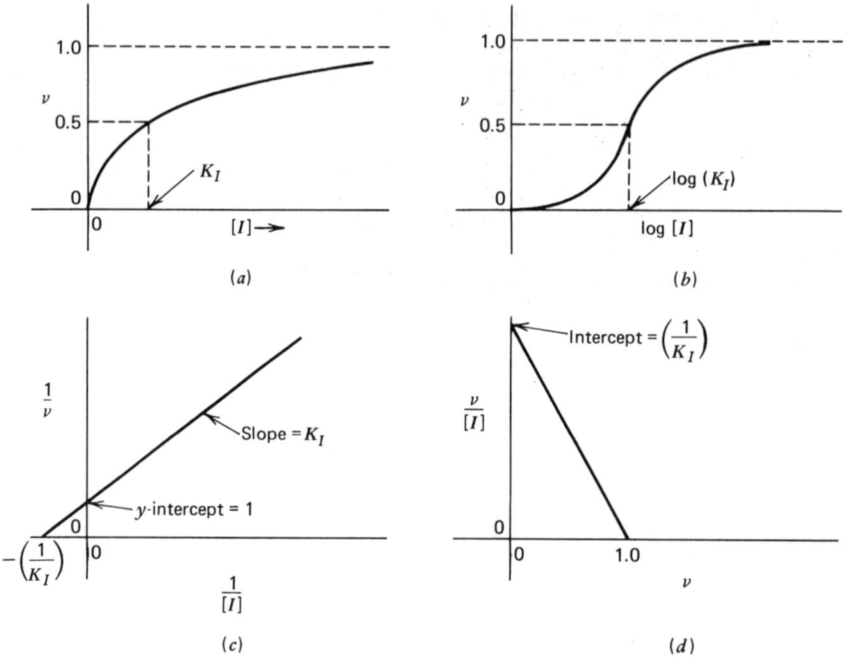

FIGURE 3-9. Comparison of various graphical means for determination of the equilibrium dissociation constant, K_I, for a one-step, one-site binding of a small molecule (or ion), I, to a macromolecule. The plots are designated as follows: (*a*) "direct" plot, based on Eq. 3-43; (*b*) "Bjerrum" or "titration" plot, based on Eq. 3-48; (*c*) "reciprocal" or "Benesi-Hildebrand" plot, Eq. 3-51; (*d*) "Scatchard" plot, based on Eq. 3-52. Although plot (*b*) is probably most familiar to most readers, plot (*d*) is preferred when there are several binding sites per macromolecule, with different binding constants to different sites (see text). A list of the relevant formulas for the various graphical data reductions, generalized to binding to n equivalent sites per macromolecule, is given in Table 3-4.

$$\mathrm{pH} = -\log_{10} a_{\mathrm{H}^+} \cong -\log_{10}[\mathrm{H}^+] \tag{3-44}$$

Thus, in this case, we take the log of both sides of Eq. 3-41 to obtain the familiar Henderson-Hasselbalch equation

$$\mathrm{pH} = pK_a + \log_{10}([E]/[EH^+]) \tag{3-45}$$

in which

$$pK_a = -\log_{10}K_a = -\log_{10}([E][H^+]/[EH^+]) \tag{3-46}$$

More generally, we could solve Eq. 3-43 for K_I,

$$K_I = [I]((1-\nu)/\nu) \tag{3-47}$$

and then take the $\log_{10}$ of both sides to obtain

$$\log(K_I) = \log([I]) + \log((1-\nu)/\nu)$$

or

$$-\log K_I = -\log[I] - \log((1-\nu)/\nu)$$

or

$$-\log[I] = -\log K_I + \log((1-\nu)/\nu)$$

to give

$$pI = pK_I + \log((1-\nu)/\nu) \quad (3\text{-}48)$$

where

$$pI = -\log_{10}[I]$$

and

$$pK_I = -\log_{10} K_I \quad (3\text{-}49)$$

Equation 3-45 (and the more general Eq. 3-49) thus leads to the idea of a "titration" plot of ν versus pH (or more generally, pI), as shown in Fig. 3-9*b*. Just as the ordinary acid dissociation constant, K_a, is determined from the pH at which the acid is half-neutralized,

$$\text{pH} = pK_a \qquad \text{when} \qquad [E] = [EH^+] \quad (3\text{-}50a)$$

we can obtain the dissociation constant for an arbitrary one-site binding equilibrium, K_I, from the small molecule (or ion) concentration at which half the macromolecular binding sites are occupied:

$$pI = pK_I \qquad \text{when} \qquad \nu = 1/2 \quad (3\text{-}50b)$$

The condition of Eq. 3-50 is readily found as the "midpoint" of the "Bjerrum" or "titration" plot of Fig. 3-9*b*.

Finally, we could obtain a straight-line plot of experimental data from Eq. 3-48, by plotting $\log((1-\nu)/\nu)$ versus pI. This plot (see discussion of "Hill" plot at the end of this section) is especially useful for detecting *cooperative* binding.

Straight-Line Plot—"Reciprocal" ("Benesi-Hildebrand") Plot There are several ways to convert the hyperbola of Eq. 3-43 to a straight line. One way is simply to take the reciprocal of both sides of Eq. 3-43 to give

$$\boxed{(1/\nu) = 1 + K_I(1/[I])} \tag{3-51}$$

The desired dissociation constant, K_I, is thus obtained directly from the slope of a plot of $(1/\nu)$ versus $(1/[I])$, as shown in Fig. 3-9c. The advantage of the straight-line plot are: (1) it is easy to obtain a quantitative measure of the precision of the result by examining the scatter of the points about the line, and (2) K_I may be determined without having to obtain data in the *particular* concentration range (i.e., $[I] = K_I$) required in the "direct" or "titration" plot. We will encounter a formally identical situation to Eq. 3-51 in the "Lineweaver-Burk" plot for enzyme kinetics in Section 3.

Straight-Line Plot—"Scatchard" Plot A second way of converting the hyperbola of Eq. 3-43 to a straight line is readily verified to be

$$\boxed{(\nu/[I]) = (1/K_I) - (1/K_I)\nu} \tag{3-52}$$

Thus, a "Scatchard" plot of $(\nu/[I])$ versus ν (see Fig. 3-9d) will produce a straight line with slope, $-(1/K_I)$, and a y-intercept of $(1/K_I)$, where $(1/K_I) = K_B$ = the *binding* constant (rather than the *dissociation* constant) for the one-site binding of Eq. 3-40. The "Scatchard" plot turns out to be especially useful in the extension of this treatment to *multi-site* binding (see below).

The preceding treatment shows that there are several ways of extracting K_I (or equivalently, $K_B = (1/K_I)$) from experimental ν and $[I]$ values. The "direct" plot requires the least data reduction, but gives a hyperbolic curve requiring that data be obtained in a particular range of $[I]$ values. The "titration" curve shares the nonlinear difficulty of the "direct" plot, but is often used either (1) when concentration data is measured directly on a log scale, as with pH, or (2) when the experimental range of $[I]$ values is very large, as with drug-binding studies (see Section 3). The "reciprocal" plot is commonly used for one-site binding, especially in enzyme kinetics (see Section 3), while the "Scatchard" plot is particularly useful in multi-site binding (see below). Finally, even for the one-site binding case, the "Scatchard" plot (Fig. 3-9d) is the graphical display of choice, because it provides optimal weighting of the data obtained at different $[I]$ concentrations. [For a brief but useful discussion of this issue, see: D. A. Deranleau, *J. Amer. Chem. Soc. 91,* 4044 (1969).]

Data Reduction for Many-site Binding: Independent Identical Binding Sites When a given small molecule (ion) may bind to any of several independent equivalent binding sites on a macromolecule,

$$K_I = \frac{[E][I]}{[EI]} \quad \text{for each of } n \text{ independent sites per macromolecule} \tag{3-53}$$

the fraction of any *single* site that is occupied will still be described by Eq. 3-42. For the present case that *each* of the sites on any one macromolecule has the *same* binding constant for small molecule, we may simply add up the number of small molecules bound at each site (Eq. 3-42) to obtain the average number of small molecules bound per macromolecule:

$$\boxed{\nu = \frac{n[I]}{K_I + [I]}} \qquad n = \text{number of identical independent binding sites} \qquad (3\text{-}54)$$

By "independent" binding sites, we mean that binding of a small molecule to any one site does not affect the binding of a second small molecule to another site on the same macromolecule. The form of Eq. 3-54 for many-site binding is similar to that of Eq. 3-43 for one-site binding, and the form of the various graphical displays of Fig. 3-9 is preserved (see dotted lines in Fig. 3-10), as long as all the binding sites on any one macromolecule are identical (see Table 3-4).

Table 3-4 List of the Equations Leading to the Various Graphs of Fig. 3-9 for Extracting the Equilibrium (Dissociation) Constant K_I from Experimental Determination of Fraction of Binding Sites Occupied, ν, as a Function of Concentration, $[I]$, of Free Small Molecule*

Graph Title	Relevant Equations	Graph Construction	Extraction of n and K_I
"Direct" Plot	$\nu = \frac{n[I]}{K_I + [I]}$	Plot ν versus $[I]$ (Hyperbola)	$n = \lim_{[I]\to\infty} \nu$; $K_I = [I]$ for $\nu = \frac{n}{2}$
"Bjerrum" ("Titration") plot	$pI = pK_I + \log\left(\frac{n-\nu}{\nu}\right)$	Plot ν versus $\log[I]$ (Sigmoidal)	$n = \lim_{\log[I]\to\infty} \nu$; $pK_I = pI$ for $\nu = \frac{n}{2}$
"Reciprocal" ("Benesi-Hildebrand") plot	$\frac{1}{\nu} = \frac{1}{n} + \frac{K_I}{n}\left(\frac{1}{[I]}\right)$	Plot $\frac{1}{\nu}$ versus $\frac{1}{[I]}$ (Straight line)	$n = (y\text{-intercept})^{-1}$; $K_I = -(x\text{-intercept})^{-1}$
"Scatchard" plot	$\frac{\nu}{[I]} = \frac{n}{K_I} - \frac{\nu}{K_I}$	Plot $\frac{\nu}{[I]}$ versus ν (Straight line)	$n = x$-intercept; $K_I = \left(\frac{y\text{-intercept}}{n}\right)^{-1}$

* *The equations have been generalized to binding of small molecules (or ions) to n independent equivalent sites per macromolecule, as discussed below.*

$$E + nI \rightleftharpoons EI_n;\ K_I = \frac{[E][I]}{[EI]}$$ *identical, independent binding for each of the n sites per macromolecule*

$$\nu = \frac{n[I]}{K_I + [I]} =$$ *average number of small molecules bound per macromolecule*

For real macromolecules, different small-molecule-binding sites in general have different binding constants. For example, protons have different affinity for different amino acids in a protein. Although we can readily generalize Eq. 3-54 to account for sites of different equilibrium dissociation constant, K_{I_i}, with n_i of the ith type of site per macromolecule

$$\nu = \sum_i \frac{n_i\,[I]}{K_{I_i} + [I]} = \text{mean number of sites per macromolecule occupied by } I \text{ molecules} \tag{3-55}$$

extraction of the various n_i and K_{I_i} from *graphical* display is no longer so easy. Now it is true that we could fit experimental data (ν as a function of $[I]$) to Eq. 3-55 using *numerical* methods, and extract the desired n_i and K_{I_i}, with the use of modern digital computers. (In fact, even when the data can be fitted with a given graphical curve, it is desirable to compute the final parameters numerically, since the numerical calculation is more readily *reproduced* by other investigators.) *Nevertheless,* a *graphical* display is always desirable as a qualitative, diagnostic tool—an "odd" data point can be found quickly from a graph and the measurement repeated, while a few "bad" data points can vitiate the most careful numerical analysis. Also, inclusion of error "bars" for each plotted point makes it easy to decide which data points should be most heavily weighted in making the final "fit" to the data. We now analyze various graphical means for extraction of binding (actually, dissociation) constants and number of each type of binding site on a macromolecule.

Figure 3-10 shows the behavior of each of the usual graphical displays of binding data, for the case of *two* types of binding sites ($K_{I_1} = 0.001$ M and $K_{I_2} = 0.04$ M, where $n_1 = 10$ and $n_2 = 30$ sites of each type). It is clear that of the four usual displays, the Scatchard is best suited for graphical determination of n_1, n_2, K_{I_1}, and K_{I_2}.

In spite of the advantages of Scatchard data reduction as seen in Fig. 3-10, there is one type of binding for which the experimental data are measured directly on a logarithmic scale, namely pH measurements. We now manipulate the "Bjerrum" ("titration") plot into a form more useful for titration data of polybasic macromolecules.

Buffer Capacity Titration curves may more easily be decomposed into the contributions from groups of particular pK_a, by plotting the *slope* of a titration curve versus pH, rather than the titration curve itself. In other words, we begin by defining the *buffer capacity* as the reciprocal of the rate of change of pH of a solution with respect to the addition of strong base to the solution:

$$\text{Buffer capacity} \equiv \beta = \frac{dB}{d(\text{pH})} \tag{3-56}$$

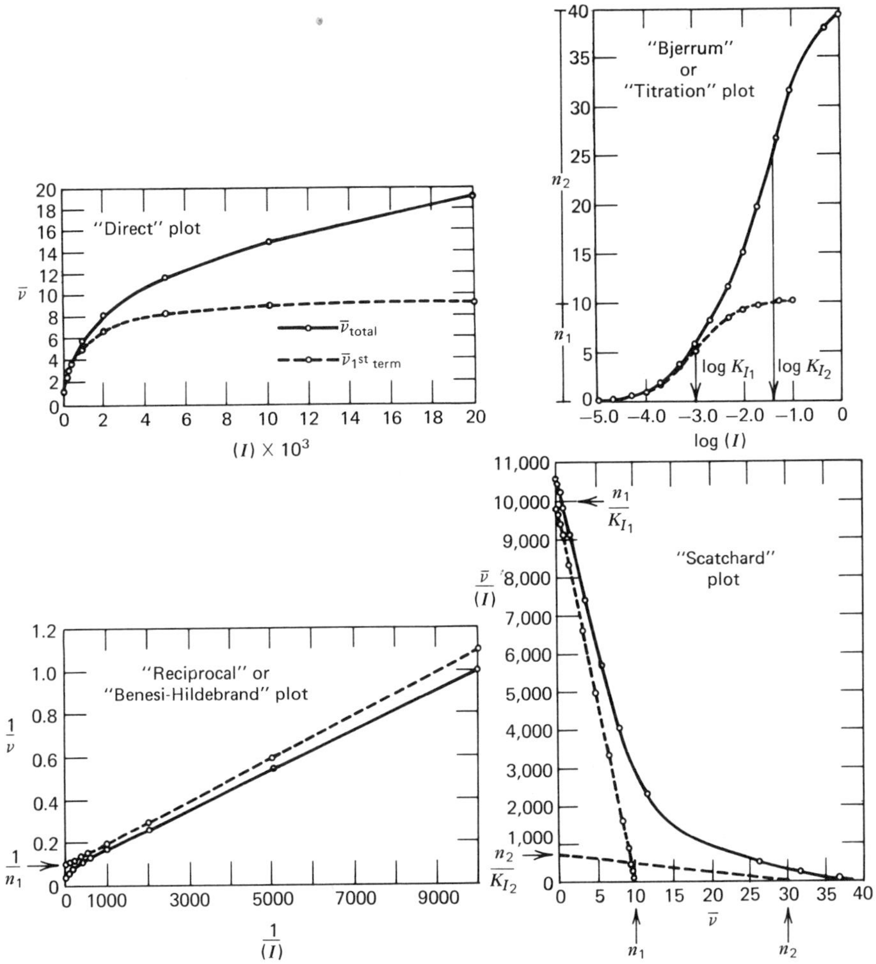

FIGURE 3-10. Comparison of various graphical means for determination of binding constants, and number of sites on a macromolecule that bind a given small molecule with that binding constant. Each dotted curve represents macromolecules with 10 identical independent binding sites, each with equilibrium dissociation constant of 10^{-3} M. The solid curve in each graph denotes macromolecules with the same 10 binding sites with $K_{I_1} = 10^{-3}$ M, and an additional 30 identical independent binding sites with $K_{I_2} = 0.04$ M. It is evident that the Scatchard plot provides the easiest means for determining n_1, n_2, K_{I_1}, and K_{I_2} from experimental data. ν in each plot is the average number of small molecules bound per macromolecule (Eq. 3-55), and $[I]$ is the small molecule concentration at equilibrium. When $I = H^+$, then the "titration" plot is easily constructed from pH measurements: this case is treated under the heading "buffer capacity" in this section. (Graphs adapted from J. T. Edsall and J. Wyman, *Biophysical Chemistry*, Vol. 1, Academic Press, N. Y., 1958, pp. 614, 616–618.)

in which $d(\mathrm{pH})$ is the pH increment induced by the addition of an increment dB of base. Equation 3-56 fits with the usual concept of a buffer: a large "buffer capacity" corresponds to a titration in which a large amount of base must be added to achieve even a small change in pH.

The buffer capacity of water itself is quickly derived from the equilibrium constant for ionization of water:

$$\boxed{K_w = [\mathrm{H^+}][\mathrm{OH^-}]} \qquad (3\text{-}57)$$

from which

$$\log K_w = \log[\mathrm{H^+}] + \log[\mathrm{OH^-}]$$

or

$$-\log[\mathrm{H^+}] = \mathrm{pH} = \log[\mathrm{OH^-}] - \log K_w \qquad (3\text{-}58)$$

For a solution of strong base, say, NaOH, $B = [\mathrm{OH^-}]$ in Eq. 3-56, and since $d(\mathrm{pH}) = d \log[\mathrm{OH^-}]$ from Eq. 3-58, we find that

$$\frac{d(\mathrm{pH})}{dB} = \frac{d \log[\mathrm{OH^-}]}{d\,[\mathrm{OH^-}]} = (1/2.303)\,\frac{d \ln[\mathrm{OH^-}]}{d\,[\mathrm{OH^-}]} = (1/(2.303)[\mathrm{OH^-}])$$

or more simply $dB/d(\mathrm{pH}) = \beta = 2.303[\mathrm{OH^-}]$ for *strong base* added to water. (3-59a)

Similarly, for a *strong acid* added to water

$$\frac{dB}{d(\mathrm{pH})} = 2.303[\mathrm{H^+}] \qquad (3\text{-}59\mathrm{b})$$

Combining Equations 3-59a and 3-59b for the buffer capacity of water containing dissolved strong acids or bases

$$\boxed{dB/d(\mathrm{pH}) = \beta = 2.303([\mathrm{H^+}] + [\mathrm{OH^-}]) \qquad \text{for water}} \qquad (3\text{-}59\mathrm{c})$$

Lastly, consider the buffer capacity of a *weak acid*, HA, in water:

$$\mathrm{HA} \rightleftharpoons \mathrm{H^+} + A^-$$

$$K_a = \frac{[\mathrm{H^+}][A^-]}{[\mathrm{HA}]} \qquad (3\text{-}60)$$

In the classic titration experiment, in which B is the amount of added base ($\mathrm{OH^-}$), we obtain one A^- for every $\mathrm{OH^-}$ added:

$$\mathrm{HA} + \mathrm{OH^-} \rightarrow A^- + \mathrm{H_2O}$$

so that Eq. 3-60 may be rewritten

$$K_a = \frac{[H^+]B}{[HA]} \tag{3-61}$$

Furthermore, $[HA] = [HA]_0 - B$, in which $[HA]_0$ is the amount of weak acid present before an amount B of base is added. Equation 3-61 thus becomes

$$K_a = \frac{[H^+]\,B}{[HA]_0 - B} \tag{3-62}$$

which may be solved for B to give

$$B = \frac{K_a\,[HA]_0}{[H^+] + K_a} \tag{3-63}$$

Differentiating Eq. 3-63,

$$dB/d\,[H^+] = -\frac{K_a\,[HA]_0}{([H^+] + K_a)^2} \tag{3-64}$$

and using the chain rule for differentiation,

$$\frac{dB}{d\text{pH}} = \frac{dB}{d\,[H^+]}\,\frac{d\,[H^+]}{d(\text{pH})} = \left(-\frac{K_a\,[HA]_0}{([H^+] + K_a)^2}\right)(-2.303[H^+]$$

$$\frac{dB}{d(\text{pH})} = \boxed{\beta = \frac{2.303\,K_a\,[H^+][HA]_0}{([H^+] + K_a)^2}} \tag{3-65}$$

A plot of buffer capacity versus pH is shown in Fig. 3-11 for the weak acid, acetic acid. The reader should verify (see Problems) that buffer capacity,

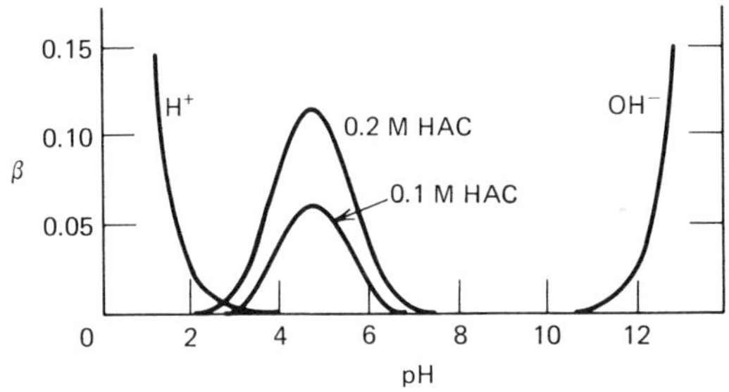

FIGURE 3-11. Buffer capacity, β, as a function of pH for solutions of strong acid (H^+), strong base (OH^-) and acetic acid solutions that are 0.1M or 0.2M in acetic acid initially. Note that for the acetic acid curves the *height* of the curve is proportional to *initial acid concentration,* while the $pH = pK_a$ at the pH for which β is a *maximum.* [See Eq. 3-65ff.] (Curves taken from H. B. Bull, *An Introduction to Physical Biochemistry,* 2nd ed., F. A. Davis Co., Philadelphia, Pa., 1971, p. 112.)

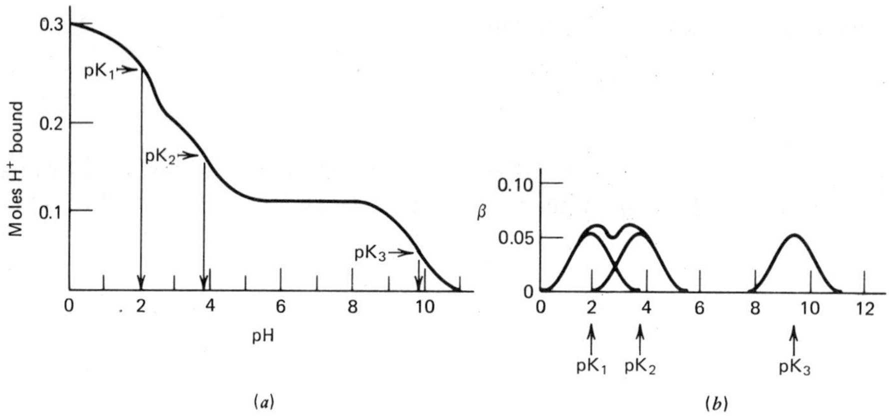

FIGURE 3-12. Comparison of two means for reducing titration data to obtain the pK_as of individual titrable groups on a molecule. [*a*] Conventional "titration" or "Bjerrum" plot for 0.1M aspartic acid, titrated with a strong base (say, NaOH). Note that the distinction between the first two pK_as is not especially well-defined. [*b*] Plot of buffer capacity, β, as a function of pH for the same system. This plot is readily broken down (as shown) into the contributions from each of the three dissociable protons of aspartic acid, to give the three desired pK_as for this molecule. Entirely analogous procedures may be used to establish the individual pK_as, and number of groups with that pK_a, for proteins with a much greater number of titrable groups, as shown in Fig. 3-13.

β, will show a *maximum* when $\text{pH} = pK_a$ for the weak acid in question Furthermore, Eq. 3-65 shows that the *maximum* buffer capacity is proportional to the *total amount* of weak acid initially present. The value of such plots may be appreciated from examination of the titration of a polybasic molecule, such as aspartic acid in Fig. 3-12. Although the locations of the various pK_as of this molecule are not especially prominent by inspection of the titration curve itself, it is easier to decompose a plot of buffer capacity versus pH into contributions from the three dissociable protons of aspartic acid.

An important feature of the plot of buffer capacity versus pH is that the *shape* of the curve is always the *same*. It is thus possible to decompose the titration curves of proteins containing a large number of dissociable protons into contributions from the various dissociable protons of the protein, by fitting the observed buffer capacity versus pH plot to an equation of the form

$$\boxed{\beta = \sum_i \frac{2.303\, K_i\, [\mathrm{H}^+]\, n_i [\mathrm{HA}]_0}{(K_i + [\mathrm{H}^+])^2}} \tag{3-66}$$

in which n_i is the number of dissociable protons per macromolecule having $K_a = K_i$. Since the shape of the buffer capacity contribution (versus pH) for

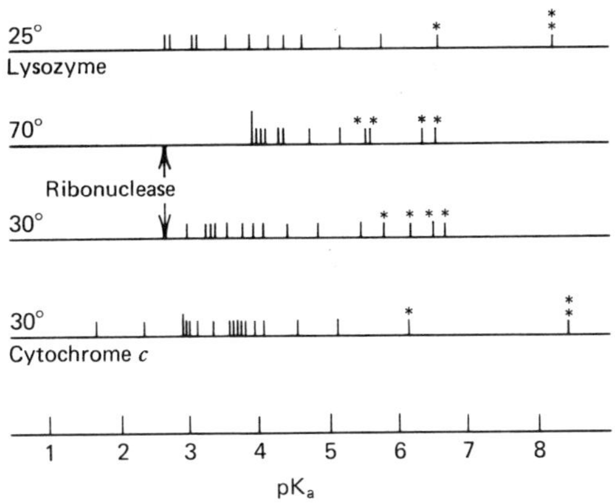

FIGURE 3-13. Individual pK_a values for the ionizable groups of three proteins (lysozyme at 25°, ribonuclease at 30° and 70°, and cytochrome *c* at 30°C), as determined by decomposition of a plot of buffer capacity versus pH as illustrated for aspartic acid in Fig. 3-12. Unmarked pK_a values refer to carboxyl ionization, one asterisk to imidazole, and two asterisks to amino. The vertical scale indicates the relative number of groups having a given pK_a. [From H. B. Bull and K. Breese, *Arch. Biochem. Biophys. 117,* 106 (1966).]

each group is the same, we need only specify n_i and K_i for each type of dissociable proton in order to fit the experimental data. Examples of such fits are shown in Fig. 3-13.

Data Reduction for Many-site Binding: Interacting Binding Sites

In the preceding discussion, we considered a macromolecule with many possible sites for binding of small molecule, *I*, where the various binding sites could differ in equilibrium dissociation constant, K_I, but where binding of an *I* to one site had *no effect* on the binding of another *I* to any other site on the same macromolecule. An experimental example of such a situation is shown by the Scatchard plot of Fig. 3-14, which clearly shows the binding of 4 NADH molecules to identical independent sites on a lactate dehydrogenase molecule.

In this section, we wish to treat the opposite extreme case for binding equilibria (Eq. 3-67), namely binding that is so highly "cooperative" that the binding of the first *I* molecule somehow *facilitates* binding of subsequent *I* molecules, with the result that the remaining sites on the macromolecule fill up immediately (i.e., as if the binding of all *n* occurred in one step):

$$\boxed{E + nI \rightleftharpoons EI_n;\; K_I = \frac{[E][I]^n}{[EI_n]}} \quad \text{one-step, } n\text{-site binding} \qquad (3\text{-}67)$$

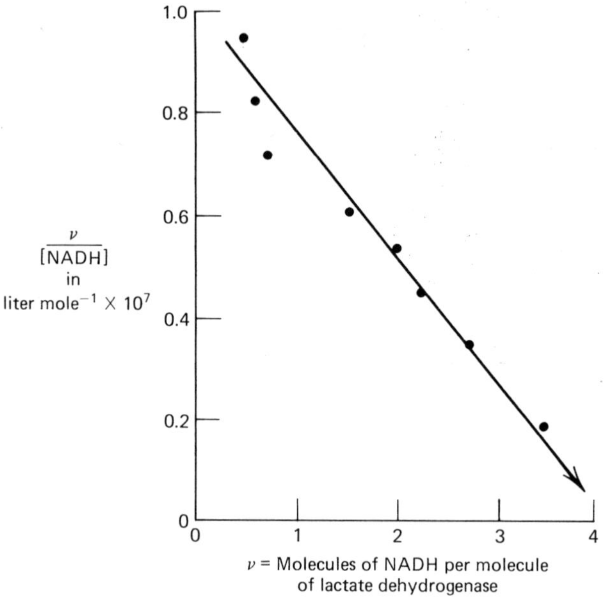

FIGURE 3-14. Scatchard plot for binding of NADH to lactate dehydrogenase. The x intercept shows that there are $n = 4$ identical NADH binding sites per macromolecule. The linearity of the plot shows that the four NADH molecules bind independently (see text). [Data from S. Anderson and G. Weber, *Biochemistry 4*, 1948 (1965).]

If we again define ν as the number of molecules of I bound per molecule of E,

$$\boxed{\nu = \frac{n[EI_n]}{[EI_n] + [E]}} \tag{3-68}$$

we can again solve Eq. 3-67 for $[EI_n]$, substitute for $[EI_n]$ in Eq. 3-68, and rearrange the result:

$$\nu = \frac{n[I]^n}{[I]^n + K_I}$$

to give

$$\frac{\nu}{n - \nu} = \frac{[I]^n}{K_I} \tag{3-69}$$

Equation 3-69 is usually simplified by noting that if θ is defined as the *fraction* of macromolecular sites occupied, then the *number* of occupied sites per macromolecule, ν, becomes

$$\nu = n\theta \tag{3-70}$$

so that

$$\frac{\nu}{n-\nu} = \frac{\theta}{1-\theta} = \frac{[I]^n}{K_I} \tag{3-71}$$

Finally, by taking the log of both sides of Eq. 3-71 (note similarity to Eq. 3-48), we find

$$\log\left(\frac{\theta}{1-\theta}\right) = n \log[I] - \log K_I \tag{3-72}$$

so that a plot of $\log(\theta/[1-\theta])$ versus $\log[I]$ should give a straight line of slope $= n$. In conclusion, when binding is independent (i.e., binding occurs at just one site at a time), $n = 1$ and the slope of such a "Hill" plot is unity; when binding is highly cooperative, the slope of the Hill plot gives the number of interacting binding sites. The classic example is given by the binding of oxygen to myoglobin ($n = 1$, binding to a single site per macromolecule) and hemoglobin ($n = 4$, binding to four interacting sites per macromolecule), as shown in Fig. 3-15. The "direct" plot for oxygen-binding to myoglobin (Fig. 3-15a) shows the usual behavior for one-site binding

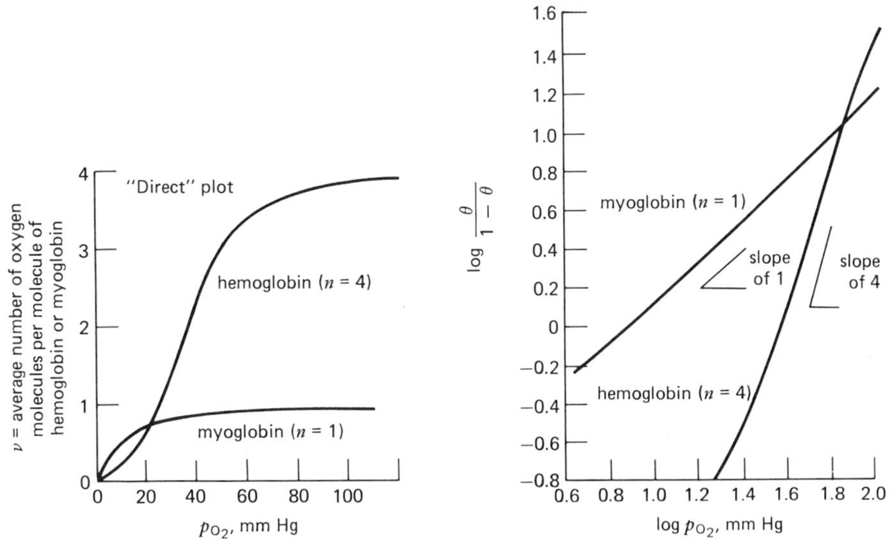

FIGURE 3-15. Cooperative binding of oxygen to myoglobin or hemoglobin, as manifested in a "direct" or in a "Hill" plot of binding data. The completely cooperative binding ($n = 4 =$ slope of Hill plot for hemoglobin) is seldom fully realized experimentally (see text).

(compare Figs. 3-9*a* or 3-10*a*), and a Scatchard plot of the same data would give a straight line that shows binding to a single site per myoglobin molecule. However, the "direct plot for oxygen-binding to hemoglobin (Fig. 3-15*a*) is of different shape and would produce a curved Scatchard (or "reciprocal") plot. The "Hill" plot for hemoglobin (Fig. 3-15*b*) shows why: when the first oxygen binds to a hemoglobin molecule, the remaining three bind so strongly that it is as if all four oxygens bind at once (Eq. 3-67 with $n = 4$). Figure 3-15*b* also shows another general effect: binding is seldom highly cooperative over the entire range of ligand concentration (the "Hill" plot slope is equal to 4 only over a certain pO_2 range, and binding is less cooperative at very high or very low pO_2). In other words, cooperativity (i.e., interaction between binding sites) is minimal when very few or almost all of the possible binding sites are occupied.

PROBLEMS

1. Explain qualitatively how you would use the Gibbs-Duhem equation to calculate the activity of sucrose as a function of sucrose concentration, using the mole fraction and solution vapor pressure data of Table 3-1. *Hint:* start with a plot of $\ln a_A$ versus X_A.

2. Confirm that for a "complex" electrolyte, M_rX_s in solution, the activity (i.e., thermodynamic "concentration") of the individual ions is related to the activity of the original dissolved salt by the equation

$$a_{M_rX_{s(\text{aq})}} = (a_{M^{+s}})^r(a_{X^{-r}})^s$$

leading us to define mean ionic activity, $a_\pm$, and mean ionic activity coefficient $\gamma_\pm$:

$$(a_{M^{+s}})^r(a_{X^{-r}})^s = (a_\pm)^{(r+s)} = (\gamma_\pm)^{(r+s)}(m_{M^{+s}})^r(m_{X^{-r}})^s$$

3. Confirm the equations, form of the graphical plots, and means for extracting the number of equivalent independent binding sites, n, and equilibrium (dissociation) constant per site, K_I, shown in Table 3-4, for the "direct," "Bjerrum," "reciprocal," and "Scatchard" plots.

4. The electron spin resonance signal for aqueous Mn^{++} changes markedly when the Mn^{++} ion binds to a macromolecule in solution. Based on ESR data for solutions containing different (known) proportions of Mn^{++} and the macromolecule, trypsin (a proteolytic enzyme of molecular weight about 25,000), the following binding data were obtained (A. G. Marshall and S. Y. Kang, unpublished data):

$\frac{[Mn^{++}]_{bound}}{[trypsin]_{total}[Mn^{++}]_{free}}$	$\frac{[Mn^{++}]_{bound}}{[trypsin]_{total}}$
2950	0.3
2100	0.5
1500	1.0
700	2.6
450	3.6
250	7.0

From a Scatchard plot based on these data, assume that there are two types of binding sites, and estimate the binding constant and number of each type of site.

5. The *isoelectric point* for a molecule may be defined as the pH at which the molecule has no net charge. Show that the isoelectric pH for a molecule with two dissociable hydrogens (and thus two pK_as) is of the form

$$\text{isoelectric pH} = (1/2)(pK_1 + pK_2), \text{ where } pK_a = -\log K_a$$

in which K_a is the acid dissociation constant for the group of interest.

6. The amino acid, lysine, usually written as $H_2N(CH_2)_4\text{—}\overset{\displaystyle H}{\underset{\displaystyle NH_2}{C}}\text{—}COOH$, has three acid dissociation constants, with $pK_1 = 2.18$, $pK_2 = 8.95$, and $pK_3 = 10.53$.

(a) What is the isoelectric point for lysine?

(b) What is the most likely formula for the zwitterion?

7. Show that the buffer capacity, β,

$$dB/d(\text{pH}) = \beta = \frac{2.303\, K_a\, [H^+][HA]_0}{([H^+] + K_a)^2}$$

has its maximum value when the $\text{pH} = pK_a$ for the weak acid in question.

REFERENCES

W. J. Moore, *Physical Chemistry,* Prentice-Hall, Englewood Cliffs, N. J., any edition. Good treatment of standard states and activities, including Debye-Hückel theory of "salting-in."

C. Tanford, *The Hydrophobic Effect,* John Wiley & Sons, N. Y. (1973). Good discussion of hydrophobicity and applications.

J. T. Edsall and J. Wyman, *Biophysical Chemistry,* Vol. 1, Academic Press, N. Y. (1958). Good treatment of ways of extracting binding constants and number of binding sites for experimental multisite binding cases.

H. B. Bull, *An Introduction to Physical Biochemistry,* 2nd ed., F. A. Davis Co., Philadelphia (1971). Good brief treatment of buffer capacity.

CHAPTER 4 Chemical Reaction Spontaneity: Temperature-Dependence of Equilibrium Constants and Reaction Rates

The simplest question about any chemical reaction is, will it "go"? According to our previous analysis of equilibrium at the usual chemical conditions of constant temperature and pressure (Table 2-1), the (Gibbs) free energy, G, is a *minimum* at equilibrium. Thus, for the general chemical reaction

$$aA + bB \rightarrow cC + dD \tag{4-1}$$

the reaction as written will be *spontaneous*, if the total change in free energy is *negative* for the conversion of a moles of A and b moles of B into c moles of C and d moles of D:

$$\boxed{\text{If } G_{\text{products}} - G_{\text{reactants}} = \Delta G_{rx} < 0, \text{ reaction is spontaneous}} \tag{4-2}$$

A simpler criterion for spontaneity comes from the previously derived relation

$$\boxed{\Delta G^{\circ}_{rx} = -RT \ln K_a} \tag{3-11}$$

in which K_a is the equilibrium constant for process 4-1 treated as a reversible reaction (Eq. 3-4), and ΔG°_{rx} represents the change in free energy on conversion of a moles of A and b moles of B into c moles of C and d moles of D, where *each* of these components is present in its *standard state* (see Chapter 3.A.2.). Thus, under standard state conditions, we can combine Equations 4-2 and 3-11 to obtain the simple criterion for equilibrium:

$$\boxed{K_a > 1 \Leftrightarrow \begin{array}{l}\text{reaction as written is spontaneous}\\ \text{for conversion of reactants in standard}\\ \text{states to products in standard states}\end{array}} \tag{4-3}$$

Similarly, for $K_a < 1$, the reaction as written would *not* be spontaneous, and for $K_a = 1$, there would be no tendency for the reaction to proceed one way or the other, where reactants and products are each in their standard states.

Equation 4-3 gives an intuitively simple picture: if equilibrium lies to the right ($K_a > 1$), the reaction is spontaneous under standard state conditions, and if equilibrium lies to the left ($K_a < 1$), the reaction is not spontaneous. The immediate problem is that the magnitude of K_a appears to depend critically on the choice of *units* for the *concentrations* of the various components of the reaction, thus confusing the spontaneity issue. We begin this chapter by resolving that apparent problem, which is critical to deciding the *sequence* (pathways) of the chemical reactions of biological metabolism. We then show how equilibrium constants (and hence chemical reaction spontaneity) depend on temperature in a predictable way. Finally, we discuss the magnitude and temperature-dependence of the individual rate constants of which any equilibrium constant is composed, since many reactions which are "spontaneous" by the thermodynamic criteria (Eq. 4-2 or 4-3) nevertheless proceed so slowly toward equilibrium that the thermodynamic criterion fails as a predictive tool.

4.A CRITERIA FOR CHEMICAL REACTION SPONTANEITY

The problem of units in equilibrium constants is immediately resolved, once we recognize that *activity* can always be thought of as a *ratio* of two *concentrations*. For example, we have already shown that activity for a *solvent* may be expressed as a *ratio* of the vapor pressure of the solution in question, P_A, to the vapor pressure of pure solvent, P_A°:

$$a_A = (P_A/P_A^\circ) = \text{activity of solvent } A,\ a_A \to 1 \text{ as } X_A \to 1 \qquad (3\text{-}9a)$$

P_A° corresponds to the "standard state" of the solvent (namely, pure solvent). Similarly, the activity of a *solute* may be expressed as a *ratio* of the molarity (or molality or mole fraction) of solute in the solution of interest to the corresponding "unit" molarity (or molality or mole fraction) of solute in an "ideal" solution:

$$a_B = \gamma_B(m_B/m_B^\circ) \qquad (4\text{-}1)$$

where

$m_B^\circ = 1$ molar (molal, mole fraction) in an ideal solution ($\gamma_B^\circ = 1$) = "standard state" of solute

For example, if we choose a standard state for solute as 1 molar solute in an ideal solution, then the activity of solute in a 2.3 molar *real* solution would be, $a_B = \gamma_B(2.3)$, where γ_B is the activity coefficient appropriate to a 2.3 molar solution, and $(m_B/m_B^\circ) = 2.3$. In other words, even though 2.3 appears to represent the molar *concentration* of solute B, we see that 2.3 is actually a *ratio* of the solute molar concentration to the (unit) molar concentration of the chosen solute "standard" state.

As a more practical example, consider a solution containing hydrogen ions. Although we could compare actual hydrogen ion concentration in a solution of interest, m_{H^+}, to a hydrogen ion standard state concentration of 1 molar in an ideal solution, the standard state would then correspond to a pH of approximately zero, which is far removed from physiological (or even ordinary chemical) reaction conditions. Biochemists therefore usually define the "standard" state of aqueous hydrogen ions as 10^{-7}M H^+ (pH 7.0) in an ideal solution. There is thus a difference of a factor of 10^7 in the *activity* of aqueous hydrogen ions, depending on the choice of "standard" state:

$$a_{H^+} = \gamma_{H^+}(m_{H^+}/m^\circ_{H^+}) = \gamma_{H^+}\, m_{H^+} \tag{4-2a}$$

where $m^\circ_{H^+} = 1$ M("chemical" convention)

or

$$a_{H^+} = \gamma_{H^+}(m_{H^+}/m^\circ_{H^+}) = 10^7\, \gamma_{H^+}\, m_{H^+} \tag{4-2b}$$

where $m^\circ_{H^+} = 10^{-7}$ M ("biochemical" convention)

A change from one concentration "unit" to another corresponds to a change from one "standard" state (i.e., concentration = 1 in the first choice of units) to another "standard" state (i.e., concentration = 1 in the second choice of units). Thus while the magnitudes of K_a and ΔG°_{rx} do depend on the choice of concentration units, Equations 3-11 and 4-3 are valid for any *single* choice of units, provided that the "standard" concentration that defines the activities in K_a is the *same* as the "standard" concentration for which ΔG°_{rx} is defined.

Having settled the issue of concentration units, we are now able to consider the spontaneity criterion for process 4-1, when the reactants and products are present at *arbitrary* (rather than *standard* or *equilibrium*) concentrations. Substituting Equations 3-2 and 3-3 into Eq. 4-2,

$$G_i = n_i\overline{G}_i = \text{total free energy of } n_i \text{ moles of } i\text{th component,} \tag{3-2}$$

and

$$\overline{G}_i = \overline{G}^\circ_i + RT\,\ln(a_i) = \text{chemical potential of } i\text{th component, having activity } a_i \tag{3-3}$$

and

$\Delta G_{rx} = G_{\text{products}} - G_{\text{reactants}}$ = change in free energy when a moles of A at activity a_A and b moles of B at activity a_B are converted to c moles of C at activity a_C and d moles of D at activity a_D, (4-2)

we obtain,

$$\begin{aligned}\Delta G_{rx} &= c\overline{G}_C + d\overline{G}_D - a\overline{G}_A - b\overline{G}_B \\ &= c\overline{G}^\circ_C + cRT \ln a_C + d\overline{G}^\circ_D + dRT \ln a_D - a\overline{G}^\circ_A - aRT \ln a_A \\ &\qquad - b\overline{G}^\circ_B - bRT \ln a_B \\ &= G^\circ_C + G^\circ_D - G^\circ_A - G^\circ_B + RT\,(\ln a^c_C + \ln a^d_D - \ln a^a_A - \ln a^b_B)\end{aligned}$$

$$\boxed{\Delta G_{rx} = \Delta G^\circ_{rx} + RT \ln \frac{a^c_C a^d_D}{a^a_A a^b_B}} \quad \text{reactants and products at } \textit{arbitrary} \text{ concentrations} \tag{4-3}$$

All reactants and products at their *standard* concentrations ($a_A = a_B = a_C = a_D = 1$) →

$$\boxed{\Delta G = \Delta G^\circ_{rx}}$$

Equilibrium at constant T,P (all reactants and products at their *equilibrium* concentrations) →

$$\boxed{\begin{aligned}&\Delta G_{rx} = 0 \\ &\Delta G^\circ_{rx} = -RT \ln \frac{a^c_C a^d_D}{a^a_A a^b_B} = -RT \ln K_a\end{aligned}}$$

Equation 4-3 gives the recipe for computing ΔG_{rx} for process 4-1, when reactants and products are present at *arbitrary* concentrations, and the respective limits for standard concentrations or equilibrium concentrations are shown below Eq. 4-3. When ΔG_{rx} has been obtained from Eq. 4-3, we can then immediately determine whether a given reaction is *spontaneous* ($\Delta G_{rx} < 0$: the reaction would proceed from left to right by itself), *nonspontaneous* ($\Delta G_{rx} > 0$: the reaction would proceed from right to left by itself), or at *equilibrium* ($\Delta G_{rx} = 0$: reaction has no tendency to proceed one way or the other). It is important to note that even when $\Delta G_{rx} > 0$ (nonspontaneous process), the reaction can be *made* to proceed by adding at least ΔG_{rx} amount of free energy to the system, as in the following example. The use of Eq. 4-3 and Eq. 3-11 is nicely illustrated by a particular reaction that occurs in association with hundreds of different biochemical reactions, namely the interconversion of adenosine diphosphate (ADP) and adenosine triphosphate (ATP).

EXAMPLE *ATP: the "Money" of the Bioenergetic Economy*

Many of the vital chemical reactions of cell metabolism are not spontaneous by the criterion of Eq. 4-3. The energy required to "drive" these reactions is obtained by coupling the reaction with another (spontaneous) reaction, chosen such that the free energy change for both reactions coupled together is negative, so that the overall (coupled) process is spontaneous. The most common "driving" reaction is the hydrolysis of ATP to ADP and inorganic phosphate (P_i). A living cell is therefore engaged in continual synthesis of ATP from ADP to provide a constant supply of ATP to help "run" the otherwise nonspontaneous metabolic reactions. The reaction shown in Eq. 4-4 is thus perhaps the most important chemical reaction in living cells (it appears in several hundred of the 1000-odd enzyme-catalyzed reactions in mammalian cells), and we will now examine the free energy for the process under various conditions.

$$\text{ADP} + P_i + \text{H}^+ \rightleftharpoons \text{ATP} + \text{H}_2\text{O} \tag{4-4}$$

for which

$$K_a = \frac{a_{\text{ATP}}\, a_{\text{H}_2\text{O}}}{a_{\text{ADP}}\, a_{P_i}\, a_{\text{H}^+}} \quad \text{at equilibrium} \tag{4-5}$$

and

$$\Delta G^\circ_{rx} = -RT \ln K_a = \text{free energy change for conversion of 1 mole of ADP to ATP, where each reactant and product is at its standard state (unit concentration in ideal sol'n)} \tag{4-6}$$

Although the actual value of ΔG°_{rx} also depends on the concentrations of Mg^{++} and Ca^{++}, a commonly accepted value for ΔG°_{rx} is

$\Delta G^\circ_{rx} = -2.2$ kcal/mole	where each reactant and product is present at unit activity	(4-7)

In other words, the reaction would appear to be spontaneous in proceeding from left to right. However, a standard state for H^+ of $a_{\text{H}^+} = 1$ when $[\text{H}^+] = 1$ molar in an ideal solution (corresponding to a pH of about zero!) is far removed from the usual $[\text{H}^+]$ found in living cells. Therefore, biochemists usually define the standard state of H^+ as $a_{\text{H}^+} = 1$ when $[\text{H}^+] = 10^{-7}$M (pH = 7.0), in order to obtain a $\Delta\overline{G}^{\circ\prime}$ value more appropriate to actual biological conditions. We can now use Eqs. 4-2 to show (see Problems) that

$$\overline{G}^{\circ\prime}_{\text{H}^+} = \overline{G}^\circ_{\text{H}^+} + RT \ln 10^{-7} \tag{4-8}$$

$\overline{G}^\circ_{\text{H}^+}$ — Free energy of H^+ at standard state of $m_{\text{H}^+} = 1$ molar in ideal sol'n

$\overline{G}^{\circ\prime}_{\text{H}^+}$ — Free energy of H^+ for standard state of $m_{\text{H}^+} = 10^{-7}$ molar in ideal sol'n (pH = 7.0)

from which we can calculate the standard free energy change, $\Delta G^{\circ\prime}_{rx}$, using our new choice of the standard state for aqueous protons (Eq. 4-2b):

$$\begin{aligned}\Delta G_{rx}^{\circ\prime} &= \bar{G}_{\text{ATP}}^{\circ} + \bar{G}_{\text{H}_2\text{O}}^{\circ} - \bar{G}_{\text{H}^+}^{\circ\prime} - \bar{G}_{\text{ADP}}^{\circ} - \bar{G}_{P_i}^{\circ}\\ &= \bar{G}_{\text{ATP}}^{\circ} + \bar{G}_{\text{H}_2\text{O}}^{\circ} - (\bar{G}_{\text{H}^+}^{\circ} + RT\ln 10^{-7}) - \bar{G}_{\text{ADP}}^{\circ} - \bar{G}_{P_i}^{\circ}\\ &= \Delta\bar{G}_{rx}^{\circ} - RT\ln 10^{-7}\\ &= -2.2 + 9.5\end{aligned}$$

$\Delta G_{rx}^{\circ\prime} = +7.3$ kcal/mole	where standard state of H^+ is 10^{-7}M, standard state of other solutes is 1M in ideal solution, and standard state of solvent is pure H_2O	(4-9)

Equation 4-9 shows that when the hydrogen ion concentration is reduced to 10^{-7}M, the reaction is now spontaneous in the *opposite* direction (i.e., from right to left in Eq. 4-4. This calculation should demonstrate that the energy required to drive a given chemical reaction is *not* constant, but depends markedly on the actual concentrations of the various components.

The reader should verify that under even more realistic physiological conditions (say, [ATP] = [ADP] = $[P_i] = 10^{-4}$M), the free energy change for the reaction of Eq. 4-4 increases to about 12.7 kcal/mole. (Present best estimates for the concentrations of all the relevant species in a typical cell suggest that the actual ΔG_{rx} for Eq. 4-4 is about +12 kcal/mole.)

One of the most basic cell "foods" is glucose, $C_6H_{12}O_6$. A large amount of (free) energy is released when glucose is broken down into CO_2 and H_2O by the cell, and a fraction of that energy is used to form ATP from ADP. Using the ΔG_{rx} for formation of ATP from ADP just obtained, it is possible to compute the *efficiency* of a cell in capturing and storing (in the form of the high-energy

Table 4-1 Free Energies of Hydrolysis (Removal of Phosphate) for Some Phosphorylated Compounds*

Compound	$\Delta G_{rx}^{\circ\prime}$
Phosphoenolpyruvate	−14.8 kcal/mole
Carbamoyl phosphate	−12.3
Acetyl phosphate	−10.3
Creatine phosphate	−10.3
Pyrophosphate	− 8.0
ATP (to ADP)	− 7.3
Glucose-1-phosphate	− 5.0
Glucose-6-phosphate	− 3.3
Glycerol-3-phosphate	− 2.2

* *$\Delta G_{rx}^{\circ\prime}$ is defined for standard states of: 10^{-7}M for H^+, 1M in ideal solution for solutes other than H^+, pure H_2O for water*

phosphate bond of ATP) the energy released from breakdown of glucose (see Problems). This calculation shows that the cell, regarded as an energy conversion device, is extremely efficient.

Although we have just seen that ΔG°_{rx} or even $\Delta G^{\circ\prime}_{rx}$ values for conversion of reactants in their standard states to products in their standard states provide a relatively crude estimate of the actual ΔG_{rx} at physiological concentrations of reactants and products, it is nevertheless useful to compare a *series* of homologous reactions based on the *same* standard states in each case, as shown in Table 4-1. The table shows the free energy change for hydrolysis (removal of phosphate) from each of several biological molecules. It is evident that some of these molecules have even higher-energy phosphate bonds than ATP. For example, by coupling the topmost reaction with the ADP–ATP reaction, phosphoenolpyruvate could be used to transfer a phosphate to ADP to form ATP—this is in fact one way in which ATP is generated biologically in the breakdown of sugars. The reader is referred to the Problems for further exercises in the dependence of ΔG_{rx} on concentrations of reactants and products.

4.B. VARIATION OF EQUILIBRIUM CONSTANTS WITH TEMPERATURE

From our original definition of Gibbs free energy

$$G = H - TS \tag{1-14}$$

and the general relation between ΔG°_{rx} and equilibrium constant, K_a,

$$\Delta G^\circ_{rx} = -RT \ln K_a \tag{3-11}$$

we quickly deduce that at any given (i.e., constant) temperature, T,

$$\Delta G^\circ_{rx} = -RT \ln K_a = \Delta H^\circ_{rx} - T\,\Delta S^\circ_{rx} \tag{4-10}$$

in which ΔH°_{rx} and ΔS°_{rx} are the respective changes in enthalpy and entropy when a mole or reactant is converted to product, with all reactants and products in their "standard" states. Rewriting Eq. 4-10 as

$$\boxed{\ln K_a = -\frac{\Delta H^\circ_{rx}}{RT} + \frac{\Delta S^\circ_{rx}}{R}} \tag{4-11}$$

it is clear that a plot of ln K_a versus $(1/T)$ will be a straight line of slope, $-(\Delta H^\circ_{rx}/R)$, and y-intercept, $(\Delta S^\circ_{rx}/R)$.* The significance of Eq. 4-11 is twofold:

* *Strictly, this is true only if ΔH°_{rx} and ΔS°_{rx} are independent of temperature. Over the usually small temperature range in which most biological reactions occur, this is approximately correct.*

first, it predicts the *temperature-variation* of the equilibrium constant for a reaction, in terms of ΔH°_{rx} and ΔS°_{rx}, and can thus be used to determine the equilibrium constant at any desired temperature, from known values of ΔH°_{rx}, ΔS°_{rx}, and K_a at any other single temperature; alternatively, it provides a means for determination of ΔH°_{rx} and ΔS°_{rx} for a given reaction from the temperature-dependence of the equilibrium constant, K_a. Since the degree of "spontaneity" of a chemical reaction depends on both ΔH°_{rx} and ΔS°_{rx}, we are now in a position to determine whether a given chemical reaction "goes" *primarily* because of a net release of energy from the breaking and reformation of chemical bonds ($\Delta H^\circ_{rx} < 0$), or because of a net increase in the "randomness" of the products over reactants ($\Delta S^\circ_{rx} > 0$). It is often particularly useful to compare *trends* in the magnitude of ΔH°_{rx} or ΔS°_{rx}, for a series of homologous reactants—such trends form the basis for discussion of "steric," "inductive," and "resonance" effects in organic reaction mechanisms.

EXAMPLE *Equilibrium Constants at "Standard" (25°C) and Body (37°C) Temperature*

The reaction of water with fumarate to form malate

$$\text{fumarate}^{=} + H_2O \rightleftharpoons \text{malate}^{=}$$

is a key metabolic step in the tricarboxylic acid cycle. From the following data, compare the equilibrium constant for this reaction at 25°C (the usual "standard" temperature for which chemical reaction data are tabulated) and 37°C (human body temperature).

$\Delta G^\circ_{rx} = -880$ cal/mole at 25°C
$\Delta H^\circ_{rx} = +3{,}560$ cal/mole, where ΔH°_{rx} may be taken as constant over the temperature range, 25 to 40°C

First of all, from Eq. 3-11, we compute $\ln K_a$ at the lower temperature:

$$\ln K_a = -\Delta G^\circ_{rx}/RT = -(-880)/(1.99 \text{ cal/mole degree})(298°\text{K}) = 1.49;$$
$$K_{a(25°C)} = e^{1.49} = \boxed{4.42} \text{ at } 25°\text{C}$$

Then, since Eq. 4-11 tells us that

$$\partial \ln K_a/\partial\left(\frac{1}{T}\right) = -\Delta H^\circ_{rx}/R \tag{4-12}$$

it is clear that so long as ΔH°_{rx} and ΔS°_{rx} are independent of temperature (an assumption that is usually valid over a small temperature range such as the 15°C in this example)

$$\boxed{(\ln K_a)_{\text{at } T_2} - (\ln K_a)_{\text{at } T_1} = -\frac{\Delta H^\circ_{rx}}{R}\left(\frac{1}{T_2} - \frac{1}{T_1}\right)} \tag{4-13}$$

or

$$(\ln K_a)_{\text{at } 37^\circ\text{C}} = 1.49 - \left(\frac{+3560}{1.99}\right)\left(\frac{1}{310} - \frac{1}{298}\right)$$
$$= 1.72$$
$$(K_a)_{37^\circ\text{C}} = e^{1.72} = \boxed{5.6} \text{ at } 37^\circ\text{C}$$

In this case, an increase in temperature shifts the equilibrium markedly toward the right-hand side of the reaction. Qualitatively, the "principle of Le Chatelier" predicts that for a reaction that absorbs heat in going from left to right ($\Delta H^\circ_{rx} > 0$, as in this case), an increase in temperature is expected to shift the equilibrium in a direction to compensate for the applied change (in this case, an input of heat). We have thus accounted for the success of the Le Chatelier principle, in terms of Eq. 4-13.

In Section 4.A and 4.B, we have tried to show that the "spontaneity" of a chemical reaction depends in a *predictable* way on the *concentrations* of the various reactants and products and on the *temperature* at which the reaction is carried out. The reader is now in a position to compute either ΔG_{rx} or the equilibrium constant, K_a, for *any* choice of standard component concentrations or temperature. Actual experimental determination of ΔG_{rx} (and thence K_a) is often most convenient from electrochemical measurements (see Chapter 5).

Differential Scanning Calorimetry

We have previously observed (Chapter 1) that determination of the heat evolved or absorbed during a chemical reaction at constant pressure, namely ΔH_{rx}, can be used to generate a useful scale of chemical bond strengths. Furthermore, in this section we have proposed a means for evaluating ΔH_{rx} from the temperature-dependence of the equilibrium constant for the reaction. Heat may also be absorbed or produced when a substance changes its *physical "phase"* (Chapter 2.A.2) at constant pressure: for example, it takes about 80 calories to melt one gram of ice, and about 540 calories to vaporize one gram of liquid water at 1 atm pressure. The larger the ΔH for a phase change, the more radical is the molecular reorganization of the material, so that changes in the *conformation* of a macromolecule may be detected from the heat evolved or absorbed during the conformational change. Since a typical phase change may occur over a relatively narrow temperature range, we might expect to detect it by measuring the amount of heat required to raise the temperature of the system through that small range. The main experimental complication is that it takes a certain amount of heat per unit temperature (C_P) to raise the temperature of *any* material under (typical) constant-pressure conditions

$$C_P = (\partial H/\partial T)_{\text{constant } P} = \text{heat capacity at constant pressure} \quad (4\text{-}14)$$

C_P is often constant for any one substance, at least over the relatively small temperature range over which biologically interesting reactions occur physiologically. Thus, we must subtract the heat-capacity contribution

$$\Delta H = \int_{T_1}^{T_2} C_P \, dT = \text{heat required to raise the temperature of a system from } T_1 \text{ to } T_2 \text{, in the absence of any phase transitions} \tag{4-15}$$

from the experimental total heat required to raise the system temperature, in order to obtain just the heat used in changing the phase of the substance. This is the principle of differential scanning calorimetry: one measures the heat required to raise the temperature of an object by a very small temperature increment, while steadily increasing the temperature. The heat-capacity contribution is subtracted, to leave a plot of heat absorbed versus temperature such as that shown in Fig. 4-2.

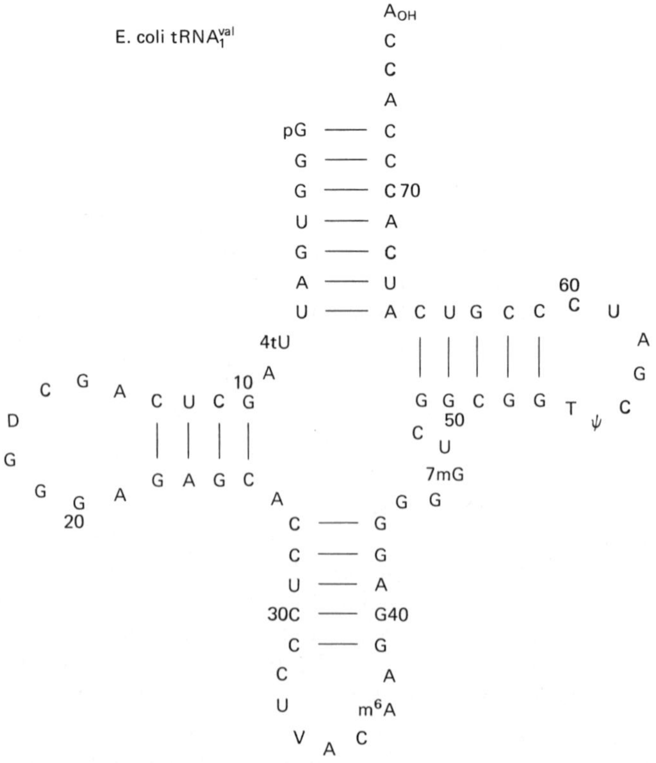

FIGURE 4-1. Nucleotide sequence for tRNA$_{val_1}$ from *E. coli* (B. G. Barrell and B. F. C. Clark, *Handbook of Nucleic Acid Sequences,* Oxford Joynson-Bruvvers, Ltd., 1974, p. 55.)

EXAMPLE *Melting of Transfer-RNA*

Since a given macromolecule in solution may have a shape described as an equilibrium between several different major conformations, it is not usually easy to detect changes between *individual* conformations by monitoring gross molecular properties (diffusion constant, sedimentation or electrophoretic velocity, optical absorption spectra, or the like). However, because heat will be absorbed as the temperature of the solution is increased past the transition temperature for each conformational ("phase") change, the individual conformational changes may sometimes be observed directly from differential scanning calorimetry. Figure 4-2 shows the enthalpy change resulting from various stages in the unfolding of a transfer-RNA molecule (Fig. 4-1) that codes for the amino acid, valine. These changes were determined by measuring the amount of heat (supplied by electrical current flow through a resistor) required to maintain the same temperature for the $t\text{RNA}_{\text{val}}$ sample as for a reference sample, as the temperature of both samples was slowly scanned upward (the reference sample thus provides an automatic correction for heat-capacity contributions to ΔH). The curves of Fig. 4-2 show that this molecule unfolds in as many as six distinct stages, with an overall total heat of melting of 370 kcal/mole. It is clear from the figure that the effect of Mg^{++} ion is to stabilize the macromolecular structure (i.e., the *t*RNA melts at a higher temperature), and to connect some regions of the molecule with other ones (since

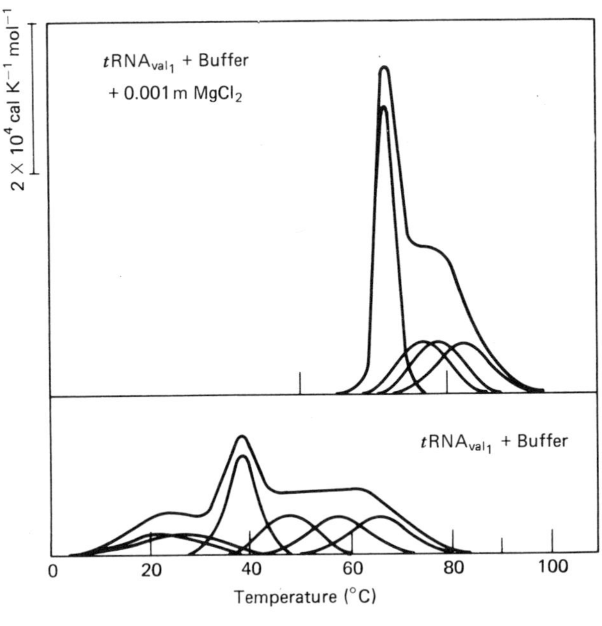

FIGURE 4-2. Decomposition of the observed calorimetric melting curves for $\text{tRNA}_{\text{val}_1}$ (3×10^{-5} M, pH 7.0, phosphate buffer) in the presence (*top*) or absence (*bottom*) of 1 mM $MgCl_2$. See text for discussion. [From P. L. Privalov, V. V. Filimonov, T. V. Benkstern and A. A. Bayev, *J. Mol. Biol. 97*, 279 (1975).]

three of the unfolding stages that are distinct in the absence of Mg^{++} merge into a single transition in the presence of Mg^{++}). Since the base sequence and base-pairing of this nucleic acid should be similar to that of $t\mathrm{RNA}_{\mathrm{phe}}$, a species whose X-ray crystal structure is known (see Section 6), it seems reasonable that the three transitions at highest temperatures correspond to "melting" of the three hydrogen-bonded base-paired regions of the molecule, and that the lower-temperature transitions correspond to disruption of the much weaker base-stacked interactions that hold the various "arms" of the "cloverleaf" structure together. With this sort of information, one can now begin to identify the sites of action of various chemical agents that bind to the *t*RNA molecule, by observing their effect (as with Mg^{++}) on the strengths of the intramolecular bonds in various parts of the macromolecule.

4.C. TEMPERATURE-DEPENDENCE OF INDIVIDUAL REACTION RATE CONSTANTS*

At equilibrium, the rates of the forward and reverse constituent chemical reactions must be equal (see Section 3). For example, for

$$A + B \underset{k_{-1}}{\overset{k_1}{\rightleftharpoons}} C + D \tag{4-16}$$

$$\text{forward rate} = k_1[A][B] = \text{reverse rate} = k_{-1}[C][D]$$

or

$$k_1/k_{-1} = \frac{[C][D]}{[A][B]} \tag{4-17}$$

But the concentration ratio on the right-hand side of Eq. 4-17 is just the equilibrium constant for the process, Eq. 4-16; therefore

$$\boxed{K_{\mathrm{eq}} = \frac{k_1}{k_{-1}}} \tag{4-18}$$

Although derived here for the particular equilibrium, Eq. 4-16, the relation of Eq. 4-18 is general.

From our recent deduction that an *equilibrium constant* varies with temperature according to an equation of the form,

$$\partial \ln K_{\mathrm{eq}}/\partial(1/T) = -\Delta H^{\circ}_{rx}/R \tag{4-12}$$

and Eq. 4-18,

* *Readers completely unfamiliar with chemical reaction kinetics may prefer to postpone coverage of this material until after completion of Section 3.*

$$\ln K_{eq} = \ln k_1 - \ln k_{-1} \qquad (4\text{-}18a)$$

it might be expected that the temperature-dependence of an individual *reaction rate constant* (k_1 or k_{-1}) might also satisfy an equation of the form

$$\partial \ln k/\partial(1/T) = -E_a/R \qquad (4\text{-}19)$$

where

$$E_{a(\text{forward } rx)} - E_{a(\text{reverse } rx)} = \Delta H^\circ_{rx} \qquad (4\text{-}20)$$

for Eqs. 4-19, 4-12, and 4-18a to be consistent. Experimentally, it is found that the component reaction rate constants for one-step equilibria do indeed satisfy the empirical equation 4-19, which is usually expressed in its integrated form

$$\boxed{k = A \exp[-E_a/RT] \qquad \text{(Arrhenius equation)}} \qquad (4\text{-}21)$$

in which the "pre-exponential" factor, A, is independent of temperature, and E_a has units of energy and is called the Arrhenius (experimental) "activation energy" for the reaction. An enormous theoretical effort over the past 60 years has been made to predict the absolute magnitudes of A and E_a for various reactions. Although *quantitative* estimates of A and E_a have largely proved inaccessible except for the very simplest gas-phase reactions, two theoretical approaches have provided *qualitative* intuitive significance to A and E_a, which may then be interpreted to furnish *mechanistic* information about the chemical reaction in question. These two theories are the "collision theory" (ca. 1917) and the "transition state" theory (ca. 1935), whose terminology we now introduce.

4.C.1. Collision Theory of Reaction Rates

The first theoretical approach is based on the idea that the rate at which two species react is proportional to the number of collisions per unit time between reactants—hence, "collision" theory. Since not all reactant molecules may possess sufficient energy to complete the reaction, one expects that

$$\begin{matrix}\text{chemical reaction} \\ \text{rate (reactions/sec)}\end{matrix} = \begin{pmatrix}\text{collisions/sec} \\ \text{between } A \text{ and } B\end{pmatrix}\begin{pmatrix}\text{fraction of} \\ \text{reactants with} \\ \text{sufficient energy}\end{pmatrix}[A][B] \qquad (4\text{-}22)$$

For ideal gas reactant molecules, it is possible to compute the collision frequency:

$$\text{collision frequency (ideal gas reactants)} = \frac{\pi N_0 d_{AB}^2}{10^6}\sqrt{\frac{8kT}{\pi\mu}}\,[A][B] \qquad (4\text{-}23)$$

in which d_{AB} is the average diameter of the reactants ($d_{AB} = (d_A + d_B)/2)$), N_0 is Avogadro's number, k is the Boltzmann constant, T is absolute temperature, and μ is the reduced mass of the reactants: $\mu = m_A m_B/(m_A + m_B)$. The fraction of reactants with sufficient energy can be shown to be $\exp[-E^*/RT]$, where E^* is the minimum energy required for the reaction to occur (see Chapter 19). Thus, since

$$d[A]/dt = -k\,[A][B] \tag{4-24}$$

comparison of Equations 4-22, 4-23, and 4-24 leads to the result

$$k = \frac{N}{1000}\pi d_{AB}^2 \sqrt{\frac{8kT}{\pi\mu}} \exp[-E^*/RT] \qquad \text{Collision theory, ideal gas reactants} \tag{4-25}$$

Equation 4-25 predicts that lighter molecules (smaller μ) will react faster, that reactants with larger cross-sectional area (πd_{AB}^2) will react faster, and that reactants for which the minimum required energy (E^*) is smaller will react faster. These qualitative predictions have been confirmed for some gas-phase reactions of simple molecules. The quantitative value for k computed from Eq. 4-25 is usually not very close to the experimental value, mainly because we have not considered any requirement that the two reactants collide in the correct relative *orientation* for reaction – clearly, as the reactant molecules become larger, so that the reaction site represents only a small fraction of the total area of the reactant, this "orientation" or "steric" requirement becomes more severe, and the experimental rate constant will in general be much smaller than predicted by Eq. 4-25. Finally, since the relative change in $\sqrt{T}$ is much less than in $\exp[-E^*/RT]$ for any usual temperature range, Eq. 4-25 predicts a temperature-dependence for reaction rate constants that has essentially the same form as the empirically correct Arrhenius equation (4-21). We will now consider a qualitative application of the collision theory, namely an explanation of the origin of first-order reactions. This example foreshadows our later treatment of steady-state enzyme kinetics, and may profitably be examined after reading Section 3.

EXAMPLE *Origin of First-order Chemical Reaction Rate Expressions*

One of the most immediate problems in describing chemical reaction rates is to explain why some reactions are *first*-order, $d[A]/dt = \pm k[A]$, since essentially all chemical elementary reactions are initiated by the collision of *two* reactant molecules. (We will not discuss photochemical reactions, which are initiated by the absorption of light by a molecule, followed by chemical reactions.) Lindemann (1922) was able to resolve this difficulty using the language of collision theory according to the mechanism outlined below.

Lindemann's intuitive leap was to suppose that out of many collisions, occasionally one of the reactants will come away with most of the energy:

$$A + A \underset{k_{-1}}{\overset{k_1}{\rightleftharpoons}} A + A^* \tag{4-26a}$$

It may then be shown that the energy of the "excited" A^* molecule quickly becomes distributed among translational, rotational, and vibrational energy. When the vibrational energy of A^* is large enough, the amplitude of the vibration(s) along one (or more) bonds becomes so large that the bond(s) break:

$$A^* \xrightarrow{k_2} \text{products} \tag{4-26b}$$

When only a small fraction of the A molecules have sufficient energy to decompose in this way (the $\exp[-E^*/RT]$ factor of the collision theory model), then the reaction mechanism of Eqs. 4-26 is formally similar to the "Michaelis-Menten" scheme for enzyme kinetics (Section 3), and may be solved using the same steady-state approximation (see Section 3 for detailed justification of this assumption):

$$d\,[A^*]/dt \cong 0 = k_1[A][A] - k_{-1}[A][A^*] - k_2[A^*] \tag{4-27}$$

or

$$[A^*] = \frac{k_1[A]^2}{k_{-1}[A] + k_2} \tag{4-28}$$

Finally, the rate of appearance of products then takes the form

$$\text{Rate of appearance of products} = k_2[A^*] = \frac{k_2k_1[A]^2}{k_{-1}[A] + k_2} \tag{4-29}$$

The physical meaning of the result, Eq. 4-29, is most easily understood from its behavior under suitable *limiting* conditions.

Low-concentration limit: $k_2 >> k_{-1}[A]$. In this limit, A^* usually goes on to products rather than back to reactants (verify from Eq. 4-27), and the rate expression, Eq. 4-29, simplifies to

$$\text{Rate of appearance of product} = (k_1k_2/k_2)[A]^2 = k_1[A]^2 \tag{4-30}$$

In this limit, the overall reaction is *second*-order, as would be expected intuitively from the initial collision of two A molecules.

High-concentration limit: $k_{-1}[A] >> k_2$. In this case, A^* usually deactivates back to reactants rather than decomposing to products, and Eq. 4-29 simplifies to

$$\text{Rate of appearance of product} = (k_2k_1/k_{-1})[A] \tag{4-31}$$

Equation 4-31 thus shows how a reaction that begins with a bimolecular collision can nevertheless show an overall *first*-order dependence on reactant concentration. It is also seen that the first-order and second-order cases are limiting situations, so that we should in general expect a reaction order in $[A]$ somewhere between 1 and 2, depending on reactant concentration—this variation has been confirmed for a variety of gas-phase reactions. Finally, the Lindemann mechanism provides a qualitative explanation for why most chemical reactions are so slow compared to collision rates between molecules, namely that only a small fraction of the total number of collisions produce products with enough energy to break chemical bonds and complete the reaction.

Collision theory thus gives a physical picture for the origin of the observed *temperature-dependence,* $\exp[-E_a/RT]$, of experimental reaction rate constants, and for the dependence of rate constants on reactant *size* and *mass.* Although the actual magnitudes of experimental rate constants will be reduced (by some "steric" factor, $p << 1$) compared to the theoretical value from Eq. 4-25, one may still compare the steric factors for a series of homologous reactants in order to deduce the relative importance of orientation in the reaction series. Finally, collision theory gives a model for the origin of *first-order* reactions and also reactions of *nonintegral order.* Such schemes are especially common for reactions involving free radicals, and free radical reactions are of great current interest in connection with atmospheric pollutant formation and also for the possible free radical scavenging action of chemical agents that increase longevity in laboratory animals.

4.C.2. Transition-State Theory of Chemical Reaction Rates

The transition-state theory begins from examination of the *potential energy* profile at various stages along the reaction "pathway" from reactants to products, as shown schematically in Fig. 4-3. The figure is a contour diagram, in which any solid curve represents a line of constant potential energy—thus, the reactants begin in a potential energy "valley" (lower right of Fig. 4-3*a*), and climb over a "saddle point," or "pass" ("activated complex" of Fig. 4-3*a*), and then descend into another "valley" after being converted to products (upper left of Fig. 4-3*a*). [By "slicing" through the potential energy contours at a fixed value of r_2 in the figure, it is clear that the products lie in a potential energy valley (Fig. 4-3*b*).] Finally, by following the potential energy along the lowest-energy path between reactants and products, we obtain Fig. 4*c*, which shows that the reactants proceed over an energy "barrier" on their way to conversion into products.

In the collision theory of reaction rates, we found it necessary to introduce a "*steric*" factor to explain why actual reactions often proceed much more slowly than expected from the fraction of molecules having sufficient energy to react. Now for ordinary equilibrium situations, changes in the extent of molecular "alignment" (i.e., "orientation," or "steric" effects) are

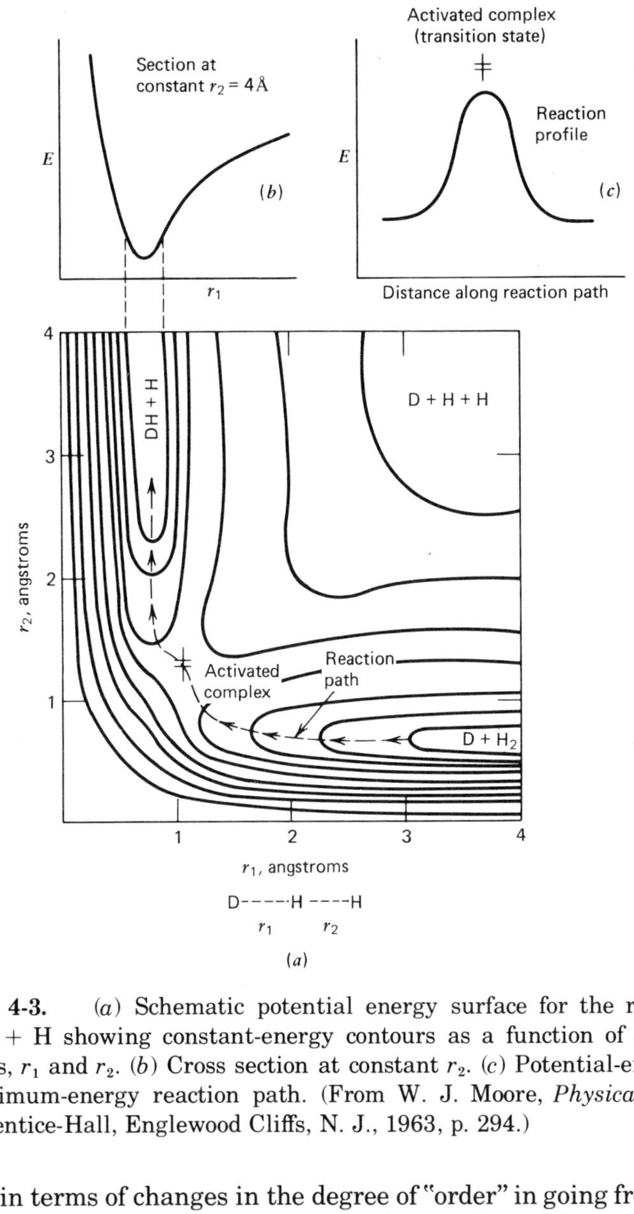

FIGURE 4-3. (*a*) Schematic potential energy surface for the reaction $D + H_2 \rightarrow DH + H$ showing constant-energy contours as a function of internuclear separations, r_1 and r_2. (*b*) Cross section at constant r_2. (*c*) Potential-energy profile along minimum-energy reaction path. (From W. J. Moore, *Physical Chemistry*, 3rd ed. Prentice-Hall, Englewood Cliffs, N. J., 1963, p. 294.)

described in terms of changes in the degree of "order" in going from one side of the equilibrium to the other, expressed as a change in *entropy* for the process:

$$\Delta G^\circ_{rx} = -RT \ln(K_a) = \Delta H^\circ_{rx} - T\Delta S^\circ_{rx} \qquad \text{at constant } T \qquad (3\text{-}11)$$

Thus, we might hope for improved understanding of chemical reaction rate constants if we could somehow inject our previous description of chemical equilibrium into the reaction path picture of Fig. 4-3. Transition-state theory is based on that connection.

In the transition-state model, it is imagined that the reactants, A and B, are in dynamic equilibrium* with the "activated complex" ("transition-state"), $AB^{\ddagger}$, of Fig. 4-3:

$$A + B \underset{}{\overset{K^{\ddagger}}{\rightleftharpoons}} AB^{\ddagger} \overset{k^{\ddagger}}{\longrightarrow} \text{products} \tag{4-32}$$

so that we may define an equilibrium constant, $K^{\ddagger}$,

$$K^{\ddagger} = \frac{[AB^{\ddagger}]}{[A][B]} \tag{4-33}$$

Since the rate of appearance of product in this scheme, Eq. 4-32, is assumed proportional to the concentration of activated complex, $AB^{\ddagger}$, we may solve for $[AB^{\ddagger}]$ in Eq. 4-33 to obtain

$$\begin{aligned} -d[A]/dt &= [AB^{\ddagger}]\cdot(\text{rate of crossing over barrier}) \\ &= [AB^{\ddagger}]\, k^{\ddagger} \\ &= k^{\ddagger}K^{\ddagger}[A][B] \end{aligned} \tag{4-34}$$

From the usual rate expression for the forward reaction

$$-d[A]/dt = k_f[A][B] \tag{4-24}$$

we can combine Equations 4-34 and 4-24 to give

$$\text{Rate constant for forward reaction} = k_f = k^{\ddagger}K^{\ddagger} \tag{4-35}$$

Using the familiar relation, Eq. 3-11, between equilibrium constant and free energy, Eq. 4-35 may be expressed

$$\begin{aligned} &\text{Rate constant for forward reaction} \\ &\qquad = k_f = k^{\ddagger} \exp\,[-\Delta G^{\circ\ddagger}/RT] \\ &k_f = k^{\ddagger} \exp\,[\Delta S^{\circ\ddagger}/R] \exp\,[-\Delta H^{\circ\ddagger}/RT] \end{aligned} \tag{4-36}$$

in which $\Delta G^{\circ\ddagger}$, $\Delta H^{\circ\ddagger}$, and $\Delta S^{\circ\ddagger}$ represent the respective changes in standard free energy, enthalpy, and entropy in going from reactants to activated complex.

* *Obviously, this cannot be a true equilibrium, since the concentrations of A and B are not constant in time. Detailed calculations however suggest that the model should nevertheless give useful results except for very small activation energies (less than about $5RT = 3000$ calories at room temperature).*

Finally, the rate of crossing over the energy barrier may be equated to the vibrational frequency, ν, of the bond(s) to be broken—in other words, if the atoms vibrate at ν oscillations per second, then there are ν opportunities per second for the bond to be broken. Since the classical energy of a vibrating bond is $h\nu$, where h is Planck's constant $= 6.62 \times 10^{-27}$ erg sec, and since the thermal energy available per vibration is kT, where k is Boltzmann's constant, we have

$$h\nu = kT \qquad (4\text{-}37)$$

so that

$$k^{\ddagger} = \nu = (kT/h) \qquad (4\text{-}38)$$

and thus

$$\boxed{\begin{aligned}\text{Rate constant for forward reaction} &= k_f \\ &= (kT/h) \exp\,[\Delta S^{\circ\ddagger}/R] \exp\,[-\Delta H^{\circ\ddagger}/RT]\end{aligned}} \qquad (4\text{-}39)$$

We can now compare the transition-state expression, Eq. 4-39, with the empirical Arrhenius expression, Eq. 4-21,

$$\frac{d \ln k_f}{dT} = \frac{E_a}{RT^2} = \frac{\Delta H^{\circ\ddagger}}{RT^2} + \frac{1}{T}$$

$$= \frac{(\Delta H^{\circ\ddagger} + RT)}{RT^2}$$

to give

$$E_a = \Delta H^{\circ\ddagger} + RT \qquad (4\text{-}40)$$

Equation 4-40 completes the connection between (theoretical) transition-state and (experimental) Arrhenius descriptions of the temperature-dependence of rate constants. Because RT is of the order of 600 calories at room temperature, while E_a for most chemical reactions is many *kilo*calories, there is only a small difference between the empirical "activation energy," E_a, and the enthalpy of activation, $\Delta H^{\circ\ddagger}$, of transition-state theory. The transition-state result, Eq. 4-39, allows us to interpret differences in chemical reaction rate constants using thermodynamic enthalpy and entropy language already familiar from our treatment of ordinary equilibria. Examples follow.

EXAMPLE *Denaturation of Proteins*

The value of Eq. 4-39, the transition state theory expression for a chemical reaction rate constant, is that it allows for separation of the *magnitude* and *temperature-dependence* of a rate constant in terms of the energy it takes to break the necessary bonds ($\Delta H^{\ddagger}$) and the change in "disorder" of the reactants ($\Delta S^{\ddagger}$). In other words, the spirit of the transition-state theory is that just as difference in free energy between *reactants* and *products* provides the best criterion for spontaneity of the *overall* reaction, the difference in free energy between *reactants* and *transition state* provides a good measure of the *progress* of the reaction along the way.

This language is particularly advantageous for describing the rate of protein denaturation reactions. For such reactions (see Problems), the activation energy can be as large as 100,000 calories/mole, since it is necessary to break a large number of (probably ionic) bonds in order to unfold a protein from its compact native conformation to a random coil chain. From Eqs. 4-39 and 4-40 it would thus appear that when $\Delta H^{\circ\ddagger}$ is a large positive number, the reaction rate constant should be very small. The "driving force" for the reaction is supplied by the very large increase in $\Delta S^{\circ\ddagger}$, the change in "disorder" in going from reactants to transition state, since there are really only a few native protein conformations but a huge number of possible denatured protein conformations. This large positive value of $\Delta S^{\circ\ddagger}$ compensates for the large positive $\Delta H^{\circ\ddagger}$ value in Eq. 4-39, and the net result is a rate constant which can be large. However, because $\Delta H^{\circ\ddagger}$ is so large, the reaction rate varies greatly over a relatively small temperature range (see Problem on denaturation of egg albumin) compared with most chemical reactions.

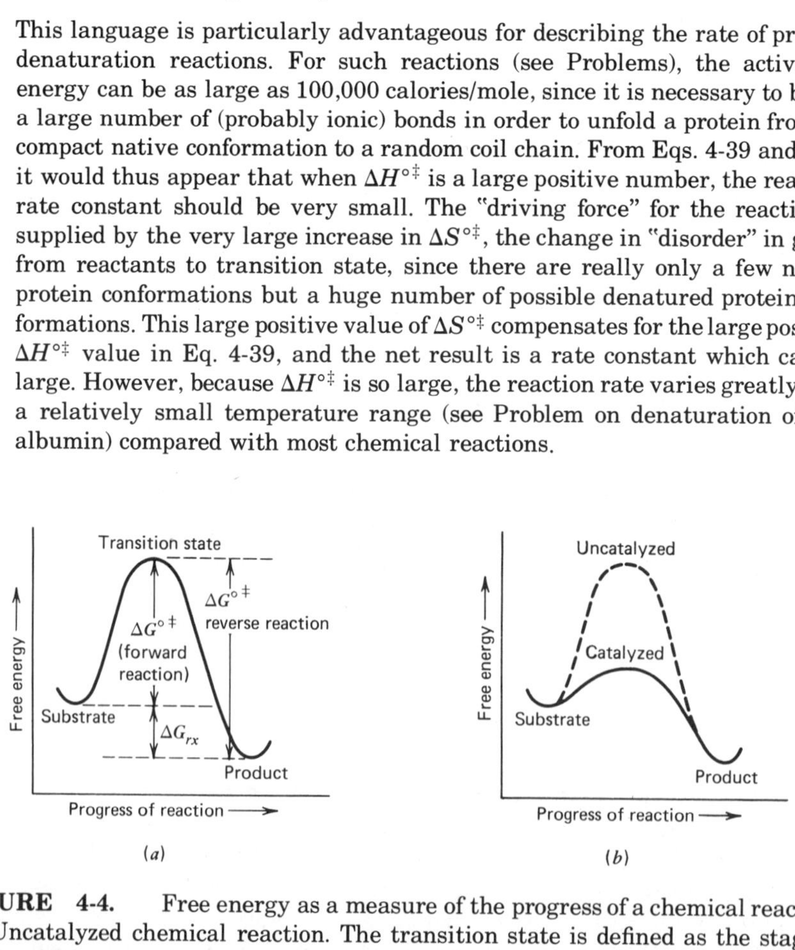

FIGURE 4-4. Free energy as a measure of the progress of a chemical reaction. (*a*) Uncatalyzed chemical reaction. The transition state is defined as the stage of maximal *free energy* between reactants and products, just as the "activated complex" was defined as the stage of maximal enthalpy between reactants and products. $\Delta G^{\circ\ddagger}$ is determined from the temperature-dependence of the reaction rate constant, according to Eq. 4-39. (*b*) Comparison of uncatalyzed and enzyme-catalyzed chemical reactions. The effect of the catalyst is to reduce the free energy of activation of the reactants.

EXAMPLE *Thermodynamic Characterization of an Enzyme-Catalyzed Reaction*

Figure 4-5 shows how it is possible to analyze the course of an enzyme-catalyzed reaction in thermodynamic terms, by assuming a mechanism of the type

$$E + A \rightleftharpoons EA \rightleftharpoons EZ \rightleftharpoons EP \rightleftharpoons E + P \qquad \text{(see Section 3)} \qquad (4\text{-}41)$$

by determining the temperature-dependence of the overall equilibrium constant (to obtain ΔH_{rx} from Eq. 4-12); the temperature-dependence of the Michaelis (pseudo-equilibrium) constant K_A for forward or K_P for reverse reaction (to obtain ΔH for binding of either A or P from Eq. 4-12); and the temperature-dependence of the forward or reverse rate constant (to obtain the "activation energy" E_a from Eq. 4-21). These ΔH values provide a full quantitative measure of the energy involved in bond breakage or bond formation for each stage of the reaction. By comparing the various ΔH values with those obtained for the uncatalyzed reaction, it is possible to make educated guesses as to the relative merits of different mechanisms for enzyme catalysis. The reader is referred to any of several current biochemistry texts for detailed discussion of enzyme catalytic mechanisms based on such data.

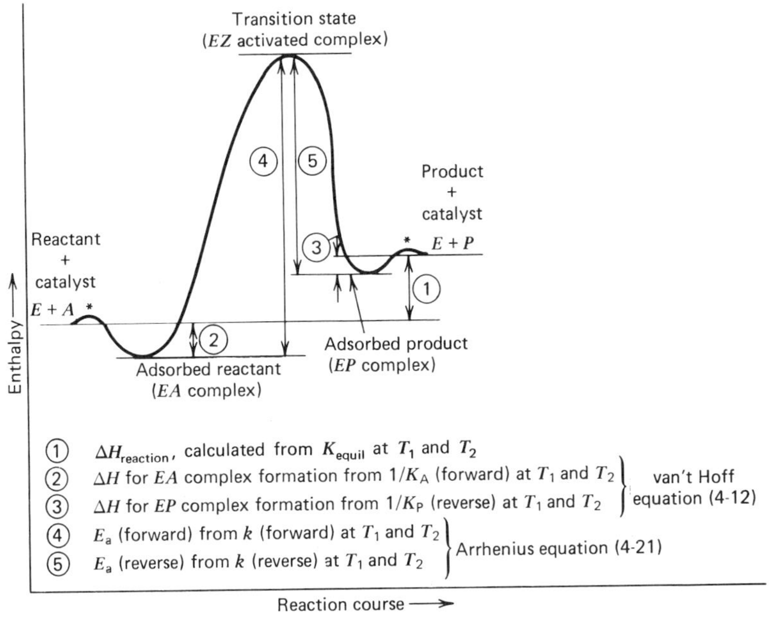

FIGURE 4-5. Progress of an enzyme-catalyzed endothermic reaction, using enthalpy as the indicator. This schematic diagram is based on the assumed mechanism given in text. (Adapted from V. R. Williams and H. B. Williams, *Basic Physical Chemistry for the Life Sciences*, 2nd edition, W. H. Freeman, San Francisco, 1973, p. 320.)

PROBLEMS

1. The free energy available from complete oxidation of one mole of glucose to carbon dioxide and water is shown below,

$$6O_2(1\text{ atm, ideal gas}) + C_6H_{12}O_6(1\text{ M, ideal solution}) \rightleftharpoons 6CO_2(1\text{ atm, ideal gas}) + 6H_2O(\text{pure } H_2O)$$

$\Delta G^\circ_{rx} = -675$ kcal/mole glucose, where the standard states of all components are as shown above

The free energy required to convert 1 mole of ADP to 1 mole of ATP using the same standard states for reactants and products is 7.3 kcal/mole ADP (pH 7.0).

(a) Compute the *maximum* number of moles of ATP that could be formed from ADP and phosphate, if *all* the energy of combustion of one mole of glucose could be utilized.

(b) The *actual* number of moles of ATP formed by a living cell from one mole of glucose is about 38. Using the result of part (a), compute the *efficiency* of energy conversion by the cell.

(c) Consider now some more typical physiological conditions:

$P_{CO_2} = 40$ mm pressure
$P_{O_2} = 100$ mm
[glucose] $= 1$ mg/cm^3
[ATP] = [ADP] = $[HPO_4^=] = 0.0001$ M
pH = 7.0

We have already shown that the energy to convert a mole of ADP to ATP under these conditions is about 12 kcal/mole ADP. Calculate the free energy for oxidation of one mole of glucose under these conditions.

(d) Finally, using the conditions of part (c), again calculate the efficiency of energy conversion by the cell.

(e) For an ordinary heat engine (i.e., a device that converts heat—in this case, heat of combustion—into mechanical work), the maximum possible efficiency is $(T_2 - T_1)/T_2$, where T_1 and T_2 are the (absolute) temperatures of the working substance in different parts of the engine. A diesel engine approaches within about 75% of this theoretical efficiency. If typical operating temperatures within the diesel engine are 1650°C and 600°C, compute the expected actual efficiency of such an engine. Compare to part (d). Suggest a reason why the biological conversion (which is isothermal) is so much more efficient.

2. (a) A common rule of thumb for many chemical reactions is that the reaction rate doubles for every 10°C increase in temperature. If the ratio of the rate constants at two temperatures that differ by 10°C is called, "Q_{10}," calculate the activation energy, E_a, corresponding to $Q_{10} = 2$ between 25°C and 35°C.

(b) Fertilization of the ovum is a process whose rate has a Q_{10}-value of about 600. Calculate the activation energy for this process.

(c) The denaturation of egg albumin is a process whose activation energy is of the order of 100,000 cal/mole. Calculate how much longer it should take to boil an egg at an altitude of 4000 feet (atmospheric pressure = 656 mm Hg, so that water boils at about 96°C) compared to sea level.

3. If $\bar{G}^\circ_{H^+}$ is the (standard) chemical potential for aqueous H^+ ions in their standard ($a_{H^+} = 1$) state of 1.00 M in an ideal solution, and if $G^{\circ\prime}_{H^+}$ is the (standard) chemical potential for aqueous H^+ ions in the (different) standard ($a_{H^+} = 1$) state of 10^{-7} M in an ideal solution, show that

$$\bar{G}^{\circ\prime}_{H^+} = \bar{G}^\circ_{H^+} - (2.303)7RT,$$ where R is the gas constant per mole, and T is absolute temperature

4. Combine the Arrhenius ($k = A\ \exp\ [-E_a/RT]$) and van't Hoff $\left(\frac{\partial \ln K_a}{\partial(1/T)} = -\frac{\Delta H^\circ_{rx}}{R}\right)$ equations to show that

$$\Delta H^\circ_{rx} = E_a(\text{forward}) - E_a(\text{reverse})$$

5. From the data plotted below, compute ΔH°_{rx} for the denaturation of chymotrypsinogen at pH 2.0 and pH 3.0. Then calculate the equilibrium constant and ΔG°_{rx} for this process at 25°C and 37°C—is the zymogen appreciably denatured at these temperatures? Finally, compute ΔS°_{rx} at both pH values and both temperatures. What do you conclude about the temperature-dependence of H°_{rx} and ΔS°_{rx}?

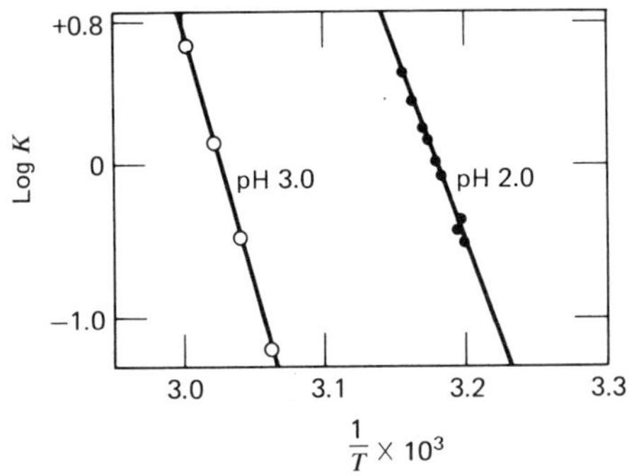

Plot of log K against $1/T$ for the denaturation of chymotrypsinogen at pH 2.0 and at pH 3.0 (M. A. Eisenberg and G. W. Schwert: *J. Gen. Physiol. 34,* 583, 1951.)

6. It is well known that most proteins are denatured by the presence of concentrated (say, 8M) urea. It is also well known that this process usually proceeds much more rapidly as the temperature of the solution is increased. Recently [R. E. Pincock and W.-S. Lin, *J. Agric. & Food Chem. 21,* 2 (1973)], it has been shown that α-chymotrypsin in the presence of dilute (0.05M urea) is actually denatured more *rapidly* as the temperature is *decreased* below 0°C! From your knowledge of the phase rule, see if you can account for this apparent *negative* activation energy.

7. For the enzyme-catalyzed conversion of fumaric acid to L-malic acid,

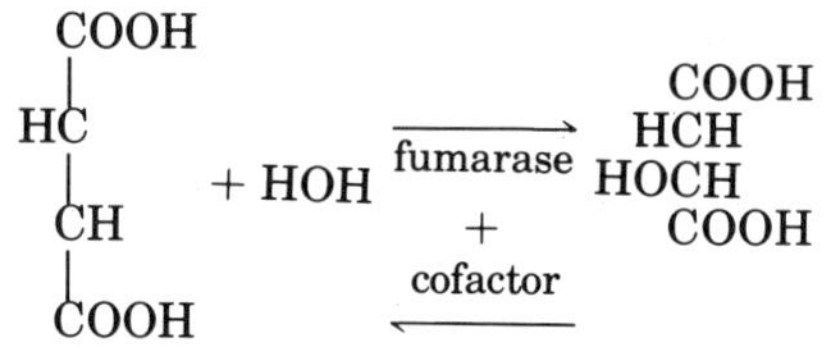

Fumaric acid Water L-malic acid

the temperature-dependence of the equilibrium binding constant (K_a), the forward and reverse rate constants (k_f and k_r), and the Michaelis constant (K_A and K_P) for both forward and reverse reactions have been determined:

$$\begin{aligned}
\partial \ln K_a/\partial(1/T) &= +1812°\text{K} \\
\partial \ln k_f/\partial(1/T) &= -3070°\text{K} \\
\partial \ln k_r/\partial(1/T) &= -7599°\text{K} \\
\partial \ln K_A/\partial(1/T) &= -2114°\text{K} \\
\partial \ln K_P/\partial(1/T) &= +604°\text{K}
\end{aligned}$$

Assuming that the Michaelis constant may be treated as an equilibrium (dissociation) constant for the binding of substrate (K_A) or product (K_P) to enzyme, sketch the enthalpy profile of this reaction as a function of reaction path. Identify: reactants, products, enzyme:substrate complex, enzyme:product complex, and activated complex (transition state).

REFERENCES

W. J. Moore, *Physical Chemistry,* any edition, Prentice-Hall, Englewood Cliffs, N. J. Good treatment of equilibrium and spontaneity.

C. Kitzinger and T. H. Benzinger, "Methods of Biochemical Analysis," *8,* 309 (1960); H. D. Brown, *Biochemical Microcalorimetry,* Academic Press, N. Y. (1969). Discussion of differential scanning microcalorimetry with applications to biological macromolecules.

W. E. Wentworth and S. J. Ladner, *Fundamentals of Physical Chemistry,* Wadsworth Publishing Company, Belmont, Calif. (1972). Good discussion of temperature-dependence of chemical reaction rate constants.

CHAPTER 5 Electrochemical Potential

So far, we have shown how a difference in chemical potential can arise between two solutions that differ in solvent concentration (osmotic pressure), charged solute concentration (Donnan equilibrium), chemical reactivity, or temperature. By suitable experimental arrangement (see below), it is possible to convert these differences in *chemical* potential to differences in *electrical* potential that may then be measured with a voltmeter (potentiometer). We shall consider electrochemical cells based on differences in solute concentration (concentration cells) and chemical reactivity (fuel cells). Concentration cells are reaching widespread use as devices for measuring the difference in concentration between an unknown and known solution, from which the unknown concentration may be determined. Fuel cells have long been used as devices for extracting ΔG_{rx}, ΔH_{rx}, and ΔS_{rx} from chemical reactions, and are gaining much consideration as practically useful means for storing energy in chemical form for later conversion to electrical power. We begin by examination of a device that can convert a difference in chemical potential (arising from a concentration difference) into a measurable voltage.

5.A. CONCENTRATION CELLS

Consider the two solutions in Fig. 5-1. A metal electrode (e.g., a metal wire) dips into the left-hand solution containing a salt of the metal, A^+X^-, at concentration, $[A^+]_L$. An identical metal electrode dips into the right-hand solution containing the same salt at a different concentration, $[A^+]_R$. Consider next what happens when a mole of electrons passes through the circuit in the direction shown (Fig. 5-1 shows the usual convention that electrons pass from left to right in the external circuit). For this to occur, electrons must be produced at the left-hand electrode by the reaction

$$A^\circ \rightarrow [A^+]_L + e^- \tag{5-1a}$$

and so that electrons do not pile up on the right-hand electrode, there must be a corresponding consumption of electrons there:

$$[A^+]_R + e^- \rightarrow A^\circ \tag{5-1b}$$

Because there are no free electrons in either solution, the electrical current must result from the migration of charged ions through the solution as

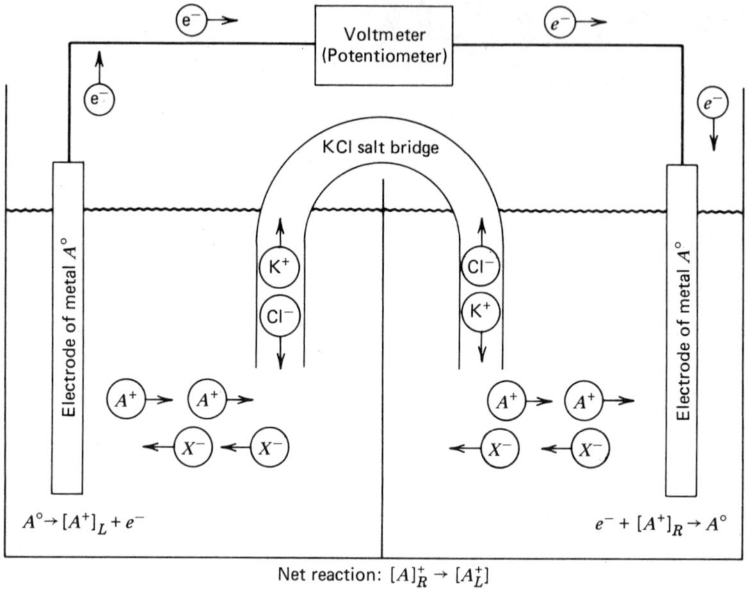

FIGURE 5-1. Electrochemical concentration cell. Operation is described in text.

shown in the figure. The salt bridge is simply a device that provides for *electrical contact* between the two solutions (i.e., electrical current, in the form of moving K^+ and Cl^- ions, can flow between the solutions), but prevents the two solutions from mixing, so that the metal ion concentrations in the two solutions remain different during the experiment. KCl is the usual choice for the salt bridge, since K^+ and Cl^- ions move at about the same speed under the influence of an electric field—when the cations and anions move at different speeds, an additional "liquid junction potential" arises as the boundary region between solution and salt bridge becomes relatively depleted of ions of a given charge (see discussion of "disc" electrophoresis in Section 2 for an experiment based on such an effect).

The electrical work done in moving n moles of *electrons* from the left-hand to the right-hand electrode is given by

$$\begin{aligned}\text{Electrical work} &= (\text{charge}) \cdot (\text{electrical potential difference}) \\ &= nF\text{E} \qquad (5\text{-}2)\end{aligned}$$

in which F is the charge of one mole of electrons (96,500 coulomb) and E is the electrical potential difference between the two electrodes as measured (assuming zero liquid junction potential) by the voltmeter of Fig. 5-1. Alternatively, we can note that the net effect of the process as far as the *solutions* are concerned is to transport n moles of A^+ ions from a solution of concentration, $[A^+]_R$ to a solution of concentration, $[A^+]_L$, for which the change in free energy is given by (refer to Equations 3-2 and 3-3):

$$\begin{aligned}\Delta G &= G_L - G_R \\ &= n(\overline{G}_{A^+})_L - n(\overline{G}_{A^+})_R \\ &= n(\overline{G}^\circ_{A^+})_L + nRT\ \ln(a_{A^+})_L - n(\overline{G}^\circ_{A^+})_R - nRT\ \ln(a_{A^+})_R\end{aligned}$$

or

$$\Delta G = nRT\ \ln([A^+]_L/[A^+]_R) \tag{5-3}$$

in which we have taken the solutions as sufficiently dilute that $a_{A+} = [A^+]$, and we have used the same standard state for A^+ ions in each solution. Recalling now our early relation

$$\Delta G = -(\text{nonPV work done } by \text{ system}) \qquad \text{(Table 1-3)}$$

we obtain,

$$\boxed{\Delta G = -nF\mathrm{E}} \tag{5-4}$$

Finally, by combining Equations 5-3 and 5-4, we have the relation

$$\boxed{\mathrm{E} = -(RT/F)\ \ln([A^+]_L/[A^+]_R)} \tag{5-5}$$

Equation 5-5 is a special case of the *Nernst equation* (see section 5.B. for the general case), and expresses the electrochemical potential, E, that corresponds to the stated concentration difference between two otherwise identical solutions. Finally, it is important to note that the electrochemical potential in this example is determined solely by the A^+ ions that are effectively transported, because we have chosen an electrode (metallic A) that responds (i.e., dissolves or grows) only to changes in the concentration of A^+ ions. If other ions could react with the electrode (see below), they would also be effectively "transported" and the measured electrochemical potential would reflect their concentrations also.

EXAMPLE *Membrane Potential in Living Cells*

A living cell actively accumulates K^+ ions within the cell and extrudes Na^+ outside the cell. For a typical ratio of K^+ concentrations inside and outside the cell of about 20:1

$$([\mathrm{K}^+]_{\text{inside}}/[\mathrm{K}^+]_{\text{outside}}) = 20$$

we can compute the corresponding electrochemical potential across the membrane from the special Nernst equation (5-5) at physiological temperature of 37°C:

$$E = \frac{(1.987 \text{ cal mole}^{-1}\text{deg}^{-1})(310°\text{K})(4.185\, joule/cal)}{(1 \text{ mole e}^{-})(96{,}500 \text{ coul/mole e}^{-})}(2.303)\log(20)$$

$$E = 0.080 \text{ Volt} = 80 \text{ millivolt}$$

This calculated potential is similar to that observed for a resting living cell. The changes in living cell membrane potential that occur during the propagation of nerve impulses or muscle contraction result from sudden fluxes of K^+ (outward) and Na^+ (inward) to produce a sudden decrease in membrane potential observable on a voltmeter using microelectrodes inside and outside the living cell.

EXAMPLE *Ion-Selective Membrane Electrodes*

The concentration cell of Fig. 5-1 can in principle be made sensitive to differences in concentration of *any* ionic solute between the left- and right-hand solutions. However, it is possible to make the current flow (and hence the potential measurement) *selective* for just a few or even just one type of ion, by replacing the (nonselective) KCl salt bridge with a substance that will bind *only* the ion of interest, as shown in Fig. 5-2. There are three common choices for the selective intermediary ion-carrier: glass, liquid, and solid.

Glass ion-selective electrode: pH measurement. The first ion-selective electrode to be invented was simply a thin (0.05mm) glass membrane. When the glass membrane is soaked for a few hours in water, the metallic cations normally present in the glass are replaced by H^+ ions at the glass surface. An electric current can then pass through the glass membrane: as protons bind on one side of the membrane, Na^+ ions move from one silicate ion to another *within* the glass and protons are eventually displaced into solution from the other side of the membrane. For suitable types of glass (quartz and pyrex will *not* work), only H^+ ions are able to bind to the membrane, and thus current will flow from one side of the membrane to the other only when there is a *proton* concentration difference across the membrane. The resulting electrochemical potential is again described by Eq. 5-5, but if the concentration of protons on one side of the membrane is fixed at a known value (say, a proton concentration of about 0.12 M, corresponding to a proton activity of 0.10), then the measured electrochemical potential provides a direct measure of the hydrogen ion activity in the sample solution (right-hand in Fig. 5-2), and the device gives a direct measurement of pH:

$$\text{pH} = -\log_{10}(a_{\text{H}^+})$$

$$E = (RT/F)\ln\left(\frac{(a_{H^+})_R}{0.1}\right) = -(RT/F)[(\text{pH})_R - 1](2.303) \qquad (5\text{-}6)$$

Most glass electrodes respond to cations other than H^+, such as Na^+, particularly at high pH, when the proton concentration may be much smaller than

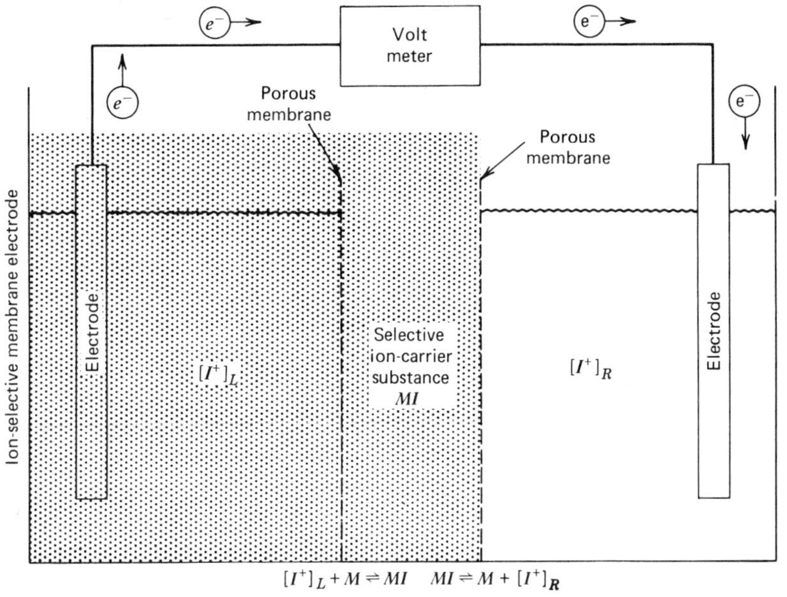

FIGURE 5-2. Highly schematic diagram of the operation of an ion-selective membrane electrode. The difference in chemical potential between the left- and right-hand solutions due to a difference in activity (concentration) of ion I^+ appears as a difference in electrical potential measured as a voltage between the two identical electrodes as in Fig. 5-1 for an ordinary concentration cell. However, in this case, the only way that current can flow from one solution to the other is by binding of I^+ ions to a carrier substance M at the left-hand boundary between solution and porous membrane, with release of I^+ ions from the MI complex at the right-hand boundary as shown at the bottom of the figure. Provided that the ion-carrier substance, M, is specific for binding of I^+ (i.e., M will bind I^+ but no other ions), the measured electrical potential will reflect the concentration only of I^+ in the sample (right-hand) compartment, since the I^+ concentration within the ion-selective electrode (left-hand) compartment is fixed and known.

that of other cations in the solution. Some glass electrodes have been designed with a composition such that the response to a particular cation is maximized: thus, it is now possible to make glass-membrane electrodes that are (relatively) specific for Na^+, Tl^+, K^+, Cu^+, NH_4^+, Rb^+, Cs^+, Li^+, or Ag^+. However, the selectivity of these electrodes does not compare to the high selectivity available with other types (see below).

Liquid ion-exchange electrode: activity coefficients. By substitut'ng a solution containing a specific ion-binding substance for the porous glass oı the previous example, it is possible to extend the response and selectivity of the electrode to include polyvalent cations and anions (the glass electrode responds only to univalent cations). For example, when the ion-binding substance is an organic phosphoric acid that binds calcium ions more strongly than most other cations, the electrode is specific for calcium. Finally, since there are two (moles of) electrons transferred per (mole of) calcium transported, Eq. 5-4 may be rewritten

$$\boxed{\Delta G = -nzF\text{E}} \tag{5-7}$$

where n now denotes the number of moles of transported ions (Ca^{++} in this case) and z is the number of moles of electrons transferred per mole of ions transported ($z = 2$ in this case). Thus, by choosing a reference solution (within the electrode, as for the pH glass electrode) of fixed calcium concentration, the liquid ion-exchange Ca^{++} electrode will give a response of the form

$$\text{E} = \text{constant} + (RT/2F)\ \ln(a_{\text{Ca}^{++}}) \tag{5-8}$$

and E provides a direct measure of calcium activity in a sample solution. Because the measured voltage varies as the logarithm of calcium activity, the electrode will respond to a very wide range of calcium concentration, as shown in Fig. 5-3. Furthermore, because the electrode response is related to calcium *activity*, knowledge of the calcium *concentration* gives a direct measure of the *activity coefficient* for calcium at that concentration. In fact, electrochemical potential measurements have provided much of the existing data on activity coefficients of ions in solution.

$$a_{\text{Ca}^{++}} = \gamma_{\text{Ca}^{++}}\, m_{\text{Ca}^{++}} \tag{3-9b}$$

Liquid ion-exchange membrane electrodes have already been prepared successfully for Ca^{++}, NO_3^-, $CO_3^=$, ClO_4^-, Cu^{++}, BF_4^-, Mg^{++}, Mn^{++}, Cl^-, and also

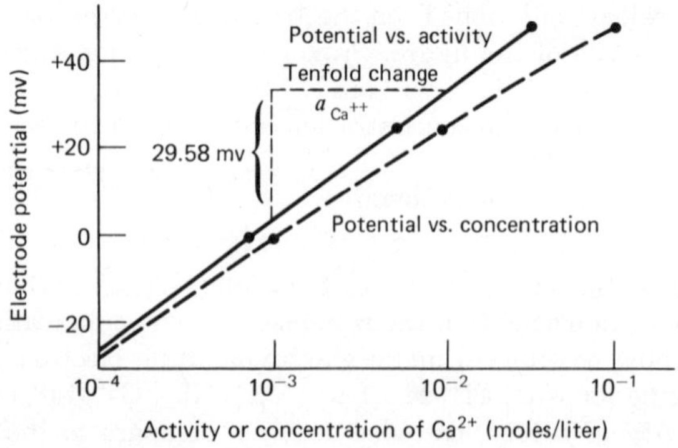

FIGURE 5-3. Response of a liquid ion-exchange calcium electrode to variations in concentration (and activity) of solutions prepared from pure calcium chloride. In accordance with Eq. 5-8, the potential changes by $(RT/2F) = 29.6$ mV for a tenfold change in calcium *activity*. Finally, since the calcium *concentration* is higher than the calcium *activity* corresponding to a given potential, the calcium activity coefficient is less than unity, and decreases with increasing calcium concentration (Eq. 3-29 and Fig. 3-1). (Data courtesy of Orion Research Incorporated, Cambridge, Mass.)

for several charged organic species. It is reasonable to expect a wide expansion in both the variety of ions detectable and the selectivity toward a given ion as these new electrodes are improved to take advantage of a wide variety of known selective ion-binding substances.

Solid-state membrane electrode. One problem with the preceding electrodes is their fragility, since it is usually necessary to encapsulate the electrode with a very thin membrane to ensure access of the sample ions to the ion-selective substance. This problem is greatly reduced by use of solid-state or crystalline materials as the ion-selective component. For example, LaF_3 is a relatively insoluble solid with much higher ionic conductivity for F^- than for any other ion, so that a solid-state membrane electrode can be constructed using a single crystal of LaF_3 as the ion-selective barrier between the sample and reference (in this case, fluoride) solution. The response of the electrode is again of the form of Eq. 5-8 with $z = 1$. By using crystals of Ag_2S (an ionic conductor for Ag^+) rather than LaF_3, a solid-state electrode will respond to Ag^+, $S^=$, Cu^{++}, Pb^{++}, Cd^{++}, I^-, Br^-, and Cl^-, since those ions either substitute for or bind to Ag^+ at the surface of the crystal. Very recently, it has been observed that the Ag_2S crystal electrode will respond to the presence of protein molecules, and will respond differently to the same concentrations of different proteins (Fig. 5-4). It seems likely that this effect is related to the number of disulfide groups on the protein, so that the Ag^+ ions react with protein cysteic residues to provide the selective reaction. Finally, it has been noted that the response of the Ag_2S electrode to a given protein can be affected by the presence of antibodies specific to that protein, so it seems that it should be possible to construct a solid-state electrode with a suitable antibody or hormone receptor bound to the membrane surface, as a means for detecting a wide range of immunochemicals in blood without any required sample preparation. It is intriguing to speculate that with such (future) methods, it may one day be possible to detect disease simply by using a pH meter with a suitable ion-selective electrode.

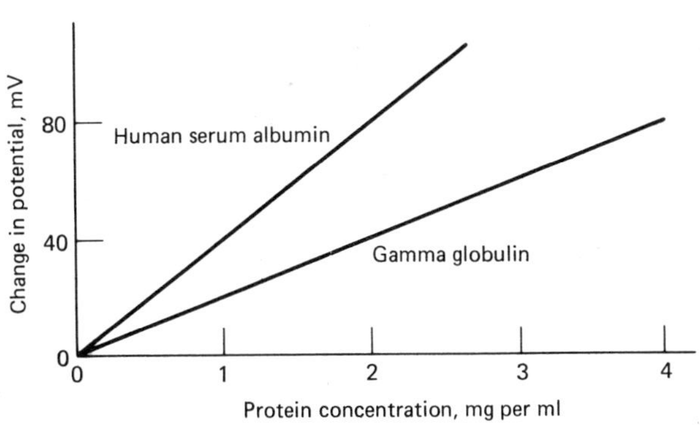

FIGURE 5-4. Response of Ag_2S crystal membrane electrode to two blood proteins. (See text.) [Data from P. W. Alexander and G. A. Rechnitz, *Anal. Chem. 46,* 860 (1974).]

5.B. FUEL CELLS

Returning to the experimental arrangement of Fig. 5-1, suppose that the two solution compartments now contain two *chemically different* reactants (e.g., Cu^{++} and Zn^{++}, in the example of Fig. 5-5), rather than the *same* species at two different *concentrations*. Figure 5-5 then shows that the net effect of the passage of 1 mole of electrons through such a circuit will be that 1/2 mole of Zn metal is oxidized to Zn^{++} in the left-hand compartment, while 1/2 mole of Cu^{++} is reduced to Cu metal in the right-hand compartment to give a net reaction:

$$\mathrm{Zn} + \mathrm{Cu}^{++} \rightarrow \mathrm{Zn}^{++} + \mathrm{Cu} \tag{5-9}$$

More generally, for the net reaction,

$$aA + bB \rightarrow cC + dD \tag{3-4}$$

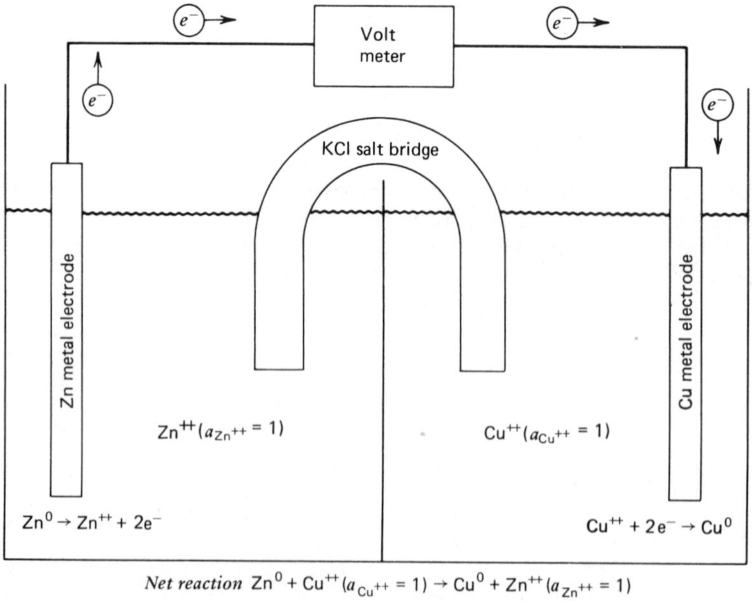

FIGURE 5-5. Electrochemical fuel cell. The passage of one mole of electrons through this circuit results in the net conversion of 1/2 mole of Zn metal to Zn^{++} ions at unit activity, and 1/2 mole of Cu^{++} ions at unit activity to Cu metal as shown. Depending on which way the electrons are permitted (or forced, using an externally applied voltage) to move around the circuit, either Zn or Cu metal will be plated out at the expense of the other.

we can combine Equations 4-3 and 5-7 to obtain the *Nernst equation* 5-10:

$$\boxed{\mathrm{E}_{rx} = \mathrm{E}^\circ_{rx} - (RT/zF)\ \ln \frac{a_C^c a_D^d}{a_A^a a_B^b}} \qquad (5\text{-}10)$$

in which E is the observed electrochemical potential corresponding to the process of Eq. 3-4, where the component solute concentrations correspond to the activities in Eq. 5-10, R is the gas constant per mole, T is absolute temperature, F is the charge of a mole of electrons, and z is the number of electrons transferred per mole of reactant converted to product. In particular, when all components are present at their standard concentrations ($a_A = a_B = a_C = a_D = 1$)

$$\mathrm{E}_{rx} = \boxed{\mathrm{E}^\circ_{rx} = -\frac{\Delta G^\circ_{rx}}{nzF}} \qquad n = \text{number of moles of reactant converted to product} \qquad (5\text{-}11)$$

where ΔG°_{rx} is the standard free energy change for the process, Eq. 3-4. Finally, since it is usual to define ΔG_{rx} and ΔG°_{rx} *per mole of reactant* converted to product, Equations 5-11 and 5-7 are often written without the factor, *n*, as in Eq. 5-12.

5.B.1. Analytical Applications: Electrochemical Determination of ΔG_{rx}, ΔH_{rx}, ΔS_{rx}, and K_a

From the preceding discussion, it is clear that a measurement of E_{rx}, using an electrochemical cell with reactant A and product D on one side and reactant B and product C on the other side, is a direct measure of ΔG_{rx} for conversion of a mole of reactant(s) to product(s), at the particular concentrations of all components present:

$$\boxed{\Delta G_{rx} = -zF\mathrm{E}_{rx}, \text{ in which } z \text{ is the number of moles of electrons that pass through the circuit per mole of reactant converted to product.}} \qquad (5\text{-}12)$$

Thus, by determination of E_{rx} for several different concentrations of components, Equations 5-10 and 5-12 can be combined to give a value for ΔG°_{rx}, from which the equilibrium constant, K_a, may be obtained (see following Example).

Because we are often interested in just *one* of the half-reactions required to define an overall redox reaction, it is convenient to choose a *particular* half-reaction as a reference, and define its E°_{rx} as zero. The most usual choice is the half-reaction,

$$(1/2)H_{2(1\ atm)} \rightleftharpoons H^{+}_{(a_{H^+}=1)} + e^-;\ [H^+] = 1.0\ M \text{ in ideal solution,}$$
$$E^\circ \equiv 0.000 \text{ for reaction in either direction.} \tag{5-13}$$

Thus, a positive E for a particular half-reaction indicates that the reaction as written would be capable of reducing H^+ to H_2 when coupled to the "standard" hydrogen electrode half-reaction of Eq. 5-13.

EXAMPLE *Reduction of Cytochrome c*

Cytochrome *c*, a heme-containing protein widely distributed in living organisms, has a central role in the electron-transfer reactions of aerobic metabolism. The mechanism of reduction and oxidation of the heme iron is therefore of general interest. For the reaction

$$\text{cytochrome } c(Fe^{+++}) + e^- \rightleftharpoons \text{cytochrome } c(Fe^{++}) \tag{5-14}$$

it happens that the optical absorbance at 550 nm is sufficiently different for the reduced Cyt^{II} and the oxidized Cyt^{III} that their respective concentrations are readily determined. Figure 5-6 is a plot of experimental E_{rx} as a function of the ratio of reduced to oxidized cytochrome *c*, where the other half-reaction is the standard hydrogen electrode of Eq. 5-13. From Eq. 5-10, we readily obtain the standard half-cell potential as +0.244 volt at 22°C, so that the overall reaction of Eq. 5-15 is spontaneous.

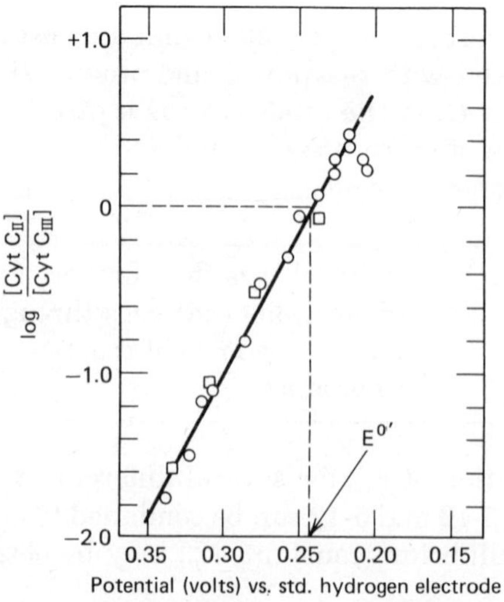

FIGURE 5-6. Equilibrium ratio (log scale) of reduced to oxidized cytochrome *c*, at various electrical potentials. The slope and zero point of this titration give the number of electrons transferred per cytochrome molecule reduced and the standard potential for the reaction. [From S. R. Betso, M. H. Klapper, and L. B. Anderson, *J. Amer. Chem. Soc. 94,* 8197 (1972).]

$$Cyt^{III}_{(pH7.0)} + \tfrac{1}{2}H_{2(1\ atm)} \rightarrow Cyt^{II}_{(pH7.0)} + H_{([H^+]=10^{-7}\ M\ \text{in ideal solution})}; \quad E^{\circ\prime}_{rx} = +0.244\ \text{volt} \tag{5-15}$$

Moreover, from the slope of the plot in Fig. 5-6, we can say that $z = 1$ electron transferred per mole of Cyt^{III} converted to Cyt^{II}, so the reaction is a "one-electron process."

The standard half-cell potentials for a variety of biologically significant redox reactions are listed in Table 5-1. Note that the convention for the potentials is that the listed reaction is for *reduction* of the listed compound, and the potential is measured against a standard hydrogen electrode (Eq. 5-13) in which hydrogen gas is oxidized to H^+ ions at unit activity. Finally, as with our previous $\Delta G^{\circ\prime}_{rx}$ values (Table 4-1), it is useful to take the standard concentration of H^+ as 10^{-7} M for the listed reaction, in order that the scale of $E^{\circ\prime}$ values more accurately reflect physiological conditions.

Table 5-1 Standard Cell Potentials for the Listed Half-reactions at pH 7.00 and Unit Activity of All Other Components, Relative to a Standard Hydrogen Electrode of $[H^+] = 10^{-7}$ M in an Ideal Solution. (Temperature = 25°C, with z the Number of Electrons Transferred Per Mole of Reactant Converted to Product)

Half-Reaction			
Oxidized Reactant	Reduced Product	z	$E^{\circ\prime}$ (volts)
α-Ketoglutarate	Succinate + CO_2	2	−0.67
Acetate	Acetaldehyde	2	−0.60
Ferridoxin (oxidized)	Ferridoxin (reduced)	1	−0.43
$2H^+$ (pH = 0.0)	H_2 (1 atm)	2	−0.41
NAD^+	NADH + H^+	2	−0.32
$NADP^+$	NADPH + H^+	2	−0.32
Lipoate (oxidized)	Lipoate (reduced)	2	−0.29
Glutathione (oxidized)	Glutathione (reduced)	2	−0.23
Acetaldehyde	Ethanol	2	−0.20
Pyruvate	Lactate	2	−0.19
Fumarate	Succinate	2	+0.03
Cytochrome *b* (+3)	Cytochrome *b* (+2)	1	+0.07
Dehydroascorbate	Ascorbate	2	+0.08
Ubiquinone (oxidized)	Ubiquinone (reduced)	2	+0.10
Cytochrome *c* (+3)	Cytochrome *c* (+2)	1	+0.22
Ferricyanide	Ferrocyanide	1	+0.36
$(1/2)O_{2(1\ atm)}$ + $2H^+$(pH 7)	H_2O	2	+0.82

EXAMPLE *Determination of Very Large or Very Small Equilibrium Constants*

When an equilibrium constant is of the order of unity, it is often possible to measure the concentration of both reactants and products directly, typically from optical absorbance. However, for many interesting reactions such as enzyme-inhibitor, hormone-receptor, antibody-antigen, and drug-receptor binding, the equilibrium binding constant may be so large that it is not feasible to detect the concentration of "un-bound" species. In such cases, electrochemical potential measurements can provide a simple means for determination of the equilibrium binding constant, as shown for the following example.

Consider the dissociation of mercuric sulfide in water:

$$HgS \rightarrow Hg^{++} + S^{=} \tag{5-16}$$

From a table of standard half-reactions similar to Table 5-1, we readily locate the relevant reactions that when combined yield the reaction of interest along with its corresponding E° value:

$$\begin{array}{rl} Hg \rightarrow Hg^{++} + 2e^- & E^0 = -0.854 \text{ volt} \\ 2e^- + HgS \rightarrow Hg + S^{=} & E^0 = -0.70 \text{ volt} \\ \hline HgS \rightarrow Hg^{++} + S^{=} & E^0 = -1.554 \text{ volt} \end{array}$$

From this standard electrochemical potential, we then obtain the standard free energy change per mole of reactant converted to product:

$$\begin{aligned} \Delta G^\circ_{rx} &= -zFE^0 \qquad (5\text{-}11) \\ &= -(2)(96{,}500)(-1.554) = +300{,}000 \text{ joule/mole HgS} \end{aligned}$$

From ΔG°_{rx}, we quickly obtain the equilibrium constant (in this case, the solubility product) for the reaction in which HgS dissolves to form the dissociated Hg^{++} and $S^{=}$ ions in water at 25°C:

$$\Delta G^\circ_{rx} = -RT \ln K_a \tag{3-11}$$

$$K_a = \exp\left[\frac{-300{,}000 \text{ joule}}{(1.987 \text{ cal mole}^{-1}\text{deg}^{-1})(298°\text{K})(4.185 \text{ joule/cal})}\right]$$

$$K_a = 2.7 \times 10^{-53} = [Hg^{++}][S^{=}]$$

in which $[Hg^{++}]$ and $[S^{=}]$ are the concentrations of Hg^{++} and $S^{=}$ in a *saturated* solution of HgS.

The solubility of HgS is therefore $(2.7 \times 10^{-53})^{1/2} = 5.2 \times 10^{-27}$ M, which corresponds to one dissociated HgS molecule in about 300 liters of solution. Obviously, there is no means for direct detection of such a small concentration! There is great current interest in mercury-sulfur complexes, since mercury has such a strong affinity for sulfur in proteins and sulfur-containing amino acids—for these cases, electrochemical potential measurements provide the simplest means for deciding which of several very strong mercury-sulfur bonds is the strongest and will determine the state of mercury in the organism.

EXAMPLE *ΔH_{rx} and ΔS_{rx} from Temperature-dependence of* E_{rx}

Since we already know the relation between ΔG_{rx} and E_{rx}, and that between ΔG_{rx} and ΔH_{rx} and ΔS_{rx}, it is straightforward to obtain ΔS_{rx} and ΔH_{rx} from the temperature-dependence of E_{rx} as follows:

$$\boxed{\Delta G_{rx} = -zF\mathrm{E}_{rx}} \quad (5\text{-}12)$$

$$\Delta G_{rx} = \Delta H_{rx} - T\Delta S_{rx} \text{ at constant temperature} \quad \text{(Table 1-3)}$$

so

$$\boxed{\Delta S_{rx} = -\,(\partial \Delta G_{rx}/\partial T)_P = zF(\partial \mathrm{E}_{rx}/\partial T)_P} \quad (5\text{-}17)$$

and

$$\Delta H_{rx} = \Delta G_{rx} + T\Delta S_{rx} \text{ at constant temperature}$$

or

$$\boxed{\Delta H_{rx} = -\,zF\ \mathrm{E}_{rx} + zFT(\partial \mathrm{E}/\partial T)_P} \quad (5\text{-}18)$$

Thus, measurement of E_{rx} and its temperature-variation provides for determination of all the thermodynamic quantities of principal interest: ΔG_{rx}, ΔH_{rx}, ΔS_{rx} and K_a. (ΔG_{rx}, ΔH_{rx}, and ΔS_{rx} in Eqs. 5-12, 5-17, and 5-18 are each defined per mole of reactant converted to product.)

5.B.2. Preparative Applications: Electroplating and Batteries

In the previous section, we considered electrochemical experiments in which the main goal was the determination of some *property* (concentration, K_a, ΔG_{rx}, ΔH_{rx}, ΔS_{rx}) of a substance or reaction mixture. In this section, we conclude with a brief look at some experiments in which the main goal is the *preparation* of a quantity of some compound or the production of a quantity of current (and thus power) using larger-scale electrochemical cells. Some relevant standard half-cell reduction potentials are collected for this purpose in Table 5-2 for a number of common electrode reactions.

EXAMPLE *Electroplating for Separation of Components of a Mixture*

In Section 2, we will discuss ion exchange chromatography as a means for large-scale separation of two or more ionic solutes in a mixture. In that case, the basis for separation is a difference in charge, size, and/or shape of the two types of ions. Electrochemical cells provide an alternative separation, based on difference in electrochemical potential between two chemically different ions.

For example, consider a solution containing approximately 0.1 M Zn^{++} and Cd^{++} ions. From Table 5-2, we find

$$Zn^{++}\ (1\ \mathrm{M}) + 2e^- \rightarrow Zn \qquad \mathrm{E}^0 = -0.76 \text{ volt}$$

and

$$Cd^{++}\ (1\ \mathrm{M}) + 2e^- \rightarrow Cd \qquad \mathrm{E}^0 = -0.40 \text{ volt}$$

Table 5-2 Standard Electrode Reduction Potentials

Electrode	Electrode Reaction	E° (volts)
	(Acid Solutions)	(25°C)
$Li\|Li^+$	$Li^+ + e \rightleftarrows Li$	−3.045
$K\|K^+$	$K^+ + e \rightleftarrows K$	−2.925
$Cs\|Cs^+$	$Cs^+ + e \rightleftarrows Cs$	−2.923
$Ba\|Ba^{++}$	$Ba^{++} + 2e \rightleftarrows Ba$	−2.90
$Ca\|Ca^{++}$	$Ca^{++} + 2e \rightleftarrows Ca$	−2.87
$Na\|Na^+$	$Na^+ + e \rightleftarrows Na$	−2.714
$Mg\|Mg^{++}$	$Mg^{++} + 2e \rightleftarrows Mg$	−2.37
$Al\|Al^{+3}$	$Al^{+3} + 3e \rightleftarrows Al$	−1.66
$Zn\|Zn^{++}$	$Zn^{++} + 2e \rightleftarrows Zn$	−0.763
$Fe\|Fe^{++}$	$Fe^{++} + 2e \rightleftarrows Fe$	−0.440
$Cd\|Cd^{++}$	$Cd^{++} + 2e \rightleftarrows Cd$	−0.403
$Sn\|Sn^{++}$	$Sn^{++} + 2e \rightleftarrows Sn$	−0.136
$Pb\|Pb^{++}$	$Pb^{++} + 2e \rightleftarrows Pb$	−0.126
$Fe\|Fe^{+3}$	$Fe^{+3} + 3e \rightleftarrows Fe$	−0.036
$Pb\|PbSO_4$	$PbSO_4 + 2e \rightleftarrows Pb + SO_4^=$	−0.35
$Pt\|D_2\|D^+$	$2D^+ + 2e \rightleftarrows D_2$	−0.0034
$Pt\|H_2\|H^+$	$2H^+ + 2e \rightleftarrows H_2$	ZERO
$Pt\|Sn^{+2}, Sn^{+4}$	$Sn^{+1} + 2e \rightleftarrows Sn^{+2}$	+0.15
$Pt\|Cu^+, Cu^{++}$	$Cu^{++} + e \rightleftarrows Cu^+$	+0.153
$Pt\|S_2O_3^=, S_4O_6^=$	$S_4O_6^= + 2e \rightleftarrows 2S_2O_3^=$	+0.17
$Cu\|Cu^{++}$	$Cu^{++} + 2e \rightleftarrows Cu$	+0.337
$Pt\|I_2\|I^-$	$I_2 + 2e \rightleftarrows 2I^-$	+0.5355
$Pt\|Fe(CN_6)^{-4}, Fe(CN)_6^{-3}$	$Fe(CN)_6^{-3} + e \rightleftarrows Fe(CN)_6^{-4}$	+0.69
$Pt\|Fe^{+2}, Fe^{+3}$	$Fe^{+3} + e \rightleftarrows Fe^{+2}$	+0.771
$Ag\|Ag^+$	$Ag^+ + e \rightleftarrows Ag$	+0.7991
$Hg\|Hg^{++}$	$Hg^{++} + 2e \rightleftarrows Hg$	+0.854
$Pt\|Hg_2^{++}, Hg^{++}$	$2Hg^{++} + 2e \rightleftarrows Hg_2^{++}$	+0.92
$Pt\|Br_2\|Br^-$	$Br_2 + 2e \rightleftarrows 2Br^-$	+1.0652
$Pt\|MnO_2\|Mn^{++}, H^+$	$MnO_2 + 4H^+ + 2e \rightleftarrows Mn^{++} + 2H_2O$	+1.23
$Pt\|Cr^{+3}, Cr_2O_7^=, H^+$	$Cr_2O_7^= + 14H^+ + 6e \rightleftarrows 2Cr^{+3} + 7H_2O$	+1.33
$Pt\|Cl_2\|Cl^-$	$Cl_2 + 2e \rightleftarrows 2Cl^-$	+1.3595
$Pt\|Ce^{+3}, Ce^{+4}$	$Ce^{+4} + e \rightleftarrows Ce^{+3}$	+1.61
$PbSO_4\|PbO_2, SO_4^=, H^+$	$PbO_2 + SO_4^= + 2e + 4H_3O^+ \rightleftarrows PbSO_4 + 6H_2O$	+1.68
$Pt\|Co^{+2}, Co^{+3}$	$Co^{+3} + e \rightleftarrows Co^{+2}$	+1.82
$Pt\|SO_4^=, S_2O_8^=$	$S_2O_8^= + 2e \rightleftarrows 2SO_4^=$	+1.98
	(Basic Solutions)	
$Pt\|Ca\|Ca(OH)_2\|OH^-$	$Ca(OH)_2 + 2e \rightleftarrows 2OH^- + Ca$	−3.03
$Pt\|H_2PO_2^-, HPO_3^=, OH^-$	$HPO_3^= + 2e \rightleftarrows H_2PO_2^- + 3OH^-$	−1.57
$Zn\|ZnO_2^=, OH^-$	$ZnO_2^= + 2H_2O + 2e \rightleftarrows Zn + 4OH^-$	−1.216
$Pt\|SO_3^=, SO_4^=, OH^-$	$SO_4^= + H_2O + 2e \rightleftarrows SO_3^= + 2OH^-$	−0.93
$Pt\|H_2\|OH^-$	$2H_2O + 2e \rightleftarrows H_2 + 2OH^-$	−0.828
$Ni\|Ni(OH)_2\|OH^-$	$Ni(OH)_2 + 2e \rightleftarrows Ni + 2OH^-$	−0.72
$Pb\|PbCO_3\|CO_3^=$	$PbCO^3 + 2e \rightleftarrows Pb + CO_3^=$	−0.506
$Pt\|OH^-, HO_2^-$	$HO_2^- + H_2O + 2e \rightleftarrows 3OH^-$	+0.88

* *W. M. Latimer, Oxidation Potentials (Englewood Cliffs, N. J.: Prentice-Hall, 2nd ed., 1952). Standard state of H^+ or OH^- is 1M in an ideal solution.*

From these E^0 values, we can predict that precipitation of metallic Cd will begin (i.e., $E_{rx} > 0$), at any (applied) potential more negative than

$$E = -0.40 - \frac{0.059}{2} \log_{10}(1/0.1)$$
$$E = -0.43 \text{ volt}$$

measured with respect to a standard hydrogen electrode. We can consider that the Cd^{++} has been quantitatively removed when the applied potential has reached a value at which $[Cd^{++}] = 10^{-6}$ M:

$$E = -0.40 - \frac{0.059}{2} \log_{10}(1/10^{-6}) = -0.58 \text{ volt}$$

Finally, the Zn^{++} will begin to precipitate out at a potential given by

$$E = -0.76 - \frac{0.059}{2} \log_{10}(1/0.1) = -0.79 \text{ volt}$$

In other words, if we maintain the cathode (the right-hand electrode at which electrons are produced in our cell diagrams) at a potential between −0.58 and −0.79 volt with regard to a standard hydrogen electrode, we should accomplish quantitative precipitation of the cadmium ions while leaving the zinc ions still present at their original concentration.

Procedures similar to that just described are used for separation and preparation of many ionic and neutral species in solution. Because the *rate* (determined by the *current* allowed to flow through the circuit), *extent* (determined by the total charge allowed to pass through the circuit), and *selectivity* toward particular species (by choice of applied potential, as in the above example) are so easily controlled experimentally by choosing the desired current, time of reaction, and applied potential, electrochemical deposition provides an attractive means for preparing a known amount of a relatively pure substance starting from an impure mixture of several components.

A potentially important preparation is the electrolysis of water to form hydrogen and oxygen gases:

$$H_2O_{(liq)} \rightarrow H_{2(1\ atm)} + (1/2)O_{2(1\ atm)} \qquad E^0_{rx} = -0.401 \text{ V}$$

Since a major problem for industrial energy production is that electric power usage varies greatly during the day and it is usually most efficient to operate the power plant at a constant rate of power production, it has been suggested that energy, in the form of hydrogen gas at high pressure, could be produced by forcing such a reaction to the right, storing the gas until power is needed, and then allowing the reaction to proceed (spontaneously) to the left to regenerate the original electric power. In order to avoid handling H_2 gas at high pressure, it appears likely that the actual solution will involve storage of H_2 gas in some other form, such as H_2 adsorbed or complexed onto some appropriate solid surface.

EXAMPLE *Batteries*

The most economically important and one of the simpler batteries is shown schematically in Figure 5-7. The cell potential is readily obtained from Table 5-2:

$$\text{CATHODE: } PbO_2 + 4H^+ + 2e^- \rightarrow Pb^{++} + 2\ H_2O$$

followed by immediate precipitation of lead sulfate

$$Pb^{++} + SO_4^= \rightarrow PbSO_4 \text{ (precipitate)}$$

$$\text{ANODE: } Pb \rightarrow Pb^{++} + 2e^-$$

followed immediately by precipitation of lead sulfate

$$Pb^{++} + SO_4^= \rightarrow PbSO_4 \text{ (precipitate)}$$

The complete cell reaction is therefore given by

$$Pb + SO_4^= \rightarrow PbSO_4 \text{ (precipitate)} + 2e^- \qquad E^0 = -0.35 \text{ volt}$$

and

$$PbO_2 + 4H^+ + SO_4^= + 2e^- \rightarrow PbSO_4 + 2\ H_2O \qquad E^0 = -1.68 \text{ volt}$$

$$Pb + PbO_2 + 4H^+ + 2SO_4^= \rightarrow 2PbSO_4 + 2H_2O \qquad E^0 = -2.03 \text{ volt}$$

Automobile batteries usually consist of either three or six of these cells coupled together so as to give a total voltage of about 6 or 12 volts. The product of both the cathode and the anode reactions is lead sulfate, $PbSO_4$, which is supposed to adhere firmly to the electrode so that if the battery runs down (all the lead and lead oxide have been converted to lead sulfate), the battery can be "recharged" by applying a voltage opposite in sign and somewhat greater in magnitude than that normally produced by the fresh battery, and the lead sulfate reconverted back to lead (anode) or lead oxide (cathode). Eventually,

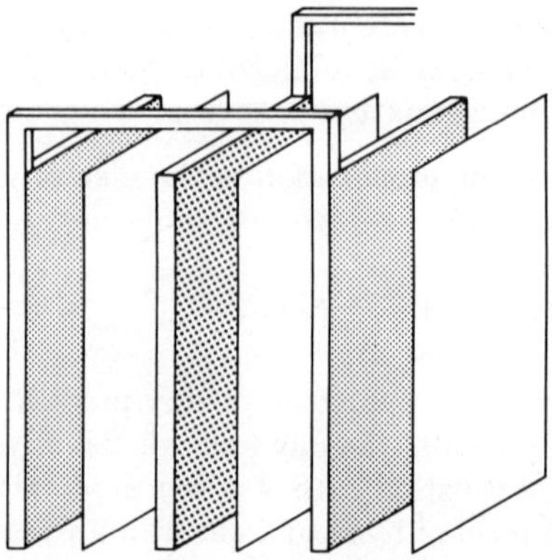

FIGURE 5-7. A schematic diagram of the lead storage cell. Lead plates are arranged in parallel sets separated by plastic spacers. Alternate lead sheets are connected to a bus bar. The anode plates are coated with finely divided metallic lead and the cathode plates are coated with finely divided lead dioxide. The whole assembly is mounted in a plastic case filled with approximately 3 M sulfuric acid. (From E. Hutchinson, *Chemistry,* W. B. Saunders, Philadelphia, 1959, p. 266.)

the lead sulfate falls off the plates and accumulates as sludge at the bottom of the battery, and the battery can no longer produce a charge even after "recharging." Because of gradual electrolysis of the water in the battery, it is necessary to add water occasionally to keep the plates covered.

The reason automobile batteries are so heavy is that a substantial amount of lead is required to produce the large current needed (see Problems) to start a car. The amount of *charge* stored in the battery is conveniently described by an "ampere-hour" rating (see Problems), while the amount of *energy* stored in the battery is rated by its watt-hours. The most serious problems in using lead-acid batteries as the power source for an electric automobile are (1) the battery is so heavy—typically, lead-acid batteries store only about 25 watt-hours per kilogram of battery, and (2) the battery cannot survive the several hundred discharge-charge cycles that would be required for use as an automobile power source.

Examination of Table 5-2 suggests that the most attractive element for use as the anode in batteries ought to be lithium, because of its extremely high reduction potential and its very light weight. However, because lithium reacts with water (!), batteries using lithium have so far been based on use of an organic solvent (if the battery must be operated near room temperature) or molten salt electrolyte (if the high 400°C operating temperature is feasible). Very recently, workers at the GTE Laboratories and at the U.S. Army Electronics Command discovered (largely by accident) a new lithium-anode battery in which the organic solvent (such as phosphorous oxyhalide, $POCl_3$) acts as a cathode reactant. Prototype versions of this battery have produced specific energies of the order of 500 watt-hours per kilogram, more than ten times the energy-per-unit-weight of conventional lead-acid batteries. However, it has been pointed out that the amount of lithium required for the energy source for 20 million urban electric cars would add up to more than the world's projected supply of lithium up to the year 2000!

Fuel cells have also been suggested as an alternative to production of electrical power from heat engines (devices that convert heat to mechanical energy and then to electrical energy), using fossil fuel (natural gas, distillate fuel oil, or gas from coal) as the ultimate energy source. The scheme would be to reform the initial fuel into hydrogen and carbon dioxide, and then use the hydrogen gas at the anode of a hydrogen-oxygen electrochemical cell (the electrolysis reaction in reverse) to produce a direct electrical current using atmospheric oxygen at the cathode. The direct current could then be changed to an alternating current using an inverter, and then stepped up to power-line voltage using a transformer. It is claimed that fuel cells could increase the conversion efficiency from roughly 30% in conventional fossil-fuel-burning power plants to well over 40%, by eliminating much of the energy lost to heat in the friction of moving parts in a conventional engine.

PROBLEMS

1. Sea water has an osmotic concentration of about 1 mole/liter of dissolved salts. Suppose that the osmotic energy in 1 cubic meter of sea water could

be released in one second (for example, by adding 1 cubic meter of fresh water to a large volume of sea water).

(a) Calculate the power (in watts) produced in this process.
(b) If the total world river flow is about 10^6 meter^3/sec, how much total power is in principle available from rapid salination of river water? Compare to the present average rate of world electric power generation of about 7×10^{11} watts.
(c) Calculate the height of a column of water corresponding to the osmotic pressure of sea water—the energy from salination of sea water can then be compared to the height of a waterfall at the mouth of a river. [A potentially practical device based on extraction of the osmotic pressure in sea water is given by R. S. Norman, *Science 186,* 350 (1974).]
(d) Alternatively, one could imagine a concentration cell with sea water on one side and fresh water on the other—we can then calculate the electrochemical potential corresponding to transport of *salt* from the sea water to the fresh water (rather than transport of *water* from the fresh water to the sea water, as in the osmotic calculation). For the system of compartments shown in the diagram, the maximal possible voltage is given by:

$$V_{\max} = (2NRT/F) \ln ((a_{\text{salt}})_{\text{sea}}/(a_{\text{salt}})_{\text{river}})$$

where N is the number of membrane pairs in the stack of the "dialytic battery." Treating sea water as 0.057M NaCl and river water as 0.0259M NaCl (assume both solutions are ideal), calculate $V_{\max}$ at 20°C for $N = 30$.

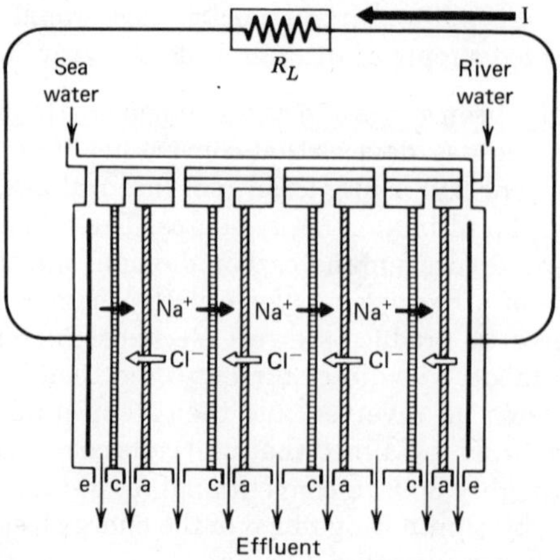

Schematic Dialytic Battery. The letters e, c, and a denote electrode, cation exchange membrane, and anion exchange membrane; R_L is the load resistance. A continuous flow of both sea and river water acts to replenish each compartment as salt ions move from sea to fresh water. [From J. N. Weinstein and F. B. Leitz, *Science 191,* 557 (1976).]

(e) Electrical power may be drawn from this system through a "load" resistor, R_L, according to the familiar relation

$$\text{Power (watts)} = W = I^2 R_L;\ I \text{ in amps}, R_L \text{ in ohms}$$

where

$$I = V/R = (V_{max}/(R_L + R_{stack}))$$

For $R_L = 8.8$ ohm, $R_{stack} = 20$ ohm, and V_{max} from part d, calculate the power produced by such a stack.

The *actual* power production by such schemes is determined by the *rate* at which water (osmotic case) or ions (battery case) can be made to go through the corresponding semipermeable membrane. However, rough calculations suggest efficiencies such that a power plant using 10% of the flow of the Mississippi River at 25% efficiency could produce 1000 megawatts, roughly comparable to the largest present hydroelectric dam plants.

2. Show that $E^{\circ\prime} = E^\circ + (RT/zF) \ln (10^{-7})^x$, in which E° is the standard half-cell reduction potential with respect to a hydrogen electrode with $[H^+] = 1$ M in an ideal solution; $E^{\circ\prime}$ is the standard half-cell reduction potential for the same reaction at pH 7.0; z is the number of electrons transferred per mole of reactant converted to product, and x is the number of moles of protons reacting per mole of reactant converted to product.

3. Calculate the electrochemical potential for the reaction

$$Zn^{++}(1\text{ M}) + H_2(1\text{ atm}) \rightarrow Zn(\text{metal}) + 2H^+\ (\text{pH } 5.76)$$

(E° for the $Zn^{++} + 2e^- \rightarrow Zn$ half-reaction is -0.763 volt.)

4. The standard half-reaction reduction potential, E°, for the dye, methylene blue, is +0.53 volt for the reaction

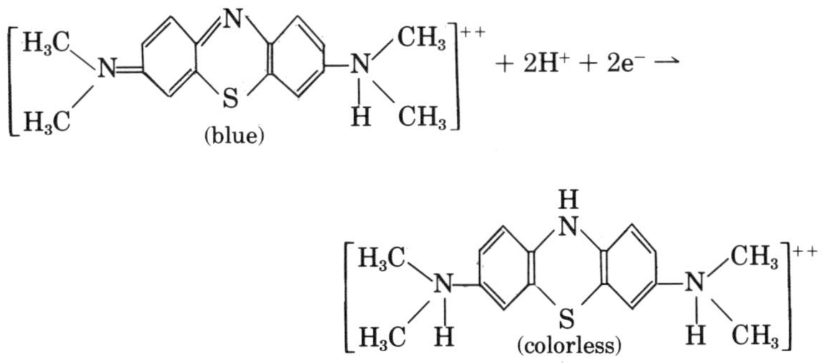

or more briefly

$$M(1\text{ molar}) + 2H^+(1\text{ M}) + 2e^- \rightarrow MH_2(1\text{ molar})$$

If measurements of the intensity of the blue color for a methylene blue solution show that the fraction of total dye in the blue form is 45%, and if the measured electrochemical potential for the solution (with respect to standard hydrogen electrode at $[H^+] = 1$ M in ideal solution) is +0.04 volt, calculate the pH of the methylene blue solution.

5. For the reaction

$$\text{Ox} + (z/2)\ \text{H}_2 \rightarrow \text{Red} + z\ \text{H}^+$$

the Nernst equation takes the form

$$\text{E} = \text{E}^\circ + (RT/zF)\ \ln\ ([\text{Ox}]/[\text{Red}])$$

Construct a plot of measured electrochemical potential, E, as a function of the fraction of molecules in the reduced form,

$$f_{\text{red}} = \frac{[\text{Red}]}{[\text{Ox}] + [\text{Red}]}$$

for $z = 1$ and for $z = 2$. Show how you could use such a plot (potentiometric titration, of which ordinary pH titrations are examples) to determine E° and z for the reaction.

6. (a) If an automobile lead-acid battery is required to deliver 150 amperes (1 ampere = 1 coulomb/sec) in order to start the engine, calculate the amount of lead oxidized on a cold morning when it takes 3 minutes to start the car.
(b) If the battery is rated for 100 amp-hour, what is the minimum number of grams of finely divided lead that such a battery must have impressed in the pores of its cathode plate?

7. For a particular electrochemical cell, for which $z = 2$, $\text{E}^\circ_{rx} = +0.425$ volt, and the overall entropy change is $\Delta S^\circ_{rx} = +17.2$ cal mole^{-1} deg^{-1} at 30°C, calculate the overall enthalpy change for this reaction and the rate of change of cell potential with temperature, $(\partial \text{E}_{rx}/\partial T)$.

REFERENCES

G. M. Barrow, *Physical Chemistry for the Life Sciences,* McGraw-Hill, New York (1974), Chapter 8. Good treatment of general electrochemistry, particularly the phosphate-transfer equilibria.

D. A. Skoog and D. M. West, *Fundamentals of Analytical Chemistry,* 2nd ed., Holt, Rinehart & Winston, New York (1969). Good general discussion of ion-selective electrodes. For a recent review of this area, see G. A. Rechnitz, *Chemical and Engineering News,* 27 Jan., 1975, pp. 29–35.

2
SUCCESS AND FAILURE

CHAPTER 6 The Random Walk Problem

Most problems in the physical or biological sciences, when stripped of specialized terminology, can be solved to a good approximation by appealing to the results from three very general and powerful styles in mathematical thought: the mathematics of *change* (the calculus), the mathematics of *chance* (the calculus of probabilities), and the mathematics of ordinary *geometry* (linear algebra). The most biologically important problems of geometry and flux are approached in later chapters; the basic problems of chance and probability are discussed in this chapter.

The story-book origin of this branch of mathematics begins from the preoccupation with gambling at the court of Louis XIV. After a few losses, the gamblers quickly learned that, on the average, they would make money by betting that at least one six would appear after four rolls of a single die. Following intuition, they expected that there should be favorable odds of obtaining a *pair* of sixes after 24 rolls of *two* dice (i.e., 4:6 = 24:36). However, in the new game, the gamblers found that the odds were reversed, and it was actually more likely that a pair of sixes would *not* appear after 24 rolls! This "gaming paradox" stimulated much intellectual activity, and was first explained by the eminent French mathematician Fermat; in solving the problem, Fermat went far toward establishing the mathematical theory of probability.

The preceding anecdote provides more than an amusing introduction to probability theory: it shows that intuition often leads in the wrong direction, therefore, in many of the highly intuitive (i.e., nonrigorous) derivations we later exhibit, it will be necessary to check the result against reality before proceeding further.

Before examining the solution of the random walk problem, one should first understand why the problem is worth doing; there are several reasons. First, the random walk problem is one of the surprisingly few interesting physical problems that has a simple solution! Second, the mechanics of the solution itself entail most of the features of statistics and probability as they appear in biophysical applications. In addition, the problem is an excellent example of the use of judicious approximations—contrary to popular belief, most of the important problems in physical science have been solved, not by intricate reasoning requiring intellectual virtuosity, but by deciding which parts of a complicated problem can be successfully ignored. Finally, the random walk problem provides a physical insight or basis for such diverse biophysical applications as diffusion, radioactivity, electrical noise, size of polymers in solution, and spectral line shapes. Fortified with these motivations, let us examine the problem.

Most of the mathematics of chance may be distilled into two simple axioms:

if

$P(A)$ is the probability* that event A will happen

and

$P(B)$ is the probability that event B will happen

and

A and B are independent events (separate tosses of the same coin, radioactive decay of two distinct unstable nuclei, etc.)

then

$$P(A \text{ and } B) = P(A) \cdot P(B) \tag{6-1}$$

where

$P(A$ and $B)$ is the probability that *both* A and B will happen (e.g., decay of both unstable nuclei during the observation period)

The other axiom is just that

$$P(A \text{ or } B) = P(A) + P(B) \tag{6-2}$$

where $P(A$ or $B)$ is the probability that *either* A or B will happen (e.g., either the first unstable nucleus will decay or the second unstable nucleus will decay). Both axioms are simply formal statements of intuition: the chance of getting a head on the first toss *and* a tail on the second toss of an ideal coin is P(head on first *and* tail on second) = (1/2)(1/2) = (1/4); similarly, the chance of a head *or* tail on a given toss is (1/2) + (1/2) = 1, i.e., it is certain that one will observe one of these two results in an ideal experiment. Our problem will be to apply these ideal cases to an *ideal* problem, and then show why the result is useful for *real* problems.

* *By "probability" of a particular result, we mean our guess for the most likely fraction of a large number of attempts that will produce that particular result. For example, the odds ("probability") of getting any particular number (say, 4) on the face of a die after one roll is 1/6, since there are six faces and thus six equivalent possible results, of which only one is desired.*

EXAMPLE *The "Gaming Paradox"*

We can now understand the gaming paradox. Because the chance of *not* getting a 6 on one throw of one die is (5/6), the chance of not having a 6 on the 1st *and* 2nd *and* 3rd *and* 4th throws of one die is the product of the probabilities for the (independent) individual throws: $(5/6)(5/6)(5/6)(5/6) = 0.48$, which is less than a 50% chance. In other words, the *other* possible result, namely of rolling at least one 6 in four throws must have a likelihood of $(1 - 0.48) = 0.52$, which is better than 50%. Similarly for 24 rolls of two dice, the chance of not getting a double six in 24 independent rolls of the two dice is $(35/36)^{24} = 0.51$, so the chance of the opposite result (namely, of at least one double six in 24 rolls) is $(1 - 0.51) = 0.49 =$ less than 50% likelihood.

To get the most (future) use from the random walk problem, we begin with the following situation. Consider N bags, each containing an equal (large) number of black and white marbles:

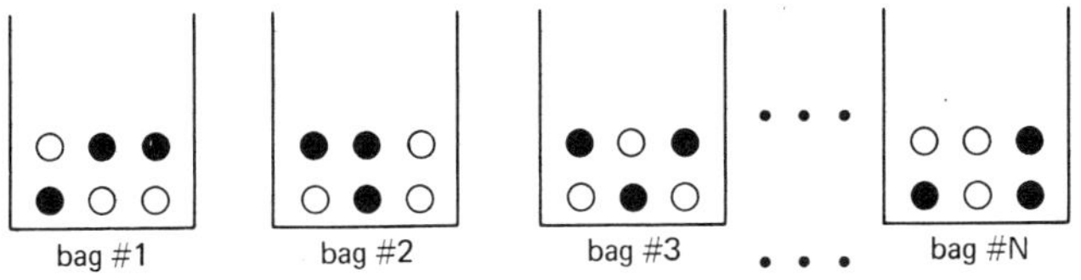

Now suppose you are blindfolded and you draw one marble from each bag (the blindfold guarantees that each draw is independent of the others). The problem then is, what is the probability that m of the marbles were black *and* that the remaining $(N - m)$ marbles were white? It is convenient to find the answer in two parts.

Bags and Marbles: Part 1 Suppose that someone has already simplified the problem by pre-selecting two groups of m bags and $(N - m)$ bags, so that all that remains is to draw a black marble from each of the m bags and a white marble from each of the $(N - m)$ bags. Since any one bag has the same number of black as white marbles, the chance of finding a black marble on any one draw is simply (1/2).

Consider first the group of m (pre-selected) bags. The probability that the draw from the first bag is black is (1/2). The probability of a black draw from the second bag is also (1/2). From axiom 6-1, one deduces that the probability that the first *and* second draws are black is $(1/2) \cdot (1/2) = (1/2)^2$. By applying axiom 6-1 successively to the third and subsequent draws, it is readily shown that the probability of drawing a black marble from each of the m bags is $(1/2)^m$.

Similarly, given the other group of $(N - m)$ bags, it is clear that the probability of drawing a white marble from each of the $(N - m)$ bags is $(1/2)^{(N-m)}$. Finally, since the draws from the two groups of bags are independent, axiom 6-1 may be invoked once more to provide the solution to half of the original problem, namely that the probability of finding a black marble in one draw from each of m pre-chosen bags *and* finding white marbles in one draw from each of the remaining $(N - m)$ bags is simply $(1/2)^m\ (1/2)^{(N-m)}$.

So far it appears that we have taken a tortuous route to reach a result that isn't very surprising—in fact we have solved half of the random walk case.

Bags and Marbles: Part 2 The second half of the solution is to figure out the number of ways that the m bags could have been chosen from the N original total bags. First, the total number of ways to pick out N bags one at a time is

$$N! = N(N-1)(N-2)\cdots 3\cdot 2\cdot 1 \tag{6-3}$$

where $N!$ (called "N factorial") arises because there are N ways to choose the first bag, $N - 1$ ways to choose the second bag, and so on, so that axiom 6-1 leads directly to Eq. 6-3. Finally, we can get to Eq. 6-3 alternatively by finding the number of ways to choose two groups of m and $(N-m)$ bags, and then choosing bags one at a time from each of the two groups:

$$N! = \begin{pmatrix}\text{number of ways to choose one group of } m \text{ bags}\\ \text{and another group of } (N-m) \text{ bags}\end{pmatrix}\cdot m!(N-m)!$$

thus, the number of ways of choosing the two groups of bags is simply

$$\frac{N!}{m!(N-m)!}$$

Putting the two parts of the solution together, let $P_N(m)$ be the probability of obtaining m black marbles in one draw from each of N bags; then

$$P_N(m) = \frac{N!}{m!(N-m)!}(1/2)^m(1/2)^{(N-m)} \tag{6-4}$$

The first factor in Eq. 6-4 gives the number of ways of selecting m bags in one group and $(N - m)$ bags in another group, and the second factor tells the probability that every draw from the m bags will be black and every draw from the $(N - m)$ bags will be white.

To allay any suspicion that may have arisen during the above derivation, it is useful to have a check on the correctness of the result. A simple test is to verify that the probability of finding either no black marbles, or one black, or two blacks, . . . , up to N blacks is unity (one is certain to obtain one of these results). Using axiom 6-2, the test may be phrased formally as

$$\text{Is} \quad P_N(0) + P_N(1) + P_N(2) + \cdots + P_N(N) = \sum_{m=0}^{N} P_N(m) = 1? \qquad (6\text{-}5)$$

The answer to this question follows directly from the familiar binomial expansion

$$(a+b)^N = a^N + Na^{N-1}b + \cdots + \frac{N!}{m!(N-m)!}a^{N-m}b^m + \cdots + b^N \qquad (6\text{-}6)$$

once one notices that Eqs. 6-5 and 6-6 are the same for the case that

$$a = b = (1/2) \qquad (6\text{-}7)$$

so that

$$(a+b)^N = 1^N = 1 \qquad (6\text{-}8)$$

and we have shown that

$$\sum_{m=0}^{N} P_N(m) = 1 \qquad \text{Q.E.D.} \qquad (6\text{-}9)$$

This sort of test is often called "normalization," and we will have occasion to try it again after tampering with the result a bit.

The molecular one-dimensional random walk problem consists of letting a molecule take only forward or backward steps in one direction with equal likelihood, and Eq. 6-4 gives the chance that m steps out of a total of N steps will be forward steps. It would therefore seem that the problem has been solved, but one now faces a situation that commonly arises in biophysical problems—the result is not expressed in a useful form. Now it is a simple matter to convert from number of steps to distance traveled, as we will later do; the present problem is the more serious one that there are factorials in the answer.

Factorials are a great embarrassment when the numbers involved are very large, as in the present case where a molecule will take billions of steps in a second, because the observed quantities (diffusion constant, sedimentation rate, electrophoretic mobility) represent an average over many molecules, and require the use of integrals for their evaluation. Plainly, an integral of Eq. 6-4 with respect to m would involve a huge number of terms and would take a long time to calculate, even for an electronic computer. However, when N is large, a short calculation leads to a very good approximation for $\ln(N!)$ (see Appendix and Problems)

$$\ln(N!) \cong N \ln(N) - N \qquad (6\text{-}10)$$

known as Stirling's approximation, and now one has only two terms in place of the N terms represented by the left-hand side of Eq. 6-10. This spectacular simplification permits the development from a dull problem one can solve (Eq. 6-4) to the molecular applications one is interested in.

In order to make use of the Stirling approximation 6-10, one must first take the log of both sides of Eq. 6-4:

$$P_N(m) = \frac{N!}{m!(N-m)!}(1/2)^N \tag{6-4}$$

$$\ln P_N(m) = \ln N! - \ln m! - \ln(N-m)! + N\ln(1/2) \tag{6-11}$$

Stirling's approximation may now be applied to $\ln N!$, $\ln m!$, and $\ln(N-m)!$ to give

$$\ln P_N(m) = N\ln N - m\ln m - (N-m)\ln(N-m) + N\ln(1/2) \tag{6-12}$$

Recourse must next be made to several manipulative tricks to prepare the previous equation for the final approximation that leads to the desired result. The first trick is to replace N in the first term of Eq. 6-12 by $[m + (N-m)]$:

$$N\ln N = m\ln N + (N-m)\ln N \tag{6-13}$$

so that

$$\ln P_N(m) = m(\ln N - \ln m) + (N-m)(\ln N - \ln(N-m)) + N\ln(1/2) \tag{6-14}$$

$$= m\ln\frac{N}{m} + (N-m)\ln\frac{N}{N-m} + N\ln(1/2) \tag{6-15}$$

The second trick is to apply the property of logarithms (see Appendix)

$$\ln(a/b) = -\ln(b/a) \tag{6-16}$$

to the first two terms of Eq. 6-15, leaving

$$\ln P_N(m) = -m\ln\frac{m}{N} - (N-m)\ln\frac{N-m}{N} + N\ln(1/2) \tag{6-17}$$

The third trick is based on an intuitive claim that will be vindicated after the derivation is completed. The claim is that the average number, $\overline{m}$, of black marbles drawn from the N bags is just

$$\overline{m} = (N/2) \tag{6-18}$$

since the chance of a black on any one draw is (1/2). Defining a new variable,

$$x \equiv (m - \overline{m}) \tag{6-19}$$

and substituting for $m = x + (N/2)$ in Eq. 6-17, one obtains

$$\ln P_N(m) = -\left(x + \frac{N}{2}\right) \ln \frac{x + (N/2)}{N} - \left(\frac{N}{2} - x\right) \ln \frac{(N/2) - x}{N} + N \ln(1/2) \tag{6-20}$$

By using the last trick of rewriting Eq. 6-20 in the form

$$\ln P_N(m) = -\left(x + \frac{N}{2}\right) \ln \left[\frac{1}{2}\cdot\left(1 + \frac{2x}{N}\right)\right] - \left(\frac{N}{2} - x\right) \ln \left[\frac{1}{2}\cdot\left(1 - \frac{2x}{N}\right)\right] + N \ln(1/2) \tag{6-21}$$

and making use of the property

$$\ln(a \cdot b) = \ln a + \ln b \tag{6-22}$$

Equation 6-21 may be simplified to

$$\ln P_N(m) = -\left(x + \frac{N}{2}\right) \ln \left(1 + \frac{2x}{N}\right) - \left(\frac{N}{2} - x\right) \ln \left(1 - \frac{2x}{N}\right) \tag{6-23}$$

When a coin is tossed, say, 1000 times, the number of heads will probably be very close to 500. Similarly, in the present case, for a large number of draws, m will be close to $\overline{m}$, in the sense that

$$\frac{m - \overline{m}}{N} = \frac{x}{N} << 1 \tag{6-24}$$

(When making approximations, it is not enough to say that a term is small; one must say that it is *small compared to something else,* as in 6-24.)

We are now in a position to dispense with both ln terms on the right side of Eq. 6-23, because for a number, a, which is much less than unity, the Taylor series (see Appendix) for $\ln(1 \pm a)$ may be approximated by just two terms:

$$\ln(1 \pm a) \cong \pm a - \frac{a^2}{2} \tag{6-25}$$

Applying the approximation, Eq. 6-25, to Eq. 6-23, one quickly shows that

$$\ln P_N(m) = -\frac{2x^2}{N} \tag{6-26}$$

or

$$P_N(m) = e^{-(2x^2/N)} \tag{6-27}$$

The troublesome factorials that appeared in Eq. 6-4 have thus been eliminated by use of two approximations (Stirling and Taylor), but it is now necessary to re-check the normalization of $P_N(m)$, which may have suffered from the approximations along the way. Because the discrete variables in the factorials have been replaced by a continuous variable, x, in Eq. 6-27, the sum in Eq. 6-5 should be replaced by an integral for the normalization test (see Fig. 6-1 and accompanying discussion). We seek to find a constant, A, such that

$$A\int_{-N/2}^{N/2} P_N(x)\,dx = 1 \tag{6-28}$$

where

x can vary from $(-N/2)$ to $(+N/2)$ corresponding to the range $0 \leq m \leq N$. Finally, since the function, $\exp[-2x^2/N]$ is so close to zero for large x, the area under that function (i.e., the integral in Eq. 6-28) will be negligible for large x, and the limits of integration may be extended to $\pm\infty$. Using the known definite integral (see Appendix)

$$\int_{-\infty}^{+\infty} e^{-ax^2}\,dx = \sqrt{\frac{\pi}{a}} \tag{6-29}$$

the constant, A, is readily evaluated:

$$A\int_{-\infty}^{+\infty} e^{-(2x^2/N)}\,dx = A\sqrt{\frac{\pi N}{2}} = 1 \tag{6-30}$$

or

$$A = \sqrt{\frac{2}{\pi N}} \tag{6-31}$$

to give the final result

$$\boxed{P_N(m) = \sqrt{\frac{2}{\pi N}}\,e^{-2x^2/N} = \sqrt{\frac{2}{\pi N}}\,e^{-2(m-(N/2))^2/N}} \tag{6-32}$$

6.A. EVEN ODDS: WALKING FOR DISTANCE

Consider a molecule that starts at the origin and takes N steps away, where each step is equally likely to be forward or backward, and each step is the same length:

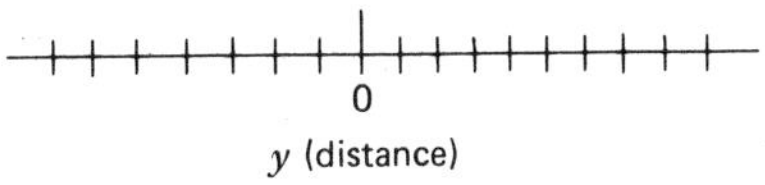

For now, the molecule will be permitted to move in only one dimension. The problem is, how far away from the origin will the molecule be after taking N total steps? This problem is formally identical to the problem we have just completed, namely the number of black marbles that result from drawing one marble from each of N bags, where any one bag contains an equal number of white and black marbles. The solution is given as Eq. 6-32, and we need only change variables from number of steps to distance traveled to complete the solution. The variables are as follows.

l = length of one step
ν = number of steps per second
t = duration of the total "walk" in seconds

Thus the total number of steps, N, is given by

$$N = \nu \cdot t \tag{6-33}$$

where

m = number of forward steps in a particular "walk" of N total steps

and

$$\overline{m} = \frac{N}{2} = \text{average number of forward steps, averaged over many walks} \tag{6-34}$$

If y is defined as the (net) distance traveled after time, t (i.e., after N steps), then

y = (m forward steps) · (length of a step)
+ (($N - m$) backward steps) · (length of 1 step) or

$$y = m \cdot l + (N - m) \cdot (-l) \tag{6-35}$$

But from Eq. 6-34, $N = 2\overline{M}$, so Eq. 6-35 may be rewritten

$$y = 2l \cdot (m - \overline{m})$$

or

$$\boxed{m - \overline{m} = \frac{y}{2l}} \tag{6-36}$$

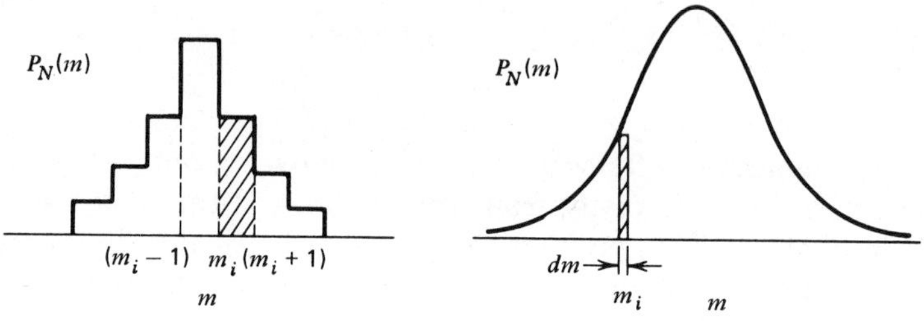

FIGURE 6-1. Probability of observing a result, m_i, when m is discrete (left plot) or continuous (right plot). See text.

At this stage, it is necessary to understand a fundamental property of probability that makes possible the calculation of average values of physical quantities for a collection of many particles, when given the properties of just one particle. What is the physical basis for proceeding from a sum (of a discrete quantity) to an integral (of a continuous quantity)? The answer is immediately apparent from scrutiny of Fig. 6-1.

For observation of a *discrete* result (number on a die, head or tail on a coin), $P_N(m)$ itself is the probability of finding that particular exact result. In other words, when there is a finite number of alternatives, there is a well-defined probability of selecting any one of them. However, when the result is a *continuously varying* quantity, such as distance or electric current, then the probability of finding a particle at any one *exact* distance is zero (since there is an infinite number of possible locations), and one must instead ask for the probability of finding the particle in a particular *region*. In Fig. 6-1 the shaded portion of the left-hand plot has an area

$$\text{Area} = P_N(m_i) \cdot [(m_i + 1) - m_i] = P_N(m_i) \tag{6-37}$$

Physically, one can identify the area of that rectangle as the probability of observing a result in the *range* $m_i \leq m < (m_i + 1)$. For the discrete example, this identification is an uninformative exercise, but, by analogy, one may now recognize the (infinitesimal) area shown for the continuous curve at the right of Fig. 6-1 as the probability of observing a result in the *range* $m_i \leq m < m_i + dm$. Therefore, in dealing with probability for continuously variable observables

$P_N(m)\,dm$ = probability that one will observe a result somewhere in the range between m and $(m + dm)$

Finally, just as the average of any *discrete* quantity is given as the sum

$$\bar{m} = m_0 \cdot P_N(m_0) + m_1 \cdot P_N(m_1) + m_2 \cdot P_N(m_2) + \cdots + m_N \cdot P_N(m_N) \tag{6-38}$$

$$\overline{m} = \sum_i m_i P_N(m_i) = \text{average value of } m \text{ when } m \text{ is discrete}$$

the average of a *continuous* quantity is given by

$$<m> = \int_{m=0}^{N} m P_N(m)\, dm = \text{average value of } m \text{ when } m \text{ is continuous} \tag{6-39}$$

EXAMPLE *Average for a Discrete Observable*

Suppose that two dice are tossed, and we are asked to predict the average result for many such experiments. There are $6 \times 6 = 36$ possible results (six on one die times six on the other independent die). The probability of two ones is (1/36); the probability of a three is (2/36), since one can throw a 2 on the first and a 1 on the second die or the other way around, and so on for the remaining choices. $P_{36}(2) = (1/36)$, $P_{36}(3) = (2/36)$, and so forth, and the average result is

$$\begin{aligned}\overline{m} &= (2)(1/36) + 3(2/36) + 4(3/36) + 5(4/36) + 6(5/36) + 7(6/36)\\ &\quad + 8(5/36) + 9(4/36) + 10(3/36) + 11(2/36) + 12(1/36)\\ &= (252/36) = 7\end{aligned}$$

The equation simply adds up each result times the probability of obtaining that result.

EXAMPLE *Average for a Continuous Variable: Average Number of Forward Steps in a Random Walk*

In a random walk with a large number of steps (so that the number of forward steps, m, may be treated as a continuously variable quantity), Eq. 6-32 may be interpreted to mean that

$$P_N(m)\, dm = \sqrt{\frac{2}{\pi N}}\, e^{-2(m-(N/2))^2/N}\, dm = \text{probability that the molecule has taken between } m \text{ and } (m + dm) \text{ forward steps after a total of } N \text{ steps} \tag{6-40}$$

The *average* number of forward steps is then defined from Eq. 6-39 as

$$<m> = \sqrt{\frac{2}{\pi N}} \int_{m=0}^{N} m\, e^{-2(m-(N/2))^2/N}\, dm \tag{6-41}$$

Since the definite integral in a math table that most closely resembles Eq. 6-41 is

$$\int_0^{\infty} x\, e^{-x^2}\, dx$$

it should seem reasonable to change variables to:

$$x = m - (N/2) \tag{6-19}$$

$$dx = dm$$

to give

$$<m> = \sqrt{\frac{2}{\pi N}}\left[\int_{-N/2}^{+N/2} x\, e^{-2x^2/N}\, dx + \frac{N}{2}\int_{-N/2}^{+N/2} e^{-2x^2/N}\, dx\right] \tag{6-42}$$

Finally, since N is large and the area under $\exp[-2(m - (N/2))^2/N]$ is negligible for large m, it is permissible to extend the limits of integration to $\pm\infty$ in Eq. 6-42. The first integral in Eq. 6-42 vanishes (see if you can see why), and the second integral has already been solved as Eq. 6-30, leaving

$$\boxed{<m> = (N/2)} \qquad \text{as expected} \tag{6-43}$$

The average *imprecision* in the number of forward steps is derived just as simply. Since the molecule is just as likely to take more than $(N/2)$ as less than $(N/2)$ forward steps

$$<(m - <m>)> = <m> - <m> = 0$$

However, what we really want is a measure of the *absolute* imprecision in number of forward steps (i.e., of observing a number of forward steps different from $<m>$, whether greater or less than $<m>$), provided by (see Problems)

$$\boxed{<(m - <m>)^2> = \frac{<m>}{2}} \tag{6-44}$$

so that $\sqrt{<m>/2}$ represents the root-mean-square deviation in the number of forward steps.

Returning to the random walk problem, if $P_N(m)\, dm$ is the probability that a molecule took between m and $(m + dm)$ steps, and if $P_t(y)$ is the probability that the same molecule ended up between y and $(y + dy)$ after time, t, then

$$P_N(m)\, dm = P_t(y)\, dy \tag{6-45}$$

because it takes m steps to travel y distance. From Eq. 6-36, Eq. 6-40, and

$$dm = (1/2l)\, dy \tag{6-46}$$

the change of variables is completed by combining Equations 6-32, 6-36, 6-45, and 6-46:

$$\boxed{P_t(y)\,dy = \sqrt{\frac{1}{2\pi N l^2}}\, e^{-(y^2/2Nl^2)}\,dy} \tag{6-47}$$

From Eq. 6-47 it is readily shown (see Problems) that

$$\langle y \rangle = 0 \tag{6-48}$$

$$\langle y^2 \rangle = N l^2$$

or

$$\sqrt{\langle y^2 \rangle} = l\,\sqrt{N} \tag{6-49}$$

Although the treatment leading to the principal result, Eq. 6-49, was based on a one-dimensional "walk," it can be shown that the same result applies if the molecule is allowed to take a three-dimensional "walk" consisting of steps of length, l, in either the x or y or z direction:

$$\boxed{\sqrt{\langle r^2 \rangle} = l\,\sqrt{N}} \tag{6-50}$$

where the result of a particular "walk" is shown in Fig. 6-2.

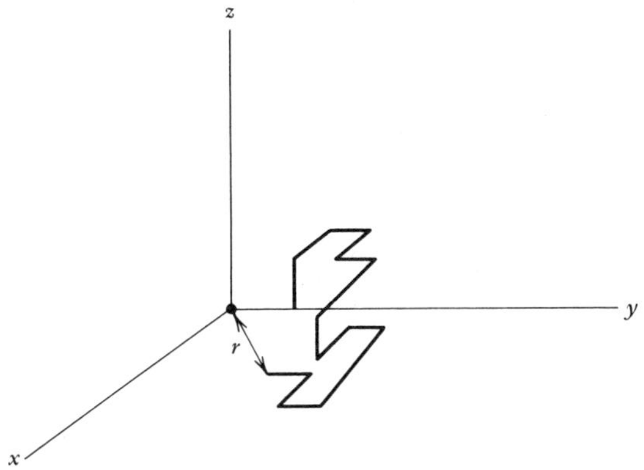

FIGURE 6-2. Three-dimensional random walk.

Equation 6-50 will now be applied to a preliminary discussion of the size of polymers in solution.

Dimensions of Polymers in Solution

The most useful simple models for the shape of a synthetic or biopolymer in solution are: sphere, ellipsoid, rigid rod, and random coil (see also the discussion under sedimentation and viscosity sections later in this Section).

The random coil shape is derived by fixing one end of the polymer at the origin, and then locating each successive link in the chain by taking a 3-dimensional random walk away from the origin: thus, the root-mean-square distance from one end of the random coil to the other is given by Eq. 6-50, in which l is now the length of one link, and N is the number of links in the polymer chain. A number of measurements, such as light-scattering, provide means for choosing among these alternatives for a particular polymer. Examples of molecules accurately described by these models are: insulin (sphere), gamma-globulin (ellipsoid), tobacco mosaic virus (rod), and natural rubber (random coil). The interest in these shapes is that biological function of most macromolecules is closely connected to the three-dimensional disposition of the long chains of which the biopolymer is constructed. Unwinding of the double-helix of DNA can be described as a rod-to-random-coil transition, for example, and shows characteristics similar to an ordinary phase change, such as the melting of ice. Inactivation of enzymes by extremes of pH, temperature, or salt concentration is often associated with an unraveling of a tightly organized structure into a more random arrangement. The random coil itself is a useful model because it offers a starting point in the determination of the solution structure of polymers: from the known molecular weight (and thus a known number of links in the chain) of a polymer, the size of the corresponding random coil can be calculated—then, by comparison of that result to the experimentally determined size, it is possible to make conclusions about the forces that hold the chain segments together.

Polyisoprene (natural rubber) consists of long chains of between 500 and 5000 isoprene units,

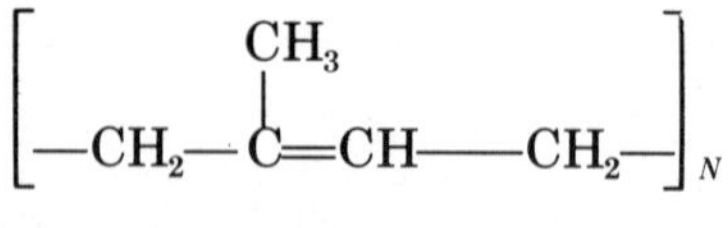

and behaves in solution as an ideal random coil. At first glance, it is difficult to see how this can occur, because for any real polymer, there are certain positions in the random walk that would cause two joints in the chain to occupy the same space, which is impossible. The situation is easily seen graphically for the simpler two-dimensional case shown in Fig. 6-3. It is clear that not all the "random walk" configurations are possible, and the resultant elimination of these structures leaves the remaining possibilities with an average end-to-end distance that is substantially larger than a truly random coil (about 20% larger from a detailed calculation). The polyisoprene anomaly was first explained by Flory, who suggested that there is a net attractive force tending to hold the chain segments to each other; for natural rubber, this force is just strong enough to reduce the actual end-to-end distance back to the value that would correspond to an ideal random coil.

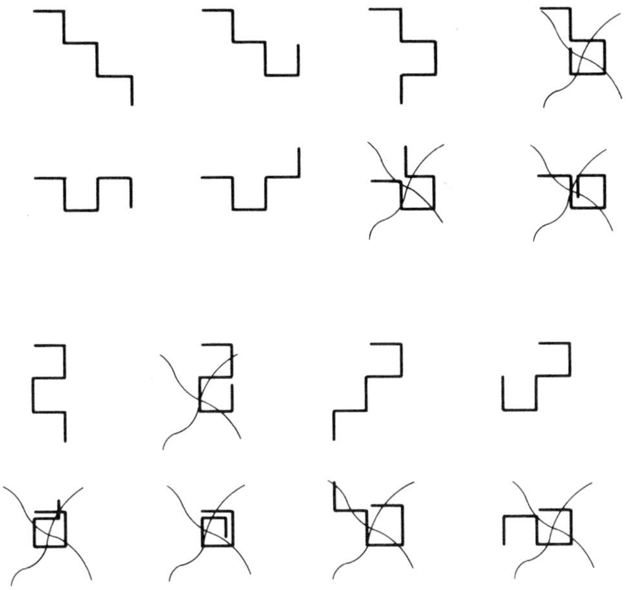

FIGURE 6-3. Demonstration of "excluded volume" for a two-dimensional random polymer. Since two atoms cannot occupy the same position in space, certain configurations from an ideal random walk are not possible. The remaining configurations show an average end-to-end distance that is appreciably longer than when all configurations are included (see text). (After *Macromolecules in Solution,* by Herbert Morawetz, Interscience, 1965, p. 123.)

Natural latex is a rather amorphous and pliant substance, but when the chains are cross-linked at wide intervals with sulfur monochloride, S_2Cl_2

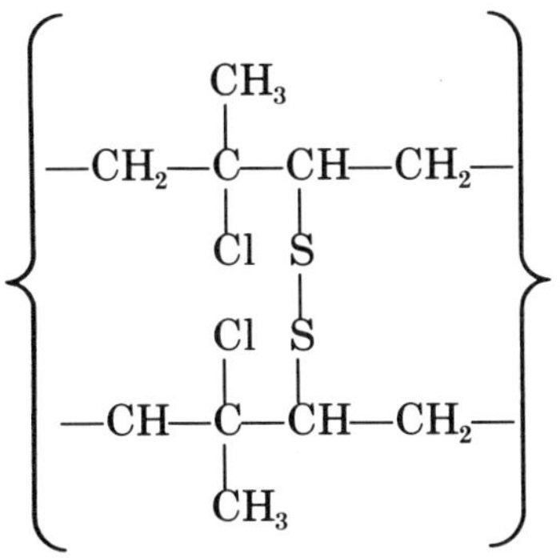

the isoprene links can no longer slide past each other very far, and the polymer can be stretched reversibly as required for use in automobile tires. The process was first attempted with sulfur and heat (hence, "vulcanization"). A similar mechanism is responsible for the "holding" quality of a hair "permanent"—in this case, heat catalyzes the formation of disulfide bonds that make hair protein more rigid.

Particularly with synthetic polymers, random coil calculations permit direct conclusions regarding the effect of solvent. In a "good" solvent that attaches itself to the polymer at many places, the polymer will swell (larger root-mean-square end-to-end distance), because the polymer chain is no longer as free to fold back on itself. Similarly, a "poor" solvent effectively compresses the polymer because of repulsive forces between solvent and external polymer surface.

Similar effects arise for the biologically more common polymers that possess many electrical charges along the chain. These charges ordinarily act to stabilize the (usually rigid) structure of enzymes and nucleic acids, through electrostatic bonds between parts of the chain. The presence of highly polar or charged species, such as urea, guanidine, salt, or detergent disrupts these electrostatic bonds and allows the biopolymer to unfold. For example, the (charged) sodium carboxymethylcellulose polymer in the presence of high salt concentration exhibits the same dimensions in solution as does the (uncharged) cellulose polymer. Detergents have recently been found of great value in the selective dismembering of biological membrane components, to permit the isolation of membrane-bound protein enzymes and drug receptors. Another use for detergents devolves from their capacity to unfold and coat most proteins into an approximate rod shape, where the length of the rod is directly related to the molecular weight of the polypeptide. The great difficulty in determination of the molecular weight of a protein is that almost all determinations depend on the shape of the protein; the addition of detergent or guanidine hydrochloride unfolds most polypeptides to a common shape (rod or random coil), and the polypeptides can then be compared on the basis of size (molecular weight) alone.

6.B. TRANSLATIONAL DIFFUSION

Translational diffusion of a substance across a *boundary* is a problem or an advantage in a great number of biophysical measurements. Analysis of diffusion is required for understanding the patterns of movement and separation of components of mixtures in the *ultracentrifuge,* in *chromatography,* and *electrophoresis.* Diffusion is the basis for analysis of the spreading of antigen on an agar plate in *immuno-assays.* Diffusion rates may be used to calculate the *molecular weight* of large molecules, one of the most basic (and most difficult) problems in biochemistry. Finally, it is essential to understand *passive* diffusion before it is possible to treat *active* transport across membrane boundaries.

In this section, we first examine an intuitive derivation of diffusion behavior, in a way that provides for a direct comparison to the rigorously derived random walk solution. The utility of the intuitive result becomes clear in the ensuing treatments of electrophoresis and sedimentation, where in addition to diffusion at a boundary, the boundary itself is forced to move under an electrical or centrifugal force.

Intuitive "Derivation" of the Diffusion Equation

Physical "laws" represent intuitive claims that cannot be derived from simpler claims, and whose consequences find no contradiction in experiment. Newton's laws of motion and the laws of thermodynamics are examples; so is the very important Fick's law of flow. Fick's intuition was that flow of substance from an area of one concentration to an area of different concentration should be proportional to the difference in concentration between the two regions, by analogy to the flow of heat, which is proportional to the difference in temperature between the two regions in question (an index finger freezes faster in liquid nitrogen than in ice water!):

$$\text{Flow} = -D\left(\frac{\partial c}{\partial y}\right)\cdot(\text{cross-sectional area across which flow occurs}) \qquad (6\text{-}51)$$

where D is the constant of proportionality, the minus sign indicates that the direction of flow is from larger to smaller concentration region, and the calculus notation is used to refine the statement of the law to incorporate an infinitesimal change in concentration in crossing the boundary. The form of Eq. 6-51 is much more general than the present applications would suggest: a similar equation holds for flow of heat proportional to a temperature gradient, flow of fluid proportional to a pressure gradient, or flow of electricity proportional to a voltage (electrical "pressure") gradient.

To test the intuitive claim of Eq. 6-51 we need merely look at a diffusion situation that compares directly to the random walk problem. Suppose, then, that at time zero, an infinitesimally thin layer of red dye is surrounded on both sides by solvent; at later times, the dye will spread out into the solvent, and the concentration of dye in any thin layer can be determined (for example) by scanning the absorption of light through a very thin slit that moves across the solution. Equation 6-51 is couched in terms of *flow* rather than *total amount* of substance, however, so we will want to consider the *rate of change* of concentration in any one thin layer (see Fig. 6-4).

Consider then the region between y and $(y + dy)$ in Fig. 6-4. For convenience, assume that the thin layer has unit cross-sectional area, so that the volume of the layer is just $(dy \cdot 1 \cdot 1) = dy$. The time rate of change in concentration in the layer is simply given by the difference between the flow of dye into the region and flow of dye out of the region:

$$\frac{\partial c}{\partial t} = \frac{\text{Flow in (moles/sec)} - \text{Flow out (moles/sec)}}{\text{Volume of region (cc)}} \text{ in moles cc}^{-1}\text{sec}^{-1}$$

$$\frac{\partial c}{\partial t} = \frac{-D(\partial c/\partial y)_{\text{at } y \text{ (entrance to region)}} + D(\partial c/\partial y)_{\text{at } y+dy \text{ (exit)}}}{dy} \qquad (6\text{-}52)$$

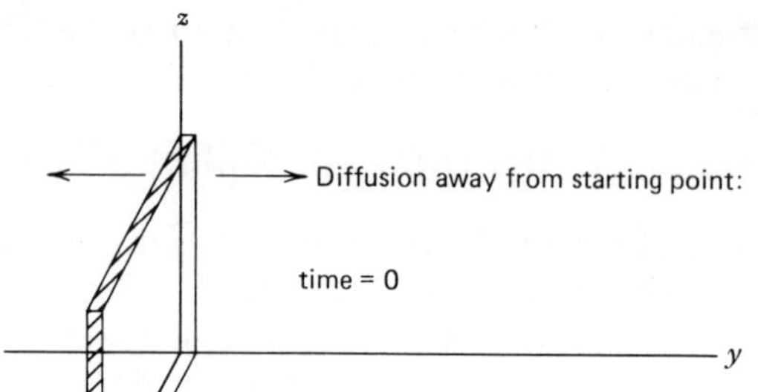

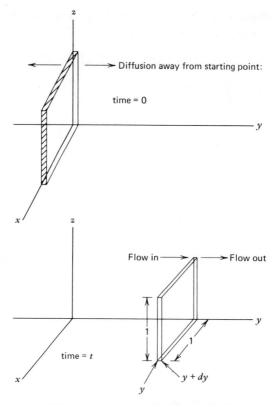

FIGURE 6-4. Progress of a diffusion experiment, starting with all the substance in a thin layer at $y = 0$ at time zero; Eq. 6-52 gives the rate of change in concentration of substance in region, y, $y + dy$ at later time, t. The layer is taken to have unit cross-sectional area.

Equation 6-52 poses the same problem that was faced early in the random walk case: the result is precise, but inconveniently phrased. The choice of an infinitesimal region may now be turned to advantage as follows. If concentration at time, t, is a function of distance, then concentration *gradient*, $(\partial c/\partial y)$, is also a function of distance, say $f(y)$. Now over any small interval, a function may be evaluated in terms of its value and the value of its derivatives at a nearby point, using a Taylor series (see Appendix):

$$f(y)_{\text{at } y} = f(y_0) + f'(y)_{\text{at } y_0}(y - y_0) + \frac{f''(y)}{2!}_{\text{at } y_0}(y - y_0)^2 + \cdots \quad (6\text{-}53)$$

Equation 6-53 may now be used to evaluate $(\partial c/\partial y)$ at $(y + dy)$, from $(\partial c/\partial y)$ at y:

$$\left[\frac{\partial c}{\partial y}\right]_{\text{at } y+dy} \cong \left[\frac{\partial c}{\partial y}\right]_{\text{at } y} + \frac{\partial}{\partial y}\left[\frac{\partial c}{\partial y}\right]_{\text{at } y} \cdot (y + dy - y) \quad (6\text{-}54)$$

Substituting the right-hand side of Eq. 6-54 for $(\partial c/\partial y)_{\text{at } y+dy}$ in Eq. 6-52 gives

$$\boxed{\frac{\partial c}{\partial t} = D\frac{\partial^2 c}{\partial y^2}} \quad \text{Diffusion Equation} \tag{6-55}$$

Equation 6-55 is the famous *diffusion equation,* which is one of the most important equations in physical science; most of the others are encountered later in this book. Solution of Eq. 6-55 entails specialized methods pioneered by Fourier, and would detract from the simplicity of the result, which is

$$\boxed{\frac{c(y)}{c(0)} = \frac{1}{2\sqrt{\pi D t}}e^{-y^2/4Dt}} \tag{6-56}$$

where $c(y)$ is the concentration of substance at distance y at time t, and $c(0)$ is the concentration of substance initially at the origin at time zero.

An example of the success of Eq. 6-56 in fitting actual experimental diffusion results is shown in Fig. 6-5. It is readily shown (see Problems) that the distance between the inflection points of the theoretical Eq. 6-56 is $2\sqrt{2Dt}$. Furthermore, since the left-hand side of Eq. 6-56 simply represents the probability of finding a given dye molecule at point y, the average absolute distance away from the origin after time t may be obtained from

$$<y^2> = \frac{1}{2\sqrt{\pi D t}}\int_{-\infty}^{+\infty} y^2 e^{-y^2/4Dt}\, dy \tag{6-57}$$

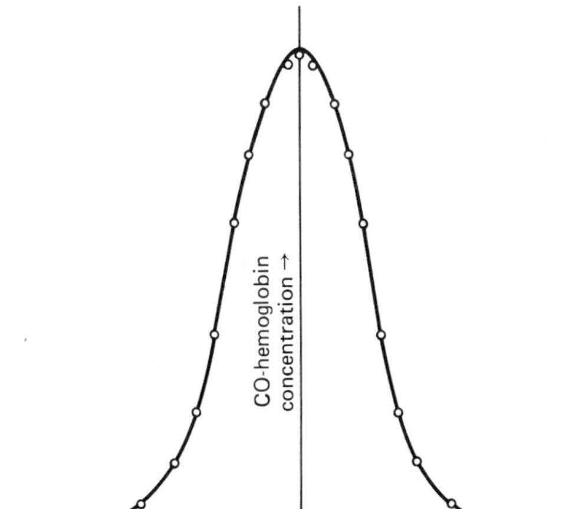

FIGURE 6-5. Diffusion of CO-hemoglobin in H_2O from a thin initial layer [O. Lamm and A. Polson, *Biochem. J. 30,* 528 (1936).] The solid line is the theoretical prediction from Eq. 6-56.

or just

$$\boxed{\sqrt{<y^2>} = \sqrt{2Dt}} \tag{6-58}$$

But the situation described by Eq. 6-56 and shown in Fig. 6-5 is identical to that treated in the random walk analysis that led to the result

$$\boxed{\sqrt{<y^2>} = \sqrt{Nl^2} = \sqrt{\nu t\, l^2}} \tag{6-59}$$

where

N = total number of steps in the walk
l = length of one "step"
ν = number of steps per unit time
t = duration of the "walk"

Correspondence of Eqs. 6-58 and 6-59 leads immediately to the important result

$$\boxed{D = \frac{1}{2}\nu\, l^2 \qquad \text{in one dimension}} \tag{6-60}$$

A treatment with more algebra but the same intellectúal content (see Problems) shows that similar results obtain when the treatment is extended to two or three dimensions:

$$D = (1/4)\nu l^2 \quad \text{for diffusion in two dimensions} \tag{6-60b}$$

$$D = (1/6)\nu l^2 \quad \text{for diffusion in three dimensions} \tag{6-60c}$$

The significance of Eqs. 6-60 is that they give the relation between the intuitive (empirical, physical) model based on continuous flow and the deductive (microscopic, statistical) model based on jerky progress by means of many small random steps. The reason physical scientists are so concerned with statistics is that almost all physical models are based on the behavior of one or two particles at a time, and some sort of averaging process is always required (Equations 6-49, 6-57) to apply the model calculation to the macroscopic experiment involving huge numbers of particles: the diffusion problem is one of the few for which a rigorous correspondence between the (microscopic) model and (macroscopic) reality can be made.

At this stage, it is reasonable to wonder why molecules should behave as if they were bouncing around very rapidly. To make the question less hypothetical, one can consider the motion of pollen particles (Brownian motion), which is readily apparent in an ordinary microscope. Detailed experiments have shown that the random motion of such particles is *not* ascribable to uneven evaporation of solvent, capillary action at the edge of the solution, cavitation (formation of small bubbles), uneven heating, attractive forces between particles, or in fact *any external* influence. The explanation for the phenomenon is that any one particle is being bombarded on all sides by solvent molecules, and while the *average* of many such collisions is to leave the particle where it started, a given particle at any one instant "feels" a net force that causes it to move. This random motion is responsible for all chemical reactions (molecules must collide before they can react), and when we consider the action of enzyme catalysts in the enhancement (speeding up) of reaction rates, we will return to a short diffusion argument to find out how often the reactants collide.

Applications The principal applications for the diffusion equation, Eq. 6-55, involve (1) diffusion from a point, (2) diffusion from one layer to another, and (3) diffusion across a membrane. Diffusion between layers occurs in chromatography, electrophoresis, and sedimentation, and is discussed in those sections as well as later in this section. Diffusion through a membrane is discussed in our examination of the Donnan equilibrium (Chapter 2), and we now look at two examples of diffusion from a point, as they are regularly applied in immunology.

EXAMPLE *Immuno-diffusion*

The immune system of the human body relies on a number of large proteins (antibodies) that circulate in the blood and may tend to localize in certain organs. The purpose of these antibodies is to recognize any strange (and presumably harmful) toxins, or antigens, and remove them from circulation by forming a very strong antigen:antibody complex. Bacteria, for example, may be recognized by characteristic oligosaccharide groups that project from the bacterial surface; protein antigens are recognized from a characteristic oligopeptide region of the protein, and so forth. Most immuno-chemistry is centered around formation, often followed by precipitation, of the antigen: antibody complex, as in the usual tests for blood "type."

To understand how the immuno-diffusion methods work, one must be aware of the dependence of the solubility of the antigen:antibody complex on the concentration ratio of the two reactants (see Fig. 6-6). As seen in Fig. 6-6, very little precipitate results when antibody *or* antigen is in excess.

One simple explanation for the behavior shown in Fig. 6-6 is illustrated graphically in Fig. 6-7. At small antibody (Ab) to antigen (Ag) ratios, soluble Ab:Ag complexes are formed, with small numbers of Ab and Ag per complex. At large Ab:Ag ratios, the polyfunctional nature of the antibody permits formation of extended networks (precipitation), with the most compact network at the largest Ab:Ag ratios.

The purpose of this introductory discussion is to show that when a solution of antibody is brought into contact with a solution of antigen, there will be a

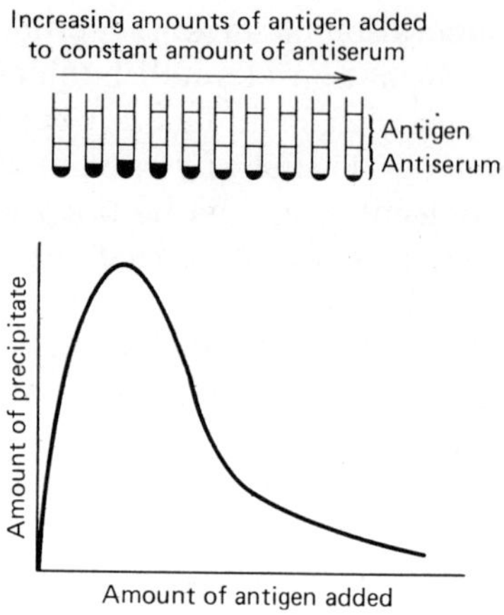

FIGURE 6-6. Typical precipitation curve for addition of antigen to a solution containing a fixed amount of antibody. (From A. H. Sehon, in *Sensitivity Chest Diseases,* M. C. Harris and N. Shure, Philadelphia, Davis, 1964.)

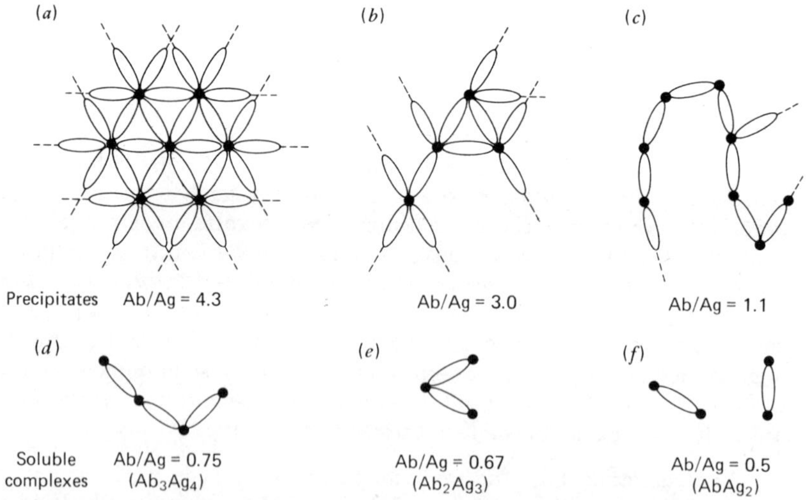

FIGURE 6-7. Schematic representation for space-filling characteristics of antibody(Ab) – antigen(Ag) complexes. Black dots are antigen molecules; oval links denote (polyfunctional) antibody molecules. The regions of Fig. 6-7 would correspond to *a* (antibody in excess), *b* (maximum precipitate formed), *c* (antigen in excess, with soluble complexes shown as *d, e,* and *f.* (From B. D. Davis, R. Dulbecco, H. N. Eisen, H. Ginsberg, and W. B. Wood, *Microbiology,* New York, Harper & Row, 1967.)

visible band of precipitation *only* for Ab:Ag ratios near the position of the maximum in the curve of Fig. 6-6.

Radial immuno-diffusion. A common diagnostic measurement in clinical immunology is determination of the quantity of individual gamma-globulins (immunoglobulins) in blood serum. In the typical assay, a small flat dish is covered with a solution containing agar (a negatively charged polysaccharide) and a specific antiserum (antibody) to the immunoglobulin (antigen) of interest; on standing, the 1% agar mixture will "set" into a relatively rigid gel. The purpose of the gel is to provide for diffusion at a slow, controlled rate. A small hole ("well") is then cut from part of the flat gel, and the immunoglobulin-containing serum is introduced into the well. As the antigen specific to the antibody in the gel diffuses radially outward, a visible "ring" appears at the boundary where the antibody:antigen concentration ratio is optimal for precipitation (see Fig. 6-8). Since the ring diameter increases with increasing concentration of antigen (see Problems), the test provides a quantitative, accurate, economical, and rapid measure of the amount of γ-globulin in the tested serum. If different antibodies are used, separate assays of the corresponding antigens may be obtained. If a mixture of antibodies is "gelled" into the gel to start with, then several rings will be observed at the same time — however, a better test for heterogeneity in the antibody fraction is provided by the Ouchterlony test.

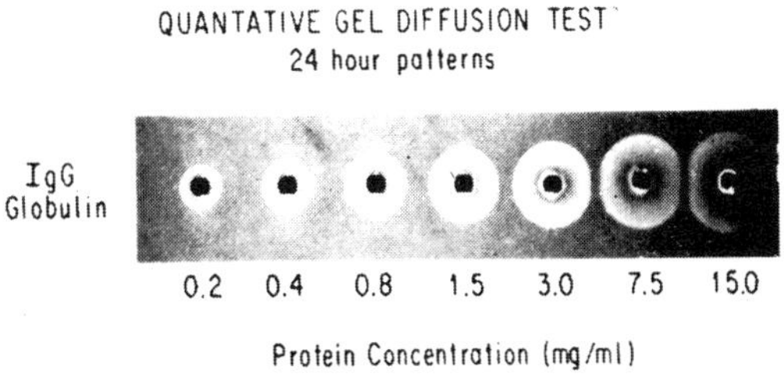

FIGURE 6-8. Photograph of the result of a radial immuno-diffusion quantitation of immunoglobulins. The diameter of the "precipitin" ring is a measure of the concentration of the immunoglobulin (see text). [From J. L. Fahey and E. M. McKelvey, "Quantitative Determination of Serum Immunoglobulins in Antibody-Agar Plates," *J. Immun. 94,* 84 (1965).]

Double diffusion — the Ouchterlony test. The other common problem for the immunologist is to determine the immunological similarity between the antigens from two different sources. In design of vaccines or antisera, it is essential to establish the Ab:Ag interactions common to both sera, to be sure that the desired combination occurs and to show that undesirable combinations do not. The standard technique again begins with formation of a thin agar layer on a glass plate, but this time the agar itself is free of antibodies. Three small wells are then cut at the points of a triangle as shown in Fig. 6-9: the lower well contains the antibodies and the upper wells contain the respective antigens of interest.

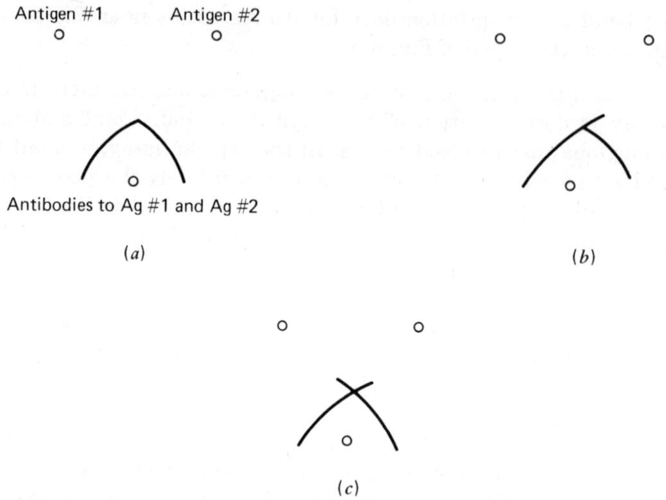

FIGURE 6-9. Typical patterns in double-diffusion gel immunochemistry. The upper two wells contain different antigens; the lower well contains immunoglobulins (antibodies). In serologic identity (*a*), the immunoglobulins react equally well with either antigen. In serologic nonidentity, each antigen reacts with an independent antibody (*c*). In serologic partial identity (*b*), one antigen reacts with both antibodies while the other antigen reacts with only one antibody (see text).

Three results describe most of the observed situations:

A *Serologic Identity.* If the antibody cannot distinguish one antigen from the other, there will be a continuous precipitin arc as seen for the upper left diagram in Fig. 6-9.

C *Serologic Nonidentity.* If the two antigens react with entirely different antibodies (i.e., the two Ab:Ag reactions are completely unrelated), then the two precipitin arcs will cross, as shown in the lower diagram of Fig. 6-9.

B *Serologic Partial Identity.* In this case (see upper right of Fig. 6-9), there are two kinds of antibody. Both antibodies form a precipitate with the antigen at upper left of diagram B in the Figure, while only one of the antibodies forms a precipitate with the antigen at upper right. Experimentally, the right-hand antigen precipitates out its antibody, while the other antibody proceeds through the precipitate region until it encounters the left-hand antigen.

A little thought shows that the precipitin arc will curve toward the well that contains the slowest-diffusing (antibody) molecule. Thus, a straight line precipitin arc is formed when the antigen and antibody have about the same molecular weight (we will later quantify the relation between diffusion rate and molecular weight), and is useful in estimating the molecular weight of the antigen, since most antibodies have about the same molecular weight of 150,000.

The immuno-diffusion methods are extraordinarily sensitive: radial immunodiffusion can detect as little as 3 micrograms/cc of antigen in serum. The techniques determine the number of antigens in a mixture and the identity of antigens in different preparations. The tests provide a criterion for purity of vaccines or other biologically active antigens, including enzymes and hormones. The gel-precipitation methods are used in sero-diagnosis of such infectious diseases as histoplasmosis, blastomycosis, and coccidiomycosis. Recently, rabbit serum containing myoglobin antibodies has been used to detect myoglobin in urine; thus, myoglobin released by a damaged heart may be detected a few hours after a heart attack to provide early diagnosis of the attack, using immunodiffusion methods.

EXAMPLE *Diffusion at a Boundary*

Diffusion at a boundary is of interest for two reasons: (1) it provides a means for determination of the diffusion constant of a macromolecule, from which we will shortly be able to find the molecular weight, and (2) it occurs in most of the preparative and analytical methods for separation of mixtures of macromolecules in the purification that must precede any study of function on a molecular level.

Measurement of diffusion constant. Consider a system consisting of an initially sharp boundary that separates a layer of solvent from a layer of solution of macromolecular concentration, $[C]_0$, as shown in Fig. 6-10. If the column of solution on either side of the boundary is sufficiently long, then the macromolecular concentration may be assumed to remain at either $[C]_0$ or zero far from the boundary—under these conditions, Eq. 6-55 may be solved to yield the concentration of macromolecule as a function of distance in the vicinity of the boundary:

$$[C]_{\text{at } y} = \frac{[C]_0}{2}\left[1 - \frac{2}{\sqrt{\pi}} \int_{u=0}^{u=(y/2\sqrt{Dt})} e^{-u^2}\, du\right] \qquad (6\text{-}61)$$

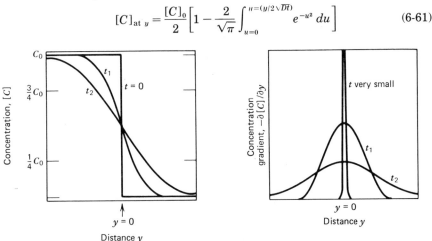

FIGURE 6-10. Progress of a diffusion experiment from an initially sharp boundary at $t = 0$. (From C. Tanford, *Physical Chemistry of Macromolecules,* John Wiley & Sons, 1961, p. 354.)

Plots of concentration near the boundary at later times are calculated from Eq. 6-6 and shown in Fig. 6-10. (The definite integral in Eq. 6-61 is the famous Gaussian or error-function, and is tabulated in many math tables.)

Both the algebra and the experiment are simplified by examination of the concentration *gradient,* $\partial[C]/\partial y$, rather than the concentration itself.

$$\left[\frac{\partial[C]}{\partial y}\right]_{\substack{\text{at } y,\\ \text{at } t}} = -\frac{[C]_0}{2\sqrt{\pi Dt}}\, e^{-y^2/4Dt} \tag{6-62}$$

The form of the concentration gradient as a function of distance at particular times is also shown in Figure 6-10.

The diffusion constant may be determined either from concentration (as measured by light absorption at a suitable wavelength, for example) or from concentration gradient (measured by refractive index methods—see Chapter 14.A.

(a) *Diffusion constant from concentration measurements.* From Eq. 6-61 it may be shown (see Problems) that the square of the distance between the points at which $[C] = [C]_0/4$ and $[C] = 3[C]_0/4$ is proportional to the time since the boundary was initially created:

$$\left(\Delta y_{\substack{\text{from } [C]_0/4\\ \text{to } 3[C]_0/4}}\right)^2 = 3.64\, Dt \tag{6-63}$$

Thus, D may be determined from the slope of an experimental plot of $(\Delta y)^2$ versus time.

(b) *Diffusion constant from concentration gradient.* From Eq. 6-62 one can readily show that the area under the plot of concentration gradient versus distance is

$$\text{Area}_{(\text{concentration gradient versus } y)} = -[C]_0 \tag{6-64}$$

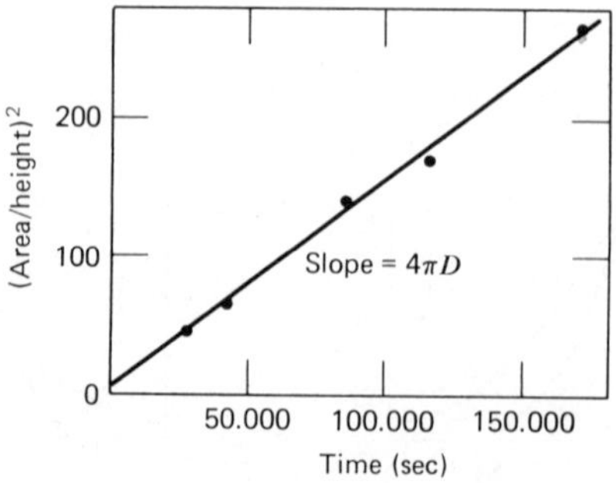

FIGURE 6-11. Determination of diffusion coefficient for ovalbumin from area and height of a plot of concentration gradient versus distance at particular times (see text). Since the boundary was not perfectly sharp at $t = 0$, the line in the graph does not pass through the origin. [From O. Lamm and A. Polson, *Biochem. J. 30,* 528 (1936).]

Table 6-1 Diffusion Coefficients for Macromolecules in Aqueous Solution*

Macromolecule	Molecular Weight	$D_{20°,\text{water}} \times 10^7$ cm²sec⁻¹
Ribonuclease	13,683	11.9
Chymotrypsinogen	23,200	9.5
Ovalbumin	45,000	7.76
Hemoglobin	68,000	6.9
Tropomyosin	93,000	2.24
Fibrinogen	330,000	2.02
Collagen	345,000	0.69
DNA	6,000,000	0.13
Tobacco mosaic virus	40,000,000	0.53

* *Diffusion coefficients have been corrected to the value that would be observed at 20°C in pure water for the same size and shape. [Data from C. Tanford,* Physical Chemistry of Macromolecules *(John Wiley, N. Y., 1961), pp. 358, 361].*

while the height of the curve at $y = 0$ is just

$$\text{height at } y = 0_{(\text{concentration gradient versus } y)} = \frac{-[C]_0}{2\sqrt{\pi D t}} \tag{6-65}$$

Thus the ratio of area to height is given by

$$\frac{\text{Area}}{\text{Height}} = 2\sqrt{\pi D t} \tag{6-66}$$

and a plot of (Area/Height)^2 versus t will give a straight line of slope, $4\pi D$, as shown in Fig. 6-11.

As evident from intuition (big molecules move more slowly) and as we will shortly show, diffusion coefficients are inversely related to molecular weight. Examples are shown in Table 6-1. The correlation between molecular weight and diffusion constant is irregular, because we have not yet accounted for the effect of macromolecular shape on the diffusion rate. This matter is explored after a discussion of viscosity.

PROBLEMS

1. Stirling's approximation, $\ln N! = N \ln N - N$, is fundamental to many statistical problems, including the Gaussian (normal) distribution on which diffusion-related phenomena are based. Derive this approximation (valid for large N) from the definition, $N! = N(N-1)(N-2) \cdots 3 \cdot 2 \cdot 1$. (*Hint:* consider a plot of $\ln N$ versus N.)

2. Show that the root-mean-square *number of steps* away from the origin in a one-dimensional random walk (Eq. 6-44), after N steps, is

$$\sqrt{<(m - <m>)^2>} = \sqrt{<m>/2} = \sqrt{(N/4)}$$

3. Show that the average *distance* away from the origin after N steps of a random walk is

$$<y> = 0$$

while the root-mean-square distance away from the origin after N steps of a one-dimensional random walk (Eq. 6-49) is

$$\sqrt{<y^2>} = l\sqrt{N}$$

where l is the length of one step.

4. It has been shown in the text that the probability that a molecule has reached the region, $y, y + dy$, starting as a random walk from the origin is

$$P_t(y)\,dy = [1/\sqrt{4\pi Dt}]\ \exp\ [-y^2/4Dt]\ dy$$

From a calculation essentially the same as for Problems 6-1 and 6-2, the "r.m.s." width of this distribution can be shown to be (Eq. 6-58)

$$\sqrt{<y^2>} = \sqrt{2Dt}$$

the most commonly quoted property of the one-dimensional random walk. In order to gain increased physical insight into this result, calculate the width of a plot of $P_t(y)$ versus y, taking as a measure of that width the separation between the points of inflection on either side of the origin. Compare to Eq. 6-58. Sketch (roughly) a plot of $P_t(y)$ versus y, showing the location of the inflection points and the r.m.s. width.

5. From light-scattering measurements (see Chapter 15.A) for a polymethylene polymer (i.e., a saturated, straight-chain hydrocarbon), the root-mean-square end-to-end distance is determined to be 85 angstrom. Assuming that the polymer is freely jointed and unbranched,
   (a) Calculate the molecular weight of the polymer (remember to count hydrogens), given that an average C—C bond is 1.54 angstrom long.
   (b) In the actual polymer, the joints are clearly not free, since a given 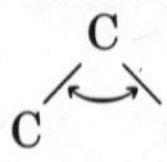 C bond angle is about 109½° (tetrahedral). Qualitatively, how will inclusion of this consideration affect the result of part (a)?

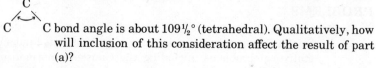

6. The diffusion constant for sucrose in water is $4 \times 10^{-6}\ cm^2sec^{-1}$.
   (a) Calculate the average (root-mean-square) distance (in centimeters) that a typical sucrose molecule moves in one hour.
   (b) Now calculate how long it will take a typical sucrose molecule to diffuse from the center to the outer edge of a blood capillary (capil-

lary diameter $= 8 \times 10^{-6}$ meter). This result should demonstrate why capillaries have to be so small in order to carry out efficient exchange of food and wastes between blood and surrounding tissue.

7. For a two- or three-dimensional diffusion case, starting with all diffusible substance at the origin at time zero, the respective probability of finding a molecule located in the two-dimensional region, x, $x + dx$ and y, $y + dy$, at time t is:

$$P_t(x,y)\,dx\,dy = \left(\frac{1}{4\pi Dt}\right) e^{-(x^2/4Dt)}\, e^{-(y^2/4Dt)}\,dx\,dy$$

and of finding a molecule located in the three-dimensional region, x, $x + dx$, y, $y + dy$, and z, $z + dy$, at later time, t, is:

$$P_t(x,y,z)\,dx\,dy\,dz = \left(\frac{1}{4\pi Dt}\right)^{(3/2)} e^{-x^2/4Dt}\, e^{-y^2/4Dt}\, e^{-z^2/4Dt}\,dx\,dy\,dz$$

Calculate the root-mean-square distance of molecules away from the origin after time, t,
(a) For a two-dimensional diffusion case
(b) For a three-dimensional diffusion case
(c) Use your result to confirm Eqs. 6-60 of the text.

8. The transitional diffusion equation in one dimension is given by Eq. 6-55: $\partial c/\partial t = D\partial^2 c/\partial y^2$ for diffusion across unit cross-sectional area. The corresponding translational diffusion equation for three-dimensional motion is given by

$$\partial c/\partial t = D\,\nabla^2 c$$

where

$$\nabla^2 = \partial^2/\partial x^2 + \partial^2/\partial y^2 + \partial^2/\partial z^2$$

(a) Confirm that a solution of the three-dimensional translational diffusion equation is:

$$c = c_0(1/4\pi Dt)^{3/2} \exp\,[-(x^2 + y^2 + z^2)/4Dt]$$

(b) Find the points of inflection of a plot of c versus r, as a measure of the "spread" of concentration over distance at a given time, t.

9. In radial immunodiffusion, suppose that all the substance of interest is concentrated at the origin (concentration, c_1) at time zero. The concentration anywhere in a plane away from the origin at later time, t, is given by the two-dimensional diffusion result

$$c(r,t) = c_1(1/4\pi Dt)\exp\,[-r^2/4Dt]$$

Suppose that antibody-antigen precipitation occurs visibly at one particular "critical" concentration, c_{critical}. Next, suppose that at time, t_1, the critical concentration occurs at $r = 1$ cm away from the origin. Finally, suppose we repeat the experiment, using a different concentration, c_2, at the origin initially. Calculate the radius of the precipitin ring as a function of the initial concentration of substance at the origin, where all immunodiffusion runs are allowed to proceed for the same length of time, t_1. (Compare your result with Fig. 6-8.)

10. Diffusion at a boundary affords two means for determination of translational diffusion constant from measurement of concentration (or concentration gradient) as a function of distance from the initial sharp boundary. Derive the equations (6-63, 6-64, and 6-65) that form the basis for these methods from Eqs. 6-61 and 6-62.

REFERENCES

N. Davidson, *Statistical Mechanics,* McGraw-Hill, N. Y. (1962), pp. 280–285. Good treatment of the random walk problem.

C. Tanford, *Physical Chemistry of Macromolecules,* Wiley, N. Y. (1961). Good discussion of random coil and translational diffusion.

CHAPTER 7 Forced March

7.A. ELECTROPHORESIS

Most large biomolecules are ampholytes; that is, the molecule has a large number of labile protons, and each proton exhibits a characteristic pK_a. From the individual pK_as, it is possible (see Chapter 3) to compute the "isoelectric point," pI, as the pH at which the macromolecule possesses an equal number of positive and negative charges, and thus behaves as a net neutral species. If the solution pH is *below* the isoelectric pH, the macromolecule will have net *positive* charge; *above pI* the molecule will have net *negative* charge. The point of this discussion is that unless the solution pH happens to coincide exactly with the unique pI, the macromolecule will be electrically charged, and will move under the influence of an electric field. Analysis of this motion proceeds in three stages. First, we must explain why a charged molecule in solution moves at a constant *velocity* proportional to the applied electric field—the applied electric field represents a constant *force* on the molecule, so one would intuitively expect the molecule to *accelerate* along the applied field. Second, we will establish the important relation between ion motion and diffusion constant, required for understanding of the effect of molecular shape on transport phenomena (diffusion, sedimentation, electrophoresis). Finally, the empirical continuous-flow model will be applied to the ion motion, to predict the result of an electrophoresis experiment.

Why Does an Ion Subjected to a Constant Force Move at Constant Velocity?

A particle with charge, Q, when subjected to an electric field (voltage gradient, electrical "pressure" gradient), E, experiences a constant net force

$$\begin{aligned} m\,a &= QE \\ m &= \text{mass} \\ a &= \text{acceleration} \end{aligned} \tag{7-1}$$

An *isolated* particle (e.g., in vacuum) would thus *accelerate* at a constant rate under the influence of such a force. However, an ion in aqueous *solution* is not free to move, but is held back by a frictional force proportional to its velocity:

$$\boxed{m\,a = QE - f\frac{dy}{dt}} \qquad (7\text{-}2)$$

where f is the frictional coefficient. As soon as the electric field is turned on, the particle will accelerate until it reaches a limiting velocity (similar to the limiting velocity of a parachute) for which the frictional force just balances the (driving) electrical force, to give a net force of zero (i.e., *constant* velocity):

$$QE = f\frac{dy}{dt}$$

or

$$\boxed{\text{limiting velocity} = \frac{dy}{dt} = \frac{QE}{f}} \qquad (7\text{-}3)$$

Microscopic Origin of Friction: Relation between Electrophoresis and Translational Diffusion

The simplest approach to this question begins from a microscopic model in which the molecule is assumed to undergo many collisions per second (random walk model); molecular motion between collisions is then calculated. To find out the average length of time between collisions, τ, use is made of a result employed in first-year chemistry courses in derivation of the ideal gas law:

$$\text{average translational kinetic energy} = \frac{1}{2}\,m\,v_{\text{thermal}}^2 = \frac{3}{2}\,kT \qquad (7\text{-}4)$$

where k is the Boltzmann constant, m is the mass of the particle, T is the absolute temperature, and v_{thermal} represents the average velocity of a gas molecule. Although the remaining argument is thus based on a property of gases, the result is nevertheless general, as is evident shortly.
Collecting definitions

τ = average time between collisions (sec/collision)
ν = frequency of collisions (collisions/sec)
l = average distance between collisions (distance/collision)
v_{th} = thermal velocity (distance/sec)

where

$$\tau = \frac{1}{\nu} \qquad (7\text{-}5)$$

Let us examine a particular molecule just after a collision. Although the molecule has a very large average *speed,* its average *velocity* is zero, since velocity is a vector quantity and the molecule is equally likely to be traveling in any direction. Under the influence of an applied electric field, the (charged) molecule will accelerate continuously until its next collision, under the force,

$$m\,a = QE \tag{7-1}$$

or

$$a = \frac{QE}{m} \tag{7-6}$$

Under that acceleration, the molecule will have traveled a net forward distance, $(a\tau^2/2)$ in the y direction; the duration of the average trip is τ, so that the average net forward velocity between collisions is given by:

$$\begin{array}{c}\text{Average net forward velocity}\\ \text{between collisions}\end{array} = \frac{QE\ \tau^2}{2m\ \tau} = \frac{QE\tau}{2m} \tag{7-7}$$

At this stage it is convenient to define *ionic mobility,* μ, as the average forward velocity per unit electric field:

$$\mu = \frac{QE\tau}{2mE} = \frac{Q\tau}{2m} \tag{7-8}$$

The key argument in this treatment is that the thermal speed (which is random in direction) is much larger than the induced velocity caused by the electric field—therefore, the direction of motion following any collision is (on the average) randomized completely, and the molecule after any collision once again has zero average velocity. Multiplying the right-hand side of Eq. 7-8 by

$$\nu\tau = 1 \tag{7-9}$$

and substituting for τ,

$$\tau = (l/v_{\text{th}}) \tag{7-10}$$

one obtains

$$\mu = \frac{Q\ \nu\ l^2}{2m\ v_{\text{th}}^2} \tag{7-11}$$

Finally substituting for v_{th} from Eq. 7-4 and using the result (Eq. 6-60c) from the three-dimensional random walk, $D = (\nu\, l^2/6)$, the desired result appears:

$$\boxed{\mu = \frac{Q\,D}{kT}} \tag{7-12}$$

The result, Eq. 7-12 relating ionic mobility in an electric field to diffusion in the absence of an electric field, is susceptible to experimental test, since μ can be studied as a function of temperature and the D value then compared with a direct measurement of D from, say, diffusion at a boundary. Equation 7-12 turns out to be a *general* result, although it was arrived at by a *specialized* derivation. This is a frequent occurrence in physical science, and furnishes additional relevance to the study of simple models.

Another general result is now available, from comparison of the continuous-flow model, Eq. 7-3, with the microscopic model, Eq. 7-12:

$$\boxed{f = \frac{kT}{D}} \tag{7-13}$$

Equation 7-13 is valid for molecules of *any shape,* and since the dependence of f on molecular shape is well-known (see viscosity section), we will soon be in a position to account for diffusion or electrophoresis behavior of neutral or charged molecules of any shape.

What Does an Electrophoresis Experiment Look Like?

Up to now, both macroscopic and microscopic treatments agree that a charged molecule in solution (or in a gel) will move at a constant velocity, given by its ionic mobility (ion velocity per unit electric field). In addition, the relations between ion mobility, diffusion constant, and friction coefficient have been derived. Experimentally, electrophoresis was first carried out by carefully layering a dilute salt solution over a solution containing the charged macromolecule of interest and observing the steady motion of the boundary on application of an electric field (Tiselius apparatus), as shown in Fig. 7-1. In the diagram, it is supposed that there are two migrating macromolecules, both negatively charged, but having different ionic mobilities.

The main problem with the apparatus of Fig. 7-1 is that it is difficult to form and maintain the sharp initial boundary required for good resolution. Modern methods therefore rely almost exclusively on use of a medium such as paper or gel to reduce convective mixing, while still permitting almost unimpeded forced motion from the electric field. These methods consist of injection of a thin band of macromolecule-containing solution into

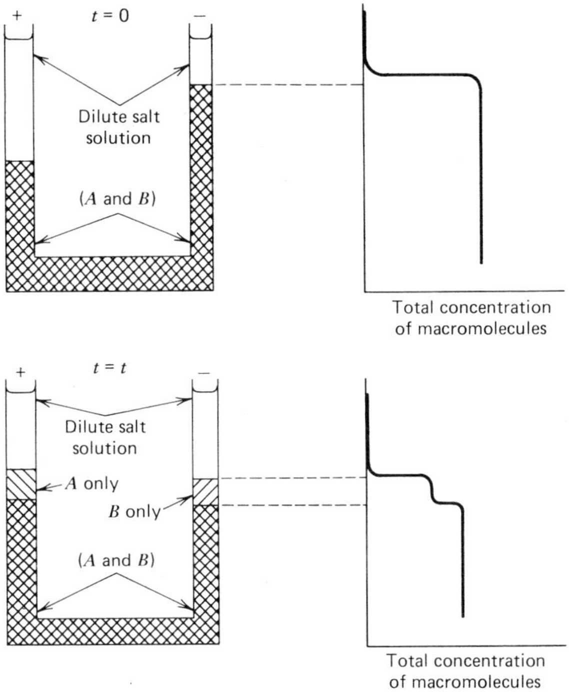

FIGURE 7-1. Separation of two macromolecules, A and B, in an early electrophoresis apparatus. Both molecules are negatively charged, and A moves faster than B under influence of an electric field. Total macromolecule concentration is plotted as a function of distance along the descending "limb" of the device.

a gel, for example, followed by migration of various ions on turning on an electric field. The result of the experiment is readily clarified from examination of the flow of substance in and out of an infinitesimal region, as shown for a positively charged species in Fig. 7-2.

From the previous treatment of diffusion from a thin layer, there will be a flow across any unit cross-sectional area of

$$\text{Diffusional flow} = -D\,\frac{\partial[C]}{\partial y} \tag{6-51}$$

In addition, because of the constant flow of ions in the presence of the electric field, there will be another contribution to flow:

$$\text{Electrophoretic flow} = (\mu E)[C] \tag{7-14}$$

where (μE) is just the ion average velocity in the electric field. The net rate of change of concentration in the region y, $y + dy$, is again obtained by the difference between the flow in and out of the region:

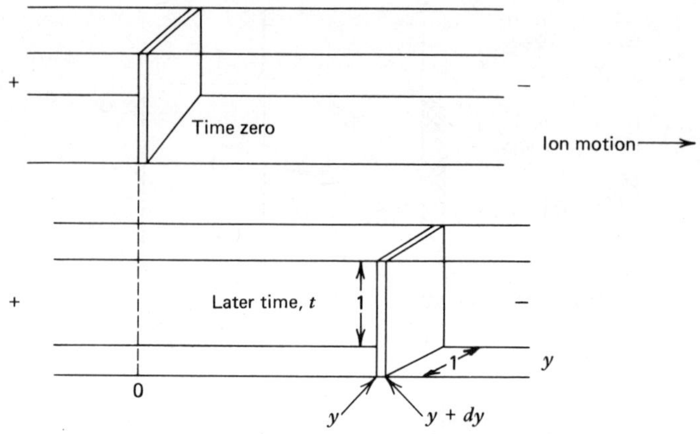

FIGURE 7-2. Diagram for analysis of gel electrophoresis experiments. Substance begins in a narrow band at $y=0$ at $t=0$; at later $t=t$, the flow of substance in and out of the region of unit cross-sectional area bounded by $y, y+dy$, is analyzed. The macro-ion is taken to be positively charged, so that it will move to the right when the electric field is turned on.

$$\frac{\partial[C]}{\partial t} = \frac{-D(\partial[C]/\partial y)_{\text{at } y} + \mu E[C]_{\text{at } y} + D(\partial[C]/\partial y)_{\text{at } y+dy} - \mu E[C]_{\text{at } y+dy}}{1 \cdot 1 \cdot dy} \tag{7-15}$$

Since both $[C]$ and $\partial[C]/\partial y$ are functions of y, recourse can be made to the Taylor series approximation for each quantity at the exit to the region of interest:

$$[C]_{\text{at } y+dy} \cong [C]_{\text{at } y} + \left[\frac{\partial[C]}{\partial y}\right]_{\text{at } y} dy \tag{6-54}$$

$$\frac{\partial[C]}{\partial y}_{\text{at } y+dy} \cong \frac{\partial[C]}{\partial y}_{\text{at } y} + \left[\frac{\partial}{\partial y}\frac{\partial[C]}{\partial y}\right]_{\text{at } y} dy \tag{7-16}$$

Substituting Equations 6-54 and 7-16 into Eq. 7-15 the final result is

$$\frac{\partial[C]}{\partial t} = D\frac{\partial^2[C]}{\partial y^2} - \mu E\frac{\partial[C]}{\partial y} \tag{7-17}$$

To visualize the significance of Eq. 7-17, it is sufficient to change variables to a coordinate frame that translates at the same rate the ions are moving due to the electric field (see Problems):

$$y' = y - \mu E t \tag{7-18}$$

The simplicity of the final result is more apparent in the new coordinates:

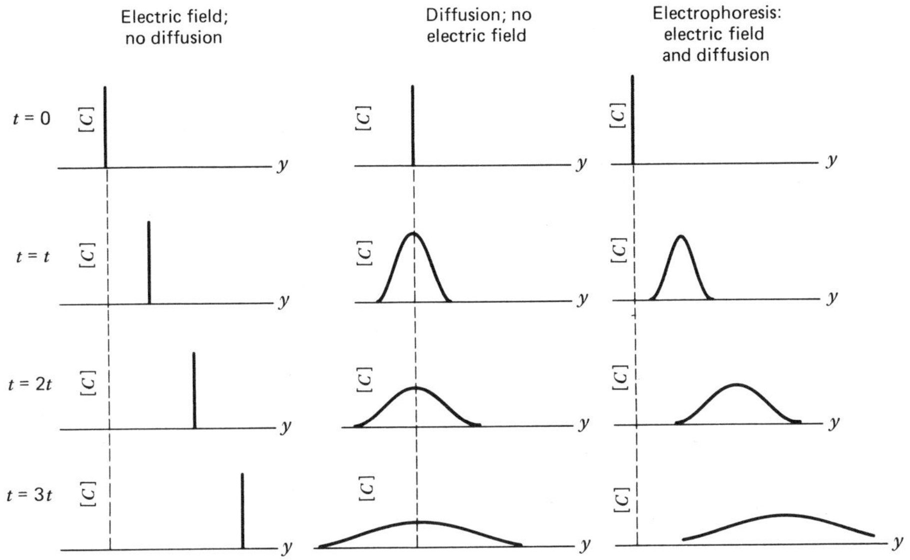

FIGURE 7-3. Plots of ion concentration versus distance under various conditions. Effect of electric field alone is given by Eq. 7-3; effect of diffusion alone is given by Equations 6-55 and 6-56; effect of diffusion plus electric field (electrophoresis) is given by Equations 7-17 and 7-19.

$$\frac{\partial[C]}{\partial t} = D\,\frac{\partial^2[C]}{\partial(y')^2} \tag{7-19}$$

Equation 7-19 thus shows that in a frame moving along with the ions in the electrophoresis experiment, diffusion proceeds as if the ions were stationary. In other words, diffusion and ion mobility are independent processes, as seen pictorially in Fig. 7-3.

Electrophoresis Techniques and Applications

Gel Electrophoresis The electrophoresis techniques provide some of the very best present means for separation, isolation, and analysis of mixtures of macromolecules, particularly proteins. Applications range from medical diagnostic information to criteria for purity of biochemical preparations, as shown below. The gel itself is most often agar, starch, cellulose acetate, or polyacrylamide, and the experiment consists simply of injection of the sample fluid into a small region of the tubular or strip-shaped gel, followed by immersion of the gel in a suitable buffer and application of an electric field by connection of electrodes at either end of the gel. The various macroions in the sample then migrate at different (constant) rates in the gel, and their location in the gel may be determined after the experiment by staining and/or measurement of light absorption at various positions along the

FIGURE 7-4. Electrophoretic pattern of normal human serum proteins for agar gel electrophoresis: anode at right, cathode at left. (After L. P. Cawley, *Electrophoresis and Immunoelectrophoresis,* Little, Brown & Co., Boston, 1969, p. 12.)

gel cylinder or strip. The result of a gel electrophoresis experiment on human serum using agar as the gel medium is shown in Fig. 7-4 (sample initially injected at point marked ⊗).

Just as photographs are made more useful by use of fine-grain film to make possible an image of greater detail and variety, both biochemical research and medical diagnosis have profited from the use of electrophoresis to separate components of biological fluids so that individual components can be analyzed separately for a better "fingerprint." The most studied and most interesting biological fluid is blood: after the cells are removed the remaining fluid is called *plasma,* and if the blood is allowed to clot, the fluid extruded is called *serum* (serum is the same as plasma, but lacks fibrinogen —see Table 7-1).

$$\text{Blood} \xrightarrow[\text{cells}]{\text{Remove}} \text{Plasma} \xrightarrow[\text{(removes fibrinogen)}]{\text{Allow to clot}} \text{Serum}$$

Table 7-1 Average Distribution of Normal Human Plasma Proteins

Albumin	50%
α_1-globulin	4
α_2-globulin	12
β-globulin	13
Fibrinogen	8
γ-globulin	13

The names of these plasma proteins derive from their position following an electrophoresis experiment; the functions of the components are described in any biochemistry text.

EXAMPLE *Gel Electrophoresis in Diagnosis of Disease*

Figure 7-5 shows the normal electrophoretic pattern for serum proteins, along with the characteristic changes that appear in three abnormal conditions. Because these experiments were conducted with cellulose acetate as the gel, the final gel strip is transparent and the proteins may be located according to their absorption of light as shown in the figure. Besides the conditions described in the figure, there have been several recent attempts to correlate the serum glycoprotein level with incidence of atherosclerosis (arteriosclerosis);

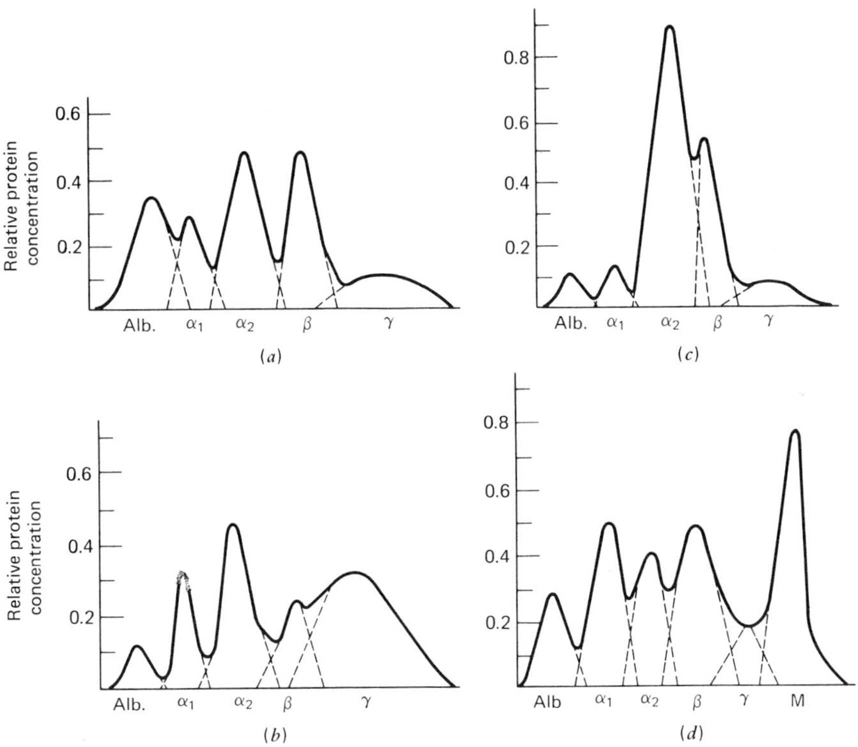

FIGURE 7-5. Cellulose acetate gel electrophoresis patterns for (*a*) normal serum, (*b*) infectious hepatitis, (*c*) lipid nephrosis, and (*d*) gamma myeloma. Protein fractions are listed in Table 7-1. Protein concentration is determined from optical absorption (see Section 4). [From *Clinica Chimica Acta,* 672 (1960).]

in this connection, it is interesting to note that Australian aborigines have a much lower incidence of arteriosclerosis and also lower serum glycoprotein level than do urban populations. With the much more selective separative capacity rendered by the recent disc-electrophoresis (see below), the diagnostic value of electrophoresis is bound to improve.

Electrophoresis is of diagnostic value in analysis of proteins of two other body fluids: cerebrospinal fluid and urine. The only difficulty is that while the protein constituents of both fluids are similar to those in serum, the respective concentrations are down by a factor of 100 and 1000. The most specific changes in the relative amounts of cerebrospinal fluid proteins are found for multiple sclerosis, meningitis, and encephalitis, with some changes also observed in degenerative diseases or neoplastic (cancerous) infiltration. Appearance of proteins in urine is a sensitive indicator of renal (kidney) malfunction—ordinarily, the kidneys filter out almost all serum protein in formation of urine, but in kidney disease, proteins can leak through into the urine. Since electrophoretic mobility is a function of molecular weight, the severity of the dysfunction can be judged in part from the size of the proteins that escape from the kidney. Detection of the relatively small "Bence Jones proteins," which correspond to the "light" chains of the immunoglobins (gamma globulins), is diagnostic for plasmocytoma and to a lesser extent primary macroglobulinemia. In all these tests, the procedure is as for serum proteins, except that electrophoresis is preceded by a 100-fold or 1000-fold concentration by dialysis under pressure (see Chapter 2.C).

The rate at which a protein can move through a solution (sedimentation, diffusion) or through a gel (chromatography, electrophoresis) is greatly affected by the macromolecular *shape*. Determination of the molecular weight of an unknown protein (with unknown shape) is thus a difficult matter. Two general methods are now available for inducing virtually any protein to take on the same shape: (1) 6M Guanidine HCl unfolds the tertiary structure of proteins so that they are to a good approximation random coils, or (2) sodium dodecyl sulfate (a common detergent) at 5×10^{-4} M or greater binds to proteins all along their length and causes them to behave as rods of constant diameter whose long axis is proportional to the molecular weight of the polypeptide chain. [In both methods, the —S—S— bonds that hold various peptide chains together must be reduced (broken) before treatment with guanidine or detergent.] Following treatment with detergent, it is found that the electrophoretic mobility correlates smoothly with molecular weight of the unraveled protein (see Fig. 7-6), in marked contrast to the irregular correlation of diffu-

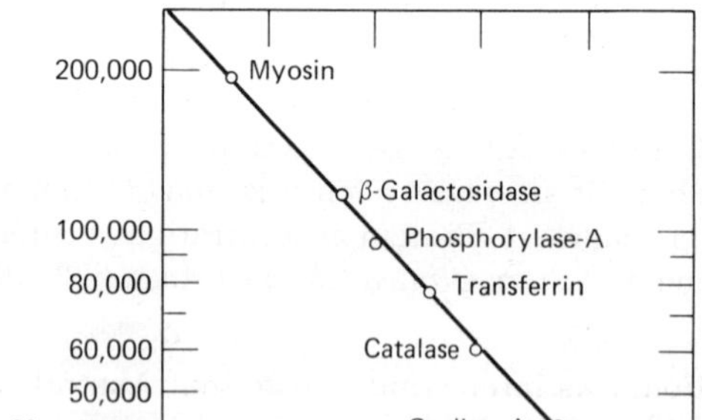

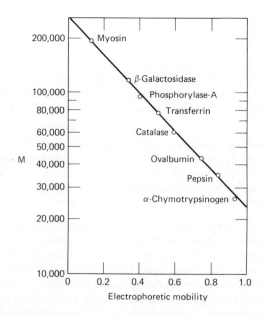

FIGURE 7-6. Correlation between molecular weight (log scale) and electrophoretic mobility for proteins that have been previously unraveled by reduction of disulfide linkages and treatment with sodium dodecyl sulfate detergent. [From H. R. Trayer et al., *J. Biol. Chem. 246,* 4486 (1971).]

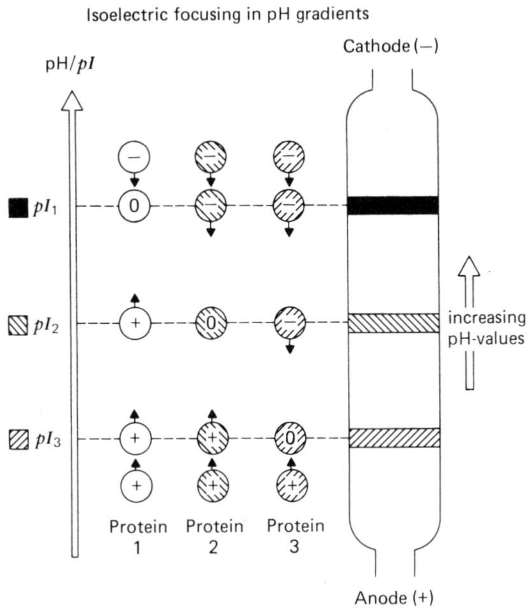

FIGURE 7-7. Three proteins of respective isoelectric points, pI_1, pI_2, and pI_3, in a schematic electrofocusing experiment. Each protein carries a negative charge above its pI (and thus is forced down the column) and a positive charge below its pI (and then is forced up the column). Each protein therefore moves in the electric field until it reaches that level in the column where the pH of the gel matches the pI of the protein. [From H. Haglund, *Methods of Biochemical Analysis 19,* 1 (1970).]

sion constant (or mobility) with molecular weight shown in Table 6-1. Confronted with an unknown protein, one need merely inject a sample containing the protein and several standard proteins of known molecular weight into a polyacrylamide gel (after prior treatment with detergent) and interpolate to find the molecular weight of the unknown protein—this is rapidly becoming the method of choice for finding the approximate molecular weight of proteins.

Isoelectric Focusing in pH Gradients The principle of isoelectric focusing is illustrated in Fig. 7-7. The technique is based on the fact that any ampholyte (a molecule with two or more dissociable protons) is positively charged at a pH below its isoelectric point and negatively charged at a pH above the isoelectric point; at the isoelectric pH, the ampholyte will be neutral and thus stationary in an electric field (see Chapter 3.C for detailed discussion of these terms). Thus, by preparing a gel (again the gel serves to minimize diffusion) with a pH-gradient throughout its length, a given protein on application of an electric field will migrate to the pH matching its isoelectric point, pI, as shown in Figures 7-7 and 7-8.

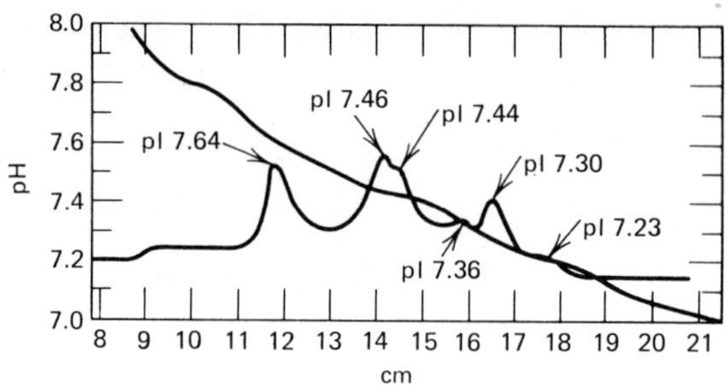

FIGURE 7-8. Separation of hemoglobin components by isoelectric focusing. The smooth curve represents the pH as a function of distance along the column; the curve with peaks indicates the appearance of various proteins (measured from optical absorbance). (From H. Haglund, *Science Tools 14,* 17 (1967), reproduced by permission from LKB Producter AB.)

EXAMPLE *Separation of Hemoglobin Components by Isoelectric Focusing*

One of the most spectacular accomplishments in isoelectric focusing is shown in Fig. 7-8. In that example, it proved possible to separate two proteins whose isoelectric points differed by only 0.02 pH unit!

Immunoelectrophoresis In this technique, electrophoresis is conducted as usual in a strip of agar (negatively charged) or agarose (neutral) gel; the strip is then removed, and trenches are placed on either side of the strip so that the edges of the strip are immersed in a solution containing various antibodies. The antibodies are chosen to be specific to certain of the proteins (antigens) that have been separated in the electrophoresis; thus a precipitin band will form for each antigen:antibody reaction. As seen in Fig. 7-9, a group of proteins with the *same* electrophoretic mobility may be resolved by reaction with antibody, because the *concentrations* of the components of the group are different, and thus their rates of radial diffusion differ.

By now it is clear that the term "gamma globulin" encompasses a large number of proteins, which are now classified as immunoglobulins IgG, IgA, and IgM, as described in any recent immunology text. The utility of immunoelectrophoresis lies in the detection of decreases or increases in the concentrations of these immunoglobulins: each immunoglobulin is thought to be produced by a group of cells (clone) deriving from a single ancestral cell, and assay of individual immunoglobulins is of great value in differential diagnosis of types of bone marrow tumors as well as deciding whether such tumors are benign (immunoglobulin level is relatively constant with time, over several months) or malignant (immunoglobulin level increases with time).

Discontinuous ("Disc"-) Electrophoresis With any conventional electrophoresis experiment, the success in resolving separate components depends

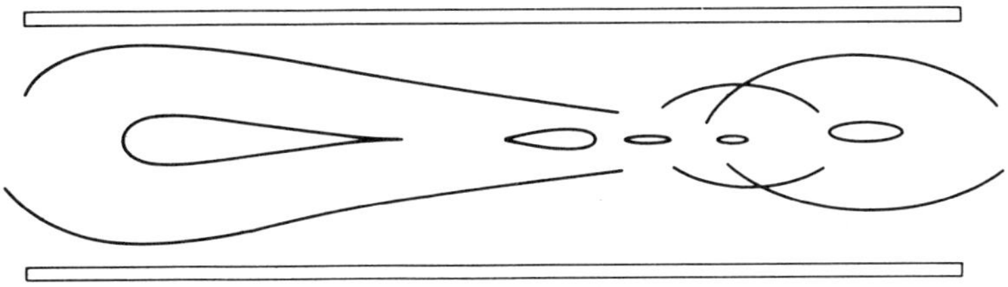

(a)

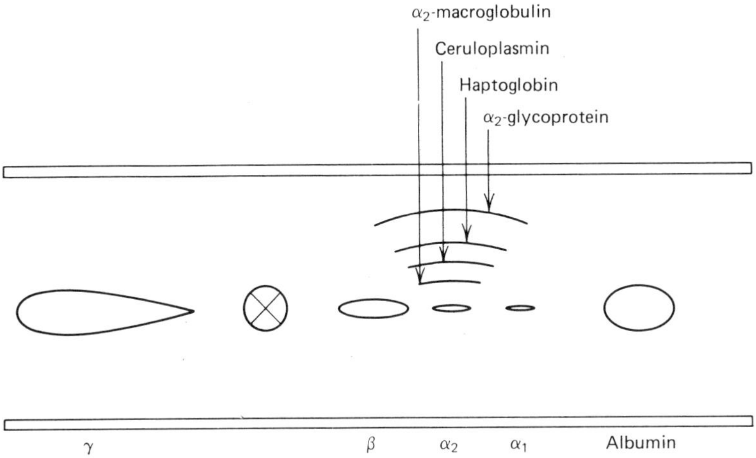

(b)

FIGURE 7-9. Immunoelectrophoresis experiments. Following normal electrophoresis, specific antibodies are placed in small trenches shown at the edges of the gel strip. As the antibody diffuses to the center of the strip, a preciptin ring is formed where the antibody reacts with its specific antigen. (*a*) Unknown antibody (in trenches) is seen to be specific toward γ-globulin, α_1-globulin, and albumin. (*b*) Using polyvalent antiserum (containing many antibodies), the α_2-globulin group is resolved into several component proteins; because these component proteins are present at different concentration, they diffuse radially outward at different rates and show distinct precipitin bands. (From L. P. Cawley, *Electrophoresis and Immuno-Electrophoresis,* Little, Brown & Co., Boston, 1969.)

on two factors: the unavoidable width of the initial band or layer of sample solution in the gel, and the "fuzzing-out" of band boundaries due to diffusion. Disc-electrophoresis has been developed theoretically by L. Ornstein and experimentally by B. J. Davis to reduce the thickness of the starting sample zone to a few microns, from an initial thickness of the order of centimeters—this leads to very sharp bands and results in great improvement in the resolving power (and thus the value) of electrophoresis. The schematic apparatus is shown in Fig. 7-10.

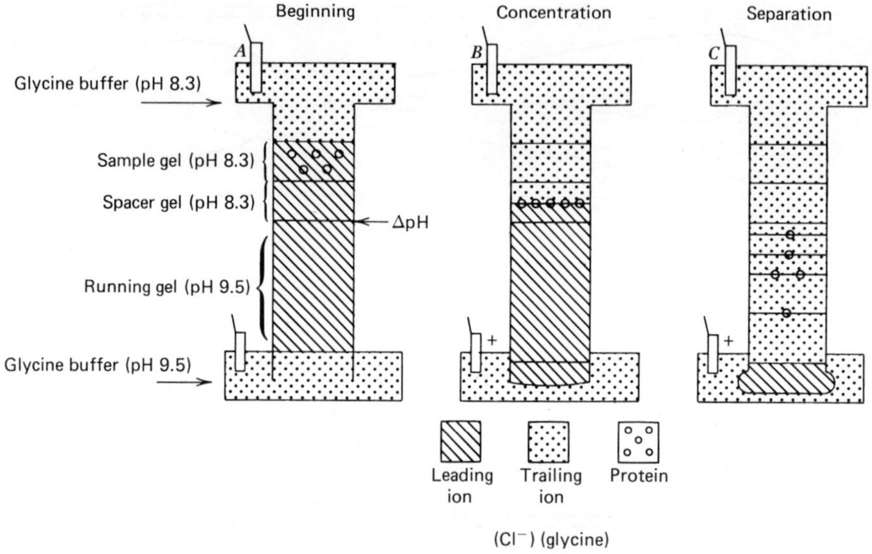

FIGURE 7-10. Schematic diagram of disc(ontinuous)-electrophoresis. Purpose of the separate gels and differences in buffer pH is indicated in the text. The result is that the various protein components in the sample are pulled along slightly behind the Cl^- boundary as Cl^- moves down the gels, so that all the proteins are stacked into a very thin layer just before they enter the "running" gel, where conventional electrophoretic separation takes place. The separate protein bands after separation are thus very thin, and it is possible to distinguish between proteins whose mobility and molecular weight differ only very slightly. [From B. J. Davis and L. Ornstein, *Ann. N.Y. Acad. Sci. 121,* 321 and 404 (1964).]

The essence of the method is the use of buffers of different pH at the two ends of the gel system. From top to bottom, the *sample* gel contains the macromolecule-containing sample mixture; the *spacer* gel separates the sample gel from the *running* gel, where the separation will ultimately take place. The sample and spacer gels have relatively large pores, so that diffusion is still minimized, but the macro-ions can move rapidly through the gels when the electric field is applied; the running gel has smaller pores, and slows down macromolecules according to their molecular weight to enhance the usual separation due to differences in electrophoretic mobility. The purpose of the different gels with different pH is to concentrate the macro-ions so that they are stacked into a narrow band just before they reach the running gel. The sample and spacer gels contain Cl^- ion; the upper reservoir and the running gels contain glycine. Glycine has two dissociable protons with $pK_1 = 2.7$ and $pK_2 = 9.8$. Thus at pH 9.5, the glycine in the running gel is composed of nearly equal amounts of $^+H_3N—CH_2—CO_2^-$ and $H_2N—CH_2—CO_2^-$ and is thus on the average negatively charged, with a correspondingly large ion mobility in an electric field (see Chapter 3.C.):

$$\mu_{Cl^-} \sim \mu_{glycine\ (pH\ 9.5)} \gg \mu_{protein} \qquad (7\text{-}20)$$

On the other hand, at pH 8.3 in the sample and spacer gels, the glycine is nearly neutral, and has an ionic mobility even slower than the macro-ions:

$$\mu_{Cl^-} > \mu_{protein} > \mu_{glycine\ (pH\ 8.3)} \qquad (7\text{-}21)$$

At the beginning of the experiment, Cl^- at the top of the *sample* gel moves rapidly downward due to the applied electric field. It would seem that Cl^- ions would rapidly leave the other negative ions behind, but a little thought shows that the result would be to create a region behind the Cl^- with a net positive charge—since separation of charge results in a voltage gradient (electric field), the trailing negatively charged protein ions are "pulled along" behind the Cl^- in much the same way that a school of fish is caught by a moving net, in order to minimize the build-up of positive charge behind the Cl^-. The reason for the pH change at the edge of the spacer gel is now clear—as soon as the Cl^- (and shortly afterward the proteins and glycine) reach the *running* gel, the pH rises and the glycine now carries appreciable negative charge and overtakes the protein, usurping its role as the ion that replaces the missing negative charge behind the Cl^-. The protein in the *running* gel thus moves at a rate dictated by its ion mobility and its ability to penetrate the smaller pores of the running gel. The net effect of this process has been to concentrate all the macro-ions in the *sample* gel into a very *thin band* as they enter the *running* gel. (One purpose of the *spacer* gel is to avoid spoiling the sharp pH boundary when the sample gel is formed just before the experiment begins.)

EXAMPLE *Improved Separation of Components of a Mixture*

Figure 7-11 shows (upper right) the result of an ordinary gel electrophoresis experiment on normal human serum. If the gel is cut into sections (shown by the shaded areas), and the protein from each of those regions is subjected to disc-electrophoresis as described in Fig. 7-10, then the resolution of each protein group into components is shown in the respective diagrams at the left of the figure. Although the gamma-globulin group has not been well-resolved, immuno-electrophoresis methods could now be applied to identify its components.

EXAMPLE *Determination of Number of Subunits of an Enzyme*

The role of enzyme subunits is described at length in Section 3. Although many enzymes have molecular weights above 100,000, it is becoming increasingly evident that most enzyme basic units have molecular weights between 15,000 and 50,000. Since the interaction between subunits can often be related to mechanisms for control of enzyme efficiency (and consequent regulation of the relative importance of different metabolic pathways), it is of great

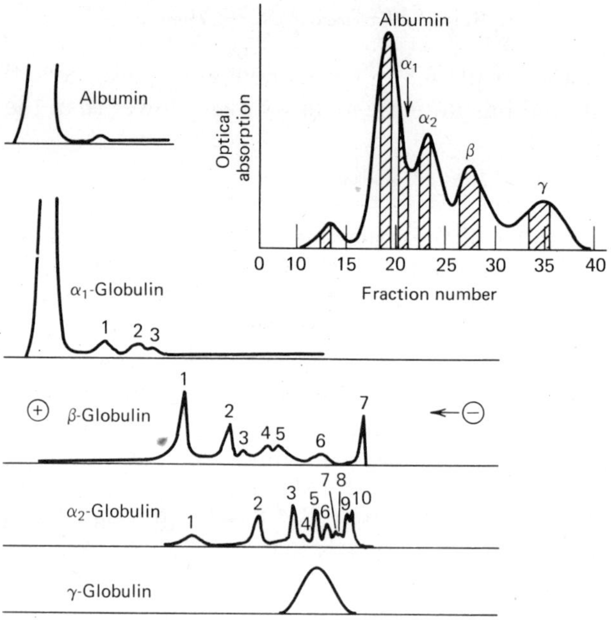

FIGURE 7-11. Disc-electrophoresis separation and identification of serum protein bands obtained from conventional gel electrophoresis of normal human serum. (From H. R. Maurer, *Disc-Electrophoresis,* Walter de Gruyter & Co., Berlin, 1971.)

interest to establish the number of subunits in an enzyme. However, while the subunits may come apart and recombine rapidly, it may be difficult to observe a subunit directly if the association constant is large (i.e., most of the subunits are aggregated). Chemical modification, with electrophoresis as the indicator, offers a clever entry into the problem.

The chemical modification consists of conversion of (positively charged) lysine residues on a protein to (negatively charged) carboxyl groups, as shown below:

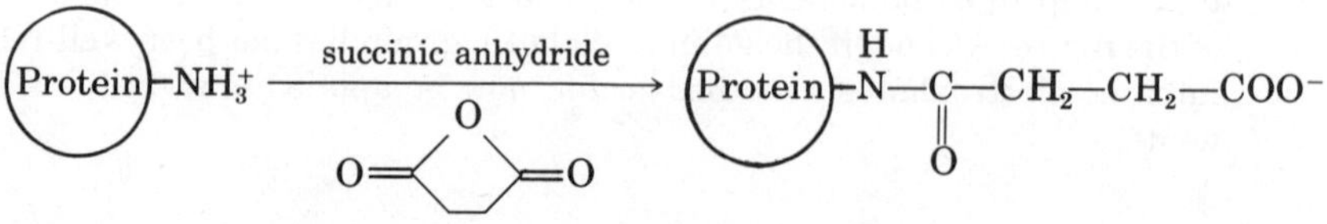

Suppose for simplicity that the enzyme in question is a dimer, and that only one lysine per subunit is affected by the succinylation reaction. An enzyme preparation is then divided into two parts, and one of the fractions is succinylated. The succinylated and native enzyme are then mixed, and if subunits are present and the dimer is dissociable, the equilibrium mixture will contain species with two, one, or zero additional negative charges than the native

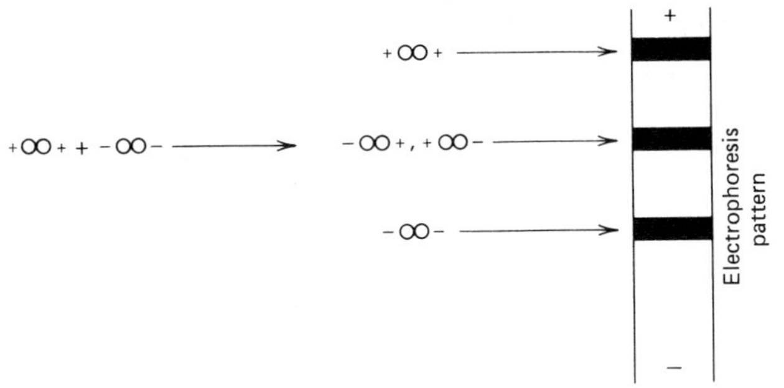

FIGURE 7-12. Schematic diagram for detection of subunits of an enzyme by mixture of succinylated and native enzyme followed by observation of additional bands in the electrophoretic pattern of the mixture. [For an actual example, see E. M. Meighen et al., *Proc. Natl. Acad. Sci. U.S.A. 65,* 234 (1970).]

enzyme. Thus, if the electrophoresis pattern for the mixture of native and succinylated enzyme shows three bands rather than just two, one has direct evidence for the presence of dimeric enzyme. Although this method has some limitations (the chemically modified subunit may not behave the same as the native subunit), the technique has been successfully applied to the determination of the tetrameric nature of aldolase and the trimeric composition of the catalytic subunit of aspartate transcarbamylase. For a schematic picture, see Fig. 7-12.

Isotachophoresis By combining the principles of the moving boundary and the discontinuous electrophoresis experiments, we obtain a new "isotachophoresis" experiment that shows great promise for analytical separation of small cations or anions such as metabolites and body electrolytes rather than the large protein ions we have considered up to now.

In the standard moving boundary electrophoresis experiment (Fig. 7-1), the faster-moving ions move ahead of the slower-moving ions in a solution, leading to a separation of the mixture into zones containing neither, one, or both ions. However, we pointed out in our discussion of disc electrophoresis (Fig. 7-10) that the faster-moving ions can only gain a limited distance ahead of the slower-moving ions, before the voltage gradient behind the fast-moving ions grows large enough to "pull" the slower-moving ions along at the same speed ("isotacho-" means "same speed") as the faster-moving ions. Thus, if we allow a moving-boundary experiment to proceed long enough, the ions in the various zones of Fig. 7-1 will eventually all be moving at the same speed.

The observed "isotachophoresis" pattern will look very similar to the bottom diagram of Fig. 7-1 but with a completely different interpretation. In conventional moving-boundary experiments, the *position* of the boundary (the "step" in Fig. 7-1, bottom diagram) gives the *qualitative* information (i.e., which type of macromolecule is present), while the *ordinate* (total macromolecule concentration in Fig. 7-1) gives the *quantitative* information (i.e., how much of each component is present). In the steady-state moving boundary ("isotachophoresis") experiment, the situation is reversed, and the ordinate gives qualitative while the abscissa gives quantitative information. Briefly, the explanation is that in the new experiment, the ordinate (usually conductance, which is a measure of the total ionic concentration) is different for chemically different ions (Na^- and K^+, for example), while the degree to which the trailing ion lags the leading ion (the distance between "steps" in Fig. 7-1) is related to their concentration difference. The presently limited number of biological examples of isotachophoresis does

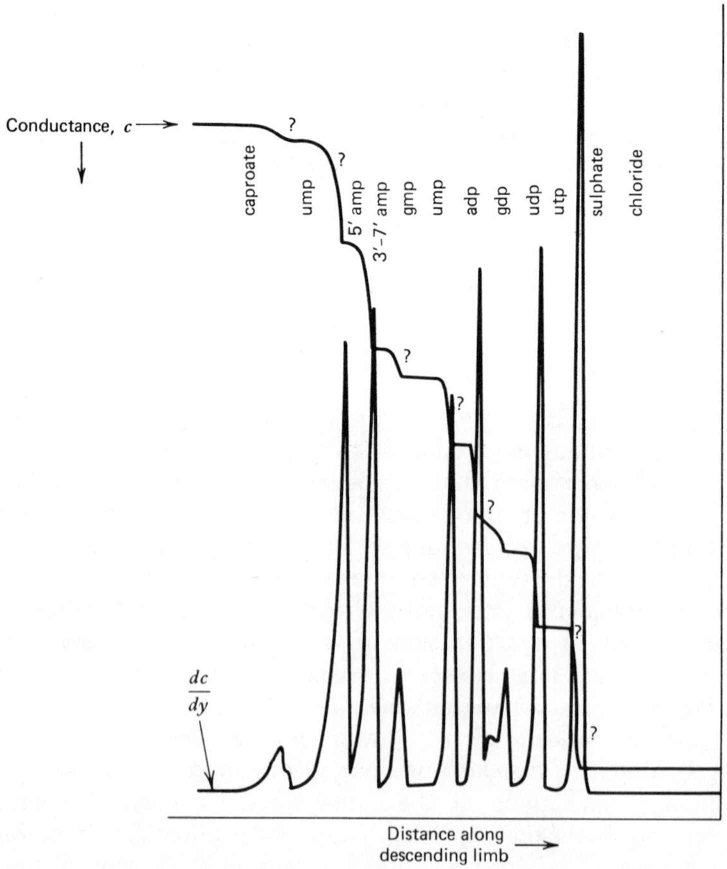

FIGURE 7-13. Isotachopherogram of some nucleotides. Conductance and its derivative are plotted as a function of distance along the descending limb of the apparatus shown in Figure 7-1. As noted in the text, this diagram is unusual in that the ordinate tells which nucleotide is which, while the position on the abscissa (distance) tells how much of each anion is present. [From F. M. Everaerts, A. J. Mulder, and Th. P. E. M. Verheggen, *American Laboratory* (Dec., 1973), p. 44.]

not justify more detailed treatment here, but the principles are treated at length in any discussion of "transference numbers" in conventional physical chemistry texts. Figure 7-13 shows a typical experimental isotachopherogram for a mixture of several biologically interesting anions. Similar procedures have been used to analyze quantitatively for potassium, sodium, calcium, magnesium, and lithium in serum, based on differences in the ionic mobility of the component ions.

7.B. SEDIMENTATION

Suspensions of very large particles in water will eventually settle out under the force of gravity, if the density of the particle is greater than the density of the solution. For biological macromolecules, however, the suspension will never settle out of its own accord, because the (randomly oriented) thermal velocity of the macromolecules is much larger than the (uniformly directed) velocity due to the force of gravity. For example, for a macromolecule of molecular weight, 100,000, the gravitational (potential) energy between two points which are 1.0 cm apart amounts to

$$\begin{aligned} E_{\text{gravity}} &= \text{mgh} \cong (100{,}000/6 \times 10^{26})\text{kg} \cdot (10 \text{ m sec}^{-2}) \cdot (10^{-2} \text{ m}) \\ &= 1.7 \times 10^{-23} \text{ joule} \end{aligned} \tag{7-23}$$

where the acceleration of gravity has been rounded off to 10 m sec^{-2}. For the same molecule, the thermal energy (see Section 5) is of the order of

$$kT = (1.38 \times 10^{-23} \text{ joule } °\text{K}^{-1}) \cdot (2.93 \times 10^{2} \text{ }°\text{K}) = 4 \times 10^{-21} \text{ joule} \tag{7-24}$$

or about 200 times bigger. The ultracentrifuge is simply a device for "amplifying" gravity to the point where the "gravitational" energy exceeds the thermal energy so that the macromolecules will settle to the bottom of the container.

Sedimentation (settling-out) *rates* allow for separation of mixtures of macromolecules according to their *size* and *density*. The sedimentation pattern at *equilibrium* furnishes an absolute determination of *molecular weight* and an indication of the heterogeneity of the mixture. Use of both methods in concert provides a definite value for macromolecular weight, and mixed information about the *shape* and *extent of hydration* (water-binding) for the macromolecule. Although newer chromatographic and electrophoretic techniques have largely supplanted the use of the ultracentrifuge as a means for preparative *separation* of proteins in solution, the great majority of protein molecular weights in the literature are determined from sedimentation methods, and sedimentation techniques are still the method of choice for characterization of nucleic acid size and shape; for study of subunit:multimer equilibria for proteins and nucleic acids; and other polymers; and for separation of biological membrane components.

7.B.1 Sedimentation Rate

Analysis of electrophoresis experiments is relatively simple because the driving force (the electric field) is *constant* in time throughout the process. In contrast, the driving force (centrifugal force) in sedimentation studies *varies* with the distance away from the center of rotation, so that the force on a macromolecule increases as the molecule sediments toward the bottom of its solution—this complication in fact renders impossible any analytical solution (i.e., an exact solution expressible in a finite number of terms) of the macromolecular motion. There are, however, two simplifying possibilities for which the mathematical solution is compact: either when there is *zero net force* on the molecule, or when there is zero *net flow* of molecules

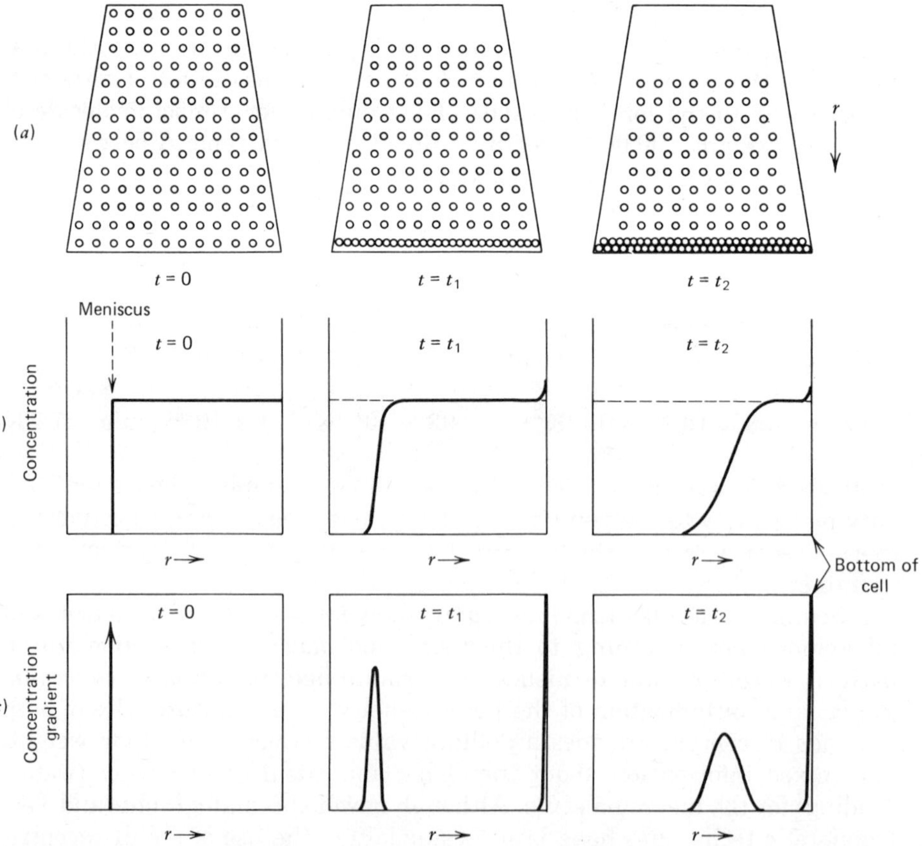

FIGURE 7-14. Progress of a sedimentation experiment at very high centrifugal force. (*a*) Distribution of macromolecules at three separate stages of the experiment. (*b*) and (*c*) show the concentration and the concentration gradient as a function of distance from the center of rotation, *r*, including the effect of diffusion at the boundary, for three separate times during the experiment. (After C. Tanford, *Physical Chemistry of Macromolecules,* Wiley, 1961, p. 366.)

across a particular region of the solution. All sedimentation techniques are designed to take advantage of one or the other of these two situations; we will consider first the case of zero net force on the macromolecule.

The progress of a sedimentation *rate* experiment is shown schematically in Fig. 7-14 for (the usual case of) a macromolecule more dense that the solution around it.

The net centrifugal force acting on a mole of macromolecules of molecular weight, M, is given by the product of the effective mass of the macromolecules and the acceleration, $\omega^2 r$, where ω is the angular velocity of the rotor:

$$F_{\text{centrifugal}} = \left[\begin{matrix}\text{mass of one mole} \\ \text{of macromolecules}\end{matrix} - \begin{matrix}\text{mass of solution displaced} \\ \text{by one mole of macromolecules}\end{matrix}\right]\omega^2 r \tag{7-25}$$

In other words, the macromolecule only tends to settle out if its density is greater than that of the solution around it. Now the mass of one mole of macromolecules is just the molecular weight, M, and

$$\begin{matrix}\text{Mass of solution displaced} \\ \text{by a mole of macromolecules}\end{matrix} = \frac{\text{g solution}}{\text{cc sol'n}} \cdot \frac{\text{g solute}}{\text{mole solute}} \cdot \frac{\text{cc solution displaced}}{\text{by 1 g solute}}$$

$$= \rho_{\text{solution}}\, M\, \bar{v}_{\text{solute}} \tag{7-26}$$

where ρ_{solution} is just the solution density, and $\bar{v}_{\text{solute}}$ (partial specific volume of solute) is the volume increase of a very large volume of solution resulting from addition of one gram of solute (the large volume of solution is just to ensure that the concentration of macromolecule will be the same before and after addition of a little more macromolecule—see Chapter 2).* The net centrifugal force may thus be written more compactly

$$F_{\text{centrifugal}} = \omega^2 r[M - M\bar{v}_{\text{solute}}\rho_{\text{solution}}] \tag{7-27}$$

As for electrophoresis, the forced motion will be opposed by a frictional force (in the opposite direction) proportional to the macromolecular velocity:

$$F_{\text{friction}} = -Nf\frac{dr}{dt} \tag{7-28}$$

where Avogadro's number, N, has been inserted into Eq. 7-28 because we are now dealing with a mole of particles rather than just one particle. A macromolecule in the ultracentrifuge will thus accelerate under the driving

* *In Chapter 2, we encountered the related quantity, partial* molal *volume,* $\bar{V}$, *where* $\bar{V} = M\bar{v}$ = *volume of solution displaced by one* mole *of solute.*

centrifugal force, until that force is exactly balanced by friction; thereafter, sedimentation will proceed at a *constant* rate, obtained by setting net force $= F_{\text{friction}} + F_{\text{centrifugal}} = 0$:

$$Nf\frac{dr}{dt} = M\omega^2 r[1 - \bar{v}_{\text{solute}}\rho_{\text{solution}}] \tag{7-29}$$

Defining a sedimentation coefficient, s, as the sedimentation rate per unit centrifugal force, and rearranging, we obtain

$$M = \frac{Nfs}{(1 - \bar{v}_{\text{solute}}\rho_{\text{solution}})}$$

where

$$\boxed{s \equiv \frac{dr/dt}{\omega^2 r}} \tag{7-30}$$

Finally, recalling that $f = (kT/D)$ (Eq. 7-13) and $(Nk) = R$, the desired result appears:

$$\boxed{M = \frac{RTs}{D(1 - \bar{v}_{\text{solute}}\rho_{\text{solution}})}} \tag{7-31}$$

Equation 7-31 represents one of the most direct determinations of molecular weight, and thus deserves some scrutiny. Experimental determination of D has been discussed in Chapter 6.B (see also section 6) and measurement of density and temperature poses no special problems. Sedimentation coefficient is obtained by first rearranging Eq. 7-30

$$\frac{dr/r}{dt} = \frac{d \log_e r}{dt} = \omega^2 s \tag{7-32}$$

Thus, a plot of log (distance from center of rotation to center of boundary) versus time gives a straight line whose slope is proportional to s, as shown in Fig. 7-15 for a glycoprotein; since s values are typically of the order of 10^{-13} sec, they are typically reported in units of Svedberg, or S, where

$$1 \text{ Svedberg} = \text{S} = 10^{-13} \text{ sec} \tag{7-33}$$

A final convention is to report all values for s and D as if they were determined at 20°C using distilled water solvent. The appropriate compensations for changes in solution density, partial specific volume, and viscosity, η, (Chapter 7.C) are given in Equations 7-34 and 7-35, assuming that f is proportional to η (Chapter 7.C). [In using these equations, it is implicitly assumed that the molecular properties of the *solute* (such as con-

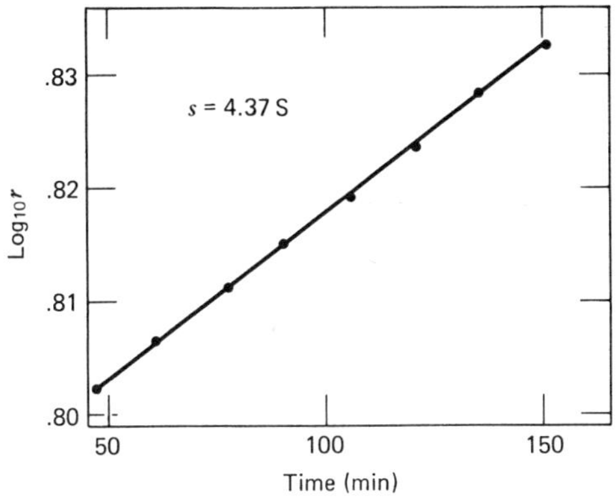

FIGURE 7-15. Determination of sedimentation coefficient from a plot of log (distance from center of rotation to middle of boundary) versus time in a sedimentation rate experiment, for a cellobiosylhydrolase enzyme from the wood-degrading fungus, *Trichoderma viride*. Rotor speed was 48,400 revolutions/min. (P. R. Griffith and A. G. Marshall.)

formation) do *not* change with temperature or concentration, often unwarranted assumptions.]

$$s_{20°\mathrm{C,water}} = s_{T°\mathrm{C,solution}} \left(\frac{\eta_{T°\mathrm{C,solution}}}{\eta_{20°\mathrm{C,water}}}\right) \frac{(1-\bar{v}\rho)_{20°\mathrm{C,water}}}{(1-\bar{v}\rho)_{T°\mathrm{C,solution}}} \tag{7-34}$$

$$D_{20°\mathrm{C,water}} = D_{T°\mathrm{C,solution}} \left(\frac{293°\mathrm{K}}{T°\mathrm{K}}\right)\left(\frac{\eta_{T°\mathrm{C,solution}}}{\eta_{20°\mathrm{C,water}}}\right) \tag{7-35}$$

As seen in the next section, viscosity is simply a measure of the frictional force that slows the sedimenting macromolecule, so the viscosity corrections to s and D are intuitively reasonable. The remaining correction to s accounts for the difference in buoyancy between a measurement at 20°C in water and a measurement at $T°$ in the macromolecule-containing solution. Finally, the temperature correction for D accounts for the fact that D is proportional to absolute temperature (see Eq. 7-13).

The *shape, size,* and *extent of hydration* of a macromolecule can be derived from s values; the details require a brief discussion of viscosity of nonspherical molecules and are left until the next section.

7.B.2. Sedimentation Equilibrium

The other sedimentation situation that is simple enough to analyze results when there is zero *net flow* of macromolecules across a given section of the spinning solution. Flow of macromolecules across a region of cross-sectional area, A, consists of a flow outward due to sedimentation and a flow inward due to diffusion. At equilibrium, concentration at any one level of the solution is constant, and there is no net flow across that region. The algebraic form of these statements may now be elaborated (from Eq. 6-51):

$$\text{Flow inward} = -DA\,\frac{\partial[C]}{\partial r} \qquad (6\text{-}51)$$

$$\text{Flow outward} = \frac{(1 - \bar{v}_{\text{solute}}\rho_{\text{solution}})M\omega^2 rA[C]}{Nf} \qquad (7\text{-}36)$$

The form of Eq. 7-36 becomes apparent from recognition that

$$[(1 - \bar{v}_{\text{solute}}\rho_{\text{solution}})M][\omega^2 r] = \begin{pmatrix}\text{net}\\ \text{mass}\end{pmatrix}(\text{acceleration})$$
$$= \text{centrifugal force (outward) on one mole of solute particles}$$

while $(1/f)$ is just the (outward) velocity per unit (outward) force per molecule: the reader can quickly be convinced that Eq. 7-36 has the correct units of flow in mass per unit time across the stated area. Because at equilibrium there can be no net accumulation or depletion of solute with time in a given region of the solution

$$\text{Flow inward} + \text{flow outward} = 0 \qquad (7\text{-}37)$$

or

$$D\,\frac{\partial[C]}{\partial r} = (1 - \bar{v}_{\text{solute}}\rho_{\text{solution}})M\omega^2 r[C]/Nf \qquad (7\text{-}38)$$

Making (again) the substitutions, $f = (kT/D)$ and $R = Nk$, and rearranging

$$\frac{d[C]}{[C]} = d\,\log_e[C] = \frac{M\omega^2(1 - \bar{v}_{\text{solute}}\rho_{\text{solution}})}{RT}\,r\,dr \qquad (7\text{-}39)$$

To cast the result in an experimentally useful form, it is necessary to integrate Eq. 7-39 between the positions r_1 and r_2, where the concentration of macromolecule is $[C]_1$ and $[C]_2$:

$$\boxed{\log_e[C]_2 - \log_e[C]_1 = \frac{M\omega^2(1 - \bar{v}_{\text{solute}}\rho_{\text{solution}})}{2RT}(r_2^2 - r_1^2)} \qquad (7\text{-}40)$$

Experimentally, one need only measure the macromolecular concentration as a function of distance away from the center of rotation, using a transparent ultracentrifuge cell and a suitable optical means for detection of concentration (such as light absorption or refraction), and construct a plot of $\log_e[C]$ versus r^2: the slope of the line will be proportional to the molecular weight of the macromolecule, where the constants of proportionality are readily measured in separate experiments. The important feature of this determination of molecular weight is that *it is not necessary to know the diffusion constant, D.*

The approach to (and the attainment of) sedimentation equilibrium are readily visualized graphically, as shown in Fig. 7-16.

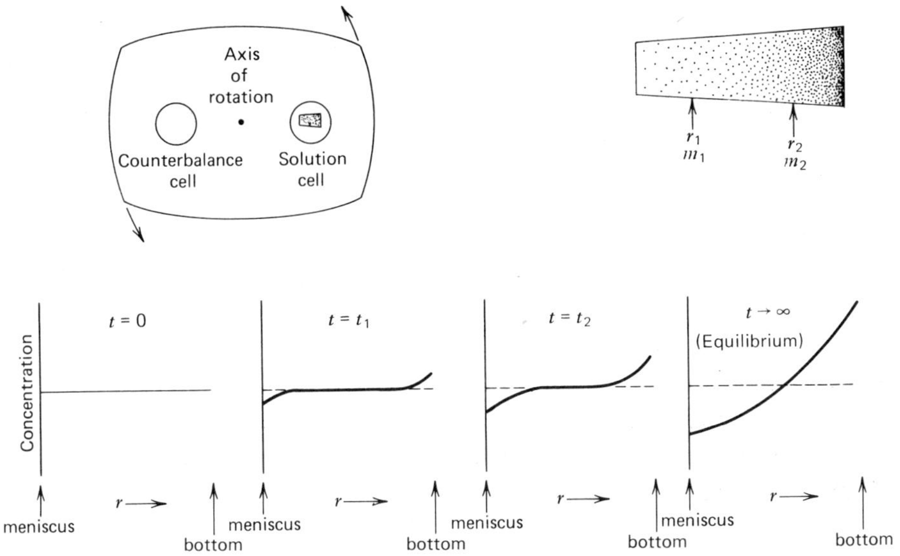

FIGURE 7-16. Schematic diagrams of the progress of a sedimentation equilibrium experiment. *Upper left:* the experimental arrangement. *Upper right:* the solution after equilibrium has been reached. *Bottom diagrams:* stages in the approach to equilibrium. r is the distance from the *center* of rotation, and m is molar concentration at r.

The remarkable accuracy of the sedimentation equilibrium technique in determination of molecular weights is evidenced by the data in Table 7-2, for molecules whose molecular weight is precisely known from the chemical formula for the molecule. The principal disadvantage of the method is that attainment of equilibrium generally requires at least 24 hours, so that measurements are time-consuming; on the other hand, sedimentation equilibrium probably provides the most accurate determination of molecular weight for large molecules in solution.

Table 7-2 Comparison of molecular weight, derived from application of Eq. 7-40 to sedimentation equilibrium data for concentration as a function of distance from the axis of rotation, with molecular weight calculated from the chemical formula for the same compound.*

Substance	Molecular Weight from Chemical Formula	Molecular Weight from Sedimentation Equilibrium
Sucrose	342.3	341.5
Ribonuclease	13,683	13,740
Lysozyme	14,305	14,500
Chymotrypsinogen *A*	25,767	25,670

* *After K. E. van Holde,* Physical Biochemistry, *Prentice-Hall, Englewood Cliffs, N. J.: 1971.*

Thermodynamic Description of Sedimentation Equilibrium In the preceding treatment of equilibrium sedimentation, equilibrium was defined as the condition of *no net flow* of solute. However, we have already noted (Table 2-1) that this is exactly the equilibrium situation for which $dG = 0$ is the appropriate criterion. Alternatively, we can use Eq. 3-3b to calculate the difference in chemical potential of macromolecular solute between the two positions, r_1 and r_2 of Fig. 7-16, at which the macromolecular molar concentration is m_1 or m_2

$$\bar{G}_{\text{solute}}\ (\text{at } r_2) - \bar{G}_{\text{solute}}\ (\text{at } r_1) = RT \ln m_2 - RT \ln m_1 = RT \ln(m_2/m_1) \tag{7-41}$$

The reader should be able to show that the work done in moving a mole of molecules of molecular weight, M, against a centrifugal force (whose acceleration is $\omega^2 r$, where ω is the angular velocity of the centrifuge) from point r_1 to point r_2 is given by

$$\text{Force} = m\omega^2 r$$

$$\text{Work} = \int_{r_1}^{r_2} F \cdot dr$$

$$\text{Work} = (1/2)M(1 - \bar{v}_{\text{solute}}\rho_{\text{solution}})\omega^2(r_2^2 - r_1^2) \tag{7-42}$$

in which we recognize $(1 - \bar{v}_{\text{solute}}\rho_{\text{solution}})$ as the buoyancy correction factor that accounts for the fact that solvent molecules must be displaced so that solute molecules can be transported through the solution. At equilibrium, the free energy difference due to a concentration gradient in the centrifuge cell (Eq. 7-41) must be equal to the free energy difference due to a difference in centrifugal energy between the same two points (Eq. 7-42). Equating Equations 7-41 and 7-42 thus leads at once to the desired expression (compare Eq. 7-40) for molecular weight, M, based on the sedimentation equilibrium experiment:

$$\boxed{M = \frac{2RT \ln(m_2/m_1)}{(1 - \bar{v}_{\text{solute}}\rho_{\text{solution}})\omega^2(r_2^2 - r_1^2)}} \tag{7-43}$$

The significance of Eq. 7-43 (which is identical to Eq. 7-40) is that it was obtained solely from thermodynamic arguments, without resorting to any particular *model* for the motion of the macromolecules in the centrifuge. Thus, it is now not so surprising that the experimental results of Table 7-2 give such good agreement with Eq. 7-40 from the simple diffusion model. Similarly, it can be shown that our one-term expression for osmotic pressure (Eq. 2-40) and the results we will later obtain for light-scattering (Section 4) represent only first-term approximations of the more general (thermodynamic) result. Although the model (diffusion, random walk, weight-on-a-spring) is intuitively satisfying and informative, the thermodynamic result is more general and precise, since it does not depend on any particular *mechanism* for the process of interest.

Archibald Method Another common sedimentation analysis is based on the fact that at *any* time during a sedimentation experiment, there can be no net flow of substance across either the *meniscus* or the *bottom* of the centrifuge cell. Thus, Eq. 7-40 will always be valid at those positions in the cell, and may be used to obtain the molecular weight from Eq. 7-43, using the concentrations at just those two points. Practically, however, the concentrations at these points are not measured directly, but are obtained by extrapolation of the concentration-distance profile (see Fig. 7-16), either to the meniscus or to the bottom of the cell. The resultant molecular weight obtained is not nearly as accurate as the measurements based on true equilibrium throughout the cell (since the latter are based on a much *larger* number of *directly* determined data points), but may be obtained *rapidly* since complete equilibrium is not required.

7.B.3. Density Gradient Sedimentation

Although the previous sedimentation techniques can give the molecular weight for a *purified* macromolecule, they are not especially suitable for measuring the molecular weights of each of several different types of macromolecules in a *mixture*. For mixtures, it is first necessary to *separate* macromolecules of different molecular weight, which may be achieved using "density gradient" techniques based on construction of a solution in which the solution density varies smoothly from a high density at the bottom of the sedimentation cell to a low density at the top.

When sedimentation *rate* experiments are conducted in a sucrose density gradient (Fig. 7-17), macromolecules sediment at a rate (approximately) proportional to their sedimentation coefficients. Thus, by introducing macromolecules of known sedimentation coefficient into an unknown sample, allowing sedimentation to proceed, and then withdrawing aliquots of solution and assaying for the presence of each macromolecular species, it is possible to determine the sedimentation coefficients of each of several unknown macromolecules in a solution (Fig. 7-17). When the sucrose gradient is arranged so that macromolecules of different molecular weight sediment at (different) *constant* rates, the method is called "isokinetic" density gradient sedimentation.

For sedimentation *equilibrium,* the density gradient may be generated during the experiment, using a solution of, for example, CsCl salt (see Fig. 7-19 for an example). A given macromolecule in such a solution will either sink or rise until it reaches the ("isopycnic") point in the cell where the density of the solution is the same as that of the macromolecule [i.e., $\rho_{\text{solution}} = (1/\bar{v}_{\text{solute}})$]. A short calculation (see Problems) then shows that if the solution density varies linearly over a short distance, the macromolecules of a given molecular weight will distribute themselves according to the now-familiar Gaussian curve, centered at the isopycnic point. The great advantage of both density gradient sedimentation techniques is that once the components of the mixture have been separated by sedimentation, aliquots of solution may be withdrawn from different levels of the centrifuge tube and assayed *chemically* (and thus very *specifically*) for their constituents.

EXAMPLE *Sedimentation of an Enzyme Mixture in an Isokinetic Sucrose Gradient*

In this example, the density gradient is prepared *before* the sedimentation experiment, by, for example, carefully layering a sucrose solution of decreasing density into the sedimentation tube. A thin layer of (less dense) macromolecule-containing solution is then layered on top of the sucrose gradient

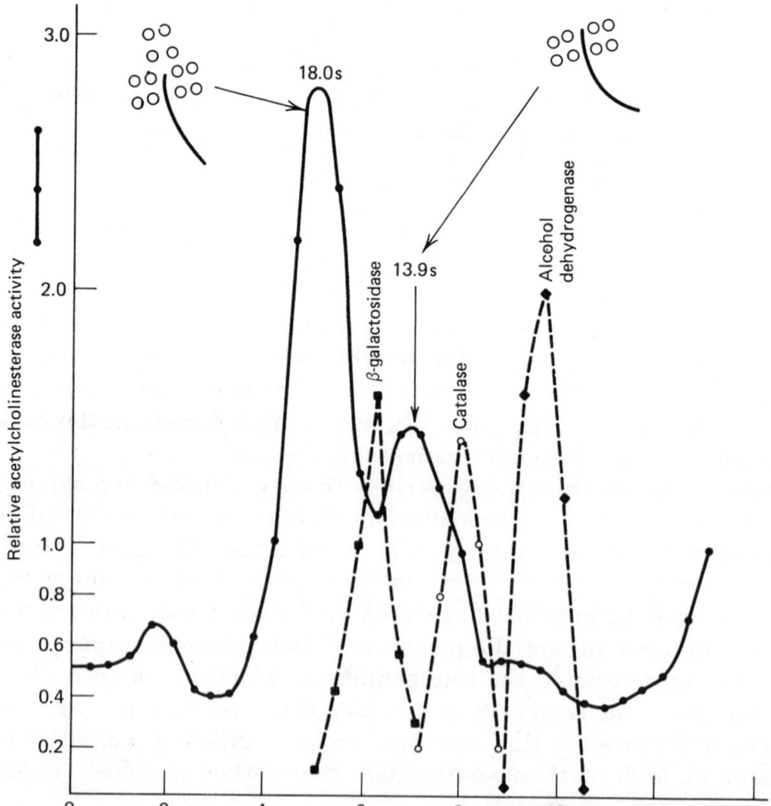

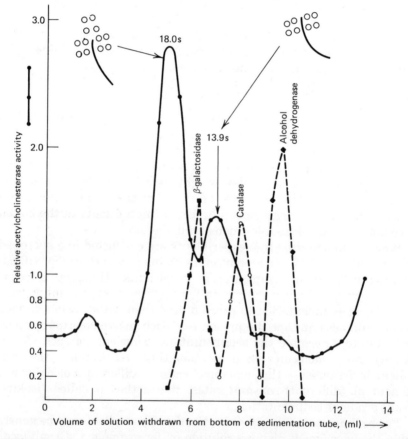

FIGURE 7-17. Sucrose-gradient sedimentation pattern for a mixture of three known enzymes (β-galactosidase, catalase, and alcohol dehydrogenase) and a sample of purified native acetylcholinesterase. Dotted lines show the activities of various aliquots toward substrates of the three known enzymes. Acetylcholinesterase activity provides an indicator of the relative amounts of the two major forms of the enzyme, whose proposed structures (based on electron micrograph evidence) are shown in the figure. [From P. J. Morrod, A. G. Marshall, and D. G. Clark, *Biochem. Biophys. Res. Commun. 63*, 335 (1975).]

solution, and sedimentation is allowed to proceed, with the various enzymes in the mixture sedimenting at different (constant) rates. Assays specific for the various enzymes are then performed on aliquots withdrawn from various depths in the centrifuge tube, with the results shown in Fig. 7-17. In this study, it was possible to determine the relative amounts of two forms of acetylcholinesterase enzyme, using three enzymes of known sedimentation coefficient.

EXAMPLE *Replication of DNA*

The most famous example of density gradient sedimentation is in study of the pairing of DNA molecules, providing direct evidence for the two-stranded

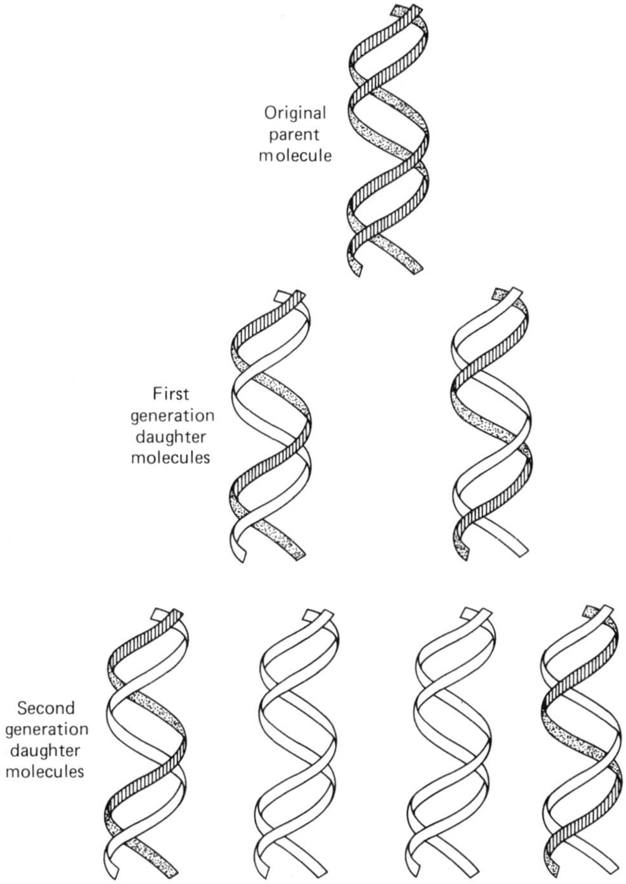

FIGURE 7-18. Mechanism of DNA replication proposed by Crick and Watson. After one replication, each daughter molecule consists of one parental chain (black) paired with one newly synthesized chain (white). Since the parental (single) chain remains intact throughout, the second and subsequent generations will each exhibit two molecules containing one parental chain. [See F. H. C. Crick and J. D. Watson, *Proc. Roy. Soc. London,* Ser. A *223,* 80 (1954).]

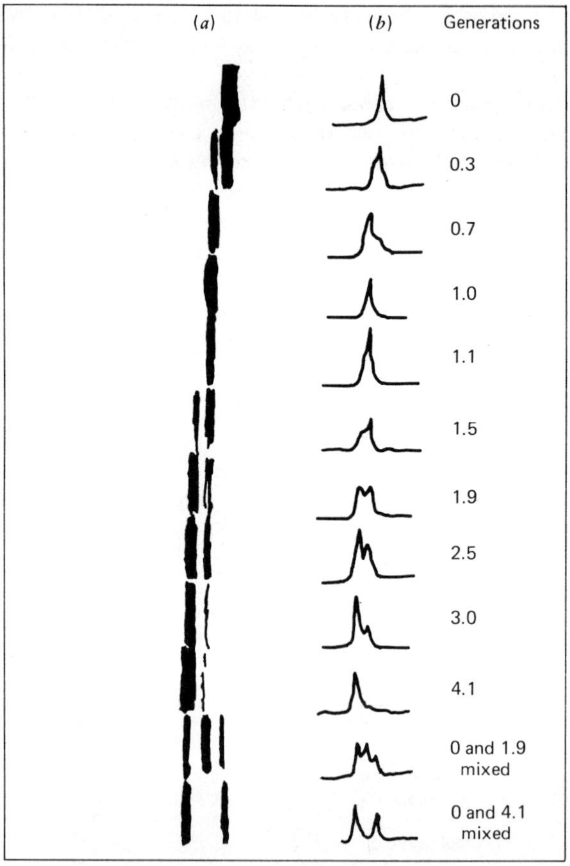

FIGURE 7-19. An application of density-gradient centrifugation. (*a*) Ultraviolet absorption photographs showing DNA bands resulting from density-gradient centrifugation of lysates of bacteria sampled at various times after the addition of an excess of ^{14}N substrates to a growing ^{15}N-labeled culture. Each photograph was taken after 20 hours of centrifugation at 44770 rpm under the conditions described in the text. The density of the CsCl solution increases to the right. Regions of equal density occupy the same horizontal position on each photograph. The time of sampling is measured from the time of the addition of ^{14}N in units of the generation time. The generation times were estimated from measurements of bacterial growth. (*b*) Microdensitometer tracings of the DNA bands shown in the adjacent photographs. The microdensitometer pen displacement above the base line is directly proportional to the concentration of DNA. The degree of labeling of a species of DNA corresponds to the relative position of its band between the bands of fully labeled and unlabeled DNA shown in the lowermost frame, which serves as a density reference. A test of the conclusion that the DNA in the band of intermediate density is just half-labeled is provided by the frame showing the mixture of generations 0 and 1.9. When allowance is made for the relative amounts of DNA in the three peaks, the peak of intermediate density is found to be centered at 50 ± 2 per cent of the distance between the N^{14} and N^{15} peaks. [From M. S. Meselson, F. W. Stahl, and J. Vinograd, *Proc. Natl. Acad. Sci. U.S.A. 44,* 671 (1958).]

nature of DNA in solution. Shortly after the DNA replication mechanism shown in Fig. 7-18 was proposed by Watson and Crick, the following experiment was carried out (see Fig. 7-19). Bacteria that had been grown in a medium containing heavy nitrogen (^{15}N) were suddenly switched to a medium containing only ^{14}N. According to the Crick-Watson replication model, by the end of the first generation, one should observe (double-stranded) DNA molecules with a molecular weight exactly half-way between the values for purely ^{15}N and purely ^{14}N DNA. After two generations, there should be an equal mixture of $^{15}N^{14}N$ and $^{14}N^{14}N$ DNA molecules. These predictions are borne out by experiment (Fig. 7-19). The spectacular resolution furnished by the density gradient experiment is apparent when it is realized that ^{15}N and ^{14}N differ by a factor of only (1/15) in mass, and since a nucleotide is only 8% nitrogen, the separation shown in Fig. 7-19 amounts to resolving between two molecules whose molecular weights differ by about 1/2 of 1 percent!

7.C. VISCOSITY

Viscosity is the parameter that characterizes the motion of molecules due to a mechanical "*shear*" force, just as ionic mobility and sedimentation coefficients characterize the response to an *electrical* or *centrifugal* driving force. Definition of viscosity leads directly to a description of the flow of a fluid through a narrow capillary, which affords a simple means for measurement of viscosity. Finally, it is possible to relate viscosity to frictional coefficients for molecules of various shapes, resulting in a number of routes toward determination of molecular shape from sedimentation, diffusion, or viscosity data.

In a description of the attractive forces between 10^{23} molecules by one parameter (viscosity), it is clear that any proposed mechanism must depend much more heavily on plausibility than rigor. The plausibility argument begins from consideration of two parallel plates, each of area A, separated by a distance, d, where the plates themselves are taken to be very thin layers of fluid. The question is, how much force must be supplied to keep the upper plate moving at velocity, v, relative to the lower plate? It is plausible

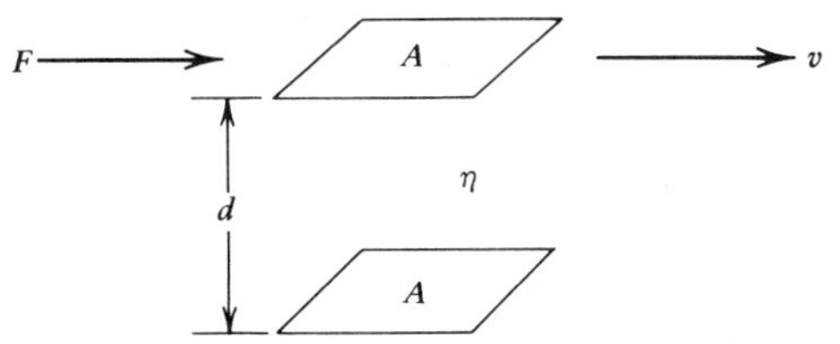

that more force will be required for plates of bigger area, for plates that are closer together, and for plates moving at larger relative velocities. Finally, it is recognized that the force required to move one plate over the other will depend on the particular fluid involved; all these intuitive suppositions may be collected in an equation:

$$\boxed{F = \frac{Av}{d}\eta} \qquad (7\text{-}44)$$

where η is defined as the viscosity of the medium, and may be thought of as the force required to maintain unit velocity difference between two plates of unit area that are one distance unit apart. Equation 7-44 typifies one of the most important aspects in analysis of any experiment, namely the separation of the *geometry* of the experiment from the *intrinsic* ("physical") property under study.

7.C.1. Flow of Fluid in a Capillary

Treatment of flow through a capillary requires no new intuition beyond Eq. 7-44, but does involve a change in the geometry of the experiment. Consider now a cylinder of radius, R_0, and length, L. We are asked to find the force required to push a smaller cylinder of fluid past the outer cylinder at velocity, v. Recognizing that the distance between the cylinders is $(R_0 - R)$ and the area of the moving cylinder is $(2\pi RL)$, Eq. 7-44 gives

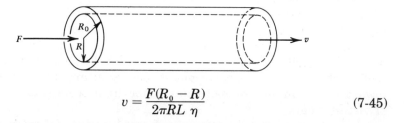

$$v = \frac{F(R_0 - R)}{2\pi RL\,\eta} \qquad (7\text{-}45)$$

An economy in notation is realized by using the relation

$$\text{Pressure} = \text{Force/area}$$

$$P = \frac{F}{\pi R^2} \qquad (7\text{-}46)$$

to substitute for F in Eq. 7-46

$$v = \frac{\pi R^2 P(R_0 - R)}{2\pi RL\eta} = \frac{PR(R_0 - R)}{2L\eta} \qquad (7\text{-}47)$$

Finally, by choosing R as infinitesimally close to R_0, so that $(R_0 - R) \rightarrow -dR$ and $v \rightarrow dv$, and integrating from the surface of the capillary to R_1 (recall that $v \rightarrow 0$ as $d \rightarrow 0$ in Eq. 7-44), the fluid velocity at any point in the interior of the capillary is obtained:

$$\int_{v\text{ at }R_0}^{v\text{ at }R_1} dv = \frac{-P}{2L\eta}\int_{R_0}^{R_1} R\,dR$$

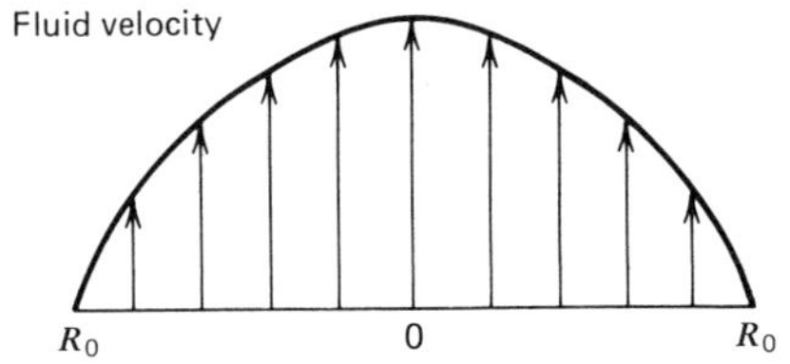

FIGURE 7-20. Fluid velocity in a capillary of radius, R_0, as predicted for laminar flow by Eq. 7-48.

$$\boxed{v_{\text{at } R_1} = \frac{P}{4L\eta}(R_0^2 - R_1^2)} \qquad (7\text{-}48)$$

A plot of fluid velocity as a function of position within the capillary is thus a simple parabola, as shown in Fig. 7-20. Capillary flow is an example of Newtonian or laminar ("layered") flow for which the lines of flow are straight, as opposed to turbulent or nonNewtonian flow in which vortices or whorls show up in the flow lines. Turbulent flow is thought to be instrumental in initiating the blood-clotting process, by introducing strong shear forces near the edges of a (jagged) cut or projection that facilitate the agglutination of platelets to the rough surface. Clotting in the normal vascular system is prevented by the smooth inner surface of blood vessels that prevents adherence of platelets and thus prevents release of pro-coagulants from the platelets. Similarly, blood flow from a sharp cut is more laminar than from a rough cut, which is why razor cuts clot so slowly. Finally, it is difficult for a clot to form in a rapidly flowing vessel (e.g., an artery), because the flowing blood carries away the pro-coagulants so quickly that their concentrations never rise high enough to sustain clot formation.

A simple measure of viscosity is provided by the Ostwald viscosimeter shown in Fig. 7-21. In the Ostwald device, one measures the length of time required for the bulb of liquid shown at upper right in the diagram to discharge between the two "fiducial" marks at the top and bottom of the bulb. In order to accommodate the result for fluid *velocity* (Eq. 7-48) to the experiment of fluid *flow* (volume/time), one need only notice that the volume of fluid passing through the (infinitesimal) cross-section bounded by R and $(R + dR)$ per unit time is given by the velocity of the fluid times the area of the infinitesimally thin ring, $v(2\pi R\, dR)$. The total flow of fluid through the total cross-sectional area of the capillary is thus obtained by integration:

$$\text{Flow (volume/time)} = \frac{\pi P}{2L\eta}\int_{R=0}^{R=R_0} [R_0^2 - R^2]\, R dR$$

$$\frac{v}{t} = \frac{\pi P R_0^4}{8L\eta} \qquad (7\text{-}49)$$

(It is interesting to note that Eq. 7-49 was first derived by Poseuille, an anatomist studying the flow of blood through blood vessels.) Since the pres-

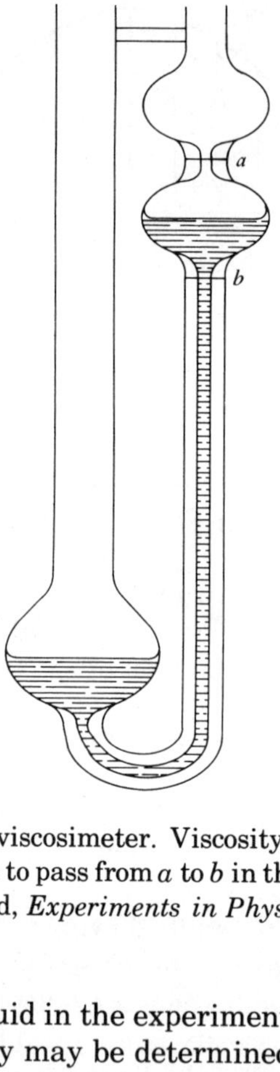

FIGURE 7-21. Ostwald viscosimeter. Viscosity is determined from the time required for the fluid meniscus to pass from *a* to *b* in the figure (see text). (From D. P. Shoemaker and C. W. Garland, *Experiments in Physical Chemistry,* McGraw-Hill, 1967.)

sure from the column of liquid in the experiment is proportional to the solution density, ρ, the viscosity may be determined from:

$$\frac{\eta}{\rho} = (\text{constant})\ (t) \tag{7-50}$$

However, the liquid issuing from the capillary at lower left has higher kinetic energy than the liquid in the bulb at upper right—when this problem is taken into account, there arises an additional term that may be made negligible by choosing the capillary diameter to be sufficiently small that the time (t) required for the experiment is sufficiently long (a few minutes):

$$\frac{\eta}{\rho} = (\text{constant})(t) + (\text{constant}')(1/t) \tag{7-51}$$

In practice, viscosity values are obtained by finding the ratio of the time required for the Ostwald bulb to discharge the unknown fluid to the time for discharge of a fluid of known viscosity:

$$\eta_{\text{unknown}} = \eta_{\text{known}} \frac{\rho_{\text{sample}}\, t_{\text{sample}}}{\rho_{\text{standard}}\, t_{\text{standard}}} \qquad (7\text{-}52)$$

The glass capillary viscosimeter (viscometer) is the simplest to build and easiest to use, but operates at a fixed (large) shearing force that is set by the height of the column of fluid. Elongated molecules, such as DNA, tend to orient with their long axes along the flow lines, when the shear rate is high, and elaborate viscometers based on mechanical or magnetic torque required to maintain a known rotation velocity between two concentric cylinders have been devised for measuring the viscosity of such molecules at the necessarily low shear rate. As seen in Section 4, the orientation of nonspherical molecules due to forced flow provides a measure of the shape of the molecule.

7.C.2. Viscosity, Friction Coefficient, and Macromolecular Shape in Solution

If we express Eq. 7-44 in the form

$$F = fv \qquad (7\text{-}53)$$

then the *frictional coefficient, f,* for the sliding of one *plate* of solution past another is simply

$$f = Fd\eta/A \qquad (7\text{-}54)$$

where η = viscosity of solvent

It is straightforward, but tedious, to further modify the geometry of the shear problem to determine the frictional resistance offered by a *sphere* or macromolecule of more complex shape. Basically, all that is involved is a refinement in the definition of viscosity (Eq. 7-44) to limit the definition to the incremental velocity difference between two plates of infinitesimal area separated by an infinitesimal distance; then with the usual procedures of calculus (complicated only by the three-dimensional nature of the problem), integration of the refined Eq. 7-44 over all the surface elements of the object in question leads to an expression of the form, Force $= f \cdot$ velocity. The form of the frictional coefficient for some simple shapes is given in Table 7-3, where the nature of a prolate (cigar-shaped) and oblate (disk-shaped) ellipsoid is shown in Fig. 7-22.

The effect of molecular shape on frictional coefficient (Table 7-3) is most directly evident from comparison to the friction coefficient that would be expected for a sphere of the same volume, as shown in Fig. 7-23. Since the force of frictional drag is directly related to the *surface area* of the ob-

Table 7-3 Friction Coefficients Corresponding to Some Simple Shapes

Shape	Frictional Coefficient	Explanation
Sphere	$f = 6\pi\eta R$	R = sphere radius
Prolate ellipsoid	$f = 6\pi\eta R_0 \dfrac{(1 - b^2/a^2)^{1/2}}{(b/a)^{2/3} \ln\{[1 + (1 - b^2/a^2)^{1/2}]/b/a\}}$	a = major axis, b = minor axis, R_0 = radius of sphere of equal volume = $(ab^2)^{1/3}$
Oblate ellipsoid	$f = 6\pi\eta R_0 \dfrac{(a^2/b^2 - 1)^{1/2}}{(a/b)^{2/3} \tan^{-1}(a^2/b^2 - 1)^{1/2}}$	a = major axis, b = minor axis, R_0 = radius of sphere of equal volume = $(a^2b)^{1/3}$
Log rod	$f = 6\pi\eta R_0 \dfrac{(a/b)^{2/3}}{(3/2)^{1/3}\{2\ln[2(a/b)] - 0.11\}}$	a = half-length, b = radius, R_0 = radius of sphere of equal volume = $(3b^2a/2)^{1/3}$

ject in question, and since a spherical object has the *least* surface area of any object having the same volume (see Problems), it is expected that frictional coefficients for nonspherical objects will in general be *larger* than for the "equivalent" sphere of the same volume (Fig. 7-23).

Experimentally, the hydrodynamic *shape* of a macromolecule (i.e., its shape in solution) may be determined from any measurement that yields

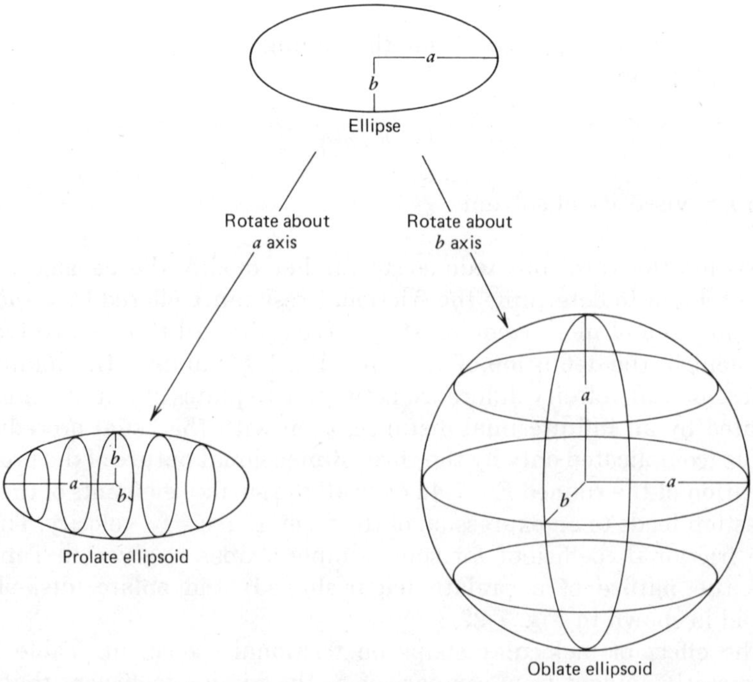

FIGURE 7-22. Prolate and oblate ellipsoids of revolution. (From C. Tanford, *Physical Chemistry of Macromolecules,* Wiley, 1961.)

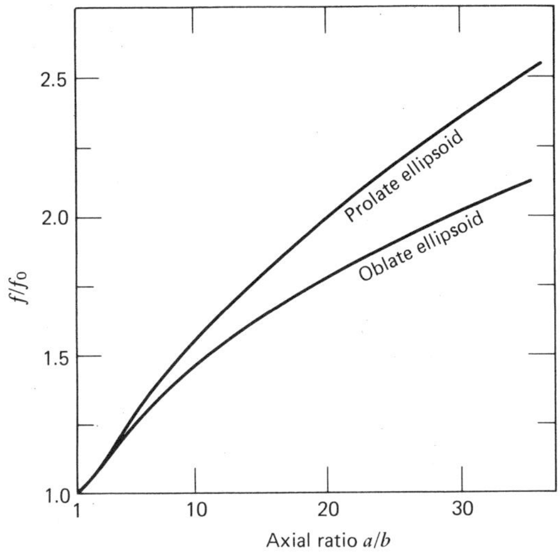

FIGURE 7-23. Frictional coefficient as a function of molecular asymmetry for a range of ellipsoids of revolution. f_0 is the frictional coefficient that would correspond to a sphere of the same volume as the ellipsoid in question. (From C. Tanford, *Physical Chemistry of Macromolecules,* Wiley, 1961.)

the *frictional coefficient,* combined with an independent measurement that yields the effective *radius* of the molecule so that f_0 may be computed from

$$\boxed{f_0 = 6\pi\eta R} \qquad \text{Stokes Law} \tag{7-55}$$

where R is the radius of the macromolecule considered as a sphere.

The molecular radius is generally obtained from (a) light scattering or low-angle X-ray scattering experiments (see Section 4), or (b) the partial specific volume of the molecule and its molecular weight. If the molecule had no water of hydration adhering to its surface, its partial specific volume would be given by

$$\bar{v}_{\text{solute}} = \frac{\text{(volume per molecule)(molecules/mole)}}{\text{grams per mole of molecules}} = \begin{matrix}\text{volume per gram}\\ \text{of solute}\end{matrix}$$

$$\bar{v}_{\text{solute}} = \frac{4\pi R^3 N}{3M};\ R = \left[\frac{3M}{4\pi N}\,\bar{v}_{\text{solute}}\right]^{(1/3)} \tag{7-56}$$

Any real macromolecule will have some amount, δ (grams solvent per gram dry solute), of solvent (with partial specific volume, $\bar{v}_{\text{solvent}}$) bound to the macromolecule and thus increases its effective size according to

$$\boxed{R = \left[\frac{3\text{M}}{4\pi N}\,[\bar{v}_{\text{solute}} + \delta\,\bar{v}_{\text{solvent}}]\right]^{(1/3)}} \tag{7-57}$$

There are two simple ways to analyze an experimental frictional coefficient, f.

1. *No Solvation; Maximum Asymmetry.* In this case, δ is set equal to zero in Eq. 7-57, and Fig. 7-23 is used to determine the corresponding axial ratio that accounts for the magnitude of f/f_0, where f_0 is calculated from Eq. 7-55, and f is determined experimentally by, say, measurement of diffusion constant, $D = kT/f$ (Eq. 7-13).

2. *Maximum Solvation; Spherical Shape.* In this limit, it is assumed that the macromolecule is a sphere, and the value of δ in Eq. 7-57 is increased until the f value calculated from Eq. 7-55 matches the experimental f value.

The molecular dimensions of a number of proteins have been calculated from their diffusion constants ($D = kT/f$), and the results are collected in Table 7-4. Because a given protein is neither a perfect sphere nor completely free of hydration, the most probable interpretation of the results is listed in the right-most column of the table, in which it has been supposed that the protein is hydrated to the extent of $\delta = 0.2$ gram water per gram of protein.

There is yet another way to obtain macromolecular shapes from hydrodynamic quantities, and that is from the magnitude of the viscosity of a macromolecule-containing solution. Up to now, we have been concerned with frictional coefficients (Table 7-3) that indicate the extent to which a *molecule* is slowed down when the molecule is forced to move through a solution; the extent of slowing-down is a function of molecular shape and *solvent* viscosity. A related (and more difficult) problem is to determine the effect of a suspension of large molecules on the pattern of flow of the *solution*, since the presence of, for example, suspended spheres means that the lines of flow of solvent molecules are distorted near the (large) spheres, with a consequent effect on the measured viscosity of the *solution*. In 1906, Einstein showed that for a suspension of rigid spheres (where the sphere is much larger than the solvent molecule so that the solvent may be treated as a continuous medium), the measured viscosity of the solution, η, is related to the viscosity of the solvent, η_0, by the formula

$$\boxed{\eta = \eta_0(1 + \nu\phi)} \tag{7-58}$$

where ϕ is the volume fraction occupied by the particles in the suspension, and $\nu = 2.5$ for spherical particles. In 1940, equations analogous to those shown in Table 7-3 were worked out by Simha for ellipsoids of arbitrary axial ratio, and the result is shown in Fig. 7-24. With final new definitions of *specific* viscosity

$$\eta_{sp} = \frac{\eta - \eta_0}{\eta_0} \tag{7-59}$$

Table 7-4 Molecular Dimensions of Proteins, Calculated from their Diffusion Coefficients. Extreme Cases Result from the Assumption that the Protein is Either a Perfect Sphere or Completely Lacks Water of Hydration (see text), Using Equations 7-55 and 7-57.

Protein	Molecular Weight	$\bar{v}^a_{\text{protein}}$	$D_{20°,w}$ x10^7	Maximum Solvation ($f/f_0 = 1$) δ^c	R^d	f/f_0^e	Maximum Asymmetry ($\delta = 0$) (a/b) for Prolate Ellipsoid	Compromise ($\delta = 0.2$) (a/b) for Prolate Ellipsoid
Ribonuclease	13,683 Dalton	0.728	11.9	0.35	18.0 Å	1.05	3.4	2.1
β-lactoglobulin	35,000	0.751	7.82	0.72	27.4	1.16	4.9	3.7
Serum albumin	65,000	0.734	5.94	1.07	36.1	1.25	6.5	4.9
Hemoglobin	68,000	0.749	6.9	0.36	31.0	1.05	3.4	2.1
Catalase	250,000	0.73	4.1	0.70	52.2	1.15	4.9	3.6
Tropomyosin	93,000	0.71	2.24	23.0	96		62	
Fibrinogen	330,000	0.710	2.02	8.4	106		31	
Collagen	345,000	0.695	0.69	218	310		300	
Myosin	493,000	0.728	1.16	49	215		100	

[a] $\bar{v}$ *has units of* cm^3g^{-1}

[b] $D_{20°,w}$ *is the diffusion constant for the stated protein, corrected to the value that would be expected at 20°C in distilled water (see Eq. 7-35), and has units of* cm^2sec^{-1}.

[c] *δ is the degree of solvation of the protein, defined as the number of grams of water per gram of dry protein.*

[d] *R is obtained by substituting the experimentally determined frictional coefficient,* $f = kT/D$, *into Eq. 7-55 (assuming a spherical macromolecular shape, and using a viscosity of 0.01 g* cm^{-1} *for water). δ is then obtained by substituting for R in Eq. 7-57.*

[e] *f is calculated as in the previous example from the experimentally measured diffusion constant,* $f = kT/D$. *δ is set equal to zero, and R is calculated from Eq. 7-57.* f_0 *is then calculated from Eq. 7-55, and the ratio,* f/f_0 *is used with Fig. 7-23 to obtain an axial ratio, a/b, for a prolate ellipsoid.*

* *From C. Tanford,* Physical Chemistry of Macromolecules, *Wiley, New York 1961.*

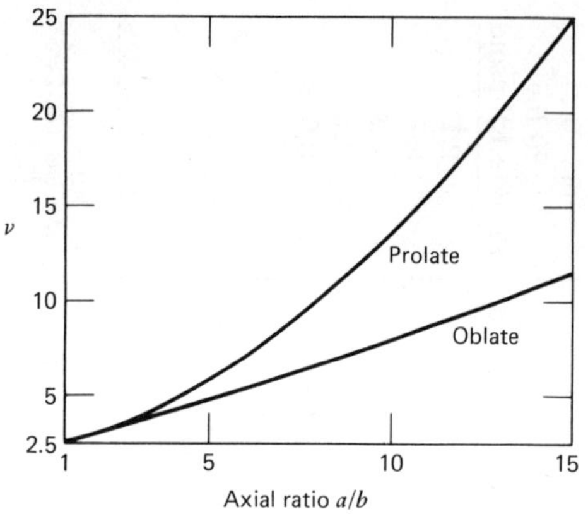

FIGURE 7-24. Viscosity factor, ν, as a function of axial ratio for ellipsoids of revolution (see Eq. 7-61). (From C. Tanford, *Physical Chemistry of Macromolecules,* Wiley, 1961.)

and *intrinsic* viscosity, $[\eta]$,

$$[\eta] = \lim_{c \to 0} (\eta_{sp}/c) \tag{7-60}$$

where c is solute concentration, the following relation may be obtained (see Problems) from Eqs. 7-58, 7-59, and 7-60:

$$[\eta] = \nu(\bar{v}_{solute} + \delta\ \bar{v}_{solvent}) \tag{7-61}$$

Equation 7-60 is simply designed to eliminate the concentration-dependence of η_{sp}, since η_{sp} is expected to be proportional to solute concentration, at least for very small solute concentration.

Table 7-5 Molecular Dimensions of Proteins, Derived from their Intrinsic Viscosities (see Eq. 7-61). Organization is as for Table 7-4.*

				Maximum Solvation ($\nu = 2.5$)		Maximum Asymmetry ($\delta = 0$)		Compromise ($\delta = 0.2$)	
	M	$\bar{v}_2$ cc/gram	$[\eta]$ cc/gram	δ grams/gram	R_0 Å	ν	a/b, Prolate Ellipsoid	ν	a/b, Prolate Ellipsoid
Ribonuclease	13,683	0.728	3.30	0.59	19.3	4.5	3.9	3.6	2.9
β-lactoglobulin	35,000	0.751	3.4	0.61	26.6	4.5	3.9	3.6	2.9
Serum albumin	65,000	0.734	3.7	0.75	33.7	5.0	4.4	4.0	3.3
Hemoglobin	68,000	0.749	3.6	0.69	34	4.8	4.1	3.8	3.1
Catalase	250,000	0.73	3.9						
Tropomyosin	93,000	0.74	52	20	91	70	29		
Fibrinogen	330,000	0.710	27	10.1	112	38	20		
Collagen	345,000	0.695	1150	460	400	1660	175		
Myosin	493,000	0.728	217	86	257	298	68		

* *From C. Tanford,* Physical Chemistry of Macromolecules, *Wiley, New York, 1961.*

Equation 7-61 may now be used in much the same way as the equations that relate friction coefficient to molecular shape. For a given measurement of solution intrinsic viscosity, one can *either* set $\nu = 2.5$ (sphere) and solve for the degree of hydration, δ, from Eq. 7-61, *or* one can set $\delta = 0$ (maximum asymmetry), calculate ν from Eq. 7-61, and then use Fig. 7-24 to obtain the axial ratio for the macromolecular solute. Results for a number of proteins (same proteins as for Table 7-4) are given in Table 7-5.

The hydrodynamic shapes of a variety of protein molecules are illustrated to scale in Fig. 7-25.

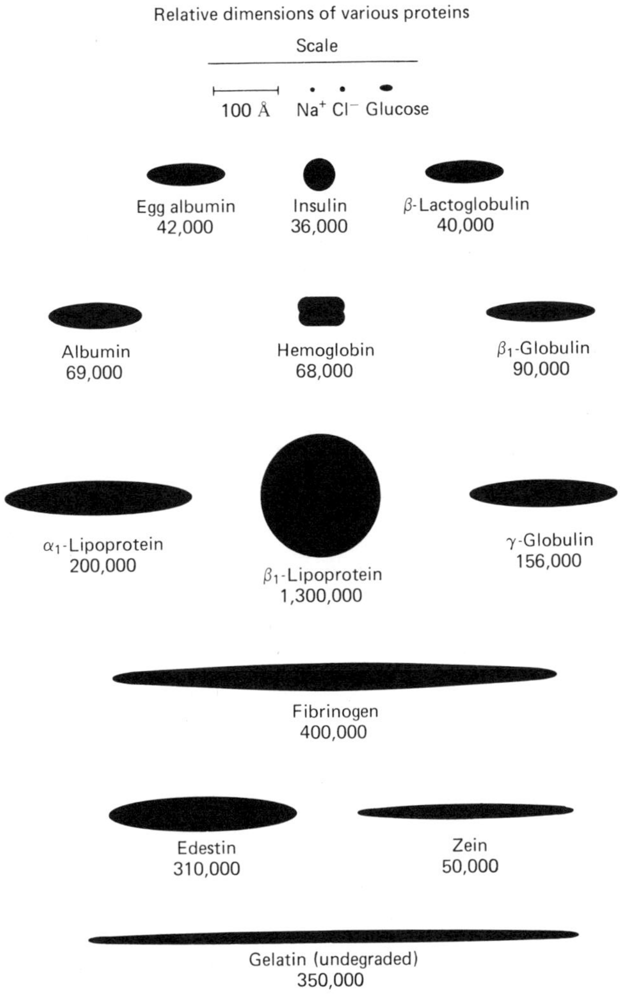

FIGURE 7-25. Dimensions of the ellipsoid of revolution that best account for the hydrodynamic properties (viscosity and frictional coefficient) of various protein molecules. [After W. J. Moore, *Physical Chemistry*, Prentice-Hall, Englewood Cliffs, N. J., 1972.)

The remaining direct uses for viscosity measurements are based on the (empirical) dependence of solution viscosity on solute molecular weight:

$$\boxed{[\eta] = KM^a} \tag{7-62}$$

where $a = 0$ for a sphere, $a = 0.5$ for an ideal random coil, and $a \cong 1.8$ for a long rigid rod. For a polymer whose K value and a value have already been determined empirically, Eq. 7-62 provides a quick estimate of the molecular weight of a particular preparation, and is especially useful for characterization of synthetic polymers. When, on the other hand, the molecular weight of, say, a protein is known, then Eq. 7-62 gives a quick indicator of the behavior of the polypeptide chain, as shown in Fig. 7-26. In Fig. 7-26, the logarithm of the intrinsic viscosity is plotted against the logarithm of molecular weight; slope of the plot yields "a" in Eq. 7-26, and shows that the proteins behave as random coils when denatured in 6M Guanidine HCl ($a = 0.66$), while DNA behaves more as a long rod ($a = 1.13$).

Intrinsic viscosity has been widely used in studies of the unfolding (denaturation) of proteins: when a globular protein denatures to a random coil under the effect of heat, acid or base, or addition of guanidine HCl or urea, the intrinsic viscosity increases (see Eq. 7-62) but when a long rod-like protein, such as collagen or myosin denatures, the intrinsic viscosity decreases because of a decrease in the asymmetry of the molecule.

Viscosity provides an indicator for the progress of enzymatic reactions that break up macromolecular substrates, as with the action of ribonuclease in depolymerizing RNA, or the action of hyaluronidase on hyaluronic acid, a highly viscous copolymer of N-acetylglucosamine and glucuronic acid

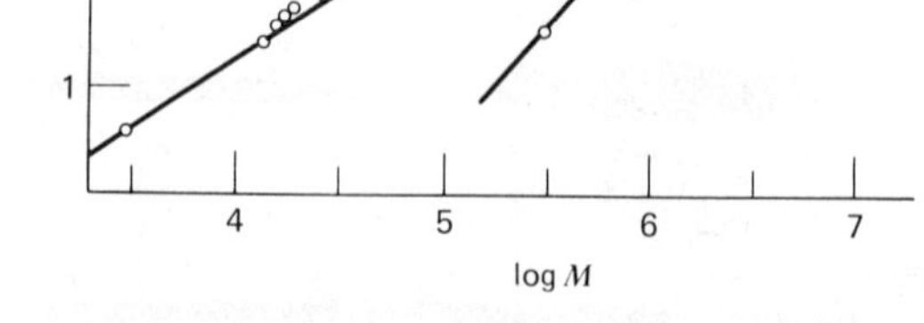

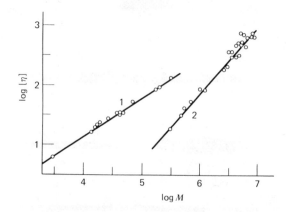

FIGURE 7-26. Plots of log $[\eta]$ versus log (molecular weight). Curve 1, various proteins that have been denatured with 6M Guanidine HCl. [From A. H. Reisner and J. Rowe, *Nature 222,* 558 (1969).] Curve 2, native DNA from a variety of sources. [From J. Eigner and P. Doty, *J. Mol. Biol. 12,* 549 (1965).] Slope of the plot is an index of molecular shape (see Eq. 7-62 ff.).

that serves to limit the spread of infectious agents by forming a cement between cells. Finally, viscosity measurements give a means of estimating the degree of cross-linking in polymers (as a random coil becomes cross-linked, the "a" value in Eq. 7-62 decreases)—viscosity evidence indicates that the polysaccharide, dextran, a branched polyglucose generated in fermentation of sucrose by certain bacteria, becomes much more highly branched as its molecular weight increases.

7.D. FORCED MARCH. POLAROGRAPHY

As ordinarily conducted (say, by mixing the reactants), most chemical reactions are irreversible; that is, they don't run backward. The great value of electrochemistry is that it can provide a means for carrying out a chemical reaction reversibly, thus allowing for determination of the equilibrium constant for the reaction, as discussed in Chapter 5. However, for dilute solutions (i.e., $a_{\text{solute}} \cong m_{\text{solute}}$) the observed reduction potential depends on a concentration *ratio* rather than on *absolute* concentration of either the oxidized or reduced species

$$\mathrm{E} = \mathrm{E}^\circ + \frac{RT}{zF} \log_e \frac{[\mathrm{Ox}]}{[\mathrm{Red}]} = \mathrm{E}^\circ + \frac{0.059}{z} \log_{10} \frac{[\mathrm{Ox}]}{[\mathrm{Red}]} \text{ at } 25^\circ\mathrm{C} \qquad (7\text{-}63)$$

where the symbols in Eq. 7-63 have the usual meanings (see Chapter 5). Polarography is a variant electrochemical technique that does give a measure of *absolute* concentration of a reducible species, even in the presence of other reducible species, and generates a number of applications ranging from inexpensive detection of trace metals in the environment to diagnosis of cancer. The technique involves measurement of electrical current (flow of "electrical fluid") as a function of applied voltage ("electrical pressure"), so our discussion begins from consideration of current in an electrical circuit.

When current flows in an ordinary electrical circuit, the current flow is proportional to the applied voltage ($i = (V/\mathbf{R})$, where $\mathbf{R}$ is electrical resistance. However, when part of the circuit involves two electrodes and a chemical solution, then virtually no current can flow until the applied voltage exceeds the minimum voltage required to start the chemical reaction (see curves A and B in Fig. 7-27). As expected from Eq. 7-63, the minimum applied voltage goes up by $(0.059/z)$ volts for every ten-fold decrease in the concentration of reducible species, [Ox], (see curves A to D of Fig. 7-27). The interesting case occurs when the [Ox] level is so small that the current voltage curve levels off to a plateau region, shown as curves C and D in Fig. 7-27. An electrode is said to be "polarized" whenever its electrical potential varies with the current passing through it, as with the cases shown in Fig. 7-27.

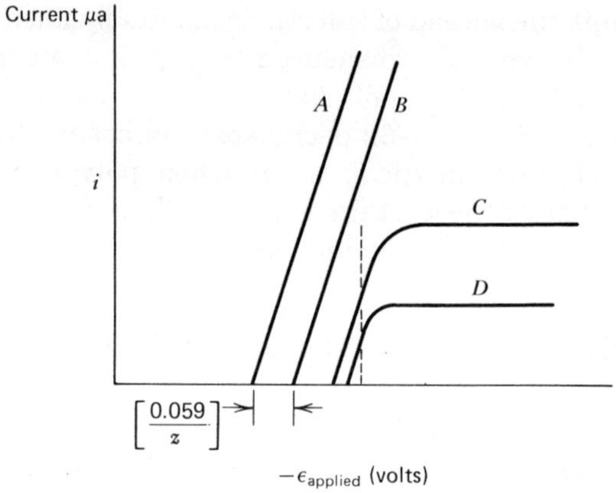

FIGURE 7-27. Plot of steady-state current flowing through the circuit shown in the following figure, as a function of applied static voltage. Individual plots correspond to different concentrations of the reducible species [Ox]. *A:* [Ox] = 0.1 M, *B:* [Ox] = 0.01 M, *C:* [Ox] = 0.001 M, *D:* [Ox] = 0.0005 M. All solutions contain 1M KCl to facilitate passage of current once the minimum applied voltage has been reached.

Resistance polarization (straight-line region in Fig. 7-27) arises from the potential drop, $i\mathbf{R}$, resulting from the resistance of the circuit. *Concentration* polarization (the "plateau"-region of Fig. 7-27) is a diffusion-related phenomenon pictured graphically in Fig. 7-28 and is the principal object of our present attention.

Figure 7-28 shows what happens to an initially homogeneous solution containing a chemically reducible species in the vicinity of a plane cathode electrode, on application of a potential sufficient to reduce the reducible

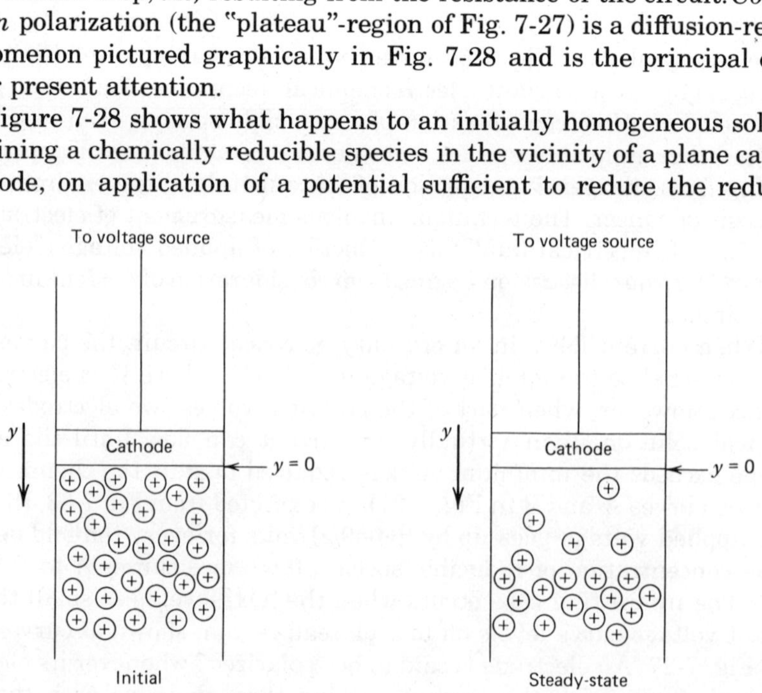

FIGURE 7-28. Concentration of a reducible species in the vicinity of a flat electrode in a stirred solution (see text).

species. It is supposed that the reducible species reaches the cathode surface by diffusion (any Coulomb attraction is ignored), and on reaching the electrode surface is reduced very rapidly (i.e., rapidly compared to the diffusional rate of reaching the surface). Thus, the reducible species will be depleted in the immediate neighborhood of the cathode, leading to a concentration gradient shown at the right of Fig. 7-28, and the steady-state current that can flow through the circuit will be limited simply by the rate at which the reducible species can diffuse to the cathode surface.

The size of the diffusion-limited current is readily calculated. First, the flow of ions across the cathode boundary (in moles per unit area per unit time) is given by Fick's first law of diffusion:

$$\text{Flow} = -D_{\text{ox}}\left(\frac{\partial[\text{Ox}]}{\partial y}\right)_{y=0} \tag{6-51}$$

where D_{ox} is the diffusion constant for the reducible species. From the number of electrons transferred per reaction, z, and the charge of a mole of electrons, F, the diffusion-limited current is immediately obtained:

$$i = -zFAD_{\text{ox}}\left(\frac{\partial[\text{Ox}]}{\partial y}\right)_{y=0} \tag{7-64}$$

From the initial condition, $[\text{Ox}] = [\text{Ox}]_0$ for all y (i.e., homogeneous solution at time zero), Fick's second law of diffusion (Eq. 6-55) may be solved to give:

$$\left(\frac{\partial[\text{Ox}]}{\partial y}\right)_{y=0} = -\frac{[\text{Ox}]_0}{(\pi D_{\text{ox}}t)^{1/2}} \tag{7-65}$$

Substitution of Eq. 7-65 into Eq. 7-64 gives the desired result:

$$\boxed{i_d = zFA[\text{Ox}]_0\left(\frac{D_{\text{ox}}}{\pi t}\right)^{1/2}} \tag{7-66}$$

where A is the area of the cathode surface in contact with the solution.

At this stage, it would appear that a similar treatment should apply to the diffusion of the reduced species to the anode, so that the resultant current would represent some complicated mixture of the cathodic and anodic processes. The innovation, which led to the development of polarography and a Nobel prize for the inventor, Jaroslav Heyrovsky, was to construct an apparatus in which the cathode is very small and the anode very large (see Fig. 7-29), so that the current through the circuit is limited solely by diffusion of the *reducible* species toward the cathode. Finally, since any small electrode is subject to adsorption and other undesirable irreversible reactions, it is desirable to refresh the cathode surface continually, as is readily accomplished by the dropping mercury electrode shown in Fig. 7-29.

It is straightforward to accommodate the previous treatment to the new geometry of a spherical cathode rather than a plane cathode surface. First, the volume of a given mercury drop before it falls is just

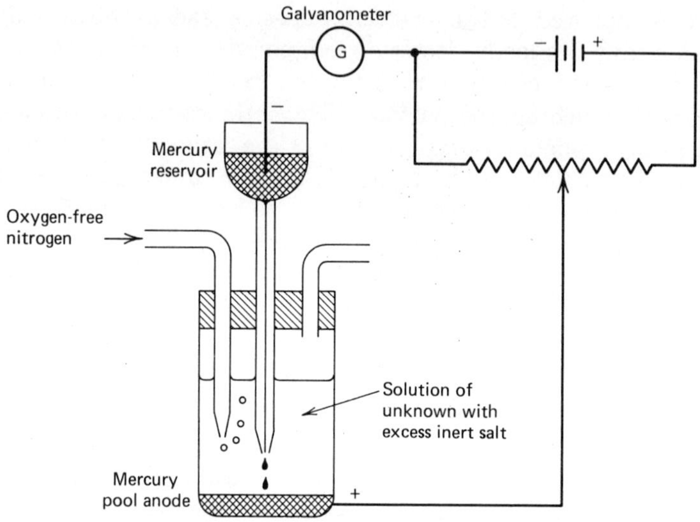

FIGURE 7-29. Schematic diagram of a polarograph. Nitrogen flushing is used to remove dissolved oxygen, since oxygen is itself a reducible species that would contribute to the diffusion-limited current under investigation.

$$(4/3)\pi r^3 = (m/\rho)t \qquad (7\text{-}67)$$

where r is the radius of the hanging drop, ρ is the density of the drop, m is the mass of mercury extruded per unit time, and t is the time that has passed since the previous drop fell off. Solving Eq. 7-67 for r, the area of the mercury drop may be obtained as

$$4\pi r^2 = A = km^{2/3}t^{2/3} \qquad (7\text{-}68)$$

where k is a constant incorporating π and the density of mercury. Then by substituting the area obtained in Eq. 7-68 into Eq. 7-66

$$i_d = Kzm^{2/3}t^{1/6}D_{\text{ox}}^{1/2}[\text{Ox}]_0 \qquad \text{(Ilkovic equation)} \qquad (7\text{-}69)$$

or just

$$i_d = (\text{constant})zD_{\text{ox}}^{1/2}[\text{Ox}]_0 \qquad (7\text{-}70)$$

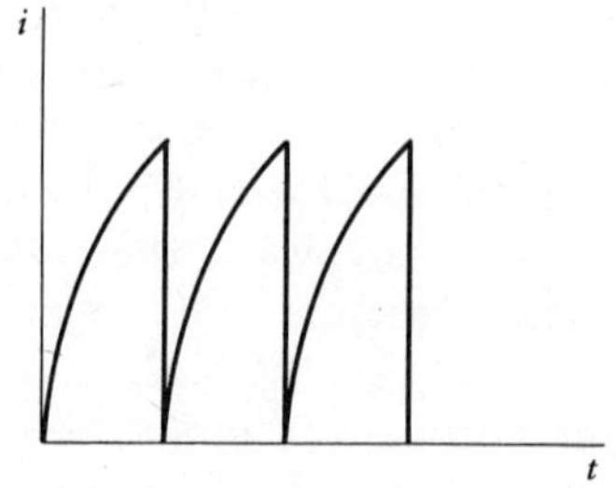

FIGURE 7-30. Diffusion-limited current versus time for a dropping mercury electrode (Eq. 7-69).

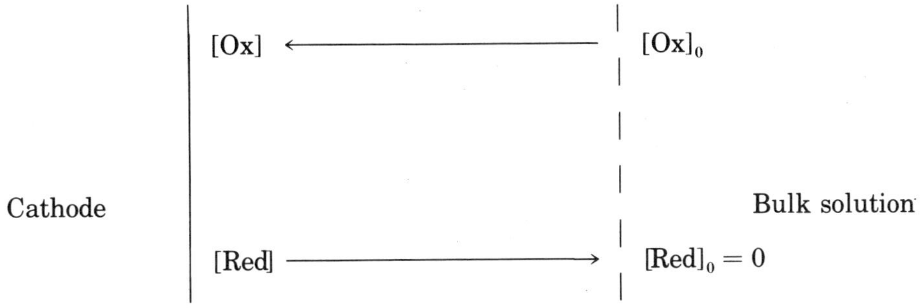

FIGURE 7-31. Schematic concentrations of relevant species, at the cathode surface (left) and in bulk solution (right).

Flow of current predicted by Eq. 7-69 is interrupted as a given drop falls, and follows a 1/6-order time-dependence as shown in Fig. 7-30. The saw-tooth pattern in Fig. 7-30 may be smoothed to a constant (average) current by using a detector (galvanometer) with a sufficiently slow time constant (see Chapter 8); the average current can be shown to be equal to 6/7 of the maximum current.

Equations 7-65 to 7-70 are based on the assumption that the current is diffusion-limited; more specifically, that the concentration of reducible species at the electrode surface is essentially zero. In general, however, the electrical "pressure" (applied potential) may not be sufficiently large to reach this limit. To generalize Eq. 7-70 to a form suitable for comparison with experiment, consider the cathode surface shown in Fig. 7-31. (If there is no reduced form present initially, then the bulk $[\mathrm{Red}]_0$ level may be taken as effectively zero.) The Ilkovic equation (7-70) may now be expressed more generally:

$$i = (\mathrm{constant}) z D_{\mathrm{ox}}^{1/2} ([\mathrm{Ox}]_0 - [\mathrm{Ox}]) \tag{7-71a}$$

$$i = (\mathrm{constant}) z D_{\mathrm{red}}^{1/2} [\mathrm{Red}] \tag{7-71b}$$

and in the diffusion-limited case for which $[\mathrm{Ox}] \to 0$

$$i_d = (\mathrm{constant}) z D_{\mathrm{ox}}^{1/2} [\mathrm{Ox}]_0, \text{ as previously derived} \tag{7-70}$$

Solving Equations 7-70 and 7-71a for [Ox], and Eq. 7-71b for [Red]

$$[\mathrm{Ox}] = \frac{(i_d - i)}{(\mathrm{constant}) z D_{\mathrm{ox}}^{1/2}} \tag{7-72a}$$

and

$$[\mathrm{Red}] = \frac{i}{(\mathrm{constant}) z D_{\mathrm{red}}^{1/2}} \tag{7-72b}$$

Substitution of Eqs. 7-72 into the Nernst equation 7-63 gives

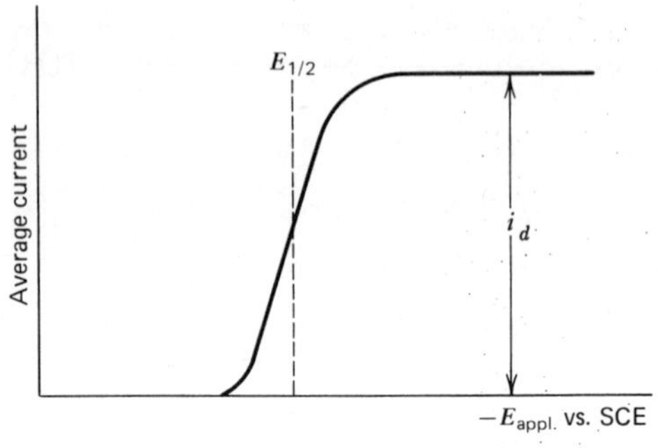

FIGURE 7-32. Schematic polarogram for cathodic reduction of a cation. (In this figure and in figure 7-33, applied voltage is linearly varied while monitoring current as a function of time.)

$$E = E^\circ + \frac{0.059}{z}\log_{10}\left(\frac{D_{\text{red}}}{D_{\text{ox}}}\right)^{1/2} + \frac{0.059}{z}\log_{10}\left(\frac{i_d - i}{i}\right) \tag{7-73}$$

The form of the original Fig. 7-27 is now explained by Eq. 7-73: the diffusion-limited current (the height of the plateau region in *C* or *D* of Fig. 7-27) is a direct measure of the bulk concentration of reducible species in solution according to Eq. 7-70; while the voltage at half-maximum current is

$$E = E_{1/2} + \frac{0.059}{z}\log_{10}\left(\frac{i_d - i}{i}\right) \tag{7-74}$$

where

$$E_{1/2} = E^\circ + \frac{0.059}{z}\log_{10}\left(\frac{D_{\text{red}}}{D_{\text{ox}}}\right)^{1/2} \tag{7-75}$$

(It is often a good approximation to take $E_{1/2} = E^\circ$, since D_{red} and D_{ox} are usually of similar magnitude.) A graphical display of the behavior predicted by Eq. 7-73 is shown in Fig. 7-32.

Applications for polarography are based either on use of Eq. 7-70 to obtain either z (number of electrons transferred per reactant molecule converted to product) or $[\text{Ox}]_0$ (concentration of a given reducible species in solution), or on use of Eq. 7-74 to obtain either $E_{1/2}$ or E° (see Chapter 5 for E° applications).

EXAMPLE *Is a Given Chemical Reaction Reversible?*

E° values provide one of the best sources for accurate equilibrium constants, as discussed in Chapter 5. However, such calculated equilibrium constants are reliable only when the electrochemical process on which they were based is *reversible*. A simple (and an accepted) criterion for reversibility is to construct a plot of E versus log $[(i_d - i)/i]$, and see if the slope agrees with $(0.059/z)$, where z is a positive integer, as predicted by Eq. 7-73. Reversibility in this context implies that the reactions at the electrodes are much faster that the rate of diffusion of either species to the electrode.

EXAMPLE *Selective Simultaneous Detection of Several Reducible Components*

Figure 7-33 shows a polarogram obtained for a solution that contained equal concentrations of three reducible species, for which the standard reduction potentials were sufficiently different to permit resolution by polarography Polarography provides one of the simplest and cheapest methods for trace

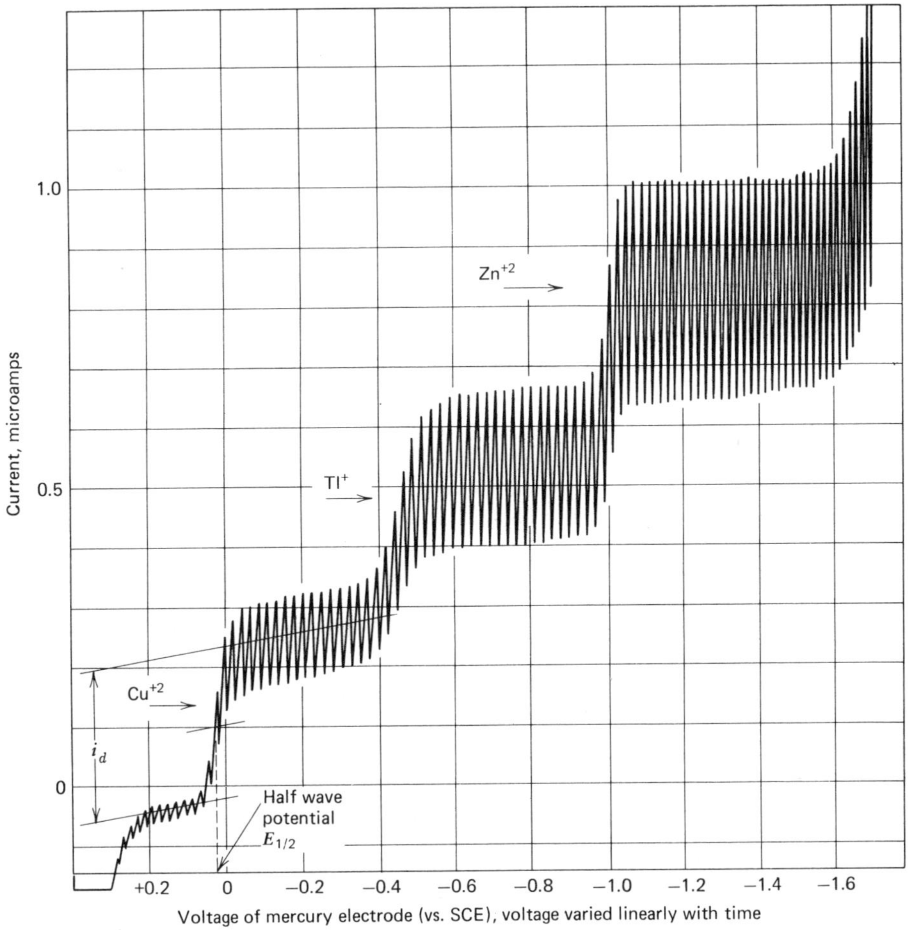

FIGURE 7-33. Polarogram of an aqueous solution containing 10^{-4} M Cu^{+2}, Tl^{+} and Zn^{+2} and 0.1 M KNO_3 as supporting electrolyte. Voltages are referred to the saturated calomel electrode. (W. B. Schaap, Indiana University, from *Physical Chemistry,* W. J. Moore, Prentice-Hall, Englewood Cliffs, N. J., 3rd ed., 1963, p. 410.)

analysis, and is particularly useful for species such as Zn^{++}, which are difficult to quantify chemically. (The diffusion-limited current, i_d, is directly proportional to concentration of the reducible species—see Eq. 7-70.)

EXAMPLE *Number of Electrons Transferred in a Given Reaction*

Because the diffusion-limited current, i_d, is proportional to the number of electrons involved in the reduction in question, it is easy to determine that number of electrons by comparison of the i_d for the unknown reaction to the i_d for the same concentration of a species (such as Zn^{++}) for which the number of electrons per reaction is known.

For example, the reduction of molecular oxygen proceeds in two steps as shown in Fig. 7-34. When the enzyme, catalase is present, the height of the first "wave" in Fig. 7-34 increases, showing that catalase catalyzes the dispro-

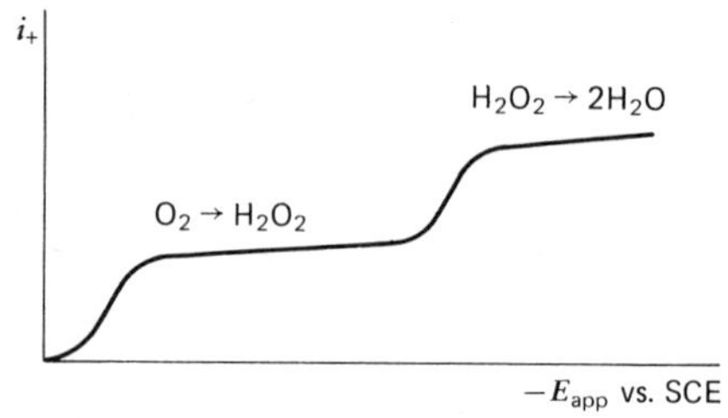

FIGURE 7-34. Polarogram for reduction of molecular oxygen.

portionation of hydrogen peroxide to water and oxygen: $H_2O_2 \rightleftharpoons (1/2)O_2 + H_2O$. Catalase thus makes it possible to convert oxygen to water without having to use the higher applied voltage (electrical "pressure") that would be required to generate the right-hand "wave" in Fig. 7-34 in the absence of catalyst.

As another example, it has been found that the reduction of the cobalt +3 in vitamin B_{12} is a *two*-electron process, leading to the formation of a stable (though unusual) +1 oxidation state for cobalt. It is becoming increasingly evident that the biological function for species containing metal ions is associated with distorted geometry of the surrounding ligands, and the B_{12} case was one of the first examples of an unusual bioinorganic complex.

EXAMPLE *Stoichiometry and Stability of Metal Complexes*

For equilibria of the type,

$$M^{+n} + pX \xrightleftharpoons{K_{eq}} MX_p \tag{7-76}$$

it can be shown that a plot of $-E_{1/2}$ versus $-\log[X]$ gives a straight line of (negative) slope, $(0.059)p/n$. Such a plot is shown in Fig. 7-35 for the complex between copper(II) and thioaspirin. From the figure, the slope is found to be

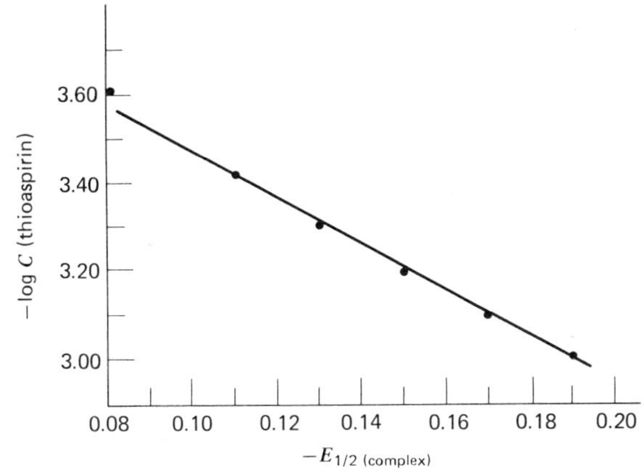

FIGURE 7-35. Graphical determination of the stoichiometry of the complex between copper(II) and thioaspirin (see text). (From W. C. Purdy, *Electroanalytical Methods in Biochemistry,* McGraw-Hill, N. Y., 1965, p. 186.)

0.181, so that $p = (0.181)/(0.059) = 3.07$. It can also be deduced that the binding constant is 2.2×10^{11}. The reader is referred to the Purdy reference for details of such procedures.

EXAMPLE *Polarographic Determination of Drugs, Vitamins, and Hormones*

A variety of drugs may be assayed polarographically, including streptomycin (structure given below), chloramphenicol, and sulfanilamide derivatives. In the case of streptomycin, the site of reduction is probably the aldehyde group:

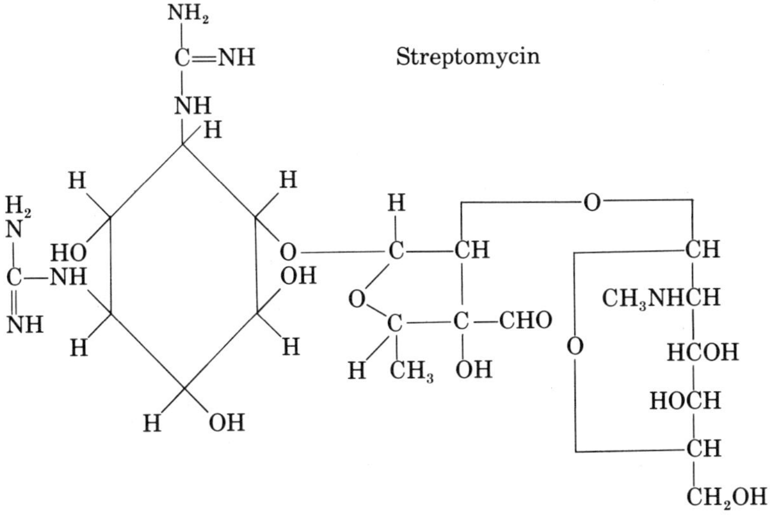

The precision of the determination compares favorably with that of the microbiological test (see section on immunodiffusion) for samples as small as 200 microgram/cc.

Polarography provides for determination of several B-vitamins, and also vitamin C (ascorbic acid). In the case of vitamin C, which is a strong reducing agent, it is the anodic two-electron *oxidation* that is observed:

$$\text{Vitamin C} \quad \underset{\text{HOH}_2\text{CHOHCH}-\overset{|}{\text{C}}}{\overset{\text{OH} \quad \text{OH}}{\underset{\diagdown\,\text{O}\,\diagup}{\overset{|}{\text{C}}=\overset{|}{\text{C}}}}\;\;\underset{\text{C}=\text{O}}{}} \quad ; \quad \underset{\text{OH} \quad \text{OH}}{-\text{C}=\text{C}-} \rightleftharpoons \underset{\text{O} \quad \text{O}}{-\overset{\|}{\text{C}}-\overset{\|}{\text{C}}-} + 2\text{H}^+ + 2e^-$$

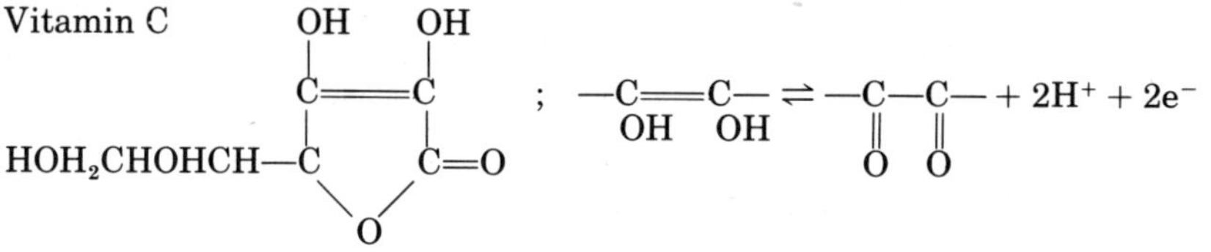

Among the steroids that can be determined polarographically are corticosterone, desoxycorticosterone, progesterone, and testosterone.

EXAMPLE Aid in Diagnosis of Cancer

This example is an exercise in serendipity. It was discovered in 1933 that a polarogram of human serum in the presence of cobaltous ion exhibited a double-wave appearance, with a peculiar maximum (shown by an arrow in Fig. 7-36) on the second wave. It was later shown that this behavior results from protein-catalyzed reduction of hydrogen ion at the electrode surface (in the absence of protein, the hydrogen ion reduction would take place at more negative potentials).

In any case, the interesting feature is that the size of the double wave differs between serum of normal and cancer patients. An empirical procedure that magnifies this difference consists of taking the ratio of the polarographic wave heights for serum that has been digested with KOH and serum that has been

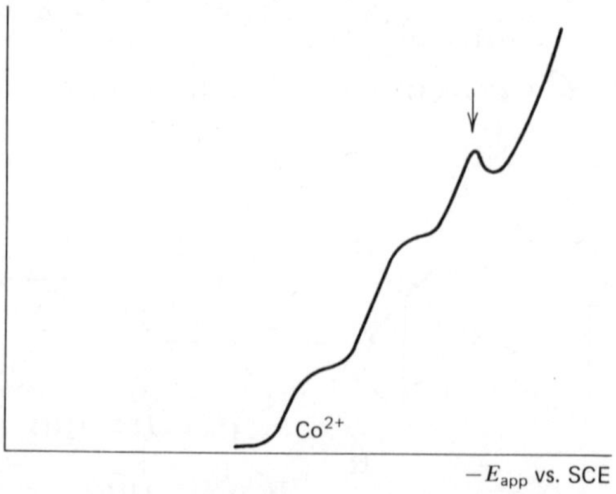

FIGURE 7-36. Polarogram of human serum in a supporting electrolyte of 0.001 M cobaltous chloride, 0.1 M ammonia, and 0.1 M ammonium chloride. (From W. C. Purdy, *Electroanalytical Methods in Biochemistry*, McGraw-Hill, N. Y., 1965, p. 163.)

precipitated with sulfosalicylic acid and filtered; this ratio is called the "protein index." Normal values for the protein index are 2.4 ± 0.17. The index increases markedly for cancer patients, and also in the presence of certain other diseases, including arthritis, gout, schizophrenia, and stomach ulcers. The importance of diagnosis cannot be overemphasized—it has been estimated that two-thirds of the people who will die from cancer could have been cured with *existing* treatments if the condition could have been recognized a few months earlier.

PROBLEMS

1. Equations 6-55 and 7-19 demonstrate the independence of diffusion and electrophoretic mobility. Obtain Eq. 7-19 from Equation 7-17 and 7-18. Hint: use "Chain Rule" of calculus.

2. In the isoelectric focussing experiment shown schematically in Fig. 7-7, what would happen if the electrodes were connected with the positive electrode at the basic end of the cell and the negative electrode at the acidic end?

3. From the data plotted in Fig. 7-15, confirm that $s = 4.37S$ for that protein.

4. Consider an equilibrium sedimentation experiment in a density gradient. For the region in the immediate vicinity of the isopycnic point, r_0 (page 189), for a given macromolecular solute, assume that the density gradient is approximately linear:

$$\rho = (1/\bar{v}_{\text{solute}}) + (r - r_0)\frac{d\rho}{dr}, \; \frac{d\rho}{dr} \cong \text{constant}$$

   Substitute this expression into the usual sedimentation equilibrium condition, Eq. 7-39, and solve for the macromolecular concentration as a function of distance from the center of rotation, r. How will the width of this distribution depend on macromolecular weight?

5. Figure 7-23 shows that nonspherical molecules offer greater frictional resistance in moving through solution than a spherical molecule of the same volume. This effect is basically due to the larger surface area of nonspherical molecules compared to a sphere of the same volume. Try to convince yourself of this fact from the following arithmetic examples.
   (a) Calculate the surface area of a spherical macromolecule of radius 10 A°.
   (b) Calculate the surface area of a cube of the same volume and compare to (a). Calculate the surface area of a right circular cylinder of the same volume, if the cylinder length is
   (c) equal to the cylinder radius
   (d) 10 times the cylinder radius

6. Determine the units (mks) for the following quantities:
   (a) ionic mobility, μ
   (b) frictional coefficient, f
   (c) sedimentation coefficient, s
   (d) translational diffusion constant, D, for one-, two-, or three-dimensional diffusion
   (e) viscosity, η

7. Given that for a particular protein, $D_{20°C,w} = 3.7 \times 10^{-7}$ cm^{2}sec^{-1}; $\bar{v} =$ 0.72 cm^{3}g^{-1}, and molecular weight = 300,000 Dalton, with viscosity of water taken as 0.01 g cm^{-1}, calculate the
   (a) protein diameter if the protein were spherical
   (b) degree of hydration, δ, corresponding to (a)
   (c) axial ratio of protein if the protein were a prolate ellipsoid having no water of hydration
   (d) axial ratio of protein as a prolate ellipsoid with $\delta = 0.2$ g water/g protein

   This problem gives a complete exercise in determination of macromolecular size, shape, and degree of hydration from diffusion data.

8. Size, shape, and degree of hydration of macromolecules may alternatively be obtained directly from viscosity data, using Eq. 7-61 (see next Problem). Derive Eq. 7-61 from Eq. 7-58 and the definitions of specific viscosity and intrinsic viscosity on page 202.

9. Using Eq. 7-61 and Fig. 7-24, obtain the last six columns of Table 7-5 from the data in the first three columns. This problem is an example of determination of molecular size and/or degree of hydration from viscosity data.

10. Show that Eq. 7-74 will generate a graph of current versus applied voltage of the type shown in Fig. 7-32.

REFERENCES

Friction

C. Tanford, *Physical Chemistry of Macromolecules,* Wiley, N. Y. (1961).

Electrophoresis

H. Haglund, "Isoelectric Focusing in pH Gradients," Methods of Biochemical Analysis **19,** 1 (1970).

R. J. Wieme, *Agar Gel Electrophoresis,* Elsevier (1961).

L. P. Cawley, *Electrophoresis and Immunoelectrophoresis,* Little, Brown & Co., Boston (1969).

H. R. Maurer, *Disc-Electrophoresis: Theory and Practice of Discontinuous Polyacrylamide Gel Electrophoresis,* Walter de Gruyter & Co (1971).

F. M. Everaerts, A. J. Mulder, and Th. P. E. M. Verheggen, "Isotachophoresis: Analytical Tool in Electrophoresis," American Laboratory (Dec., 1973), pp. 37–45.

Sedimentation

J. H. Coates, "Ultracentrifugal Analysis," in *Physical Principles and Techniques of Protein Chemistry,* Part B, ed. S. J. Leach, Academic Press, N. Y. (1970).

C. H. Chervenka, "A Manual of Methods for the Ultracentrifuge," Spinco Division, Beckman Instruments Inc., Stanford Industrial Park, Palo Alto, Calif. (1969).

Viscosity

J. H. Bradbury, "Viscosity," in *Physical Principles and Techniques of Protein Chemistry,* part B, ed. S. J. Leach, Academic Press, N. Y. (1970).

Polarography

W. C. Purdy, *Electroanalytical Methods in Biochemistry,* McGraw-Hill, N. Y. (1965).

CHAPTER 8 Bad Odds: Poisson Distribution

8.A. RADIOACTIVE COUNTING

The origin of radioactivity and its interaction with matter is a topic for advanced quantum mechanics; for now, we are concerned with applications in which the number of radioactive counts serves as an indicator of the location and the concentration of a species of chemical or biological interest. The only facts required are (1) the probability, a, that any one nucleus will emit radiation during a given time interval of observation (say, one second) is very small compared to unity, and (2) for a sample containing a large number of radioactive nuclei, the observed average number of counts is proportional to time.

This situation is closely related to the bags and marbles problem we have previously solved, except that now the desired event (observing a radioactive decay) has a probability, $a \ll 1$, rather than the case of $a = (1/2)$ that led to Eq. 6-4. In other words, observation of radioactivity is like looking for a black marble in a bag containing nearly all white marbles: the

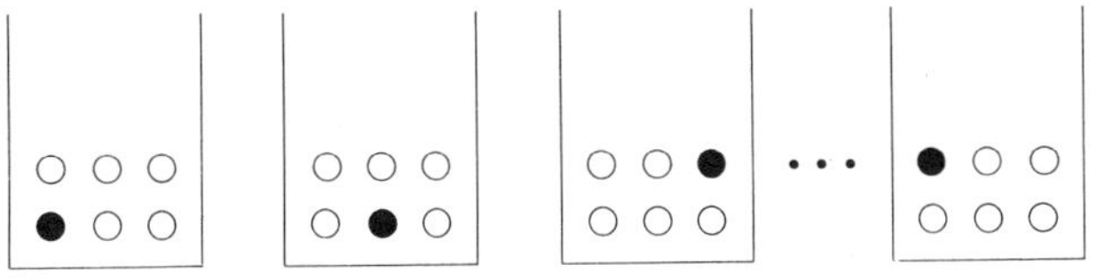

probability of finding the desired black marble (radioactive decay) in a given bag (nucleus) is much less than one, and one would like to know how many black marbles (decays) one is likely to find in examination of many bags (nuclei). From intuitive reasoning precisely analogous to that which led to Eq. 6-4, the solution is clearly

$$P_N(m) = \frac{N!}{m!\,(N-m)!}\, a^m b^{(N-m)} \tag{8-1}$$

where $P_N(m)$ is the probability of obtaining m black marbles in one draw from each of N bags, a is the probability that any single draw will yield a black marble, and b is the probability that any single draw will yield a white marble. In order to apply Eq. 8-1 to the radioactive counting problem (and others in this section), it will now be supposed that the probability of obtaining a black marble on a single draw (i.e., the probability that any one unstable nucleus will decay during the observation period) is small:

$$a \ll 1 \tag{8-2}$$

but the total number of draws (nuclei) is very large,

$$N \ll 1 \tag{8-3}$$

such that the average number of black marbles drawn in a given experiment is

$$aN = \bar{m} \tag{8-4}$$

where $\bar{m}$ ranges from zero to 1000 or more.

Solving Eq. 8-4 for $a = \bar{m}/N$, and substituting into Eq. 8-1

$$P_N(m) = \frac{N!}{m!\,(N-m)!}\left(\frac{\bar{m}}{N}\right)^m \left(1 - \frac{\bar{m}}{N}\right)^{(N-m)} \tag{8-5}$$

where it has been recognized that

$$a + b = 1 \tag{8-6}$$

Equation 8-5 may now be rearranged more conveniently:

$$\begin{aligned} P_N(m) &= \frac{N!}{m!\,(N-m)!}\left(\frac{\bar{m}}{N}\right)^m \left(1 - \frac{\bar{m}}{N}\right)^N \left(1 - \frac{\bar{m}}{N}\right)^{(-m)} \\ &= \left(\frac{\bar{m}^m}{m!}\right)\left(1 - \frac{\bar{m}}{N}\right)^N \left(\frac{N!}{(N-m)!\,N^m\left(1 - \frac{\bar{m}}{N}\right)^m}\right) \\ &= A \cdot B \cdot C \end{aligned} \tag{8-7}$$

Letting N now approach infinity, the three factors of Eq. 8-7 reduce to:

$$\lim_{N\to\infty} A = \frac{\bar{m}^m}{m!} \tag{8-8a}$$

$$\lim_{N\to\infty} B = e^{-\bar{m}} \qquad \text{(definition of } e^x\text{)} \tag{8-8b}$$

and

$$\lim_{N\to\infty} C = 1 \qquad \text{(see Problems)} \tag{8-8c}$$

Equation 8-1, under the restrictions of 8-2 to 8-4 thus reduces to:

$$\boxed{P_N(m) = \frac{\overline{m}^m}{m!} e^{-m}} \tag{8-9}$$

which is the celebrated Poisson distribution.

As a check on the derivation, we will evaluate the *average* number of successes, $<m>$ using the definition of an average of a discretely varying quantity (Eq. 6-38):

$$\begin{aligned}
<m> &= \lim_{N\to\infty} \sum_{m=0}^{N} m \frac{\overline{m}^m}{m!} e^{-\overline{m}} \\
&= e^{-\overline{m}} \lim_{N\to\infty} \left[0 + \frac{\overline{m}}{1!} + \frac{2\overline{m}^2}{2!} + \frac{3\overline{m}^3}{3!} + \cdots + \frac{N\overline{m}^N}{N!}\right] \\
&= \overline{m}\, e^{-\overline{m}} \lim_{N\to\infty} \left[1 + \overline{m} + \frac{\overline{m}^2}{2!} + \cdots + \frac{\overline{m}^N}{N!}\right] \\
&= \overline{m}\, e^{-\overline{m}}\, e^{\overline{m}}
\end{aligned}$$

$$\boxed{<m> = \overline{m} = Na} \tag{8-10}$$

in agreement with the intuitively derived Eq. 8-4. It is left as an exercise (see Problems) to show the principal useful property of the Poisson distribution:

$$\boxed{<(m - <m>)^2> = m = Na} \tag{8-11}$$

In other words, the result of many separate radioactive counting experiments will be to give an average result of $\overline{m} = Na$, with a root-mean-square deviation of $\pm\sqrt{\overline{m}} = \pm\sqrt{Na}$.

Just as the algebraic equation for an ellipse is complicated, while a

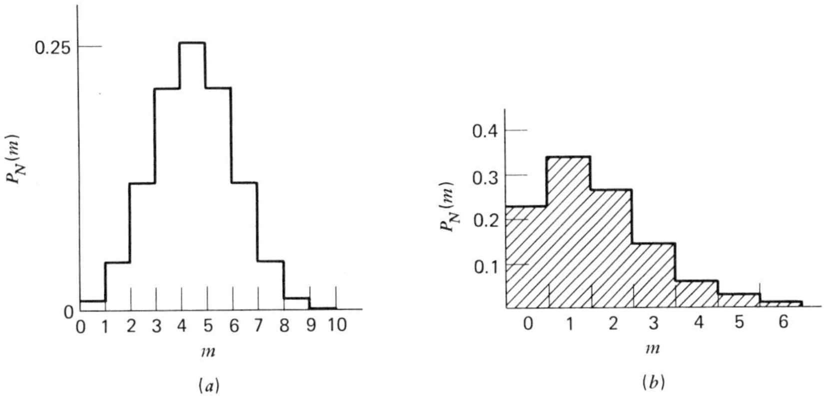

FIGURE 8-1. Two binomial distributions that approach the Gaussian and Poisson limits shown in Table 8-1. (*a*), $(0.5+0.5)^{10}$; (*b*), $(0.95+0.05)^{30}$. Note the skew in the right-hand plot.

Table 8-1. Illustration of the Common Origin of the Poisson and Normal (Gaussian) Distributions as Special Cases of the Binomial Distribution.*

BINOMIAL DISTRIBUTION

$$P_N(m) = \frac{N!}{m!\,(N-m)!}\,a^m b^{(N-m)}$$

$a = b = (1/2);\ N \to \infty$	$a \ll 1;\ N \to \infty$
NORMAL (GAUSSIAN) DISTRIBUTION	POISSON DISTRIBUTION
$P_N(m) = \sqrt{\frac{2}{\pi N}}\,e^{-2(m-<m>)^2/N}$	$P_N(m) = \frac{m^m}{m!}\,e^{-\bar{m}}$
$<m> = Na = \frac{N}{2}$	$<m> = \bar{m} = Na$
and $\sqrt{<(m-<m>)^2>} = \sqrt{\frac{<m>}{2}} = \sqrt{\frac{N}{4}}$	and $\sqrt{<(m-<m>)^2>} = \sqrt{<m>} = \sqrt{Na}$

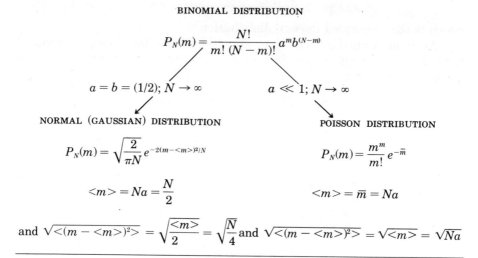

* *Note that although the algebraic form of the Poisson is different from that of the Gaussian distribution, the principal property (the standard deviation from the average result) is similar in both cases.*

sketch is simple, the meaning of Equations 8-9 and 8-11 is most readily apparent from a graph (see Fig. 8-1). Table 8-1 summarizes the basic mathematics of this chapter.

As far as radioactive counting experiments are concerned, the most important result from the Poisson treatment is that if the average number of counts observed during a given observation period is $\bar{m}$, then in any particular observation period, the observed number of counts, m, will generally be within $\pm\sqrt{\bar{m}}$ of the average value, $\bar{m}$. Now since the number of counts increases as the length of the observation period ($\bar{m} \propto t$), while the imprecision in the measurement increases as $\sqrt{\bar{m}}$, it is necessary to count for *four* times as long in order to decrease the *fractional* error, $\Delta m/\bar{m} = \sqrt{1/\bar{m}}$, by a factor of *two*, as shown in Fig. 8-2.

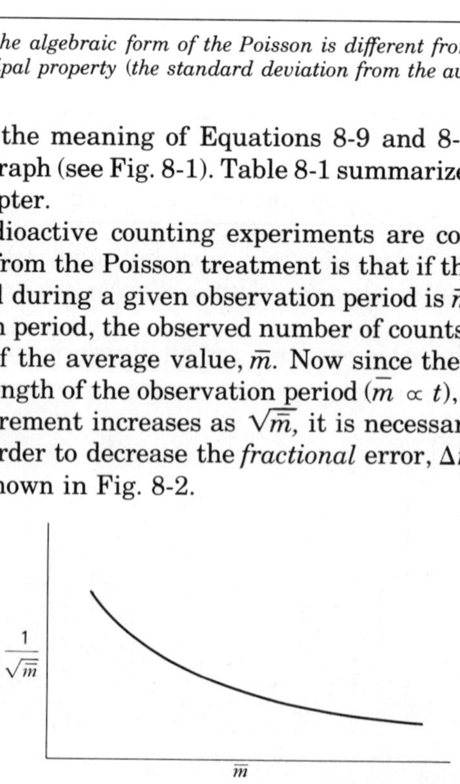

FIGURE 8-2. Counting imprecision as a function of the average number of counts.

EXAMPLE How many radioactive counts must be accumulated in order that the per cent error in the result be ±2%?

Solution $(1/\sqrt{\overline{m}}) = 0.02$
$\overline{m} = 2500$ counts, with imprecision of ±50 counts

Before proceeding to some applications, it is necessary to dispense with two practical difficulties connected with the counting process: background radiation (from cosmic rays, natural radioactivity of the surroundings) and the finite resolving time of the counter itself.

It would seem simple to find the *net* activity (observed counts per unit time)

$$A = \frac{\overline{m}}{t} \tag{8-12}$$

of a radioactive sample from the difference between the total (sample + background) activity, A_{total}, and the activity of the background (determined in the absence of the sample), $A_{\text{background}}$:

$$A = A_{\text{total}} - A_{\text{background}} \tag{8-13}$$

However, the *imprecision,*

$$\sigma = \frac{\sqrt{\overline{m}}}{t} \tag{8-14}$$

in the sample activity is determined by the imprecision in both A_{total} and $A_{\text{background}}$, from the usual rule for the imprecision of a sum:

$$\sigma = \sqrt{\sigma^2_{\text{total}} + \sigma^2_{\text{background}}} \tag{8-15}$$

$$= \sqrt{\frac{\overline{m}_{\text{total}}}{t^2_{\text{total}}} + \frac{\overline{m}_{\text{background}}}{t^2_{\text{background}}}} \tag{8-16}$$

$$= \sqrt{\frac{A_{\text{total}}}{t_{\text{total}}} + \frac{A_{\text{background}}}{t_{\text{background}}}} \tag{8-17}$$

If some of the available time is spent in counting the background ($t_{\text{background}}$) and the remainder is spent in counting the sample (t_{total}), then it may be shown that the time available should be divided according to

$$\frac{t_{\text{total}}}{t_{\text{background}}} = \sqrt{\frac{A_{\text{total}}}{A_{\text{background}}}} \tag{8-18}$$

Thus if the sample were nine times as active as the background, for example, then the sample should be counted for $\sqrt{9} = 3$ times as long.

EXAMPLE A sample (including background) has an activity of 900 counts per minute with a background activity of 225 counts per minute. How much time should be spent in counting the sample and in counting the background to achieve an imprecision of 2% in the net activity of the sample?

Solution The net activity of the sample is 900 − 225 = 675 cpm. An imprecision of 2% in that number corresponds to (0.02)675 = 14 cpm. The time should be divided according to

$$\frac{t_{\text{total}}}{t_{\text{background}}} = \sqrt{\frac{900}{225}} = 2$$

or

$$t_{\text{total}} = 2\, t_{\text{background}}.$$

Substituting into Eq. 8-17

$$14^2 = \frac{900}{t_{\text{total}}} + \frac{225 \cdot 2}{t_{\text{total}}}$$

to give

$$t_{\text{total}} = 6.9 \text{ minutes}$$

and

$$t_{\text{background}} = 3.5 \text{ minutes}$$

For any radioactivity counter, there is a short period (100 μsec for geiger counters, less than 10 μsec for scintillation counters) following registry of a given count, during which the counter cannot record any further counts. If this period of insensitivity, or "dead time," is τ, and the observed number of counts per second is A_{obs}, then the total length of time that the counter is inoperative during one second is $A_{\text{obs}}\,\tau$. During that period, the number of counts that will be missed is $A \cdot A_{\text{obs}}\,\tau$, where A is the true number of counts per second from the sample. Thus

$$A = \text{observed counts} + \text{unobserved counts},$$

or

$$A = A_{\text{obs}} + A \cdot A_{\text{obs}}\,\tau \tag{8-19}$$

and

$$A = \frac{A_{\text{obs}}}{1 - A_{\text{obs}}\,\tau} \tag{8-20}$$

To determine the dead time, τ, of the counter, one need merely measure the count rate for three situations: sample #1 alone, sample #2 alone, and sample #1 and #2 together. The corrected (true) count rates for #1 and #2 alone must then sum to the combined corrected count rate for #1 and #2 together:

$$\frac{A_1}{1 - A_1\tau} + \frac{A_2}{1 - A_2\tau} = \frac{A_{(1\text{ and }2)}}{1 - A_{(1\text{ and }2)}\,\tau} \tag{8-21}$$

Since $\tau \ll 1$, and $\tau^2 \ll \tau$, Eq. 8-21 may be simplified to yield

$$\tau = \frac{A_1 + A_2 - A_{(1\text{ and }2)}}{2A_1A_2} \tag{8-22}$$

EXAMPLE A geiger counter gave a count rate of 1242 counts/sec for sample #1 alone, 1371 counts/sec for sample #2 alone, and 2209 counts/sec for samples #1 and #2 together. Calculate the dead time of the counter and the error introduced when it is not corrected for. Background is 3 counts/sec and is negligible.

Solution

$$\tau = \frac{1242 + 1371 - 2209}{2(1242)(1371)} = 1.19 \times 10^{-4}\text{ sec}$$

The true count rate for source #1 is given by

$$A_1 = \frac{A_{\text{obs}}}{1 - A_{\text{obs}}\,\tau} = \frac{1242}{1 - 1242(1.19 \times 10^{-4})}$$

$A_1 =$ 1460 counts/sec, so that neglect of the correction for dead time leads to an error of about 15% in the counting rate

Isotopic Dilution and Tracer Methods

Modern medical diagnosis often relies in part on the determination of the *amount* of a particular metabolite in a particular body fluid or organ. However, while it is generally possible to isolate the desired substance in *high purity,* it is generally not possible to obtain *high yield.* The technique of isotopic dilution is a clever solution to this problem, and has led to the common use of radioisotopes in medicine (see examples).

In isotopic dilution, radiation *activity* (count *rate*) is used as a measure of the *concentration* of radioactive material per gram of substance. If a sample of some pure compound consisting of n_0 grams of activity, A_0 per gram, is mixed with an additional n_{unk} grams of the same inactive compound, then the count rate for the mixture will be "diluted" to the value,

$$A = A_0 \left[\frac{n_0}{n_0 + n_{\text{unk}}}\right] \tag{8-23}$$

just as the concentration of an ordinary chemical is decreased on dilution with solvent. The amount of the inactive compound may then be found by rearrangement of Eq. 8-23:

$$\boxed{n_{\text{unk}} = n_0 \left[\frac{A_0 - A}{A}\right]} \tag{8-24}$$

EXAMPLE *Calculation of Human Blood Volume*

a. *Labeling the plasma.* In this method, human serum albumin (a large protein, MW 68,000, normally present in human blood) is covalently iodinated with ^{131}I to some of the tyrosine residues at a position ortho to the hydroxyl group, prior to the test. Then 10cc of labeled albumin with activity of about 1 μCi/cc (1 Curie amounts to 3.7×10^{10} counts/sec) is injected into the elbow vein; it takes about ten minutes for the injected albumin to become equilibrated with the total blood volume. A blood sample of 3 cc is then withdrawn from the individual, and the radioactivity measured.

   If the activity of the 3cc blood sample is 3287 counts/min, where the background count for 3cc of blood withdrawn from the same individual *before* the injection of labeled albumin is 175 counts/min, and the activity of the labeled albumin (3 cc) used for injection is 5702 counts/min at a dilution of 1:200 (to make the count rate more similar to that to be measured for the blood), then the plasma volume is simply

$$\text{Plasma volume} = \frac{200\ (5702 - 175) - (3287 - 175)}{(3287 - 175)}\ 10\ \text{cc}$$
$$= 3550\ \text{cc} = 3.5\ \text{liter}$$

   Finally, the relative volume of the blood cells themselves may be determined by centrifuging a blood sample and determining the packed cell volume. For the example given, the blood cell volume might be 2.8 liters, to give a total blood volume (plasma plus cells) of 6.3 liter.

b. *Labeling the red blood cells.* For this experiment, ^{51}Cr is incubated with the patient's own blood *in vitro,* until most of the $CrO_4^{=}$ has become bound to the blood *cells.* Then a reducing agent is added (say, sodium ascorbate) to reduce any unreacted $CrO_4^{=}$ to Cr^{+3}. The Cr^{+3} binds almost exclusively to plasma proteins. Finally, following injection, equilibration, and extraction of blood as before, one can determine the count rate for the labeled blood before injection, the count rate for whole blood after injection, and the count rate for plasma (obtained by centrifugation) after injection, to find the volume of red blood cells.

Determination of blood volume is of most immediate use in deciding whether blood or plasma transfusions are required in cases of bleeding, burns, or surgical shock, and has been responsible for saving a great number of lives of accident victims.

For lean individuals, blood volume varies nearly in direct proportion to body weight, at about 80 cc/kg for males. When a person puts on weight, however, the ratio of blood volume to body weight drops, because fat tissue requires less associated volume of blood vessels. Women, on the average, have a blood volume about 20% less than men (about 65 cc/kg) because of their greater ratio of fat-to-lean tissue.

Although the *total* blood volume may change somewhat when the body malfunctions, a more sensitive diagnostic indicator is the *ratio* of red blood cell volume to total blood volume, the so-called hematocrit. In severe anemia, the hematocrit may fall from its normal value (40% for males, 36% for females) to as low as 15, due to a shortage of red blood cells. In contrast, excess red blood cells (an a hematocrit of up to 70%) result from continued exposure to high altitude or from a tumor of the blood cell-producing organs. Finally, since the kidneys are largely responsible for returning the blood volume to normal following a relatively sudden change (such as drinking a lot of water), measurement of blood volume as a function of time following a deliberate change in blood volume can provide a probe of renal (kidney) function. In all these cases, the radioisotope methods provide quick and accurate results for diagnosis.

EXAMPLE *Thyroid Function*

Although the *physical* properties of (stable) ^{127}I and (radioactive) ^{131}I are different, their *chemical* properties are the same, so that ^{131}I provides a natural tracer for the metabolism of iodine and the iodinated compounds connected with thyroid function. The thyroid gland is unique in its capacity to selectively concentrate *and* retain iodide, for conversion to iodo-tyrosine and then thyronine.

a. *Iodide uptake.* Although the *total amount* of iodide in an individual is roughly constant, and also relatively unaffected by thyroid malfunction, the *rate of uptake* of iodide is a good diagnostic indicator of thyroid activity: the thyroid gland extracts from 0.5 to 6.8% of the circulating iodide pool per hour in a normal case, rising to 2.5 to 6.8% in hyperthyroidism, or 0 to 1.3% in myxedema. Uptake of ^{131}I is readily measured by counting the activity at a position on the patient's neck just above the thyroid, with calibration using the un-diluted ^{131}I sample at the same distance from the counter.

b. *Protein-Bound Iodine.* A difficulty with the iodide-uptake measurement is that while a hyperactive thyroid gland concentrates iodide rapidly, it also releases it more rapidly than normal back to the blood stream as protein-bound iodine (PBI). Radioactive ^{131}I is injected into the patient as before, but this time (after equilibration) a blood sample is withdrawn, the protein-containing fraction is removed from the remainder (by precipitation, say), and both fractions are counted; the ratio of PBI to total circulating ^{131}I then gives a good measure of thyroid activity.

EXAMPLE *Location of Brain Tumors*

Certain isotopes, such as ^{74}As or ^{64}Cu, emit positrons. Immediately after emission a positron decays to two gamma rays that travel in exactly opposite directions. Brain tumors tend to concentrate these ions; by putting counters on opposite sides of the patient's head and recording only those events which register simultaneously on both counters, it is possible to localize the tumor position rather closely.

EXAMPLE *Pernicious Anemia*

In this disease, the level of vitamin B_{12} in the blood falls to a fraction of normal. The biochemical origin appears to be the absence of a critical mucoprotein in the gastrointestinal tract; this protein normally combines with B_{12} to make it more soluble and more readily absorbed into the blood. Since vitamin B_{12} contains cobalt, the simplest test for the condition is to have the patient ingest some ^{58}Co-labeled B_{12} and then assay the level of ^{58}Co in the blood after 8 hours or so.

EXAMPLE *Tests for Blood Circulation*

Obstructions (clot, thrombosis) or constrictions in various blood vessels may readily be detected by injection of $^{24}NaCl$ at a point preceding the suspected problem, and monitoring the radioactivity at various places along the vessel in question. A similar test is used to measure heart output by injecting ^{51}Cr-labeled red blood cells into a vein and monitoring the counts above the aorta.

EXAMPLE *Survival Time of Red Blood Cells*

In many types of anemia, the red blood cells fail to last their normal lifetime of about 120 days. By preparing a sample of the patient's blood with the cells labeled with ^{51}Cr, and then injecting one aliquot back into the patient and another into a normal individual, two (unfavorable) results may appear. If the red blood cell lifetime is too short in both cases, then the fault lies in some abnormality of the cells themselves. But if the survival time is short for the patient and normal for the normal person, then the patient's blood contains some substance that is destroying the red cells.

The above examples represent some of the more common applications of the techniques of radioactive counting and isotopic dilution in medical diagnosis. Tracer methods are also widely used in biochemistry: by labeling a metabolic precursor and determining the extent of enrichment in subsequent metabolic products, one can clarify the chemical mechanisms by which the molecules are transmuted from one form into another. Examples may be found in biochemistry texts.

8.B. ELECTRICAL NOISE

In most biophysical measurements, the data appear on some sort of electrical device, ranging from pH meters to the recorders that monitor body function. Contrary to what one might expect, however, the quality of the response from such devices is sometimes determined by the intrinsic *noise* of the device, rather than the weakness of the *signal* being displayed—it is *signal-to-noise ratio* that is the critical quantity. Since the biological concentrations of most substances are rather low compared to laboratory optimum, it is clear that direct study of interesting substances will generally involve weak responses, so that one should have a feeling for the origin of the noise one will observe in the electrical device and what can be done about it. Fortunately, the nature of some kinds of electrical noise can be understood quite simply from the random walk model.

8.B.1. Shot Noise

Any circuit containing vacuum tubes or transistors contains one or more discontinuities: the cathode-anode separation in a vacuum tube or the junction in a transistor. A measurement of electric current consists of counting the electrons that cross such a junction. However, as with radioactive counting, the probability that any one electron will cross the junction is much less than one, while there are a large number of electrons that could potentially go across. We have already done the statistics for this situation: if the number of electrons crossing such a junction per unit time is (I/e), where I is the current (charge units/sec) and e is the number of charge units per electron, then the total number of electrons that will pass through the circuit in some time, τ, is just

$$\bar{m} = \frac{I\tau}{e} \tag{8-25}$$

and the imprecision in measurement of that number is exactly analogous to the imprecision in the number of radioactive counts from a sample of many nuclei, and is given by

$$\Delta\bar{m} = \sqrt{\frac{I\tau}{e}} = \sqrt{\bar{m}} \tag{8-26}$$

The signal-to-noise ratio, S/N, is thus

$$\boxed{S/N = \frac{\bar{m}}{\sqrt{\bar{m}}} = \sqrt{\frac{I\tau}{e}}} \tag{8-27}$$

Equation 8-27 shows that in order to increase the signal-to-noise ratio by a factor of *two*, it is necessary to increase the signal (input) by a factor of *four* (rather than by a factor of *two*, as one might expect from intuition). To understand the factor τ in Eq. 8-27, one must ask the purpose of the circuit in question. Typically, one desires that a particular circuit respond as quickly as necessary to a particular change: a recorder in a spectrophotometer must be able to register a signal by the time the monochromator has scanned the desired wavelength region; a pH meter must be able to register the pH-reading in the time it takes to deliver titrant, and so forth. The time constant, τ, represents the *shortest time during which the circuit must be able to register a response.* In electrical vernacular, it is more common to refer to the *bandwidth* of a circuit, defined as

$$\text{Bandwidth} = \frac{1}{\tau} \tag{8-28}$$

The bandwidth then represents the fastest (*largest*) *frequency* at which the circuit can respond to a change in signal. Shot noise may thus be reduced by "filtering" out the higher frequencies—that is, reducing the "bandwidth," or (equivalently) increasing the "time constant" of the circuit.

The results of the preceding paragraph, while quantified in Eq. 8-27,

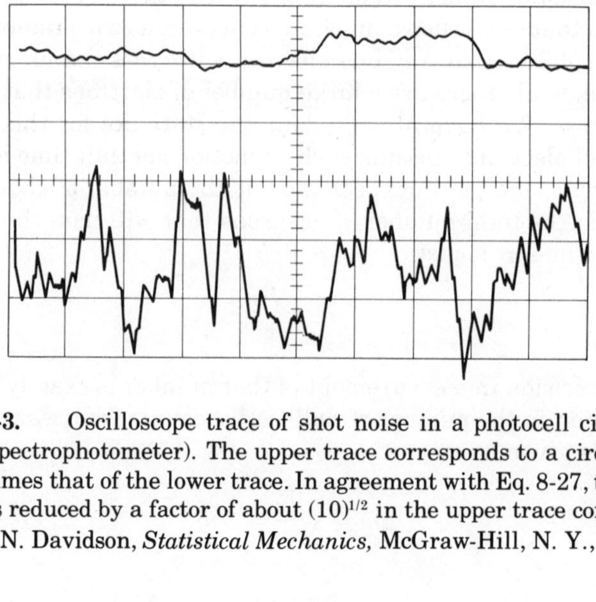

FIGURE 8-3. Oscilloscope trace of shot noise in a photocell circuit (the detector for a spectrophotometer). The upper trace corresponds to a circuit time constant of ten times that of the lower trace. In agreement with Eq. 8-27, the amplitude of the noise is reduced by a factor of about $(10)^{1/2}$ in the upper trace compared to the lower. (From N. Davidson, *Statistical Mechanics,* McGraw-Hill, N. Y., 1962, p. 300.)

are actually just statements of experience. It is clear that one may improve signal-to-noise by increasing signal strength, such as by widening the slit that passes incident light through a spectrophotometer. However, in doing so, one lets through a wider spectrum of colors and loses "resolution," or the capacity of distinguishing between absorption of light of slightly different frequency. This example is typical of experimental situations involving electronic detection—in order to obtain the most *information* (best "resolution" in this case), one is forced to give up *signal-to-noise ratio.* A similar compromise is involved in curing the "bandwidth" problem; Fig. 8-3 shows that it is simple to reduce the *noise* from a recorder pen by making it respond more slowly (then the pen no longer "sees" the rapid noise flickering), but in order for the pen to respond to the *desired signal* (say, an absorption peak as the recorder sweeps through an optical spectrum), it is necessary to slow the recorder sweep rate by a corresponding factor, making the measurement require much more time for completion. This represents the other major sort of compromise with electronic detection—if noise is to be *reduced,* the experiment takes a lot *longer* to do.

8.B.2. Resistor Noise (Johnson Noise, Nyquist Noise, Thermal Noise)

The recognition that this type of electrical noise was directly related to the Brownian motion (unavoidable random motion) of electrons in resistors was reached by Nyquist shortly after discovery of the phenomenon by Johnson in 1928. Electrons in any ordinary conductor are free to roam about according to their thermal energy (see Section 5). Now, because this motion is random, there may be more electrons at one end of a resistor than at the other at a given instant, resulting in a potential difference (voltage) between the two ends of the resistor. At a later instant, there will likely be a different voltage as the electrons redistribute themselves. It is possible to treat this situation by a model in which electrons are allowed to take steps either forward or backward in the resistor, using the bags and marbles solution given at the beginning of this chapter in much the same way as it was applied to the random walk and the "random coil" (Chapter 6.A.). The imprecision in voltage may be expressed as

$$\Delta \mathbf{V} = 2\sqrt{kT \cdot \mathbf{R} \cdot B} \tag{8-29}$$

where k is the Boltzmann constant, T is absolute temperature, $\mathbf{R}$ is the resistance of the resistor, and B is the frequency bandwidth under consideration.

The most significant feature of either shot noise or resistor noise is that it is constant per unit bandwidth. This is called "white" noise, by analogy to "white" light, which has equal intensity at all frequencies. The final type of noise to be discussed does not have this property.

8.B.3. (1/f) Noise (Flicker Noise)

Particularly in solid-state devices, a serious source of noise at low frequencies is so-called "flicker" noise, which varies inversely with frequency (see Fig. 8-4). "Drift" in direct current readings—a slow meandering of the meter or recorder—is especially troublesome. The origin of this sort of noise is obscure, but may have to do with the degree of granularity in the conductor. In any case, flicker noise may be made negligible by conducting all measurements at frequencies of 1000 Hz or higher—this is one reason that many devices that must record a d.c. ("direct current") or slowly varying response have mechanical "choppers" (as in spectrophotometers, to interrupt the light beam many times per second) or electronic rapid switching to convert a d.c. signal to an a.c. ("alternating current") signal before detection.

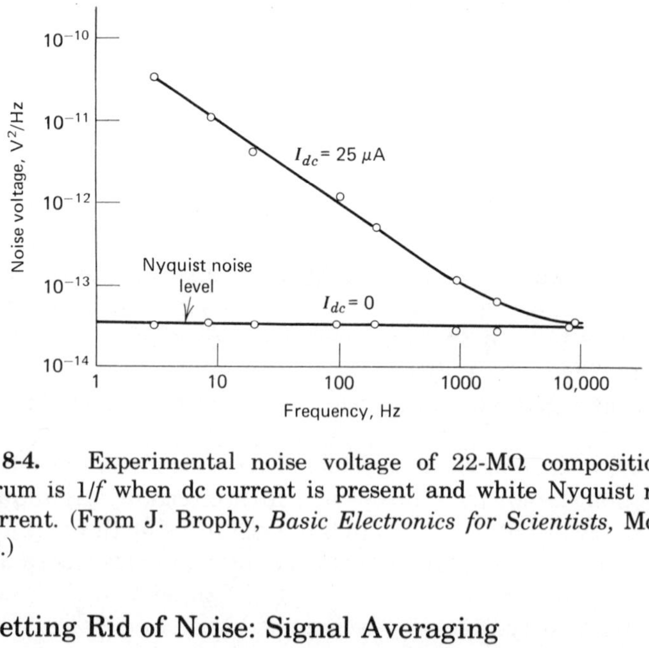

FIGURE 8-4. Experimental noise voltage of 22-MΩ composition resistor Note spectrum is $1/f$ when dc current is present and white Nyquist noise in absence of current. (From J. Brophy, *Basic Electronics for Scientists,* McGraw-Hill, N. Y., 1972.)

8.B.4. Getting Rid of Noise: Signal Averaging

The basic means for elimination of noise are filtering, modulation and phase-sensitive detection, and signal averaging. Of these, the first has been mentioned, and the second is basically the "chopping" idea of the preceding paragraph. Signal averaging is another example of the random walk at work. A simple example is provided by measurement of an absorption spectrum of a molecule.

The typical spectroscopy experiment (Section 5) is to measure the absorption of radiation by matter as a function of frequency, usually by vary-

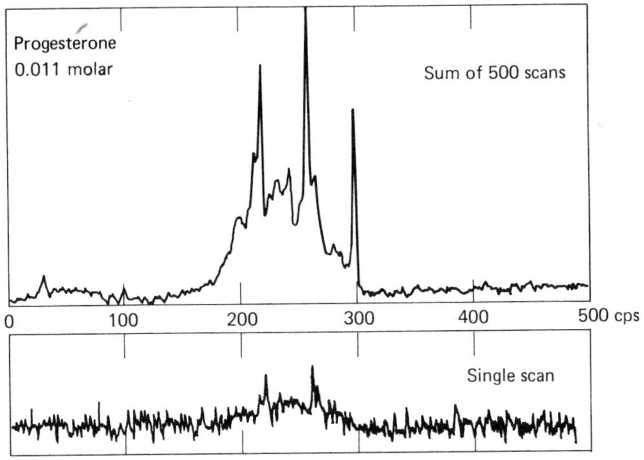

FIGURE 8-5. Proton magnetic resonance spectrum of the female sex hormone, progesterone, showing the increase in signal-to-noise that results when many individual scans are summed together to make a single composite result. (The vertical scale of the upper scan has been compressed to facilitate comparison with the lower.) [R. R. Ernst and W. A. Anderson, *Rev. Sci. Instrum. 37,* 93 (1966).]

ing the frequency of the radiation as a recorder pen moves along its chart paper. In any single such experiment, the noise in the baseline may be objectionable. However, if many scans are performed, an interesting feature appears: if a particular frequency is chosen for scrutiny, the recorder pen position in a given scan be treated as if it could move only up or down by the same amount away from the baseline. Thus (Chapter 6.A), the random walk model predicts that the average distance moved after summing many steps (many scans in this case) will increase as the square root of the number of steps (scans). This result is demonstrated graphically in Fig. 8-5, which shows the comparison between the appearance of a single scan and the sum of many scans.

EXAMPLE In a study of a dilute solution of the cell-breaking enzyme, lysozyme, by proton magnetic resonance spectroscopy, it was found that the result of adding together 30 scans of the same spectrum was to yield a single spectrum with a signal-to-noise ratio of 2:1. Assuming that the noise was "white" (i.e., random in frequency), calculate the number of scans that would have to be summed to yield a spectrum with a signal-to-noise ratio of 7:1.

Solution. The signal increases proportional to the number of scans, while the noise increases proportional to the square root of that number. Thus

$$(S/N) = (\text{constant}) \cdot \sqrt{n}$$

where n is the number of scans, so that

$$2 = (\text{constant}) \cdot \sqrt{30}$$

Now one could stop to evaluate this constant but it's quicker just to write

$$7 = (\text{constant}) \cdot \sqrt{n}$$

and take the ratio of the two equations

$$(7/2) = \sqrt{(n/30)}$$
$$n = (7/2)^2 \cdot 30 = 368 \textit{ scans}$$

In this particular example, it took 500 sec for one scan, so the S/N of 7 in this problem would require about 51 hours, or more than two days of continuous scanning! Fourier transform methods (Chapter 6) have reduced this time by about a factor of 1000, so that it is now feasible to obtain such spectra of biologically interesting molecules.

8.C. DID THE TREATMENT HELP THE PATIENTS?

The fundamental problem in medical *diagnosis* is deciding whether the patient is normal or not, and in medical *treatment* is in deciding whether or not an improvement resulted from the treatment. Although no single indicator is unequivocal, it is useful to have a basis for quantitative evaluation of the diagnostic test. In such cases, as in comparing a patient's blood count to the average for normal patients, or in comparing the fatality rate of a new operation to that of an existing procedure, Poisson probability paper (Fig. 8-6) is a useful yardstick, particularly when the number of meaningful data points is small as in many medical examples.

Figure 8-6 gives the probability that an event will occur *at least C*

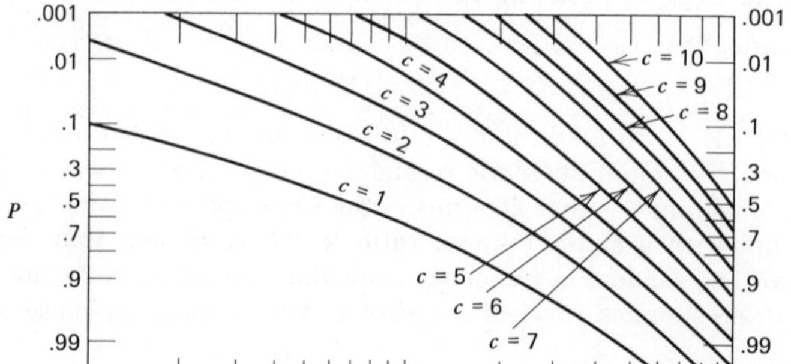

FIGURE 8-6. Poisson probability graph: probability, P, that an event will occur at least c times when the usual (expected) number of occurrences is $\bar{m}$. (From M. J. Moroney, *Facts from Figures,* Penguin Books, Baltimore, Md., 1956, p. 105.)

times, when the expected number of occurrences is $\bar{m}$. For example, when the expected number of occurrences is 3, it is only 20% likely that a particular observation will show 5 or more occurrences.

EXAMPLE *Success of a New Treatment*

Experience with a certain disease shows that it has an untreated mortality rate of 10%. When a new treatment was tested on 40 patients, 7 deaths resulted. The Poisson probability graph shows that when the expected number of events (in this case, fatalities) is $(0.10)(40) = 4$, the probability that 7 or more deaths would have occurred without treatment is only about 10%. Thus, it is 90% probable that the new treatment was worse than the original treatment.

The applicability of the Poisson distribution is extremely general. The first successful fit of experimental data to theory was for data collected by Bortkewitch (see Problems), consisting of the chance that a cavalryman was killed by a horse-kick in the course of a year—the data resemble the plot in Fig. 8-1b. Other examples include: number of calls from public phones, number of wrong telephone numbers dialed, number of blood cells, bacteria, or viruses in a given counting experiment, number of cancer deaths per household, number of vacancies on the Supreme Court, number of labor strikes per week, number of bullets hitting a target, and so on (see Problems).

8.D. CHROMATOGRAPHY

The Poisson distribution arises in a natural way in the theory for certain kinds of chromatography (liquid-liquid partition and the more recent and useful gas-liquid, gel, ion-exchange, and affinity chromatography). As usual, the derivation is based on an intuitive, artificial model: it is supposed that the chromatograph column can be taken to consist of a stack of hypothetical disks, or plates, each containing a volume, V, of solvent (see Figure 8-7).

From Fig. 8-7 and its legend, it is clear that after a total of N infinitesimal increments of solvent have been added, with equilibration between each successive addition, the resultant concentration in a given plate (say, the mth plate counting down from the top) will follow a binomial distribution:

$$[C(m)] = [C]_0 \frac{N!}{m!(N-m!)} \left(\frac{dV}{V}\right)^m \left(1 - \frac{dV}{V}\right)^{(N-m)} \tag{8-30}$$

where $[C(m)]$ is the concentration of solute in the mth plate. Inspection of Eq. 8-30 quickly shows that since $(dV/V) << 1$, and $N >> 1$ (in a typical

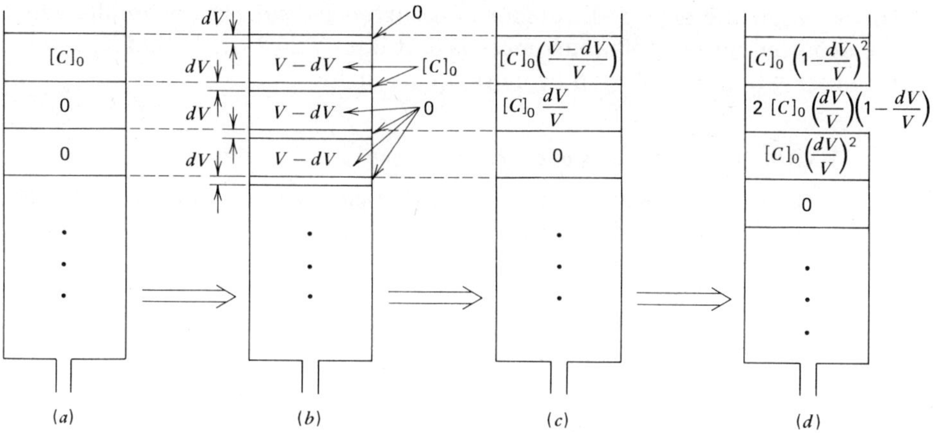

FIGURE 8-7. Schematic chromatography with intermittent flow of solvent. At the beginning of the experiment (*a*), the column is treated as a stack of many separate hypothetical "plates," with the sample occupying the topmost plate. Then (*b*) a small volume of solvent, dV is added, with a corresponding dV of solvent allowed to issue from the bottom of the column. The first plate then contains volume dV of solute concentration zero and volume $(V - dV)$ of concentration, $[C]_0$. On equilibration (*c*), the first plate assumes a homogeneous composition of solute concentration, $[C]_0[(V - dV)/V] = [C]_0[1 - (dV/V)]$, while the second plate contains concentration $[C]_0(dV/V)$ of solute. When the same process is repeated, and equilibration again allowed to proceed, the distribution of solute among the various hypothetical plates of the column is shown in (*d*). The binomial nature of the distribution can now be demonstrated by induction.

gas chromatograph, there may be several hundred thousand plates), it is possible to apply the Poisson limit to Eq. 8-30 by the same reasoning used to obtain Eq. 8-9, with the identifications,

$$a = (dV/V) \ll 1 \tag{8-31}$$

$$Na = \overline{m} = N(dV/V) = \frac{v}{V} \tag{8-32}$$

where v is the volume of solvent that has passed through the column, to give

$$\text{Fraction of initial material residing in the } m\text{th hypothetical plate} = P_N(m) = \frac{[C(m)]}{[C]_0} = \frac{(v/V)^m e^{-(v/V)}}{m!} \tag{8-33}$$

In Eq. 8-33, $\overline{m} = v/V$ represents the number of plate volumes of solvent that has passed through the column. A graphical illustration of Eq. 8-33 is shown in Fig. 8-8, in which each solid curve is a plot of solute concentration

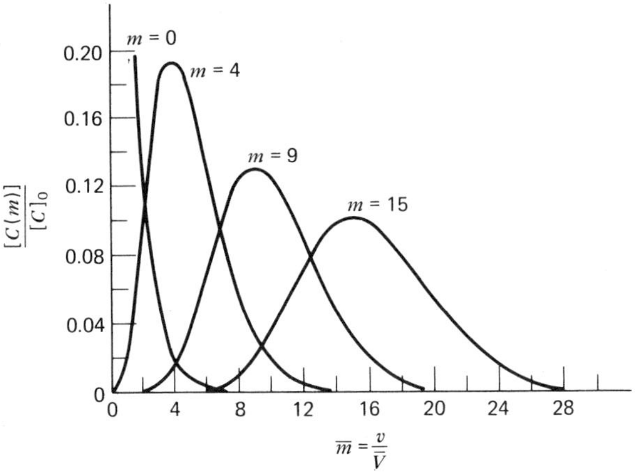

FIGURE 8-8. Relative solute concentration, $[C(m)]/[C]_0$, in the mth hypothetical "plate" of the chromatograph column of Fig. 8-7, as a function of the number of plate volumes of solvent, v/V, that has passed through the column. If m is the last plate of the column, the corresponding curve will give the concentration of solute in successive volume fractions issuing from the column, as in a typical experiment (see Fig. 8-9). (From E. W. Berg, *Physical and Chemical Methods of Separation*, McGraw-Hill, N. Y., 1963, p. 110.)

in the m*th plate,* as a function of the number of plate volumes of solvent that has passed through the column. It is to be noted that the concentration profile *on the column* (i.e., concentration as a function of plate number for a fixed amount of eluent passed through the column) begins with a skewed shape (Fig. 8-1b) and gradually becomes more symmetric as the solute spreads out while the concentration "peak" moves down the column. However, in the usual column chromatography experiment, solute concentration is monitored as the solution *leaves* the column; thus, while the solute profile *on the column* is at first a narrow "peak" as the first solute fractions issue from the last plate of the column, the concentration profile *on the column* will have broadened considerably by the time the later fractions from the same "peak" have left the column. In other words, the expected solute concentrations for fractions collected as they *leave* the column will be skewed, showing a "tail" as the final fractions leave the column, as seen in Fig. 8-8. A practical example is shown in Fig. 8-9, which is a gas-liquid chromatogram of various amino acids, each of which has been derivatized to make it easier to vaporize. The gas-liquid-chromatography ("glc")column itself may consist simply of a glass capillary tube (say, 0.02 cm inner diameter and a few hundred feet long), coated with solvent, while the (vaporized) sample is forced through the column by an inert "carrier" gas such as nitrogen. Alternatively, the column may be packed beforehand with some nonvolatile substance on which the liquid coating can be introduced as described above. *Separation* of different solutes is based on differences in

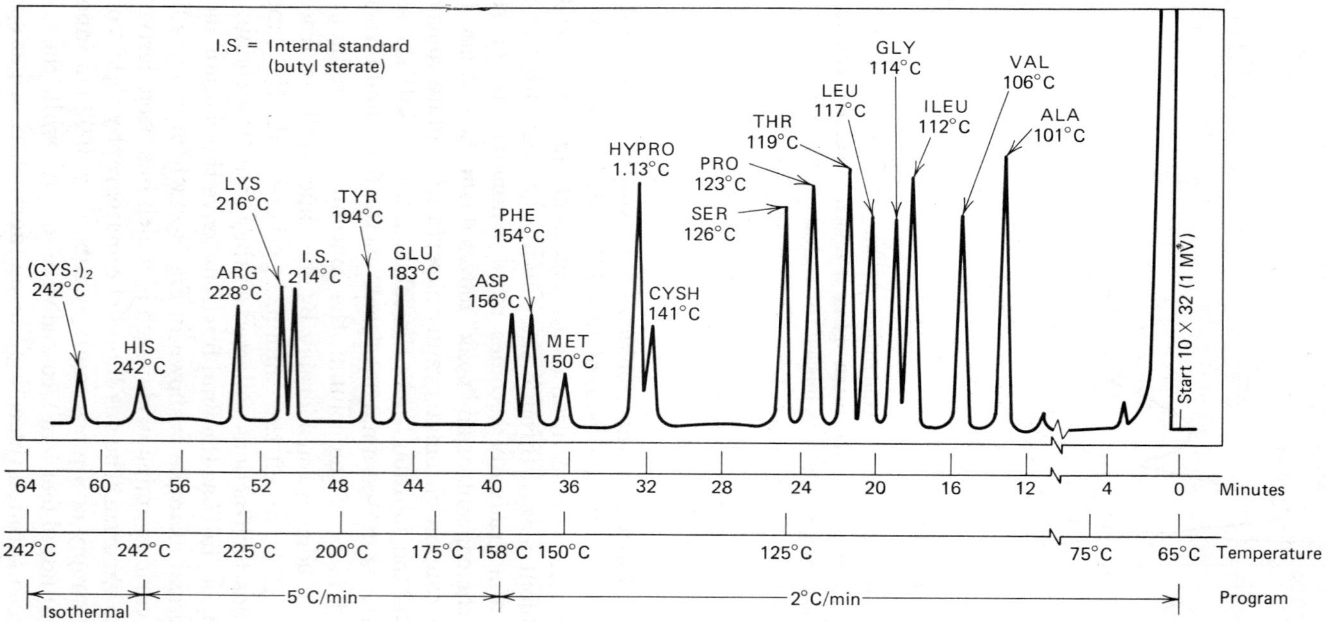

FIGURE 8-9. Separation of amino acid derivatives with neopentyl glycol sebacate. Each peak represents 5 μg amino acids. Column: 1.5 m $\times$ 3.7 mm I.D. of 0.5 w/w % neopentyl glycol sebacate coated on 80/100 a.w. Chromosorb G, Microtek Model 220 gas chromatograph with flame-ionization detectors; 1-mv potentiometric recorder. Carrier gas flow: 64 ml/min N_2. (From C. W. Gehrke and D. L. Stalling, in *Separation Techniques in Chemistry and Biochemistry*, ed. R. A. Keller, Marcel Dekker, Inc., N. Y., 1967, p. 44.)

their affinity for the liquid on the column, and the elution of more strongly-bound solutes is facilitated by increasing the temperature of the column during the elution (see Fig. 8-9).

Gas-liquid chromatography has become one of the most universal tools for separation of mixtures of small (less than 1000 molecular weight or so) molecules, primarily because the column behaves as if it contained a huge number of hypothetical plates, allowing for many successive equilibrations as the "solute" passes down the column—fractional distillation is superior to single-stage distillation for a similar reason. The great disadvantage of the technique is that it is necessary to volatilize the substance of interest; since many interesting chemicals are not sufficiently volatile, it is necessary to derivatize them with a functional group that increases their vapor pressure, and even this procedure becomes infeasible for biological macromolecules, whose chromatographic separation is now discussed.

Gel Chromatography

Although the preceding treatment for the *width* of the solute distribution on a chromatograph column was derived from an artificial model involving intermittent flow of solvent, the result is quite general and can be derived without recourse to the special scheme shown in Fig. 8-7. In this section, another simple model is offered to account for the *position* of the maximum of the solute distribution (i.e., the mechanism by which chromatography is able to separate mixtures of molecules based on molecular size). In gas-liquid chromatography, the basis for separation of mixtures of solutes is simply that a given gaseous species has a different affinity for the liquid stationary phase than do other, chemically different species.

For gel chromatography, on the other hand, solutes are separated largely according to their *size* rather than their chemical or solubility properties, and this method generates a large family of popular methods for isolation, purification, and characterization of the macromolecules of life. Figure 8-10 shows the appearance of a commonly used gel, Sephadex, a dextran (poly-glucose) cross-linked with epichlorhydrin. The principal structural feature of a chromatograph column composed of tightly packed beads of such a gel is that solute molecules that are sufficiently *small* can enter the pores or interstices of the gel beads, and thus have access to a much larger percentage of the total volume of the column than *large* molecules that are too big to enter the pores. In other words, if a mixture of large (compared to gel pore diameter) and small molecules are charged onto the top of a gel chromatograph column, and solvent is forced down the column, it is the *largest* molecules that progress the *farthest* down the column, since the effective volume of the column is smaller for the large molecules that are excluded from the interior of the gel beads.

Consider a gel bead whose interior is assumed to consist of a collection

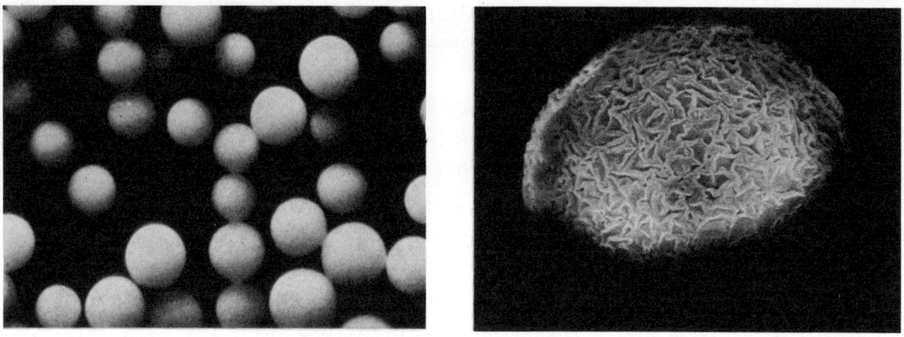

FIGURE 8-10. Low-magnification (left) and high-magnification (right) pictures of the bead polymerized dextran, Sephadex. Note the porous surface of the gel beads, providing for entry of sufficiently small solutes. (From *Pharmacia*, "Sephadex, Gel Filtration in Theory and Practice," 1970.)

of tiny cone-shaped cavities into which a molecule of effective radius, r, is able to penetrate to a depth, h, as shown in Fig. 8-11, in which the conical cavity itself has a base radius of R and height of H. The volume of the conical cavity is

$$\text{Volume of cavity} = \pi H R^2/3 \tag{8-34}$$

Study of Fig. 8-11 shows that the volume accessible to the center of the molecule is given by

$$\text{Volume available to molecule} \cong \pi h(R - r)^2/3 \tag{8-35}$$

for a long, narrow cavity ($H >> R$).

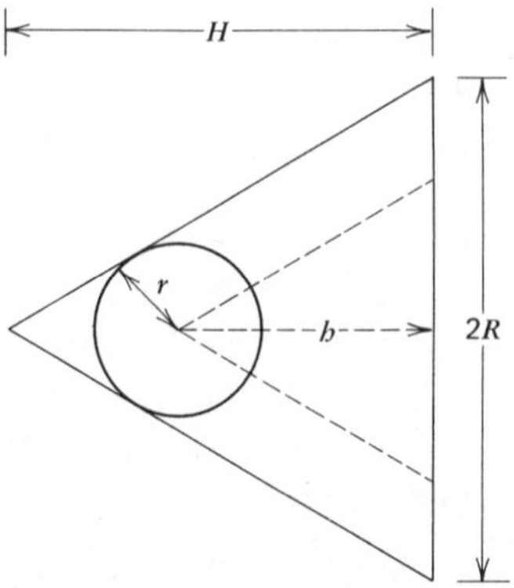

FIGURE 8-11. Schematic diagram of a conical cavity in the interior of a gel bead. A molecule of effective radius, r, penetrates to a depth, h, of a cavity whose own height is H and whose radius at its base is R. ($H \gg R$ in the present example.)

The fraction of the total gel bead interior volume accessible to the molecule is then

$$\text{Fraction of bead volume available} = \frac{h(R-r)^2}{HR^2} = \frac{h}{H}(1-(r/R))^2 \quad (8\text{-}36)$$

But since

$$(h/H) \cong (R-r)/R \quad (8\text{-}37)$$

$$\begin{aligned}\text{Fraction of bead volume available} &\cong (1-(r/R))^3,\ r < R \\ &\cong 0,\ r > R\end{aligned} \quad (8\text{-}38)$$

Equation 8-38 predicts that solute molecules whose effective diameter is *larger* than the pore diameter should all progress down the column in the volume between (i.e., outside) the beads, at a rate essentially *independent* of molecular size. Similarly, molecules that are very much *smaller* than the pore diameter ($r << R$ in Eq. 8-38 are expected to proceed at a slow rate (because these molecules have access to essentially all the volume both inside and outside the beads) that is again *independent* of molecular size. Finally, for molecules whose effective diameter is of the order of, but smaller than the pore diameter, the rate of migration down the column will vary (nonlinearly) with molecular weight (i.e., size), as seen in Fig. 8-12.

When confronted with a nonlinear, but regular curve for experimental data, it is a common practice to re-plot the data on semi-log paper to straighten the curve more toward a straight line for facilitating interpolation between two known points to estimate an unknown value. Such a data reduction is shown in Fig. 8-13, for the rates of migration of a variety of proteins whose shapes are approximately the same (spherical). Figure 8-13

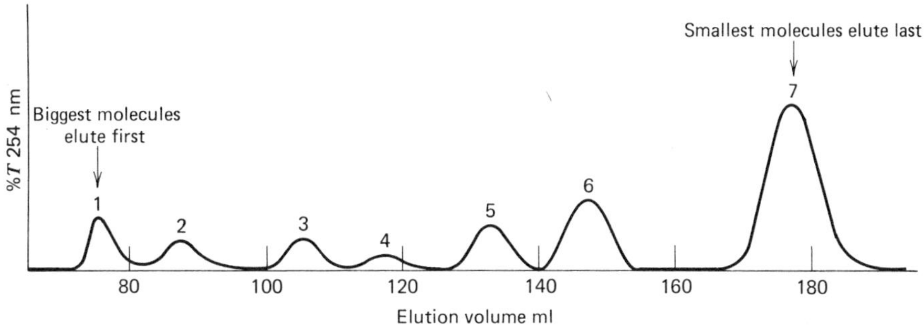

FIGURE 8-12. Separation of a mixture of proteins on Sephadex G-200 Superfine. Pharmacia column K 16/100, bed height 90.6 cm. Sample volume 1 ml, flow rate 1.5 cm/h. Sample contained thyroglobulin (1), IgG (2), serum albumin (3), ovalbumin (4), chymotrypsinogen (5), cytochrome *c* (6) and vitamin B_{12} (7). (Work from Pharmacia Fine Chemicals.)

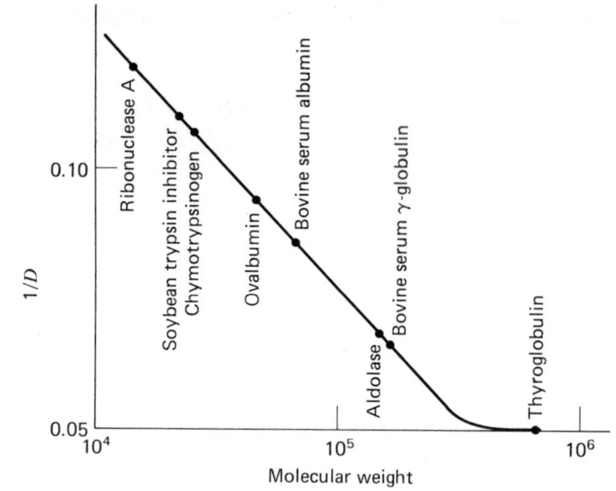

FIGURE 8-13. Relationship between the reciprocal of the migration distance (1/*D*) and the molecular weight of proteins separated by thin-layer gel filtration on Sephadex G-150 Superfine. Thin-layer plate: 20 × 30 cm. Layer thickness: 0.5 mm. Time: 7 hrs 40 min. Angle: 15°. The proteins were transferred from the gel layer to a Whatman 3 MM paper which was stained in 0.1% bromophenol blue methanol/glacial acetic acid (9:1, v/v). Migration distances were measured on the paper replica after rinsing with 5% acetic acid. (From Pharmacia Fine Chemicals.)

provides one of the most widely used methods for estimating the molecular weight of an unknown protein; by monitoring the elution volume for each of the components of a mixture of the unknown with several known proteins, the molecular weight of the unknown protein is readily obtained by interpolation from Fig. 8-13. The method may be extended to nonglobular (nonspherical) proteins simply by reducing all proteins in the mixture to a common (random-coil) shape by carrying out the experiment in the presence of a denaturant such as guanidinium hydrochloride. The method may be extended to macromolecules of much larger molecular weight, such as nucleic acids and viruses, by use of a gel with no cross-links at all (and thus having much larger pores), such as agarose, a linear polysaccharide consisting of alternating units of D-galactose and 3,6 anhydro-L-galactose; an example is shown in Fig. 8-14.

Apart from molecular weight determination and the obvious corollary of preparative separation of mixtures of macromolecules according to molecular size, a major application for gel chromatography is the de-salting or removal of other low molecular weight compounds from solutions of macromolecules. As discussed in Chapter 3.B.1, one of the simplest means for isolating a given protein from a mixture of proteins is selective precipitation with a salt such as ammonium sulfate; when the desired protein is then re-dissolved, it is necessary to remove the salt to proceed with characterization of the protein. Because of the great difference in molecular

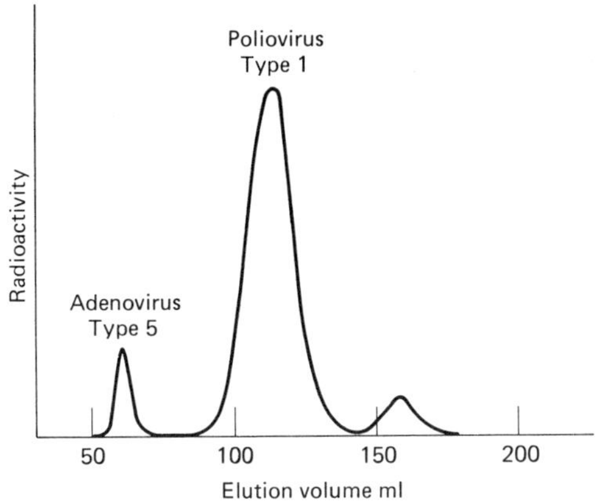

FIGURE 8-14. Separation of ^{32}P-labelled adenovirus and poliovirus on Sepharose 2B. Bed dimensions: 2.1 × 56 cm. Eluant: 0.002 M sodium phosphate buffer, pH 7.2 and 0.15 M sodium chloride. Flow rate: 2 ml/cm^2 h. Sample volume: 1 ml. (From Pharmacia Fine Chemicals.)

weight between the salt and the protein, gel chromatography is an easy procedure for salt removal.

The arguments based on a gel bead interior composed of conical cavities that led to Eq. 8-38 are not the only ones to account for an approximately linear dependence of elution volume on log(molecular weight). In an alternative treatment, the gel interior may be represented by uniform cylindrical channels; on entering such a channel, a molecule may then be imagined to encounter increased frictional resistance, manifested as a smaller apparent diffusion constant and thus slower progress through the bead than through the bulk solution outside the bead. Finally, a proper theory for gel chromatography ought to incorporate the effect of differential adsorption of various solutes to the gel surface itself—it has recently been shown that glycoproteins (proteins with covalently bound sugar in their structures) exhibit apparent molecular weights by gel chromatography that are too small, because the sugar moiety of the protein has affinity for the polysaccharide "backbone" of the gel itself and is thus retarded more than an ordinary protein of the same molecular weight.

In closing this section, it may be noted that it is not necessary to use a gel made from a natural polymer; polystyrene, chlorobutyl rubber, and acrylamide synthetic polymers have been used, and even porous glass beads will accomplish the same differential migration of macromolecules of different molecular weight, as shown in Fig. 8-15. In that figure, linear plots of log(molecular weight) versus relative elution volume are linear over a molecular weight range of 17,000 to 385,000, using glass with a pore size of about 500 A. Sodium dodecyl sulfate was used to unfold and coat the various proteins so that they would assume a common (rod) shape.

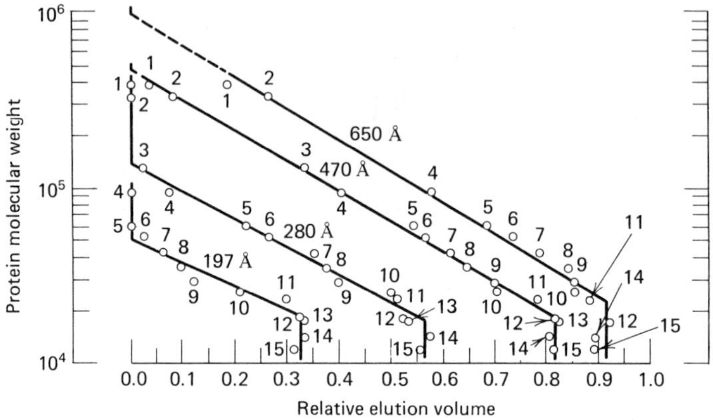

FIGURE 8-15. Protein subunit molecular weight versus relative elution volume of SDS-protein complexes chromatographed on controlled pore glass of narrow pore size distribution. Proteins are: (1) hemocyanin, (2) thyroglobulin, (3) β-galactosidase, (4) phosphorylase, (5) catalase, (6) glutamate dehydrogenase, (7) ovalbumin, (8) pepsin, (9) carbonic anhydrase, (10) α-chymotrypsinogen, (11) trypsin, (12) β-lactoglobulin, (13) myoglobin, (14) lysozyme, (15) cytochrome *c*. [From R. C. Collins and W. Haller, *Anal. Biochem. 54,* 47 (1973).]

Ion-Exchange Chromatography

In gel chromatography, macromolecules may be separated according to their *size;* by introducing charged groups on the gel, macromolecules may be further separated according to their *charge,* in what is called ion-exchange chromatography. The most common ion-exchange resins (gels) are prepared by addition of —SO_3H or —COOH (to form a negatively charged resin) or —NR_2, —NHR (to form a positively charged resin) to a gel composed of either dextran (Sephadex) or a copolymer of styrene and divinylbenzene (Dowex, Amberlite).

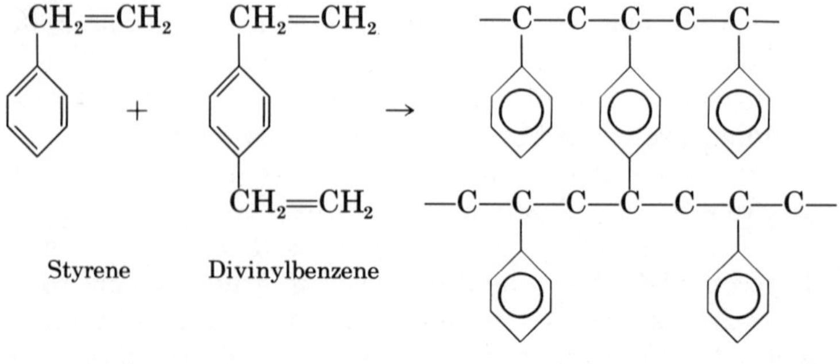

As discussed at greater length in Chapter 3.C, the net charge of a protein depends on pH, as shown in Fig. 8-16. Figure 8-16 suggests that it should be possible to prepare a protein at a pH at which the protein is, say,

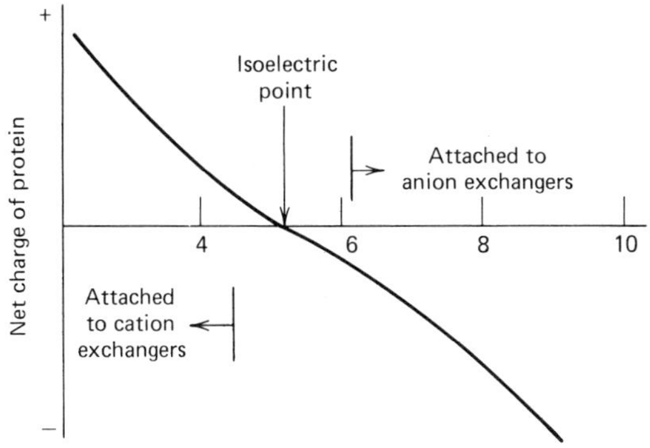

FIGURE 8-16. The net charge of a protein as a function of pH. The pH ranges in which the protein is bound to anion or cation exchangers are shown.

negatively charged, then pour the solution through an anion exchange resin to bind the protein to the gel, then elute the protein off the column by (in this case) lowering the pH until the protein is no longer sufficiently negatively charged to remain attached to the anion exchange gel. By eluting with a solution whose pH changes as more eluent is added (pH gradient), various macromolecules in a mixture may be separated if their isoelectric points (Fig. 8-16) are different, as shown schematically in Fig. 8-17 and specifically in Fig. 8-18.

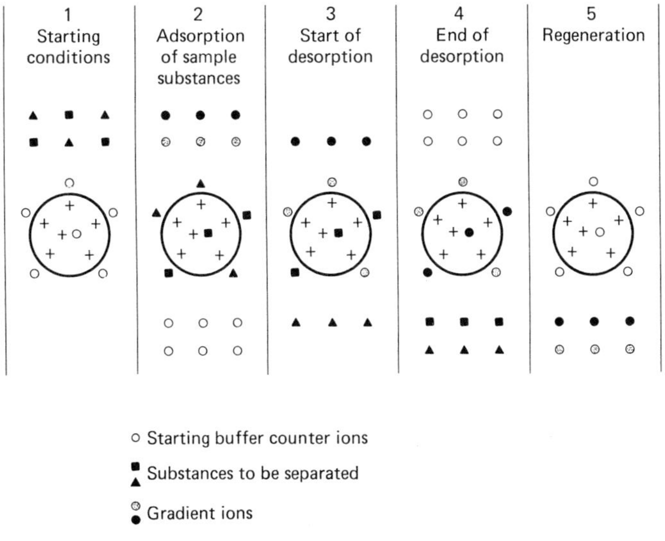

Figure 8-17. Schematic diagram for separation of a mixture of two charged proteins by use of an ion-exchange chromatograph column and an eluent with a gradient of either pH or salt concentration for selective removal of the two macromolecules from the ion-exchange column. (From Sephadex Ion Exchangers, a guide to ion exchange chromatography, from Pharmacia Fine Chemicals.)

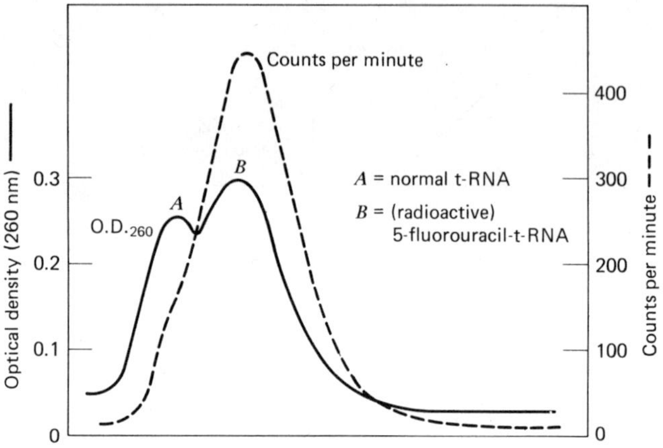

FIGURE 8-18. Ion-exchange chromatography using salt gradient elution to separate normal transfer-RNA from transfer-RNA containing radioactive 5-fluorouracil in place of normal uracil. $O.D._{260}$ gives a measure of the amount of t-RNA present, while radioactivity serves to identify the 5-fluorouracil-t-RNA. The two RNA's were detached from the *anion*-exchange column (DEAE-cellulose) using eluent containing increasing concentration of NaCl (0.3M to 0.5M). Since 5FU has a lower pK_a (8.1) than normal uracil (9.5), the 5FU-RNA carries the larger negative charge, is bound more strongly to the column, and requires the higher salt concen tration for elution. (5FU is an anti-tumor agent, that is incorporated into the t-RNA of bacteria that have been fed the drug.) (Data from J. L. Smith and A. G. Marshall.)

EXAMPLE *Amino Acid Analysis*

In this case, the column of choice is a *cation* exchanger made from sulfonated polystyrene. The pH of the amino acid mixture is lowered to about pH 3 to ensure that all the amino acids are positively charged; the mixture is then put onto the column, where the amino acids, A^+, bind according to

$$A^+ + RSO_3^-Na^+ \rightleftharpoons RSO_3^-A^+ + Na^+$$

Since the most basic amino acids are, on the average, more positively charged, they will bind most strongly to the column. As the pH (and salt concentration) of the column are then gradually increased (see Fig. 8-19), the acidic amino acids will be neutralized first (and thus detached from the column), then the neutral, and finally the basic amino acids. In contrast to the gas-liquid chromatograph method, there is no need to derivatize the amino acids to achieve separation by ion-exchange chromatography; on the other hand, much more material is required for the ion-exchange separation.

EXAMPLE *Analysis for Trace Constituents*

Because of the high affinity of ions for an ion exchange resin, it is possible to *concentrate* the desired ions by passing a dilute solution through an ion ex-

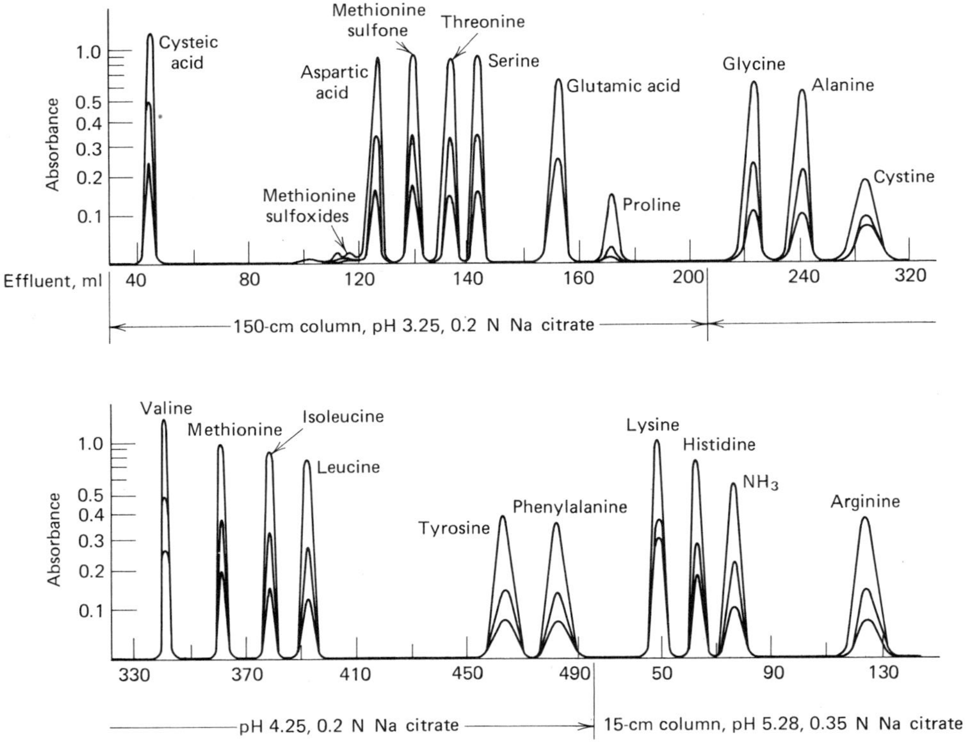

FIGURE 8-19. Automatically recorded chromatographic analysis of a synthetic mixture of amino acids on a sulfonated polystyrene resin. [Data from D. H. Spackman, W. H. Stein, and S. Moore, *Anal. Chem., 30,* 1190 (1958).]

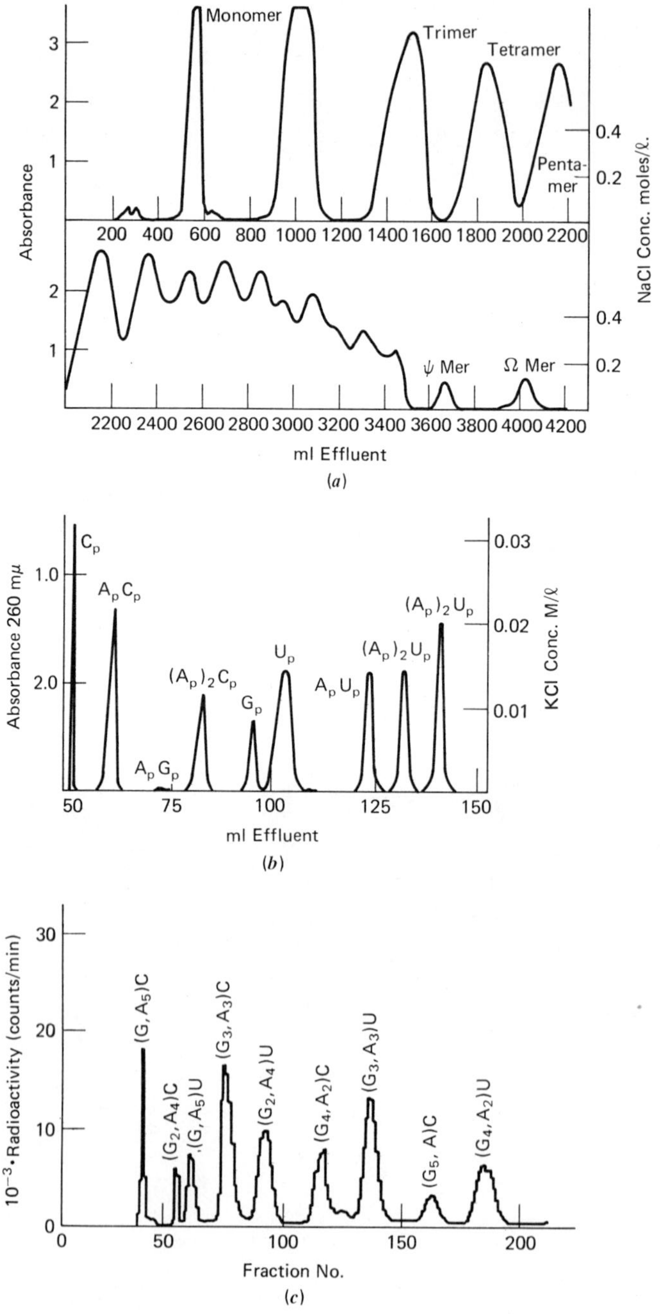

FIGURE 8-20. Examples of separations of oligonucleotides by anion-exchange chromatography on DEAE-Sephadex. [Plots *a* and *b* are from D. A. Lloyd and S. Mandeles, *Biochemistry 9*, 932 (1970); plot *c* is from H. R. Matthews, *Eur. J. Biochem. 7*, 96 (1968).]

change resin (to hold the ions) and then eluting with a small volume of concentrated acid (or base, or salt) to dislodge the ions of interest:

$$\begin{aligned} \text{Adsorption of ions:} \quad & RSO_3H + M^+ \rightleftharpoons RSO_3M + H^+ \\ \text{Desorption of ions:} \quad & RSO_3M + Na^+ \rightleftharpoons RSO_3Na + M^+ \\ \text{Regeneration of resin:} \quad & RSO_3Na + H^+ \rightleftharpoons RSO_3H + Na^+ \end{aligned}$$

This procedure is particularly useful for estimation of fluoride ion in drinking water, of metal ions such as Pb, Fe, and Cu in wine and other goods, and for detection of (especially radioactive) metals in urine. Ions such as Cu^{++}, Ni^{++}, Mg^{++}, Ca^{++}, Na^+, and K^+ are essentially 100% retained by the column, and may thus be concentrated from extremely dilute solutions up to a level at which a variety of conventional assays are possible.

EXAMPLE *Timed-release Drugs*

Ion-exchange resins may be used to form complexes with a variety of drugs, to yield a tablet with adequate mechanical strength, but with slow dissociation of drug from resin after the complex reaches the intestinal tract. An anion-exchange resin has been used to provide an oral preparation of penicillin; cation-exchange resins have been employed similarly to provide orally administered forms for morphine as an analgetic (pain-killer) or histamine as an anti-tussive (anti-cough) agent. In a related application, an anion-exchange resin has been used to remove radioactively labeled tri-iodothyronine from human serum for tests of thyroid function (see also Chapter 8.A).

EXAMPLE *Separation of Oligonucleotides*

Figure 8-20 shows that oligonucleotides may be separated on an anion-exchange resin (in this case, diethylaminoethyl dextran), according to differences in charge resulting from differences in chain length, size, and base composition.

EXAMPLE *Desalinization of Water*

As the student is probably well aware, it is possible to de-salt water by passing the water through successive (or combined) anion- and cation-exchange resins, as in a number of commercial resins used for preparing water for steam irons from tap water. However, this method is relatively expensive for large-scale desalinization because of the need to interrupt the process for periodic regeneration of the resin. It is possible to prepare ion-exchange resins in the physical form of a flexible sheet, on the other hand, which may be used to remove mineral ions from the center compartment to the cathode or anode compartments of the apparatus shown in Fig. 8-21. Diffusion of positive ions from the center compartment to the right-hand compartment (for example) is thus prevented by the presence of the (positively-charged) anion-exchange membrane. The whole procedure therefore succeeds because the ion-exchange membrane is essentially impermeable to ions of the same charge as the membrane.

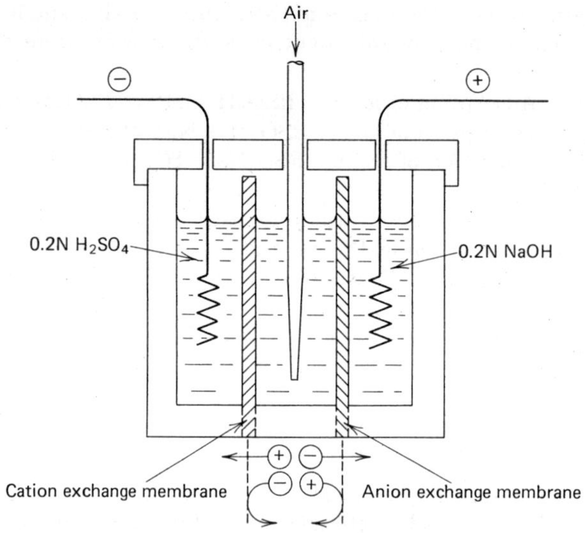

FIGURE 8-21. Desalinization apparatus using electrodialysis with ion-exchange resin membranes. Salt water is introduced into the center compartment. As the electrolysis proceeds, cations move from center to left and anions move from center to right compartments; any diffusion of cations into the right, or anions into the left compartment is prevented by the one-way ion entry property of the ion-exchange membrane. When sufficient salt has been removed, de-ionized water is taken from the center compartment and the process is repeated. (From "Ion Exchange Resins," BDH Chemicals Ltd., Poole, England.)

It may be noted in passing that the de-salting process in some survival kits is rather different. In that case, cations and anions are removed simultaneously according to a reaction of the type,

$$RSO_3Ag + Na^+ + Cl^- \rightarrow RSO_3Na + AgCl \downarrow$$

Affinity Chromatography

Affinity chromatography is a recent development that offers extraordinarily selective isolation and purification of a desired macromolecule from the usual mixture of macromolecules that results from any early stage of isolation. The method simply takes advantage of the inherent functional specificity of binding by a biological macromolecule: a specific enzyme inhibitor, for example, is covalently attached to a gel column whose pores are sufficiently large to admit the enzyme of interest; when enzyme is added to the column, the enzyme will then bind to the column-bound inhibitor while other unrelated proteins will pass through the column unaffected; to remove the column-bound enzyme afterwards, it suffices to elute with a potent *soluble* enzyme inhibitor, which detaches the desired enzyme from the column.

As a result of this procedure, a particular biomacromolecule may be selectively removed and concentrated from a crude preparation by use of a *single* chromatograph column. In practice, it has been found that attaching the affinity ligand (the small molecule that shows specific binding to the macromolecule of interest) directly to the column gives some binding, but that much stronger binding is possible when a "spacer" chain is introduced between the "backbone" of the gel and the affinity ligand itself. Because of the large pore sizes required, the gel employed is generally one of the natural linear agarose polymers. Details of the chemical linkages involved may be found in the Cuatrecasas reference at the end of this chapter and in many of the newer "bio-organic" chemistry texts. An example of the spectacular one-step purification possible with this technique is shown in Fig. 8-22. Using a column consisting of agarose to which the illustrated ligand was covalently attached, it is seen that most of the desired acetylcholin-

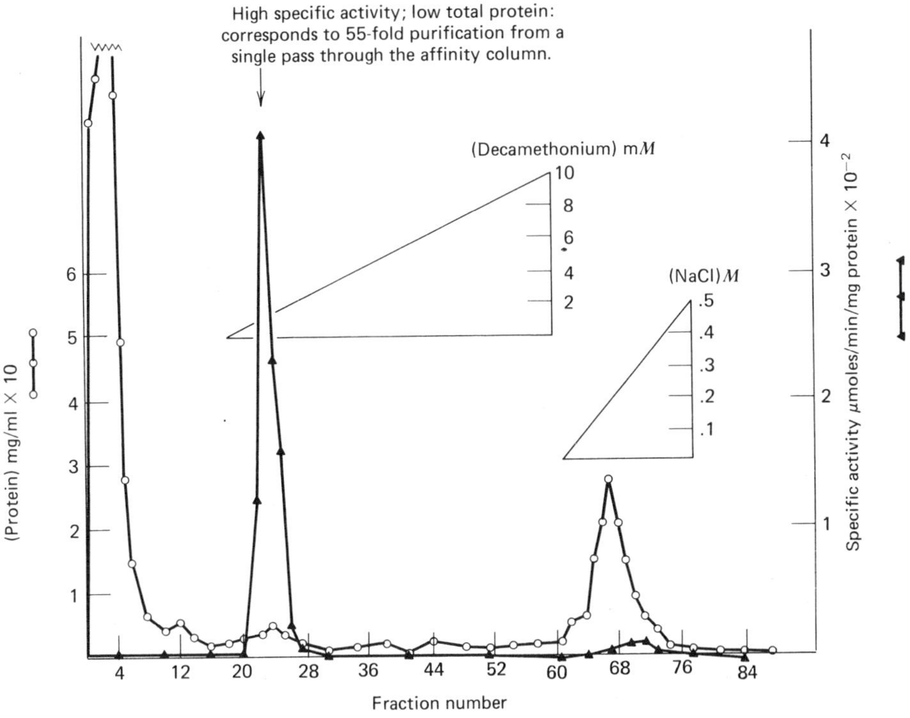

FIGURE 8-22. Elution profile for *Torpedo* Acetylcholinesterase using the affinity ligand Sepharose—2B—N(H)—$(CH_2)_5$—C(=O)—N(H)—C_6H_4—$\overset{+}{N}(CH_3)_3$. Decamethonium is a strong inhibitor of the enzyme, and is used to detach the acetylcholinesterase from the affinity column. (Data from P. J. Morrod, D. G. Clark, and A. G. Marshall.)

Table 8-2 Summary of Techniques (Chapters 1 to 8) for Distinguishing between Different Macromolecules in a Mixture in Solution

Technique	Separation Based on Differences in	Separation Used Primarily for Analytical or Preparative Applications?	Examples in Chapter
Crystallization	solubility	preparative	2.A.1
Salting-Out	solubility in salt solution	preparative	3.B.1
Dialysis	solute ability to pass through semipermeable membrane	preparative	2.C
Equilibrium Dialysis	solute ability to pass through semipermeable membrane	analytical	2.C, 3.C
Immunodiffusion	macromolecular size, shape, and immunospecificity	analytical	6.B
Electrophoresis; "Disc"-electrophoresis; isotachophoresis	macromolecular charge, size, and shape	analytical	7.A
Isoelectric focussing	macromolecular isoelectric point	analytical	7.A
Immunoelectrophoresis	macromolecular charge, size, shape, and immunospecificity	analytical	7.A
Sedimentation rate	macromolecular weight, buoyant density, and shape	analytical	7.B.1
Equilibrium sedimentation	macromolecular weight and buoyant density	analytical	7.B.2
Density-gradient Sedimentation	macromolecular buoyant density (isopycnic gradient)	preparative	7.B.3
	macromolecular weight (isokinetic gradient)	preparative	7.B.3
Viscosity	macromolecular size, shape, and density	analytical (crude)	7.C.2
Polarography	standard electrochemical redox potential	analytical	7.D
Column Chromatography:			
gas-liquid	solubility and volatility	both	8.D
gel	macromolecular size, shape	both	8.D
ion-exchange	macromolecular charge, size and shape	both	8.D
affinity	macromolecular affinity for specific column-bound ligand	preparative	8.D

esterase enzyme was retained on the column, while enzymatically inactive protein passed on through (fractions 0 to 15 or so). Then, on elution with a gradient of a soluble, potent inhibitor of acetylcholinesterase, the enzyme was dislodged from the column to give a protein peak of very high specific activity. Finally, a salt gradient was applied to remove any remaining protein from the affinity column, and it is evident that very little active enzyme is released. In this particular example, it would have required at least four separate chromatographic and other purification procedures to obtain an enzyme of comparable specific activity. Affinity methods have also been used for isolation of specific nucleic acids, a variety of enzymes, drug receptors, hormone receptors, and the like. If current attempts to couple antibodies to agarose gels are successful, it may even prove possible to selectively remove disease-carrying cells, viruses, or antigens from human blood.

Table 8-2 summarizes the techniques for distinguishing between different macromolecules in a solution mixture, according to the basis for separation in each technique. The table incorporates most of the techniques discussed in Chapters 1 to 8, and the appropriate chapter is given for reference to examples.

PROBLEMS

1. Complete the derivation of the Poisson distribution by proving Eq. 8-9. *Hint:* Begin by applying the more accurate form of Stirling's approximation,

$$N! = \sqrt{2\pi N}\, N^N e^{-N}$$

to the factorials in term "C", and rearrange the result to obtain

$$C = e^{-m}\left[\left(1-\frac{m}{N}\right)^{(m-1/2)}\right]\left[\frac{1}{\left(1-\frac{m}{N}\right)^{N}}\;\frac{1}{\left(1-\frac{\bar{m}}{N}\right)^{m}}\right]$$

then take the limit as $N \to \infty$ of each of the four factors just shown.

2. Obtain the principal useful property of the Poisson distribution (Eq. 8-11):

$$\sqrt{<(m-<m>)^2>} = \sqrt{\bar{m}}$$

3. Bortkewitch collected the following data showing the chance of a cavalryman being killed by a kick from a horse in the course of a year. The data derive from the records of ten army corps for twenty years (200 readings):

Number of deaths/year	Number of corps-years in which this number of deaths occurred
0	109
1	65
2	22
3	3
4	1
5 or more	0

(a) Show that the average number of deaths per corps-year is $\bar{m} = 0.61$.
(b) Compare the observed number of corps-years in which a given number of deaths occurred with that predicted from a Poisson distribution, with $\bar{m} = 0.61$.
(c) As an additional test, calculate the probability that there is at least one death per corps-year, at least two, at least three, and so on, and place each point on the appropriate line ($C = 1$, $C = 2$, etc.) of the Poisson paper, Fig. 8-6, and see if the points lie on a vertical straight line.
(d) If this problem seems irrelevant, pretend that the data are derived from the number of times a given microscope quadrant contains zero, 1, 2, and so on number of normal red blood cells from a diluted blood sample!

4. If radioactive counting has been carried out for a period, t, how much longer will the experiment have to be continued to improve the precision of the measurement by a factor of 35%?

5. A radioactivity counter gives a count rate of 1500 counts/sec for sample A alone, 735 counts/sec for sample B alone, and 2082 counts/sec for samples A and B together.
   (a) Calculate the dead time of the counter.
   (b) Obtain the true count rates for samples A and B alone.

6. Suppose you added 2.0 grams of a test sample of pure X of known radioactivity of 1000 c.p.m. to a mixture of substances known to include some nonradioactive X. Further suppose that after mixing thoroughly, some X is isolated and purified from this mixture, and this purified X has an activity of 150 c.p.m. per gram of pure X. How much nonradioactive X much have been present in the original unknown mixture?

7. A certain photomultiplier tube has a gain of 10^6 – that is, 10^6 electrons are produced for every photon that enters the front of the tube. If the number of photons per second hitting the photomultiplier is 4, and the time constant of the subsequent electronic circuit is 0.1 millisec, calculate the signal-to-noise ratio for such a signal. (You may assume that the noise is determined by shot noise from the photomultiplier).

8. A group of 20 patients is subjected to a new medical treatment for a condition in which 20% of similar patients succumb after the conventional treatment. If 7 of the new group die, how strong is the evidence that the new treatment was better or worse than the old?

9. Show that chromatographic resolution (the ratio of the distance that a given solute "peak" moves down the column to the "width" of such a peak) is proportional to the square root of the height of the gel bed of the column.

10. What is the signal-to-noise ratio for a circuit that carries 5 microamp with a bandpass of 20 kHz?

REFERENCES

Poisson Distribution

M. J. Moroney, *Facts from Figures,* Penguin Books, Baltimore (1956).

Radioactive Counting

S. Silver, *Radioactive Isotopes in Medicine and Biology: Medicine,* 2nd ed., Lea and Febiger (1962).

E. W. Phelan, *Radioisotopes in Medicine,* U. S. Atomic Energy Commission, Division of Technical Information (1966). Obtain free from USAEC, P.O. Box 62, Oak Ridge, Tennessee.

Electronic Noise

J. Brophy, *Basic Electronics for Scientists,* 2nd ed., McGraw-Hill (1972).

T. Coor, "Signal-to-Noise Optimization in Chemistry," *J. Chem. Educ. 45,* A533, A583 (1968).

Gel Chromatography

R. E. Pecsok and D. Saunders, in *Separation Techniques in Chemistry and Biochemistry,* ed. R. A. Keller, Marcel Dekker (1967).

Sephadex: Gel Filtration in Theory and Practice, Beaded Sepharose 2B-4B-6B: Agarose Gel Filtration, both from Pharmacia Fine Chemicals AB, Box 175, S-751 04 Uppsala 1, Sweden.

Ion-Exchange Chromatography

"Ion Exchange Resins," BDH Chemicals Ltd, Poole, England. "Sephadex Ion Exchangers," Pharmacia Fine Chemicals AB, Box 175, S-751 04, Uppsala 1, Sweden.

Affinity Chromatography

P. Cuatrecasas, "Protein Purification by Affinity Chromatography," *J. Biol. Chem.* *245,* 3059–65 (1970).

3

GROWTH AND DECAY

CHAPTER 9 First-Order Rate Processes

Bacterial growth, radioactive decay, pharmacokinetics, and all of chemical reaction kinetics are based on the general relation,

$$d[\quad]/dt = \pm k[\quad] \tag{9-1}$$

in which [ ] represents the amount or concentration of material remaining at time, *t;* the *sign* preceding *k* indicates whether the amount of substance is increasing (positive), or decreasing (negative), and the *magnitude* of *k* is a measure of the *rate* of the process. Equation 9-1 leads immediately (see below) to the exponential growth or decay inherent in the processes mentioned above, and we are now in a position to account for the form of Eq. 9-1, using statistical arguments developed in Section 2.

Macroscopic changes in amount or concentration of substance tend toward *exponential* time-behavior, simply because the behavior of any *one* bacterium (atom, pair of reactive molecules) is governed by *chance.* There is a well-defined *probability* that any *one* bacterium (atom, pair of reactive molecules) will change during a given observation period, and that probability is independent of the prior history of that system. (Stated more simply, no one unstable atom "knows" whether or not any of the other unstable atoms in the sample have decayed.) To see how this condition leads to exponential *decay,* for example, consider a room containing 100 people, each of whom tosses a coin once a minute. Suppose that anyone who obtains "heads" is asked to leave the room. After the first toss, there will be approximately 50 people remaining; after two tosses, about 25 people; and the number of people in the room will vary approximately exponentially with time. As the number of coin-tossers (bacteria, unstable atoms, pairs of chemically reactive molecules) becomes larger, it was shown in Chapter 8 that the *relative* precision in such a measurement is greatly improved. When the sample is small (see Fig. 9-1, left), a smooth exponential curve is a relatively poor representation of the stepwise decrease in the quantity of interest; however, when the sample is much larger (Fig. 9-1, right), the disappearance is very accurately described by an exponential curve.

Bacterial Growth; Radio- and Chemical Dating

From the preceding example, we might expect that the number of coin-tossers (bacteria, unstable atoms, pairs of reactive molecules) will vary in time according to

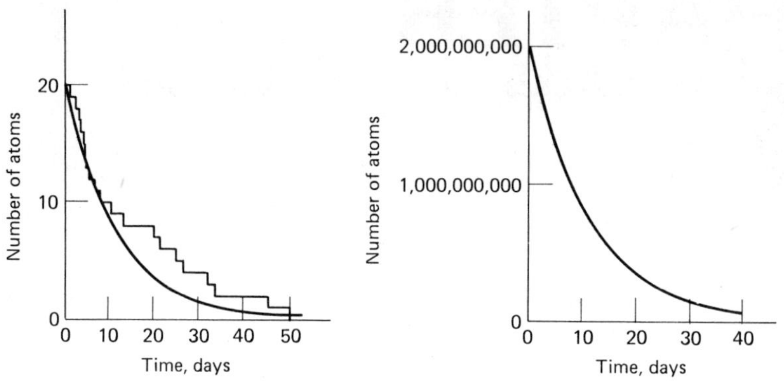

FIGURE 9-1. Disappearance of ^{131}I atoms with time. *Left:* initial number of ^{131}I atoms = 20. *Right:* initial number of ^{131}I atoms = 2,000,000,000.

$$\boxed{[\ \ ] = [\ \ ]_0\, 2^{\pm(t/T)}} \quad \begin{cases} \xrightarrow{\text{growth}} [\ \ ]_0\, 2^{(t/T)} \\ \xrightarrow{\text{decay}} [\ \ ]_0\, 2^{-(t/t)} \end{cases} \tag{9-2}$$

T represents the doubling time (growth) or halflife (decay) for the process; that is, when $t = T$,

$$[\ \ ] = [\ \ ]_0\, 2^{\pm 1} \quad \begin{cases} \xrightarrow{\text{growth}} 2[\ \ ]_0 \\ \xrightarrow{\text{decay}} [\ \ ]_0/2 \end{cases} \tag{9-3}$$

and the population doubles (halves) every T seconds.

It will now be shown that a different intuitive starting point, Eq. 9-1, leads to the same result. Equation 9-1 simply implies that the number of observed bacterial divisions (atomic decays, chemical reactions) ought to be proportional to the number of bacteria (unstable atoms, pairs of reactive molecules) present. Rearranging Eq. 9-1,

$$\frac{d[\ \ ]}{[\ \ ]} = d \ln [\ \ ] = \pm k dt \tag{9-4}$$

and integrating from initial time zero to final time, t, during which the amount of substance has changed from its initial value $[\ \ ]_0$ to its final value, $[\ \ ]$,

$$\ln [\ \ ] - \ln [\ \ ]_0 = \pm kt \tag{9-5}$$

or

$$\boxed{[\ \ ] = [\ \ ]_0\, e^{\pm kt}} \tag{9-6}$$

Defining a "lifetime," τ, for growth or decay,

$$\boxed{\tau \equiv (1/k)} \tag{9-7}$$

then by analogy to Eq. 9-3, it is clear that when $t = \tau$, the population has grown (decayed) to e times ($1/e$th of) its initial value:

$$[\ \ ] = [\ \ ]_0\, e^{\pm 1} \begin{cases} \xrightarrow{\text{growth}} e[\ \ ]_0 \\ \xrightarrow{\text{decay}} [\ \ ]_0/e \end{cases} \tag{9-8}$$

In chemical reaction kinetics, the *rate constant, k,* is the usual parameter; in bacterial growth and radioactive decay problems, the *doubling time (halflife), T,* is the usual parameter; and in fast reaction kinetics and relaxation problems (Section 4), the *lifetime,* τ, is the parameter of choice. These parameters are related as follows: combining Eqs. 9-2 and 9-6

$$\frac{[\ \ ]}{[\ \ ]_0} = e^{\pm(t/\tau)} = 2^{\pm(t/T)} = e^{\pm kt} \tag{9-9}$$

Taking the logarithm to base e

$$\pm(t/\tau) = \pm(t/T)\log_e 2 = \pm k$$

or

$$\boxed{k = \frac{1}{\tau} = \frac{\ln 2}{T} \cong \frac{0.693}{T}} \tag{9-10}$$

The results of this section are now applied to some growth or decay problems.

EXAMPLE *Bacterial Growth*

The exponential nature of the growth of microorganisms is illustrated in the data of Fig. 9-2 for a yeast culture. It is now generally believed that unicellular organisms are potentially immortal in the sense that cultures of many microorganisms have been kept alive and dividing for years. In fact, a similar property is true of certain plant, animal, or human tissue cultures, and if the cells of human tissue from a tissue culture are dissociated, then so-called "clones" of daughter cells can be grown in colonies similar to those from microorganisms, provided that the cell culture is continually supplied with nutrients and kept washed free of metabolic wastes.

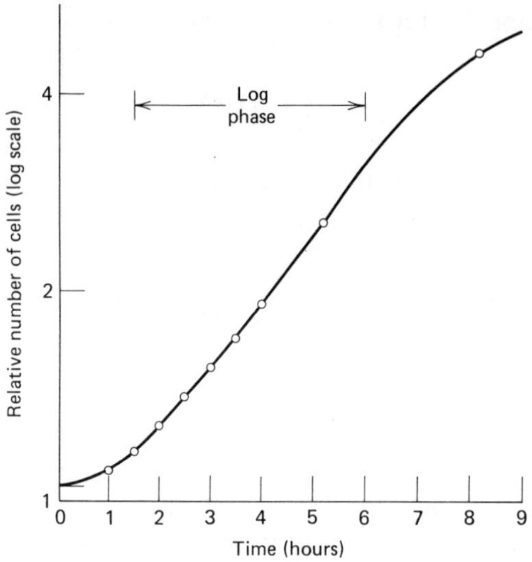

FIGURE 9-2. Population curve for a *Saccharomyces cerevisiae* (yeast) culture. Following an initial "lag" period, the yeast cells enter a "logarithmic phase," during which the cells are dividing at their maximal rate, with number of cells increasing exponentially with time. As the nutrients in the initial mixture are depleted, and metabolic wastes accumulate (upper right of figure), the number of cells eventually levels off (and ultimately decreases). (Data from G. Luoma, J. L. Smith, and A. G. Marshall.)

One reason for interest in the growth of microorganisms is that cancer tumors appear to exhibit exponential cell growth of very similar nature (see Fig. 9-3). All present cancer treatments are based on means for suppression of tumor growth once it has started [ionizing radiation and incorporation of radioactive salts (Section 5), and various inhibitors of the synthesis of metabolites required for the continued growth of the tumor (Chapter 11)]. The study of cancer is largely a study of the origin of *normal* cell stasis (neither growing or dividing) versus growth or division.

EXAMPLE *Radio-Dating from Remaining Amount of Parent Nuclide: ^{14}C-Dating*

Natural carbon consists of about 99% ^{12}C and 1% ^{13}C. In addition, the nitrogen in the atmosphere is continually bombarded by cosmic neutrons to produce minute amounts of radioactive ^{14}C; this ^{14}C is oxidized to $^{14}CO_2$ and eventually incorporated into plants and later into animals. Since the biological ^{14}C disappears with a half-life of 5668 years, there is a steady-state balance between ^{14}C formation and ^{14}C decay, such that living organisms reach a concentration of about 10^{-12}g of ^{14}C for each gram of ^{12}C. The resultant ^{14}C radioactivity in modern organisms, A_0, thus amounts to about

$$A_0 = 15.3 \pm 0.1 \text{ disintegrations/minute per gram of carbon}$$

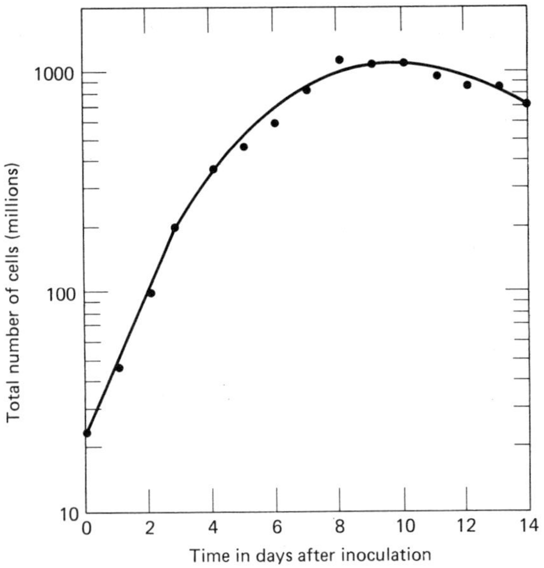

FIGURE 9-3. Population growth curve for cells of the Ehrlich ascites tumor. This tumor is made up of individual cells growing in the body (ascitic) fluid. The initial injection on "0" days consisted of 22×10^6 cells. Note the initial logarithmic rise for almost four days, the later decline in rapid growth, and then the decrease in number of cells as some of them die, the whole curve resembling that in Fig. 9.2. The mean survival time for mice with the tumor so inoculated is 14 days. Each point is the average of six determinations. (From A. C. Giese, *Cell Physiology,* W. B. Saunders Co., Philadelphia, 1962, p. 521.)

In fact, radioactivity measurements are the most accurate means for detection of ^{14}C. Provided that the rate of production of ^{14}C in the atmosphere has remained relatively constant, and that exchange of carbon between the object and the atmosphere ceases when the object dies, comparison of the remaining ^{14}C radioactivity, A, with the presumed initial activity, A_0, yields a measure of the age of the object from Eqs. 9-6 and 9-10:

$$A/A_0 = e^{-(t \ln 2/T)}$$

or

$$\boxed{t = \frac{T}{\ln 2} \ln (A_0/A)} \qquad (9\text{-}11)$$

Figure 9-4 shows some representative dates derived from the carbon-dating method. By comparing the radio-carbon apparent dates with the actual dates corresponding to individual tree ring samples from a very old bristlecone pine

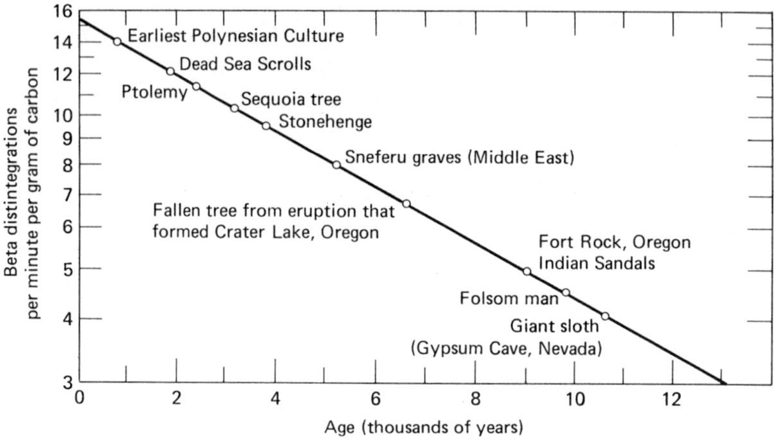

FIGURE 9-4. Beta-disintegration rate as a function of age for some objects dated by C^{14} activity. (After B. H. Mahan, *University Chemistry*, 2nd ed., Addison-Wesley, Reading, Mass., 1969, p. 800.)

(Fig. 9-5), the variations in atmospheric ^{14}C production are seen to be measurable, particularly for the older dates. Other errors in age determination by carbon-14 dating derive from the recent extensive burning of fossil fuels (coal, petroleum) that have essentially no ^{14}C-activity and thus dilute the modern atmospheric ^{14}C, and the very recent increase in the production of ^{14}C as a result of synthetic nuclear explosions (see Fallout example). Finally, it should be pointed out that the (already low) ^{14}C-activity in a carbon sample has dropped to about 0.1% of its original value after 10 half-lives have elapsed, so that carbon-14 dating is pretty well limited to dates from about 2000 to 20,000 years before present.

EXAMPLE *Radio-Dating from Daughter:Parent Ratio: Potassium-Argon Dating*

In spite of its short half-life (relative to the age of the earth), ^{14}C is evident today because it is being formed continuously in the upper atmosphere; the situation differs in two important ways for most other isotopes of paleontological interest. First, since modern radioactive ^{40}K, for example, represents the remnant of whatever ^{40}K was present when the earth was formed, the isotopic abundance of ^{40}K is about the same in all modern rocks (0.0119 atom %). Second, when a ^{40}K-containing sample is suddenly isolated from its surroundings, as by solidification of a rock, it is the exchange of the *daughter* isotope with its surroundings that is interrupted, rather than the exchange of the *parent* isotope as with ^{14}C-dating. These two differences lead to a rather different procedure for age determination using radioisotopes other than ^{14}C.

Since most useful rocks and minerals contain between 0.1% and 10% potassium, *total* potassium is readily determined gravimetrically or by flame photometry; ^{40}K amount in the sample may then be determined since ^{40}K is

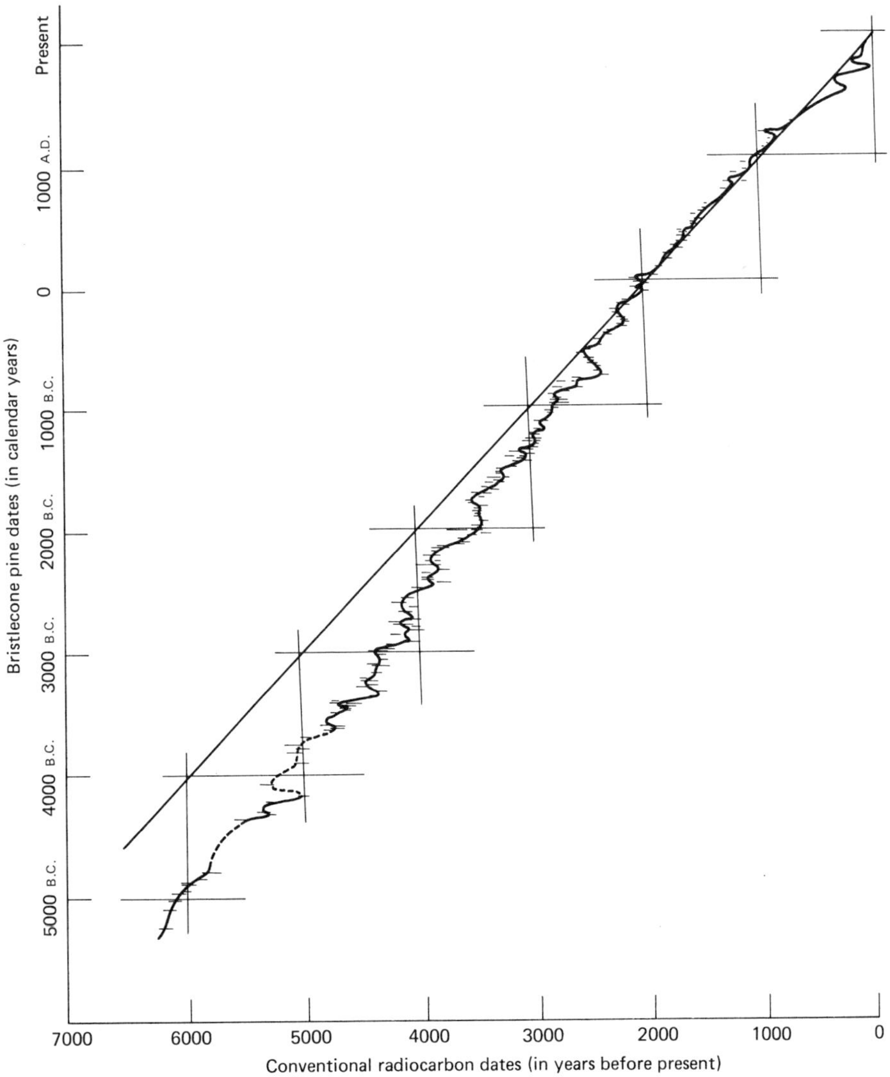

FIGURE 9-5. Plot of tree-ring date versus conventional radiocarbon age ("present" is 1950 A.D.), for the period, 5300 B.C. to present, using wood from *Pinus aristata* (bristlecone pine) and *Sequoia gigantea* (giant sequoia). If atmospheric ^{14}C levels had been constant over this period, the data points would all fall on the 45° solid line; deviations thus reflect variations in atmospheric ^{14}C. [From "Radiocarbon Variations and Absolute Chronology," ed. I. U. Olsson, Wiley Interscience, 12th Nobel Symposium, 1970.]

present as a constant proportion of the total potassium. The daughter isotope, ^{40}Ar, is determined by melting the sample under vacuum, and then analyzing for ^{40}Ar with a mass spectrometer. The accuracy of the method is limited by the precision of the (necessary) correction for the presence of modern atmospheric ^{40}Ar. The analysis of the data is based on use of Eqs. 9-6 and 9-10, in which N_D, N_0, and N are now defined as the amount of daughter isotope present today, the amount of parent isotope present at the time the rock solidified, and the amount of parent isotope present today, respectively:

$$N_D = N_0 - N \tag{9-12}$$

or

$$= Ne^{+t \ln 2/T} - N = N(e^{t \ln 2/T} - 1) \tag{9-13}$$

$$1 + \frac{N_D}{N} = e^{t \ln 2/T}$$

to give

$$t = \frac{T}{\ln 2} \ln \left[1 + (N_D/N)\right] \tag{9-14}$$

One final procedural complication is due to the dual radioactive decay pathway for ^{40}K:

$$^{40}\text{K} \begin{array}{l} \nearrow\ ^{40}\text{Ca} \quad T_a = 1.47 \times 10^9 \text{ years} \\ \searrow\ ^{40}\text{Ar} \quad T_b = 1.19 \times 10^{10} \text{ years} \end{array}$$

When the dual nature of the ^{40}K decay is taken into account (see Problems), Eq. 9-14 is modified to the form

$$t = \frac{1}{\ln 2}\left[\frac{1}{\frac{1}{T_a} + \frac{1}{T_b}}\right] \ln \left[1 + \left(\frac{T_b}{T_a} + 1\right)\left(\frac{N_D}{N}\right)\right] \tag{9-15}$$

One of the more famous examples of potassium-argon dating is shown in Fig. 9-6, which showed that humans as tool-making animals are at least 1.75 million years old. (The various "KA" numbers in the figure refer to separate samples: KA1047 came from 12 inches above the *Zinjanthropus* specimen; KA850 from one inch above the living floor; and KA1180 from 81 inches below.) Potassium-argon dating is useful for dates ranging from the age of the oldest known rocks (about 3.5×10^9 years) up to rocks as young as about 50,000 years. For further applications, consult the end of chapter references.

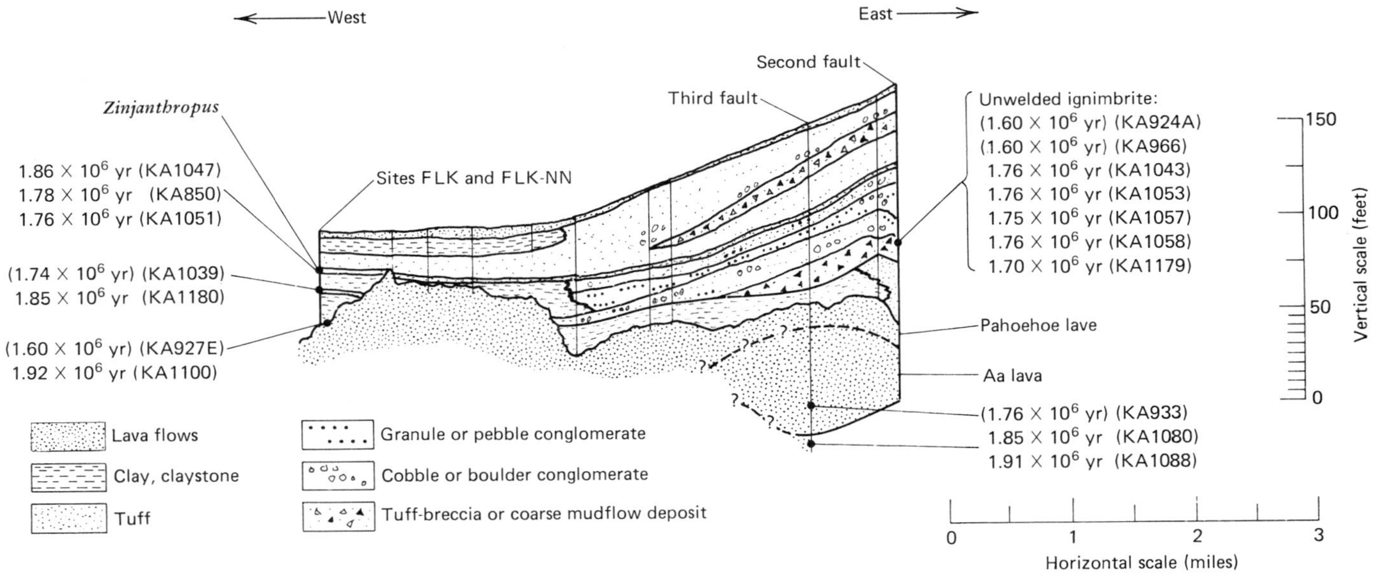

FIGURE 9-6. Stratigraphy of Bed I, Olduvai Gorge, Tanzania, showing the locations of the fossil hominid *Zinjanthropus* and of radiometrically dated samples. Vertical lines indicate positions of measured sections. The sequence is reconstructed as it would have appeared at the end of Bed I time. Ages in parentheses are considered unreliable by Evernden and Curtis (1965a). After C. S. Grommé and R. L. Hay, *Earth Planet. Sci. Letters,* v. 2, p. 111–115, 1967.

EXAMPLE *Chemical Dating*

A very recent and promising method for dating of objects that are 5000 to 150,000 years old is based on two facts: (1) only L-amino acids are commonly found in living organisms, and (2) over a period of tens of thousands of years, there is substantial racemization (interconversion of L- and D-amino acid),

$$L \underset{k}{\overset{k}{\rightleftarrows}} D; \ K_{eq} = \frac{k}{k} = 1 \tag{9-16}$$

where k is the (first-order) rate constant for interconversion of L- and D-isomers. From Eq. 9-1, the rate of change in the amount of D-isomer will be governed by its rate of formation (from L-isomer) and its rate of disappearance (to L-isomer):

$$\frac{d[D]}{dt} = k[L] - k[D] \tag{9-17}$$

With the use of three manipulative tricks, Eq. 9-17 can be developed for determination of ages of organic objects from measurement of the present ratio of $[D]/[L]$ isomers. Defining the variable, x, by the relations

$$\begin{aligned} [L] &= [L]_0 - x \\ [D] &= [D]_0 + x \end{aligned} \tag{9-18}$$

where $[L]_0$ and $[D]_0$ denote the *initial* concentrations of L- and D-isomer present at the time of death of the organism (i.e., the time at which racemization begins), and it is assumed that $[L]_0 > [D]_0$ since L-isomers predominate in living organisms,

$$d[D]/dt = k([L]_0 - x) - k([D]_0 + x) \tag{9-19}$$

Using the first trick, it may be noted from Eq. 9-18 that

$$d[D]/dt = dx/dt \tag{9-20}$$

so that Eq. 9-19 may be written

$$dx/dt = k([L]_0 - [D]_0) - 2kx \tag{9-21}$$

With a second trick, a new variable, u, may be defined as

$$u \equiv \frac{[L]_0 - [D]_0}{2} - x \tag{9-22}$$

so that

$$-dx/dt = du/dt = -2ku \tag{9-23}$$

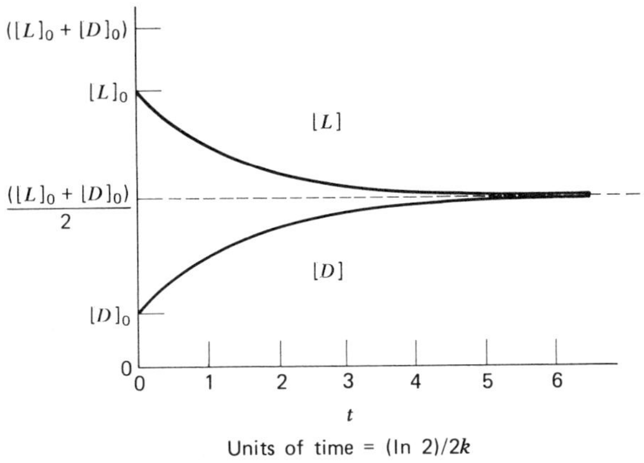

FIGURE 9-7. Concentration of L-isomer (top curve) and D-isomer (bottom curve) as a function of time after the onset of racemization of a mixture having initial isomer concentrations of $[L]_0$ and $[D]_0$ at time zero. The algebraic form of this graph is given by Eq. 9-26. Note that the asymptotic composition (dotted line) consists of a 1:1 mixture of D- and L-isomers.

Equation 9-23 is now of the same form as Eq 9-1, and may be integrated to give

$$\ln u - \ln u_0 = -2kt \tag{9-24}$$

or

$$\ln\left(\frac{[L]_0 - [D]_0}{2}\right) - \ln\left(\frac{[L]_0 - [D]_0}{2} - x\right) = 2kt \tag{9-25}$$

Using Eqs. 9-18 again, Eq. 9-25 may be simplified to

$$\boxed{\ln\left(\frac{[L]_0 - [D]_0}{[L] - [D]}\right) = 2kt} \tag{9-26}$$

Equation 9-26 was used to construct Fig. 9-7. This figure shows that the concentration *difference*, $([L]_0 - [D]_0)$, decays exponentially, with a time constant of $(-1/2k)$. The final trick in the derivation is to note that if the dead organism has been isolated from its surroundings (as by burial), then the total amino acid level should be constant in time:

$$[L] + [D] = [L]_0 + [D]_0 \tag{9-27}$$

and Eq. 9-26 may be rewritten as

$$\ln \frac{([L]_0 - [D]_0)}{([L] - [D])} \frac{([L] + [D])}{([L]_0 + [D]_0} = 2kt$$

or finally

$$t = (1/2k)\left(\ln\left[\frac{1 + \dfrac{[D]}{[L]}}{1 - \dfrac{[D]}{[L]}}\right] - \ln\left[\frac{1 + \dfrac{[D]}{[L]}}{1 - \dfrac{[D]}{[L]}}\right]_0\right) \tag{9-28}$$

From Eq. 9-28, the age of an organic object may be determined from measurement of the present ($[D]/[L]$)-ratio for the object and the ($[D]/[L]$)-ratio for a modern organic object that is taken to represent ($[D]_0/[L]_0$). In practice, the amino acid of interest, say aspartic acid, is first isolated as a *D,L* mixture by ion exchange chromatography, as discussed in Chapter 8.D, and then combined with L-leucine to form a mixture of two diasteromeric di-peptides that may then be resolved by another ion-exchange chromatography step. Some typical ($[D]/[L]$)-ratios are shown in Fig. 9-8 (the ninhydrin color yield is

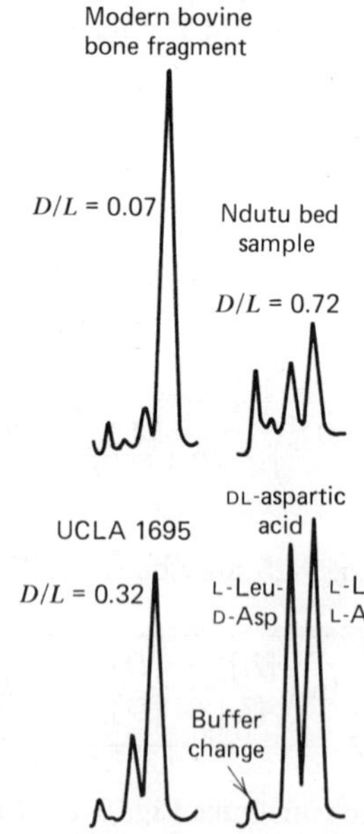

FIGURE 9-8. Part of the amino-acid analyzer printouts for various bone samples showing the diastereomeric dipeptides L-leucyl-D-aspartic acid and L-leucyl-L-aspartic acid. A 56 × 0.9-cm column filled with Beckman-Spinco AA-15 resin was used for the separation. The column was eluted with pH 3.24 buffer for 40 min, then the buffer was switched to pH 4.25 for the remainder of the run. The buffer change, L-Leu-D-Asp, and L-Leu-L-Asp peaks come at 72, 79, and 83 min with this elution sequence. The small peak that elutes between the buffer peak and L-Leu-D-Asp is glycine that was not completely separated from aspartic acid on the Dowex 50 (H^+) column. [From J. L. Bada and R. Protsch, *Proc. Natl. Acad. Sci. U.S.A. 70,* 1331 (1973).]

lower for the *L-D* dipeptide than for the *L-L* dipeptide, leading to the unequal apparent amounts of *L-D* and *L-L* in the chromatogram for racemic dipeptide at the bottom right of the figure—this disparity is readily corrected using a suitable conversion factor). Sample UCLA 1965 in Fig. 9-8 had a known radiocarbon age of 17,550 ± 1000 years, and could thus be used to determine the racemization rate constant, k_{asp}, which was found to be

$$k_{asp} = 1.48 \times 10^{-5}\ \text{yr}^{-1} \text{ at } 24°\text{C}$$

With this calibration, the aspartate racemization rate could then be used to date other objects from the same location (the uppermost part of the Olduvai Gorge, Tanzania). Such procedures yielded an age for the Eyasi I Hominid of 34,000 years.

Radiocarbon dating accuracy depends on the constancy of *^{14}C production* in the atmosphere; *chemical* dating accuracy depends on the constancy of *temperature* of the object, since chemical reaction rates vary markedly with temperature. However, chemical dating requires only a few grams of sample, whereas radiocarbon dating should have hundreds of grams for reliable results; moreover, by using an amino acid (such as alanine) that racemizes more slowly than aspartic acid, it should be possible to date bones as old as 100,000 to 150,000 years:

$$k_{asp} \cong 7\ k_{isoleucine};\ k_{asp} \cong 3\ k_{alanine}$$

EXAMPLE *Fallout and Other Radioactive Waste*

With the certain upcoming decline in supply of fossil fuel, the economic pressure for increasing construction of nuclear power plants is inevitable, along with the resultant controversy concerning the risks associated with increased public exposure to radioactive wastes from those power plants. In this brief section, the various sources of radiation are listed, along with some quantitative indications of their relative significance.

The largest present (and likely future) source of radiation is natural radiation from cosmic rays and their products and from radioactive isotopes occurring naturally in the earth's crust (and thus in building materials and food). The second largest source (about half the average per capita dose of natural radiation) is diagnostic radiology (X rays, radioisotope tracer location and metabolic studies). The first of these sources is unavoidable, and the benefits of the second very greatly outweigh the harm arising from the radiation exposure. The largest source of radiation dosage that is in principle avoidable arises from atmospheric testing of nuclear weapons, as seen in Table 9-1. The dose unit, *rem,* is defined as the quantity of any ionizing radiation which has the same biological effectiveness as 1 *rad* of X rays, where a *rad* amounts to 100 ergs absorbed per gram of absorbing material. As a biological calibration of the numbers in Table 9-1, it may be noted that doses as small as 80 millirad may produce a 40% increase in the risk of cancer if the radiation is given to a fetus during the first three months of pregnancy (based on study of 700,000

Table 9-1 Per Capita Dose Rates and Integral Doses in the United States for Years 1960, 1970, and 2000[a]

Source of Exposure	1960 183×10^6 [b]		1970 205×10^6 [b]		2000 321×10^6 [b]	
	Per Capita Dose Rate (mrem/year)	Integral Dose (million man-rem/year)	Per Capita Dose Rate (mrem/year)	Integral Dose (million man-rem/year)	Per Capita Dose Rate (mrem/year)	Integral Dose (million man-rem/year)
Natural	130	24	130	27	130	42
Occupational	0.75	14×10^{-2}	0.8	16×10^{-2}	0.9	27×10^{-2}
Nuclear power	0.0001	18×10^{-6}	0.002	41×10^{-5}	0.2	64×10^{-3}
Fuel reprocessing	–	–	0.0008	17×10^{-5}	0.2	64×10^{-3}
AEC activities other than open-air weapons testing	0.01	18×10^{-4}	0.01	20×10^{-4}	0.01	32×10^{-4}
Open-air weapons testing	13.0[c]	2.4	4.0	0.82	4.9	1.6
TV, consumer products, air travel	1.6	29×10^{-3}	2.6	40×10^{-3}	1.1	6×10^{-2}
Diagnostic radiology	72	13.3	72	14.8	72	23.1
	217	40	209	43	208	66

[a] *From Environmental Protection Agency (1972); M. Eisenbud,* Environmental Radioactivity, *2nd ed. (Academic Press, N. Y., 1973),*
[b] *Population of United States.*
[c] *1963.*

Table 9-2 Approximate Yields of the Principal Nuclides per Megaton of Fission*

Nuclide	Half-life	MCi
^{89}Sr	53 days	20.0
^{90}Sr	28 years	0.1
^{95}Zr	65 days	25.0
^{103}Ru	40 days	18.5
^{106}Ru	1 year	0.29
^{131}I	8 days	125.0
^{137}Cs	30 years	0.16
^{131}Ce	1 year	39.0
^{144}Ce	290 days	3.7

From M. Eisenbud, Environmental Radioactivity, *2nd ed. (Academic Press, N. Y., 1973), p. 321.*

children by Dr. Brian MacMahon, Harvard School of Public Health, 1962). It should also be noted that a 40% increase in a very small number is still a very small number.

Radionuclides from atmospheric nuclear explosions originate from the immediate products of the fission process itself (Table 9-2), from radioactive products of the interaction of the bomb radiation with air (Table 9-3), and from radioactive products of the interaction of bomb radiation with soil (Table 9-4). (One Curie represents the quantity of radioactive material that will comprise 3.7×10^{10} disintegrations per second.) It may be noted in connection with Tables 9-2, 9-3, and 9-4, that there has been a total of more than 400 *mega*tons yield of atmospheric nuclear weapons tests alone since 1945. Ranking just below nuclear weapons testing in Table 9-1 is the contribution from television (mostly X rays from color TV sets), consumer products (mostly radioactive paint on wristwatch dials), and air travel (by raising the individual above much of the protective atmospheric radiation shield). Except for people who actually work with radioisotopes ("occupational category" in Table 9-1), nuclear power plants offer a contribution whose magnitude is rapidly becoming significant, though still very small. Nuclear (civilian) power plants have only recently overtaken military production of plutonium in the United States as a source of high-level radioactive waste; the accumu-

Table 9-3 Principal Radionuclides Induced in Air*

Isotope	Half-life (years)	Ci/megaton
^{3}H	12.3	<1
^{14}C	5600	3.4×10^{4}
^{39}A	~260	59

From M. Eisenbud, Environmental Radioactivity, *2nd ed. (Academic Press, 1973) p. 325.*

Table 9-4 Principal Radionuclides Induced in Soil*

Isotope	Half-life	Ci/megaton
^{24}Na	15 hr	2.8×10^{11}
^{32}P	14 days	1.92×10^{8}
^{42}K	12 hr	3×10^{10}
^{45}Ca	152 days	4.7×10^{7}
^{56}Mo	2.6 hr	3.4×10^{11}
^{55}Fe	2.9 yr	1.7×10^{7}
^{59}Fe	46 days	2.2×10^{6}

* *From M. Eisenbud,* Environmental Radioactivity, *2nd ed. (Academic Press, N. Y., 1973), p. 325.*

lated liquid waste would amount to some 60 million gallons by 2000 A.D. without further processing. It has been AEC policy since 1969 that high-level *liquid* wastes must be converted into *solid* form for ultimate storage. An attractive storage location may turn out to be salt beds, since salt has a radiation-shielding characteristic similar to concrete, a natural plasticity that should effectively seal the waste container, and the highest capacity to dissipate heat of any type of rock. (It is already possible to remove about 99.9% of radioactive noble *gases* from reactor effluents.)

Because of its long half-life (28 years) and its tendency to replace calcium (as in bone), ^{90}Sr has been suggested as the limiting factor in the extent to which human beings can tolerate global contamination by fission products. Figures 9-9 and 9-10 show the close correlation between ^{90}Sr in the environment and in human bone—the variations with calendar year correspond to the level of atmospheric nuclear testing, accounting for the maximum effect about 1963.

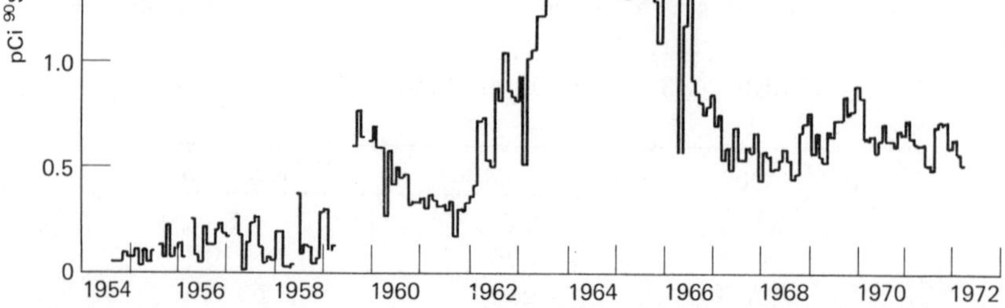

FIGURE 9-9. ^{90}Sr in tap water in New York City, 1954–1971. From M. Eisenbud, *Environmental Radioactivity,* 2nd ed., Academic Press, N. Y., 1973, p. 371.

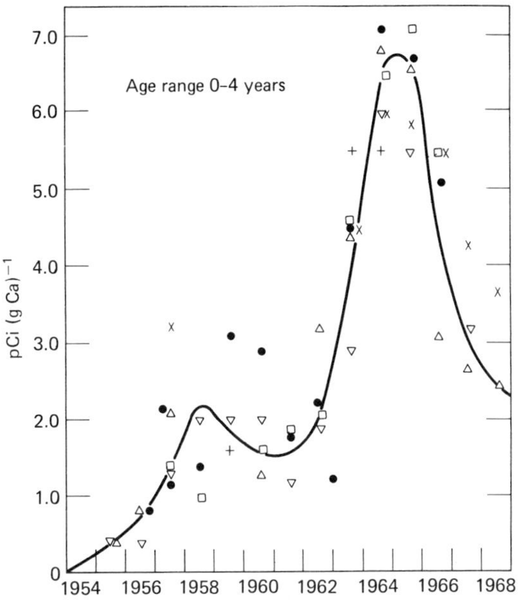

FIGURE 9-10. ^{90}Sr/Ca ratios in human bone samples in North Temperate Zone (age range 0–4 years). Legend: □, Canada; △, Denmark; ▽, Federal Republic of Germany; ◇, France; +, Poland; ×, Soviet Union (Moscow); ●, United Kingdom (adults, West London; other ages, national); ○, United States (New York City). From M. Eisenbud, *Environmental Radioactivity,* 2nd ed., Academic Press, N. Y., 1973, p. 378.

Following the heavy nuclear atmospheric testing schedules of 1961 and 1962, tropospheric ^{14}C content had risen to more than 500% above natural levels in the northern hemisphere, with an increase of 20 to 30% (above 1950 values) in human tissue as early as 1964. Apart from concern about its direct radiation hazard, ^{14}C which decays after it has been incorporated into nucleic acids represents a direct danger of mutations from damage to the gene structure itself. A final comment on radioisotopes concerns ^{131}I: because of the human thyroid gland's ability to concentrate this element (and its copious production from fission), radiation doses to the thyroid can be relatively high, even when the concentration of ^{131}I is relatively small: following a test on May 19, 1953, exposures for St. George, Utah, are estimated to have been as high as 84 rad.

PROBLEMS

1. For a typical dog, an intravenous dose of 30 mg of phenobarbital per kg body weight will usually produce surgical anesthesia. The drug is m tabolically eliminated by a first-order process ($[\ \ ] = [\ \ ]_0 \exp(-kt)$) with a half-life of 4.5 hr. If a 14-kg dog is anesthetized, how many mg of pheno-

barbital must be administered two hours later (when the anesthesia is beginning to lighten) to restore the full original depth of anesthesia?

2. Suppose that the true age of a biological sample is 57,000 years, and that the sample is contaminated with 1% by weight of modern carbon. Calculate the (apparent) age that would be deduced from ^{14}C-dating, if the experimenter were unaware of the existence of the contaminant.

3. Derive an expression for the age of an object (Eq. 9-15) determined from the dual decay scheme of radioactive ^{40}K, and show that the overall half-life

$$T = (\ln 2)/(k_a + k_b) = 1.31 \times 10^9 \text{ years}$$

4. For a chemical-dating experiment based on racemization of L-aspartic acid,

$$L_{\text{asp}} \underset{}{\overset{k_{\text{asp}}}{\rightleftharpoons}} D_{\text{asp}}$$

$$k_{\text{asp}} = 1.48 \pm 0.09 \times 10^{-5} \text{ yr}^{-1} \text{ at } 24°\text{C}$$

calculate the age of a biological specimen whose measured $[D]/[L]$-ratio is 0.72, if a modern specimen would have a $[D]/[L]$-ratio of 0.07.

REFERENCES

F. Hodgson, *Dating by Radioisotopes,* William Clowes & Sons, Ltd., London (1970). General coverage of radio-dating.

G. B. Dalrymple and M. A. Lanphere, *Potassium-Argon Dating,* W. H. Freeman, San Francisco (1969).

J. L. Bada and R. Protsch, "Racemization Reaction of Aspartic Acid and Its Use in Dating Fossil Bones," *Proc. Nat. Acad. Sci.* U. S. A. *70,* 1331 (1973). Principles of chemical-dating: for recent application toward dating of living specimens, see *Nature 262,* 279 (1976).

M. Eisenbud, *Environmental Radioactivity,* 2nd ed., Academic Press, N. Y. (1973). Facts about fallout.

CHAPTER 10 Catalysts

10.A. MICHAELIS-MENTEN (STEADY-STATE) KINETICS

10.A.1. Need for a Steady-State Hypothesis

Contrary to common opinion, the difficulty with many biophysical problems is not that they cannot be *solved* (often exactly), but rather that the result is of such algebraic complexity as to afford little insight into the process in question. The development of simplified enzyme kinetics provides one of the best examples of the use of judicious *approximations* to render a complicated problem tractable and intuitively understandable, and the treatment finds widespread utility in enzyme and drug assays, as well as for determination of the mechanisms of enzyme-catalyzed reactions and drug responses.

The very existence of an enzyme:substrate complex was first deduced (indirectly!) from a study of the reactant concentration-dependence of the rate of inversion of sucrose by the enzyme known as "invertase" by Brown in 1902. Evidence had been accumulating for the preceding twenty years that the rates of chemical reactions involving enzymes appeared to differ in a fundamental way from ordinary chemical reaction rates. Although most chemical reaction rates are *proportional* to the concentration of reactant present (Eq. 9-1, or just plain intuition, sometimes known as the "mass action" law), the rates of enzyme-mediated reactions are often *independent* of the concentration of reactant present. A simple explanation of such behavior is that successful completion of the reaction requires the combination of reactant ("substrate") with some catalyst ("enzyme"); one can then imagine that a sufficiently high substrate concentration will "saturate" the enzyme, so that virtually all the enzyme at any given instant is combined with substrate, and the rate of the overall chemical process thus reaches a limiting value proportional to the concentration of enzyme present, but independent of the concentration of (excess) substrate present. We now pursue the algebraic formulation of the problem, in order to present the mathematical justification for the above reasoning.

If combination of substrates with enzyme is to be an essential feature of an enzyme-catalyzed reaction, it would be expected that the simplest general model from which to begin a formal analysis would be that shown below:

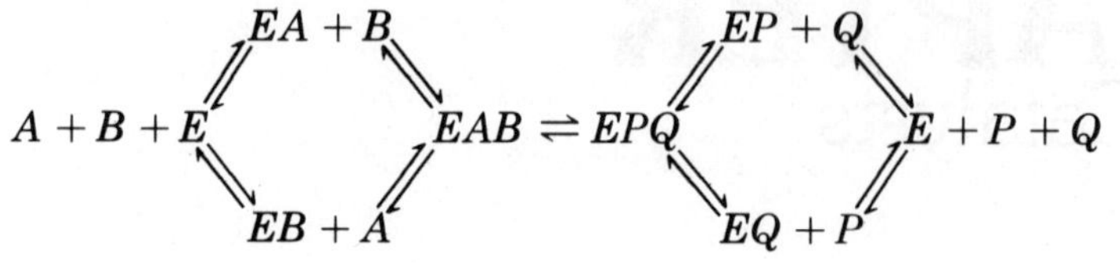

$$A+B+E \begin{matrix} \nearrow EA+B \searrow \\ \\ \searrow EB+A \nearrow \end{matrix} EAB \rightleftharpoons EPQ \begin{matrix} \nearrow EP+Q \searrow \\ \\ \searrow EQ+P \nearrow \end{matrix} E+P+Q \qquad (10\text{-}1)$$

in which A and B are the two reactants, P and Q the products, E the free catalyst, and EA, EB, EAB, EPQ, EP, and EQ the various possible combinations between enzyme and substrate(s) or product(s). *The algebra corresponding to this model is unmanageable for all practical purposes.* One immediate simplification is to ignore the second substrate, as we will later be able to justify, leaving

$$E + A \rightleftharpoons EA \rightleftharpoons EP \rightleftharpoons E + P \qquad (10\text{-}2)$$

in which A and P now denote substrate and product in the usual literature notation. A further reduction is to ignore the presence of all but one chemical intermediate:

$$E + A \underset{k_{-1}}{\overset{k_1}{\rightleftharpoons}} EA \underset{k_{-2}}{\overset{k_2}{\rightleftharpoons}} E + P \qquad (10\text{-}3)$$

Further progress may be made by limiting the observation period to include only the *initial* rate of the reaction, before any significant amount of product has been formed—the back-reaction of E with P to form EA may then be ignored:

$$\boxed{E + A \underset{k_{-1}}{\overset{k_1}{\rightleftharpoons}} EA \overset{k_2}{\longrightarrow} E + P} \qquad (10\text{-}4)$$

The final simplifying assumption is that the concentration of enzyme is small compared with the (initial) concentration of substrate and also with the (ultimate equilibrium) concentration of product. The simplification provided by the approximations that led to Eqs. 10-2, 10-3, and 10-4 is intuitively evident; the reason for the final condition

$$\boxed{[A]_0 \gg [E]_0} \qquad (10\text{-}5)$$

is that it leads to a *steady-state* concentration of the EA complex. Note that the restrictions of single substrate, single intermediate, and irreversible formation of product can readily be removed later on without disrupting the basic form of the result, but the steady-state approximation is absolutely essential to a compact and intuitively satisfying treatment of enzyme kinetics. (Details of the *mechanisms* suggested by steady-state kinetics may be tested using *transient* kinetics methods to be discussed in Section 4.) Since the steady-state condition, Eq. 10-5, is really the only critical assumption, its origin is now demonstrated.

Having gone to the trouble of reducing a proposed general enzyme kinetics scheme (Eq. 10-1) to the simplest useful form (Eq. 10-4), it is now necessary to return to an elementary level to build up the background needed to handle the solution of Eq. 10-4. From the next five illustrations (Figs. 10-1 to 10-5), the reader can trace the development of a kinetic model that leads to a steady-state concentration of an intermediate molecule—the algebra accompanying those figures is included for completeness, but should not detract from their simplicity. The *biochemically* useful results follow from the algebra in Chapter 10.A.6.

10.A.2. Single Forward Reaction

$$\boxed{A \xrightarrow{k_1} B} \tag{10-6}$$

The notation of Eq. 10-6 is shorthand for an equation that describes the rate of disappearance of reactant, A:

$$d[A]/dt = -k_1[A] \tag{10-7}$$

the familiar "first-order" decay already encountered in Chapter 9. This equation has already been solved (see Eqs. 9-1 to 9-6), with the result shown pictorially in Fig. 10-1.

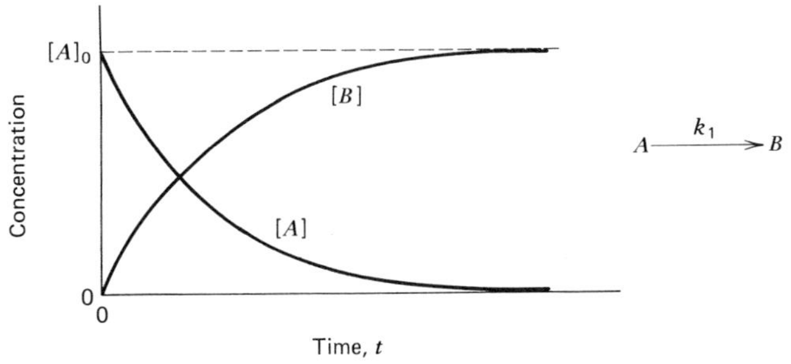

FIGURE 10-1. Plot of reactant concentration, $[A]$ and product concentration, $[B]$, for the reaction scheme of Eq. 10-6, for the initial condition that no product was present originally: $[B]_0 = 0$.

10.A.3. Forward and Back Reaction

$$\boxed{A \underset{k_{-1}}{\overset{k_1}{\rightleftarrows}} B} \tag{10-8}$$

Equation 10-6 describes a very simple chemical reaction rate expression; to progress toward the desired Eq. 10-4, the next stage is to include the possibility of back reaction:

$$d[A]/dt = -k_1[A] + k_{-1}[B] = -d[B]/dt \tag{10-9}$$

In other words, the amount of A present decreases at a rate proportional to the amount of A and increases according to the amount of B present. We have already solved the slightly simpler case for which $k_1 = k_{-1}$ in our treatment of chemical dating (Chapter 9.A, Equations 9-16 to 9-28). Because the present problem is quite analogous, it is left as a homework problem to show that

$$[A] = [A]_0 - \frac{k_1[A]_0}{k_1 + k_{-1}}(1 - \exp[-(k_1 + k_{-1})t]) \tag{10-10}$$

$$[B] = \frac{k_1[A]_0}{k_1 + k_{-1}}(1 - \exp[-(k_1 + k_{-1})t]) \tag{10-11}$$

where it has been assumed that no product was present initially: $[B]_0 = 0$.

The principal content of Equations 10-10 and 10-11 is seen in Fig. 10-2. $[A]$ and $[B]$ again approach limiting values in an exponential way,

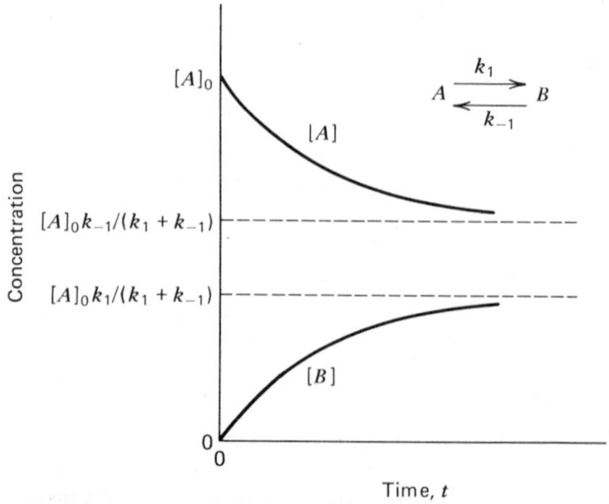

FIGURE 10-2. Plot of concentrations of components A and B for the reaction scheme of Eq. 10-8; the initial concentration of B has been assumed to be zero. The special case for which $k_1 = k_{-1}$ has already been illustrated (see Fig. 9-7).

but because of the back reaction, $[A]$ never decays to zero, and the final limiting concentrations of $[A]$ and $[B]$ are reached more rapidly than in the preceding example. In both Figures 10-1 and 10-2, it should be noted that the reaction is essentially *complete* before any of the concentrations reaches a *constant* level.

10.A.4. Consecutive Reactions

$$\boxed{A \xrightarrow{k_1} B \xrightarrow{k_2} P} \tag{10-12}$$

By now it is becoming clear that even very simple kinetic models can generate extensive algebraic effort in their solution! The last case that we solve explicitly is that indicated in the usual "shorthand" of Eq. 10-12:

$$d[A]/dt = -k_1[A] \tag{10-13}$$

$$d[B]/dt = k_1[A] - k_2[B] \tag{10-14}$$

$$d[P]/dt = k_2[B] \tag{10-15}$$

We already know how to solve Eq. 10-13:

$$\boxed{[A] = [A]_0 \exp[-k_1 t]} \tag{10-16}$$

Substituting for $[A]$ in Eq. 10-14, one obtains

$$d[B]/dt = k_1[A]_0 \exp[-k_1 t] - k_2[B] \tag{10-17}$$

for which the solution is readily verified as

$$\boxed{[B] = \frac{k_1[A]_0}{k_2 - k_1}(\exp[-k_1 t] - \exp[-k_2 t])} \tag{10-18}$$

Substituting Eq. 10-18 for $[B]$ in Eq. 10-15, one finds

$$d[P]/dt = \frac{k_2 k_1[A]_0}{k_2 - k_1}(\exp[-k_1 t] - \exp[-k_2 t]) \tag{10-19}$$

that the reader should be able to solve to give:

$$\boxed{[P] = \frac{k_1[A]_0}{k_2 - k_1}(\exp[-k_2 t] - 1) - \frac{k_2[A]_0}{k_2 - k_1}(\exp[-k_1 t] - 1)} \tag{10-20}$$

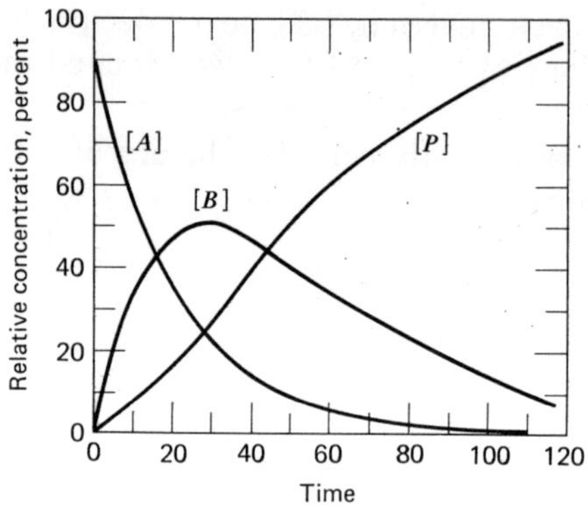

FIGURE 10-3. Concentration changes in consecutive first-order reactions. After W. J. Moore, *Physical Chemistry,* 3rd ed., Prentice-Hall, Englewood Cliffs, N. J., 1962, p. 267.

The calculated concentrations, $[A]$, $[B]$, and $[P]$ are shown in Fig. 10-3, for the case, $k_1 = 2k_2$. Figure 10-3 shows that there is an instant (at a middle stage of completion of the reaction) during which the concentration of the middle species, $[B]$, levels off.

10.A.5. Uncatalyzed Reaction with One Intermediate

$$A + B \underset{k_{-1}}{\overset{k_1}{\rightleftharpoons}} X \xrightarrow{k_2} P \qquad (10\text{-}21)$$

The scheme, Eq. 10-21, resembles the desired enzyme-catalyzed reaction sequence, Eq. 10-4, except that the reactant, B, is not regenerated at the conclusion of the reaction. The mathematical equations corresponding to Eq. 10-21 are:

$$d[A]/dt = -k_1[A][B] + k_1[X] = d[B]/dt \qquad (10\text{-}22)$$

$$d[X]/dt = k_1[A][B] - k_{-1}[X] - k_2[X] \qquad (\text{"rate" equations}) \qquad (10\text{-}23)$$

$$d[P]/dt = k_2[X] \qquad (10\text{-}24)$$

and

$$[A]_0 - [A] = [B]_0 - [B] \qquad (\text{"conservation" equation}) \qquad (10\text{-}25)$$

Equations 10-22 to 10-24 give the rates of change of concentration of each of the components; for example, in Eq. 10-22, component A disappears at a rate proportional to the amount of each reactant present, and is formed

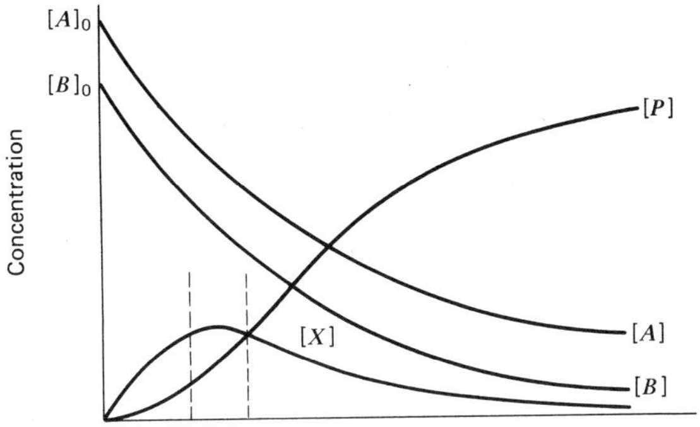

FIGURE 10-4. Progress curve for the reaction sequence

$$A + B \underset{k_{-1}}{\overset{k_1}{\rightleftharpoons}} X \xrightarrow{k_2} P$$

with $k_1 \cong k_{-1} \cong k_2$ and $[A]_0 \cong [B]_0$. Adapted from W. W. Cleland by H. R. Mahler and E. H. Cordes, *Biological Chemistry,* Harper & Row, N. Y., 1966, p. 224.

at a rate proportional to the amount of X present. Equation 10-25 simply states that for every A molecule that reacts, a molecule of B also reacts.

With the above four independent equations in the desired four unknowns ($[A]$, $[B]$, $[X]$, $[P]$), a solution exists—unfortunately, it is not in general possible to express that solution explicitly, and the resultant concentrations must be approximated (to any desired accuracy) by numerical integration using a digital computer. One typical solution is shown in Fig. 10-4, for $k_1 \cong k_{-1} \cong k_2$ and $[A]_0 \cong [B]_0$. The qualitative behavior of the system in Fig. 10-4 is similar to that of the simpler case of consecutive reactions in Fig. 10-3.

10.A.6. Catalyzed Reaction with One Intermediate

$$\boxed{A + E \underset{k_{-1}}{\overset{k_1}{\rightleftharpoons}} EA \xrightarrow{k_2} P + E} \qquad (10\text{-}26)$$

Equation 10-26 is the kinetic scheme that is the usual starting point for enzyme-catalyzed reaction kinetics, and differs from the preceding example (Eq. 10-21) only in that the catalyst, E, is regenerated on completion of the reaction sequence. The corresponding rate equations are:

$$d[A]/dt = -k_1[A][E] + k_{-1}[EA] \qquad (10\text{-}27)$$

$$d[EA]/dt = k[A][E] - k_{-1}[EA] - k_2[EA] \qquad (10\text{-}28)$$

$$d[P]/dt = k_2[EA] \qquad (10\text{-}29)$$

$$d[E]/dt = -k_1[A][E] + k_{-1}[EA] + k_2[EA] \qquad (10\text{-}30)$$

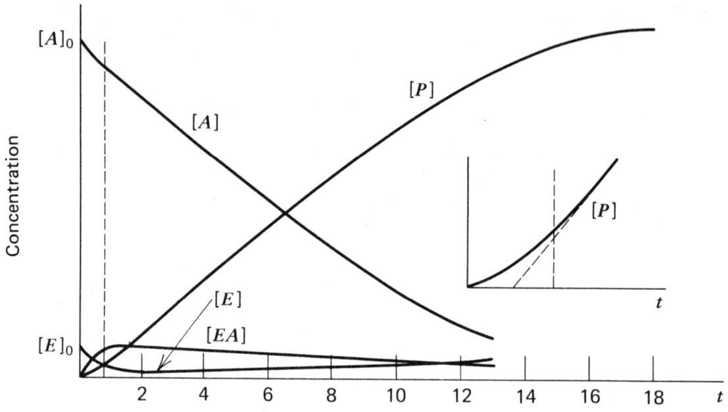

FIGURE 10-5. Progress curve for the reaction sequence

$$A + E \underset{k_{-1}}{\overset{k_1}{\rightleftharpoons}} EA \xrightarrow{k_2} P + E$$

with $k_1 \cong k_{-1} \cong k_2$ and $[A]_0 >> [E]_0$. Adapted from W. W. Cleland by H. R. Mahler and E. H. Cordes, *Biological Chemistry*, Harper & Row, N. Y., 1966, p. 226.

along with

$$[E]_0 = [E] + [EA] \qquad \text{("conservation" equation)} \tag{10-31}$$

Although no explicit solution of Equations 10-27 to 31 is possible in general, computer-assisted numerical integration again provides a result (Fig. 10-5). The new and critical features of Fig. 10-5 are that following an initial "induction" period (region to the left of the dotted vertical line in Fig. 10-5), the *concentration* of the *intermediate species,* $[EA]$, is approximately *constant,* and there is also an approximately *constant rate of appearance of product* (slope of the $[P]$ versus time curve at the right of the dotted line in the figure). This portion of the reaction progress curve is called the "*steady-state*" region: the larger the initial excess concentration of reactant over enzyme, $[A]_0 >> [E]_0$, the shorter the "induction" period during which the steady-state condition (constant $[EA]$) is invalid.

The steady-state condition, $[EA]$ = constant, or equivalently, $d[EA]/dt \cong 0$, provides for great simplification in the rate equations—in fact, it is no longer necessary to solve any differential equations at all! Consider Eq. 10-28:

$$d[EA]/dt = k_1[A][E] - (k_{-1} + k_2)[EA] = 0$$

or

$$(k_{-1} + k_2)/k_1 = [E][A]/[EA] \tag{10-32}$$

Because the right-hand side of Eq. 10-32 has a constant value and because its form resembles that of a simple equilibrium constant, it is usual to define a "Michaelis" constant, K_A

$$\boxed{K_A = (k_{-1} + k_2)/k_1} \tag{10-33}$$

K_A may be thought of as an "equilibrium" constant for dissociation of EA, *either* back to reactants E and A *or* on to products E and P. $(1/K_A)$ is thus a measure of the *strength of binding* of substrate to enzvme: a smaller value of K_A corresponds to stronger binding. Rewriting Eq. 10-32,

$$[EA] = [E][A]/K_A \tag{10-34}$$

Now the rate of appearance of product (i.e., the desired observable) is proportional to $[EA]$, but it is desirable to have a more useful expression for $[EA]$ than Eq. 10-34. Therefore, noting that the total amount of enzyme initially present must be in one of the two forms, $[E]$ and $[EA]$,

$$\boxed{[E]_{\text{total}} = [E] + [EA]} \tag{10-35}$$

Equation 10-34 may be rewritten as

$$[EA] = ([E]_{\text{total}} - [EA])[A]/K_A$$

or

$$[EA] = [E]_{\text{total}}[A]/(K_A + [A]) \tag{10-36}$$

Finally, the rate of appearance of product, $d[P]/dt = v$, the velocity of the overall reaction, may now be obtained from Equations 10-36 and 10-29:

$$d[P]/dt = v = k_2[E]_{\text{total}}[A]/(K_A + [A]) \tag{10-37}$$

The graphical behavior of reaction velocity as a function of substrate concentration is shown in Fig. 10-6: v is proportional to $[A]$ at relatively low concentrations of A, but becomes independent of $[A]$ when there is a sufficient excess amount of A present. The simple kinetic model, Eq. 10-4, is thus able to account for the puzzling (lack of) concentration-dependence sometimes shown by enzyme-catalyzed reactions discussed at the beginning of this section. Since the actual enzyme concentration, $[E]_{\text{total}}$, is often not known (i.e., most enzyme preparations are impure), it is useful to observe that for very high substrate concentration, $[A] >> K_A$, v reaches a maximum value, V_{max}, as seen in Fig. 10-6 from Eq. 10-37:

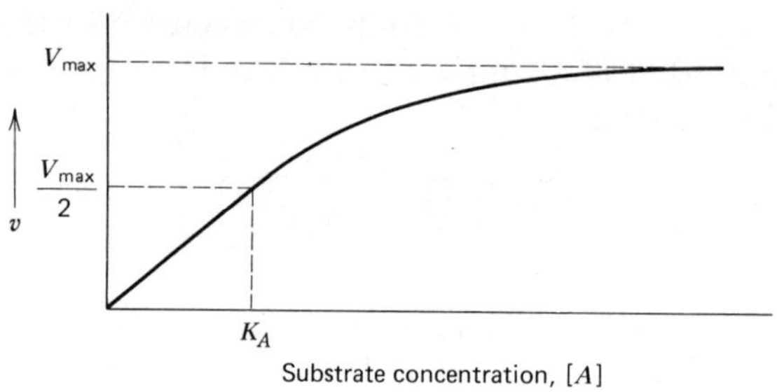

FIGURE 10-6. Plot of steady-state enzyme-catalyzed reaction velocity, v, versus substrate concentration, $[A]$, at constant enzyme concentration. At high substrate concentration, v reaches the limiting value, V_{max}. K_A is equal to the substrate concentration at half maximal rate.

$$\boxed{v = V_{max}[A]/(K_A + [A])} \tag{10-38}$$

Figure 10-6 affords another simple interpretation for K_A: K_A is the substrate concentration at which the (steady-state) reaction velocity, v, reaches half its limiting value:

$$v = V_{max}/2 = V_{max}[A]/(K_A + [A])$$

Solving for $[A]$,

$$[A] = K_A$$

when

$$v = V_{max}/2 \tag{10-39}$$

Equation 10-38 is the equation of a (rectangular) hyperbola—successful extraction of V_{max} and K_A from a graph of that equation (Fig. 10-6) depends on collection of rate data at substrate concentrations large enough to approach accurately the limiting velocity, V_{max}. One is therefore led to look for a determination of V_{max} and K_A that will give reliable results at *arbitrary* choices for substrate concentration. One idea, by analogy to the well-known titration situation, pH $= pK_a + \log([A^-]/[HA])$, is to define $pA = -\log[A]$, and to consider

$$\boxed{pA = pK_A + \log((V_{max} - v)/v)} \tag{10-40}$$

which can be derived from Eq. 10-38 (see Problems). Enzyme kinetic data can then be analyzed in the same way that one usually analyzes pH titra-

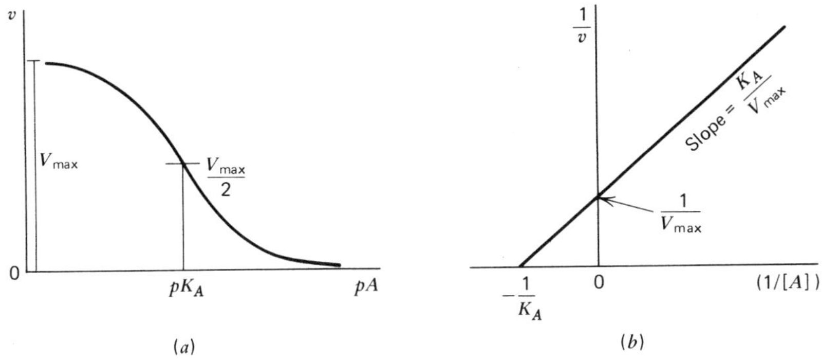

FIGURE 10-7. Alternative plots of enzyme-catalyzed reaction rate data, for use in determination of $V_{\max}$ and K_A. (*a*) Plot of v versus pA, according to Eq. 10-40. The plot resembles a titration curve, with "endpoint" given by $V_{\max}$ and "midpoint" at $[A] = K_A$. (*b*) Lineweaver-Burk plot of $(1/v)$ versus $(1/[A])$. The y intercept is $(1/V_{\max})$ and the x intercept is $-(1/K_A)$. In this plot, it is not necessary to include rate data for very high substrate concentration to obtain accurate estimates for $V_{\max}$.

tion data, as shown in Fig. 10-7*a*. The "endpoint" of the "titration" curve gives $V_{\max}$, and the "midpoint" gives $pK_A = -\log K_A$. Again with Fig. 10-7*a*, the problem is that it is necessary to collect data at very high $[A]$ to reach the "endpoint" to determine $V_{\max}$.

The difficulties with the methods of Figures 10-6 and 10-7*a* can readily be resolved by reducing the basic equation of a hyperbola (Eq. 10-38) to an equation of a straight line. Among several possible straight-line equations, the one most universally popular in enzyme kinetics is that of Lineweaver and Burk,* derived as follows. Take the reciprocal of both sides of Eq. 10-38:

$$1/v = (K_A + [A])/V_{\max}[A]$$

or

$$\boxed{1/v = (K_A/V_{\max})(1/[A]) + (1/V_{\max})} \qquad \begin{pmatrix}\text{Lineweaver-Burk}\\ \text{equation}\end{pmatrix} \qquad (10\text{-}41)$$

Equation 10-41 is plotted in Fig. 10-7b. One reason that *the Lineweaver-Burk plot is the method of choice* for obtaining $V_{\max}$ and K_A from enzyme-catalyzed reaction rate data is because it is possible to determine the x

** The alert reader will by now recognize the striking similarity between the "direct," "titration," and "Lineweaver-Burk" equations 10-38, 10-40, and 10-41 and the "direct," "Bjerrum," and "reciprocal" equations for ordinary binding equilibria given by equations 3-43, 3-48, and 3-51. The algebraic parallel is quite close, and we shall return to explore it after we have completed discussion of the diagnostic value of the Lineweaver-Burk plot.*

and y intercepts (and thus K_A and V_{max}) from data taken at *arbitrary* values of substrate concentration, $[A]$.

The principal parameters available from a Lineweaver-Burk plot are: V_{max}, K_A, and k_2.

1. V_{max}. Since $V_{max} = k_2[E]_{total}$, V_{max} can be used as a measure of the *total amount* of enzyme present in an unknown solution (enzyme "assay"). The usual convention for the magnitude of V_{max} is in terms of enzyme "units," where, for example, 1 enzyme unit = 1 μmole product formed per minute, for substrate-saturated enzyme. Enzyme activity measurements are becoming increasingly useful in medical diagnosis of disease or injury. A few examples are illustrative.

EXAMPLE *Diagnosis of Heart Disease*

The concentration of the enzyme, glutamic oxaloacetic transaminase (GOT), is particularly high in heart muscle (myocardium). Following a myocardial "infarction" (local tissue damage in the heart due to obstruction of circulation by a blood clot), there is a sharp rise in GOT activity in serum, as shown in Fig. 10-8. Severity of the attack is directly related to the level of increase in SGOT level. The significance of SGOT activity as a diagnostic index is that it

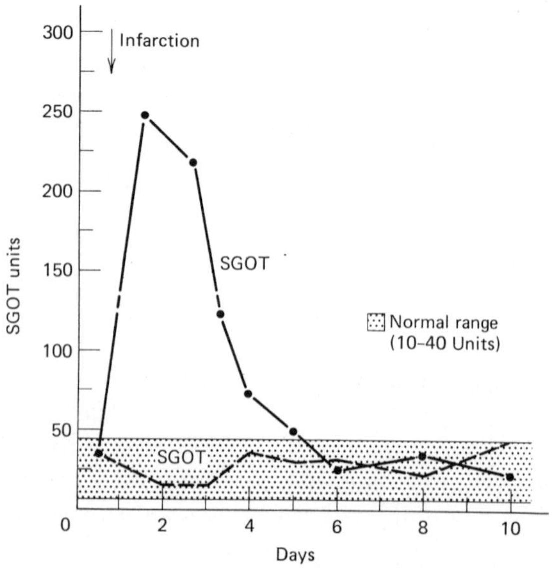

FIGURE 10-8. Concentration of serum glutamic oxaloacetic transaminase (SGOT), determined by V_{max} measurement, following acute myocardial infarction. (From E. L. Coodley, in *Diagnostic Enzymology*, ed. E. L. Coodley, Lea and Febiger, Philadelphia, 1970, p. 41.)

detects up to 96% of myocardial infarctions, while other techniques, such as the electrocardiograph (EKG) fail to detect 25 to 30% of these conditions. In particular, elevation of serum enzyme concentration is often apparent when there is complete absence of any electrocardiographic change: enzyme diagnosis thus provides what may be the earliest indication of sudden tissue damage in the heart.

EXAMPLE *Differential Diagnosis of Liver and Biliary Disease*

Additional useful diagnostic information may be obtained by determination of the concentrations of several *different* enzymes from the same serum sample. Although no one enzyme concentration may be diagnostic for a particular condition, it may still be possible to narrow down the number of possibilities by simultaneous comparison of several different enzyme concentrations. In the examples shown below (Fig. 10-9), for instance, primary biliary cirrhosis (bottom right picture) is the only condition in which there is a high level of

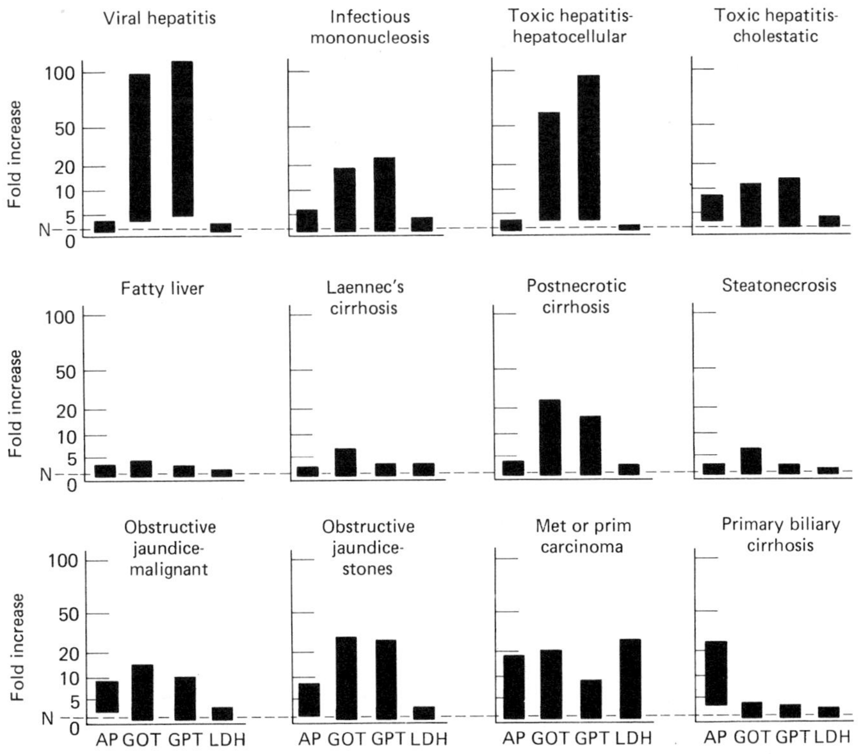

FIGURE 10-9. Relative enzyme concentrations (1-fold increase corresponds to a normal physiological condition) typically encountered in the various pathological conditions listed. The enzymes are listed in the text. (From H. J. Zimmerman and L. B. Seeff, in *Diagnostic Enzymology,* ed. E. L. Coodley, Lea and Febiger, Philadelphia, 1970, p. 29.)

alkaline phosphatase (AP, enzyme classification 3.1.3.1) *and* simultaneously low levels of glutamic oxaloacetic transaminase (GOT, 2.6.1.1), glutamic pyruvic transaminase (GPT, 2.6.1.2) and lactate dehydrogenase (LDH, 1.1.1.27).

2. K_A. The K_A parameter is primarily useful in establishing enzyme *specificity,* because K_A is a measure of the tendency of the enzyme-substrate complex to dissociate, either on to product or back to reactant. By comparing the K_A values for a number of *different* substrates, it is possible to determine the sorts of molecules that bind best to the enzyme; generally the "natural" substrate (the one on which the enzyme acts *in vivo*) is among the substrates that bind most strongly (i.e., have smallest K_A). By measurements such as these, it has been possible to classify, for example, various proteolytic (protein-breaking) enzymes according to their preferred sites of action along a polypeptide chain, as shown in Fig. 10-10. Chymotrypsin (E.C. 3.4.4.5) is shown to cleave the chain by attacking the carboxyl end of the bond between tyrosine and the adjacent amino acid; the specificity is, however, not absolute, and chymotrypsin can also bind phenylalanine, tryptophan, and methionine residues. Such specificity is often useful in cleaving a protein at known sites, prior to amino-acid analysis (see Chapter 8.D)—by cleaving at several different (known) places and analyzing

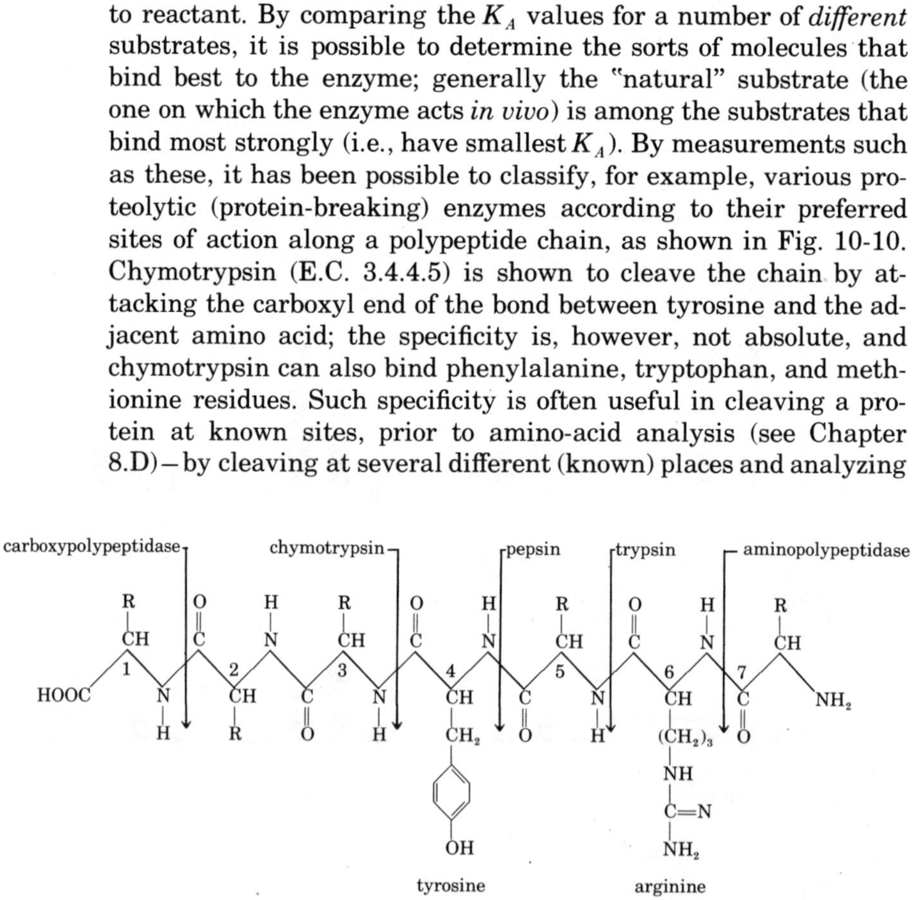

FIGURE 10-10. A diagram to indicate the specificity of various proteolytic enzymes. The numbers refer to the amino acid residues only two of which, tyrosine and arginine, are labeled below. The polypeptidases are specific, one to the free carboxyl end (left) of a protein molecule or peptide, the other to the free amino end (right) of such molecules. Pepsin is specific to the amino side of tyrosine (or phenylalanine) residues inside a protein molecule; chymotrypsin is specific to the carboxyl side of such residues; and trypsin is specific to the carboxyl side of arginine or lysine residues. (From A. C. Giese, *Cell Physiology,* W. B. Saunders Co., Philadelphia, 1962, p. 307.)

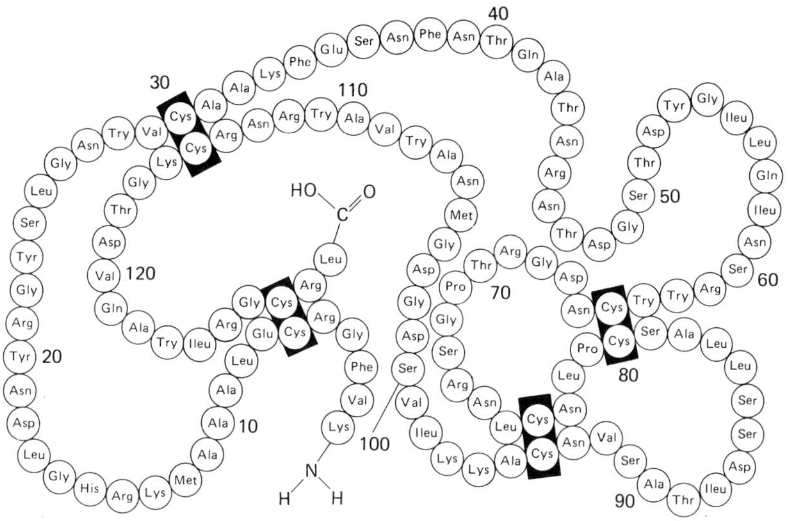

FIGURE 10-11. Amino-acid sequence of the enzyme, lysozyme, as determined by P. Jolles, *Proc. R. Soc. 167B,* 350 (1967) and by R. E. Canfield, *J. Biol. Chem. 238,* 2698 (1963).

the products, it is possible to deduce the *amino acid sequence* of a protein, an example of which appears in Fig. 10.11.

3. k_2. When the enzyme concentration is known, as for example with a purified enzyme preparation, it is possible to calculate k_2 from V_{max}:

$$\boxed{V_{\text{max}} = k_2[E]_{\text{total}}} \qquad (10\text{-}42)$$

The k_2 parameter is a measure of the *efficiency* of enzyme catalysis, and is useful in obtaining information about enzyme catalytic *mechanism.** For example, the deacylation process (k_2-step) for chymotrypsin has been shown to be 2.5 times slower in D_2O than in H_2O, suggesting that proton-transfer is involved in this rate-limiting step as would be expected for a "general base-catalysis" mechanism. From the pH-dependence of k_2 for chymotrypsin, it appears that the proposed basic group in the enzyme active site

* *k_2, the "catalytic constant," can be thought of as the number of moles of product formed per unit time by one mole of pure* enzyme *saturated with substrate. A related quantity, "turnover number," is obtained by:*

$$\textit{turnover number} = k_2/(\textit{number of active sites per enzyme molecule}) \qquad (10\text{-}43)$$

and is thus a measure of the number of molecules of product formed per unit time by one enzyme active site *saturated with substrate.*

has a pK_a of 6.8 (see pH effects later in this chapter). Typical values of k_2 range from about 10^1 to 10^6, as shown in Table 10-1.

Table 10-1 Maximum Turnover Numbers for Several Enzymes

Enzyme	Maximum Turnover Number (maximum number of molecules product formed per second per enzyme active site)
Chymotrypsin	1000
Ribonuclease	100
Carboxypeptidase	100
Kinases	1000
Catalase	5,000,000
Peroxidase	10
Carbonic anhydrase	36,000,000

The enzyme, catalase, which catalyzes the disproportionation of hydrogen peroxide, H_2O_2, to water and oxygen, has one of the highest turnover rates of any well-characterized enzyme. It is interesting to note that the catalase system is the basis for the unique defense mechanism of the bombardier beetle. This beetle possesses a "reactor gland" at the tip of its abdomen containing about 25% (wt/vol) H_2O_2. When the animal is disturbed or attacked, the catalase reaction proceeds so rapidly that the water in the gland is heated from ambient to about 100°C temperature in about a second, which in combination with the high pressure from the generated oxygen propels boiling hot water at the attacker!

10.B. REMOVAL OF MICHAELIS-MENTEN RESTRICTIONS ON ENZYME-CATALYZED REACTIONS

The necessity for a steady-state condition, Eq. 10-5, has been extensively justified in Chapter 10.A.1.–6. In this section, the other three restrictions (no back-reaction, single intermediate, single substrate) are removed, and it becomes clear that the basic character of the problem remains almost the same and the result almost as simple.

10.B.1. Back-reaction Permitted

When back-reaction between enzyme and product is permitted, the Michaelis-Menten scheme is generalized to:

$$\boxed{E + A \underset{k_{-1}}{\overset{k_1}{\rightleftharpoons}} EA \underset{k_{-2}}{\overset{k_2}{\rightleftharpoons}} E + P} \qquad (10\text{-}3)$$

The corresponding equation for rate of change of the critical component, EA, is:

$$d[EA]/dt = k_1[E][A] - (k_{-1} + k_2)[EA] + k_{-2}[E][P] \qquad (10\text{-}44)$$

and at steady-state conditions, $[E]_0 \ll [A]_0$ and/or $[P]_0$, we may set $d[EA]/dt = 0$ and solve for $[EA]$:

$$[EA] = \frac{k_1[A] + k_{-2}[P]}{k_{-1} + k_2}[E] \qquad (10\text{-}45)$$

Solving the conservation equation

$$[E]_0 = [E] + [EA] \qquad (10\text{-}46)$$

for $[E]$ and substituting for $[E]$ in Eq. 10-45,

$$[EA] = \frac{(k_1[A] + k_{-2}[P])[E]_0}{k_{-1} + k_2 + k_1[A] + k_{-2}[P]} \qquad (10\text{-}47)$$

Finally, substituting for $[E]$ and $[EA]$ in the rate equation

$$d[P]/dt = v = k_2[EA] - k_{-2}[E][P] \qquad (10\text{-}48)$$

one obtains the overall steady-state rate equation

$$\boxed{v = \frac{(k_1k_2[A] - k_{-1}k_{-2}[P])[E]_0}{k_{-1} + k_2 + k_1[A] + k_{-2}[P]}} \qquad (10\text{-}49)$$

In the absence of product ($[P] = 0$), Eq. 10-49 reduces to the usual Michaelis-Menten result

$$v_{\text{forward}} = \frac{V_{\text{max}}^{A}\,[A]}{K_A + [A]} \qquad (10\text{-}50)$$

similarly, if the concentration of "reactant" goes to zero ($[A] = 0$), the Michaelis-Menten equation again appears, but this time with regard to "product":

$$v_{\text{reverse}} = \frac{V_{\text{max}}^{P}\,[P]}{K_P + [P]} \qquad (10\text{-}51)$$

in which V_{max} and K_A have their usual meanings. These limits have two obvious implications. First, if the enzyme-catalyzed reaction is truly reversible, then the reverse reaction is also catalyzed, and if we begin with enzyme and "product," we can determine V_{max} and K_P for the "reverse" reaction as easily as we ordinarily measure the "forward" catalyzed rate.

Second, the back-reaction complication can in general be avoided simply by limiting measurements to experiments in which $[P] \cong 0$.

Direct inspection of Eq. 10-49 shows that as an enzyme-catalyzed reversible reaction proceeds, the reaction velocity slows down for two reasons. First, some product is always being removed by back-reaction (negative factor in numerator of Eq. 10-49). Second, an increasing proportion of the available free enzyme becomes tied up as EP complex, and is thus unavailable for combination with substrate ($k_{-2}[P]$ term in denominator of Eq. 10-49). The reversible enzyme-catalyzed reaction, Eq. 10-49, thus provides a simple example of biological regulation at a molecular level—product is formed rapidly when only a little product is present, but product is formed more slowly when $[P]$ is large. We consider other (more complicated) mechanisms for enzyme-catalyzed reaction rate regulation in the next chapter.

10.B.2. More Than One Intermediate

In the ordinary Michaelis-Menten scheme, Eq. 10-4, it is tacitly assumed that substrate and product bind to the enzyme to give the same complex; a more reasonable idea might be that the nature of the complex changes as the reaction proceeds:

$$E + A \underset{k_{-1}}{\overset{k_1}{\rightleftharpoons}} EA \underset{k_{-3}}{\overset{k_3}{\rightleftharpoons}} EP \xrightarrow{k_2} E + P \tag{10-52}$$

The relevant rate equations are:

$$d[EA]/dt = k_1[E][A] + k_{-3}[EP] - (k_{-1} + k_3)[EA] = 0 \text{ at steady state} \tag{10-53}$$

$$d[EP]/dt = k_3[EA] - (k_{-3} + k_2)[EP] = 0 \text{ at steady state} \tag{10-54}$$

and

$$d[P]/dt = v = k_2[EP] \tag{10-55}$$

plus the conservation equation

$$[E]_0 = [E] + [EA] + [EP] \tag{10-56}$$

The solution proceeds in the following steps. Solve Eq. 10-56 for $[E]$ and substitute for $[E]$ in Eq. 10-53; then solve Eq. 10-53 for $[EA]$ and substitute for $[EA]$ in Eq. 10-54 and solve for $[EP]$; finally, substitute for $[EP]$ in Eq. 10-55 to obtain the steady-state reaction velocity, v (see Problems)

$$v = \frac{k_3k_2k_1[A][E]_0}{k_{-1}k_{-3} + k_{-1}k_2 + k_3k_2 + [A]k_1(k_3 + k_{-3} + k_2)} \tag{10-57}$$

Except for the identity and meaning of the constants, Eq. 10-57 is thus exactly the same functional form as the single-intermediate result, Eq. 10-38. (It is readily shown that adding the possibility of back-reaction to Eq. 10-52 gives a result of the same form as the corresponding single-intermediate case, Eq. 10-49.)

In other words, the addition of a second intermediate to the basic Michaelis-Menten kinetic scheme gives a result *experimentally indistinguishable* (at least by steady-state reaction kinetics) from the simpler one-intermediate situation. Although the apparent K_A and $V_{\max}$ have entirely altered meaning in terms of the rate constants for individual elementary kinetic steps, an experimental plot of $(1/v)$ versus $(1/[A])$ will still be a straight line of slope $K_A/V_{\max}$ and y-intercept $(1/V_{\max})$. This calculation is more than another algebraic exercise—it points up one of the most serious weaknesses of steady-state enzyme kinetics, namely the *inability to detect multiple intermediates in a reaction.* It will shortly be seen in Section 4 that an entirely different experiment based on the *transient* response to a sudden impulse (rather than *steady-state* response to a sustained chemical change) provides a means for solving this important problem inaccessible to steady-state methods.

10.B.3. Two Substrates: Why Does the One-substrate Scheme Ever Fit the Data?

The calculations of the previous section illustrate a typical problem in designing scientific experiments, namely that it is useless to construct a more elaborate and detailed theory unless those details can be extracted by reduction of suitable experimental data! In this section, we consider the reaction rate laws for the three simplest cases in which two substrates are explicitly involved: the "random," the "sequential" ("ordered"), and the "ping-pong" mechanisms. It will prove possible to distinguish the "ping-pong" from the other two by the usual Lineweaver-Burke plot of $(1/v)$ versus $(1/[A])$, while more elaborate experiments are required to distinguish the "random" from the "ordered" mechanism. The three mechanisms are illustrated below.

RANDOM

$$\begin{array}{ccccc} E + A \rightleftharpoons EA & & & & EP \rightleftharpoons E + P \\ & \searrow\!\!\!\nwarrow & & \nearrow\!\!\!\swarrow & \\ & & EAB \rightleftharpoons EPQ & & \\ & \nearrow\!\!\!\swarrow & & \searrow\!\!\!\nwarrow & \\ E + B \rightleftharpoons EB & & & & EQ \rightleftharpoons E + Q \end{array} \tag{10-58}$$

ORDERED

$$E + A \rightleftharpoons EA \rightleftharpoons EAB \rightleftharpoons EPQ \rightleftharpoons EQ \rightleftharpoons E + Q \tag{10-59}$$

$$\left.\begin{array}{l} E + A \leftrightarrows EA \rightleftharpoons EP \rightleftharpoons E' + P \\ E' + B \rightleftharpoons EB \rightleftharpoons EQ \rightleftharpoons E + Q \end{array}\right\} \quad \text{PING-PONG} \tag{10-60}$$

In both the random and ordered mechanisms, both reactants must combine with the enzyme before reaction can take place; in the ping-pong mechanism, the enzyme combines with and converts first one substrate and then the other. The algebraic solution of the various mechanisms, Eqs. 10-58, 10-59, and 10-60 is straightforward, and the solutions are:

$$v = \frac{V[A][B]}{C_0 + C_1[A] + C_2[B] + [A][B]} \quad \text{Random or ordered} \tag{10-61}$$

and

$$v = \frac{V[A][B]}{C_1[A] + C_2[B] + [A][B]} \quad \text{Ping-pong} \tag{10-62}$$

in which C_0, C_1 and C_2 are combinations of various rate constants in the respective mechanisms.

Consideration of Equations 10-61 and 10-62 quickly produces some immediately useful and interesting results. First of all, if one of the reactants (say, B) is present in great *excess* over the other reactant ($[B]_0 \gg [A]_0$), then Equations 10-61 and 10-62 each reduce to the same form

$$v = \frac{V[A]}{C_2 + [A]} \tag{10-63}$$

which is exactly the same form as initially obtained in the Michaelis-Menten scheme by ignoring the presence of the second substrate completely (compare to Eq. 10-38). This situation is the case for any enzyme-catalyzed reaction (such as esterolysis) in which one of the reactants is water, as for many digestive enzymes (trypsin, chymotrypsin, pepsin, etc). The second important feature of the two-substrate rate laws is that when the concentration of the second substrate (say, B) is held *constant* while the concentration of A is varied, then a plot of $(1/v)$ versus $[1/[A]$ will still give a straight line. However, if another plot of $(1/v)$ versus $(1/[A])$ is constructed from data taken at a *different* (constant) concentration of B, then the Lineweaver-Burk plots will take on one of the two forms shown in Fig. 10-12. Equations 10-61 and 10-62 and the corresponding plots in Fig. 10-12 show that it should be possible to distinguish between either the ping-pong mechanism on the one hand (Lineweaver-Burk plots from several fixed $[B]$ form a set of parallel straight lines), and the random or ordered mechanisms on the other hand (Lineweaver-Burk plots from several fixed $[B]$ form a set of straight lines that intersect at a negative value of $(1/[A])$ and a value of $(1/v)$ that may be positive, negative, or zero). Examples of the ordered mechanism include reactions catalyzed by NAD- and NADP-requiring dehydrogenases; a typical random mechanism is followed by creatine kinase (a phosphotransferase); a ping-pong example is pyridoxal phosphate-

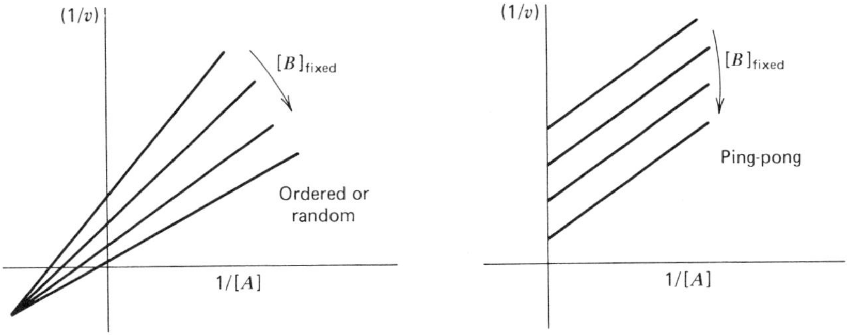

FIGURE 10-12. Plots of reciprocal enzyme-catalyzed reaction velocity versus reciprocal concentration of first substrate, $[A]$, for several different (fixed) concentrations of second substrate, B. The ping-pong mechanism produces a family of parallel lines, while both the ordered and the random mechanisms produce a family of lines that always intersect to the left of the y axis at a point that may lie above, on, or below the x axis.

requiring transaminase. Further (inhibition or isotopic-exchange) measurements are required to distinguish the random from the ordered mechanism.

In this section, we have shown that it is possible to generalize the basic Michaelis-Menten kinetic scheme, Eq. 10-4, to allow for back-reaction of enzyme with product, presence of multiple intermediates, and presence of two substrates that may react separately (ping-pong) or together (random or ordered). These complications are readily avoided or ignored in construction of a Lineweaver-Burk plot of $(1/v)$ versus $(1/[A])$ to give a straight line: (1) product inhibition effects are minimized by limiting the experimental measurements to *initial* reaction rates (so that $[P]$ is effectively zero); (2) since the form of the rate equation is the same for multiple intermediates as for one, there is nothing to be gained by including additional intermediate enzyme-substrate or enzyme-product complexes; (3) the effect of the second substrate is suppressed by restricting experiments to cases in which the second substrate concentration is constant and/or in excess. Surprisingly, the simple Michaelis-Menten scheme not only provides the *simplest* mechanism that can account for the dependence of enzyme-catalyzed reaction rates on concentration of substrate and enzyme, but relatively little new knowledge is gained by attempting to generalize the formulation to more realistic (and thus more complicated) situations. An important exception to the preceding claim is the information available from steady-state enzyme kinetics experiments regarding the mechanism by which drugs, poisons, hormones, and natural metabolites regulate the rates of specific enzyme-catalyzed reactions. However, even in analyzing the effects of inhibitors, it will suffice to consider an enzyme that combines with just one substrate to form just one intermediate, and to ignore back-reaction from product. The value of Chapter 10.A and 10.B has thus been *to establish the coarsest degree of approximation to enzyme-catalyzed reaction reality that yields readily extracted useful information.*

A final technique of major importance not treated in mathematical detail here is that of radioisotopically labeled reactants. For example, if only one reactant is labeled, the ping-pong mechanism is readily identified, since only one of the products will be labeled and the labeling will occur even in the absence of the second substrate. It is easy to imagine (but tedious to chronicle) the possible experiments using labeled reactants or products in various combinations that might be used to unravel the ordering in relatively complicated kinetic schemes—the reader should consult the Cleland reference for a wealth of examples.

PROBLEMS

1. For the simple equilibrium kinetic scheme

$$A \underset{k_{-1}}{\overset{k_1}{\rightleftharpoons}} B$$

compute expressions for $[A]$ and $[B]$ as functions of time, for the initial condition that $[B] = 0$ at time zero. (*Answer:* Eqs. 10-10, 10-11.)

2. For the case of two consecutive first-order reactions

$$A \xrightarrow{k_1} B \xrightarrow{k_2} P$$

compute expressions for $[A]$, $[B]$, and $[P]$ as functions of time, for the initial condition that $[B] = [P] = 0$ at time zero. (*Answer:* Eqs. 10-16, 10-18, 10-20.)

3. For the usual Michaelis-Menten enzyme-catalyzed reaction scheme

$$E + A \underset{k_{-1}}{\overset{k_1}{\rightleftharpoons}} EA \xrightarrow{k_2} E + P$$

show that

$$pA = pK_A + \log\left[\frac{V_{\max} - v}{v}\right] \qquad \text{(Eq. 10-40)}$$

where:

$$pA = -\log[A];\ K_A = (k_{-1} + k_2)/k_1,\ V_{\max} = k_2[E]_{\text{total}}$$

and

$$v = d[P]/dt$$

4. Derive the equation for the steady-state rate, for a Michaelis-Menten scheme with back-reaction permitted (Eq. 10-49, Chapter 10.B.1).

5. Derive the equation for the steady-state rate, for a Michaelis-Menten scheme with two intermediates (Eq. 10-57, Chapter 10.B.2).

6. Derive the equation for the steady-state rate for a Michaelis-Menten scheme with two substrates in a ping-pong mechanism (Eq. 10-62, Chapter 10.B.3).

7. Derive an equation for the steady-state rate, for a Michaelis-Menten scheme with two substrates in a simplified "sequential" sequence shown below. (Eq. 10-61, Chapter 10.B.3.)

$$E + A \underset{k_{-1}}{\overset{k_1}{\rightleftharpoons}} EA \underset{k_{-3}}{\overset{k_3}{\rightleftharpoons}} EAB \xrightarrow{k_2} E + \text{products}$$

8. Take the reciprocals of Equations 10-61 and 10-62 (the two-substrate rate equations) to show that the corresponding Lineweaver-Burk plots of $(1/v)$ versus $(1/[A])$ indeed have the form shown in Fig. 10-12.

REFERENCES

Almost all published treatments of uncatalyzed (or catalyzed) chemical reaction rates are very similar, because almost all are based on classic work by Frost and Pearson (or Cleland). The references given below contain the most readable detailed summaries of these treatments.

H. R. Mahler and E. H. Cordes, *Biological Chemistry,* Harper & Row, N. Y. (1966). Good treatment with modern examples.

M. Dixon, *Enzymes,* Academic Press, N. Y. (1964). Old, but still good.

W. W. Cleland, in *The Enzymes,* Vol. II, ed. P. D. Boyer, Academic Press, N. Y. (1970). The most formal, most general, most detailed, and least quickly readable treatment.

CHAPTER 11 Regulation of Enzyme-catalyzed Reaction Rates

In Chapter 10.A, we found that the rate of an enzyme-catalyzed reaction increases directly with the amount of substrate present. Thus, if the function of the enzyme is to *remove substrate,* the Michaelis-Menten system is already self-regulated: the more substrate present, the faster the rate of its removal. However, if the function of the enzyme is to *produce product,* then the Michaelis-Menten scheme does *not* provide regulation, since the rate of production of product is independent of product concentration. In Chapter 10.B.1, we pointed out that inclusion of back-reaction in the Michaelis scheme does provide for regulation of the amount of product, since the rate of production of product decreases as more product is formed. However, both these types of regulation are natural to the single main function of the enzyme, which is to convert substrate to product. In this chapter, we shall consider a variety of types of regulation, based on the presence of chemicals *other than* the substrates or products themselves, and based on binding of chemicals to sites not necessarily near the actual catalytic site on the enzyme. The effects of such agents provide both for "finer" (in the sense of more sensitive) control of the enzyme-catalyzed reaction rates, and also provide means for biochemical compensation of externally introduced environmental changes (such as change in pH) by altering the rates of appropriate enzyme-catalyzed rates.

11.A. DRUGS, POISONS, AND HORMONES: TYPES OF ENZYME INHIBITION AND ACTIVATION

Some drugs are *structurally nonspecific:* that is, their potency correlates with some physicochemical property such as surface tension, solubility, or degree of ionization, rather than with the presence of some specific chemically functional group. Such drugs include general anesthetics (chloroform, ethyl ether) whose physiological potency correlates with molecular size and lipid solubility; another group comprises certain hypnotics and analeptics (picrotoxin, dimefline), which act by binding nonspecifically to the outer layer of membranes and modifying the permeability of synaptic (nerve junction) membranes to ions. However, a large and increasing number of drugs and poisons are found to have their effect by changing the efficiency by which specific enzymes catalyze metabolic reactions. Because uncatalyzed reaction rates are much slower than catalyzed rates, the selectivity of such structurally specific agents provides a "chemical amplifier" effect in which just a few drug molecules can have a very large physiological effect (the potent heart drug, ouabain, is maximally effective when only ten drug molecules per cell are present). The interest in study of enzyme inhibitors is thus twofold: first, by finding which sorts of molecules are "good" inhibitors, one can begin to predict which structural features of the inhibitor are essential, and thus provide a rational basis for design of drugs and antidotes; and second, the same information can often be used to deduce the nature of the natural function of the enzyme being disrupted by the inhibitor.

As in our initial approach to enzyme kinetics, we will begin from the simplest general model that might be expected to account for the action of inhibitors. We will then make simplifying assumptions that render the model tractable and suitable for comparison with experiment, and then apply the knowledge to some examples of the action of various types of inhibitors.

Types of enzyme inhibition are usually classified as to the effect of the inhibitor on a Lineweaver-Burk plot of $(1/v)$ versus $(1/[A])$—that is, whether the presence of inhibitor changes the slope, or y intercept, or both. In order to understand the basis for this classification, consider first the general mechanism shown below:

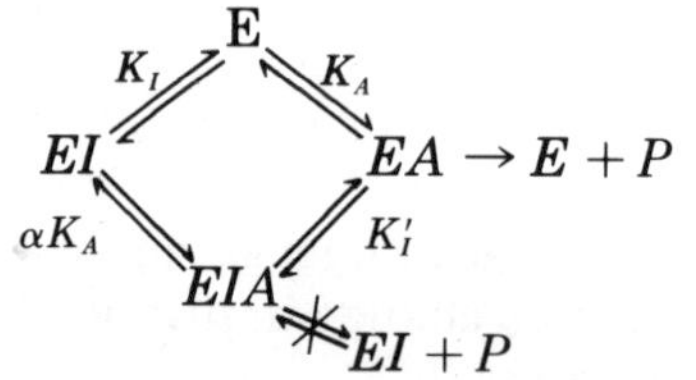

(11-1)

in which K_I = dissociation constant for dissociation of EI to E and I, K_A = dissociation constant for dissociation of EA to E and A, and so on. For example, for the step

$$E + I \underset{k_{\text{back}}}{\overset{k_{\text{forward}}}{\rightleftharpoons}} EI$$

$$K_I = k_{\text{back}}/k_{\text{forward}}$$

When we were faced with a similarly complex kinetic scheme in the previous Chapter (Eq. 10-1), it proved expedient to consider some special *limiting* cases. There are three simple limits for the scheme of Eq. 11-1, as shown in Table 11-1, of which only two are of major experimental importance. First, when binding of inhibitor, I, completely *prevents* binding of substrate, A, ($\alpha = \infty$ in Eq. 11-1), I is called a "competitive" inhibitor. In the opposite limit ($\alpha = 1$) that binding of I has *no effect* on binding of A, I is called a "noncompetitive" inhibitor. A final limit called "uncompetitive"

Table 11-1 Limiting Cases of Generalized Enzyme Inhibition (Eq. 11-1), for which the Graphical Data Reduction (Lineweaver-Burk plot of $(1/v)$ versus $(1/[A])$) Has a Simple Interpretation

Limit of Eq. 11-1	Name of Limit	Description of Limit	Experimental Examples
$\alpha \to \infty$	Competitive Inhibition	Binding of inhibitor prevents binding of substrate	Numerous (see text)
$\alpha = 1$	Noncompetitive Inhibition	Binding of inhibitor has no effect on binding of substrate	Numerous (see text)
$K_I = \infty$	Uncompetitive Inhibition	Inhibitor can bind only after binding of substrate	Rare
$\alpha \neq 1, \infty$ $K_I \neq \infty$	Mixed Inhibition	General case of Eq. 11-1	This is the usual experimental result

inhibition occurs when the inhibitor can bind to enzyme only when substrate A is already present ($K_I = \infty$ in Eq. 11-1). For the general case (α has arbitrary magnitude), inhibition is said to be "mixed." These limits are summarized in Table 11-1 (and visualized mechanistically in Fig. 11-4), and we next treat the two (competitive and noncompetitive) limits of principal experimental interest.

11.A.1. Competitive Inhibition

The kinetic scheme for competitive inhibition is obtained simply by deleting the species, *EIA*, from the general scheme, Eq. 11-1, to leave

$$\left.\begin{array}{l} E + A \underset{k_{-1}}{\overset{k_1}{\rightleftharpoons}} EA \xrightarrow{k_2} E + P \\ E + I \underset{k_{-3}}{\overset{k_3}{\rightleftharpoons}} EI \\ EI + A \to \text{no reaction} \end{array}\right\} \quad EI \overset{K_I}{\rightleftharpoons} E \overset{K_A}{\rightleftharpoons} EA \xrightarrow{k_2} E + P \tag{11-2}$$

Even though the *EI* complex is a kinetic "dead-end" in the sense that no products are formed from it, the inhibitor effectively removes some of the free enzyme E that would otherwise be available for binding of substrate A. Therefore, the concentration of EA complex is less than in the absence of inhibitor, and the reaction rate (which is proportional to $[EA]$) is decreased, resulting in inhibition of the reaction. The necessary rate equations at steady state ($[E]_0 << [A]_0, [I]_0$) are:

$$d[EA]/dt = -(k_{-1} + k_2)[EA] + k_1[E][A] = 0 \text{ at steady state} \quad (11\text{-}3)$$

$$d[EI]/dt = -k_{-3}[EI] + k_3[E][I] = 0 \text{ at steady state} \quad (11\text{-}4)$$

It is convenient to regroup the terms of Equations 11-3 and 11-4 to define K_A and K_I

$$\frac{[E][A]}{[EA]} = \frac{k_{-1} + k_2}{k_1} = K_A, \text{ the Michaelis-Menten constant for A} \quad (10\text{-}33)$$

and

$$\frac{[E][I]}{[EI]} = \frac{k_{-3}}{k_3} = K_I, \text{ the dissociation constant for } EI \text{ complex} \quad (11\text{-}5)$$

Denoting the total enzyme concentration initially present as $[E]_0$, the conservation equation 11-6 is obtained:

$$[E]_0 = [E] + [EA] + [EI] \quad (11\text{-}6)$$

Finally, the ultimate immediate goal is to determine the enzyme-catalyzed reaction rate,

$$d[P]/dt = v = k_2[EA] \quad (11\text{-}7)$$

The solution begins by solving Eq. 11-6 for $[E]$, and substituting for $[E]$ in Equations 10-33 and 11-5:

$$K_A = \frac{([E]_0 - [EA] - [EI])[A]}{[EA]} \quad (11\text{-}8)$$

$$K_I = \frac{([E]_0 - [EA] - [EI])[I]}{[EI]} \quad (11\text{-}9)$$

Then, solving Eq. 11-9 for $[EI]$,

$$[EI] = \frac{([E]_0 - [EA])[I]}{[I] + K_I} \quad (11\text{-}10)$$

and substituting $[EI]$ in Eq. 11-10 into Eq. 11-8 and solving for $[EA]$,

$$[EA] = \frac{K_I[E]_0[A]}{[I]K_A + K_I[A] + K_I K_A} \quad (11\text{-}11)$$

which when substituted into the desired rate Eq. 11-7, gives

$$v = \frac{k_2[E]_0 K_I[A]}{K_I[A] + K_A[I] + K_I K_A} \quad (11\text{-}12)$$

Equation 11-12 shows that when $[A]$ is sufficiently large, the reaction velocity v approaches a limiting maximal rate, $V_{max} = k_2[E]_0$:

$$v = \frac{V_{max} K_I[A]}{K_I[A] + K_A[I] + K_I K_A} \tag{11-13}$$

Taking the reciprocal of both sides of Eq. 11-13 yields the desired result:

$$\text{Competitive inhibition} \quad (1/v) = (1/V_{max}) + \left(1 + \frac{[I]}{K_I}\right)\frac{K_A}{V_{max}}(1/[A]) \tag{11-14}$$

Equation 11-14 is to be compared to the Michaelis-Menten result in the *absence* of competitive inhibitor:

$$\text{No inhibitor} \quad (1/v) = (1/V_{max}) + \frac{K_A}{V_{max}}(1/[A]) \tag{10-41}$$

Comparison of Equations 11-14 and 10-41 shows that the effect of the presence of a constant concentration of competitive inhibitor on a Lineweaver-Burk plot of $(1/v)$ versus $(1/[A])$ is to increase the *slope* while leaving the *y-intercept* unchanged (see Fig. 11-1).

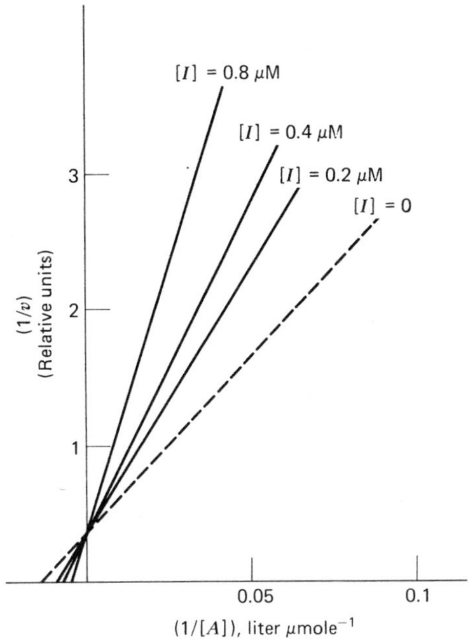

FIGURE 11-1. Competitive inhibition of acetylcholinesterase by eserine. Plots of $(1/v)$ versus $(1/[A])$ at various (fixed) concentrations of eserine. Substrate A was acetylthiocholine. From these data, it was found that $K_I = 4.6 \times 10^{-7}$ M. Since $K_A = 7.6 \times 10^{-5}$ M, it is clear that eserine is a very strong competitive inhibitor. (From J. A. Benbasat and A. G. Marshall, unpublished data.)

The qualitative consequences of the results are readily explained. Since inhibitor I and substrate A effectively compete for the same binding site on the enzyme, and since binding is in both cases reversible, it is clear that for sufficiently large excess of $[A]$ over $[I]$, essentially all the enzyme will be present as EA complex, resulting in the same limiting maximal enzyme-catalyzed rate, V_{max}. At lower concentrations of A, however, the inhibitor I effectively removes some of the enzyme as EI complex, so that the concentration of EA (and hence the rate, v) is reduced.

EXAMPLE *Eserine Inhibition of Acetylcholinesterase*

As an impulse travels along a nerve, it ultimately encounters a nerve-nerve or nerve-muscle "synaptic" junction. When the impulse reaches the "presynaptic" end of the original nerve fiber, a chemical "transmitter" substance is released from hundreds of tiny bags ("vesicles"), and a new impulse is produced at the "post-synaptic" nerve or muscle fiber as soon as the transmitter molecule can diffuse across the synaptic "cleft" to a "receptor" protein. For the postsynaptic nerve or muscle fiber to recover for receipt of other impulses later on, the transmitter molecules must be removed from the cleft once their function has been completed; this removal is accomplished by an enzyme located at the outer edges of the cleft. When the chemical transmitter is acetylcholine, as for the "parasympathetic" nervous system (whose effects include slowing of heart rate, constriction of coronary arteries, and contraction of eye

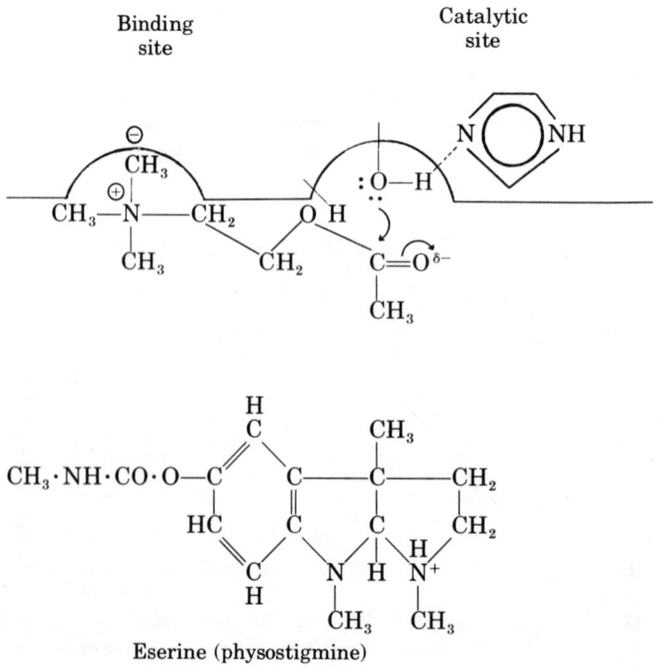

FIGURE 11-2. Active site of the enzyme, acetylcholinesterase.

pupils), the transmitter-destroying enzyme is acetylcholinesterase. When the transmitter is nor-epinephrine (nor-adrenalin), as for the "sympathetic" nervous system (whose effects include increase in heart rate, dilation of coronary arteries, and dilation of eye pupils), the transmitter-destroying enzyme is catechol-o-methyltransferase.

Acetylcholinesterase is thought to possess well-defined and distinct catalytic and binding sites (see Fig. 11-2). Eserine possesses a positively charged quaternary ammonium group that competes very effectively with the natural acetylcholine substrate for the same binding site, and thus prevents catalysis by preventing binding of substrate. The physiological effect of eserine is that acetylcholine is no longer destroyed after its stimulation of the desired neuron or muscle fiber and therefore accumulates in the synaptic cleft and continues to stimulate the next neuron or muscle. This effect is called "parasympathomimetic," since the effect mimics the normal stimulation of parasympathetic nerves. This is an example of how a physiological *stimulant* effect can result from a chemical reaction *inhibition*. The competitive inhibitor behavior of eserine is seen from the behavior of Lineweaver-Burk plots in the presence of varying (fixed) amounts of eserine in Fig. 11-1.

EXAMPLE *Sulfa Drugs*

For more than 100 years, it has been known that various dyes have particular affinity for certain types of bacteria. Reasoning that some dyes might somehow kill those bacteria, the I. G. Farbenindustrie in Germany tested its new dyes on bacterial cultures, and by 1935 had discovered the first sulfonamide ("sulfa") drug that would prevent further multiplication of hemolytic streptococci in mice. From the literally thousands of chemical derivatives synthesized since then, it has become clear that the sulfonamide moiety is the critical functional component (see examples below).

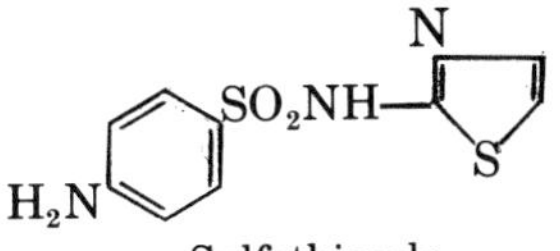

Sulfathiazole
(for severe infections)

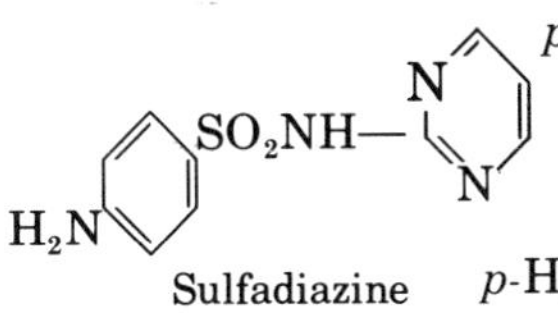

Sulfadiazine
(has low toxicity)

p-$H_2NC_6H_4SO_2NHCONH_2$
Sulfacetamide
(for local application)

p-$H_2NC_6H_4SO_2NHC(NH_2){=}NH$
Sulfaguanidine
(for bacillary dysentery)

The mechanism of action of the sulfonamides is to compete with p-aminobenzoic acid in the biosynthesis of the coenzymatically active form of folic acid in bacterial metabolism:

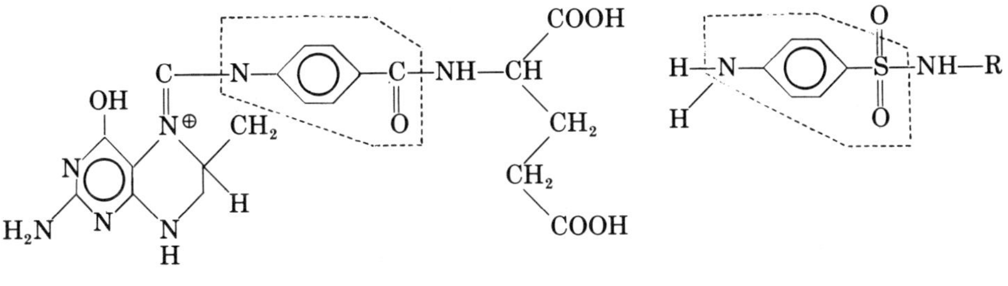

The sulfa drugs are bacteriostatic (i.e., prevent multiplication of bacteria) but not bacteriocidal (i.e., do not kill bacteria) toward pneumococci, streptococci, staphylococci, coliform bacteria of the intestine, and the organisms that cause gangrene and plague. Nowadays, many of the sulfa drugs have been replaced by antibiotics of much more complicated chemical structure derived from microbial metabolites (penicillin, terramycin, aureomycin). One of the main interests in determination of the metabolic pathways in a wide variety of organisms is to establish differences between pathways in other organisms and in humans, in order to design therapeutic drugs that can then be directed toward inhibition of critical enzymes in the disease organism that are absent or less essential in humans (see cancer chemotherapy at the end of Chapter 11.A).

11.A.2. Noncompetitive Inhibition

In noncompetitive inhibition, the inhibitor does not affect *binding* of substrate to enzyme, but does prevent reaction of substrate to form products when inhibitor is present in the enzyme-substrate complex:

$$\left.\begin{aligned} E + A &\underset{k_{-1}}{\overset{k_1}{\rightleftharpoons}} EA \xrightarrow{k_2} E + P \\ E + I &\underset{k_{-3}}{\overset{k_3}{\rightleftharpoons}} EI \\ EI + A &\underset{k_{-1}}{\overset{k_1}{\rightleftharpoons}} EIA \rightarrow EI + P \\ EA + I &\underset{k_{-3}}{\overset{k_3}{\rightleftharpoons}} EIA \rightarrow EI + P \end{aligned}\right\}$$

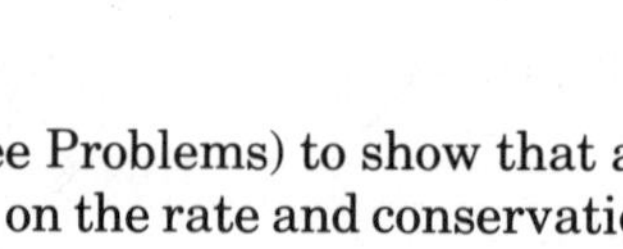

(11-15)

It is left as an exercise for the reader (see Problems) to show that algebraic reduction of the system, Eq. 11-15, based on the rate and conservation Equations 11-16a to 11-16e:

$$d[EA]/dt = -(k_{-1} + k_2)[EA] + k_1[E][A] - k_3[EA][I] + k_{-3}[EIA] \tag{11-16a}$$

$$d[EI]/dt = k_3[E][I] + k_{-1}[EIA] - k_{-3}[EI] - k_1[EI][A] \tag{11-16b}$$

$$d[EIA]/dt = k_1[EI][A] + k_3[EA][I] - (k_{-1} + k_{-3})[EIA] \tag{11-16c}$$

$$d[P]/dt = k_2[EA] \tag{11-16d}$$

$$[E]_0 = [E] + [EA] + [EI] + [EIA] \tag{11-16e}$$

yields a rate equation that can be rearranged to give ($[E]_0 << [A]_0, [I]_0$):

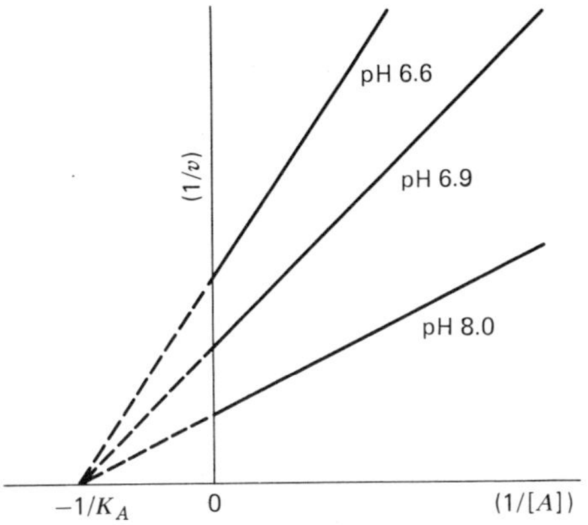

FIGURE 11-3. Noncompetitive inhibition of the α-chymotrypsin-catalyzed hydrolysis of acetyl-L-tryptophan amide by hydrogen ion. (From S. Bernhard, *Structure and Function of Enzymes,* W. A. Benjamin, Inc., New York, 1968, p. 88.)

Table 11-2 Limiting Types of Enzyme Inhibition and their Effects on a Lineweaver-Burk Plot

Type of Inhibition	Mechanistic Description	Properties of Lineweaver-Burk Plot of $(1/v)$ versus $(1/[A])$		
		y Intercept	Slope	x Intercept
No inhibitor	Michaelis-Menten	$\frac{1}{V_{\max}}$	$\frac{K_A}{V_{\max}}$	$-\frac{1}{K_A}$
Competitive inhibition	Binding of inhibitor to enzyme prevents binding of substrate and vice versa	$\frac{1}{V_{\max}}$	$\left(1+\frac{[I]}{K_I}\right)\frac{K_A}{V_{\max}}$	$-\frac{1}{K_A\left(1+\frac{[I]}{K_I}\right)}$
Noncompetitive inhibition	Binding of inhibitor to enzyme has no effect on binding of substrate, but the EIA complex cannot react to give EI + product	$\left(1+\frac{[I]}{K_I}\right)\left(\frac{1}{V_{\max}}\right)$	$\left(1+\frac{[I]}{K_I}\right)\frac{K_A}{V_{\max}}$	$-\frac{1}{K_A}$

$$(1/v) = \left(1 + \frac{[I]}{K_I}\right)\left(\frac{1}{V_{\max}} + \left(\frac{K_A}{V_{\max}}\right)\left(\frac{1}{[A]}\right)\right) \qquad \text{Noncompetitive inhibition} \qquad (11\text{-}17)$$

$$(1/v) = (1/V_{\max}) + \left(\frac{K_A}{V_{\max}}\right)(1/[A]) \qquad \text{No inhibition} \qquad (10\text{-}41)$$

Comparison of Equations 11-17 and 10-41 shows that the effect of the presence of a constant concentration of a noncompetitive inhibitor on a Lineweaver-Burk plot of $(1/v)$ versus $(1/[A])$ is to increase both the slope and the y intercept so as to leave the x intercept unchanged (see Fig. 11-3 for an experimental example). Although purely noncompetitive inhibition is uncommon, a similar mechanism can be used to account for activated binding and cooperative binding of substrates by enzymes (see "allosterism" later in this chapter).

A summary of the principal properties of competitive and noncompetitive inhibition is given in Table 11-2.

11.A.3. Mixed Inhibition

Figure 11-4 gives a schematic representation of the two extreme cases of enzyme inhibition, in which binding of inhibitor either prevents binding of substrate (competitive inhibition) or has no effect on binding of substrate (noncompetitive inhibition). It might be expected (and is in fact found) that most reversible inhibitors have an effect that is intermediate between these two extremes, and is classified as "mixed" inhibition. Although the algebraic formulation of the rate equation becomes unwieldy, the graphical treatment remains simple, and gives curves that lie between those expected for the competitive and noncompetitive limits, as shown in Fig. 11-5.

A particularly useful means for analyzing enzyme-catalyzed reaction rate data in the presence of inhibitors is obtained by re-grouping Equations 11-14 and 11-17:

$$(1/v) = \frac{1}{V_{\max}}\left(1 + \frac{K_A}{[A]}\right) + \left(\frac{K_A}{K_I V_{\max}[A]}\right)[I] \qquad \text{(competitive inhibition)} \qquad (11\text{-}18)$$

and

$$(1/v) = \frac{1}{V_{\max}}\left(1 + \frac{K_A}{[A]}\right) + \frac{1}{K_I V_{\max}}\left(1 + \frac{K_A}{[A]}\right)[I] \qquad \text{(noncompetitive inhibition)} \qquad (11\text{-}19)$$

Equations 11-18 and 11-19 form the basis of the "Dixon" plot of $(1/v)$ versus $[I]$ at constant $[A]$. Such plots (see Fig. 11-5 and discussion) are a convenient means for experimental determination of K_I.

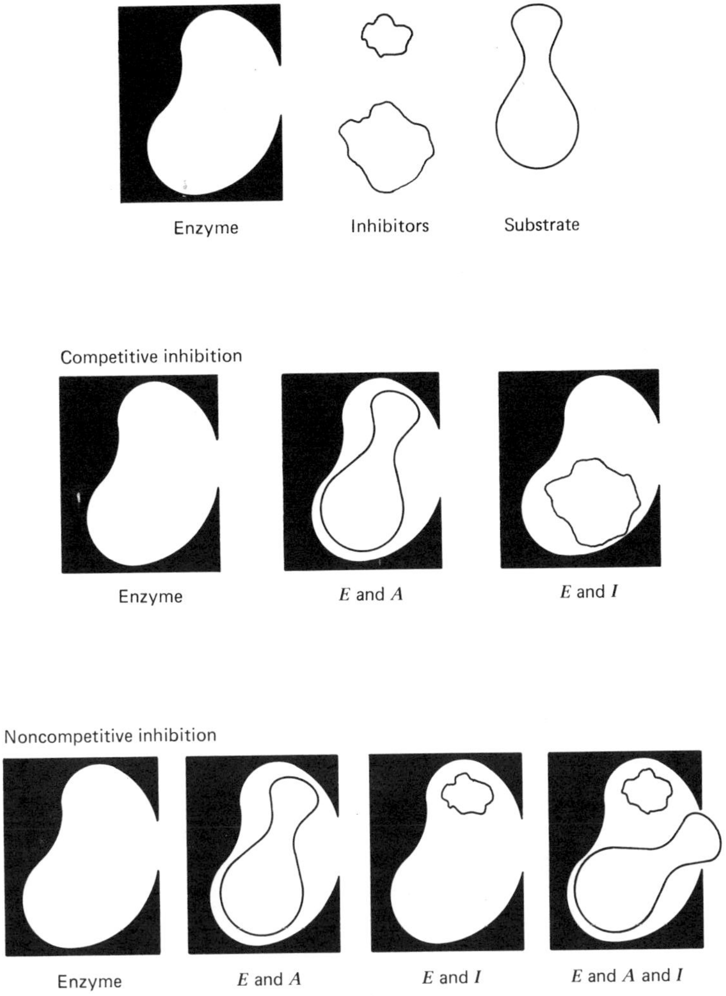

FIGURE 11-4. Schematic diagrams for competitive and noncompetitive enzyme inhibition. For competitive inhibition, substrate A can no longer bind to the enzyme once the inhibitor-enzyme complex has been formed. In noncompetitive inhibition, the substrate binds just as well whether inhibitor is present or not, but the enzyme-catalyzed reaction cannot occur when inhibitor is bound to the enzyme.

As for the equilibrium binding of small molecules to macromolecules (Chapter 3.C), there are many possible choices for graphical data reduction to extract the parameters of interest. A side-by-side comparison of the usual methods is given in Table 11-3, for binding equilibria and for enzyme kinetics in the *absence* of inhibitors. As for equilibrium binding, the "direct" and "titration"-type data reductions for enzyme kinetics suffer from the problem that it is necessary to collect rate data over a huge range of sub-

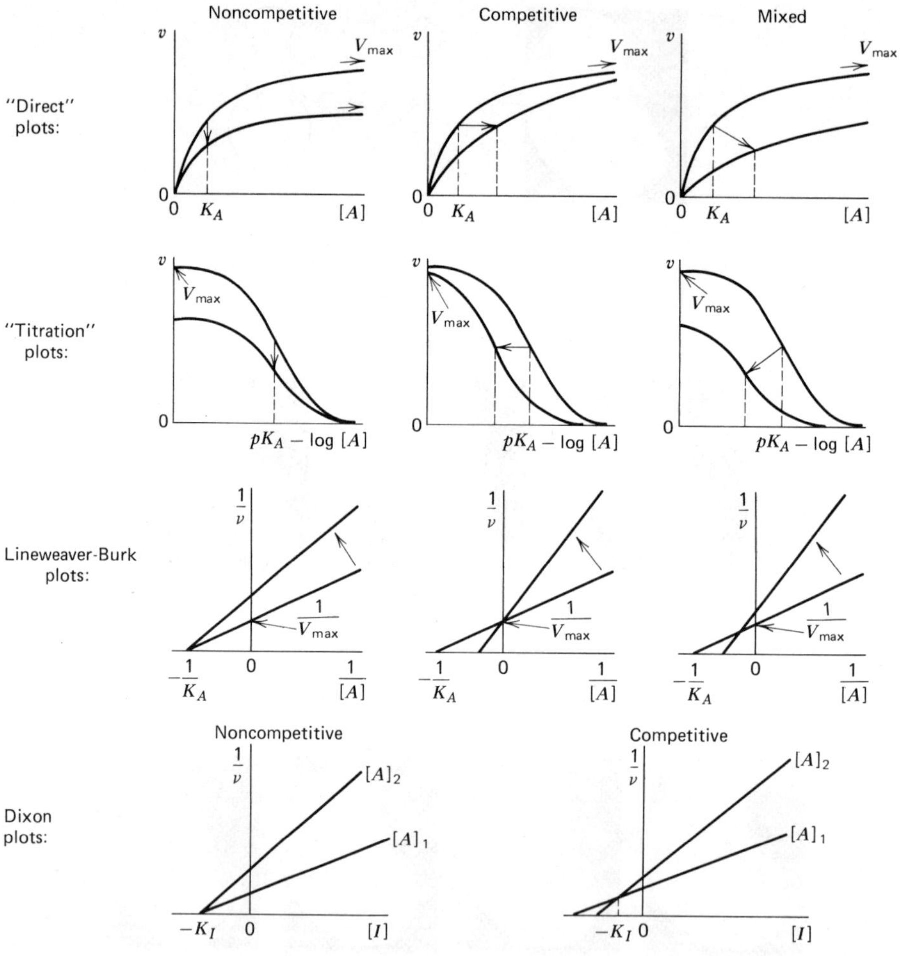

FIGURE 11-5. Various graphical displays of rate data for enzyme-catalyzed reactions in the presence of competitive, noncompetitive, or mixed-type inhibitor. In the upper nine plots, the enzyme-catalyzed reaction rate as a function of substrate concentration, $[A]$, is determined in the presence or absence of a *fixed inhibitor concentration,* $[I]$. In the two lower-most "Dixon" plots, rate data is determined at two different choices of *fixed substrate concentration,* as a function of concentration of inhibitor. Uses for the various displays are discussed in the text.

strate concentration to extract V_{max} and K_A graphically from experimental data. Among the two straight-line plots (Lineweaver-Burk and Eadie), the "Eadie" plot (analogous to the "Scatchard" plot for ordinary binding equilibria) offers the same advantages for enzyme-catalyzed rate data reduction (optimal "weighting" of data for different concentrations, simpler graphical display for multi-site catalysis), but the Lineweaver-Burk plot (analogous to the "reciprocal" plot for binding equilibria) is almost universally em-

Table 11-3 Side-by-side comparison of the equations and graphs that are used to extract either the equilibrium dissociation constant K_I (or the Michaelis-Menten constant K_A) from measurements of the fraction of binding sites occupied (or steady-state enzyme-catalyzed reaction rate), for the equilibrium binding of small molecule(s) to a macromolecule (or enzyme catalysis of a substrate reaction).*

	Binding Equilibrium		Steady-State Enzyme Kinetics
	$E + {}_nI \rightleftharpoons EI_n$; $K_1 = \frac{[E][I]}{[EI]}$ for each of n sites		$E + A \underset{k_{-1}}{\overset{k_1}{\rightleftharpoons}} EA \xrightarrow{k_2} E + P$; $K_A = \frac{[E][I]}{[EI]} = \frac{k_{-1} + k_2}{k_1}$ $V_{\max} = k_2[EI]$
Direct plot	$\nu = \frac{n[I]}{K_I + [I]}$; plot ν versus $[I]$ $n = \lim_{[I]\to\infty} \nu$; $K_I = [I]$ at $\nu = \frac{n}{2}$	Direct plot	$v = \frac{V_{\max}[A]}{K_A + [A]}$; plot v vs $[A]$ $V_{\max} = \lim_{[A]\to\infty} v$; $K_A = [A]$ for $v = \frac{V_{\max}}{2}$
Bjerrum or Titration plot	$pI = pK_I + \log\left(\frac{n - \nu}{\nu}\right)$; plot ν versus $\log[I]$ $n = \lim_{\log[I]\to\infty} \nu$; $pK_I = pI$ at $\nu = \frac{n}{2}$	Log-dose plot	$pA = pK_A + \log\left(\frac{V_{\max} - v}{v}\right)$; plot v versus $\log[A]$ $V_{\max} = \lim_{\log[A]\to\infty} v$; $pK_A = pA$ at $v = \frac{V_{\max}}{2}$
Reciprocal or Benesi-Hildebrand	$\frac{1}{\nu} = \frac{1}{n} + \left(\frac{K_I}{n}\right)\left(\frac{1}{[I]}\right)$; plot $\frac{1}{\nu}$ versus $\frac{1}{[I]}$ $n = (y \text{ intercept})^{-1}$; $K_I = -(x \text{ intercept})^{-1}$	Lineweaver-Burk plot	$\frac{1}{v} = \frac{K_A}{V_{\max}}\left(\frac{1}{[A]}\right) + \frac{1}{V_{\max}}$; plot $\frac{1}{v}$ versus $\frac{1}{[A]}$ $V_{\max} = (y \text{ intercept})^{-1}$; $K_I = -(x \text{ intercept})^{-1}$
Scatchard plot	$\frac{\nu}{[I]} = \frac{n}{K_I} - \frac{\nu}{K_I}$; plot $\frac{\nu}{[I]}$ versus ν $n = x$ intercept; $K_I = \left(\frac{y \text{ intercept}}{n}\right)^{-1}$	Eadie plot	$\frac{v}{[A]} = \frac{V_{\max}}{K_A} - \frac{1}{K_A}(v)$; plot $\frac{v}{[A]}$ versus v $V_{\max} = x$ intercept; $K_A = -(\text{slope})^{-1}$
	ν = number of sites occupied per (average) macromolecule		$\frac{v}{V_{\max}} = \frac{[EA]}{[E]_{\text{total}}} =$ average fraction of combining sites occupied by substrate

* *Notation is as in Chapter 3.C, and Chapters 10 and 11. (The equilibrium binding equations have been generalized to the binding of small molecules (or ions) to n independent equivalent sites per macromolecule, as discussed in Chapter 3.C.)*

ployed in the literature. The reasons are partly historical, and also because the Lineweaver-Burk plot provides especially simple *diagnosis* of the type of *inhibition,* when inhibitors are present. Finally, it should be noted that any of the above plots may be used to extract the inhibition parameter, K_I, provided that the substrate concentration, $[A]$, is known. However, when (as is often the case) the substrate is polymeric (polynucleotide, polypeptide, polysaccharide), there may be many potential sites for enzymatic attack along the polymer chain, and it is not easy to estimate the concentration of potentially cleavable substrate bonds. In such cases, the Dixon plot provides a simple graphical means for determination of K_I (equilibrium constant for dissociation of the enzyme:inhibitor complex), without ever having to know the *absolute* substrate concentration.

In Chapter 11.A, we have generated a number of kinetic schemes, equations, and graphs designed to provide *quantitative* criteria for the effectiveness and mechanism of various inhibitors in slowing enzyme-catalyzed reactions. One of the more recent and most important applications of these methods has been in the identification and mechanism of action of the most potent inhibitors of nucleic acid biosynthesis, because these inhibitors often show anti-tumor action, as discussed in the following example.

EXAMPLE *Cancer Chemotherapy*

The primary forms of cancer therapy are surgery and radiation. Surgery can be remarkably successful when the tumor is localized and relatively accessible: it is principally responsible for extending survival for 5 years or more in 85% of patients with skin cancer, 60% of women with breast cancer, and 40% of cases of cancer of the colon. Even when the tumor is less accessible, radiation is particularly useful in extending survival for 5 years or more in 80% of children with retinoblastoma (a cancer of the eye), in 75% of patients with Hodgkin's disease (a cancer of the lymph system), and in 50% of cases of cancer of the nasopharynx. However, the most difficult feature of cancer treatment is caused by the ability of many tumors to metastasize (i.e., detach and move through the blood or lymph circulation to other locations in the body). Because tumors are not usually clinically detectable until they contain at least 10^9 cells, many undetected metastases may remain after the primary tumor has been removed by surgery or radiation. It is in these situations that chemical treatment is increasing in popularity.

A characteristic feature of tumor cells is rapid growth and replication; chemotherapy is thus largely directed at inhibition of growth. One class of such inhibitors is the antimetabolites, which interfere with biosynthesis of nucleic acids by substituting as competitive inhibitors in specific enzymes of that biosynthetic pathway (see Fig. 11-6); an example is methotrexate, which blocks folic reductase early in the purine biosynthesis pathway. Other examples are shown in Fig. 11-6. Of the other agents listed in that figure, procarbazine (a methyl hydrazine derivative) acts by depolymerizing cellular DNA; the antibiotics (e.g., actinomycin D) bind nonspecifically by intercalating between successive turns of the DNA helix and interfering with

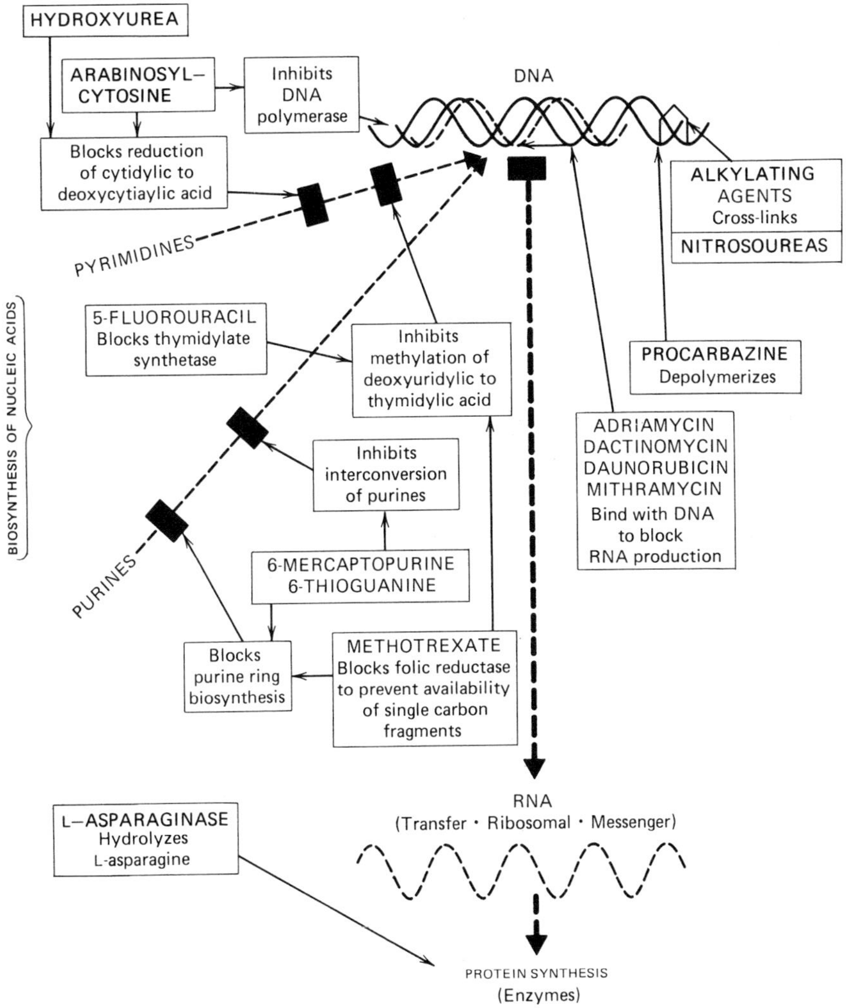

FIGURE 11-6. The mechanism of action of several antitumor agents that interfere with the replication of cells. [From Irwin H. Krakoff, Memorial Hospital for Cancer and Allied Diseases, New York City; reprinted from T. H. Maugh, II, *Science 184,* 972 (1974).]

transcription; and L-asparaginase simply hydrolyzes asparagine in the blood and thus helps to starve certain types of tumor cells that require an external source of asparagine. A recent example of progress in cancer chemotherapy is in post-operative treatment of osteogenic sarcoma, a children's bone tumor in which the 5-year survival after primary (amputation of the affected limb) therapy is less than 20%, generally due to lung metastases appearing in less than a year after surgery. Massive doses of methotrexate, followed by administration of "citrovorum factor" (another folic acid analog, but one that can serve as a substitute for the normal product of folic acid reductase, thus bypassing the inhibited enzyme) have shown an absence of metastases for up to 26 months after surgery in a large majority of cases. Multi-drug regimes made up from a combination of several of the agents shown in Fig. 11-6 have increased the incidence of long-term survival in acute leukemia from 0 to more than 50%.

One of every four cancer patients in 1954 could expect to survive 5 years or more by then-available therapy. The proportion has risen to 1 in 3 by 1975, and a substantial fraction of this 30% increase in "cure" rate (amounting to some 55,000 lives per year in the U.S. alone) can be attributed to advances in chemotherapy.

The importance of detailed knowledge of enzyme inhibitor mechanism in clinical treatment is well-illustrated by closer examination of two of the major anti-tumor drugs: methotrexate and 5-fluorouracil. The drug 5-fluorouracil is a competitive inhibitor of thymidylate synthetase, the enzyme that catalyzes the synthesis of thymidylic acid from deoxyuridylic acid; methotrexate inhibits folic reductase, as previously noted. It might be expected that better anti-tumor activity should be obtained by *combining* these two drugs, and in fact many chemotherapy programs against cancer do consist of mixtures of drugs. However, it turns out that the product of the methotrexate-inhibited reaction is a necessary *cofactor* (see Chapter 11.B.4) for binding of 5-fluorouracil to thymidylate synthetase. Thus, the action of 5-fluorouracil would be expected to be *reduced* on addition of methotrexate, and the effect of the two drugs combined should be less than the sum of the separate effects of either drug alone! This theory has been verified in cultured mouse tumor cells and may be true for humans. With this realization, one might hope to eliminate the antagonism between the two drugs by finding some *third* chemical that will replace 5-10-methylenetetrahydrofolic acid (the product of the reaction inhibited by methotrexate) as a cofactor for binding of 5-fluorouracil to thymidylate synthetase—a promising candidate seems to be the naturally occurring vitamin, folinic acid (5-formyltetrahydrofolic acid). Here is a case where unraveling the detailed enzymatic inhibitor mechanisms provided data of direct potential clinical impact. [See T. H. Maugh II, *Science 194,* 310 (1976).]

11.B. FURTHER TYPES OF ENZYME REGULATION

In Chapter 10.B.1, it was shown that accumulation of product is itself inhibitory to enzyme-catalyzed reaction rates, and thus automatically helps to stabilize the steady-state level of product (product is formed fastest when there is only a little product present, but is formed more slowly once appreciable product appears). However, since most enzyme-catalyzed reactions function to produce product rather than to speed up the attainment of equilibrium, most such reactions are "one-way" in practice, so that binding of product to enzyme is necessarily much weaker than binding of substrate to enzyme—thus, it would require a very high concentration of product to inhibit most enzyme-catalyzed reactions. In Chapter 11.A, it was seen that a more sensitive regulation is provided by the presence of external agents (inhibitors) that are themselves not chemically changed on binding to enzyme, but whose presence on the enzyme may prevent either binding or reaction of substrate (or both).

Even more sensitive control of enzyme-catalyzed rates is possible when (as we shall show in this section) the enzyme-catalyzed reaction rate is accelerated ("activated") by the presence of substrate or external agents—an increasing number of such cases appear to involve multimeric enzymes composed of several subunits, and the subunit structure may be the simplest general biological solution to "fine-control" of metabolic reaction rates. Finally, the variation of enzyme-catalyzed reaction rate with pH provides a means by which the component steps in metabolism can respond so as to counteract a variation in the external environment (pH), and we will begin this section by providing a simple mechanism for pH control of enzyme reactions.

11.B.1. pH Control of Enzyme-Catalyzed Reaction Rates

Enzyme activity typically exhibits a bell-shaped profile as a function of pH, as exemplified in Fig. 11-7 for the enzyme, histidase. Since different enzymes have optimal activity at different pH, metabolic paths can obviously be shifted from one direction to another upon change in pH, whether introduced by the organism itself or as a response to an external change. It is possible to account for these pH effects by supposing (in the simplest case) that the enzyme can exist in three states of ionization, EH_2, EH, and E, of which only one state, EH, is capable of binding substrate and catalyzing the reaction. It is therefore logical to begin by finding the relative concentrations of these various enzyme forms as a function of pH.

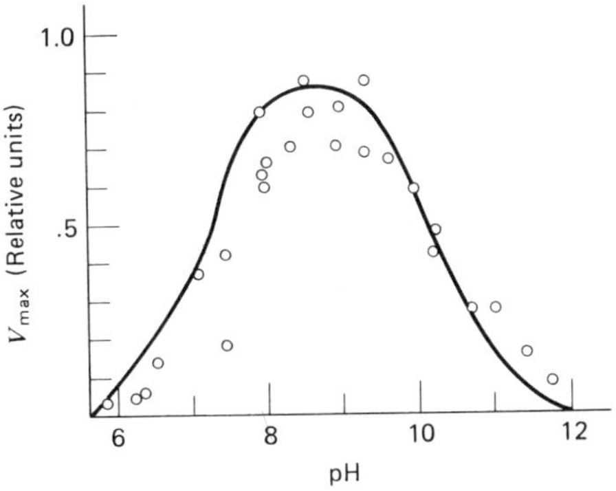

FIGURE 11-7. Enzyme activity as a function of pH, for the enzyme histidase acting on the substrate, histidine. Circles are experimental data; solid line is a plot of Eq. 11-23. [From A. C. Walker and C. L. A. Schmidt, *Arch. Biochem. Biophys.* 5, 445 (1944).]

$$E\mathrm{H}_2^{++} \underset{}{\overset{K_1}{\rightleftharpoons}} E\mathrm{H}^+ \underset{}{\overset{K_2}{\rightleftharpoons}} E \tag{11-20}$$

in which

$$K_1 = \frac{[E\mathrm{H}^+][\mathrm{H}^+]}{[E\mathrm{H}_2^{++}]}$$

$$K_2 = \frac{[E][\mathrm{H}^+]}{[E\mathrm{H}^+]} \tag{11-21}$$

From Eq. 11-21, it may be shown (see Problems) that the relative fraction of enzyme present in each of the various states is

$$[E\mathrm{H}_2^{++}]/[E]_{\mathrm{total}} = \frac{[\mathrm{H}^+]^2}{[\mathrm{H}^+]^2 + K_1[\mathrm{H}^+] + K_1K_2} \tag{11-22a}$$

$$[E\mathrm{H}^+]/[E]_{\mathrm{total}} = \frac{K_1[\mathrm{H}^+]}{[\mathrm{H}^+]^2 + K_1[\mathrm{H}^+] + K_1K_2} \tag{11-22b}$$

$$[E]/[E]_{\mathrm{total}} = \frac{K_1K_2}{[\mathrm{H}^+]^2 + K_1[\mathrm{H}^+] + K_1K_2} \tag{11-22c}$$

Since it is supposed that only the $E\mathrm{H}^+$ enzyme form is active in binding and catalysis of substrate, and since V_{max} is proportional to the concentration of active enzyme present, the bell-shaped pH profile of enzyme activity is thus accounted for by Eq. 11-23 (from Eq. 11-22b) as seen in the graphs of Fig. 11-8:

$$V'_{\mathrm{max}} = \frac{V_{\mathrm{max}}K_1[\mathrm{H}^+]}{[\mathrm{H}^+]^2 + K_1[\mathrm{H}^+] + K_1K_2} \tag{11-23}$$

in which V_{max} is the maximum enzyme-catalyzed rate that would be observed if all the enzyme were in the active $E\mathrm{H}^+$ form, and the remaining factor simply gives the fraction of total enzyme in the $E\mathrm{H}^+$ state. Equation 11-23 provides a good fit to experimental data, as seen in Fig. 11-7.

It is readily shown (see Problems) that the maximum of the bell-shaped curve in Fig. 11-8 occurs at a pH halfway between $\mathrm{p}K_1$ and $\mathrm{p}K_2$, or

$$[\mathrm{H}^+]_{\text{at maximum }[E\mathrm{H}^+]} = \boxed{\sqrt{K_1K_2} = [\mathrm{H}^+]_{\mathrm{optimal}}} \tag{11-24}$$

The *range* of pH over which the enzyme is at least half maximally active can be computed from the $[\mathrm{H}^+]$ values at which the $E\mathrm{H}^+$ fraction is equal to half its maximum value (see Problems), namely the roots of the equation

$$\boxed{[\mathrm{H}^+]^2 - (K_1 + 4[\mathrm{H}^+]_{\mathrm{opt}})[\mathrm{H}^+] + [\mathrm{H}^+]_{\mathrm{opt}}^2 = 0} \tag{11-25}$$

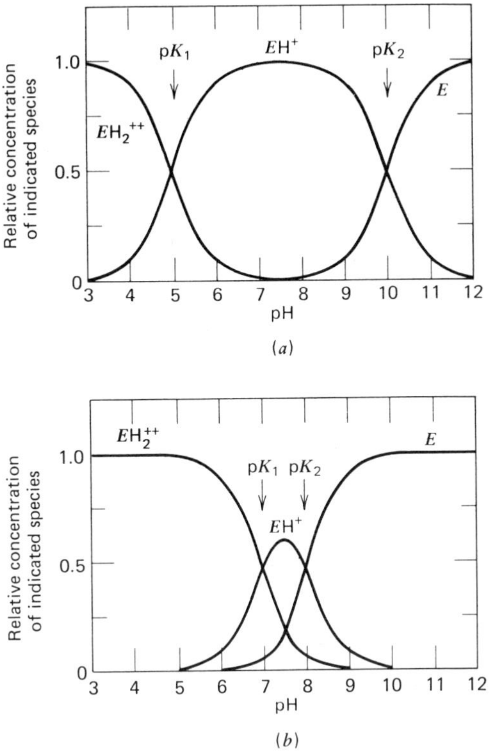

FIGURE 11-8. Relative concentration of enzyme forms, EH_2^{++}, EH^+, and EH as a function of pH. (*a*) pK_a values of the two ionizing groups are 5 and 10. (*b*) pK_a values of the two ionizing groups are 7 and 8. For a fit of such theoretical curves to experimental data, see Fig. 11-7. (From M. Dixon and E. C. Webb, *Enzymes*, 2nd ed., Academic Press, New York, 1964, p. 120.)

The equations derived so far account for the limiting rate behavior (V_{max}) under conditions of a great excess of substrate. To account for rate behavior at lower (i.e., ordinary) substrate concentrations, it is necessary to construct a slightly more elaborate mechanism, in which it is recognized that the equilibrium constant for protonation of free enzyme is not necessarily the same as for enzyme-substrate complex. It will still be assumed that only the EH^+ enzyme form is capable of binding substrate and catalyzing the reaction:

$$\begin{array}{ccccc} & E & & EA & \\ & {\scriptstyle +H^+}\big\downarrow\!\big\uparrow{\scriptstyle K_{e2}} & & {\scriptstyle +H^+}\big\downarrow\!\big\uparrow{\scriptstyle K_{ea2}} & \\ A + & EH & \underset{k_2}{\overset{k_1}{\rightleftharpoons}} & EHA & \xrightarrow{k_3} EH + \text{products} \\ & {\scriptstyle +H^+}\big\downarrow\!\big\uparrow{\scriptstyle K_{e1}} & & {\scriptstyle +H^+}\big\downarrow\!\big\uparrow{\scriptstyle K_{ea1}} & \\ & EH_2 & & EH_2A & \end{array} \tag{11-26}$$

It may be shown that the mechanism, Eqs. 11-26, leads to a result of the usual Michaelis-Menten form

$$(1/v) = (1/V'_{max}) + (V'_{max}/K'_A)(1/[A]) \qquad (11\text{-}27)$$

in which the (apparent) V'_{max} and K'_A are now functions of pH:

$$V'_{max} = \frac{V_{max}}{1 + [H^+]/K_{ea1} + K_{ea2}/[H^+]} \qquad (11\text{-}28)$$

$$K'_A = K_A \frac{1 + [H^+]/K_{e1} + K_{e2}/[H^+]}{1 + [H^+]/K_{ea1} + K_{ea2}/[H^+]} \qquad (11\text{-}29)$$

and V_{max} and K_A represent the maximal velocity and Michaelis-Menten constant that would be obtained if all the enzyme were present in EH^+ ionization state. Again, as in previous examples for two substrates, multiple intermediates, inhibitors, and so on, the result of the present complication (pH-dependence of enzyme binding and catalysis) is to produce a result of the same functional form as the simple Michaelis-Menten scheme.

11.B.2. Why do enzymes have subunits?

Part of the nature of biological control at the molecular level is based on enzyme *specificity,* which ensures that the rates of just one or a few chemical reactions are adjustable by changes in concentration of substrate or inhibitor. If the biological machine is pictured as a device with many control "knobs," then greater enzyme specificity allows for the inclusion of a greater *number* of knobs, which therefore provide very elaborate control combinations for regulation of the system. However, another easily visualized "knob" property is its "gain"—in other words, how large a perturbation is required before the system responds to a given level? The subunit structure of many enzymes effectively acts to increase the *sensitivity,* or "gain" of the system, so that a much smaller applied concentration change results in the same effect on enzyme-catalyzed reaction rate. For example, Fig. 11-9 shows the enzyme-catalyzed reaction velocity as a function of substrate concentration for a reaction (top curve) catalyzed by an isolated "catalytic" subunit of the enzyme, aspartate transcarbamylase, compared to the velocity of the same reaction (bottom curve) when catalyzed by the whole enzyme ("catalytic" subunit combined with "regulatory" subunit). It is obvious that the reaction rate varies much more sharply with substrate concentration for the assembled multimeric enzyme than for the isolated catalytic subunit—since this reaction is the first step unique to the biosynthesis of pyrimidine nucleotides, it is clearly desirable to have "fine control" of this reaction rate; that control is provided by the shift in the enzyme rate versus substrate concentration curve from its usual hyperbolic shape to a sigmoidal shape that we now attempt to explain mechanistically.

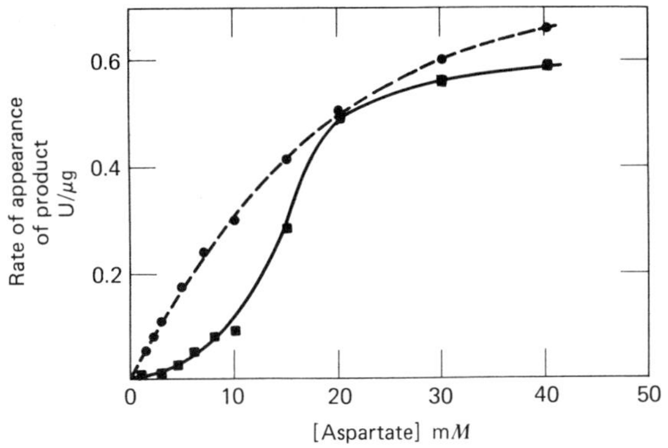

FIGURE 11-9. Plot of rate of appearance of product versus concentration of (aspartate) substrate. (Enzyme activity is expressed as units per microgram of catalytic polypeptide.) *Dotted line:* catalytic enzyme subunit alone. *Solid line:* assembled multimeric enzyme, consisting of two catalytic subunits and three "regulatory" polypeptide subunits. [From J. S. Mort and W. W.-C. Chan, *J. Biol. Chem. 250,* 653 (1975).]

Sigmoid enzyme kinetic plots, such as that in Fig. 11-9, are observed in many critical enzyme-catalyzed reactions, and there are at least half a dozen distinct types of mechanisms to account for this behavior. An increasing number of cases seem to rely on mechanisms in which there is more than one substrate-binding site per enzyme molecule (often with the additional site(s) located on separate subunit(s)), such that the reaction rate goes faster, the more substrate molecules bound to a given enzyme molecule. For example, sigmoid kinetics arise if the enzyme can form the complex EA_2 that undergoes reaction much faster than the initial EA complex. In the especially simple limit that EA is completely inactive, the steady-state appearance of product takes the form

$$E + A \underset{k_{-1}}{\overset{k_1}{\rightleftharpoons}} EA \xrightarrow{k_2[A]} EA_2 \xrightarrow{k_3} E + A + P \tag{11-30}$$

$$v = \frac{k_1 k_2 k_3 [E]_0 [A]^2}{k_3(k_1 + k_2) + k_1 k_3 [A] + k_1 k_2 [A]^2} \tag{11-31}$$

Thus at very large $[A]$, the velocity levels off to a V_{max} determined by the amount of enzyme present, $[E]_0$; at very small $[A]$, v increases (quadratically) as the square of $[A]$, and the resultant plot of v versus $[A]$ is sigmoidal. The key feature of this mechanism is that the enzyme-substrate complex becomes *increasingly* reactive as more substrate molecules bind to a given enzyme molecule; that is, there must be *interaction* between the two substrate-binding sites on the enzyme. (One can show that if binding of one substrate molecule is independent of binding of a second substrate molecule to the same enzyme molecule, ordinary hyperbolic behavior is seen in a plot of v versus $[A]$.)

In current biochemical jargon, *allosterism* refers to any increase (or decrease) in enzyme-catalyzed reaction rate resulting when a (substrate or inhibitor) molecule binds at an enzyme site remote from the active site. The previous example, Eq. 11-30, would be called allosteric *activation* by substrate. Noncompetitive inhibition (Chapter 11.A.2) is an example of allosteric *inhibition*. We next examine the two most popular current mechanisms that show how allosteric activation by substrate can result in a natural way from some simple supposed properties of a multimeric enzyme; allosteric inhibition may be explained on similar grounds.

In 1965, Monod, Wyman, and Changeux published a paper that has become the most widely quoted theoretical biochemistry paper ever written because the authors were able to account for the sigmoidal binding curves that had long been seen for binding of oxygen to the well-studied tetramer, hemoglobin, and which are becoming increasingly evident for substrate or inhibitor binding to a variety of important multimeric enzymes at control or "branch" points of metabolism. The model is based on an enzyme that consists of identical subunits; a given subunit may exist in one of two conformations; but if one of the subunits in a particular enzyme molecule has a stated conformation, then all the other subunits in that enzyme molecule must have the same conformation (see Fig. 11-10). In other words, the multimeric enzyme can exist in just two conformational forms, E_R and E_T. To account for sigmoidal kinetics, it is sufficient to suppose that the free enzyme exists predominantly in, say, the E_T form; that is, that the equilibrium constant, L,

$$L \equiv \frac{[E_T]}{[E_R]} >> 1 \tag{11-32}$$

and that the substrate has greater affinity for, say, the E_R form:

$$K_T \equiv \frac{[E_T][A]}{[E_TA]} >> \frac{[E_R][A]}{[E_RA]} \equiv K_R \tag{11-33}$$

As substrate concentration is gradually increased, substrate will first bind to the available E_R molecules, but because of the $E_T \rightleftharpoons E_R$ equilibrium, more E_R will then be formed at the expense of E_T molecules. Thus, the concentration of $[E_RA]$ complex molecules will actually increase faster (i.e., increase with $[A]$ to a higher power than first) than proportional to A concentration. This results in a sigmoidal plot of enzyme-catalyzed reaction velocity versus $[A]$, since the reaction velocity is proportional to $[E_RA]$. The mathematics behind this argument is surprisingly short, and is left as an exercise (see Problems). With the definitions

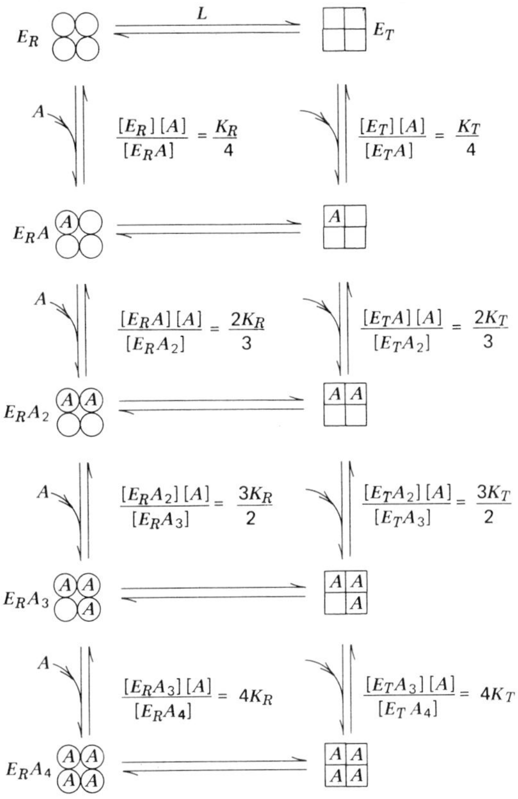

FIGURE 11-10. Schematic model for binding of substrate to a multimeric enzyme. [From J. Monod, J. Wyman, and J.-P. Changeux, *J. Molec. Biol. 12*, 88 (1965).] The enzyme itself can exist in only two forms, shown as four circles or four squares for the above tetrameric enzyme example. Each subunit is assumed to bind one substrate molecule with the same dissociation constant, except for statistical factors—for example, there are two ways for E_RA_2 to lose an A but there are three ways for an A to become attached to E_RA, leading to a $\frac{2}{3}$ statistical factor in the expression for $[E_RA][A]/[E_RA_2]$. The degree of "cooperativity" (S-shaped binding curve) in binding of substrate by enzyme is determined by the relative magnitudes of the various equilibrium constants (see Eq. 11-34 ff.)

$$\alpha = [A]/K_R$$

and

$$c = K_R/K_T \tag{11-34}$$

the fraction of enzyme sites occupied by substrate, $\overline{Y}_A$, becomes (see Problems):

$$\overline{Y}_A = \frac{L\alpha c(1+c\alpha)^{n-1} + \alpha(1+\alpha)^{n-1}}{L(1+c\alpha)^n + (1+\alpha)^n} \tag{11-35}$$

in which n is the number of subunits per multimeric enzyme molecule.

Figure 11-11 shows some representative plots of $\overline{Y}_A$ (remember that reaction velocity is proportional to the fraction of enzyme sites that are occupied by substrate, $\overline{Y}_A$) versus α (remember that α is proportional to $[A]$, Eq. 11-34), for several choices of L (the equilibrium constant for $E_R \rightleftharpoons E_T$) and c (the substrate relative preference for binding to one enzyme form over the other). The theoretical curves in Fig. 11-11 may be compared favorably with the experimental results of Fig. 11-9. It may also be noted that this model provides for ordinary hyperbolic v versus $[A]$ plot shape in the limit that (1) $c = 0$ and $L = 1$ or (2) $c =$ finite and $L =$ large.

It was pointed out at the beginning of this section that the key element in explanations of sigmoidal enzyme kinetics for multimeric enzymes is some change in enzyme conformation on binding of substrate, so that binding of subsequent substrate molecules to the same enzyme molecule is facilitated. The model of Fig. 11-10 clearly represents an extreme situation in which binding of substrate is related to a *simultaneous* change in conformation of *all* the subunits of the enzyme at once. A modified model is that binding of substrate changes the conformation of the subunit involved, and that change affects the conformational preference of adjacent subunits, as shown in Fig. 11-12; the new feature here is that the enzyme can exist

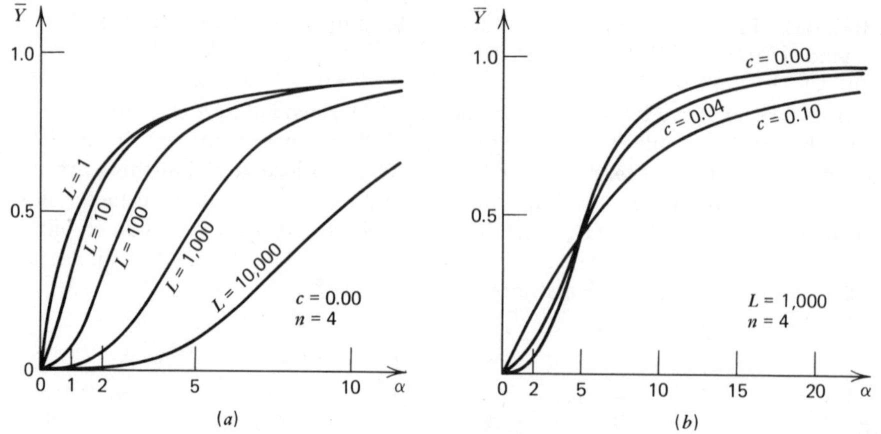

FIGURE 11-11. Theoretical plots of the fraction of enzyme sites occupied by substrate, $\overline{Y}$, versus a parameter (α) which is proportional to substrate concentration, $[A]$, for an enzyme consisting of four identical subunits ($n = 4$). Parameters L and c are defined in Equations 11-32 and 11-34. Note the ability of the Monod-Wyman-Changeux model to account for sigmoidal enzyme rate plots. [From J. Monod, J. Wyman, and J.-P. Changeux, *J. Mol. Biol. 12*, 88 (1965).]

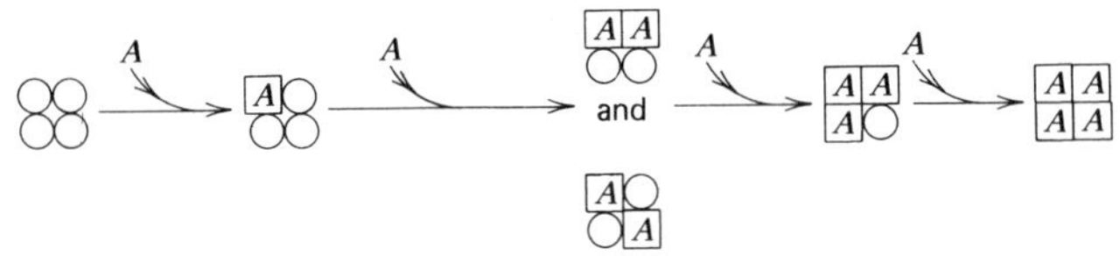

FIGURE 11-12. Alternative model for cooperative binding of substrate A, to a tetrameric enzyme. Substrate molecules are supposed to have greater affinity for the □ subunit than for the ○ subunit, and change in conformation of one subunit favors change in conformation of adjoining subunits. [See A. J. Cornish-Bowden and D. E. Koshland, Jr., *Biochemistry 9,* 3337 (1970).]

in more than two forms, according to how many of its subunits exist in a given conformation.

The model of Fig. 11-12 is also capable of generating sigmoidal kinetic curves and is not discussed further here. It is in principle possible to distinguish between the models of Figures 11-10 and 11-12 (see Section 4 on transient enzyme kinetics), but there are presently few convincing experimental examples.

As a final topic in this section, it is worth noting that the preceding models for subunit conformational change on binding of *substrate* are readily extended to subunit conformational change on binding of *other* "effector" molecules whose effect is to increase ("activator") or decrease ("inhibitor") the multimeric enzyme-catalyzed reaction rate. Figure 11-13 shows the theoretical effect of the presence of a fixed concentration of activator (top curve), or inhibitor (bottom curve) on the plot of $\overline{Y}_A$ versus α for a system in which $c = 0$ (i.e., substrate itself has affinity for only one of the two possible enzyme forms of the Monod-Wyman-Changeux model).

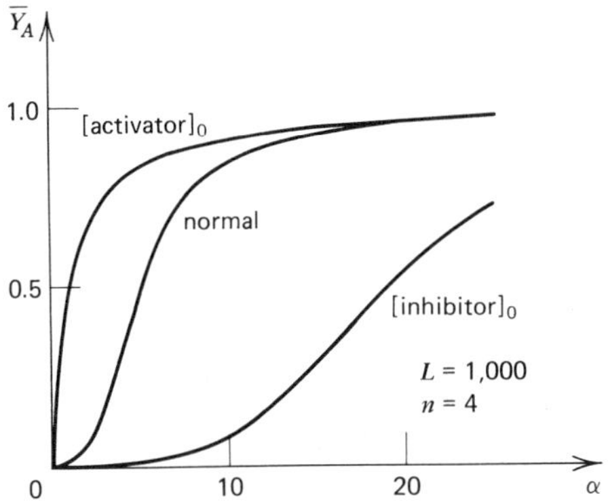

FIGURE 11-13. Extended Monod-Wyman-Changeux model of allosteric activation or inhibition of a multimeric ($n = 4$) enzyme-catalyzed reaction. Theoretical plots of fraction of enzyme sites occupied by substrate in the presence of fixed concentration of activator (*top*), inhibitor (*bottom*), or no added agents (*middle curve*), as a function of α, where α is proportional to substrate concentration, $[A]$. [From J. Monod, J. Wyman, and J.-P. Changeux, *J. Mol. Biol. 12,* 88 (1965).]

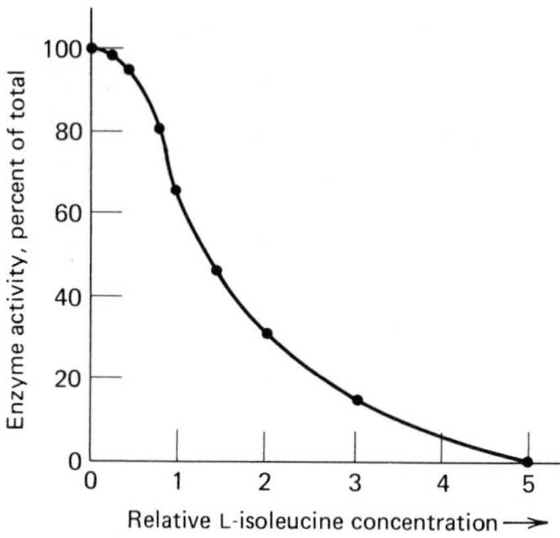

FIGURE 11-14. Effect of L-isoleucine concentration on the catalytic activity of L-threonine deaminase. [From "The Control of Biochemical Reactions" by J. P. Changeux. Copyright © April 1965 by Scientific American, Inc. All rights reserved.]

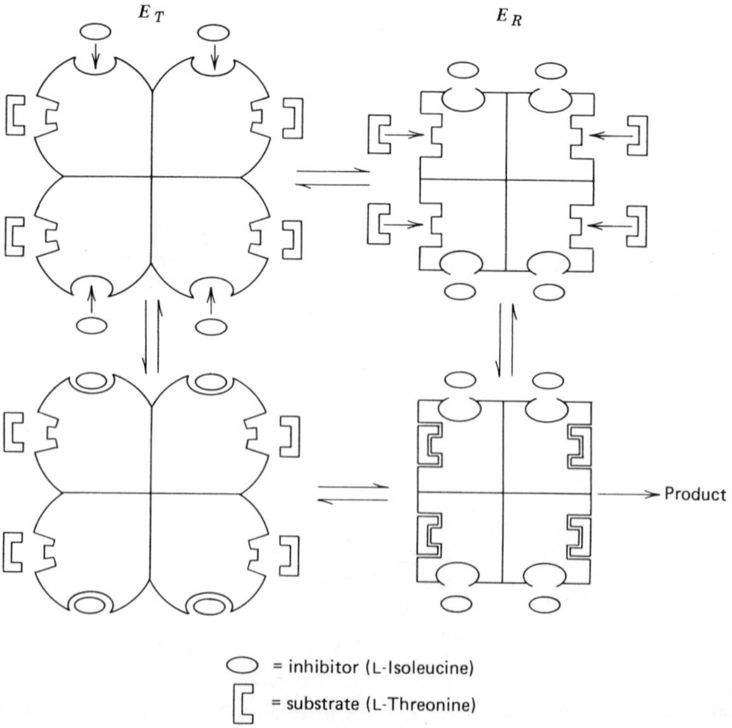

FIGURE 11-15. Schematic representation of the two conformational states of L-threonine deaminase—one that binds inhibitor (E_T state) and one that binds substrate and catalyzes its reaction (E_R state). (From M. L. Bender and L. J. Brubacher, *Catalysis and Enzyme Action*, McGraw-Hill, N. Y., 1973, p. 196.)

EXAMPLE *Allosteric Inhibition of* L*-threonine Deaminase*

Figure 11-14 shows the sigmoidal *decrease* in enzyme-catalyzed rate produced by addition of the allosteric inhibitor, L-isoleucine, for the enzyme, L-threonine deaminase. As with the sigmoidal *increase* in enzyme-catalyzed rate on addition of *substrate*, this sigmoidal *decrease* in the presence of *inhibitor* can be accounted for from an interacting subunit model.

The sigmoidal substrate-binding behavior and inhibition by isoleucine of the tetrameric enzyme, L-threonine deaminase, can be explained by the allosteric inhibition model of Fig. 11-13 as shown schematically in Fig. 11-15. In Fig. 11-15, it is supposed that there are two enzyme conformations, E_R and E_T. If substrate binds only to the E_R form, then by binding to the enzyme, the substrate will shift the $E_T \rightleftharpoons E_R$ equilibrium to the right, and initial substrate binding will be higher than first order (i.e., sigmoidal binding curve). Similarly, if the inhibitor is assumed to bind only to the E_T form, addition to inhibitor will shift the $E_T \rightleftharpoons E_R$ equilibrium to the left, leaving less available active enzyme, and resulting in allosteric inhibition.

EXAMPLE *Activation and Inhibition of Aspartate Transcarbamylase Enzyme*

Figure 11-16 shows some plots of experimental data for activation of the enzyme, aspartate transcarbamylase, by ATP, and inhibition by CTP.

These effectors are known to bind to a subunit different than the subunit which binds the substrate, and thus form classic examples of allosteric effectors.

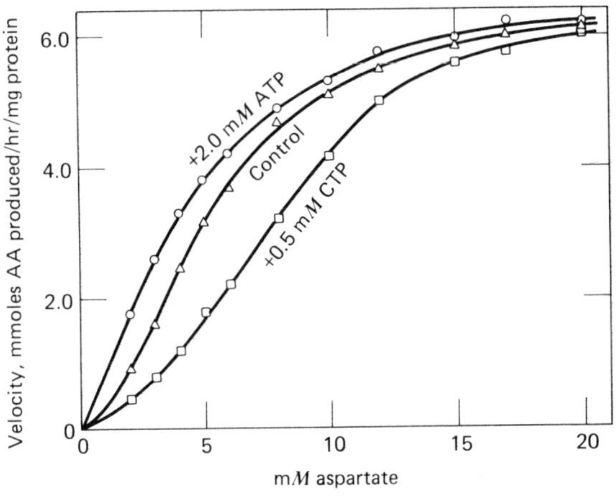

FIGURE 11-16. Plot of enzyme-catalyzed reaction velocity versus substrate concentration for reaction of aspartate with carbamyl phosphate (control) in the presence of either ATP activator (*top curve*) or CTP inhibitor (*bottom curve*). These plots should be compared with the theoretical predictions from the Monod-Wyman-Changeux model of Fig. 11-15 and it is apparent that the model can successfully account for the experimental effects. [From Fig. 3 of J. C. Gerhart, *Curr. Top. Cell. Regul. 2*, 275 (1970).]

In summary, it is possible to achieve "finer control" of enzyme-catalyzed reaction rates by shifting to a mechanism that produces sigmoid rather than hyperbolic plots of reaction velocity versus substrate concentration. One means for accomplishing this end is the development of multimeric enzymes having the properties of the models of Figures 11-10 or 11-12. In both cases, the essential feature is that binding of substrate favors a change in conformation of part or all of the enzyme.

11.B.3. Self-Inhibition by Substrate

A rather direct means for regulation of enzyme-catalyzed rates is provided by inhibition of the reaction at high substrate concentration, as shown in Fig. 11-17 for urease.

A particularly simple model for this behavior is that binding of a second substrate molecule to a given enzyme molecule effectively prevents the reaction:

$$\boxed{\begin{array}{l} E + A \underset{k_{-1}}{\overset{k_1}{\rightleftharpoons}} EA \overset{k_2}{\longrightarrow} E + P \\ \\ EA + A \underset{k_{-3}}{\overset{k_3}{\rightleftharpoons}} EA_2 \rightarrow \text{no reaction} \end{array}} \tag{11-36}$$

It is left as an exercise (see Problems) to show that the resultant steady-state rate expression takes the form

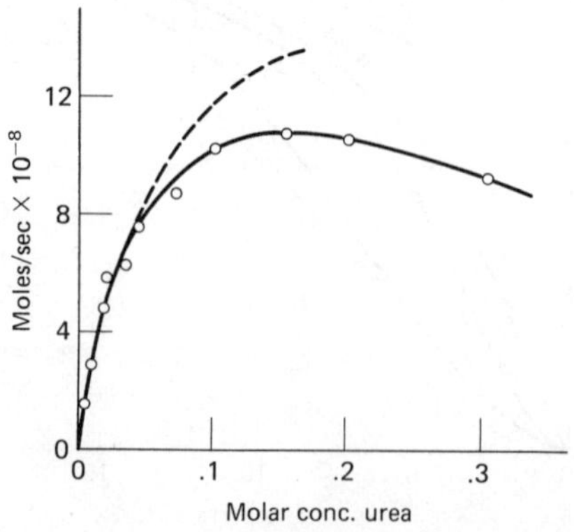

FIGURE 11-17. Self-inhibition by substrate of the decomposition of urea by urease. *Dotted line:* expected hyperbolic behavior in the absence of any inhibition. *Solid line:* experimental data showing increasing inhibition at higher substrate concentrations. [From data of K. J. Laidler and J. P. Hoare, *J. Amer. Chem. Soc. 71*, 2699 (1949) in H. B. Bull, *Introduction to Physical Biochemistry*, F. A. Davis (Philadelphia, 1971), p. 415.]

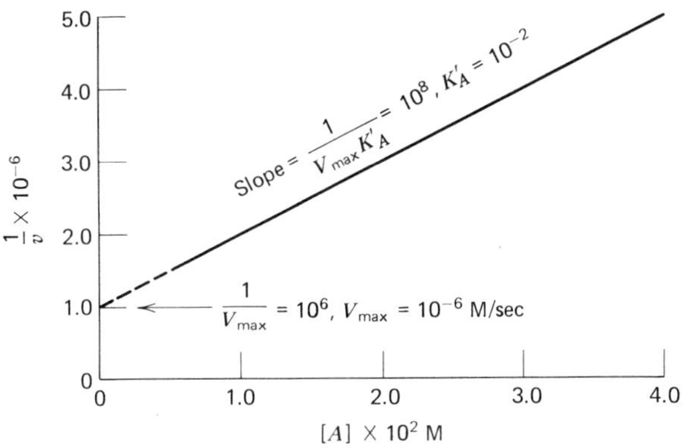

FIGURE 11-18. A typical plot of $1/v$ versus $[A]$ for an enzyme-catalyzed reaction that is inhibited by high substrate concentrations. K_A is 10^{-4} M, K'_A is 10^{-2} M, and V_{max} is 10^{-6} M/sec. (From C. Walter, *Steady-State Applications in Enzyme Kinetics,* The Ronald Press Co., N. Y., 1965, p. 80.)

$$d[P]/dt = v = \frac{V_{max}K'_A[A]}{[A]^2 + K'_A[A] + K_AK'_A} \tag{11-37}$$

By rearranging Eq. 11-37, it is evident that

$$v = \frac{V_{max}}{1 + \frac{K_A}{[A]} + \frac{[A]}{K'_A}} \cong \frac{V_{max}K'_A}{K'_A + [A]} \text{ for large } [A] \tag{11-38}$$

Thus, K'_A is readily extracted from a plot of $(1/v)$ versus $[A]$ at high $[A]$ levels (see Fig. 11-18), while K_A and V_{max} are evaluated from the usual Lineweaver-Burk plot of $(1/v)$ versus $[A]$ at very low $[A]$ levels.

At this stage, the reader may recall the strong similarity between Eq. 11-37 for self-inhibition by substrate and Eq. 11-23 for variation of enzyme-catalyzed rate with pH. Indeed, both mechanisms produce an enzyme activity that is bell-shaped with respect to log of concentration of substrate or hydrogen ion, and both mechanisms predict a maximal rate at a substrate (or hydrogen ion) concentration that occurs (see Problems) at the geometric mean of the two dissociation constants for binding of substrate (or hydrogen ion) to the enzyme (see Fig. 11-19). With the present substrate self-inhibition mechanism, the reaction rate increases as substrate is made available, but the rate eventually decreases in the presence of excess substrate, providing a direct and simple means for homeostatic regulation of the rate of production of product.

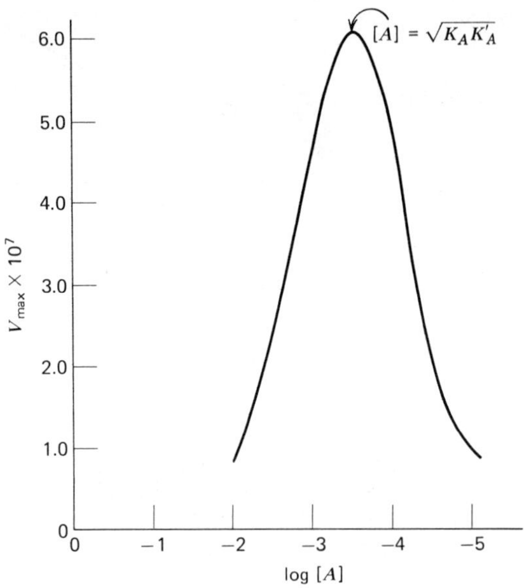

FIGURE 11-19. A theoretical plot of rate versus log of substrate concentration for an enzyme-catalyzed reaction that is inhibited by high substrate levels. The $K_A = 10^{-4}$ M, $K'_A = 10^{-3}$ M, and $V_{max} = 10^{-6}$. The substrate concentration giving the optimal rate is 3.15×10^{-4} M, and the optimal rate is 6.2×10^{-7} M, less than two-thirds the V_{max}. (From C. Walter, *Steady-State Applications in Enzyme Kinetics,* The Ronald Press Co., N. Y., 1965, p. 81.)

11.B.4. ACTIVATION BY METAL IONS

Of the 1000-odd mammalian enzymes now known, about one-fourth have metals either built into the permanent enzyme structure, require added metal ions for activity, or are further activated by metal ions. Since enzymes with metal ions firmly bound are kinetically indistinguishable from enzymes with no metal, we will concentrate in this section on enzymes for which the metal ion is relatively weakly bound and can go on and off the enzyme rapidly. The simplest general kinetic mechanism takes a form similar to that for ordinary enzyme inhibition (compare Eq. 11-1):

$$\begin{array}{ccccc} & & & & E + P \\ & & & \nearrow & \\ & & EA & & \\ & {}^{+A}\!\!\nearrow\!\!\swarrow & & \searrow\!\!\nwarrow^{+M} & \\ \mathrm{E} & & & & \mathrm{EMA} \rightarrow E + P + M \\ & {}_{+M}\!\!\searrow\!\!\nwarrow & & \nearrow\!\!\swarrow_{+A} & \\ & & EM & & \end{array} \qquad (11\text{-}39)$$

As usual, the general solution of this scheme is so complicated as to be of little practical or intuitive value. However, a large number of enzymes can be treated as if they behave according to some simple limiting cases. Supposing first that there is no reaction unless both M and A are bound to the enzyme, one can imagine two limits listed below:

1. No EA complex in absence of M (reaction proceeds by bottom path of Eq. 11-39):

$$\boxed{E + M \underset{k_{-m}}{\overset{k_m}{\rightleftharpoons}} EM \underset{k_{-A}}{\overset{k_A}{\rightleftharpoons}} EMA \overset{k}{\longrightarrow} E + P + M} \qquad (11\text{-}40)$$

for which the steady-state rate law takes the form

$$v = \frac{V_{\max}\,[A][M]}{[A][M] + K_A[M] + K_A K_M} \qquad (11\text{-}41)$$

in which $K_M = k_{-m}/k_m$ = dissociation constant for EM complex, and $K_A = (k_{-A} + k)/k_A$ = Michaelis-Menten constant for binding of substrate.

2. No EM complex in absence of A (reaction proceeds by top path of Eq. 11-39):

$$\boxed{E + A \underset{k_{-A}}{\overset{k_A}{\rightleftharpoons}} EA \underset{k_{-m}}{\overset{k_m}{\rightleftharpoons}} EAM \overset{k}{\rightarrow} E + P + M} \qquad (11\text{-}42)$$

for which the steady-state rate law is

$$v = \frac{V_{\max}[(K_M + [M] + [A]) - \sqrt{(K_M + [M] + [A])^2 - 4[M][A]}]}{(K_M + [M] + [A]) - \sqrt{(K_M + [M] + [A])^2 - 4[M][A]} + 2K_A} \qquad (11\text{-}43)$$

Metal participation in enzyme-catalyzed reactions of the types just described is most common in reactions involving phosphate transfer:

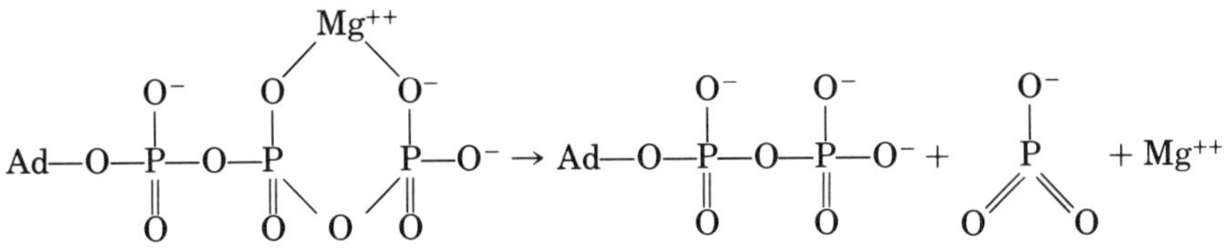

ATP ADP (= adenosine diphosphate)

in which the metal ion is usually Mg^{++} and occasionally Mn^{++}. The two mechanisms, Equations 11-40 and 11-42, are readily distinguished from plots of enzyme-catalyzed reaction velocity versus $[M]$, for various fixed $[A]$, as shown schematically in Fig. 11-20. This figure shows that when experimental rate data is gathered under conditions such that $[A] > [M]$,

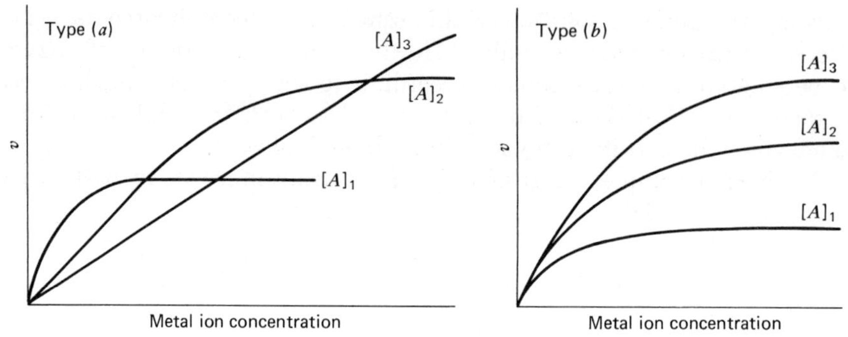

FIGURE 11-20. Steady-state graphical method for distinguishing metal-activation mechanisms (*a*) and (*b*) shown in Equations 11-40 and 11-42, by plotting enzyme-catalyzed reaction velocity, v, versus metal concentration $[M]$, for three (fixed) concentrations of substrate A. $[A]_1 < [A]_2 < [A]_3$ for both graphs. (After H. Gutfreund, *Enzymes: Physical Principles,* Wiley-Interscience, London, 1972, p. 152.)

then at constant $[M]$ (i.e., draw a vertical line through the left region of the plots in Fig. 11-20), the reaction velocity varies *inversely* with $[A]$ for mechanism (a) but is either *constant* or varies *directly* with $[A]$ for mechanism (b). Experiments such as these have shown that yeast hexokinase, which catalyzes the reaction

$$\text{Glucose} + \text{ATP} + \text{Mg}^{++} \rightleftharpoons \text{Glucose-6-phosphate} + \text{ADP} + \text{Mg}^{++} \quad (11\text{-}45)$$

proceeds by mechanism (a), while pyruvate kinase, which catalyzes the reaction

$$\text{Pyruvate} + \text{ATP} + \text{Mn}^{++} \rightleftharpoons \text{Phosphoenolpyruvate} + \text{ADP} + \text{Mn}^{++} \quad (11\text{-}46)$$

proceeds via mechanism (b). Mechanism (b) is followed by a much larger group of enzymes than mechanism (a).

11.C CHEMICAL OSCILLATIONS: QUIRK OR CHEMICAL BASIS FOR BIOLOGICAL CLOCKS?

One of the most startling consequences of feedback activation and feedback inhibition of chemical reactions is provided by a handful of chemical systems in which the concentration of a given intermediate in the reaction rises and falls many times before equilibrium is reached. The limited present understanding of these systems is illustrated by the fact that all of the known reactions were discovered by accident, and no one has yet succeeded in producing a chemically oscillating system by *a priori* choice of

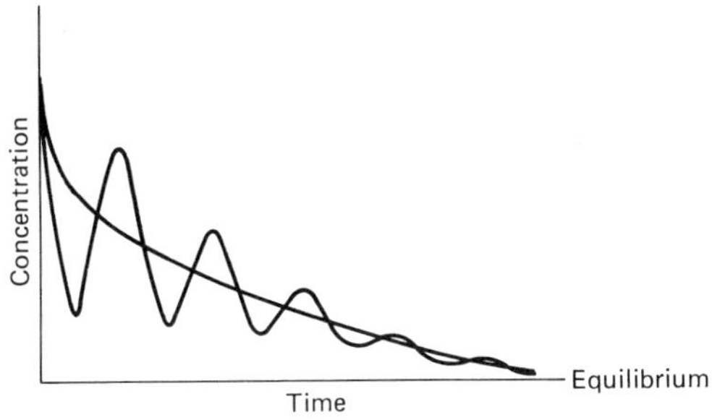

FIGURE 11-21. Variation of concentration of chemical intermediate with time on the way to chemical equilibrium, for a system subject to chemical oscillation. [After H. Degn, *J. Chem. Educ. 49,* 302 (1972).]

suitable reactants. Nevertheless, most explanations of the behavior are based on elaborations of the following simple model, first suggested by Lotka in 1920, which is most easily understood in terms of animal populations in a predator-prey context:

$$\begin{array}{rll} G + R \rightarrow 2R & \text{Grass plus rabbits = more rabbits} & \\ R + F \rightarrow 2F & \text{Rabbits plus foxes = more foxes} & (11\text{-}47) \\ F \rightarrow D & \text{Foxes die} & \end{array}$$

It is intuitively clear that for a constant supply of G (grass), the concentrations (populations) of R and F (rabbits and foxes) will eventually reach a steady-state constant level. However, if there is a sudden change in climate, and the grass supply changes suddenly, then the animal populations will oscillate on the way to a new steady state: rabbits will multiply rapidly until the fox supply reduces the number of rabbits, and the foxes eventually overpopulate until the rabbit supply is small again and the whole cycle repeats (Fig. 11-21).

At this stage, one might next think of trying to establish the critical feature of the mechanism, 11-47, which seems to lead to oscillatory behavior. From comparison of many such schemes proposed since Lotka's original idea, there seems to be a quite small number of general mechanisms:

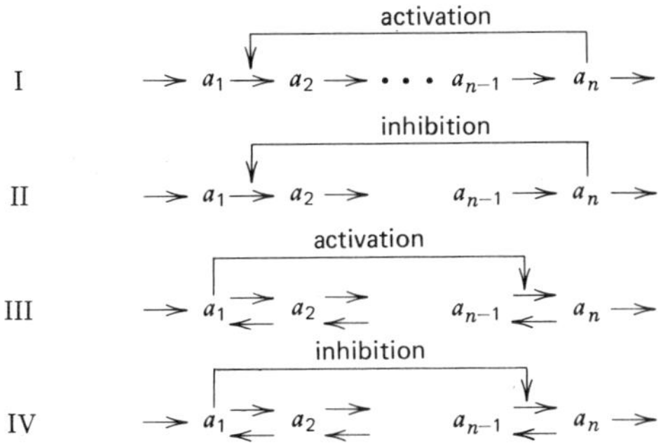

(11-48)

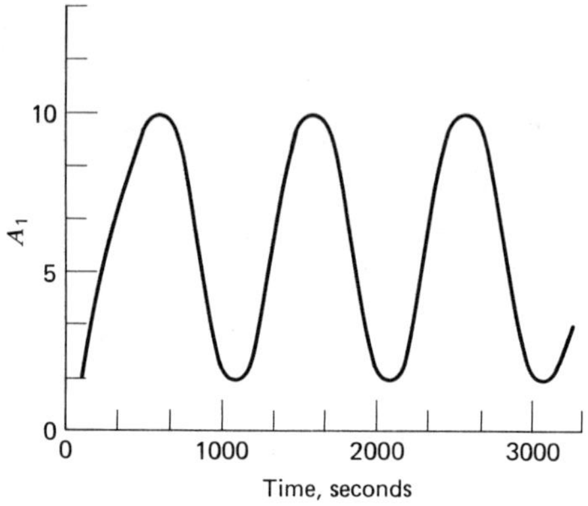

FIGURE 11-22. Variation of concentration of A_1 versus time in the mechanism of Eq. 11-49. It is supposed that reactants A_1 and A_2 are externally supplied at a constant rate. [After R. A. Spangler and F. M. Snell, *Nature 191,* 457 (1961).]

The key feature in common between these mechanisms is feedback activation or inhibition of prior rates in the scheme. No hypothetical reaction scheme showing oscillatory solutions is known that does not contain at least one of the above four types of feedback control. The simple Lotka scheme, Eq. 11-47, is now seen as a version of case I in Eq. 11-48, in which there are only two intermediates (rabbits and live foxes) and each reaction is at most second-order.

Although chemical reaction schemes involving three or more species, like the well-known celestial three-body problem in physics, are not rigorously solvable, it is generally possible to search for oscillating solutions using an iterative computer procedure. The result for a relatively simple particular mechanism for the model enzyme system, Eq. 11-49, is shown graphically in Fig. 11-22, in which the oscillation is clearly evident.

$$
\begin{array}{rcl}
A_1(\text{exterior}) \rightleftharpoons & A_1 + X \rightleftharpoons A_1X \xrightarrow{k} X + P_1 & \rightleftharpoons P_1(\text{exterior}) \\
 & P_1 + Y \rightleftharpoons P_1Y & \\
 & P_2 + X \rightleftharpoons P_2X & \\
A_2(\text{exterior}) \rightleftharpoons & A_2 + Y \rightleftharpoons A_2Y \rightarrow Y + P_2 & \rightleftharpoons P_2(\text{exterior})
\end{array}
\qquad (11\text{-}49)
$$

The basic system of Eq. 11-49 is simply two enzyme-catalyzed reactions, in which the product of each reaction inhibits the other reaction.

EXAMPLE *Oscillating Reactions in Glycolysis*

The glycolytic pathway is replete with feedback loops, as shown in Fig. 11-23:

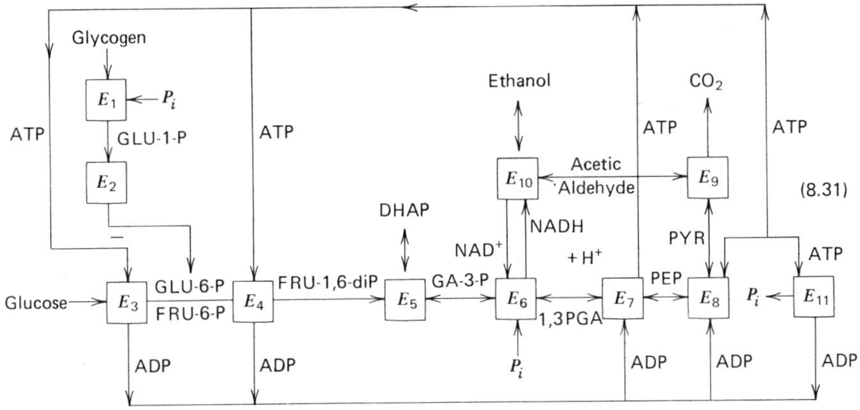

FIGURE 11-23. Metabolic reactions involved in glycolysis.

where

ATP = adenosine-5′-triphosphate
ADP = adenosine-5′-diphosphate
GLU-1-P = glucose-1-phosphate
P_i = inorganic phosphate
GLU-6-P = glucose-6-phosphate
FRU-6-P = fructose-6-phosphate
FRU-1,6-diP = fructose-1,6-diphosphate
DHAP = dihydroxyacetone phosphate
GA-3-P = glyceraldehyde-3-phosphate
NAD^+ = oxidized pyridine nucleotide
NADH = reduced pyridine nucleotide
1,3PGA = 1,3-diphosphoglyceric acid
PEP = phosphoenolpyruvate
PYR = pyruvate
E_1 = phosphorylase
E_2 = phosphoglucomutase
E_3 = hexokinase
E_4 = phosphofructokinase
E_5 = aldolase
E_6 = glyceraldehyde-3-phosphate dehydrogenase
E_7 = phosphoglycerate kinase
E_8 = pyruvate kinase
E_9 = pyruvate decarboxylase
E_{10} = alcohol dehydrogenase
E_{11} = adenosine-5′-triphosphatase

Based on the criteria for chemical oscillations given in Eq. 11-49, this system might be considered as a prime candidate for experimental evidence of oscillatory behavior—the evidence was provided in 1964 by Chance and coworkers, and some of their data is shown in Fig. 11-24. In the experiment

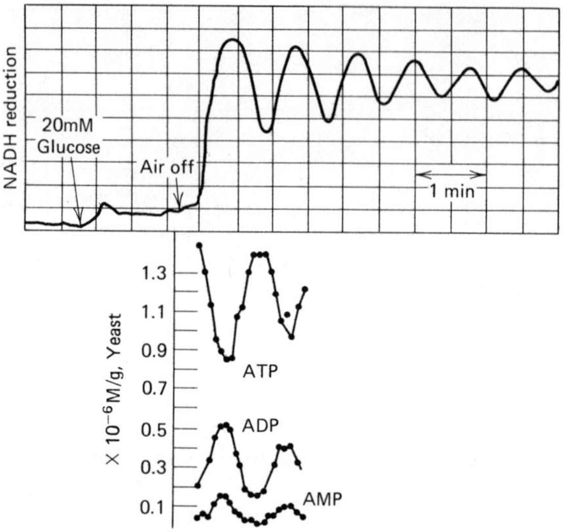

FIGURE 11-24. Oscillating concentrations of various metabolic components in the glycolytic cycle of suspended yeast cells. Components shown include reduced pyridine nucleotide (NADH), adenosine triphosphate (ATP), adenosine diphosphate (ADP), and adenosine monophosphate (AMP). [From A. Betz and B. Chance, *Arch. Biochem. Biophys. 109,* 585 (1964).]

shown, an aerobic suspension of yeast cells was first exposed to a supply of glucose, and, after a short wait, the air supply was turned off and the system proceeded to anaerobic metabolism. The relative concentration of reduced pyridine nucleotide (NADH) is readily monitored from its characteristic fluorescence. In addition, samples were removed at the indicated times, and analyzed for ATP, ADP, and AMP as shown in the lower traces. It is interesting to note that the ADP and AMP and NADH vary together ("in phase") with time, while ADP and ATP oscillate about 180° out-of-phase with each other. Although the oscillatory behavior is readily explained by invoking any or all of the feedback loops (the most detailed treatment is based on computer solution of 57 simultaneous differential equations representing 101 chemical equations!), the more useful approach is to try to identify the critical feedback segment. The most likely enzyme candidate appears to be phosphofructokinase, which is inhibited by adenosine triphosphate (ATP, a substrate of the enzyme), and activated by the products ADP and fructose-1,6-diphosphate.

One must be careful in ascribing any direct biological clock significance to the glycolysis oscillations just described. Since clock behavior will be expected for any system with many feedback loops, and since the feedback loops may have evolved solely for regulatory functions described in Chapter 11.A and 11.B, the oscillatory consequences of feedback may simply be accidental. (The damping, or dying out, of the oscillations in Fig. 11-24 can be greatly reduced by more direct simulation of biological homeostatic conditions by a relatively constant supply of glucose rather than the sudden surge in glucose used to start the rhythm shown in the figure; the glycolytic system can be made to oscillate for several hours.) However, an intriguing speculation follows from the analogy to dropping a stone in a quiet pool of water. The water where the stone was dropped rises and falls, just as the concentration of ATP in the glycolysis example rises and falls at the site of chemical perturbation from addition of glucose—it thus seems reasonable to suppose that the ATP fluctuations will propagate in *space,* just as the water wave spreads out from its starting point. Thus, such "clock" behavior may offer a means for *synchronizing* the behavior of spatially separated cells (as in a tissue, such as the heart) according to a trigger signal (pacemaker?) that originates elsewhere. Oscillating chemical "waves" that spread out in space have in fact been observed in a number of experimental model systems exhibiting chemical oscillation.

One serious problem with oscillating chemical reactions as a proposed basis for biological rhythms is the relatively *slow* period of *biological* rhythms (seconds to a day or so) compared with the relatively *fast* rates of *chemical* reactions. However, it should first be noted that the period of the chemical *oscillation* (about 1000 seconds in Fig. 11-22) can be very much longer than the period of the constituent chemical *reactions* ($1/k = 5$ sec in the same example). Second, a well-known property of oscillators is that by "mixing" or "coupling" two fast oscillators, one can readily extract an oscillation at the "difference" frequency that can be much slower than either of the component oscillations.

Still, a simpler explanation for the slower biological clocks, including the "circadian," or daily rhythm, might be based on an oscillation resulting from a molecular process that is itself inherently much slower than most metabolic reactions. A recently suggested model is based on the relatively slow (see Chapter 14.D) lateral (translational) diffusion of proteins in membranes. Since circadian rhythms are preserved even in many enucleate cells (cells without nuclei), and even in the presence of a variety of nucleic acid and protein synthesis inhibitors, the membrane model provides an explanation that applies almost independently of precise enzyme concentrations and thus accounts for the preceding observations.

EXAMPLE *A Membrane Model for the Circadian Clock*

Figure 11-25 shows how a fluid mosaic membrane [see Chapter 2.A.2 and Chapter 14.D, or S. J. Singer and G. L. Nicolson, *Science 175,* 720 (1972)] might provide an oscillation in the transmembrane ion gradient to serve as the basis for the circadian rhythm. The membrane is (for simplicity) supposed to exist in just two states, which differ by having the membrane proteins either aggregated (and thus able to actively pump K^+ ions across the membrane) or spread out (and thus unable to cause active transport). Light might then act

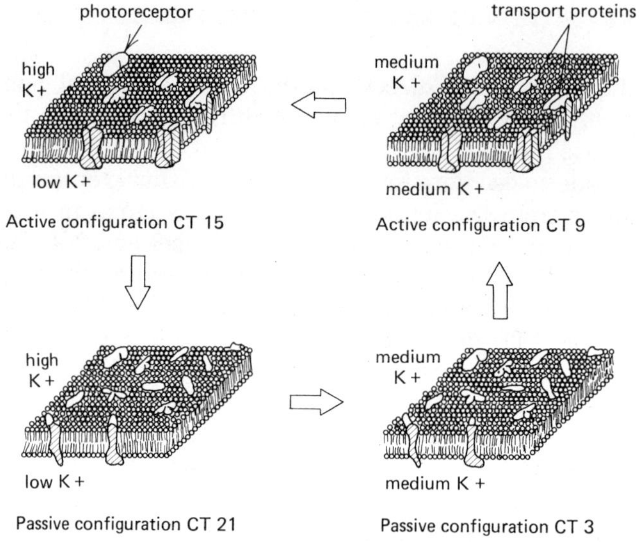

FIGURE 11-25. Schematic representation of the way in which a fluid mosaic membrane might keep time. Ion concentrations and ion fluxes (transport) act as a feedback system to produce self-generated circadian oscillations. (See Fig. 11-26.) Light acts on the membrane, either directly or via hormonal coupling, to open an ion gate (or cause the membrane transport proteins to spread out to an inactive configuration, thus preventing active transport of ions) and depleting the transmembrane ion gradient to move the clock forward or backward (see text and Fig. 3-47). [From D. Njus, F. M. Sulzman, and J. W. Hastings, *Nature 248,* 116 (1974).]

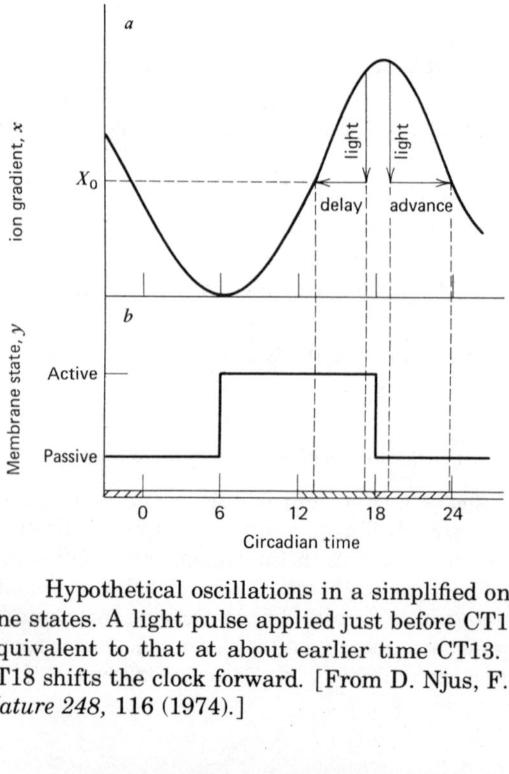

FIGURE 11-26. Hypothetical oscillations in a simplified one-ion system with only two membrane states. A light pulse applied just before CT18 moves the clock back to a state equivalent to that at about earlier time CT13. Similarly, a light pulse just after CT18 shifts the clock forward. [From D. Njus, F. M. Sulzman, and J. W. Hastings, *Nature 248,* 116 (1974).]

to "open a K^+ ion gate," perhaps by inducing the proteins to spread out (see rhodopsin example of Section 4 for a similar case). Then as light alternately appears and disappears in the daily cycle, ion flux and thus ion gradient would follow suit. Since many circadian rhythms persist even when no light is present, there may be an inherent mechanism unconnected with light that changes the state of the membrane. However, it is interesting to note that the model of Fig. 11-25 accounts also for the phase change ("re-setting the clock") effects of a light pulse applied to real systems. For example, applying a light pulse just before time CT18 in Fig. 11-26 would open the K^+ ion gate and shift the ion concentrations back to what they would have been at time CT13 ("moving the clock backward"), while a light pulse just after CT18 would shift the ion concentrations to a point forward in time.

PROBLEMS

1. For noncompetitive inhibition, obtain the steady-state enzyme-catalyzed reaction rate expression and show that the Lineweaver-Burk plots at different (fixed) inhibitor concentration intersect in a point on the $(1/[A])$ axis. In other words, obtain Eq. 11-17 from the rate and conservation Eqs. 11-16 for the mechanism of Eq. 11-15.

2. For the limiting cases of pure competitive or noncompetitive inhibition (Eqs. 11-14 or 11-17), show that Dixon plots of $(1/v)$ versus $[I]$ at several (fixed) $[A]$ values will intersect in a point whose $[I]$ value is $-K_I$.

3. For the pH-control model of enzyme-catalyzed reactions

$$EH_2^{++} \underset{}{\overset{K_1}{\rightleftharpoons}} EH^+ \underset{}{\overset{K_2}{\rightleftharpoons}} E$$

in which only form EH^+ is capable of binding substrate and catalyzing the desired reaction, and

$$K_1 = [EH^+][H^+]/[EH_2^{++}] \text{ and } K_2 = [E][H^+]/[EH^+]$$

(a) Obtain the fraction of enzyme in each of the respective states of protonation, EH_2^{++}, EH^+, and E (Eqs. 11-22)

(b) Show that the maximum enzyme-catalyzed reaction rate will occur for

$$[H^+]_{opt} = \sqrt{K_1K_2}$$

(c) Show that the *range* of pH over which the enzyme is at least half-maximally active is given by the roots of the equation

$$[H^+]^2 - (K_1 + 4[H^+]_{opt})[H^+] + [H^+]_{opt}^2 = 0$$

4. One of the simplest kinetic schemes that accounts for a sigmoidal behavior of enzyme-catalyzed reaction velocity as a function of substrate concentration is the two-stage binding shown below:

$$E + A \underset{k_{-1}}{\overset{k_1}{\rightleftharpoons}} EA \underset{A}{\overset{k_2}{\longrightarrow}} EA_2 \overset{k_3}{\longrightarrow} E + A + P$$

Find a general expression for steady-state rate of appearance of product for this scheme, and analyze its limiting behavior to obtain the expected graphical form of (a), a plot of v versus $[A]$, and (b), a plot of $(1/v)$ versus $(1/[A])$.

5. In this problem, the reader will work out the complete derivation of the Monod-Wyman-Changeux model for cooperative binding of substrate by a multimeric enzyme. Referring to the model shown in Fig. 11-10, with the definitions

$$L = [E_T]/[E_R], \ \alpha = [A]/K_R, \text{ and } c = K_R/K_T$$

where

$$K_T = [E_T][A]/[E_TA] \qquad \text{and} \qquad K_R = [E_R][A]/[E_RA]$$

show that the fraction of enzyme sites occupied by substrate, $\bar{Y}_A$, is given by

$$\bar{Y}_A = \frac{\alpha L\, c(1 + c\alpha)^{n-1} + \alpha(1 + \alpha)^{n-1}}{L(1 + c\alpha)^n + (1 + \alpha)^n}$$

for an enzyme of n subunits.

Hints. First solve the case for $n = 4$ shown in Fig. 11-10, and obtain the final result by induction. Begin by showing that

$$\bar{Y}_A = \frac{\begin{array}{c}[E_RA] + 2[E_RA_2] + 3[E_RA_3] + 4[E_RA_4] \\ + [E_TA] + 2[E_TA_2] + 3[E_TA_3] + 4[E_TA_4]\end{array}}{\begin{array}{c}4\{[E_R] + [E_RA] + [E_RA_2] + [E_RA_3] + [E_RA_4] \\ + [E_T] + [E_TA] + [E_TA_2] + [E_TA_3] + [E_TA_4]\}\end{array}}$$

$$= \frac{[E_RA](1 + \alpha)^3 + [E_TA](1 + c\alpha)^3}{4\{[E_R](1 + \alpha)^4 + [E_T](1 + c\alpha)^4\}}$$

Then show that

$$\frac{[E_TA]}{[E_RA]} = c\, L$$

so that the previous result simplifies to

$$\bar{Y}_A = \frac{\alpha c L(1 + c\alpha)^3 + \alpha(1 + \alpha)^3}{L(1 + c\alpha)^4 + (1 + \alpha)^4}$$

6. For the self-inhibition by substrate mechanism of Eq. 11-36,
(a) Derive a steady-state rate expression (Eq. 11-37)
(b) Show that the maximal enzyme-catalyzed rate will occur for

$$[A]_{\text{opt}} = \sqrt{K_A K_A'}$$

in which $K_A = (k_{-1} + k_2)/k_1$ and $K_A' = k_{-3}/k_3$ for the scheme of Eq. 11-36

(c) predict the shape of a plot of $(1/v)$ versus $(1/[A])$ for this case.

REFERENCES

H. R. Mahler and E. H. Cordes, *Biological Chemistry,* Harper & Row, N.Y. (1966). Good general treatment.

M. Dixon, *Enzymes,* Academic Press, N. Y. (1964). Good critical comparison of various means of graphical reduction of enzyme-catalyzed rate data.

K. Laidler and P. S. Bunting, *The Chemical Kinetics of Enzyme Action,* Clarendon Press, Oxford (1973), Chapter 11. Good discussion of allosterism.

A. S. Mildvan, in *The Enzymes,* ed. P. D. Boyer, Vol. II, Academic Press, N. Y. (1970). Good overview of mechanisms of metal-ion-assisted enzyme catalysis.

H. Degn, "Oscillating Chemical Reactions in Homogeneous Phase," *J. Chem. Educ. 49,* 302 (1972). Discussion of nonenzymatic chemical oscillations.

C. Walter, *Enzyme Kinetics: Open and Closed Systems,* Ronald Press Co. (N. Y., 1966), Chapter 8; B. Hess and A. Boiteux, "Oscillatory Phenomena in Biochemistry," *Ann. Rev. Biochemistry 40,* 237 (1971). Biochemical examples of chemical oscillations.

I. H. Segel, *Enzyme Kinetics,* John Wiley & Sons, Inc. (New York, 1975). Encyclopedic catalog of the various special limiting cases of steady-state enzyme-catalyzed rate expressions.

CHAPTER 12 Pharmacokinetics: Chemical Reaction Kinetics with Renamed Variables

12.A. TIME COURSE OF DRUG ACTION: INTAKE AND ELIMINATION OF DRUGS

The mathematical basis for the mechanism and time course of action of drugs forms a close parallel with our prior treatments of statistics and the kinetics of uncatalyzed and enzyme-catalyzed chemical reactions. It is convenient to consider first the rates at which a drug reaches its site of action via intake and elimination (Chapter 12.A), and then to treat the molecular mechanisms by which the drug acts once it has arrived at the proper place (Chapter 12.B). Because the parallel with previous chemical reaction kinetics is so close, we will illustrate the time course of drug action with a single example from common experience.

EXAMPLE *Sobering Up—Time Evolution of Ethanol in the Body*

Following the administration of a single dose of a drug, the drug level at various stages may in general be described by the scheme

$$A \xrightarrow{k_1} B \xrightarrow{k_2} C \qquad (12\text{-}1)$$

in which $[A]$ denotes the drug concentration at the site of ingestion (say, the stomach or gastrointestinal tract for an orally taken drug), $[B]$ is the drug concentration in the blood, and $[C]$ represents the amount of drug eliminated via metabolism, secretion, or excretion. The kinetic scheme of Eq. 12-1 has already been solved (Chapter 10.A.4, consecutive first-order reactions), with the typical concentration-versus-time behavior shown in Fig. 10-3. The only conceptual difference here is that the volume of the "reaction vessel" changes as the reaction proceeds, since the drug may become diluted or concentrated as it enters the blood or other organs—the main effect of this change is that the vertical scales for $[A]$, $[B]$, and $[C]$ in Fig. 10-3 will in general be different. We now consider some special cases of Eq. 12-1, and then apply them to the concentration of ethanol in human blood.

Instantaneous ingestion and instantaneous absorption. When, as is often the case, ingestion and absorption of a drug are rapid compared with the rate of elimination, it is convenient to suppose that drug elimination follows a simple exponential decrease:

$$[B] = [B]_0 \exp[-k_2 t] \qquad (12\text{-}2)$$

in which $[B]_0$ now represents the (instantaneously ingested and absorbed) drug concentration in the blood just after administration of the first dose. If second and succeeding doses of the same initial amount are now administered at regular intervals, T seconds apart, then it is readily shown (see Problems) that the drug level in the blood increases exponentially to a plateau level after many doses:

$$f = \frac{\text{fraction of eventual plateau}}{\text{level reached after } n \text{ equal doses}} = 1 - \exp[-k_2 nT] \qquad (12\text{-}3)$$

The actual drug level in the blood as a function of time is shown schematically in Fig. 12-1. Once k_2 is known, the more usual question becomes, what number of doses of a given size are required to attain a given fraction, f, the plateau level for an infinite number of doses of that size? From Eq. 12-3 (see Problems), the answer is found to be

$$n = [\ln(1 - f)]/(-k_2 T) \qquad (12\text{-}4)$$

A final common question is, how large a dose must be given at a stated interval (say $T = 4$ hours) for a given fractional blood level to be maintained once it has been reached? Clearly, just before the next dose, the amount, $[B]_0(1 - \exp[-k_2 T])$ has been eliminated and must be replaced.

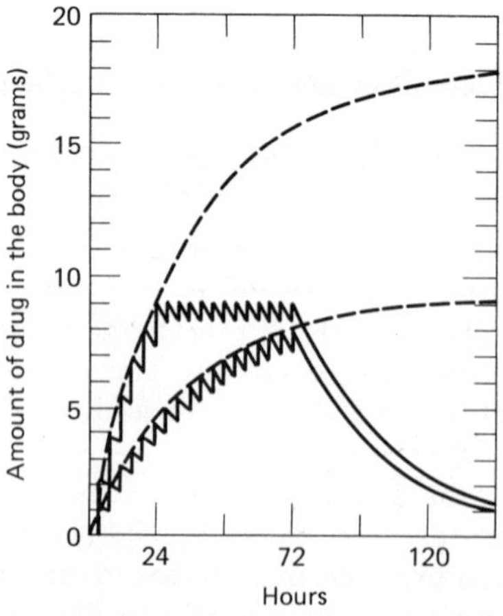

FIGURE 12-1. Drug accumulation with repeated dosage at regular intervals for a drug that is absorbed instantaneously. Elimination rate constant $k_2 = 0.0289$ hr^{-1} (elimination half-time = 24 hr), and dosing interval $T = 4$ hr. Lower curve: dose = 1 g. Upper curve: dose = 2 g for first six doses, then 1 g. Broken curves indicate course of continued accumulation if drug administration were continued (drug stopped at 72 hr). Solid curves computed from Equations 12-3 and 12-4. Elimination curve is seen to be identical to accumulation curve, but inverted. [After J. H. Gaddum, *Nature 153*, 494 (1944).]

First-order absorption; zero-order elimination. With ethanol and many other drugs, the body is capable of eliminating a fixed maximal amount of drug per unit time, no matter how much drug is present. In the case of ethanol in particular, the elimination occurs by enzymatic oxidation to acetaldehyde (and ultimately to acetic acid) in the liver. Because the rate of the enzymatic reaction is determined by the *enzyme* concentration (when ethanol is in great excess), the elimination of ethanol is essentially *zero*-order with respect to ethanol level. The mechanism, Eq. 12-1, thus reduces to

$$d[B]/dt = k_1[A] - k_2 \tag{12-5}$$

which may be integrated (initial condition, $[B]_0 = 0$) to give

$$[B] = [A]_0(1 - \exp[-k_1 t]) - k_2 t \tag{12-6}$$

Two features of the result, Eq. 12-6, are of immediate interest in predicting degree of intoxication. First, the maximal ethanol level reached after ingestion may be found by setting $d[B]/dt = 0$ from Eq. 12-6 to find

$$[\text{Ethanol}]_{\text{max}} = [A]_0 - (k_2/k_1) - (k_2 t_{\text{max}}) \tag{12-7}$$

in which t_{max} is the time it takes to reach that maximum ethanol level. However, since we shall soon see that $k_2 << k_1$ (Table 12-1), and $t_{\text{max}} << (1/k_2)$, Equation 12-7 shows that the maximal blood level of ethanol is very nearly the total amount ingested, as seen in Fig. 12-2.

Table 12-1 Conversion Factors and Values of Constants Used in Analysis of Time Evolution of Ethanol in the Body

Ethanol (g)	Concentration of Ethanol in Body Fluid (g/l)	Equivalent Volume of 80 Proof (40% v/v ethanol) Liquor ml	oz	Qualitative Physiological and Psychological Response
20	0.5	63	2.1	sense of euphoria
40	1.0	127	4.3	mild to medium disturbance in motor coordination, *legally intoxicated.*
60	1.5	190	6.4	overt lack of coordination and slurring of speech
80	2.0	254	8.6	amnesia
100	2.5	317	10.7	induced sleep or unconsciousness

$k_1 = 10$ hr^{-1}; $k_2 = 0.192$ g ethanol/l body fluid hr = 10 ml ethanol/hr; volume of body fluid = 40l.
From G. V. Calder, J. Chem. Educ. 51, *19 (1974).*

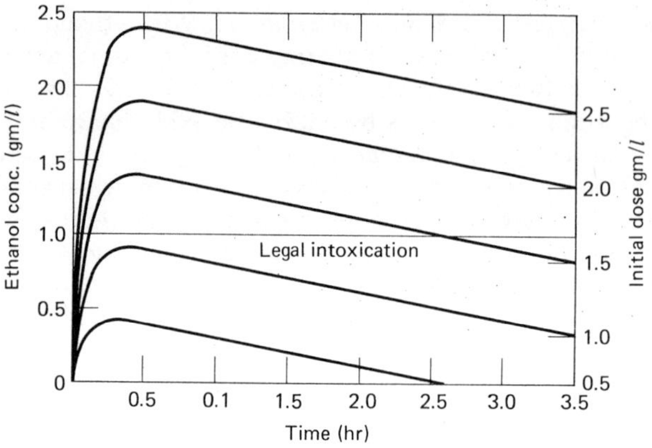

FIGURE 12-2. Time evolution of ethanol in the body, assuming instantaneous ingestion and absorption. (See Table 12-1 for description of units.) [From G. V. Calder, *J. Chem. Educ. 51,* 19 (1974).]

The second interesting calculation from Eq. 12-6 is the time it takes for the blood level of ethanol to drop below an arbitrary (e.g., legal) limit, once that limit has been exceeded. It is left as an exercise to show that this "sobriety time," t_{sobriety}, may be expressed in more familiar units (Table 21-1) as

$$t_{\text{sobriety}} = 1.22(\text{ounces EtOH ingested} - 4.3) \text{ hr} \tag{12-8}$$

while the time required to eliminate all the alcohol ingested becomes

$$t_{\text{hangover}} = 1.22(\text{ounces EtOH ingested}) \text{ hr} \tag{12-9}$$

Specifically, for every ounce of 80 proof liquor ingested in excess of 4.3 oz, about 1 1/4 hr are required to reach the legally sober limit; also, it takes about 5 hr to remove all the ethanol from a person intoxicated just to the legal limit.

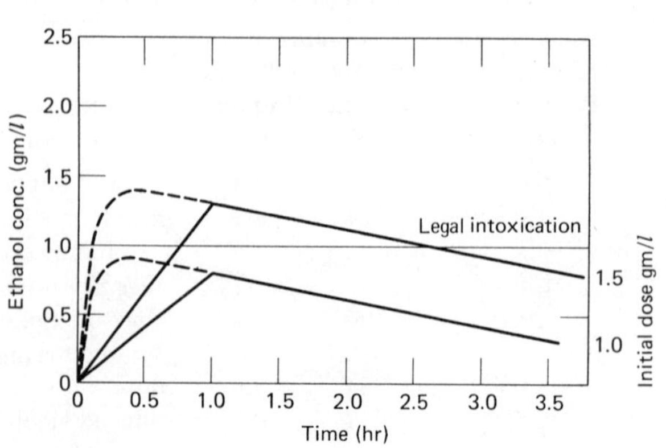

FIGURE 12-3. The time evolution of ethanol in the body assuming a constant rate of ingestion for 1 hr and subsequent termination. [From G. V. Calder, *J. Chem. Educ. 51,* 19 (1974).]

Zero-order ingestion; first-order absorption; zero-order elimination—the cocktail party. Obviously no one usually ingests ethanol in a single sudden large dose. A more reasonable model might therefore be a constant rate of ingestion for a given length of time, followed by stoppage of ingestion. It is again left as an exercise for the interested reader to show that the blood level of ethanol in this case follows the course shown in Fig. 12-3.

12.B. THEORIES OF DRUG-EFFECT CONNECTION

12.B.1. Graded Response: Occupancy theory; Rate theory

Although some drugs act nonspecifically, such as the general anesthetics (Chapter 11.A), general theories of drug action are based on the principle that a given drug acts by binding to a specific "*receptor*" (often a membrane-bound protein, as in the action of insulin) to cause a *biochemical* response, such as speeding or slowing of some metabolic reaction or change in membrane permeability to specific ions or molecules, leading to a *physiological* effect, such as change in breathing or heart rate. In one of the simplest theories, it is imagined that a given organism possesses a fixed number of possible equivalent independent receptors, as shown schematically in Fig. 12-4, to which a given drug molecule may bind.

From the model of Fig. 12-4, one can construct the kinetic description,

$$D + R \underset{k_{-1}}{\overset{k_1}{\rightleftharpoons}} DR \longrightarrow \text{Effect (response)} \tag{12-10}$$

in which D, R, and DR denote drug, receptor, and drug-receptor complex and $K_D = k_{-1}/k_1$ is the equilibrium dissociation constant for dissociation of

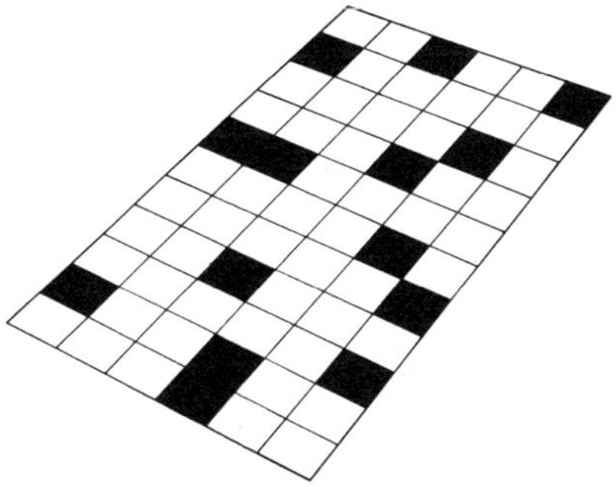

FIGURE 12-4. Schematic diagram of the array of possible drug receptors (grid) for an organism. Receptor sites occupied by bound drug molecules are shaded. In this "occupancy" theory of drug action, it is assumed that the degree of physiological response is proportional to the fraction of receptor sites occupied by drug molecules. Sites are assumed independent, so that a drug molecule binds equally well to any site, whether or not any adjacent sites are already occupied.

the DR complex. If the (graded) response is assumed proportional to the fraction of receptor sites occupied by drug molecules, then the maximum effect is expected when $[D] >> [R]$, so that essentially all the receptor sites are occupied. The reader should by now note a pronounced similarity between Eq. 12-10 (and its implications) and the Michaelis-Menten enzyme kinetic scheme (Eq. 10-4). Pressing the algebra to completion

Effect = constant · (fraction of receptor sites occupied by drug molecules)

$$\text{Effect} = \frac{\text{constant} \cdot [DR]}{[R] + [DR]} = \text{constant} \frac{[DR]}{[R]_{\text{total}}} \tag{12-11}$$

Recalling that

$$K_D = \frac{k_{-1}}{k_1} = \frac{[D]\,[R]}{[DR]} = \frac{([D]_{\text{total}} - [DR])([R]_{\text{total}} - [DR])}{[DR]} \tag{12-12}$$

and proceeding to the limit that drug is in great excess over receptor

$$[D]_{\text{total}} >> [R]_{\text{total}} \tag{12-13}$$

Equation 12-12 simplifies to

$$K_D \cong \frac{[D]_{\text{total}}([R]_{\text{total}} - [DR])}{[DR]} \tag{12-14}$$

Solving Eq. 12-14 for $[DR]$ and substituting into Eq. 12-11,

$$\text{Effect} = \text{constant} \frac{[D]_{\text{total}}}{K_D + [D]_{\text{total}}} \tag{12-15}$$

Finally, since

$$\text{Effect}_{\text{maximum}} = \lim_{[D]_{\text{total}} >> K_D} (\text{Effect}) = \text{constant} \cdot 1 \tag{12-16}$$

Equation 12-15 may be written

$$\boxed{\text{Effect} = \frac{\text{Effect}_{\text{max}}\,[D]_{\text{total}}}{K_D + [D]_{\text{total}}}}$$

which may be compared with the analogous result from Michaelis-Menten kinetics:

$$v = \frac{V_{\text{max}}[A]}{K_A + [A]} \tag{10-38}$$

Table 12-2 Analogous Terms, Symbols, Assumptions, Equations, and Graphs for Enzyme-Substrate and Drug-Receptor Relationships

	Enzyme-substrate Case	Drug-receptor Case
Symbols	A = substrate E = enzyme EA = enzyme-substrate complex P = products v = initial stebdy-state reaction velocity I = inhibitor	D = drug(agonist) R = receptor DR = drug-receptor complex Eff = effect (response) Ant = antagonist
Reaction scheme	$E + A \underset{k_{-1}}{\overset{k_1}{\rightleftharpoons}} EA \xrightarrow{k_2} P + E$	$D + R \overset{K_D}{\rightleftharpoons} DR \longrightarrow \text{Eff}$
Assumptions	One substrate per enzyme catalytic site Initial rate only (no back reaction) Monomeric enzyme (no cooperative binding effects) $[A]_{\text{total}} >> [E]_{\text{total}}$ v proportional to $[EA]$ $v \rightarrow V_{\max}$ when $[EA] \rightarrow [E]_{\text{total}}$	One drug per receptor site No analogous assumption (but see "rate" theory later in this section) Binding of drug to a given receptor is independent of whether or not adjacent receptors are already occupied $[D]_{\text{total}} >> [R]_{\text{total}}$ Eff proportional to $[DR]$ Eff $\rightarrow$ $\text{Eff}_{\max}$ when $[DR] \rightarrow [R]_{\text{total}}$
Graphs	v versus $[A]$ at constant $[E]$ [a] $(1/v)$ versus $(1/[A])$ at const $[E]$ v versus log $[A]$ at constant $[E]$	Eff versus $[D]$ at constant $[R]$ (1/Eff) versus $(1/[D]$ at const $[R]$ [a] Eff versus $\log[D]$ at constant $[R]$
Formulae	$v = \dfrac{V_{\max}[A]}{K_A + [A]}$ $K_A = \dfrac{[E][A]}{[EA]} = \dfrac{k_{-1} + k_2}{k_1}$ $v = (V_{\max}/2)$ at $[A] = K_A$	$\text{Eff} = \dfrac{\text{Eff}_{\max}\,[D]}{K_D + [D]}$ $K_D = \dfrac{[R][D]}{[DR]} = \dfrac{k_{-1}}{k_1}$ $\text{Eff} = \text{Eff}_{\max}/2$ at $[D] = K_D$
	Formulas for competitive and noncompetitive enzyme inhibition are entirely analogous to competitive and noncompetitive antagonism of the action of agonist drugs.	

[a] *Denotes preferred graphical display.*

A convenient comparative collection of terms, assumptions, and formulae for drug-receptor versus enzyme-substrate combination is given in Table 12-2. From Table 12-2, it is seen that the principal difference (other than different names for the variables) between enzyme kinetics and drug-receptor situations is in the preferred mode of graphical presentation of data. Since the pharmacologist is most interested in the effects in the vicinity of the "ED_{50}" $= K_D =$ dose at which the effect is half the maximum possible effect, the "log-dose-response" curve of Eff versus $\log[D]$ is usually employed. An example is shown in Fig. 12-5, in which the effect of epi-

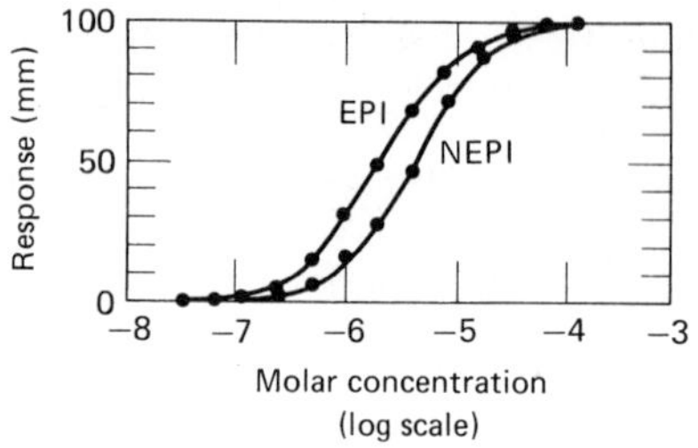

FIGURE 12-5. Log-dose-response curves for action of epinephrine (EPI) and norepinephrine (NEPI) on isolated cat spleen. Epinephrine has the smaller 50%-effective-dose, or ED_{50}, and is thus the more potent drug by about a factor of two in ED_{50} compared to norepinephrine. [From R. K. Bickerton, *J. Pharmacol. Exp. Therap. 142*, 99 (1963).]

nephrine (EPI) is compared to the effect of norepinephrine (NEPI) in contraction (in mm) of an isolated strip of cat spleen tissue. From the midpoints of these "titration"-type plots, it is seen that the more potent drug (left-hand curve) is epinephrine, having about twice the affinity (i.e., half the K_D) for binding compared to norepinephrine. Because the two curves have the same shape, it seems reasonable to suppose that both drugs act in the same way. (One reason that overdoses of the "psychedelic" drug LSD are so common among users is that this drug has an extraordinarily small ED_{50} of about 1 *micro*gram drug per kg body weight.)

The analogy between drug-receptor and enzyme-substrate interactions can be extended to include situations of competitive and noncompetitive inhibition, and most of the other extensions already discussed in Chapters 10 and 11. As a typical example, Fig. 12-6 shows the competitive inhibition ("antagonism") by the antihistamine, diphenhydramine, against the normal agonism of histamine, as determined by a Lineweaver-Burk-type plot of (1/Eff) versus (1/[D]), in which the measured effect was the fall in blood pressure induced by histamine. The plots for different concentrations of diphenhydramine intersect in a point that lies on the y axis, a result diagnostic for competitive binding.

In the preceding model, a drug molecule is assumed to exert its effect continuously during the time it resides at the receptor site. An alternative view might be that a drug exerts its effect only at the moment that it combines with the receptor (as for production of an action potential spike following arrival of neurotransmitters at their receptors in the postsynaptic membrane); this model forms the basis of so-called "rate" theory of drug-receptor interactions. Algebraically,

$$\text{Effect} = k_1(1 - f)\ [D](\text{constant}) \tag{12-18}$$

so that the effect is proportional to the rate of formation of the DR complex, following the same basic kinetic scheme of Eq. 12-10, in which $(1 - f)$ is the

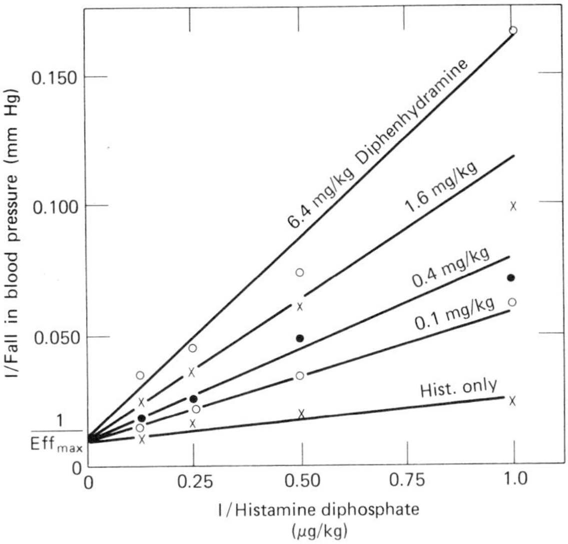

FIGURE 12-6. Double-reciprocal (Lineweaver-Burk type) plot of reciprocal response versus reciprocal agonist concentration for blood pressure fall caused by histamine in the dog. Response has been measured for histamine alone (lowermost line) and in the presence of various (fixed) concentrations of the antagonist drug, diphenhydramine (upper plots), with all doses in mg drug per kg body weight of subject animal. Competitive antagonism is indicated (see text). [From G. Chen and D. Russell, *J. Pharmacol. Exp. Therap. 99*, 401 (1950).]

proportion of receptor free to bind with drug, D. Since at equilibrium, the rates of association and dissociation of the DR complex are equal,

$$\text{Effect} = (\text{constant})(k_{-1}\, f) \tag{12-19}$$

and using the fact that $f = [D]/(K_D + [D])$ when $[D] >> [R]_{\text{total}}$

$$\boxed{\text{Effect} = \frac{(\text{constant})k_{-1}\,[D]}{K_D + [D]}} \tag{12-20}$$

The "rate" theory result, Eq. 12-20, clearly exhibits the same functional relation between $[D]$ and Eff as the "occupancy" theory result, Eq. 12-17. Thus, we reach the interesting conclusion that the drug-receptor model leads to a relation between $[D]$ and Eff that is entirely analogous to the relation between $[A]$ and v in enzyme kinetics, whether the effect of the drug is related simply to the fraction of occupied receptor sites or to the rate at which drug molecules combine with receptor.

The biological relevance of the drug-receptor model depends on demonstration of the existence of the *receptor* entity, just as the relevance of the substrate-enzyme model depends on the isolation of the *enzyme*. Recently, an increasing number of hormonal receptors have been demonstrated and a number have actually been isolated, including the acetylcholine receptor, various sex hormone receptors, and the cholera toxin receptor. In many cases the "receptor" is itself an enzyme, and the "effect" of the drug is simply to act as an inhibitor of the enzyme with subsequent physiological consequences. One of the most dramatic examples of both kinds of effects is provided by an examination of the mode of action of various drugs that affect mood and mental health.

EXAMPLE *Psychopharmacological Agents and Physiology of Catecholamines*

The overwhelming significance of psychopharmaceuticals is clearly evident in the data of Fig. 12-7, showing the great reduction in number of patients in mental hospitals (in spite of increasing general population and diagnosis of mental conditions) following the introduction of the first antipsychotic phenothiazine derivative (chlorpromazine). The mode of action of this and a large variety of other types of antipsychotic drugs is most readily appreciated from a basic knowledge of the components of the physiology of sympathetic nerve terminals, shown schematically in Fig. 12-8.

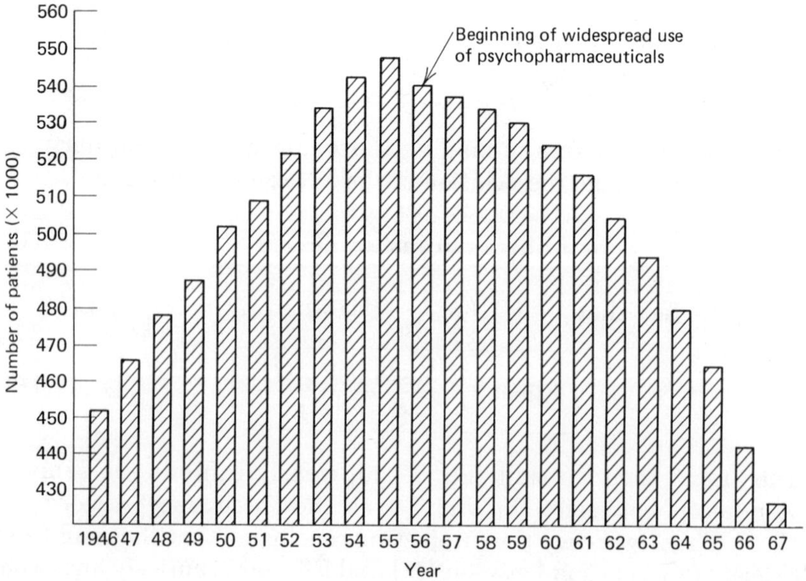

FIGURE 12-7. Number of resident patients in state and local government mental hospitals in the United States, 1946–1967 (based on USPHS figures). A further confirmation of the effectiveness of treatment is given by the twofold increase in the live-discharge rate recorded since the introduction of these drugs. [From D. H. Efron, ed., *Psychopharmacology. A Review of Progress*, U.S. Dept. of HEW (Washington, D.C., 1968), p. 2.]

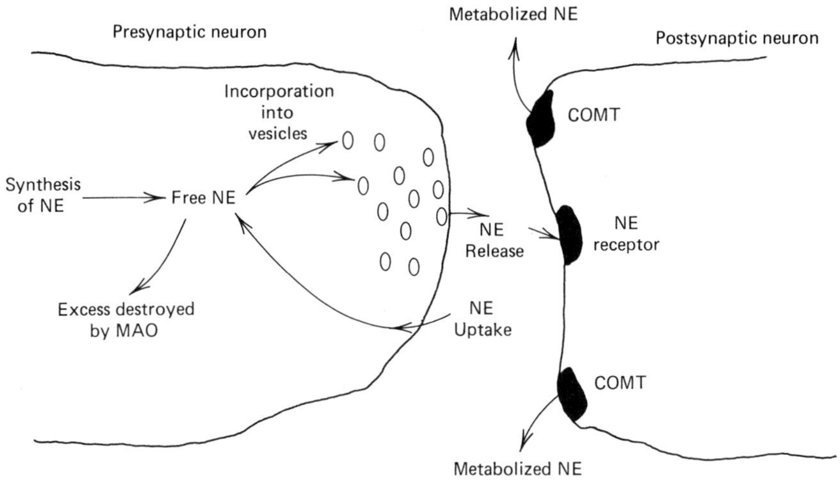

FIGURE 12-8. Schematic diagram of a sympathetic nerve junction. The neurotransmitter norepinephrine (NE) is synthesized and incorporated into presynaptic vesicles that are released in great numbers on arrival of an action spike at the nerve terminus. Once released, the transmitter may bind to the receptor, causing formation of a new action spike that then travels down the postsynaptic neuron, or the NE may be reabsorbed into the presynaptic vesicle for destruction by monoamine oxidase enzyme (MAO) or reincorporated into vesicles again, or may be metabolized by catechol-o-methyltransferase (COMT) as the transmitter diffuses toward the edges of the synaptic cleft. Action of various psychogenic drugs on this system is described in the text.

Chlorpromazine and other phenothiazine derivatives have proved very useful in alleviation of such psychoses as schizophrenia and mania, as well as in relief of chronic hallucinations and confusion. This class of tranquilizer drugs seems to act by blocking the receptor itself [in this case, the transmitter involved is adrenaline (epinephrine) rather than noradrenaline (norepinephrine)]. Another useful antipsychotic tranquilizer drug is reserpine, whose principal effect is to deplete norepinephrine by causing its release from presynaptic storage vesicles to permit its quick destruction by monoamine oxidase (MAO). Another group of "antidepressant" drugs includes a variety of tricyclic compounds such as imipramine, which block the re-uptake of NE transmitter and thus increase postsynaptic nerve stimulation by causing a buildup of transmitter at the synapse. (Amphetamine also acts as a stimulant by blocking NE uptake.) A final class of antidepressant drugs is the monoamine oxidase inhibitor group, such as iproniazid (no longer used) and phenelzine. By blocking MAO, these drugs prevent destruction of the transmitter in the presynaptic neuron and thus cause stimulation by the extra NE thus allowed to accumulate.

12.B.2. All-or-none response: sleep and death

The theories of the preceding section apply when the response can vary over a *range,* as with degree of muscle contraction or range of blood pressure or

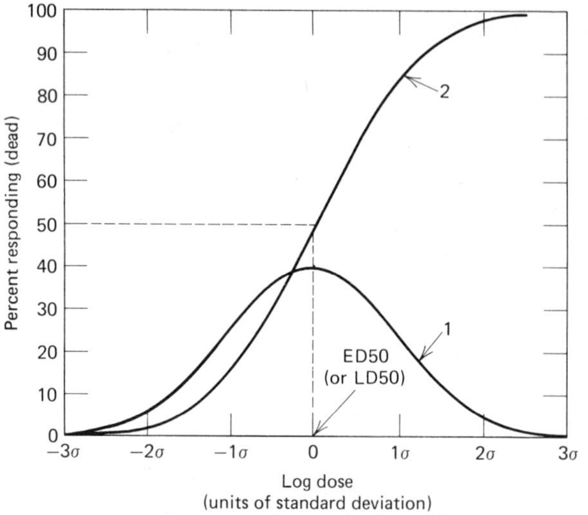

FIGURE 12-9. The Quantal (all-or-none) log dose-response curve. The horizontal axis shows log(dose) in units of standard deviation. Curve 1 represents the log-normal distribution of threshold sensitivities of individual subjects to the drug; curve 2 is the cumulative distribution. σ is the standard deviation of sensitivities. The median (lethal) dose is called LD50. (After A. Goldstein, L. Aronow, and S. M. Kalman, *Principles of Drug Action,* Harper & Row, N. Y., 1969, p. 353.)

heart rate. A somewhat different description is required when the response is of the *all-or-none* type (did the subject die or not?). For a large number of all-or-none response examples, the number of subjects showing greater than threshold response versus the logarithm of drug dose follows a normal (Gaussian) distribution, sometimes called the "log-normal" distribution. (A log-normal distribution occurs whenever the response is proportional to the size of the stimulus *and* to the size of the animal.) When the *cumulative* response from such a population is plotted versus log(dose), the graph shown in Fig. 12-9 results. This time the vertical axis represents the *fraction* of *many* subjects that showed threshold behavior (sleep, death, a given degree of intoxication, etc.), rather than the *degree* of response from a *typical* individual as in the preceding section. The principal point of Fig. 12-9 is that the cumulative fraction of responding individuals shows the same sigmoidal behavior as a function of log(dose) as did the degree of response versus log(dose) for the graded response case. All quantal log(dose)-response curves have the same shape when the x axis is expressed in units of standard deviation of the original log-normal curve, as in Fig. 12-9. In place of the ED_{50} (dose at which a graded response is half the maximum possible response), we have the LD_{50} (dose for which half the subject population would show greater than threshold response).

It is always desirable to reduce any plot of experimental data to a straight line, because the quality of the data can then be evaluated using universally acceptable criteria. With the sigmoidal log dose-response data for the graded response case, the appropriate straight-line display is obtained from a Lineweaver-Burk type double-reciprocal plot. With the present log-normal distribution, a straight-line display results from a plot of "probit" versus log(dose), as now described.

If the y axis of Fig. 12-9 is changed from percent responding to the number of standard deviations corresponding to that percentage

Cumulative Percent Response	Number of S.D.
2	−2.0
7	−1.5
16	−1.0
31	−0.5
50	0.0
69	+0.5
84	+1.0
93	+1.5
98	+2.0

then curve 2 of Fig. 12-9 becomes a straight line. (Pharmacologists typically add +5.0 to the number of standard deviations so that all the numbers are

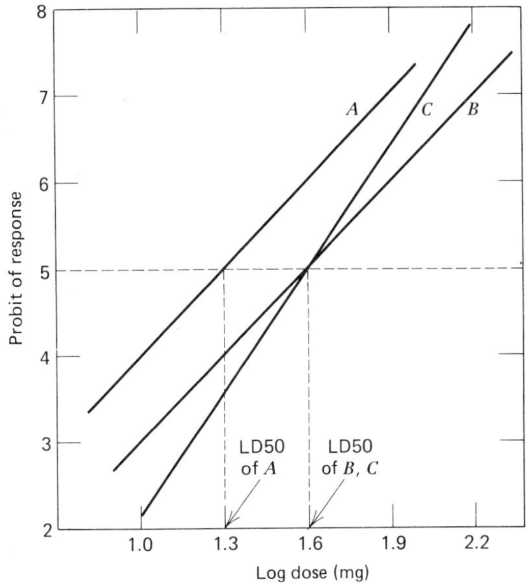

FIGURE 12-10. Plots of probit versus log(dose) for three theoretical drugs, *A*, *B*, and *C*. Drug *A* is more potent than drugs *B* and *C*, since *A* has the smaller LD50. Drug *C* has the steeper slope, and thus represents a more homogeneous population (with respect to variation in response to the drug) than drugs *B* and *A* (see text). (From A. Goldstein, L. Aronow, and S. M. Kalman, *Principles of Drug Action*, Harper & Row, N. Y., 1969, p. 356.)

positive over the dose range of interest—the resultant quantity is called "probit.") Finally, suppose the x axis is now plotted directly as log(dose), rather than in standard deviations as in Fig. 12-9. A plot of probit versus log(dose) will then be a straight line in which the intersection with the x axis indicates drug potency, as shown in Fig. 12-10. (See Problems for a numerical example.)

A little thought should convince the reader that the slope of a probit versus log(dose) curve is $(1/\sigma)$, where σ is a measure of the variation in drug sensitivity shown by a given animal population. Thus, a steeper slope corresponds to a smaller σ and a population whose sensitivities cluster around a smaller range of drug dosage.

EXAMPLE *Safety Margin in Choice of Barbiturates*

Any drug will have some undesirable side effects, and one therefore wants to choose a drug in which there is a large difference between the smallest dose that will be *threshold*-effective for (say) 99% of subjects, and the smallest dose at which undesirable threshold effects for (say) 1% of subjects become significant. Figure 12-11 shows that a probit versus log(dose) plot furnishes a simple and direct means for such evaluation. In the illustrated example, the "therapeutic ratio," LD50/ED50 was greater for aprobarbital (5.3) than for

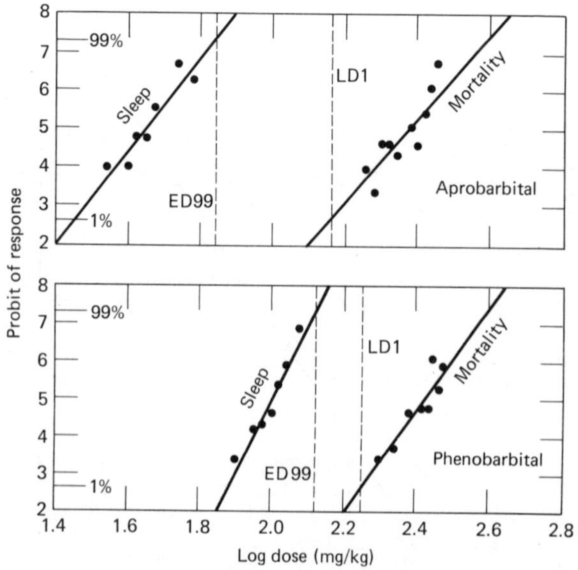

FIGURE 12-11. Comparison of therapeutic ratios of two barbiturates in mice. Groups of 20 mice were injected with each dose subcutaneously. Sleep was defined as the loss of righting reflex. Deaths were recorded at 24 hours. Probit versus log(dose) curves were calculated, using appropriate weighting factors for the points at various probit values (see Problems for a similar calculation). [From R. H. K. Foster, *J. Pharmacol. Exp. Therap. 65,* 1 (1939).]

phenobarbital (2.6). Since the LD50 for the two drugs is similar while the ED50 for aprobarbital is much lower than for phenobarbital, it would appear that aprobarbital offers a larger margin of safety between effective and lethal doses. However, because of the differences in slope of the effective dose curves, the safety margin, LD1/ED99 is almost the same (2.0 and 1.3 for aprobarbital versus phenobarbital). These data show that it is necessary to consider the behavior of the *whole* subject population, not just the behavior of the "average" subject (ED50, LD50) in evaluating drug dosage and safety.

PROBLEMS

1. (a) Calculate the concentration of drug, $[B]$, in the blood after n doses, $[B]_0$, each of which is instantaneously ingested and absorbed. Assume that the drug is eliminated by a first-order kinetic process of rate constant, k_2. (Ans. Eq. 12-3).

(b) How many doses, each of amount, $3[B]_0$, would be required to reach a blood concentration of *half* the eventual plateau level produced by an infinite number of doses, each of amount $[B]_0$? (Ans. Eq. 12-4).

2. Calculate the "sobriety" time, the time it takes for the blood alcohol level to drop below the (arbitrary) limit of 1.0 g/l, for a case of zero-order ingestion rate of 1.5 g/l/hr for a period of 1 hr (after which consumption ceases). Assume that absorption is a first-order process with rate constant, $k_{in} = 10\ hr^{-1}$, and that elimination is a zero-order process of rate constant $k_{out} = 0.192$ g ethanol/liter body fluid/hr. (Volume of body fluid = 40 liter.)

3. 120 rats were divided into 8 groups, and each group exposed to the dosage of a toxic drug shown in the table below (hypothetical data).
(a) Calculate and plot the cumulative percent of subjects killed as a function of the logarithm of the dose (compare to Figure 12-9).
(b) Convert cumulative percent to probit (refer to Fig. 12-9 for normal (Gaussian) curve values as a function of standard deviation from the mean value), and plot probit versus log(dose). [Compare to Figure 12-10.]

Toxicity test (hypothetical data), showing relation between dosage and mortality.

Dose (mg/kg)	*Proportion Killed*
1.6	0/15
2.0	1/15
2.5	2/15
3.2	7/15
4.0	10/15
5.0	12/15
6.3	14/15
8.0	15/15

REFERENCES

G. V. Calder, "The Time Evolution of Drugs in the Body," *J. Chem. Educ. 51,* 19 (1974).

W. C. Bowman, M. J. Rand, and G. B. West, *Textbook of Pharmacology,* Blackwell Scientific Publications, Oxford (1968), chapter 18.

A. Goldstein, L. Aronow, and S. M. Kalman, *Principles of Drug Action,* Harper & Row, N. Y. (1969).

4
WEIGHT ON A SPRING

CHAPTER 13 The Driven, Damped Weight on a Spring: One of the Most Important Problems in Physical Science

Everything we perceive about our surroundings is based on the *forces* that connect us to those surroundings. The content of physics might be defined as a study of the origin and consequences of forces in nature. As for any quantity that varies with distance, one can in general expand the force (function) in a power series about some convenient point in space, say, the origin:

$$\text{Net force} = m(d^2x/dt^2) = A_0 + A_1x + A_2x^2 + \cdots \qquad (13\text{-}1)$$

in which x represents the displacement from some arbitrary origin (say, the center of an atom). The first thing to recognize is that even a relatively simple force, such as the Coulomb force between two charged particles, $m(d^2x/dt^2) = qq'/x^2$, may be represented by an *infinite* number of terms in the power series of Eq. 13-1. Fortunately, one can understand much of what is presently known about common forces by making two simplifications in Eq. 13-1.

First, we can often dispense with the constant term, A_0, by choosing a suitable "reference" frame. For example, we successfully neglected the (constant) force of gravity when we considered electrophoresis experiments. Second, if the force is so *weak* that the observed displacement, x, is very *small*, then we may neglect all the higher order terms, leaving

$$\text{Net force} = m(d^2x/dt^2) = A_1x \qquad (13\text{-}2)$$

Finally, if the constant, A_1, is now re-labeled as a new constant, $-k$, then Eq. 13-2 is immediately recognized as the equation for the (restoring) force for a weight (of mass, m) on a spring:

$$\text{Net force} = \boxed{m(d^2x/dt^2) = -kx} \qquad (13\text{-}3)$$

For example, even though we may know that the force binding an electron to an atom or molecule is a *Coulomb* force, we may still treat the electrons in a substance as if the electrons were bound to their respective atoms by *springs*, provided that whatever forces we apply do not displace the electrons very far from their equilibrium positions. Thus, because the electric and magnetic forces from electromagnetic waves (light, X rays, radiofrequencies, microwaves, infrared, etc.) on atoms and molecules are indeed weak, we can successfully describe many effects of electromagnetic radiation on molecules using the same mathematics that describe the response of a simple mechanical weight on a spring to some externally applied "jiggling." Before we consider the motion of a "jiggled" *spring,* it is useful to review the vocabulary of *wave* motion, since electromagnetic radiation can be described by the same language as for ordinary water waves.

13.A. VOCABULARY FOR WAVE MOTION

We are interested in two types of waves: longitudinal and transverse. A *longitudinal* wave (e.g., a sound wave) may be defined as a disturbance whose displacement oscillates *along* the direction of propagation of the disturbance. A *transverse* wave (e.g., a water wave or electromagnetic wave) is a disturbance whose displacement oscillates in a direction *perpendicular* to the direction of propagation of the wave. In other words, air molecules move forward and backward as a sound wave propagates forward, while a cork floating on water moves up and down vertically as the water wave moves along horizontally. Both types of waves are illustrated in Fig. 13-1.

The *amplitude* of a wave is the *maximum* displacement from the equilibrium position. The *wavelength,* λ, for a monochromatic (see below) wave is the distance (along the direction of propagation) between two successive points of maximal amplitude *at a given instant in time.* The *velocity, c,* of a monochromatic wave is defined as the distance a given wave crest moves per unit time. The *frequency,* ν, of a monochromatic wave is the number of times per second that the wave amplitude *at a given point in space* passes through the maximum value. The *period, T,* is the time required to complete one cycle of the oscillation of a monochromatic wave. Finally, the *phase* (or "phase angle"), ϕ, of a wave at a given instant of time, *t,* can be defined as the number of radians of oscillation (there are 2π radians per oscillation, and thus $\omega = 2\pi\nu$ radians per second for a wave of frequency, ν oscillations/sec) which have accumulated since some arbitrary zero time. The relations between the previous definitions are given by the following equations:

Distance/sec = (distance/cycle) (cycles/sec)

$$c = \lambda\nu \tag{13-4}$$

$$T = 1/\nu \quad \text{sec/cycle} \tag{13-5}$$

$$\phi = \phi_0 + \omega t \quad \text{radians accumulated since time zero} \tag{13-6}$$

$$\omega = 2\pi\nu \quad \text{radians/sec} \tag{13-7}$$

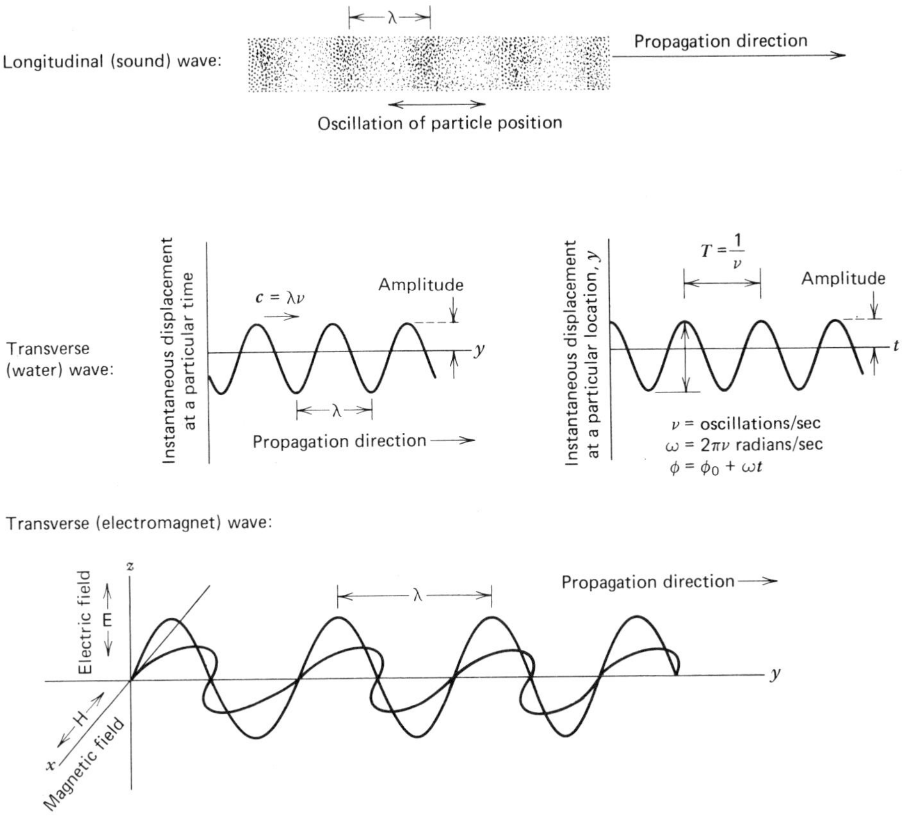

FIGURE 13-1. Schematic diagrams of a longitudinal (sound) wave, a transverse (water) wave, and a transverse (electromagnetic) wave. In each case, the wave is taken to be monochromatic (see text). The electromagnetic wave is also taken to be plane-polarized (see text), for simplest display. All waves are generated continuously at the left of the figure, so that propagation is to the right. The parameters associated with the wave motion are described below. It will prove useful in later discussion to switch back and forth between the "snapshot" picture of wave displacement as a function of propagation distance at a particular time (middle left diagram), and the equivalent alternative picture of wave displacement as a function of time at a given propagation distance (middle right diagram).

The principle of *superposition* states that when two waves travel through the same region of space, their displacements *add*, as, for example, for two sound waves of different frequency (pitch) or two light waves of different frequency (color). A wave is said to be *monochromatic* if all its components have the same frequency (and thus the same wavelength). The *intensity* of a wave is the energy flow across unit area perpendicular to the direction of propagation. *Intensity* is proportional to the *square* of the wave *amplitude,* as may be seen by analogy to the water wave: if the amplitude of a water wave doubles, then the water molecules must travel twice as far in a given length of time, and therefore have on the average twice as much velocity. Since kinetic energy is proportional to the square of velocity, intensity must be proportional to the square of wave amplitude.

For the remaining terminology, and for most of this Section, we shall concentrate on the electromagnetic (transverse) wave. One must first appreciate that most interactions of the *magnetic* field component of an electromagnetic wave with matter are much *weaker* than interactions of the *electric* field component with matter. Thus, except when we are considering exclusively magnetic effects (Chapters 14.D and 16.B.3), we ignore the magnetic field component completely.

Electromagnetic radiation is said to be *plane-polarized* when the directions of the electric field vectors at various positions along the direction of propagation at a given instant all lie in a plane. Plane-polarized radiation need not be monochromatic, as evident from the diagram below. When many

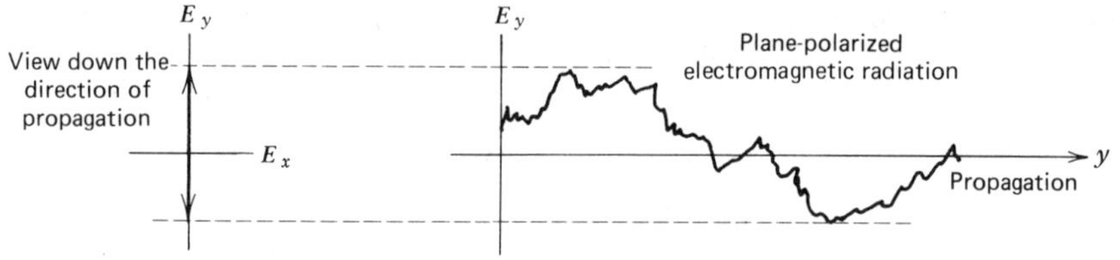

electromagnetic wave components are traveling through the same region of space, they may add in various ways as shown in Fig. 13-2. When the components are all of the same frequency (wavelength) *and* all of the same phase (i.e., each component wave goes through zero displacement at the same point in space at a given instant), then the addition of amplitudes is said to be *coherent,* and the resultant is a sine wave of the same phase as its individual components, with greatly increased amplitude. On the other hand, when there is no common phase relation between the component waves, addition is said to be *incoherent,* and the resultant wave is either a sine wave of the same frequency but different amplitude and phase, or a periodic (but not sine) wave, depending on whether the component waves have the same or different frequency (see Fig. 13-2). Lasers (Chapter 19.B.2) are devices in which component monochromatic waves are added coherently to give a resultant of amplitude several orders of magnitude larger than possible with an ordinary incoherent source, such as a glowing lamp filament. It can be shown (see Problems) that incoherent monochromatic addition always produces a sine wave, but the amplitude of that sine wave will always be smaller than for coherent addition. Many applications of lasers are based on their extremely high intensity (see Chapter 19.B.2).

Circularly (*elliptically*) *polarized* light exhibits an electric field whose direction moves around a *circle* (*ellipse*) when viewed down the direction of propagation. Although circularly (elliptically) polarized light need not be

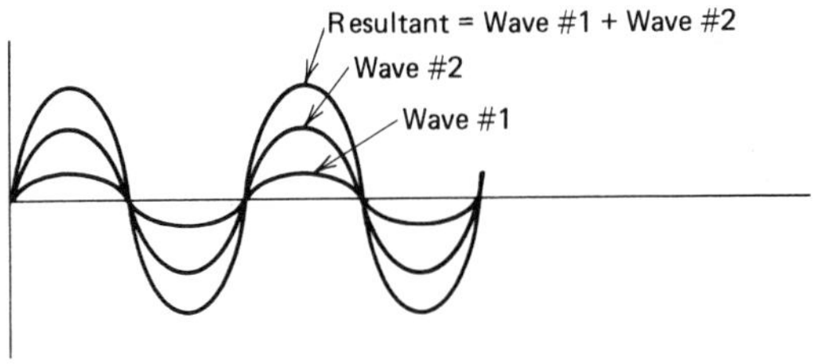

coherent addition of magnitudes (displacements) of two waves of equal frequency and equal phase, to give a resultant sine wave of increased amplitude.

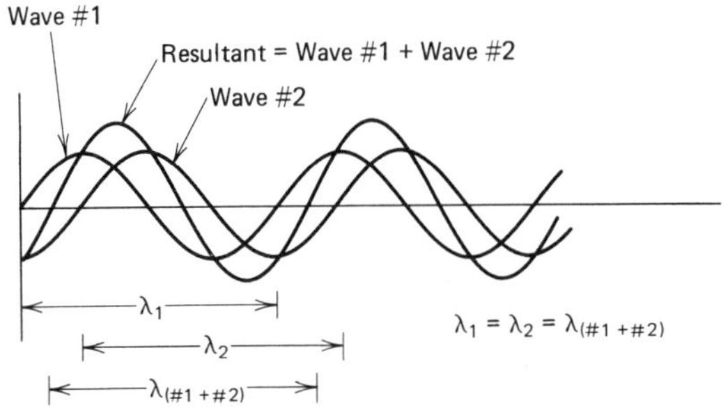

incoherent addition of magnitudes of two waves of equal frequency but different phase, to give a resultant sine wave of different amplitude and phase than either of its two components.

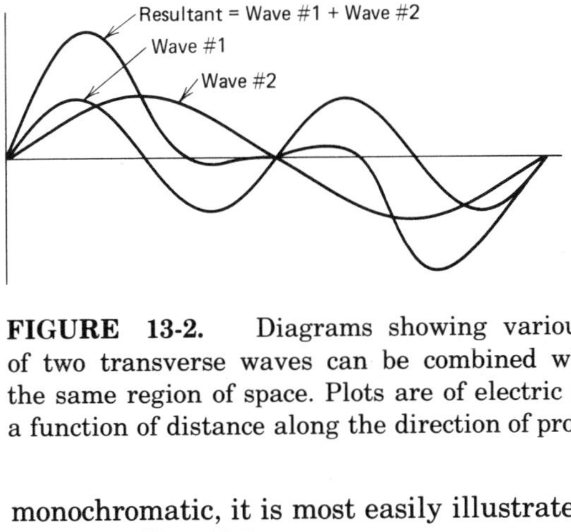

incoherent addition of magnitudes of two waves of different frequency to give a resultant which is periodic, but which is no longer a simple sine wave.

FIGURE 13-2. Diagrams showing various ways in which the magnitudes of two transverse waves can be combined when the two waves are traversing the same region of space. Plots are of electric field of an electromagnetic wave as a function of distance along the direction of propagation at a given instant in time.

monochromatic, it is most easily illustrated using monochromatic light, as shown in Fig. 13-3.

Finally, from the definition,

$$\omega = 2\pi\nu \tag{13-7}$$

in which ω is called the *angular frequency* of the wave in radians $\sec^{-1}$, it is possible to express the electric field magnitude, E, of an electromagnetic monochromatic wave as a function of time at a given (arbitrary) point in space as:

$$\mathbf{E} = E_0 \cos(\omega t) \qquad \text{at } y = 0 \tag{13-8}$$

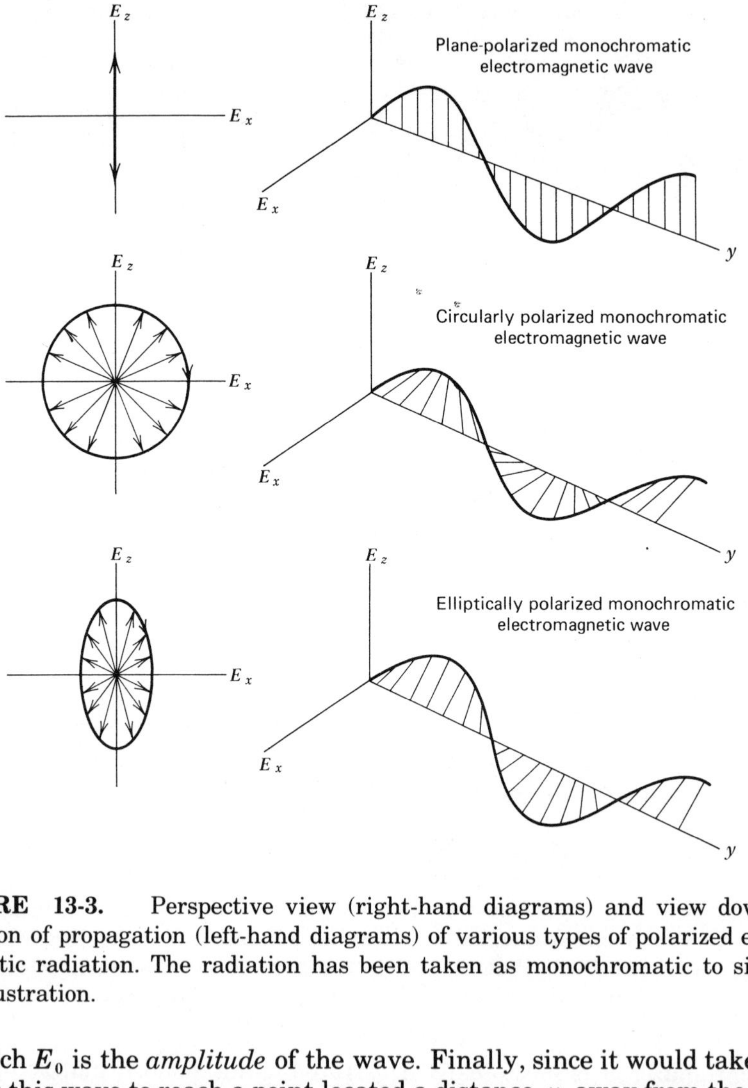

FIGURE 13-3. Perspective view (right-hand diagrams) and view down the direction of propagation (left-hand diagrams) of various types of polarized electromagnetic radiation. The radiation has been taken as monochromatic to simplify the illustration.

in which E_0 is the *amplitude* of the wave. Finally, since it would take (y/c) sec for this wave to reach a point located a distance, y, away from the given point, where c is the velocity of electromagnetic radiation, the reader should convince him or herself that the magnitude of the electric field at point, y, will obey the equation

$$E = E_0 \cos(\omega(t - y/c)) \tag{13-9}$$

(Note that E has the same value at $t = 0$ for $y = 0$ as at $t = y/c$ for $y = y$.)

Equations 13-8 and 13-9 show why *angular* frequency is almost always used in the mathematical description of spring motion, since the electromagnetic wave frequency, ν, appears in the equations as $2\pi\nu = \omega$. Armed with the necessary vocabulary, we are now prepared to discuss the motion of a weight on a spring, first in the absence (and later in the presence) of additional driving and damping forces.

"Natural" Motion of the Weight on a Spring Itself

The equation of motion of a weight of mass, m, attached to a spring of stiffness (spring constant), k, in the *absence* of any damping or driving force is

$$\text{restoring force} = -kx$$

or

$$md^2x/dt^2 + kx = 0 \qquad (13\text{-}3)$$

It is readily verified that Eq. 13-10 is a solution of this equation of motion. (Since the mathematical procedures for solving differential equations, such as Eq. 13-3, are rather specialized, we will generally proceed directly to the solution without indicating the route by which it was obtained—the differential equations in this chapter are solved in most books on differential equations.)

$$x = x_0 \cos(\omega_0 t) \qquad (13\text{-}10)$$

where

$$\omega_0 = \sqrt{(k/m)} = \text{"natural" frequency for weight-on-a-spring} \qquad (13\text{-}11)$$

for a maximum initial displacement, $x = x_0$ at time, $t = 0$. Since there is no frictional "damping," the displacement (position) of the weight continues to oscillate sinusoidally indefinitely at the "natural" frequency, $\nu_0 = (\omega_0/2\pi)$ cycles/sec.

Experiments for Extracting Weight-on-a-Spring Parameters

In order to determine which *mathematical* problem to solve, it is first necessary to examine briefly the types of *experiments* from which properties of the weight on a spring can be evaluated. The best experiments are of two types: (a) hit the weight with a sudden blow (i.e., displace the weight suddenly) and then monitor the spring position as a function of *time*, or (b) apply a steady "jiggling" to the weight (at a driving frequency that may be different from the "natural" frequency of the isolated weight on a spring itself), wait until the system settles into a steady-state motion, and measure the amplitude of the motion as a function of *driving frequency*. These two approaches are shown schematically in Fig. 13-4.

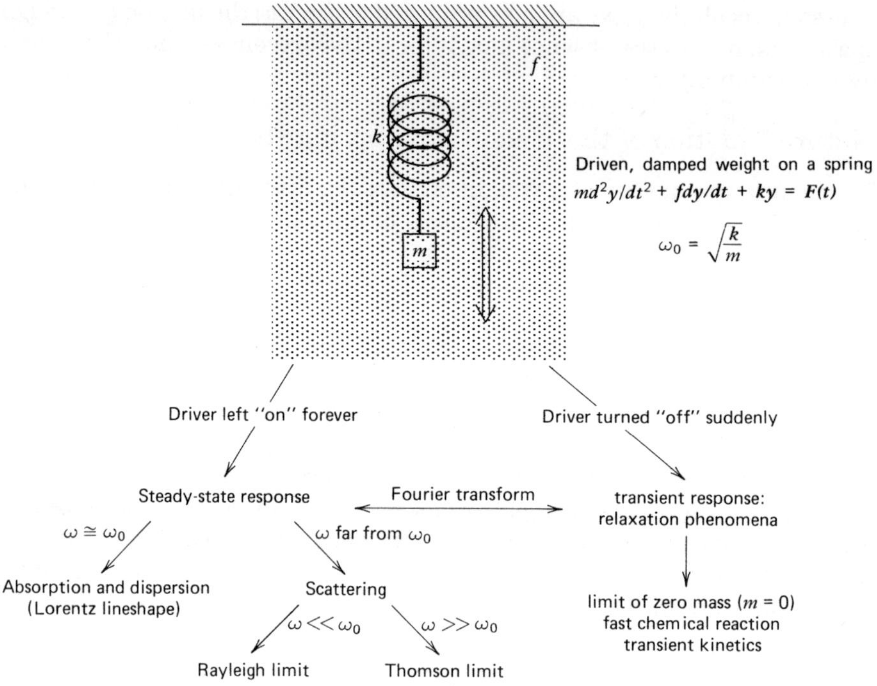

FIGURE 13-4. Interrelations between various scattering, spectroscopy, and transient experiments, based on analogy to the motion of a sinusoidally driven (or suddenly displaced), damped weight on a spring. The weight has mass, m, bound to a spring of force constant, k, immersed in a medium to give frictional coefficient, f, and subjected either to a sudden displacement or to a sinusoidally time-varying force, $F(t) = F_0\cos(\omega t)$. Steady-state phenomena are discussed in Chapters 14 and 15, transient phenomena in Chapter 16, and Fourier methods in Section 6.

If we consider that electrons in matter behave as if they were bound by springs to their respective atoms, then the effect of the oscillating electric field of an incident electromagnetic wave (such as a light wave) is to drive the electron up and down against its restraining spring. The electron, once set into motion will emit electromagnetic radiation in various directions, just as a tuning fork once set vibrating will emit sound in various directions. It is then possible to *locate* the vibrating electron on a spring by measuring the amplitude (or more commonly, the intensity, which is proportional to the square of the amplitude) of the re-radiated ("scattered") radiation from the driven electron as a function of direction. Such experiments form the basis for location of atoms in crystals by X-ray and neutron diffraction, for example. In addition, the *natural frequency*, ν_0,

$$\nu_0 = (1/2\pi)\sqrt{(k/m)} \quad \text{oscillations/sec} \tag{13-12}$$

can be determined either (a) by counting the number of oscillations per second following a sudden blow (pulse), or (b) by determining the externally applied driving frequency at which the steady-state spring response is maximal (see below). Since both the sudden blow (pulse) and steady-state jiggling experiments give the same information, it is not surprising to find that they are related by a well-defined mathematical procedure, the Fourier transform (Section 6). The mathematical development for the driven, damped weight on a spring is most directly appreciated from Fig. 13-4, which shows a schematic flow chart for calculations and experiments.

Now if the driven spring were completely free to move, then when the driving frequency is set equal to (in "resonance" with) the natural spring frequency, ν_0, we would expect the spring to absorb power continuously and execute larger and larger amplitude motion without limit. However, any real spring (including our electron bound to a molecule) is hindered in its motion by frictional resistance, or drag, expressed by a "damping" force proportional to the velocity of the moving mass (electron):

$$\text{damping force} = -f(dx/dt) \tag{13-13}$$

The damping constant, f, can again be determined using either of our prior experiments: following a sudden blow (pulse), the amplitude of the mass (electron) motion will decrease at a rate related to f; alternatively, as we jiggle the spring continuously with a sinusoidal driving force, the maximum displacement is inversely related to f, because as friction increases, more of the driving power is dissipated as heat, leaving less power available for moving the weight on the spring.

In the spectral examples mentioned later, the natural frequencies of various springs correspond to the frequencies of power absorption "*peaks*" in various types of spectra; the damping constant of a given spring is related to the *width* of a spectral line; and the spatial distribution of springs (electrons) on a molecule is reflected in the "scattered" (re-radiated) *radiation-versus-direction* pattern discussed above. The practical interest in these parameters is that the *natural frequencies* indicate the strength of the spring of interest, and are related to color, chemical bond strength, optical activity, and local environment near the spring; *spectral line widths* can be used to measure the rates of processes too fast to study directly, such as molecular collisions, translations, and rotations, and very fast chemical reactions; radiative *scattering* can be used to locate the spatial distribution of atoms within a single molecule (i.e., the molecular three-dimensional structure) in crystals or ribosomes, or even molecules in solution.

13.B. THE DRIVEN, DAMPED WEIGHT-ON-A-SPRING: STEADY-STATE RESPONSE

Our present (brief) mathematical treatment follows the flow chart of Fig. 13-5. When subjected to a continuous, sinusoidally time-varying driving force, $F_0 \cos(\omega t)$, a weight on a spring will eventually settle into a steady

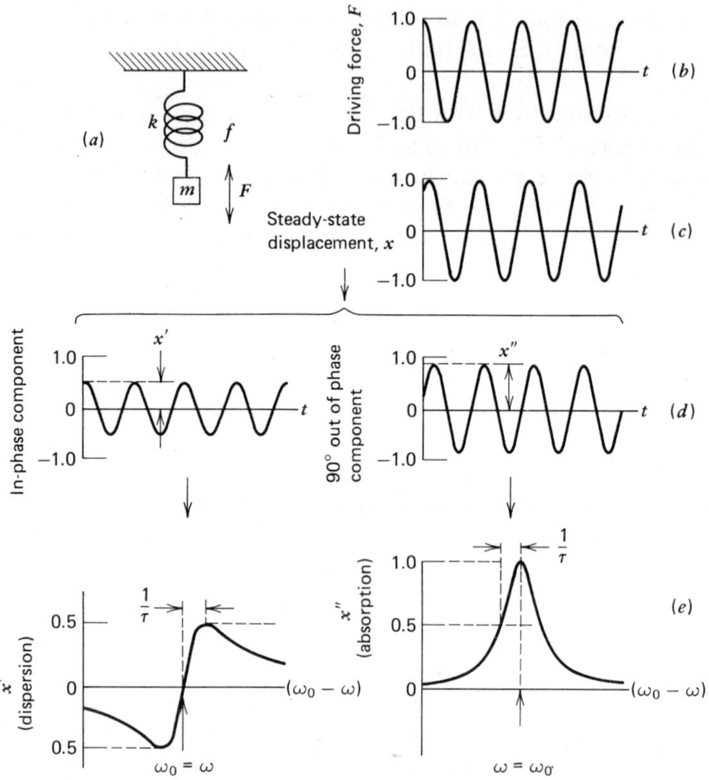

FIGURE 13-5. Dispersion and absorption, as derived from the steady-state response of a driven, damped weight on a spring. When subjected to a sinusoidally time-varying driving force (*b* in the figure), the mass on a damped spring (Fig. 13-5*a*) will eventually settle into a steady-state oscillation at the *same frequency* as the driver, but with *different phase* (see Fig. 13-5*c*). The magnitude of the displacement (Fig. 13-5*c*) may be decomposed into one component that is exactly *in-phase* with the driver (Fig. 13-5*d*, left diagram) and a component exactly *90°-out-of-phase* with the driver (Fig. 13-5*d*, right diagram). The variation with (driving) frequency of the in-phase amplitude is called the "dispersion" spectrum (Fig. 13-5*e*, left diagram) and the frequency-variation of the 90°-out-of-phase amplitude is called the "absorption" spectrum. See text for examples from scattering and spectroscopy. $\omega_0 = (k/m)^{1/2}$, and $(1/\tau) = f/2m$ in the mechanical analog.

vibration *at the same frequency as the driver,* in contrast to the undriven spring (Equations 13-10 and 13-11) that vibrates at its own *natural* frequency. When there is no friction (damping), the driven spring oscillates exactly *in-phase* with the driver (see Fig. 13-5*b*), but when friction is present, the driven spring displacement will in general lag behind the driver, as shown in Figures 13-5*b* and 13-5*c*. It is mathematically convenient (and physically useful) to "decompose" the steady-state displacement, x, into two *components* (each of which still oscillates at the same frequency as the driver) that are respectively either exactly *in-phase* (amplitude $= x'$) or exactly *90°-out-of-phase* (amplitude $= x''$) with the driver, as shown in Fig. 13-5*d*. The variation of the component amplitudes, x' and x'', with *driving* frequency is shown in Fig. 13-5*e*. These plots are known as "dispersion" and "absorption" for reasons we soon explain. The overwhelming advantage of using this mechanical model to generate the mathematical form of x' and x'' is that once the treatment is completed, we will be able to identify x' and x'' with each of several types of spectroscopic signals without any additional mathematical effort, simply by changing the names of the variables. With this motivation, we will now proceed to trace the mathematical steps that describe the graphs of Fig. 13-5.

The equation of motion for the position, x, of a weight of mass, m, on a spring of force constant, k, driven by an oscillating force, F,

$$F = F_0 \cos(\omega t) \tag{13-14}$$

and subject to a frictional (damping) force proportional to the velocity of the moving weight, is given by:

$$\boxed{m\, d^2x/dt^2 = \text{net force} = -k\, x - f\, dx/dt + F_0 \cos(\omega t)} \tag{13-15}$$

Here $-kx$ is the restoring force from the spring, $-f\, dx/dt$ is the frictional force opposing the motion, and $F_0 \cos(\omega t)$ is the applied driving force. It is left to the reader to verify that Eq. 13-16 is a solution of Eq. 13-15, where Eq. 13-12, $\omega_0 = \sqrt{(k/m)}$, has been used to simplify the final expression:

GENERAL CASE

$$x = x' \cos(\omega t) + x'' \sin(\omega t) \tag{13-16a}$$

where

$$x' = F_0 \frac{m(\omega_0^2 - \omega^2)}{m^2(\omega_0^2 - \omega^2)^2 + f^2\omega^2} \tag{13-16b}$$

and

$$x'' = F_0 \frac{f\,\omega}{m^2(\omega_0^2 - \omega^2)^2 + f^2\omega^2} \tag{13-16c}$$

Equation 13-16 simply states mathematically what is shown pictorially in Fig. 13-5, namely that the steady-state displacement, x, may be broken down into two components, $x' \cos(\omega t)$ and $x'' \sin(\omega t)$, which may be thought of as the parts of the response either exactly *in-phase* with the driver [i.e., vary in time as $\cos(\omega t)$], or exactly *90°-out-of-phase* with the driver [since $\cos(\omega t - (\pi/2)) = \sin(\omega t)$]. The terms x' and x'' will henceforth be referred to as the amplitudes of the in-phase or 90°-out-of-phase components of the response, and are the quantities of physical interest.

NOTE The following short section is not essential for understanding any of the topics treated in this chapter. However, it is highly desirable for any reader who hopes to follow most of the literature in these areas, and it provides a much simpler mathematical description of absorption and dispersion, particularly as encountered again by way of Fourier transform methods in Section 6.

Up to now, all formal treatment has been in terms of *real* (as opposed to *complex*) variables, which accounts for all the phenomena we treat in this Section. However, a considerable saving in effort is realized, particularly for further applications in Section 6, by introducing *complex* notation. A reasonable question is, what is the physical meaning of the imaginary part of the complex numbers involved? The answer is that we will only ever associate *real* quantities, or real parts of complex quantities with the result of a physical measurement.

Stated most simply, the *first* reason complex quantities are useful is that the time derivative of $\exp[i\omega t]$ is just the same function again, multiplied by a constant: $i\omega\exp[i\omega t]$. In contrast, the derivative of $\sin(\omega t)$ is a different function, $\cos(\omega t)$; similarly the derivative of $\cos(\omega t)$ is again a different function, $-\sin(\omega t)$. Thus, for any problem (such as the driven, damped, harmonic oscillator) involving time derivatives of sines and cosines, it is easier to employ the function, $\exp[i\omega t]$

$$\exp[\pm i\omega t] = \cos(\omega t) \pm i \sin(\omega t) \qquad (13\text{-}17)$$

The *second* reason that complex variables are useful is that they provide an automatic way of keeping two parts of a problem separated (as the real and imaginary parts of a complex number). Complex numbers thus serve the same sort of purpose as using two perpendicular axes in ordinary geometry to keep two components of a vector separated in a plane. Both the above advantages of complex notation begin to become clear when we start over again to solve the driven, damped, harmonic oscillator problem, this time with complex numbers. The problem we actually want to solve is

$$m(d^2x/dt^2) + f(dx/dt) + kx = F_0 \cos(\omega t) \qquad x \text{ is real} \qquad (13\text{-}15)$$

Suppose, however, that we add the term, $iF_0 \sin(\omega t)$ to the right-hand side of Eq. 13-15, and then solve for the (complex) x solution:

$$m(d^2x/dt^2) + f(dx/dt) + kx = F_0 \exp[i\omega t], \qquad x \text{ is complex} \qquad (13\text{-}18)$$

In Eq. 13-18, it is understood that x is a complex number, but we will only be interested in the real part of x when we have finished solving the problem. If we now suppose that there is a steady-state solution of the form

$$\text{complex } x = \chi \exp[i\omega t] \tag{13-19}$$

then the value of the (complex) amplitude, χ, is readily found by substituting Eq. 13-19 into 13-18:

$$\chi = F_0 \frac{1}{(k - m\omega^2) + if\omega} = F_0 \frac{1}{m(\omega_0^2 - \omega^2) + if\omega} \tag{13-20}$$

Using the identity

$$\frac{1}{a + ib} = \frac{a - ib}{a^2 + b^2} \tag{13-21}$$

Equation 13-20 may be rewritten

$$\chi = F_0 \frac{m(\omega_0^2 - \omega^2) - if\omega}{m^2(\omega_0^2 - \omega^2)^2 + f^2\omega^2} \tag{13-22}$$

Now let us evaluate the real part of the complex x solution of Eq. 13-19:

$$\text{Re(complex } x) = \text{Re}(\chi \cdot \exp[i\omega t])$$

$$= \frac{F_0 m(\omega_0^2 - \omega^2)}{m^2(\omega_0^2 - \omega^2)^2 + f^2\omega^2} \cos(\omega t) + \frac{F_0 f\omega}{m^2(\omega_0^2 - \omega^2)^2 + f^2\omega^2} \sin(\omega t) \tag{13-23}$$

$$= x' \cos(\omega t) + x'' \sin(\omega t) \tag{13-24}$$

In other words, the real part of the solution of the complex Eq. 13-18 is identical to the (real) solution of the (real) equation 13-15 that we wanted in the first place. Furthermore, it is now evident that χ can be expressed as

$$\boxed{\chi = x' - i x''} \tag{13-25}$$

so that x' and x'' can be written as the real and imaginary components of the complex amplitude of the complex solution of Eq. 13-18. Although x'' thus appears at first glance to be associated with a mathematically imaginary term, the apparent anomaly is immediately resolved when it is recognized that the (real) physical response is obtained from the real part of $(\chi \cdot \exp[i\omega t])$, so that x'' ultimately is expressed as the amplitude of the

90°-out-of-phase response in Eq. 13-23. The complex amplitude, χ, thus shows the advertised property of conveniently keeping x' and x'' separated as the real and imaginary parts of a complex number; moreover, the mathematical difficulty in going from Eq. 13-18 to 13-20 is much less than the difficulty in going from Eq. 13-15 to 13-16 (see Problems), completing the claimed advantages for using complex notation.

One of the common approaches for analyzing a complicated formula (such as Eq. 13-16) is to examine its behavior in various physically meaningful limits. For example, in the limit of no damping ($f \to 0$), the 90°-out-of-phase response component, x'', reduces to zero

$$\lim_{f \to 0} x'' = 0 \tag{13-26}$$

so that the system can respond only *in-phase* with the driver. We will now generalize this result to the physical situations in which we probe the system with a driving frequency that is either very near or very far from the natural frequency of the spring.

13.B.1. Scattering Limit: Negligible Damping and/or Far from Resonance

Here we simply note that in reaching the limit of Eq. 13-26 most generally, we require just that the second term in the denominator of Eq. 13-16c be much smaller than the first term in the denominator, namely that $f\omega << |m(\omega_0^2 - \omega^2)|$. Thus, either if the damping is negligible (small friction coefficient, f) *or* if we choose a driving frequency far from the natural frequency of the spring (large value of $|(\omega_0 - \omega)|$), the 90°-out-of-phase component is negligible and the in-phase component reduces to the limiting value

SCATTERING LIMIT

$$\lim_{f << \left|\frac{m(\omega_0^2-\omega^2)}{\omega}\right|} (x') = \frac{F_0}{m(\omega_0^2 - \omega^2)} \tag{13-27a}$$

$$\lim_{f << \left|\frac{m(\omega_0^2-\omega^2)}{\omega}\right|} (x'') = 0 \tag{13-27b}$$

For present purposes, electromagnetic radiation (such as visible light or X rays) consists of an oscillating electric field that, on reaching our electron-on-a-spring model of matter, drives the (charged) electron at the frequency of the applied radiation. Maxwell's equations predict that such a moving electron will produce (re-radiate, "scatter") new radiation with an electric field amplitude proportional to the second derivative of the electron position:

$$E_{\text{scatt}} \propto \frac{d^2}{dt^2}(x' \cos(\omega t)) = \frac{-F_0\,\omega^2}{m(\omega_0^2 - \omega^2)} \cos(\omega t)$$

or

$$E_{\text{scatt}} \propto \frac{-F_0}{m((\omega_0/\omega)^2 - 1)} \cos(\omega t) \qquad (13\text{-}28)$$

Finally, noting that radiation *intensity* is proportional to the square of radiation *amplitude* (as for any transverse wave), we obtain the physically useful result

$$I_{\text{scatt}} = \frac{\text{constant}}{((\omega_0/\omega)^2 - 1)^2} \qquad (13\text{-}29)$$

As shown in many first-year undergraduate chemistry texts, the "natural" frequency (frequencies) of an electron bound to a hydrogen (or hydrogen-like) atom fall at the blue end of the optical spectrum (in fact, in the ultraviolet). It is thus useful to examine Eq. 13-29 in the *further* limit that the probing radiation is of much *lower* or much *higher* frequency that this natural electron motion.

13.B.1.a. *Rayleigh Limit: Driving Frequency Smaller than Natural Frequency.* In the limit that the driving radiation frequency (say, visible light) is much smaller than the natural frequency (say, frequency corresponding to an "electronic" transition in an atom or molecule, in the ultraviolet) of a bound electron on a spring, Eq. 13-29 reduces to the form

$$\boxed{\lim_{\omega \ll \omega_0} I_{\text{scatt}} = \text{constant} \cdot (\omega^4/\omega_0^4)} \qquad \text{RAYLEIGH SCATTERING (Visible Light)} \qquad (13\text{-}30)$$

Equation 13-30 establishes the well-known property that blue light is scattered much more strongly than red light ($\omega_{\text{blue}} >> \omega_{\text{red}}$), since scattering of visible light is proportional to the *fourth power* of the frequency of the driving radiation. In particular, this explains why the sky is blue, since light reaching an observer in directions other than in a straight line from the sun must arise from driven oscillating electrons on molecules in the atmosphere. Furthermore, since the wavelength of visible light is of the order of the dimensions of very large macromolecules in solution, such as synthetic or biological polymers, the light waves scattered from different parts of a big molecule will interfere destructively (Fig. 13-2) to an extent that depends on the macromolecular size and shape. Thus, light scattering intensity measurements can give precise though not particularly detailed information (because of the complicated process of adding many waves together to obtain the final scattered intensity) about macromolecular size and shape in solution (Chapter 15.A). Finally, since the frequency of the scattered light will be slightly shifted according to whether a given molecule was moving toward or away from the observer at the instant of scattering (Doppler effect), light scattered from molecules (or even bacteria) exposed to a very monochromatic source (such as a laser) will be observed to exhibit a range of frequencies, and the extent of that range can be used to calculate the average speed and crude estimates of the type of motion for that molecule (or bacterium), as discussed in Chapter 21.

13.B.1.b. *Thomson Limit: Driving Frequency Larger than Natural Frequency.* The other simple limiting case of Eq. 13-29 is when the driving frequency is much larger than that of the electrons on springs, as when X-rays or fast-moving electrons (remember that a electron has wave character, where the wavelength is calculated from the electron momentum by the deBroglie relation, $\lambda = h/mv$, where h is Planck's constant) impinge on matter:

$$\boxed{\lim_{\omega >> \omega_0} I_{\text{scatt}} = \text{constant}} \qquad \text{(13-31)}$$

THOMPSON SCATTERING (X rays, fast-moving electrons)

At the very high frequencies associated with X-rays or fast-moving electrons the radiation wavelength has become small compared to the size of a typical molecule, and scattering from individual atoms or molecules can be detected. Since there are no X-ray lenses, we cannot make an X-ray microscope, and are forced to rely on diffraction experiments to study molecular structure using X-radiation (Chapter 15.B). However, because electrons are *charged* particles, they can be deflected (focussed) by applied electric or magnetic fields, as any television set owner can testify, so it has proved possible to construct electron microscopes, based on the images resulting when fast-moving electrons (regarded here as electromagnetic radiation of very short wavelength) strike a sample specimen (Chapter 15.C). Equation 13-31 shows that the electron-microscope must however be "color-blind"; that is, the scattering is independent of the "color" (frequency) of the incident radiation.

13.B.2. Lorentz Limit: Driving Frequency Near Natural Frequency

In the opposite extreme that the driving frequency is very close to the natural frequency, as in the vicinity of a spectroscopic absorption line, the general case (Eq. 13-16) reduces to a different simple limit. First, if

$$|\omega_0 - \omega| << (\omega_0 + \omega) \qquad \text{(13-32)}$$

then we may approximate $\omega \cong \omega_0$ to give

$$(\omega_0^2 - \omega^2) = (\omega_0 - \omega)(\omega_0 + \omega) \cong 2\omega_0(\omega_0 - \omega) \tag{13-33}$$

Substituting Eq. 13-33 into Equations 13-16b and 13-16c, we obtain

$$\lim_{\frac{|\omega_0-\omega|}{\omega_0+\omega} << 1} (x') = \frac{2m\omega_0(\omega_0 - \omega)F_0}{4m^2\omega_0^2(\omega_0 - \omega)^2 + f^2\omega_0^2} \tag{13-34a}$$

$$\lim_{\frac{|\omega_0-\omega|}{\omega_0+\omega} << 1} (x'') = \frac{f\omega_0 F_0}{4m^2\omega_0^2(\omega_0 - \omega)^2 + f^2\omega_0^2} \tag{13-34b}$$

Finally, recognizing that the friction (damping) coefficient, f, has units of mass·time^{-1}, we may simplify Eq. 13-34 by defining a characteristic "relaxation" time, τ, according to Eq. 13-35:

$$\tau = 2m/f \tag{13-35}$$

to give the final "dispersion" and "absorption" functions of Eq. 13-36. (The terms enclosed in brackets in Eqs. 13-36 are plotted in Figure 13-5e)

$$\text{Dispersion} = \lim_{\frac{|\omega_0-\omega|}{\omega_0+\omega} << 1} (x') = \frac{F_0}{2m\omega_0}\left[\frac{(\omega_0 - \omega)\tau^2}{1 + (\omega_0 - \omega)^2\tau^2}\right] \tag{13-36a}$$

(spectroscopy)

$$\text{Absorption} = \lim_{\frac{|\omega_0-\omega|}{\omega_0+\omega} << 1} (x'') = \frac{F_0}{2m\omega_0}\left[\frac{\tau}{1 + (\omega_0 - \omega)^2\tau^2}\right] \tag{13-36b}$$

The bracketed portions of Eq. 13-36 are often referred to as the "Lorentz" line shape, and occur in many forms of spectroscopy. The 90°-out-of-phase component is also known as the "absorption" line shape, since a plot of x'' versus driving frequency can often be related to the absorption of power when radiation strikes matter (Chapter 14). As shown in Fig. 13-5*e*, the driving frequency at which the absorption signal is maximal provides a direct measure of the *natural frequency* of the sample—this condition is sometimes referred to as "resonance," by analogy to the maximum transfer of energy that occurs when a tuning fork is driven at its natural frequency. Similarly, the width at half-maximum height of the absorption-versus-frequency plot gives a direct measure of the "relaxation rate," $(1/\tau)$, which is proportional to the *friction coefficient* of the system, and thus gives information about the *surroundings* or *environment* in which the spring tries to move.

Table 13-1 Summary of phenomena whose amplitude-versus-frequency responses are related to the in-phase or 90°-out-of-phase components of the amplitude of displacement of a sinusoidally driven, damped weight on a spring

Phenomenon	Natural Motion	Direction of Natural Motion	Relation Between Driving, ω, and Natural Frequency, ω_0.	Type of Field to Drive Motion	Range for Natural Frequencies	In-Phase Component of Response	90°-Out-of-Phase Component of Response	Information Available
Scattering			$m(\omega_0^2 - \omega^2) >> f\omega$			$E_{scatt} \propto \frac{1}{\left(\frac{\omega_0}{\omega}\right)^2 - 1}$	0	Locate scattering center(s), and determine their distribution from interference patterns or focusing.
Light scattering	Electron-on-a-spring	Linear	$\omega << \omega_0$	Electric	Visible Light	$E_{scatt} \propto (\omega/\omega_0)^2$ (Rayleigh Limit)	0	Obtain size, shape, and average speed of very large molecules (or bacteria) in solution.
X-Ray scattering	Electron-on-a-spring	Linear	$\omega >> \omega_0$	Electric	X-Ray	E_{scatt} indep. of ω (Thomson Limit)	0	Determine three-dimensional structures of crystalline molecules from diffraction patterns.
Electron scattering	Electron-on-a-spring	Linear	$\omega >> \omega_0$	Electric	X-Ray	E_{scatt} indep. of ω (Thomson Limit)	0	Focus transmitted or scattered electrons to obtain electron microscope image.
Spectroscopy			$\omega \cong \omega_0$			$x' \propto \frac{(\omega_0 - \omega)\tau^2}{1 + (\omega_0 - \omega)^2\tau^2}$	$x'' \propto \frac{\tau}{1 + (\omega_0 - \omega)^2\tau^2}$	Obtain spring strength from ω_0; learn about spring environment from τ.
Electronic	Electron-on-a-spring	Linear	$\omega \cong \omega_0$	Electric	Visible, Ultraviolet	Dispersion $= n - 1$; n = refractive index	Absorption $\rightarrow \epsilon$	Learn about π-electron bonding patterns in conjugated systems from electronic absorption spectrum.
Electronic	Electron-on-a-spring	Linear	$\omega \cong \omega_0$	Electric	Visible, Ultraviolet	$\eta_{\parallel} - \eta_{\perp}$ (Birefringence)	$\epsilon_{\parallel} - \epsilon_{\perp}$ (Dichroism)	Detect and quantify extent of linear order in molecular arrays using either dichroism or birefringence.
Electronic	Electron-on-a-spring	Spiral	$\omega \cong \omega_0$	Electric	Visible Ultraviolet	$\eta_{\circlearrowleft} - \eta_{\circlearrowright}$ (Optical Rotation)	$\epsilon_{\circlearrowleft} - \epsilon_{\circlearrowright}$ (Circular Dichroism)	Detect and quantify "handedness" in organic molecules and inorganic complexes.
Vibrational	Nuclei-on-springs	Linear	$\omega \cong \omega_0$	Electric	Infrared	Dispersion $= n - 1$	Absorption	Determine chemical bond strengths from apparent force constant, k, of spring (from determination of ω_0).
Pure rotational	Molecular rotation at discrete frequencies	Circular	$\omega \cong \omega_0$	Electric	Microwave	Dispersion	Absorption	Determine chemical bond lengths and bond angles from ω_0 values obtained from location of absorption spectral peaks.

Ion cyclotron resonance	Ion in fixed magnetic field	Circular	$\omega \cong \omega_0$	Electric	Radio-frequency	Dispersion	Absorption	Obtain mass spectrum of ionized molecules, from location of absorption spectral peaks.
Electron spin resonance	Precession of electron magnetic moment	Circular	$\omega \cong \omega_0$	Magnetic	Microwave	Dispersion	Absorption	Identify and distinguish between unpaired electrons from metals or free radicals, from location and shape of absorption spectrum.
Nuclear magnetic resonance	Precession of nuclear magnetic moment	Circular	$\omega \cong \omega_0$	Magnetic	Radio-frequency	u-mode	v-mode	Identify and distinguish between magnetic nuclei (such as ^{13}C) in a single molecule from pattern of peaks in absorption spectrum.

* *Phenomena are classified according to several listed criteria, with listed applications.*

f = *frictional (damping) coefficient for moving weight on a spring*

m = *mass of weight on spring*

$\omega_0 = (k/m)^{1/2}$ = *"natural" frequency of oscillation of undriven, undamped weight on spring of force constant, k*

E_{scatt} = *amplitude of (in-phase) scattered radiation from electron on a spring, driven at a frequency far from resonance*

$\tau = 2m/f$ = *"relaxation" time*

ϵ = *absorption coefficient for the radiation in question*

Alternatively, Fig. 13-5*e* shows that the same information (namely, ω_0 and f) may be obtained from examination of the in-phase component versus driving frequency (dispersion signal). The name, dispersion, comes from the fact that a plot of $(n - 1)$ versus incident radiation frequency has the same shape as a plot of our in-phase component versus driving frequency, where n is the refractive index of the sample (i.e., the parameter describing the "dispersion" of white light into its component colors by a prism).

It thus appears (as can be justified in a more rigorous treatment) that *power absorption* and *refractive index* are closely related properties of matter (Chapter 14.A). The dividends of our formal approach now begin to become evident, as shown in Table 13-1. For example, if the "springs" that bind an electron to a molecule have different strengths in different directions, and if the molecules are all lined up in similar directions, then we would expect that the electrons would vibrate with different amplitude when driven in different directions (using plane-polarized radiation as a means for exciting vibration in a particular direction). We would thus expect *different* absorption (*dichroism*) and *different* refractive index (*birefringence*) for incident light plane-polarized in different directions: experimental examples are given in Chapter 14.B.

The above examples describe resonant processes in which the natural motion is represented by *electrons* on springs. However, different *nuclei* in a molecule are connected by chemical bonds, which may also be represented as springs whose force constants give a measure of the bond strengths of those bonds. The frequencies corresponding to these spring force constants and nuclear masses fall in the infrared part of the spectrum. Infrared power absorption and refractive index thus give information about *bond strengths* in molecules (Chapter 14.A).

The natural motion under discussion need not be *linear;* it may be *circular.* For example, free molecules (in the gas phase) tend to rotate at fixed natural frequency (quantum mechanical calculations tell us which natural frequencies will occur) determined by the moments of inertia about various special axes in the molecule. Those moments of inertia can in turn be related to bond lengths and bond angles in the molecule. The natural frequencies for such "pure rotational" spectra fall in the microwave region, so that microwave spectroscopy provides a very accurate measure of *bond lengths* and *bond angles* in isolated small molecules, based on microwave power absorption.

The *cyclotron* is a device based on the fact that ions of a given charge-to-mass ratio will rotate in the presence of a static magnetic field at a fixed natural frequency determined by the charge-to-mass ratio of the ion. For ordinary ionized molecules at convenient applied magnetic field strength, such frequencies fall in the radiofrequency region. Thus, measurement of radiofrequency power absorption as a function of incident radiation frequency will give an "ion cyclotron resonance" (I.C.R.) spectrum with peaks corresponding to each distinct charge-to-mass ion present. The device thus functions as a *mass spectrometer,* and has several unique advantages in studying ion-molecule reactions in the gas phase (Chapter 20.B.2).

Finally, since a fixed magnetic moment will precess in a circular path about a fixed magnetic field direction, and since it is possible to "drive" that motion by using the oscillating (or rotating) *magnetic* field component of electromagnetic radiation, a plot of "*magnetic resonance*" power absorption versus incident radiation frequency will give a spectrum whose peaks serve to identify particles of *different magnetic moment.* When the particles are paramagnetic electrons or free radicals, the natural frequencies fall in the microwave, and the "*electron spin resonance*" (E.S.R.) spectrum serves to identify and distinguish between various chemical species containing *unpaired electrons* (such as paramagnetic metal ions or free radicals in solution). When the particles are *paramagnetic nuclei* (e.g., ^{1}H, ^{2}H, ^{13}C, ^{15}N, ^{19}F, ^{31}P, etc.), the natural frequencies fall in the radiofrequency range, and the "*nuclear magnetic resonance*" (N.M.R.) spectrum can be used to observe, for example, distinct ^{13}C signal peaks corresponding to chemically distinct carbon atoms in an organic molecule (Chapter 14.D).

The generality of the weight-on-a-spring model is further demonstrated by showing that the absorption and speed of sound waves (Chapter 14.G) or electrical oscillations (Chapter 14.E and 14.F) may be described by analogy to the motion of a *massless* "weight" on a spring. The relevant equations may be obtained simply by setting $m = 0$ on the left-hand form of Eq. 13-20, and are discussed in Chapter 14.

As summarized in Table 13-1, the plots of Fig. 13-5 appear in almost the same form in a variety of different chemically useful contexts, and are seen to be qualitatively understandable from a single simple model, in the limit that the observing ("driving") frequency is close to the natural frequency for the motion in question. In the opposite extreme, the table shows that the in-phase component of the amplitude of the displacement of the driven spring, whose square is proportional to *scattered* radiation intensity, can provide useful information when the observing frequency is much larger (Thomson scattering) or much smaller (Rayleigh scattering) than the natural frequency in question. It is emphasized that no *quantum mechanical* calculations are required to apprehend these examples: a *classical* weight-on-a-spring model suffices to account for the observed frequency-dependencies listed here. (The value of quantum mechanics is in predicting the *number* and *magnitudes* of the "natural" frequencies that will be encountered for a given type of natural linear or circular motion.)

13.C. THE DAMPED WEIGHT-ON-A-SPRING: TRANSIENT RESPONSE

As presented in Fig. 13-4 and accompanying discussion, there is a second type of experiment that gives information about the mass (m), force constant (k), and frictional coefficient (f) for a damped weight on a spring. In this second experiment, we displace the mass from its equilibrium position, then let it go at time zero, and monitor the (transient) displacement as a function of *time*. The problem is then stated by the equation for a damped (undriven) weight on a spring:

$$\boxed{m\, d^2x/dt^2 + f\, dx/dt + kx = 0} \tag{13-37}$$

It is left as an exercise (see Problems) for the reader to verify that Eq. 13-38 is a solution of Eq. 13-37:

$$x = x_0 \exp[-t/\tau] \cos(Wt) \tag{13-38}$$

in which

$$(1/\tau) = f/2m \tag{13-35}$$

and

$$W = \sqrt{\omega_0^2 - (1/\tau)^2} \tag{13-39}$$

This result becomes especially simple in the usual physical case that

$$(1/\tau) << \omega_0 \qquad \text{(slight damping)} \tag{13-40}$$

for which

$$\boxed{\lim_{\frac{1}{\tau} << \omega_0} x = x_0 \exp[-t/\tau] \cos(\omega_0 t)} \tag{13-41}$$

Equation 13-41 indicates that when the mass is initially displaced by x_0 at time zero, it will then oscillate at its *natural* frequency, $\omega_0 = \sqrt{k/m}$, with an amplitude that *decreases exponentially with time,* with exponential time constant, τ. In more familiar terms, striking a tuning fork will cause it to vibrate at its natural pitch, with a loudness that decreases exponentially with time. It turns out to be experimentally more convenient to extract the weight-on-a-spring parameters, ω_0 and τ, from such "transient" experiments than from the previously discussed steady-state experiments, for the cases treated in Chapter 16.B (magnetic relaxation, gamma-ray correlations, and fluorescence depolarization).

The displacement, Eq. 13-41, for an undriven, damped, weight on a spring is shown in Fig. 13-6. The displacement oscillates at the natural frequency, ω_0, of the undamped weight on a spring, and the envelope of the oscillations decays exponentially with time constant, $\tau = 2m/f$.

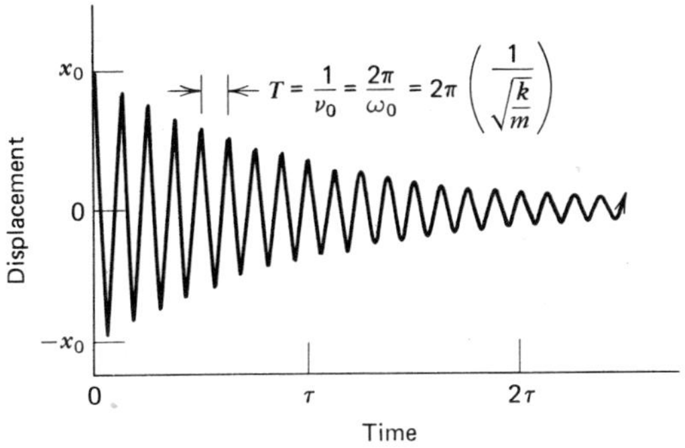

FIGURE 13-6. Diagram of the location of a damped weight on a spring as a function of time, where the mass has been initially displaced by distance, x_0, at time zero, and then released. (See Eq. 13-40 ff.)

13.D. ZERO-MASS ON A DAMPED SPRING: RELAXATION PHENOMENA

A final limiting situation is the case in which $m = 0$ (massless "weight" on a spring). The equation of motion can now be thought of as:

$$f\,dx/dt + kx = F_0 \cos(\omega t) \qquad \text{Steady-state experiment} \qquad (13\text{-}42)$$

or

$$f\,dx/dt + kx = 0 \qquad \text{Transient experiment} \qquad (13\text{-}42)$$

The solutions to Equations 13-42 and 13-43 can be verified (see Problems) as

$$x = x' \cos(\omega t) + x'' \sin(\omega t) \qquad (13\text{-}44)$$

where

$$x' = F_0 \frac{k}{k^2 + f^2\omega^2} \qquad \text{Steady-state} \qquad (13\text{-}45a)$$

and

$$x'' = F_0 \frac{f\omega}{k^2 + f^2\omega^2} \qquad (13\text{-}45b)$$

or

$$x = x_0 \exp[-kt/f] \qquad \text{Transient} \qquad (13\text{-}46)$$

Applications of the *steady-state* massless damped spring response are given in Chapter 14.E to G. Applications of the *transient* response of a massless damped spring are listed in Chapter 16.A, and are based on the fact that when a system at chemical equilibrium is subjected to a sudden, small shift in equilibrium, the approach of the component concentrations to their new values can be described by Eq. 13-43. The chemical reaction rate constants involved are often related in a simple way to the exponential time constant of the response (Eq. 13-46). One of the big advantages of the "transient" chemical kinetics method is that it is possible to determine rate constants for reactions too fast to be determined simply by mixing the reactants and monitoring the variation of the component concentrations with time. Table 13-2 shows that the rate constants for most of the chemical and motional rates of chemical and biochemical interest are too fast to be measured directly. Table 13-3 shows the range of such rates that may be determined either from the line width of the steady-state response or from the

Table 13-2. Schematic diagram of rate constants for various chemical and physical processes (typical rates are shown).

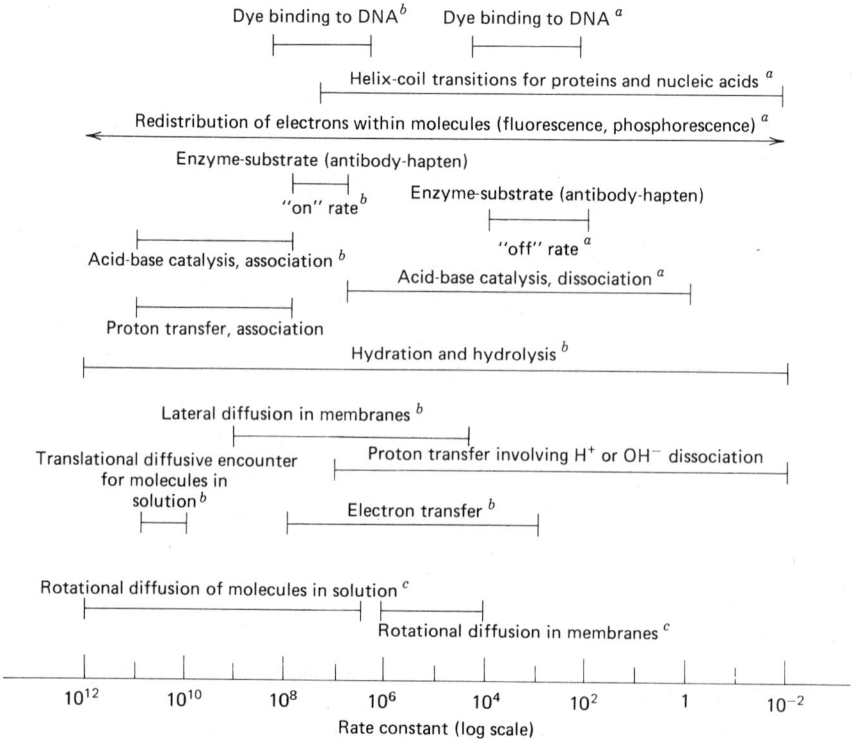

[a] *rate constant* = (*1/lifetime*) *for first-order process;* [b] *rate constant* = *2nd-order rate constant in* M^{-1} *sec*$^{-1}$; [c] *rate constant* = (*1/(rotational correlation time)*).

Table 13-3. Schematic diagram of the range of rates typically accessible by means of each of the techniques listed.

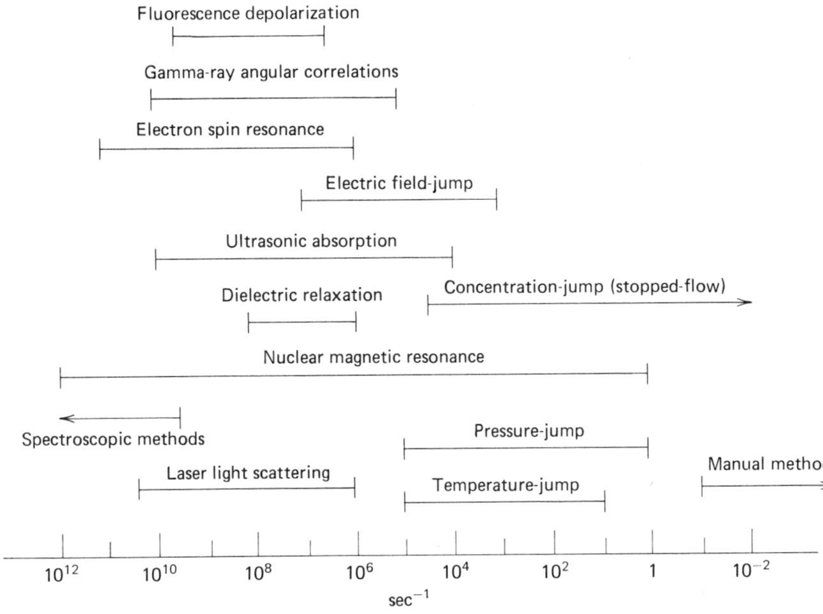

exponential decay constant of the transient response for the appropriate weight on a spring.

In this chapter, we have calculated the steady-state and transient responses for a damped weight (of zero or finite mass) on a spring. Major formulae are collected in Table 13-4 for reference. These calculations provide a basis for discussion of an extremely wide range of physical measurements. Experiments based on the *steady-state* response at driving frequencies *near* the "natural" undamped weight-on-a-spring frequency are collected in Chapter 14. Experiments based on the *steady-state* response at driving frequencies *far from* resonance are treated in Chapter 15. *Transient* experiments are discussed in Chapter 16, and the special (but again generally useful) case of *two or more springs* connected together is analyzed in Chapter 17.

Table 13-4 Formulas for the displacement, x, of a weight (mass $= m$) on a damped (frictional coefficient $= f$) spring (force constant $= k$) set into motion, either by a nonequilibrium displacement ($x_0 \neq 0$) at time zero ("transient" response), or by a continuous oscillatory driving force, $F_0 \cos(\omega t)$ ("steady-state" response).

STEADY-STATE

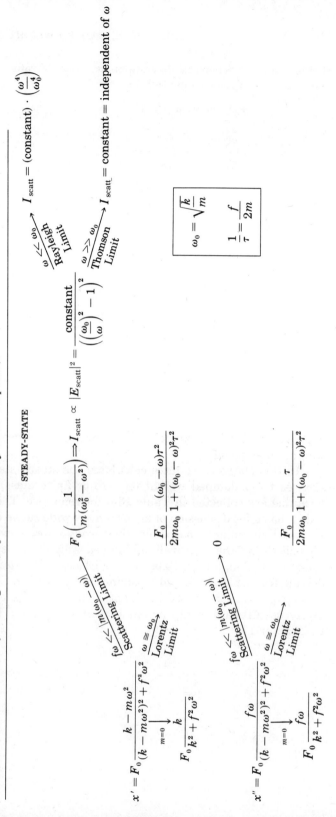

$$x' = F_0 \frac{k - m\omega^2}{(k - m\omega^2)^2 + f^2\omega^2} \xrightarrow{m=0} F_0 \frac{k}{k^2 + f^2\omega^2}$$

Scattering Limit ($f\omega << |m(\omega_0 - \omega)|$):

$$x' \to F_0 \left(\frac{1}{m(\omega_0^2 - \omega^2)} \right) \Rightarrow I_{\text{scatt}} \propto |E_{\text{scatt}}|^2 = \frac{\text{constant}}{\left(\left(\frac{\omega_0}{\omega} \right)^2 - 1 \right)^2}$$

Rayleigh Limit ($\omega << \omega_0$):

$$I_{\text{scatt}} = (\text{constant}) \cdot \left(\frac{\omega^4}{\omega_0^4} \right)$$

Thomson Limit ($\omega >> \omega_0$):

$$I_{\text{scatt}} = \text{constant} = \text{independent of } \omega$$

Lorentz Limit ($\omega \cong \omega_0$):

$$x' \to \frac{F_0}{2m\omega_0} \frac{(\omega_0 - \omega)\tau^2}{1 + (\omega_0 - \omega)^2\tau^2}$$

$$x'' = F_0 \frac{f\omega}{(k - m\omega^2)^2 + f^2\omega^2} \xrightarrow{m=0} F_0 \frac{f\omega}{k^2 + f^2\omega^2}$$

Scattering Limit ($f\omega << |m(\omega_0 - \omega)|$):

$$x'' \to 0$$

Lorentz Limit ($\omega \cong \omega_0$):

$$x'' \to \frac{F_0}{2m\omega_0} \frac{\tau}{1 + (\omega_0 - \omega)^2\tau^2}$$

$$\omega_0 = \sqrt{\frac{k}{m}}$$

$$\frac{1}{\tau} = \frac{f}{2m}$$

	TRANSIENT
$m \neq 0$	$x = x_0(e^{-(t/\tau)}) \cos\left[t\sqrt{\omega_0^2 - \left(\frac{1}{\tau}\right)^2}\right] \xrightarrow{\frac{1}{\tau} << \omega_0} x_0 e^{-(t/\tau)} \cos(\omega_0 t)$
$m = 0$	$x = x_0 e^{-(kt/f)}$

* *For the special "scattering" limit (upper right corner), the intensity of scattered radiation from a driven electron-on-a-spring is proportional to the square of the second time-derivative of the displacement, as discussed in the text. Expressions for the "natural" frequency of oscillation of the undamped, undriven weight-on-a-spring (ω_0), and the "relaxation time" (τ) are included in the box at upper right.*

PROBLEMS

1. Show that the sum of two sine waves of the same frequency (but different phase and different amplitude) gives a resultant sine wave of the same frequency. In other words, show that there exists a ψ, such that

$$A \sin(\theta) + B \sin(\theta + \phi) = C \sin(\theta + \psi)$$

[This formula is the algebraic basis for the middle plot of Fig. 13-2. More important, a special case of this formula ($\phi = \pi/2$) is the basis for breaking down an arbitrary sine wave into sine and cosine components (Fig. 13-5), the process that forms the basis for Chapters 14 and 15.]

2. The equation of motion for the displacement, x, of a weight of mass, m, on an undriven, undamped spring (force constant, k) is given by Eq. 13-3:

$$m(d^2x/dt^2) + kx = 0 \qquad (13\text{-}3)$$

(a) Show that $x = x_0 \cos(\omega_0 t)$ is a (real) solution of Eq. 13-3, and determine the form of ω_0.
(b) Show that $x = x_0 \exp[i\omega_0 t]$ is a (complex) solution of Eq. 13-3, and again determine ω_0. Note that the real part of the complex displacement is the same as the (real) solution of the original (real) Eq. 13-3.

3. The equation of motion for the displacement, x, of a weight of mass, m, on a sinusoidally driven (driving force = $F_0 \cos(\omega t)$), damped (frictional constant = f) spring (force constant = k) is:

$$m(d^2x/dt^2) + f(dx/dt) + kx = F_0 \cos(\omega t) \qquad (13\text{-}15)$$

(a) Show that a (real) solution of Eq. 13-15 is $x = x' \cos(\omega t) + x'' \sin(\omega t)$, and determine the form of x' and x''.
(b) Beginning this time from the complex form of Eq. 13-15:

$$m(d^2x/dt^2) + f(dx/dt) + kx = F_0 \exp[i\omega t] \qquad (13\text{-}18)$$

show that a (complex) solution of Eq. 13-18 is:

$$x = \chi \exp[i\omega t]$$

where

$$\chi = x' - ix''$$

and

$$\text{Re}[\chi \exp(i\omega t)] = x' \cos(\omega t) + x'' \sin(\omega t)$$

and x' and x'' are as in part a. In other words, show that the real part of the complex displacement is the same as the real solution of the original real equation. Note the simpler algebra involved in the "complex" manipulations. These calculations are the basis of this chapter.

4. Confirm the text expressions for x' and x'' ("dispersion" and "absorption") in the Lorentzian limit that the driving frequency is near the "natural" frequency of the weight on a spring. In other words, obtain Eq. 13-36 from Eq. 13-16. These two expressions describe the behavior of refractive index and power absorption as a function of incident radiation frequency in the spectroscopy applications of Chapter 14.

5. The Lorentz line shape (Chapter 13.B.2) may be written:

$$A(\omega) = \frac{\tau}{1 + (\omega_0 - \omega)^2\tau^2} = \text{"absorption" line shape}$$

and

$$B(\omega) = \frac{(\omega_0 - \omega)\tau^2}{1 + (\omega_0 - \omega)^2\tau^2} = \text{"dispersion" line shape}$$

(a) Plot $A(\omega)$ versus ω, and compute the peak height (maximum of $A(\omega)$), and the full width of the curve at half its maximum peak height.
(b) Plot $B(\omega)$ versus ω, and compute the extrema (maximum and minimum values) of $B(\omega)$. Then compute the frequency separation between the two extrema. The above calculations comprise the principal properties of a Lorentzian spectral line shape (Chapter 14).
(c) In certain spectroscopy experiments, it is convenient to detect the first derivative (with respect to driving frequency) of $A(\omega)$, as for electron spin resonance spectroscopy (Chapter 14.D). The resulting plot of $dA(\omega)/d\omega$ resembles the "dispersion" line shape of part 5.b. Calculate the locations, magnitudes, and frequency separation between the extrema of the $dA(\omega)/d\omega$ versus ω plot, and compare to the "dispersion" line shape of part 5.b.

6. The equation of motion for the displacement, x, of a weight of mass, m, on an *undriven*, damped (frictional constant $= f$) spring (force constant $=$ k) is given by Eq. 13-37:

$$m(d^2x/dt^2) + f(dx/dt) + kx = 0 \qquad (13\text{-}37)$$

(a) Show that a (real) solution of Eq. 13-37 is $x = x_0 \exp[-t/\tau] \cos(Wt)$, and find the form of τ and W. Then find the limiting behavior of the solution when damping is weak $[(1/\tau) << \omega_0]$.
(b) Show that a (complex) solution of Eq. 13-37 is $x = x_0 \exp[ict]$, and find the form of c. Show that the real part of this complex solution is the same as the (real) solution of the original (real) equation.

7. The equation of motion for the displacement, x, of a damped (frictional constant $= f$), *massless* ($m = 0$) spring (force constant $= k$) is given by Equation 13-42 and 13-43:

$$f(dx/dt) + kx = F_0 \cos(\omega t) \quad \text{Sinusoidally driven} \qquad (13\text{-}42)$$

$$f(dx/dt) + kx = 0 \quad \text{Undriven} \qquad (13\text{-}43)$$

(a) Show that a (real) solution of Eq. 13-42 is $x = x' \cos(\omega t) + x'' \sin(\omega t)$, and determine the form of x' and x''.
(b) Show that a (real) solution of Eq. 13-43 is $x = x_0 \exp[-kt/f]$.
(c) Show that a (complex) solution of the complex form of Eq. 13-42

$$f(dx/dt) + kx = F_0 \exp[i\omega t] \qquad (13\text{-}42a)$$

is

$$x = \chi \exp[i\omega t]$$

where

$$\chi = x' - ix''$$

and x' and x'' are as in part a. Again note the briefer algebra required when using complex notation. These calculations are the basis for all the "transient" ("relaxation") phenomena of Chapter 16.

REFERENCES

Almost any elementary physics text with a section on "mechanics" will give some sort of treatment of the weight-on-a-spring problem. However, the most lucid and most general treatment is found in R. P. Feynman, R. B. Leighton, and M. Sands, *The Feynman Lectures on Physics,* Addison-Wesley, Reading, Mass. (1964).

CHAPTER 14 Absorption and Dispersion: Steady-state Response of a Driven, Damped Weight on a Spring

Quantum mechanics (see Section 5) predicts that the energy of a given molecule can take on only discrete (as opposed to a continuous range of) values ("levels"). An upward "transition" from a lower to a higher energy "level" can occur when a photon of that energy ($h\nu = E_{\text{upper}} - E_{\text{lower}}$) reaches that molecule, provided that certain "selection" rules (see Section 5) are satisfied. Since energy is absorbed by the molecule for such a transition, we thus anticipate a "peak" in a plot of power (i.e., energy per unit time) absorption versus photon frequency, ν, whenever the photon energy is approximately equal to the energy of an "allowed" transition. Although it is true that *quantum mechanical* calculations are required to determine the *frequencies* and *intensities* of the peaks in the (power absorption) "spectrum" of such molecules, we can use our previous *classical* treatment of the steady-state displacement of a driven, damped weight on a spring to account for the *shape* of any one of the "peaks" in the absorption spectrum. In particular, we will often be able to extract useful information about the molecular *environment* from the *width* of a particular spectral line, since the width is determined by the particular "frictional" forces that couple the molecule to its surroundings.

We have already shown that the steady-state displacement of a sinusoidally driven, damped weight on a spring may be separated into two components, x' and x'', which are either exactly *in-phase* (x') or *90°-out-of-phase* (x'') with respect to the driver (see Fig. 13-5). In this chapter, we show that the "absorption," x'', may be identified with: power absorption of optical (ultraviolet, visible, infrared) radiation, dichroism, circular dichroism, dielectric loss, ultrasonic absorption, and magnetic resonance absorption; while the "dispersion," x', may be identified with: $(n - 1)$, where n is refractive index, and with birefringence, optical rotation, dielectric constant, ultrasonic dispersion, and magnetic resonance dispersion. We begin with examination of absorption and refractive index of light, the phenomena that gave the names "absorption" and "dispersion" to x'' and x'.

14.A. ABSORPTION AND REFRACTIVE INDEX: BASIS FOR SPECTROSCOPY AND MICROSCOPY

Suppose that monochromatic radiation is directed through a thin slab of matter of thickness, Δy (see diagram below). In this section, we are interested in the electric field amplitude of the emerging wave, at a distance y from the slab. First, since the observed electric field at y was produced

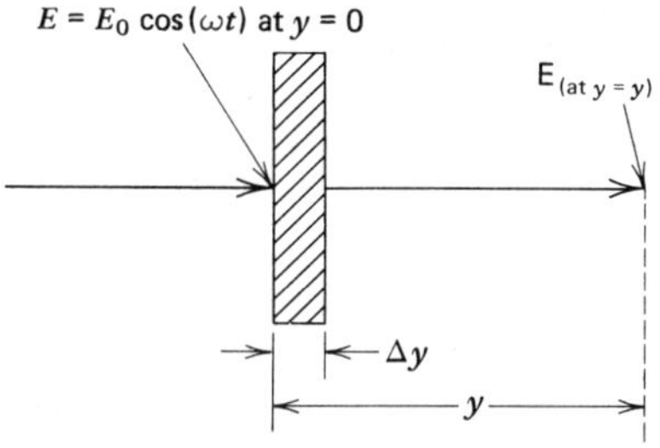

by oscillating electrons in the slab at a time $(t - (y/c))$ ago, where c is the speed of light (in a vacuum), it is readily seen that even if the slab had no effect whatever on the incident radiation, the electric field at y would be of the form

$$E_{(\text{at } y)} = E_0 \cos\left[\omega\left(t - \frac{y}{c}\right)\right] \tag{13-9}$$

$$= \text{Re}\left[E_0 \exp\left[i\omega\left(t - \frac{y}{c}\right)\right]\right] \tag{14-1}$$

However, if the wave slows down while passing through the slab, then from the usual definition of refractive index, n'

$$\boxed{\text{Refractive index} = n' = \frac{(\text{speed of light in vacuo})}{(\text{speed of light in slab})}} \tag{14-2}$$

the emerging wave will have been slowed down by a time, $t = (n' - 1)(\Delta y)/c$, so that the electric field at y now becomes

$$E_{(\text{at } y)} = E_0 \cos\left[\omega\left(t - \frac{(n' - 1)(\Delta y)}{c} - \frac{y}{c}\right)\right] \tag{14-3a}$$

* *Throughout this section, the results are displayed in both real and complex notation where appropriate to familiarize the reader with the intuitively understandable ("real") description as well as with the mathematically compact ("complex") format.*

$$= \text{Re}\left[E_0 \exp\left[\left(i\omega\left(t - \frac{(n'-1)(\Delta y)}{c} - \frac{y}{c}\right)\right]\right]$$

$$= \text{Re}\left[\underbrace{\exp[-i\omega(n'-1)(\Delta y)/c]}_{\substack{\text{Effect of}\\ \text{refractive index} \neq 1}} \underbrace{E_0 \exp\left[i\omega\left(t - \frac{y}{c}\right)\right]}_{\substack{\text{Electric field of wave}\\ \text{with no slab present}}}\right] \quad (14\text{-}3b)$$

Equation 14-3b shows that with complex notation it is easy to separate the behavior of the wave in the absence of any refraction from a term that consists solely of the effect due to slowing down of the radiation while passing through the matter.

In addition to the reduction in *velocity* (Eq. 14-3) while passing through the slab, there will be a *loss. dI, in intensity, I*, on passing through the slab. For a sufficiently thin slab, we would expect the rate of decrease in intensity to be proportional to the intensity, to the thickness of the slab, *dy*, and to the concentration of absorbing molecules, *m* (in moles/liter), as in Eq. 14-4.

$$dI = -kImdy \quad (14\text{-}4)$$

in which k is a proportionality constant characteristic of the molecule in question. Equation 14-4 in its integrated form is known as "Beer's Law":

$$\log_e(I/I_0) = -km\ \Delta y$$

$$\boxed{\begin{aligned}\text{"Transmittance"} = I/I_0 &= 10^{-(km\Delta y/2.303)} = 10^{-(\epsilon m \Delta y)}\\ &= 10^{-A} = e^{-(2{,}303\epsilon\Delta yN/N_0)}\end{aligned}} \quad (14\text{-}5)$$

where ϵ is called the "molar absorbancy index" (formerly called the "molar extinction coefficient"), I_0 is the intensity incident on the slab, and A is called the "absorbance," or "optical density," N is the number of molecules per cm^3, and N_0 is Avogadro's number. Thus, since electromagnetic radiation intensity is proportional to the square of electric field amplitude of the transmitted wave, and since $\sqrt{e^{-x}} = (e^{-x})^{1/2} = e^{-(x/2)}$, Eq. 14-3 can now be extended to incorporate the absorption of energy described by Eq. 14-5:

$$E_{(\text{at } y)} = \exp[-2{,}303\epsilon\, N(\Delta y)/2N_0]\, E_0 \cos\left[\omega\left(t - \frac{(n'-1)\Delta y}{c} - \frac{y}{c}\right)\right] \quad (14\text{-}6a)$$

$$= \text{Re}[\underbrace{\exp[-2{,}303\epsilon\, N\Delta y/2N_0]}_{\substack{\text{Effect of finite}\\ \text{absorption}}}\ \underbrace{\exp[-i\omega(n'-1)\Delta y/c]}_{\substack{\text{Effect of refractive}\\ \text{index} \neq 1.0}}\ \underbrace{E_0 \exp\left[i\omega\left(t - \frac{y}{c}\right)\right]}_{\substack{\text{Electric field of wave}\\ \text{with no slab present}}}] \quad (14\text{-}6b)$$

Examination of the first two factors of Eq. 14-6b shows that considerable economy in notation can be achieved by defining a *complex* "refractive index," n,

$$\boxed{n = n' - in''} \tag{14-7}$$

so that Eq. 14-6b simplifies to

$$E_{(\text{at } y)} = \text{Re}[\exp[-i\omega(n-1)(\Delta y)/c]\, E_0 \exp\left[i\omega\left(t - \frac{y}{c}\right)\right]] \tag{14-8}$$

in which

$$n'' = \frac{2303cN}{2\omega N_0}\epsilon \tag{14-9}$$

The importance of Eq. 14-8 is that the "imaginary" part, n'', of the complex "refractive index", n, is directly proportional to the molar extinction coefficient for *absorption* of power by the matter (Eq. 14-9), while the "real" part, n', of the complex "refractive index" is the ordinary *refractive index* of Eq. 14-2. Again, we recognize that n' and n'' are "real" in both the physical and mathematical sense—we simply save space by using Eq. 14-7 in Eq. 14-8.

The previous derivation was based on the assumptions that an electromagnetic wave slowed down and that its energy was partly lost on passing through a thin slab of matter. If we now suppose only that electrons in matter behave as damped, driven springs when subjected to the oscillating electric field of an incident electro-magnetic wave, it is easy to see why power is absorbed by matter, since energy is dissipated into friction in driving the damped springs. It is less easy to see why the incident wave should appear to slow down, and the answer is slightly beyond the scope of this discussion. Basically all that is involved is a calculation of the electric field produced at a remote observer, due to the simultaneous coherent (driven) motion of a uniformly distributed collection of damped electrons on springs spread over the slab diagrammed above. It turns out that the resultant electric field from all those springs (in the limit of a very thin slab of infinite lateral dimension) can be written very simply as a sinusoidal wave that is 90° out of phase with the incident wave. Thus, by an argument identical to that rehearsed in Fig. 13-2, the sum of the electric field of this scattered wave and the incident (driving) electric field must also be a sine wave whose phase is in general different from that of the incident wave (see Fig. 14-1). The net effect is that the electric field seen by an observer at the far right of Fig. 14-1 appears to have been slowed down compared to the behavior expected for the incident wave itself.

The mathematics corresponding to the preceding discussion and dia-

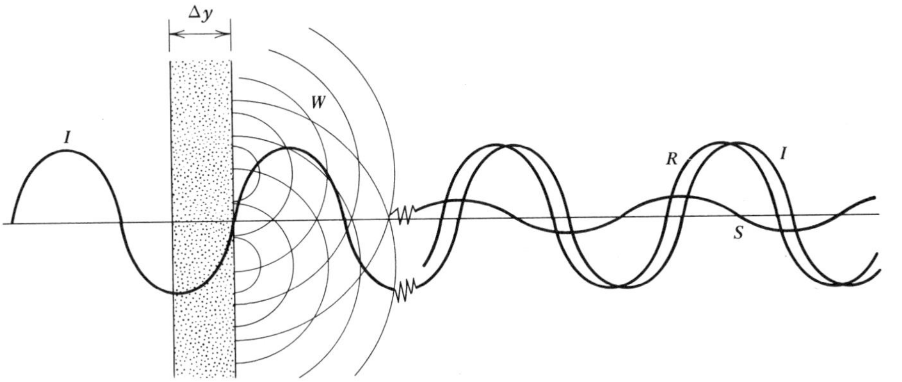

FIGURE 14-1. The relationship between forward scattering and refraction. The incident wave, *I*, after passing through a thin slab of matter, is accompanied by a forward scattered wave, *S*, which is 90° out of phase with it. This forward-scattered wave can be considered the composite of all the wavelets, *W*, scattered by the particles in the slab. The resultant, *R*, of the incident and forward-scattering waves is the observed emergent wave. It is retarded a little in phase. (From K. E. van Holde, *Physical Biochemistry*, Prentice-Hall, Englewood Cliffs, N. J., 1971, p. 187.)

gram serve only to provide the proportionality constant in Eq. 14-10, and are omitted here.

$$n - 1 = \frac{N}{(\omega_0^2 - \omega^2) + if\omega}\,\frac{q^2}{2\epsilon_0 m}, \; n = n' - in'' \qquad (14\text{-}10)$$

in which ω_0, ω, and f are as in Eq. 13-16, q is electronic charge, N is the number of molecules per unit volume, ϵ_0 is dielectric constant, and m is electron mass. Strictly speaking, as we learn in the next section, even a single atom can have many discrete "natural" frequencies, ω_0, so that in practice it is necessary to evaluate $(n - 1)$ in Eq. 14-10 from a sum of many terms of the type shown on the right-hand side of the equation. It is clear from comparison of Eq. 14-10 with 13-20 that the refractive index, n', and the molar extinction coefficient, ϵ (which is proportional to n''), will have a frequency-dependence of the form shown at the bottom of Fig. 13-5 (see also Fig. 14-4). The origin of the "absorption" and "dispersion" terminology is now apparent: n'' is proportional to molar absorbency index, and n' is refractive index, whose variation with frequency is responsible for the "dispersion" (spreading out) of the component colors of white light by a prism. Most of the applications for refractive index or absorbance are based either on the variation of n' and n'' (or more familiarly, ϵ) with (driving) frequency of the incident light, or on use of n' or n'' as a measure of the number of scattering centers per unit volume, N (or more familiarly, concentration of the absorbing molecules).

Refractive Index and Its Applications

EXAMPLE *Concentration-dependence of Refractive Index: Schlieren Optics*

In the analytical ultracentrifuge (see Chapter 7B), it is desirable to be able to monitor the progress of sedimenting macromolecules as they move toward the "bottom" (outer edge) of the sedimentation cell. Now in our later treatment of Rayleigh light scattering, it will be argued that at low concentrations, the refractive index of a solution should vary linearly with solute concentration:

$$n \cong n_0 + (\partial n/\partial[c])[c] \quad (14\text{-}11)$$

in which n is the refractive index of the solution, n_0 the refractive index of the solvent, and $[c]$ the (molar) concentration of solute. [Since the right-hand side of Eq. 14-11 represents the first two terms in a Taylor series for n, expanded about n_0, it is clear that the approximation will be valid whenever $(n - n_0) << 1$.] In other words, refractive index of a dilute solution varies linearly with solute concentration. Figure 14-2 shows the behavior of a "plane" electromagnetic wave (i.e., one in which the electric field has the same value at any point along the vertical axis shown in the figure) on passing through a solution (as in the sedimentation cell) in which the concentration increases in going from the top of the solution to the bottom. Since the incident waves that pass through the more dilute part of the solution are slowed down less than those passing through the more concentrated part, the emerging wave front is partly bent, as shown in Fig. 14-3.

The schlieren (from German, "streaked") optical system, shown in Fig. 14-3, is simply a device for removing image intensity from that region of the solution within which the concentration *gradient* is large (i.e., the concentration changes from one part of the region to the other), and is based on Fig. 14-2. The image thus resembles a collection of intensity "peaks" that have an ap-

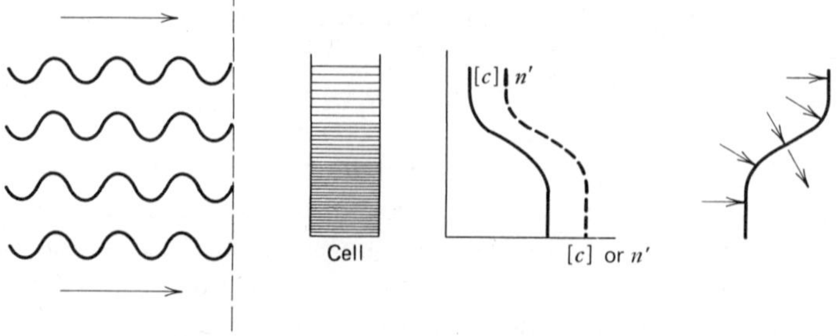

(a) (b)

FIGURE 14-2. Behavior of a plane wave front on passing through a solution in which there is a concentration gradient. Since the refractive index varies linearly with solute concentration (a), the emergent wave front has the form shown in (b), in which the arrows indicate the direction of the normals (perpendiculars) to the wavefront.

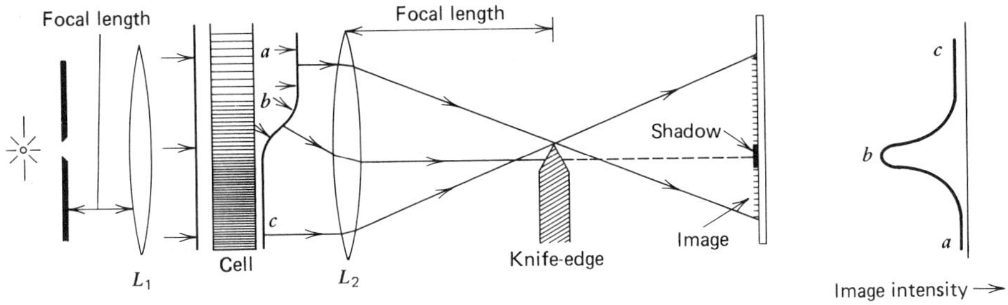

FIGURE 14-3. Schlieren optical system to detect regions of high *concentration gradient,* based on the linear relation between refractive index and *concentration.* The lens L_2 is positioned to bring the cell image to focus on a screen, and the knife-edge is placed so as to intercept light passing through a region of large concentration gradient. (After R. B. Setlow and E. C. Pollard, *Molecular Biophysics,* Addison-Wesley, Reading, Mass., 1962, p. 102.)

pearance similar to the concentration peaks of electrophoresis or chromatography.

EXAMPLE *Frequency-dependence of Refractive Index: Phase-Contrast Microscope*

Although the electron microscope (Chapter 15.C) is capable of resolving very fine detail, its high-vacuum operating conditions render study of most living biological tissue impossible. The light microscope can obviously be used for study of living tissue, but contrast is usually based on a difference in *absorption* of light by the specimen and the background. Unfortunately, as we see in the next section, the "natural" frequencies of electrons bound to matter, ω_0, tend to lie in the ultraviolet, so that unstained samples often show very poor contrast since visible light (say green light at about 5×10^{-5} cm) is so far from "resonance" that absorption is weak (see Fig. 14-4). On the other hand, refractive index exhibits significant effect $(n' - 1) \neq 1$, over a much wider fre-

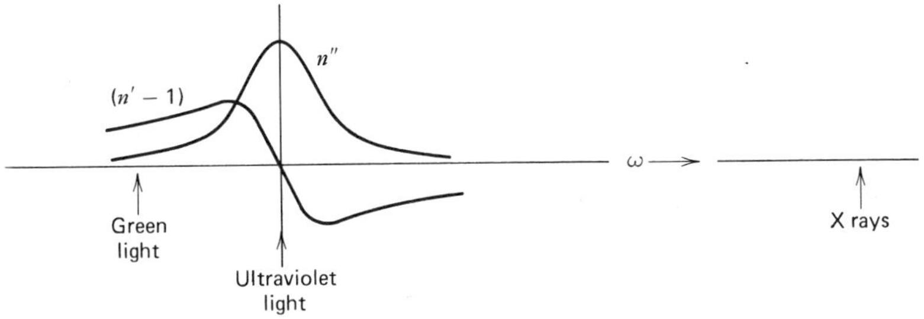

FIGURE 14-4. Diagram showing that refractive index can be appreciably different from unity even when absorption is negligibly small (green light) for visible light. This fact is basic to the advantage of phase-contrast microscopy over ordinary absorption microscopy (see text).

quency range (see Fig. 14-4), and thus potentially provides better contrast for microscopy with visible light than ordinary absorption.

For example, a bacterium is essentially nonabsorbing for most visible light, say green light of wavelength, 5×10^{-5} cm. However, the same bacterium has a refractive index of 1.34, compared to the surrounding medium (water) refractive index of 1.33. If the bacterium thickness is about 10^{-4} cm, then light passing through the bacterium will be slowed down more than light passing through the medium, and the optical path difference between the two rays will be $10^{-4}(1.34 - 1.33) = 10^{-6}$ cm. For the green light in question, this path difference amounts to $10^{-6}/(5 \times 10^{-5}) = 0.02$ of one wavelength. In the phase-contrast microscope, light passing through the sample is combined with a reference light beam whose path is adjusted so that there is a 180° phase difference (i.e., a path difference of exactly one-half of one wavelength) between the light passing through the background and the light from the reference beam. Light passing through the background is thus completely removed by destructive interference. However, light passing through an object, such as the bacterium above, will have traveled a path whose length is *not* exactly one-half wavelength different from the reference beam, and which therefore is not completely canceled by the reference beam. Objects in a phase-contrast microscope thus appear bright on a dark background. The advantage of the technique is that no staining is necessary, so that the possibility of artifacts in the image is reduced.

A final bonus from Fig. 14-4 is in explanation of the nonexistence of X-ray microscopes. Since the refractive index is essentially unity at frequencies either much higher (X-rays) or much lower than the "natural" ultraviolet frequencies of electrons bound to molecules, there is no form of matter that can be used as a lens to refract (bend) X-rays, and with no lenses, there can be no microscopes. That is why the interaction of X-rays with matter must be observed indirectly by scattering experiments rather than directly by microscopy (see Chapters 15.B and 22.A).

Light Absorption and Its Applications: visible and ultraviolet frequencies

EXAMPLE *Characteristic Visible and Ultraviolet Spectra for Macromolecules in Solution*

Figure 14-5 shows the absorption as a function of frequency for some species of biological interest. Detailed theoretical justification for the values of the resonant frequencies, ω_0, is unwarranted on two grounds. First, except for very small molecules, present theory does not predict those frequencies with useful accuracy; and second, because the many absorption peaks are relatively broad compared to the spacing between peaks even for a single molecule, the resolution of detail in the spectra is not particularly good. (It is possible to obtain some rough predictions of absorption maxima from a simple quantum mechanical model of an electron confined to a container whose dimensions are approximately those of the molecule in many cases.) For our purposes, it suffices to note that absorption peaks of appreciable height ($\epsilon > 2000$ or so) are

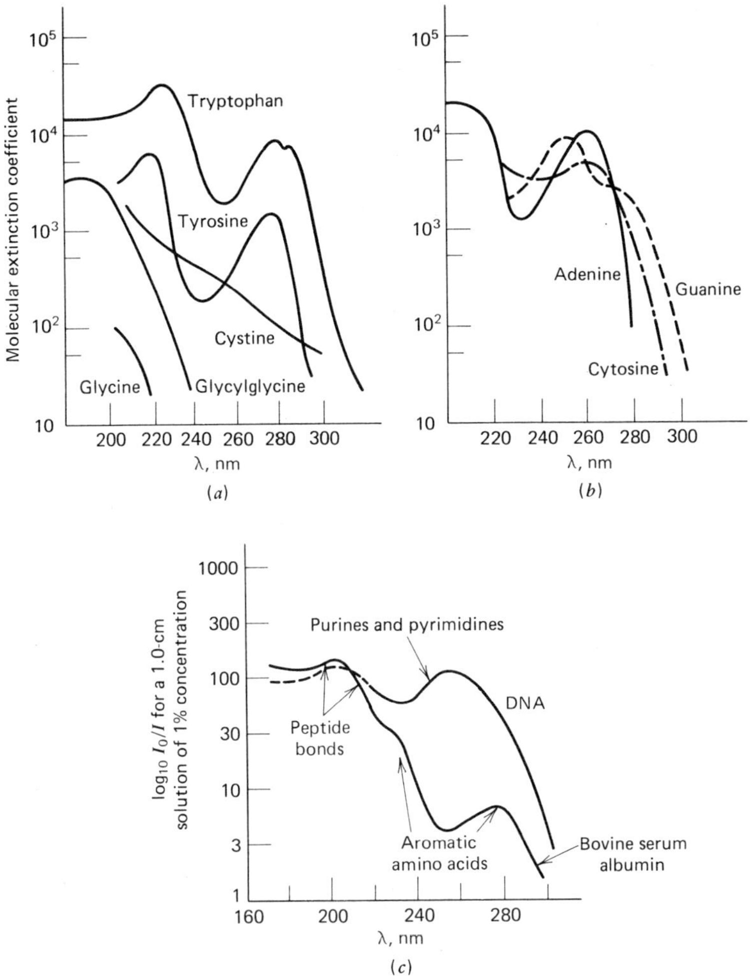

FIGURE 14-5. Ultraviolet-visible absorption spectra. (*a*) Representative amino acids; (*b*) representative nucleic acid bases; (*c*) representative protein (bovine serum albumin) and nucleic acid (DNA), at neutral pH. (From R. B. Setlow and E. C. Pollard, *Molecular Biophysics,* Addison-Wesley, Reading, Mass., 1962, pp. 225–226.)

generally associated with *metal ions,* or with *double bonds* or *conjugated bonding* in molecules. Most nucleic acids (see Fig. 14-5*c*) show characteristic absorption at about 260 nm, while proteins which contain aromatic amino acids typically show absorption at about 280 nm. Below about 200 nm, almost all types of molecules show large absorption, so that little distinguishing information is available from an absorption spectrum (moreover, since water and oxygen in particular absorb strongly at those wavelengths, experiments must be conducted with dry samples in vacuo, further decreasing practical interest in that part of the spectrum). Virtually all applications of ultraviolet-visible absorption spectroscopy in study of macromolecules are based on small *differences* in absorption introduced by changes in molecular shape (and thus change in environment near the electron on a spring), change in bulk environment (changing the solvent), or change in local environment near a particular electron on a spring on binding of some small molecule nearby, as shown in the following examples.

EXAMPLE *Ultraviolet-difference Spectra and Macromolecular Conformational Change*

Figure 14-6*a* illustrates the often substantial change in the ultraviolet absorption spectrum of a macromolecule (in this case, the synthetic polymer poly-L-glutamic acid) on changing from one conformation to another. The result suggests that it should be possible to follow the conformational change quantitatively by monitoring the *difference* in ultraviolet absorption between a solution containing exclusively one conformation and a solution containing a mixture of the two conformations—an example of this idea is shown in the accompanying Fig. 14-6*b* for the (different) macromolecule, ribonuclease. The change in physical state of the ribonuclease enzyme exhibits an S-shaped curve as a function of temperature, characteristic of a cooperative process of the same type we encountered in allosteric enzyme kinetics in Chapter 11.B.2. Such cooperative behavior characterizes most "phase-transitions" associated with melting or boiling of molecules: once the process has started, it

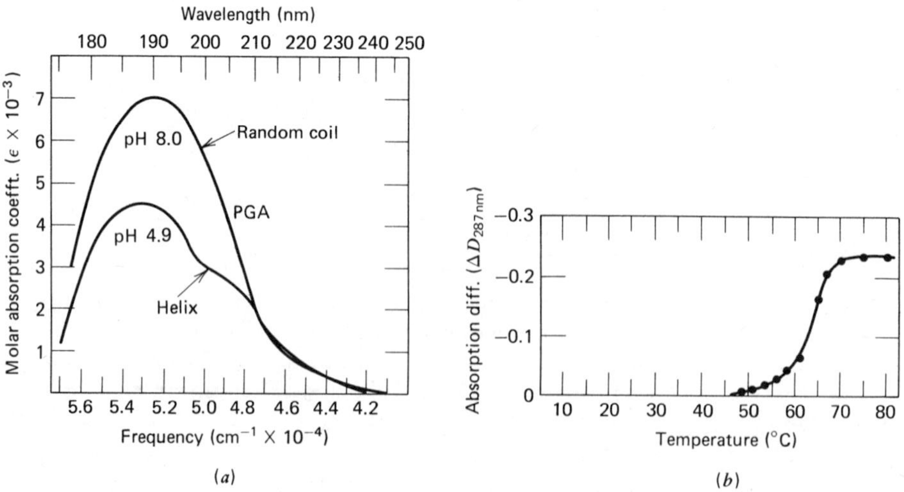

FIGURE 14-6. (*a*) absorption spectrum (ultraviolet) for poly-L-glutamic acid in its helical (pH 4.9) and random coil (pH 8.0) forms. (*b*) difference ultraviolet optical density ($\log(I_0/I)$) at 287 nm for ribonuclease solution as a function of temperature, measured against a reference solution maintained at the same pH (6.83) at a low temperature (< 43°). [(*a*) from I. Tinoco, Jr., A. Halpern, and W. T. Simpson, in *Polyamino Acids, Polypeptides and Proteins,* ed. M. A. Stahmann, Univ. of Wisconsin Press, Madison, Wisconsin, 1962, p. 147; (*b*) from J. Hermans, Jr., and H. A. Scheraga, *J. Amer. Chem. Soc. 83,* 3283 (1961).]

proceeds more and more readily as the temperature is raised. We have discussed phase transitions in Chapter 2A, where they were related to the fluidity of (and thus the rate of transport across) biological membranes.

EXAMPLE *Ultraviolet-difference Spectra and Buried Aromatic Amino Acid Residues in Proteins*

One of the most useful applications of ultraviolet spectroscopy is based on the fact that a change in solvent (most usually from water to 20% v/v ethylene

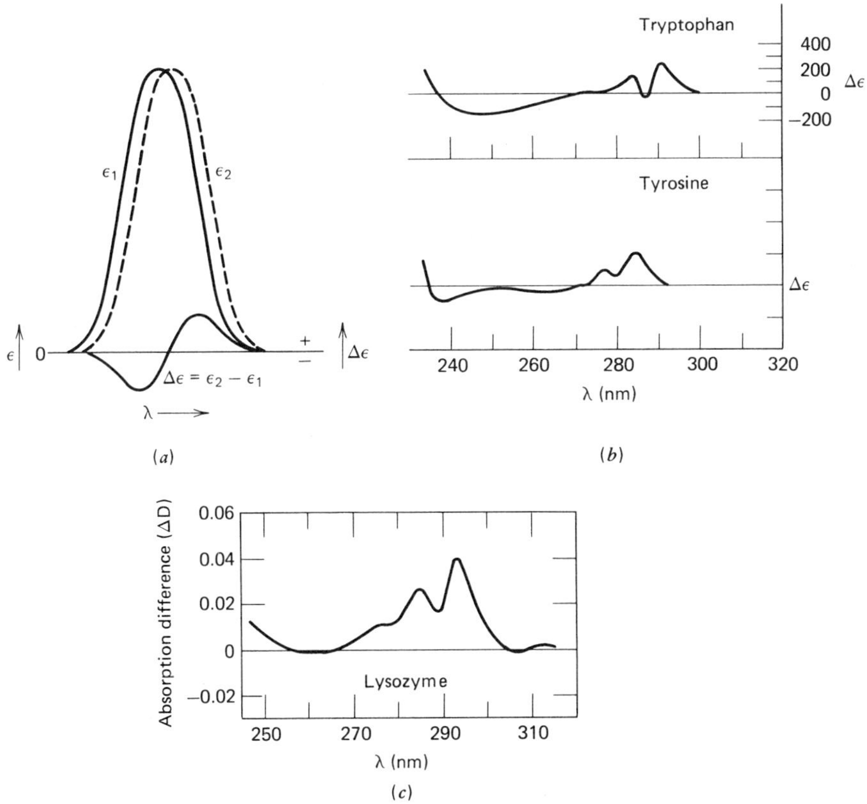

FIGURE 14-7. Origin and appearance of ultraviolet difference spectra. (*a*) Plot of the difference, $\Delta\epsilon = \epsilon_2 - \epsilon_1$, between the ultraviolet absorbances for normal (ϵ_1) and solvent-shifted (ϵ_2) ultraviolet spectra. When the shift is small, the difference spectrum has approximately the same shape as a plot of the *slope* of the original spectrum versus wavelength. (*b*) Examples of difference spectra for tryptophan and tyrosine in presence and absence of 20% v/v ethylene glycol—since the change in solvent causes a slight increase (as well as a shift) in ϵ, the negative part of the difference spectrum is moved upward on the vertical scale. (*c*) Difference spectrum for lysozyme in presence and absence of glycol chitin substrate. [(*a*) from J. W. Donovan et al., *J. Amer. Chem. Soc. 83,* 2686 (1961); (*b*) from J. W. Donovan, in *Physical Principles and Techniques of Protein Chemistry,* ed. S. J. Leach, Academic Press, New York, 1969, Part A, p. 125; (*c*) from K. Hayashi et al., *J. Biochem.* (Tokyo) *54,* 381 (1963).]

glycol) will lead to a small shift (usually to higher wavelength) of the ultraviolet absorption peak(s) of aromatic amino acids. Figure 14-7*a* shows that when the shift is small, the difference spectrum obtained by subtracting the original spectrum from the shifted spectrum closely resembles a plot of the *slope* of the original absorption curve versus wavelength, as may be seen by comparing the solvent-shifted difference spectra for tryptophan and tyrosine in Fig. 14-7*b* with the original absorption spectra of those amino acids in Fig. 14-5*a* over the same wavelength region. Since only those aromatic amino acid residues accessible to the solvent will be shifted, the solvent-shifted difference spectrum thus provides a measure of the number of "exposed" (as opposed to "buried") aromatic amino acids for a protein in solution. One of the more definitive examples in which such information was useful is shown in Figure 14-7*c*, which gives the difference spectrum between lysozyme in the presence of its substrate and lysozyme alone. Clearly, the spectrum resembles that which is obtained by using a change in solvent to shift the spectrum of free tryptophan (Fig. 14-7*b*). By comparing solvent-shifted difference spectra for lysozyme in the presence and absence of substrate, it was observed that there was indeed one less tryptophan accessible to solvent in the presence of substrate. Several years later, X-ray crystallographic data for lysozyme in presence and absence of substrate showed that the active site of the enzyme consists of a cleft, with a tryptophan residue (#62 on the polypeptide chain) located at the bottom of the crevice, and that tryptophan is masked on binding of substrate. Because the amino acid sequence of a protein is much more easily determined than its full tertiary structure, it is clearly useful to know which amino acids are exposed in attempting to limit the number of possible alternative ways in which the polypeptide chains can be folded to form the tertiary structure.

EXAMPLE *Ultraviolet Isosbestic Point: Evidence for a Two-state Chemical Equilibrium*

To determine the mechanism of any biological process, it is necessary to discover the number of steps, or stages, through which the reaction proceeds. When the reactions are sufficiently slow, the various "jump" techniques described in Chapter 16.A.1 provide that information. A simpler and more general method is based on observation of an "isosbestic point" in a visible-ultraviolet spectrum. Suppose that the absorption spectrum of a molecule in one chemical state happens to intersect the absorption spectrum of the same molecule in a different chemical state, as in the two extreme cases shown in Fig. 14-8 for two chemically different states of the invertebrate respiratory protein, hemerythrin. Now there is nothing unusual for two different species to have the same optical absorbance at one or more frequencies, but suppose that we now examine the optical spectra of the same molecule when both species are present in dynamic equilibrium. If all the spectra still intersect in a single point at a particular frequency, then that point is called an *isosbestic point,* and its existence is evidence that the molecule likely exists in only two distinct forms. In other words, although it is conceivable that two forms of the molecule might happen to have the same absorbance at a particular fre-

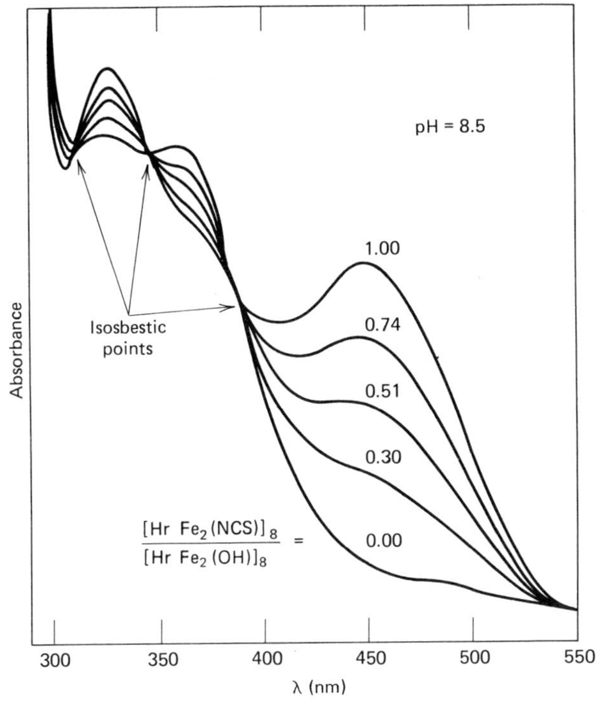

FIGURE 14-8. Ultraviolet-visible absorption spectra for hemerythrin, for varying degrees of replacement of OH^- ligand by NCS^- ligand. Isosbestic points are indicated by arrows, suggesting that this process may be described as a two-state equilibrium. (Courtesy of Prof. A. W. Addison, University of British Columbia.)

quency, it is unlikely that three or more forms of the molecule would all have the same absorbance at that frequency.

EXAMPLE *Infrared Absorption and Chemical Bonding*

Ultraviolet and *visible* absorption may be thought of as arising from the driven motion of an *electron* vibrating with respect to an atomic or molecular backbone. *Infrared* absorption may be identified with driven vibration of *atoms* with regard to the rest of a molecule. Fortunately, among the many possible ways in which the atoms of a molecule might be expected to vibrate, we can often identify specific infrared absorption associated with the vibration of a single chemical group, such as a carbonyl or hydroxyl function (see Chapter 17.B for further discussion). Since the strength of the "spring" for a given chemical bond (and thus the resonant absorption frequency for that spring) depends on the electronegativity at the bonded atoms, one would expect to see a variety of infrared absorption frequencies corresponding to, for example, carbonyl groups in different kinds of molecules. The most important applications for infrared spectroscopy are based on this "fingerprint"

Table 14-1 Vibrational Frequency of the O—O Stretch for Some di-Oxygen Compounds.

Compound	Frequency (cm^{-1})
O_2	1555[a]
O_2^-	1107[a]
H_2O_2	878
Hemoglobin:O_2	1107
Hemerythrin:O_2	844[a]

[a] *These frequencies were determined by Raman methods rather than by direct infrared absorption (see Chapter 17.C). [From W. S. Caughey, C. H. Barlow, J. C. Maxwell, J. A. Volpe, and W. J. Wallace, "The Biological Role of Porphyrins and Related Substances,"* Annals New York Acad. Sci. 244, *1 (1975).*]

spectral profile. Because of the wide variety of number and kinds of chemical bonds, almost no two molecules have identical infrared absorption spectra, and infrared spectra are therefore extremely useful for identifying unknown molecules or comparing a suspected structure with a known one based on their respective infrared absorption (see Chapter 17.B for examples).

An example of the sort of empirical way in which infrared spectra are commonly useful is shown in Table 14-1. The vibrational frequency associated with the stretching of an oxygen-oxygen bond is directly related to the strength of that chemical bond. Thus oxygen molecule O_2, "superoxide" O_2^-, and peroxide, $O_2^=$, show substantially different vibrational frequencies at 1555, 1107, and 878 cm^{-1} ($\nu/c = 1/\lambda$). Comparing these numbers with the vibrational frequency for oxygen bound to hemoglobin (1107 cm^{-1}), it is clear that oxygen in hemoglobin behaves (spectroscopically) as superoxide ion. In contrast, oxygen bound to the invertebrate respiratory protein, hemerythrin, shows vibrational behavior (844 cm^{-1}) that more closely resembles peroxide.

EXAMPLE *Infrared Absorption and Hydrogen-bonding*

Because a hydrogen-bond tends to change the chemical bond strength of the hydrogen-bonded atom, one would expect to find a change in the vibrational frequency associated with the hydrogen-bonded atom, manifested as a shift in the infrared absorption "peak" corresponding to that vibration. The most important hydrogen-bonding in biology is that which stabilizes the double-helix configuration of nucleic acids. Direct evidence for hydrogen-bonds in the double-helix is provided by the largely self-explanatory spectra in Fig. 14-9. Although UMP and AMP do not pair up in solution, and thus show no hydrogen-bonding, poly-A and poly-U show an infrared spectrum markedly different from the sum of the absorption spectra of the two separate components. Identification of the infrared absorption bands with specific carbonyl and ring vibrations has shown that the inter-chain interaction involves the $C_4{=}O$ rather than the $C_2{=}O$ group of the uracil bases.

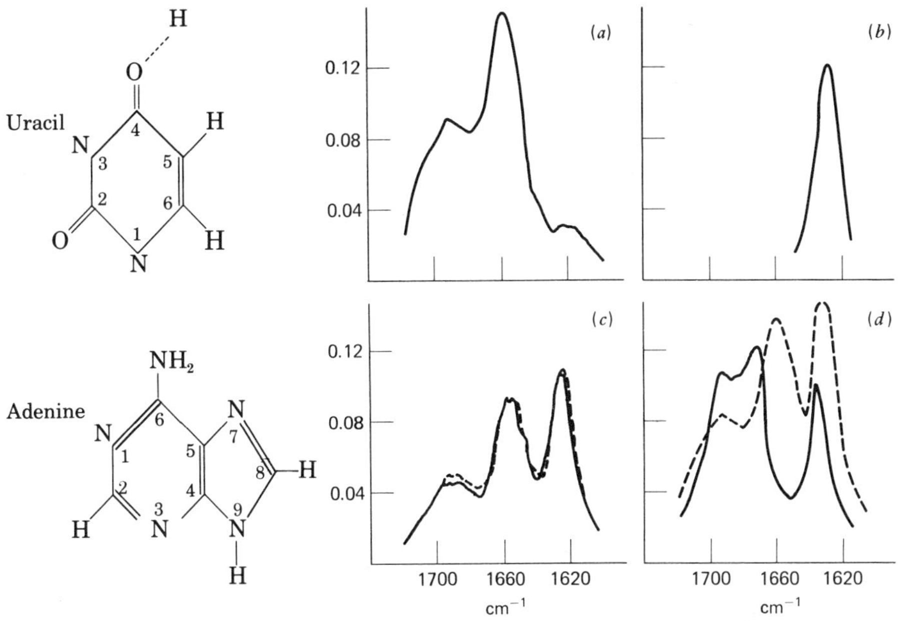

FIGURE 14-9. Direct infrared evidence for hydrogen-bonding in base-paired polynucleotides. All plots are infrared spectral absorption (O.D.) as a function of frequency, for buffered solutions in D_2O, pD = 7.6. (*a*), poly-U (0.68 M); (*b*), poly-A (0.62 M); (*c*), equimolar mixture (0.33 M each) of Na_2UMP and Na_2AMP, in which solid line is the observed spectrum for the mixture, and dotted line is the sum of the two separate absorption spectra of the individual components; (*d*), poly-A + poly-U (0.62 M), in which solid line is for the mixture and dotted line for the sum of the spectra for the two individual components. [From H. T. Miles, *Biochim. Biophys. Acta 30,* 324 (1958).]

14.B. DICHROISM AND BIREFRINGENCE: DETECTION OF LINEAR ORDER IN MOLECULAR ARRAYS

In all our previous electron-on-a-spring examples, it has been supposed that a given electron is bound by just one kind of spring. However, an interesting class of phenomena is based on the more realistic case shown schematically in Fig. 14-10, in which a given electron is bound by springs of different properties in different directions. For example, if we shine plane-polarized incident radiation on one such system (Fig. 14-11) such that the plane of polarization is chosen to lie in the yz plane of Fig. 14-11, then the electron will be driven against the spring on the z axis, while for incident light plane-polarized along the xy plane, the driven electron will be forced to stretch the spring on the x axis. Since the two perpendicular springs are assumed to be of different strength, the absorption and refractive index for light plane-

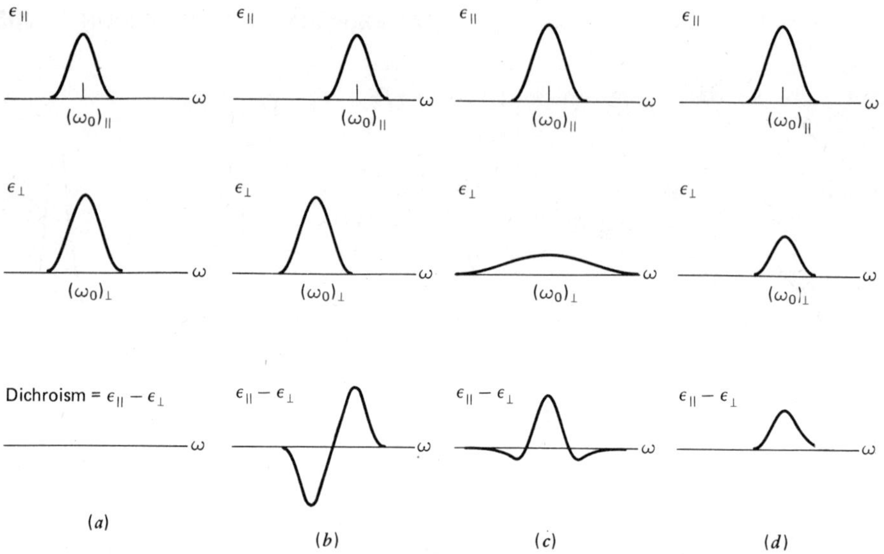

FIGURE 14-10. Types of optical absorption for light plane-polarized in either of two mutually perpendicular directions. (*a*): identical absorption for light plane-polarized in either direction ("isotropic" absorption), to give zero dichroism. (*b*): electrons driven in different directions are bound by springs of different stiffness ($(\omega_0)_{\parallel} \neq (\omega_0)_{\perp}$). (*c*): electrons driven in different directions are slowed by different frictional resistance ($f_{\parallel} \neq f_{\perp}$). (*d*) electrons driven in different directions have different maximum displacement. "Anisotropic" situations (*b*), (*c*), or (*d*) (or any combination thereof) will produce nonzero dichroism. (*b*) and (*c*) can be understood in classical terms; (*d*) requires quantum mechanical treatment.

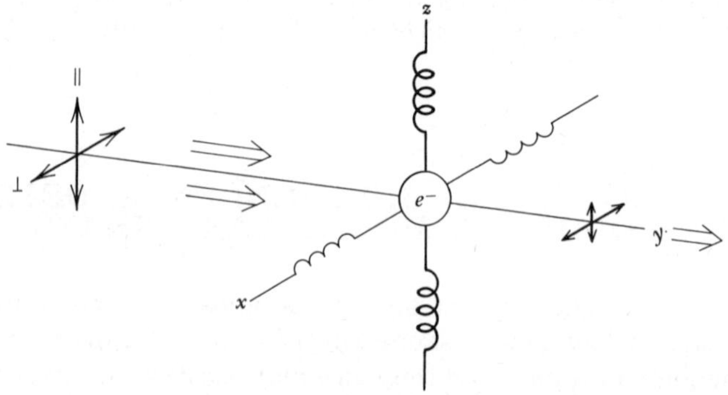

FIGURE 14-11. Schematic diagram of one basis for dichroism and birefringence (Fig. 14-10*b*). It is imagined that electrons in the optically "anisotropic" substance are bound by springs of different strengths in different directions. Light plane-polarized along the direction of one of the springs will be absorbed and slowed down to a different extent than light plane-polarized along the direction of the other type of spring (note difference in absorption in Fig. 14-10*b* for the two plane-polarized components after passing through the substance). The difference in absorbancy is called *dichroism* and the associated difference in refractive index is called *birefringence*.

polarized in two perpendicular directions will be different at any given frequency. The above argument will apply *only* when a majority of such molecules are somehow *aligned* with their "strong" (and "weak") springs pointing in the *same direction;* for molecules oriented randomly in solution, for example, there will be the same number of strong springs pointing along z as along x and no difference in absorption or refractive index for light plane-polarized along two perpendicular directions will be observed (Fig. 14-10a). *Dichroism* ("two-colors") refers to a difference in *absorption* (or extinction coefficient) for light plane-polarized in two perpendicular directions with respect to the incident direction. If the *absorption* at a given frequency is different in different directions, the molecule will appear to have a different color when viewed with the two kinds of plane-polarized light (hence, "di"-chroism). *Birefringence* refers to a difference in *refractive index* for incident light plane-polarized in two perpendicular directions. Applications of these phenomena are based on introduction or detection of partial alignment of such "optically anisotropic" molecules. Means for aligning molecules include orientation by use of thin films, by capillary flow, and with electric fields, as illustrated in the following examples.

EXAMPLE *Infrared Dichroism and Polypeptide Configuration*

The data in Fig. 14-12 show that for a synthetic polypeptide, poly-L-alanine, light plane-polarized in one direction relative to the oriented fiber axis is absorbed to a different extent than light polarized in a plane perpendicular to the first. More important, the *sign* of the difference is reversed on going from

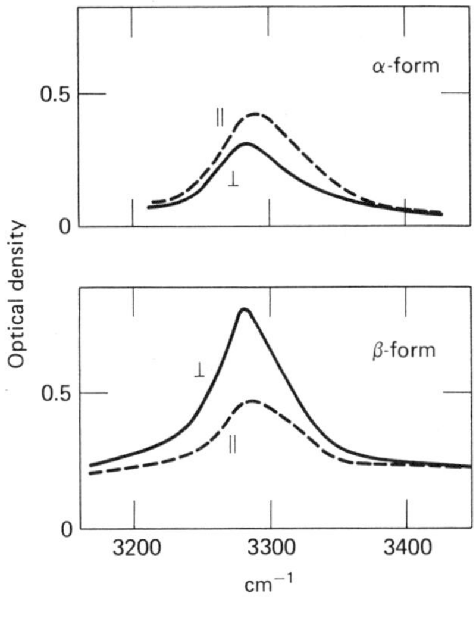

FIGURE 14-12. Dichroism in the infrared absorption for an oriented film of poly-L-alanine. Full line: electric vector of plane-polarized incident light is perpendicular to direction of oriented fiber axis; broken line: electric vector parallel to direction of the fiber axis. *Top:* α-helix polypeptide configuration. *Bottom:* β-extended pleated sheet structure. [From A. Elliott, *Proc. Roy. Soc. A226,* 408 (1954).

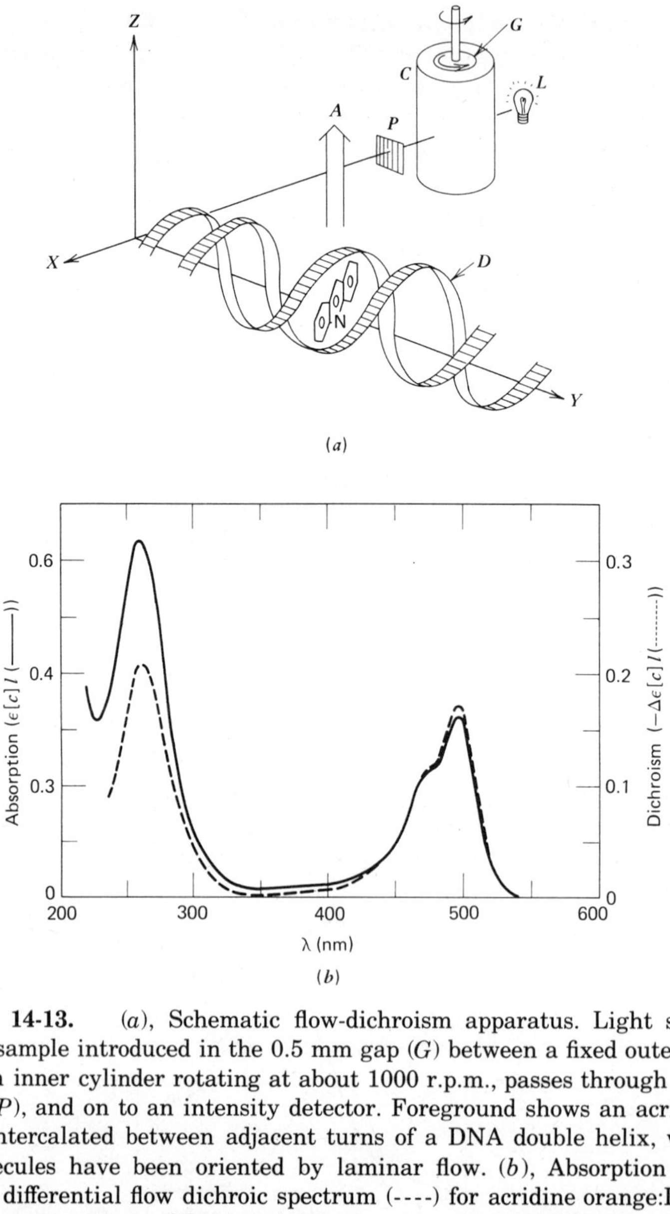

FIGURE 14-13. (*a*), Schematic flow-dichroism apparatus. Light source (*L*) shines on sample introduced in the 0.5 mm gap (*G*) between a fixed outer cylinder (*C*) and an inner cylinder rotating at about 1000 r.p.m., passes through a (plane-) polarizer (*P*), and on to an intensity detector. Foreground shows an acridine ring skeleton intercalated between adjacent turns of a DNA double helix, where the DNA molecules have been oriented by laminar flow. (*b*), Absorption spectrum (——) and differential flow dichroic spectrum (----) for acridine orange:DNA complex. The concentrations of DNA and AO were 1.5×10^{-3} M and 1.87×10^{-4} M, respectively. DNA and AO were dissolved in 0.001 M phosphate buffer containing 0.0001 M EDTA and 0.0025 M NaCl (pH 7.0). Under these conditions, more than 99.6% of AO cations are bound to DNA and the ratio of nucleotide to bound dye is about 8. [From H. Takesada, E. Saito, H. Fujita, K. Suzuki, and A. Wada, *Bull. Chem. Soc. Japan 43,* 181 (1970).]

α-helix to β-pleated sheet configuration (see Section 6 for further discussion of those terms), so that dichroism in the infrared may be used to distinguish between these two principal polypeptide configurations.

EXAMPLE *Ultraviolet Visible Flow Dichroism and Dye Intercalation in DNA*

It is possible to align molecules in solution, provided that the molecular shape is appreciably nonspherical, simply by producing laminar flow (see Chapter 7.C.1) so that a rod-shaped molecule will tend to line up with its long axis along the flow lines. We have previously treated laminar flow in a capillary; for optical measurements it is more convenient to introduce laminar flow in the narrow gap (see Fig. 14-13*a*) between fixed and rotating concentric cylinders. Then, by measuring the absorption of light plane-polarized along or perpendicular to the direction of flow, it is possible to determine whether the absorbing moiety is oriented parallel or perpendicular to the long axis of the molecule in question. For example, Fig. 14-13*b* shows the result of a measurement of flow-induced dichroism for the dye, acridine orange (AO), bound to DNA. Since acridine orange is a flat, aromatic molecule, the π-electrons in this case effectively move in the plane of the aromatic ring, so the increased absorption observed in a direction perpendicular to the direction of flow (and hence perpendicular to the long axis of the DNA molecule) shows that acridine orange must bind to DNA so that the plane of the acridine rings is perpendicular to the axis of the DNA helix. In other words, the acridine lies parallel to the planes of the DNA bases. It has been suggested that mutagenic or antibiotic agents such as acridine orange may disrupt transcription by "intercalating" (i.e., squeezing) between adjacent base pairs in the DNA helix. Further evidence in this case is provided from viscosity measurements, which indicate that DNA becomes longer after acridine orange is bound to it.

EXAMPLE *Birefringence and Order in Rod Outer Segments of Retinal Cells*

A birefringent substance has different refractive indices for light polarized in different planes: birefringent calcite crystals can be used to disperse an unpolarized light beam into two plane-polarized components that emerge in different directions, and are a convenient means for producing or analyzing polarized light. (Newer and cheaper "polaroid" is based on the strong dichroism shown when dichroic molecules such as molecular iodine are oriented in vinyl or other polymeric films.) In biological lipid bilayer membranes, in which hydrocarbon chains are preferentially aligned perpendicular to the plane of the membrane, there is pronounced dichroism (and thus according to Fig. 14-14, an associated birefringence) when many such membranes are stacked on top of each other, as in a rod cell outer segment in the retina of the eye. It is important to recognize that birefringence can result *either* from alignment of molecules that are birefringent individually ("intrinsic" birefringence) or from a nonuniform arrangement (for example, spherical beads on a string) of molecules that are individually isotropic optically ("form" birefringence).

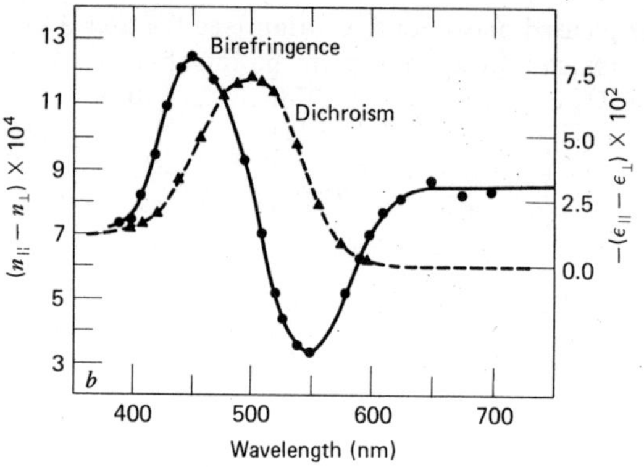

FIGURE 14-14. Birefringence and dichroism spectra for dark-adapted rod outer segments from frog retina. The dichroism shows that there is greater absorption of light polarized perpendicular than parallel to the *rod axis* (light is thus absorbed more strongly when plane-polarized parallel rather than perpendicular to the *membranes* themselves). [From P. A. Liebman, W. S. Jagger, M. W. Kaplan, and F. G. Bargoot, *Nature 251,* 31 (1974).]

In the case of rod outer segments, the *sign* of the (form) birefringence corresponding simply to a stacking up of many layers of membranes is opposite to the sign of the net birefringence actually observed, which must therefore be dominated by intrinsic birefringence arising from optical anisotropy of the lipid molecules themselves. The dichroism and birefringence for rod outer segments are shown in Fig. 14-14. An interesting recent observation is that there is a 1% *change* in the observed birefringence almost immediately after irradiation of the rod cells with strong light. It is interesting to note that there are about 100 phospholipid molecules per rhodopsin molecule in the rod outer segment, so that this small birefringence change may correspond to the disorientation of a single phospholipid per rhodopsin absorbing light; in any case, the birefringence measurement appears to be extraordinarily sensitive to small changes in molecular order.

14.C. CIRCULAR DICHROISM AND OPTICAL ROTATION: OPTICAL ACTIVITY AND "HANDEDNESS" OF MOLECULES

In the previous two subsections, we have been able to account for absorption and refractive index by treating electrons in matter as if they were suspended on springs driven by an oscillating force from the electric field of incident electromagnetic radiation, and we accounted for dichroism and birefringence as the *difference* between absorption and refractive index for light plane-polarized in two perpendicular directions. In this section, we try to account for circular dichroism and optical rotatory dispersion ("circular birefringence") using the same language and equations already covered,

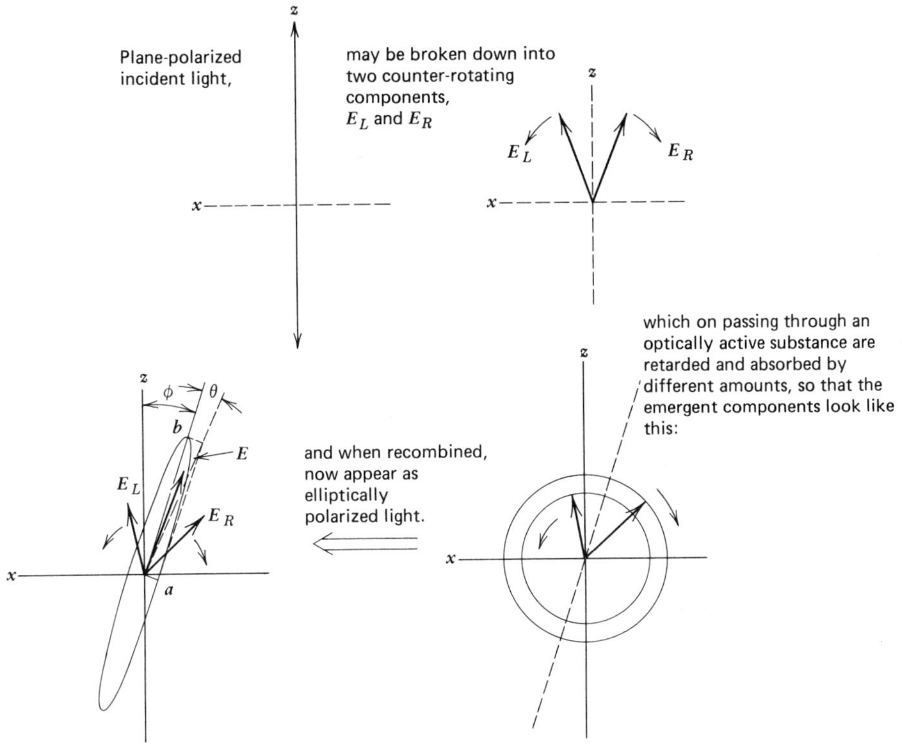

FIGURE 14-15. Schematic diagram that accounts for the apparent rotation, by angle, ϕ, of plane-polarized incident light by passage through an optically active substance (optical rotation), and also the fact that the emergent beam is now elliptically polarized to a slight extent usually quantified according to the "ellipticity," θ.

then indicate how naturally occurring optical activity may be used to determine helical content of macromolecules, and how the presently strongly biased "handedness" of biological molecules might have arisen.

To provide a simple description of these phenomena, it is useful to express plane-polarized light as the (vector) sum of two counter-rotating components, each of which rotates at the same constant velocity, as shown in Fig. 14-15. On entering an "optically active" substance, these two circularly polarized (see Fig. 13-3 for review) components may be *absorbed* to different extents (*circular dichroism*) and *slowed down* to different extents (*optical rotation*). If we now examine the light emerging from the substance using a plane-polarizer that passes only light plane-polarized in one direction, then we will observe a maximum intensity at a direction rotated by angle, ϕ, away from the initial direction, so that it would appear on first glance that the only effect is that the plane of the incident plane-polarized light has been rotated by angle, ϕ ("optical rotation").* However, if we next measure the intensity of emergent light in a direction perpendicular to the direction of maximal intensity, we will in general observe a nonzero intensity, because the emergent light is actually *elliptically* polarized since one of its electric field components is now smaller than the other. Because the difference in refractive index (or absorbance) between left and right circularly polarized radiation is so small, experimental determinations of the circular dichroism, $E_L - E_R$, and circular birefringence $\eta_L' - \eta_R'$, are based on measurement of the respective "ellipticity," θ, and "optical rotation," ϕ, as summarized in Equations 14-11 and 14-12 (see Problems).

Qualitative Mechanism for C.D. and Optical Rotation for "Handed" Molecules (Optional)

Figure 14-15 presents a complete phenomenological picture that accounts for circular dichroism (C.D.) and optical rotation, but does not explain why a right-handed molecule, say, should treat left and right circularly polarized light differently. An introductory explanation is provided from the plots in Fig. 14-16.

In Fig. 14-16, incident plane-polarized electric field is broken down into two circularly polarized components shown in detail in the figure. The reader should realize that the two lower plots follow directly from the two middle plots. Also shown is the magnetic field, H_2, associated with electric field component, E_2. The effect of the electric fields, E_1 and E_2, is to force electrons to move either up and down, or in and out of the plane of the paper. The effect of the magnetic field H_2 is to drive the electrons in a circular path in a plane perpendicular to the main axis of the helix. However, because the electrons

* *Optical rotation is commonly reported in at least the following forms:*

$\phi = (n_L' - n_R')\pi/\lambda$ *radians per centimeter, where n_L' and n_R' are the respective refractive indices for left and right circularly polarized light and λ is wavelength* (14-11a)

$\delta = \phi(l)$ *radians, in which l is the path length in cm* (14-11b)

$\alpha = 180\delta/\pi$ *degrees* (14-11c)

$[\alpha] = \alpha/d[C]_0$ *deg cm² dekagram⁻¹, $[C]_0$ is concentration in gram/cc and d is optical path length in decimeters* (14-11d)

= *"specific rotation"*

$[\phi] = 100\alpha/l[c]$, *deg cm² decimole⁻¹, in which $[c]$ is in moles/liter* (14-11e)

= *"molar rotation"*

Similarly, for ellipticity, $[\theta] = 3298\,(\epsilon_L - \epsilon_R)$, *where ϵ_L and ϵ_R are the molar extinction coefficients for left and right circularly-polarized light,* (14-12a)

= *molar ellipticity in degree cm² decimole⁻¹* (14-12b)

$= 100\psi/l[C]$ *where* $\psi = \dfrac{1800}{\pi}$; $[C]$ = *concentration (moles liter⁻¹)*

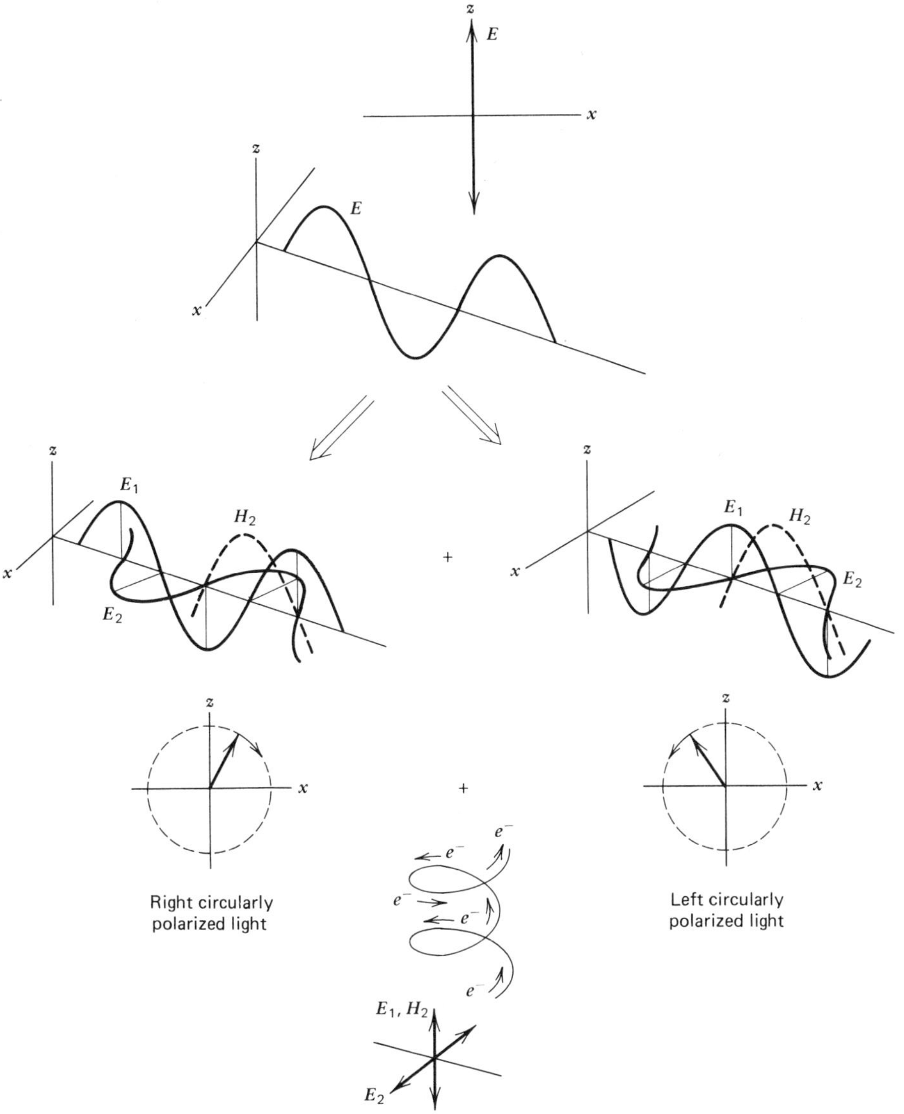

FIGURE 14-16. Detailed decomposition of plane-polarized light into the sum of two circularly polarized components. The effect of the electric and magnetic fields for each component on a system consisting of electrons that are constrained to move on a helical path is discussed in the text.

are constrained to move along a *helical* path, H_2 has the effect of forcing electrons to move *up and down* the helix, which is similar to the effect of E_1. Moreover, because the force due to H_2 depends on the time-derivative of H_2, and because the derivative of a sine wave gives the same function displaced by 90°, the forces from E_1 and H_2 for *right* circularly polarized light will actually be *in phase* (i.e., always in the same direction), while the forces from E_1 and H_2 for *left* circularly polarized light will be *180° out-of-phase* (i.e., always opposed). Thus, we conclude that an electron on a helix will be driven by a different net force when subjected to right or left circularly polarized light, and should therefore be expected to exhibit different absorption and refractive index with respect to those types of radiation.

The reader should understand that the same sort of reasoning can be used to argue that there should be an equal but opposite effect when right and left circularly polarized components encounter a *left*-handed rather than a *right*-handed helix. Careful study of the above reasoning shows that the key feature is that the electrons must be able to move up and down at the same time as they move sideways. For example, in a planar molecule, the electrons are constrained to move in the molecular plane, so that the induced electric and magnetic dipole moments are at right angles to each other, and the above argument does not apply—planar molecules do not in fact show optical activity. It might seem that the argument about the helix breaks down if we turn the helix upside down, but it turns out that a right-handed helix always looks different from a left-handed one, no matter what the viewing angle. A common chemical (sufficient) criterion for optical activity is that a molecule contain an "asymmetric" carbon—that is, one with four different substituents.

Frequency-dependence of Circular Dichroism and Optical Rotation

If, as argued in Fig. 14-15, C.D. and O.R. arise as the difference between two kinds of *absorption* and two kinds of *refractive index,* respectively, then the frequency-dependence for these phenomena might be expected to resemble the frequency-dependence for ordinary absorption and refractive index, as shown in Fig. 14-17. The frequency-dependence of O.R. is called optical rotatory dispersion (O.R.D.). It soon becomes clear (Chapter 15.A) that the "Drude" approximation bears the same relation to O.R.D. plots as the "scattering" or "no-damping" approximation bears toward the in-phase response of an electron on a spring driven by a sinusoidal driving force (see Chapter 13.B.1). The Drude plot clearly fails near resonance, because as resonance is approached, the contribution of the 90°-out-of-phase response (C.D. in this case) begins to dominate. As with absorption and refractive index, C.D. and O.R.D. are two manifestations of the same molecular phenomenon, and provide essentially the same information (i.e., the resonant frequency and the "damping constant" related to the width of the spectral plot). O.R.D. was historically observed and characterized first, but comparison of the two lowermost plots of Figures 13-5 and 14-4 again shows that the "absorption" (in this case, C.D.) is concentrated over a smaller frequency range than is the "dispersion" (in this case, O.R.D.), so that C.D. provides the more selective observation, with less overlap between "absorption" occurring at two closely spaced "natural" frequencies $(\omega_0)_A$ and $(\omega_0)_B$. The only remaining justification for use of O.R.D. measurements is that since the O.R.D. does spread over a wider frequency range, it is possible

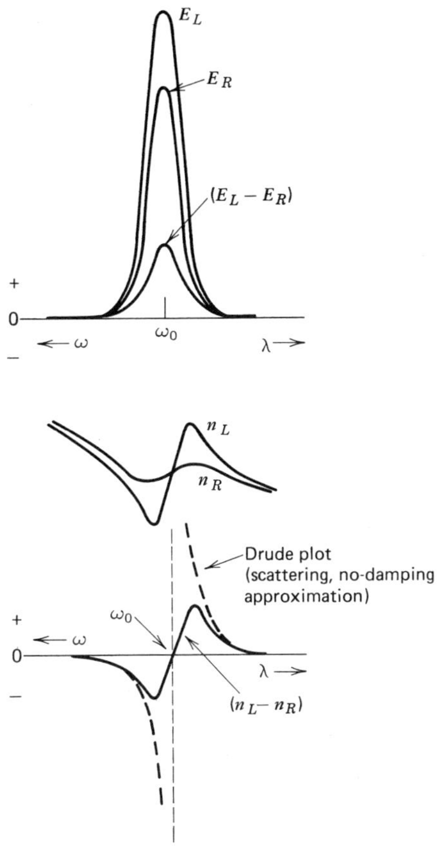

FIGURE 14-17. Frequency-dependence of circular dichroism (C.D.) and optical rotation (O.R.), obtained as the difference between extinction coefficients or refractive indices for left- and right-circularly polarized light. Optical rotatory dispersion is sometimes approximated by a "Drude" plot, whose functional form is

$$[\phi] = (\text{constant}) \frac{\lambda_0^2}{\lambda^2 - \lambda_0^2} \tag{14-13}$$

in which λ_0 is the wavelength corresponding to the "natural" or "resonant" maximal absorption frequency. The Drude plot approximation to optical rotatory dispersion is shown as the dotted lines in the lowermost diagram."

to obtain information about "resonances" that may be beyond the spectral range (usually at ultraviolet frequencies) of the spectrometer. Since C.D. and O.R.D. occur only when there is a "handedness" or helicity to a molecule or part of a molecule, it would seem desirable to use these measurements quantitatively to determine the fraction of a given macromolecular chain that is helical—information otherwise obtainable only from X-ray crystallography. This is a major application for C.D. and O.R.D. in biochemistry, and the following examples are representative of the semi-empirical treatment used.*

EXAMPLE *C.D. and O.R.D. as Measures of the Degree of Helicity in Macromolecules*

It is clear from Fig. 14-17 that there can only be appreciable C.D. or O.R.D. at frequencies at which there is appreciable ordinary absorption in the first place. Thus, although naturally occurring polypeptides and proteins have asymmetric centers at each tetrahedral alpha-carbon atom, that carbon atom and its connecting bonds do not exhibit any strong ultraviolet-visible absorption, so it might seem that C.D. and O.R.D. of proteins would be very weak. However, the asymmetric nature of the alpha-carbon is partly communicated to nearby amide groups that do absorb in the ultraviolet-visible, and lead to pronounced C.D. and O.R.D. Furthermore, an additional contribution to the optical activity can arise from the secondary structure of the protein (both the alpha-helix and beta-structures are optically active since neither is superimposable on its mirror image). Optical activity of peptides thus arises from both "point" asymmetry of the individual amino acids and also "form" asymmetry of the peptide backbone. Figure 14-18 shows the C.D. and O.R.D. spectra obtained for three different conformations of the polypeptide, poly-L-lysine: alpha-helix, beta-antiparallel structure (see Section 6), and random coil, in the far ultraviolet spectral region. It is clear that both C.D. and O.R.D. are very sensitive to changes in polymer shape.

It is logical to think of using the characteristic C.D. or O.R.D. spectra for the model polypeptide of Fig. 14-18 in attempting to assign the quantitative relative proportions of the three common secondary structures (α-helix, β-structure, and random coil) by comparison of Fig. 14-18 with C.D. or O.R.D. spectra from actual proteins in solution. The results are tabulated in Table 14-2, and show that C.D. is in general more accurate than O.R.D. when the results are compared with structures determined by X-ray diffraction, which is what we have learned to expect from the generally sharper and better-resolved features of C.D. spectra compared to O.R.D. spectra.

Evolutionary Origin of Optical Activity in Nature

Anyone who thinks seriously about handedness in nature must wonder how it could have reached the present near-total preponderance of chemical isomers of one handedness rather than the others (the most familiar example is the great preference for L-amino acids over D-amino acids in most living organ-

* *Optical activity (nonzero C.D. and O.R.D.) is induced in all matter, "handed" or not, by applying a static large magnetic field. "Magnetic" C.D. is a relatively new technique as applied to biochemical problems, and there are relatively few cases where "M.C.D." spectra have been used to correlate macromolecular structure and function. One promising avenue is that the M.C.D. spectra of tyrosine and tryptophan are measurably different, even though the ordinary optical absorption spectra of the two are very similar (Fig. 14-5a).*

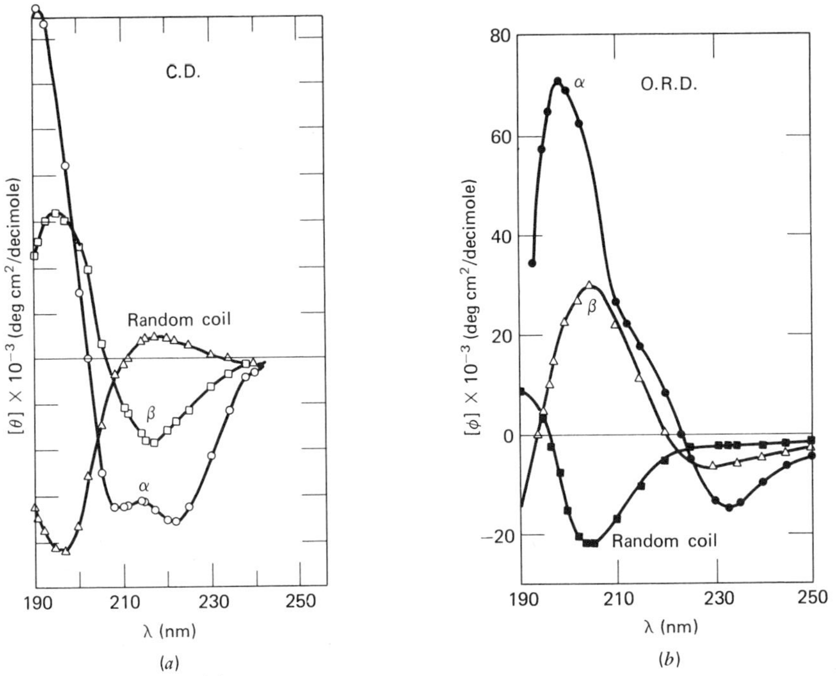

FIGURE 14-18. C.D. (left) and O.R.D. (right) spectra for poly-L-lysine in three distinct structural forms: alpha-helix (pH 11), antiparallel beta-structure (pH 11, heated at 50°C for 12–25 min, then cooled to 22°C), and random coil (pH 5). [C.D. data from N. Greenfield and G. D. Fasman, *Biochemistry 8,* 4108 (1969); O.R.D. data replotted from N. Greenfield, B. Davidson, and G. D. Fasman, *Biochemistry 6,* 1630 (1967).]

Table 14-2 Structural content for various proteins, determined by X-ray diffraction and by comparison of the C.D. or O.R.D. spectrum of the native protein with the C.D. or O.R.D. spectra of several proteins in each of the three listed conformations (α-helix, β-form, and disordered or random coil form). Compare Figure 14-18 for typical reference spectra.

	X Ray			C.D.[a]			O.R.D.[b]		
Protein	% α	% β	% coil	% α	% β	% coil	% α	% β	% coil
Myoglobin	77	0	23	77	2	21	77	5	18
Lysozyme	29	16	55	29	16	55	25	0	75
Lactate dehydrogenase	29	20	51	31	6	63	30	8	62
Papain	21	5	74	21	10	69	22	19	59
Ribonuclease	19	38	43	18	44	38	20	51	29
Insulin	22–45	12	66–43	31	18	51	25	7	68
Nuclease	24	15	61	27	10	63	27	8	65
Cytochrome c	11–39	0	89–61	27	6	67	26	8	66
α-Chymotrypsin	9			8	10	82	11	0	89
Chymotrypsinogen A	6			9	36	55	8	26	66

[a] *Based on reference values between 205 and 240 nm at 1 nm intervals.*
[b] *Based on reference values between 225 and 240 nm at 1 nm intervals.*
[*From Y.-H. Chen, J. T. Yang, and H. M. Martinez,* Biochemistry 11, *4120 (1972).*]

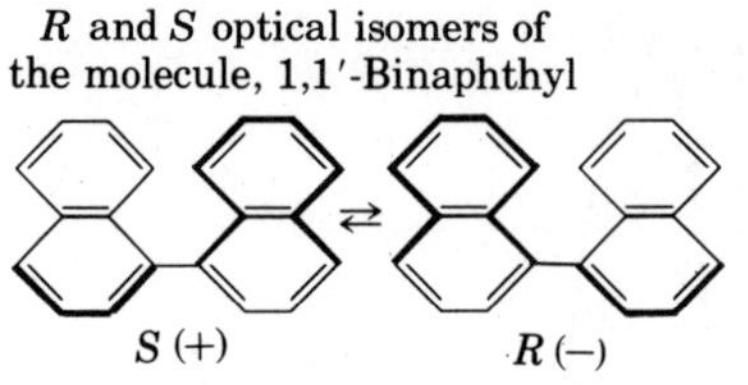

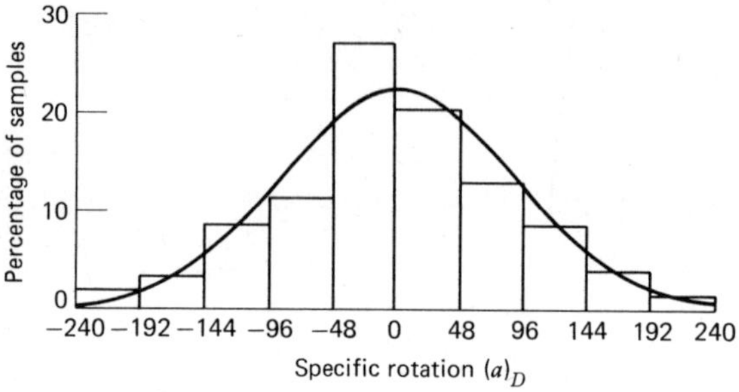

FIGURE 14-19. Spontaneous resolution of a racemic mixture into crystals that when redissolved sometimes show pronounced optical activity. Two hundred samples were crystallized from a purely racemic melt at 150°C. The smooth curve shows the distribution of the relative number of samples expected to exhibit a particular optical rotation, based on a Gaussian curve with a mean of +0.14 (i.e., ≅0) and a standard deviation of 86.4. [From R. E. Pincock, R. R. Perkins, A. S. Ma, and K. R. Wilson, *Science 174,* 1018 (1971).]

isms). This question has proved difficult: traditional explanations have been based on an introduction of handedness either from a supposed preponderance of, say, left circularly polarized light striking the primordial atmosphere (since it has been shown that circularly polarized light can selectively destroy molecules of a given handedness), or from similar effects based on interactions of one of the inherently "handed" particles of nature, such as neutrinos, muons, or natural β-rays (left circularly polarized electrons). Although there is no evidence that sunlight arriving at the earth has ever been appreciably circularly polarized, recent experiments [W. A. Bonner et al., *Nature 258,* 419 (1975)] have shown that (natural) left circularly polarized electrons bring about slightly greater degradation of D-leucine rather than L-leucine. [As a control, it was found that (unnatural) right circularly polarized electrons produce slightly greater degradation of the L-isomer.]

However, it is not necessary to invoke any *external* influence to explain the origin of optical activity. A fascinating recent experiment that demonstrates the *spontaneous* appearance of optical activity from an initially "racemic" mixture (i.e., a 50–50 mixture of the two "handed" isomers) having no optical activity to begin with is shown in Fig. 14-19. Using a synthetic molecule (to avoid prior problems of contamination by naturally occurring optically active material that then provided a "seed" mechanism for preferential production of a particular optical isomer), the authors observed that the optical activity of crystals obtained from an initially racemic mixture sometimes showed a great preference for one optical isomer, and sometimes a great preference for the other, with an approximately statistical (Gaussian) distribution of results in between. In experiments not illustrated here, it was further shown that addition of small amounts of an optically active compound could effectively control the stereospecificity (i.e., choice of right- or left-handed product) of the reaction. These experiments thus suggest a possible mechanism for the production of the first preponderantly optically active compound from a previously optically inactive mixture, and also show how the presence of the first optically active species might have led to the preferential selection of a given optical isomer of many *other* types of molecules.

Finally, it is interesting to note that in another recent study [Kovacs and Garay, *Nature 254,* 538 (1975)], crystallization of racemic D,L-sodium ammonium tartrate resulted in marked and systematic preference for L-salt crystal formation when the experiment was conducted in the presence of (natural) ^{32}P-phosphate radiation. Thus, it may be that while optical activity could have arisen spontaneously, it may have been guided toward a preference for L-forms by the presence of naturally occurring (but "handed") radiation.

14.D. MAGNETIC RESONANCE ABSORPTION AND DISPERSION: NUCLEAR TETHERBALL; SPIN LABELS AS PROBES OF MOLECULAR FLEXIBILITY: SELECTIVE pH METER

Electrons and nuclei can be characterized according to a number of static properties, including mass, charge, and parity (a spatial symmetry property). The electron and a number of interesting nuclei (including ^{1}H, ^{2}D, ^{13}C, ^{19}F, ^{31}P, ^{23}Na, ^{35}Cl, etc.) exhibit an additional property known as *magnetic moment.* We have already treated the *linear* motion of a *charged* particle due to its interaction with the *electric* component of electromagnetic radiation; we now deal with the *circular* motion of a *magnetic* particle due to its interaction with the *magnetic* component of electromagnetic radiation. The nature of the magnetic interaction can be readily appreciated by analogy to the motion of a continuously driven tetherball (see Fig. 14-20).

As shown in the figure, the magnitude of the magnetic moment, M_0, which results on application of a static magnetic field, H_0, is given by Eq. 14-14:

$$M_0 = \chi_0 H_0 \tag{14-14}$$

in which χ_0 is called the "magnetic susceptibility." Finally, in the absence of friction, a tetherball (or a magnetic moment) will precess at some "natural" frequency, ω_0, which depends on the length of the rope (or the magnitude of M_0); it turns out that the magnetic precession frequency, ω_0, is proportional to H_0:*

$$\omega_0 = \gamma H_0 \tag{14-15}$$

in which γ is called the magnetogyric ratio, and is very much different for isotopically different nuclei.

* *Equation 14-15 requires quantum mechanical justification—see Chapter 18.A.*

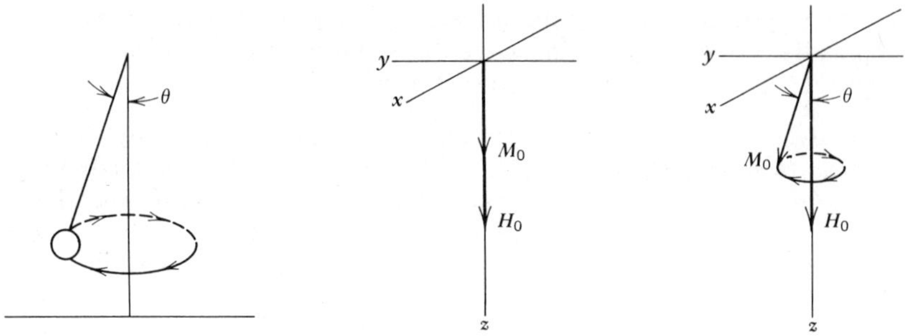

FIGURE 14-20. Magnetic precession, illustrated by tetherball analogy. When the tetherball (far left) is subjected to a steady tangential force (and a frictional force proportional to the tetherball velocity), the ball traces out a steady-state circular path at angle, θ, from vertical. A group of magnetic nuclei or electrons in a fixed magnetic field exhibits a magnetic moment (i.e., behaves like a bar magnet) whose magnitude, M_0, is proportional to the magnitude of the applied magnetic field, H_0 (Eq. 14-14), as shown, in the middle diagram. The magnetic moment may be pushed in a circular path just like the tetherball, by applying a tangential force as shown at the far right. (The tangential force is derived by applying additional small magnetic fields oscillating in the x-y plane.) For small tangential force, both systems precess at angle, θ, and for sufficiently small θ, the amplitude of the (experimentally observed) displacement along the x axis is proportional ($\sin\theta \cong \theta$) to the amplitude of the applied tangential force. In other words, the system is "linear" (response is proportional to driving force), and behaves just like the driven, damped weight on a spring of Chapter 13, except that the "natural" motion is circular rather than straight-line.

The second important feature of the figure is that the magnetic system is "linear"; that is, the amplitude of the observable response (which turns out to be the x-component of the precessing M_0) is proportional to the amplitude of the driving force, at least when the driving force is small enough that the precession is held to small angles. Looking back over our previous treatment of a driven, damped weight on a spring, it is apparent that the same features are present here in the case of magnetic precession. There is a natural motional frequency, ω_0; the motion would continue indefinitely in the absence of friction, and would be exactly in-phase with the driver; in the presence of friction, the motion will in general lag the driver and the response can be expressed in terms of components that are rotating exactly in-phase or 90° out-of-phase (one-quarter of a revolution behind) the driver. In fact, the only real difference is that the "natural" motion is circular rather than linear, and that aspect is completely nonessential to the argument. Without working through the algebra, which is in any case very similar to our earlier treatment, we can express the magnetic moment response M_x along the x-axis, to a driving magnetic field of amplitude H_1 rotating in the same sense as the magnetic precession, as

$$M_x = (\chi' \cos\omega t + \chi'' \sin\omega t)(H_1) \quad (14\text{-}16)$$

in which

$$\chi' = \frac{\chi_0\omega_0}{2} \frac{(\omega_0 - \omega)T_2^2}{1 + (\omega_0 - \omega)^2 T_2^2} \quad (14\text{-}17a)$$

and

$$\chi'' = \frac{\chi_0\omega_0}{2} \frac{T_2}{1 + (\omega_0 - \omega)^2 T_2^2} \quad (14\text{-}17b)$$

and $(1/T_2)$ may be thought of as a measure of the friction that pulls the magnetic moment back toward its "resting" position along H_0 on the nuclear tetherpole. Equations 14-17a and 14-17b are immediately recognized as having the Lorentz frequency-dependence of Fig. 13-5. χ'' is called the "absorption-mode" or "v-mode" magnetic resonance signal, and χ' is called the "dispersion-mode" or "u-mode" magnetic resonance signal. The resonant ("natural") frequencies of a variety of magnetic nuclei and the electron are shown in Table 14-3; the *nuclear* magnetic resonance (NMR) frequencies are seen to lie in the *radiofrequency* range, while the *electron* paramagnetic resonance (EPR) frequency is in the *microwave* region, for a typical choice of applied magnetic field, H_0, of about 23.5 kGauss.

The first major feature of Table 14-3 is that the magnetic resonance signals for different magnetic isotopes differ by very much more than the width of an absorption signal; since radiofrequencies can be measured easily to 0.1 Hz, it is obviously very easy to resolve the magnetic resonance signals from, say, protons from those of carbon-13, for example. Even more important, there are small natural or induced magnetic fields in any mole-

Table 14-3 Properties of Various Magnetic Isotopes

Magnetic Isotope	Magnetic Resonance Frequency, ν_0, (MHz) for Applied Magnetic Field of 23.5 kGauss	Typical Absorption Line Width (Hz)	Natural Abundance of Magnetic Isotope	Typical Range of ν_0 Values for Different Compounds, in Multiples of Typical Linewidth
^{1}H	100.0	0.5	0.9998	2000 linewidths
^{2}D	15.4	2	0.0002	
^{13}C	25.1	0.5	0.01	20,000
^{15}N	10.1	0.5	0.004	
^{19}F	94.1	0.5	1.00	40,000
^{23}Na	26.5	15[a]	1.00	
^{31}P	40.5	0.5	0.07	40,000
^{35}Cl	9.8	15[a]	0.75	
^{199}Hg	17.9	0.5	0.17	
free electron	65,749.2	2×10^6	—	10 (nitroxides)

[a] *Aqueous ion.*

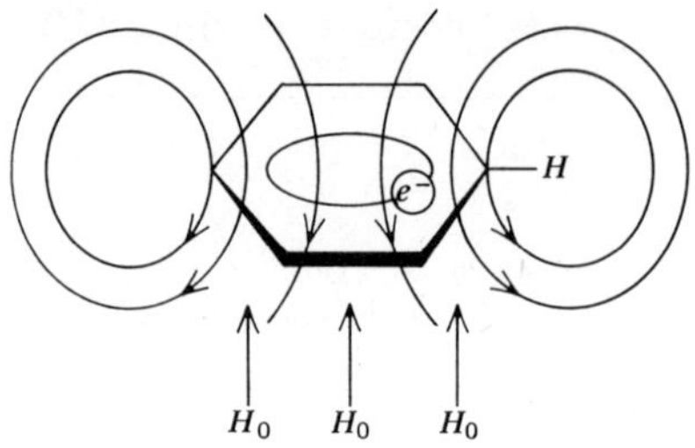

FIGURE 14-21. Application of a static external magnetic field, H_0, to an aromatic carbon ring leads to a circulation of π-electrons and that electron current is then associated with a small opposing ("induced") magnetic field whose effect is to *add* to the applied magnetic field at a proton located outside the edge of the ring, as shown at the right of the figure. That proton thus exhibits a slightly *larger* magnetic resonance frequency than would be the case in the absence of the aromatic ring.

cule, and those fields are slightly different for magnetic nuclei attached to or near to different chemical groups. Consider the effect of an applied magnetic field on the π-electrons of an aromatic carbon ring. The applied magnetic field will produce a force on those electrons that causes them to circulate in the plane of the ring, producing a small *induced* magnetic field in a direction opposed to that of the applied field (see Fig. 14-21). Following the magnetic field direction to the edge of the aromatic ring, it is clear that a proton just outside the edge of the ring will be subjected to a slightly *larger* magnetic field (and thus resonate at slightly higher frequency, ν_0) than if no aromatic ring were present. Although the effect is small (only about 0.000007 of the strength of the applied magnetic field, H_0), the magnetic resonance signals for protons are so narrow (only about 0.00000001 of the resonant frequency), it is easy to distinguish protons located in different chemical bonds or near different chemical functional groups. Figure 14-22 shows a carbon-13 nuclear magnetic resonance spectrum with the different resonant frequencies corresponding to several chemically different carbons in an organic molecule. Nuclear magnetic resonance thus provides an outstandingly *selective* tool for characterizing several different parts of a molecule at once—compare Fig. 14-22, for example, to the relatively featureless ultraviolet-visible or infrared spectra of Figures 14-5 or 14-9. However, as we discover in Chapter 19.B.4, the weakness of the magnetic resonance technique is its poor *sensitivity;* sample concentration must typically exceed 10^{-3}M, while the optical techniques can often yield information at concentrations less than 10^{-7} M. When sufficient *amount* and/or *solubility* of sample is possible, nuclear magnetic resonance is of great value.

The accessible information about any coherently driven damped spring is the resonant frequency, ν_0 (from the driving frequency at the steady-state absorption maximum) and the damping constant (from the spectral linewidth). The "fingerprint" advantage of the high selectivity of NMR is of

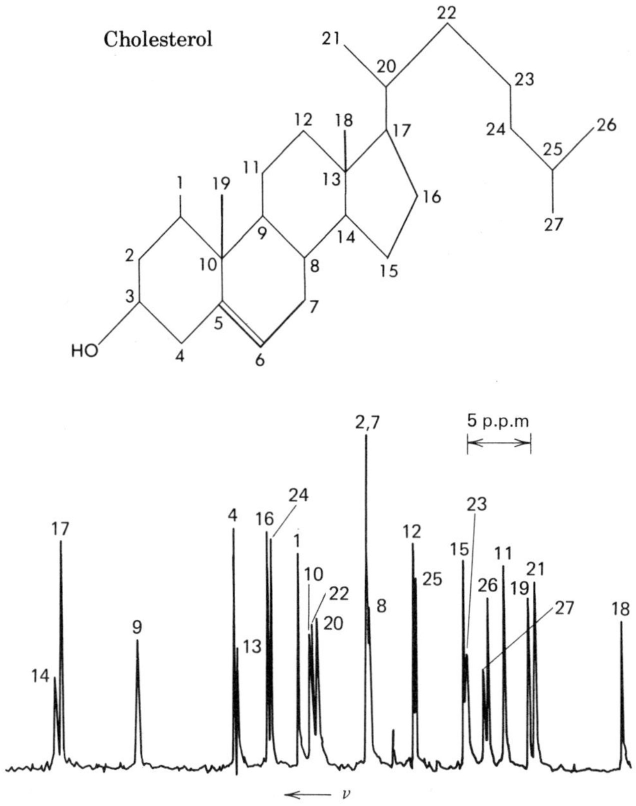

FIGURE 14-22. Carbon-13 nuclear magnetic resonance (NMR) spectrum (absorption versus irradiating frequency) for cholesterol. Individual carbon resonances are resolved and identified. Frequency scale is listed in parts per million of the resonant frequency. [From H. J. Reich, M. Jautelat, M. T. Messe, F. J. Weigert, and J. D. Roberts, *J. Amer. Chem. Soc. 91,* 7445 (1969).]

obvious value in assigning the structures of *small* organic molecules; in the following examples, we see how the same characteristic can be useful in obtaining information about particular small parts of a biological *macromolecule.*

EXAMPLE *The Selective pH Meter: Ribonuclease Proton NMR*

The imidazole ring of histidine plays a critical role in the function of enzymes that break down proteins and nucleic acids. By the same arguments given in Fig. 14-21, one can explain why the proton NMR signals for the ring protons of histidine occur at higher resonant frequency (at fixed observing magnetic field strength), or, as shown in Fig. 14-23, at lower applied magnetic field strength (when the irradiating frequency is fixed). It is possible to resolve all four histidines of the enzyme ribonuclease (RNase) as separate absorption

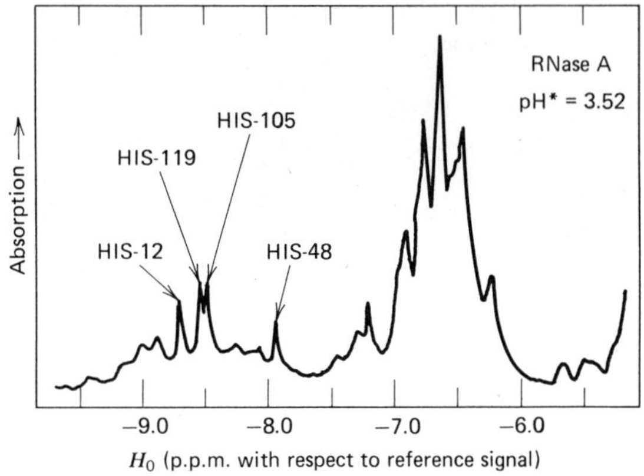

FIGURE 14-23. ^{1}H NMR spectrum ($H_0 = 58.75$ kGauss) of the aromatic proton region for bovine pancreatic ribonuclease A. The identified peaks correspond to the C-2 protons of histidines at positions 12, 119, 105, and 48 of the amino acid sequence. pH* refers to the reading of a pH meter for a D_2O solution (see text) [From J. L. Markley, *Accounts Chem. Res. 8,* 70 (1975)].

peaks, since the environment for each of the four histidines is sufficiently different from that of the others.

The proton NMR spectrum of RNase thus provides a means for monitoring *simultaneously* and *separately* the environment of four parts of the protein. For example, since histidine is the only amino acid whose side chain can serve either as an acid or as a base in the physiological pH range

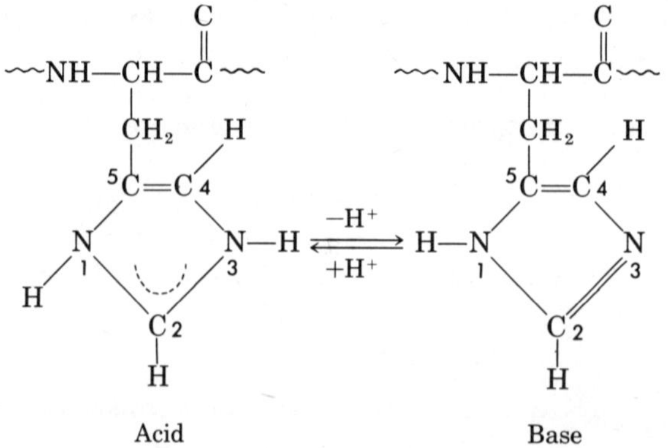

it is logical to think of using the *proton NMR resonant frequency* (absorption peak position) for each histidine C-2 proton as the *indicator* for a titration of the protein over a range of pH. (While the response of an ordinary glass electrode differs by about 0.4 pH unit for pH measured in D_2O compared to H_2O, there is also approximately equal (see Problems) change in the pK_a for dissociation of a histidine deuteron in D_2O compared to a histidine proton in H_2O. Thus, a given pH meter reading in D_2O corresponds to approximately

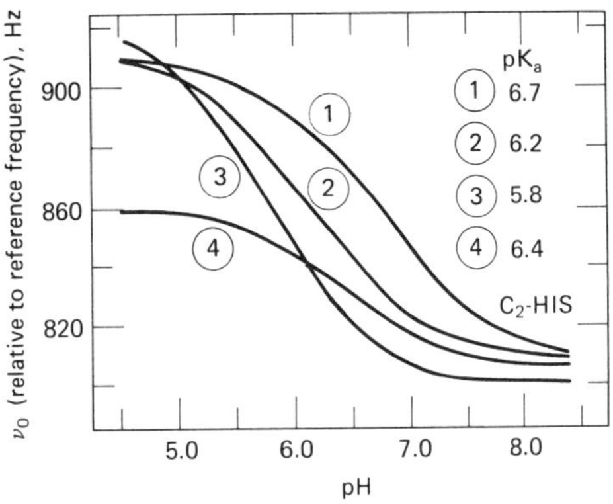

FIGURE 14-24. Titration curves (with C-2 histidine proton NMR resonant frequency as indicator) for various histidines of Rnase A in D_2O. Assignment of curves: (1) His-105, (2) His-119, (3) His-12, (4) His-48. [From D. H. Meadows, O. Jardetsky, R. M. Epand, H. H. Ruterjans, and H. A. Scheraga, *Proc. Natl. Acad. Sci.* U. S. A. *60,* 766 (1968).]

the *same* degree of dissociation of histidine C-2 deuterons as for C-2 protons at the same pH meter reading in H_2O. In other words, the buffer should be adjusted to give the *same* pH meter reading in D_2O as in H_2O, if the protein is to have the *same* degree of dissociation in both solvents.) Figure 14-24 shows the result of such a titration. It is clear that the various histidines exhibit different pK_as. What is more interesting is that when, say, His-12 is chemically modified (as verified by amino acid analysis), the pK_a of His-119 is affected, while the pK_as for the other two histidines are unaffected; similarly, when His-119 is chemically modified, only the pK_a of His-12 is changed. These results suggest that His-12 and His-119 are probably close together in space in the intact enzyme, even though the two residues are widely separated in the amino acid sequence. This sort of information is valuable in deciding how the polypeptide chain must be folded together in the native enzyme. Moreover, the pK_a of His-105 is 6.72 ± 0.02, which is essentially the same as for a small tripeptide model compound, suggesting that His-105 is in an environment relatively exposed to the bulk solution, while His-12 and His-119 are in a more special interior environment, which turns out to be the active site of the enzyme.

Similar experiments with other proteins have produced histidine pK_a values ranging from less than 4.0 (α-lytic protease) to 8.1 for two of the histidines of hemoglobin. Chymotrypsin and trypsin, for example have histidine pK_a values of about 4.5 to 5.5, more than a full pH unit smaller than "normal," almost certainly reflecting the catalytic participation of histidine in the function of those enzymes.

EXAMPLE *NMR and Fast-exchange: A Chemical Amplifier*

Consider a molecule that may be found at one of two sites (such as a small molecule existing either free in solution or bound to a macromolecule). Further suppose that there is a sufficient difference between the two environments so that at least one of the proton NMR signals for the molecule occurs at a different resonant frequency at the two sites: $\Delta = \omega_B - \omega_A$, where ω_A and ω_B are the proton resonant frequencies for the two sites for that proton. Finally, suppose that the molecule can jump back and forth rapidly between the two locations. If we look at the behavior of the proton on just one molecule, its nuclear magnetic moment will precess at angular frequency ω_A for as long as the molecule is at site *A*, then at frequency ω_B as soon as the molecule jumps to site *B*, and so forth. Provided that the "exchange rate," or rate of jumping between the two sites, is fast compared to the difference in precessional frequency between the sites, it should seem reasonable that the proton magnetic moment will precess at an average frequency, ω_{ave} given by Eq. 14-18.

$$\omega_{\text{ave}} = f_A \omega_A^{\circ} + f_B \omega_B^{\circ} \tag{14-18a}$$

$$f_A = \frac{[A]}{[A] + [B]}; \; f_B = \frac{[B]}{[A] + [B]} \tag{14-18b}$$

in which f_A and f_B simply represent the relative fractions of total sites that are *A* or *B*. To summarize: when there is no chemical exchange between sites *A* and *B*, there will be two distinct NMR signals; when there is rapid chemical exchange between the two sites, there will be a single NMR signal (see Fig. 14-25). NMR thus provides an obvious access to study of chemical exchange rates.

The poor sensitivity of NMR has already been mentioned as a primary drawback to the generality of its application to study of (dilute) macromolecules in

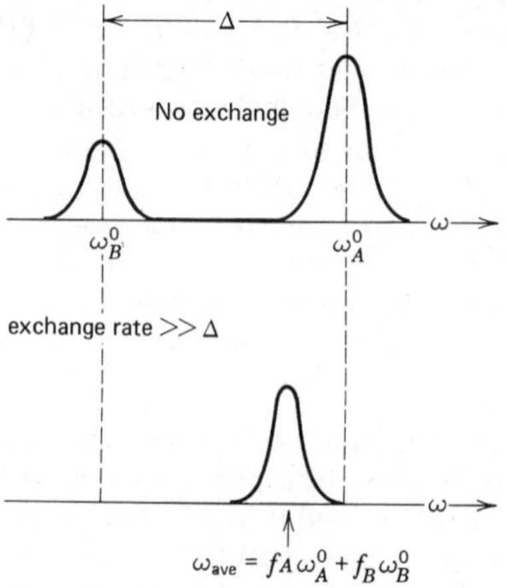

FIGURE 14-25. Schematic diagram of NMR spectra for a molecule that may reside in two different sites *A* and *B*. *Top:* no chemical exchange between sites *A* and *B*. *Bottom:* chemical exchange between *A* and *B* is fast compared to the *difference* in resonant frequencies between sites *A* and *B* ("fast-exchange" NMR limit). An average signal is observed, whose position is weighted according to the relative fractions of nuclei at the two kinds of sites.

solution. Fast chemical exchange provides a clever circumvention of this difficulty in many cases. For example, suppose that the two sites differ in frequency by 100 Hz (remember that NMR frequencies for a given nucleus can vary over several thousand Hz), and suppose that one site (say, A) is 100 times more concentrated than the other site (B). In this case, Eq. 14-18b suggests that $f_A \cong 1$ and $f_B \cong [B]/[A] = 0.01$, so that we would expect to see an average signal shifted about 1 Hz (see Eq. 14-18) from the A signal. Since proton NMR line widths are typically 0.5 Hz or less, it would be easy to detect such a frequency shift. The significance of the calculation is that we have obtained information about a *dilute* site (B) by observation of a small change in the NMR frequency for the strong signal from the more *concentrated* site A. Because proton NMR signals are readily obtained at concentrations as low as 10^{-3}M, the above calculation indicates that one might hope to learn about the NMR properties of a macromolecule whose concentration could be as low as 10^{-5}M, a much more realistic concentration from both preparative and physiological standpoints. One of the earliest and most striking applications of this type of arithmetic is based on NMR study of the binding of small sugar inhibitors to an enzyme (lysozyme) that breaks down polysaccharide cell walls (see Fig. 14-26).

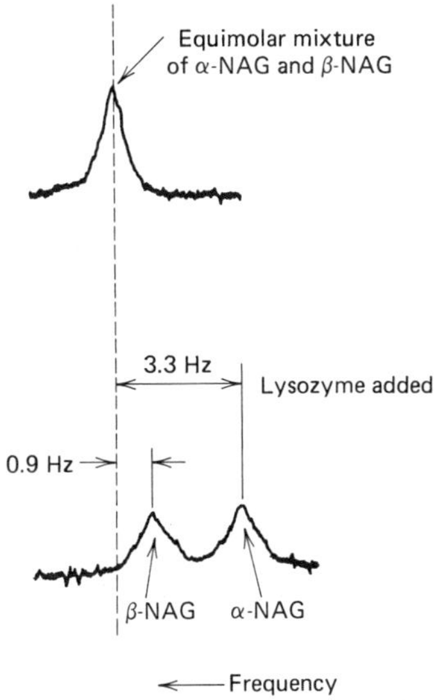

FIGURE 14-26. Proton magnetic resonance spectra of 0.05 M racemic NAG (top spectrum) and of 0.05 M racemic NAG in the presence of 0.003 M lysozyme (bottom spectrum). Both sugar anomers bind to lysozyme in the "fast exchange" NMR limit, and the greatest frequency shift is observed for the α-anomer. (The NMR signal is from the acetamido methyl proton in all cases.) Because of the fast-exchange situation, it is possible to obtain the proton magnetic resonance frequencies for both α-NAG and β-NAG *bound to lysozyme,* with a signal strength associated with free sugar at 0.05 M, even though the enzyme concentration is more than ten times smaller. [Adapted from M. A. Raftery, F. W. Dahlquist, S. I. Chan, and S. M. Parsons, *J. Biol. Chem. 243,* 4175 (1968).]

Lysozyme is an enzyme that breaks the sugar-sugar linkages of cell wall polysaccharides consisting largely of α-linked polymers of N-acetyl-D-glucosamine/(NAG), shown below. For the NAG monomers here, both of the free anomers show the same proton magnetic resonance frequency (top spectrum of Fig. 14-26). However, when lysozyme is added to the solution, there is rapid chemical exchange of both α-NAG and β-NAG with the enzyme, and the resonances of both anomers are shifted (by different amounts) on binding to the enzyme (bottom spectrum, Fig. 14-26).

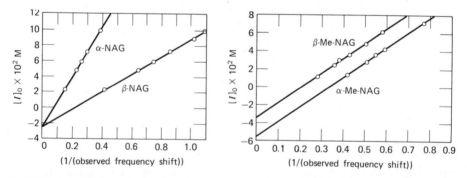

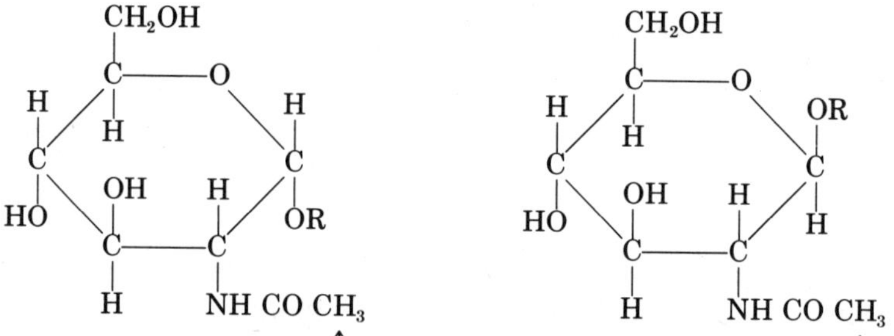

α-NAG: R = H   Proton NMR
Me-α-NAG: R = CH_3 signal chosen for monitoring

β-NAG: R = H   Proton NMR
Me-β-NAG: R = CH_3 signal chosen for monitoring

Using a form of data reduction (Fig. 14-27) common to many sorts of equilibria in which one component is in excess, it is possible to determine the binding constant and "bound" NMR frequency (relative to the free sugar NMR frequency) by plotting $[I]_0$ versus [1/(observed frequency shift at that concentra-

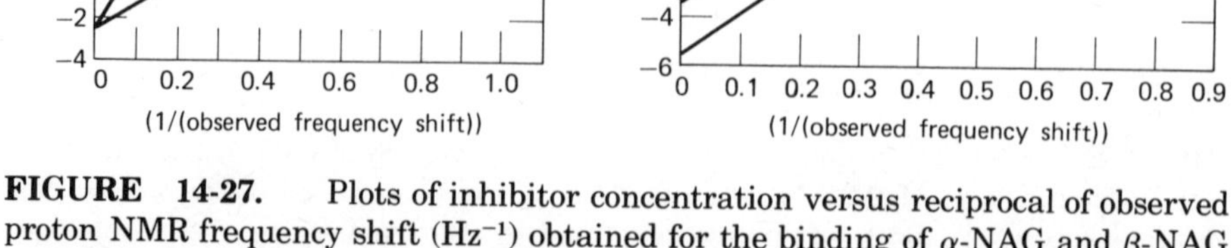

FIGURE 14-27. Plots of inhibitor concentration versus reciprocal of observed proton NMR frequency shift (Hz^{-1}) obtained for the binding of α-NAG and β-NAG to lysozyme (left-hand plot) or binding of α-Me-NAG and β-Me-NAG to lysozyme (right-hand plot). The *y-intercept* of the plot gives the dissociation constant for the enzyme:inhibitor complex, and the *slope* is proportional to the "bound" NMR frequency shift for the enzyme:inhibitor complex (see Problems); see text for discussion. [Data for α-NAG and β-NAG from F. W. Dahlquist and M. A. Raftery, *Biochemistry 7,* 3269 (1968); data for α-Me-NAG and β-Me-NAG from M. A. Raftery, F. W. Dahlquist, S. I. Chan, and S. M. Parsons, *J. Biol. Chem. 243,* 4175 (1968).]

tion)] for a series of experiments at constant enzyme concentration, $[E]_0$, and varying inhibitor concentration, $[I]_0$ (see Problems). Results for two pairs of inhibitors are shown in Fig. 14-27.

A number of conclusions follow from the NMR data plotted in Fig. 14-27. First, although α-NAG and β-NAG have similar binding constants for binding to lysozyme, the α-anomer shows the much larger "bound" frequency shift (0.68 ppm versus 0.51 ppm for the β-anomer). It seems highly likely that this frequency shift arises from proximity of the acetamido methyl group to the aromatic ring of tryptophan #108 in the amino acid sequence of lysozyme, so the acetamido methyl group of the α-anomer must lie closer to the TRY 108 than that of the β-anomer. (Independent experiments have shown that α-NAG and β-NAG compete for the same binding site.) In contrast, the "bound" frequency shifts for both α-Me-NAG and β-Me-NAG are the same, even though their binding constants are different. It would thus appear that both the substituent and linkage at the anomeric position affect the binding of monosaccharides by lysozyme, but by different mechanisms.

All examples so far in this section are drawn from *nuclear* magnetic resonance; however, Table 14-3 shows that an unpaired *electron* also has a magnetic moment and thus exhibits magnetic resonance in an applied static magnetic field, when driven at its precession frequency in the microwave range. As we see in Chapter 19, the *intensity* of electron magnetic resonance absorption [usually called "electron spin resonance" (ESR) or "electron paramagnetic resonance" (EPR)] is several orders of magnitude stronger than for nuclear magnetic resonance, so that ESR signals are observed without difficulty at concentrations down to 10^{-6}M. Another difference between ESR and NMR is in the nature of the spectral display: because ESR spectra are typically obtained by low-amplitude low-frequency modulation (see Chapter 17.C for a related case), the experimentally observed absorption line shape appears as the *first derivative* (with respect to frequency) of the usual Lorentzian shape of Fig. 13-5, as shown in Fig. 14-28.

An additional feature of ESR spectra is that there may be more than one absorption peak for a single unpaired electron – this aspect has a simple quantum mechanical explanation (Chapter 18A) for which there is also a

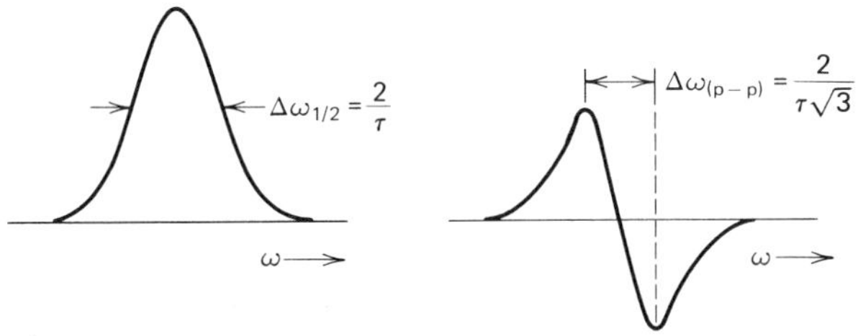

FIGURE 14-28. Lorentzian spectral absorption line shape (left), and first derivative of Lorentzian line shape (right). ESR spectra are typically reported as the first derivative of absorption line shape. (Compare with Fig. 13.5.)

classical analogy (Chapter 17.A), and is not discussed further here. The majority of biochemical applications for ESR have evolved from the family of compounds called nitroxides, RN $\dot{\rightarrow}$ O, in which the dot over the NO bond indicates the presence of an unpaired electron whose ESR absorption is to be observed. The nitroxide electron is magnetically anisotropic in that there is a different absorption of energy when the N—O bond is aligned parallel than when the bond is perpendicular to the applied static magnetic field, H_0. Thus, it is possible to determine the direction and extent of *linear order* in nitroxide-labeled solids by observing a difference in ESR absorption for different sample orientations, just as we have already used dichroism and birefringence to the same end.

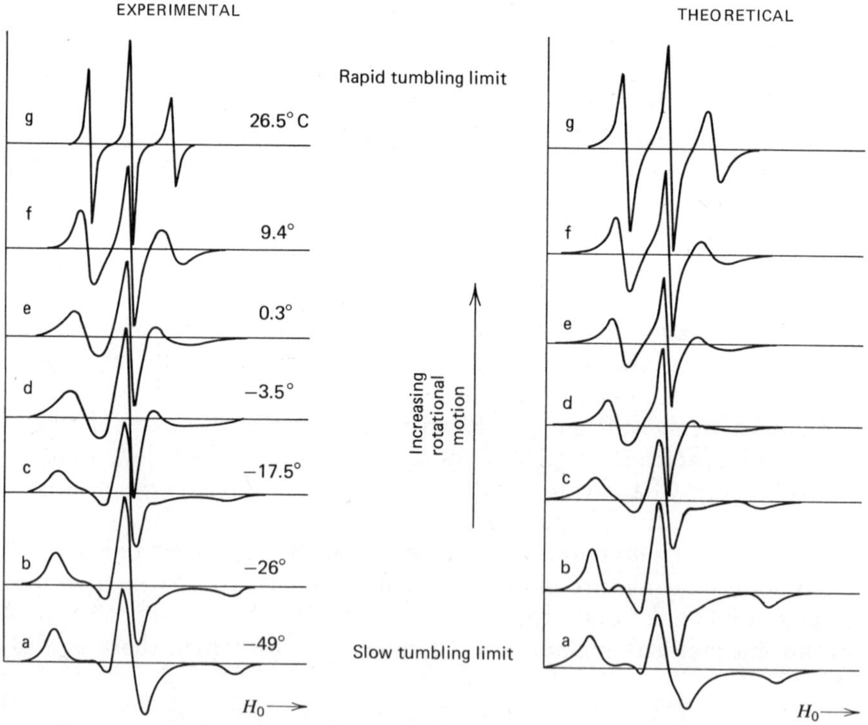

FIGURE 14-29. Experimental (left) and theoretical (right) ESR spectra for a di-tertiary butyl nitroxide derivative in glycerol, for a variety of rotational motional rates. At −49°C, the nitroxide solution is essentially frozen, and the spectrum represents a sum of the individual ESR spectra corresponding to many different randomly oriented molecules. At 26.5°C, the (much simpler) spectrum represents an average result, since each nitroxide molecule samples many rotational positions in a time that is short compared with the time it would take to distinguish between the corresponding different ESR spectral frequencies. [Experimental data from P. Jost, A. S. Waggoner, and O. H. Griffith, in *Structure and Function of Biological Membranes,* ed. L. Rothfield, Academic Press, N. Y. (1971), p. 83; theoretical plots from R. G. Gordon and T. Messenger, in *Electron Spin Relaxation in Liquids,* ed. L. T. Muus and P. W. Atkins, Plenum Press, N. Y. (1972), pp. 371–72.]

Since ESR line shape changes as the molecule containing, say, the N—O bond is rotated, it might be anticipated that ESR line shape should be sensitive to the *rate of rotational reorientation* of nitroxide-labeled (so-called "spin-labeled") molecules in solution. From our previous analysis of fast exchange in NMR spectra, we might expect that for very slow rates of rotational motion (slow compared to the difference in ESR spectral absorption peak positions for different angles between the N—O bond and the magnetic field), the ERR spectrum should represent a *sum* of all the individual spectra corresponding to all possible molecular orientations; on the other hand, for rapid molecular tumbling, we would expect some sort of *averaging* of all the possible ESR peak positions to give a much simpler spectrum. Although the calculations in this case are well beyond the scope of this book (and as yet imperfect in any case), the general features of the results are apparent in Fig. 14-29, showing ESR line shapes for spin-labeled molecules tumbling at various rates in solution. The effect of rotational motion (especially rotational diffusion) is treated later in Chapter 21 for some cases that are computationally simpler than the ESR problem.

EXAMPLE *Fluidity in the Interior of Biological Membranes*

Nitroxide spin-labeled analogs of stearic acid (see structures below) have recently been synthesized and introduced into erythrocyte ghost membranes (red blood cells with the hemoglobin removed). Comparison of the two ESR spectra in Fig. 14-30 with those in Fig. 14-29 shows that when the nitroxide label is near the polar "head group" of the molecular chain, the label is fully immobilized, but when the label is located at the other (nonpolar) end of the

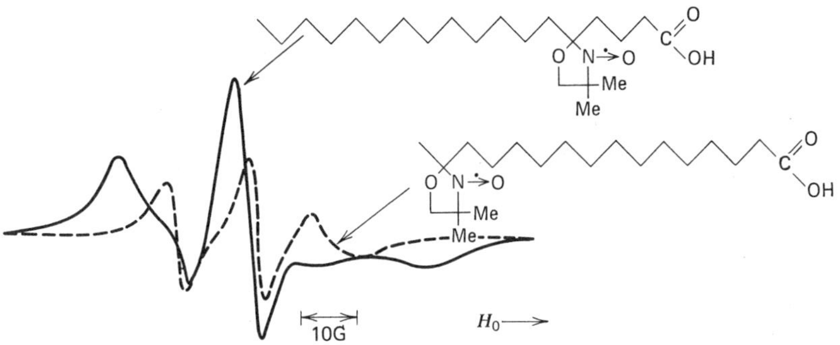

FIGURE 14-30. ESR spectra for two spin-labeled derivatives of stearic acid, which have been incorporated (separately) into erythrocyte ghost membranes. The spectrum for the derivative labeled near the carboxyl group resembles the "polycrystalline" ("rigid glass," "powder") spectrum characteristic of highly immobilized nitroxide groups (see bottom plots of Fig. 14-29), while the spectrum for the derivative labeled at the other end of the hydrocarbon chain resembles that of a nitroxide with a high degree of rotational freedom (compare with upper plots of Fig. 14-29). [Adapted from C. F. Chignell, in "*Aldrichimica Acta*" *7,* 1 (1974), from Aldrich Chemical Company, 940 W. St. Paul Ave., Milwaukee, Wisc.]

chain, there is a high degree of rotational freedom. A series of such experiments have produced a number of interesting conclusions about the structure and function of biological membranes. First, it has been shown that the relatively rigid order in the membrane persists from the surface down through about eight carbons of the chain into the membrane interior. Second, various drugs such as general anesthetics act to increase the fluidity within the membrane bilayer, while other agents such as DDT and cholesterol act to decrease membrane interior fluidity. Perhaps most interesting is that the functions of biological membranes (such as transport of glucose) have now been shown to correlate closely with the relative membrane fluidity. Membranes can be treated as multi-phase systems (see Chapter 2.A.2), and there are pronounced changes in transmembrane transport rates of, say, glucose at the temperature at which the membrane undergoes a phase transition from "solid" (very restricted rotational mobility) to "solid" plus "fluid" (patches of the membrane exhibit a large degree of rotational freedom), as deduced from the appearance of ESR spectra of spin-labeled phospholipid bilayer vesicles.

EXAMPLE *ESR Spin-labels as a Molecular Dipstick*

Carbonic anhydrase is an enzyme whose active site is a deep crevice, at the bottom of which is a single zinc atom. An aromatic sulfonamide inhibitor will bind to the active site by coordinating directly to the zinc atom. By comparing the ESR spectra for a series of spin-labeled sulfonamide derivatives (see Fig. 14.31) having varying distance, *d*, between the sulfonamide site of attachment to the enzyme and the pyrrolidine ring of the spin-label, it was found that the rotational mobility of the nitroxide group could be used to estimate the dimensions of the active site. Assuming a funnel-shaped active site, the

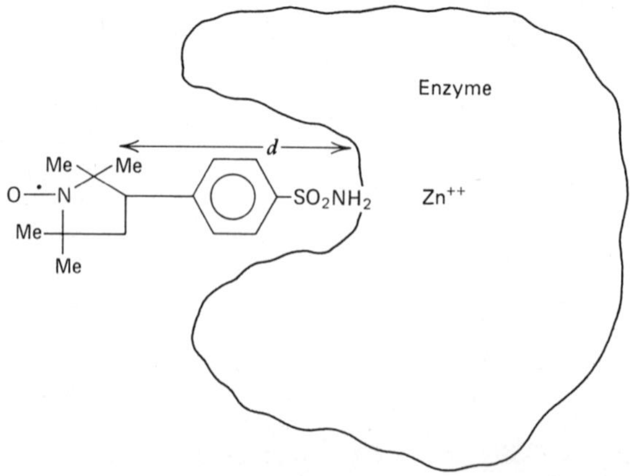

FIGURE 14-31. Schematic diagram of the binding of a sulfonamide spin-label to the active site of the enzyme, carbonic anhydrase. As the length of the "molecular dipstick" is increased, there is increased rotational mobility as the nitroxide moiety begins to protrude from the enzyme surface. [Based on experiments of R. H. Erlich, D. K. Starkweather, and C. F. Chignell *Mol. Pharmacol. 9*, 61 (1973).]

active site of carbonic anhydrase C from human erythrocytes was found to be about 14 Å deep.

EXAMPLE *ESR Spin-labels in a Test for Morphine Addiction*

The mammalian immune system will readily manufacture antibodies against macromolecular antigens. This property can be used to produce an antibody against morphine, where the morphine is first coupled to bovine serum albumin protein (BSA) and then given to rabbits (see structures below). Next, a spin-labeled morphine derivative is prepared. When the spin-labeled morphine binds to the anti-morphine antibody, there is a high degree of immobilization with a broad and asymmetric ESR spectrum. When morphine itself is added, it displaces the spin-labeled morphine, whose ESR spectrum promptly reverts to the sharp symmetric three-line pattern of a small molecule in solution (top plot of Fig. 14-29). The concentration of free spin-labeled morphine is readily determined from the ESR absorption peak height of the low-field peak, and the same experiment may now be used to determine the concentration of morphine in any biological sample (such as the blood of a heroin addict). Since morphine substitutes, such as methadone and propoxyphene (as well as unrelated drugs such as barbiturates or amphetamines) are not recognized by the anti-morphine antibody, the test is highly specific, simple, and cheap, and has been used on a wide scale.*

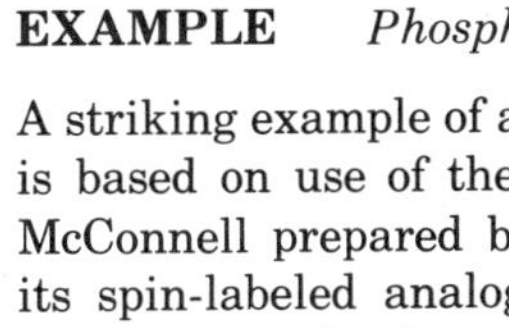

R = H (Morphine)
R = —CH_2CO—BSA (Morphine coupled to bovine serum albumin)
R = —CH_2CO

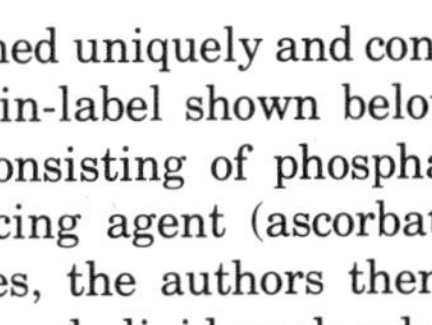

(Spin-labeled morphine)

EXAMPLE *Phospholipid Flip-Flop in Membrane Vesicles*

A striking example of a conclusion reached uniquely and convincingly by ESR is based on use of the phospholipid spin-label shown below. Kornberg and McConnell prepared bilayer vesicles consisting of phosphatidylcholine and its spin-labeled analog. Using a reducing agent (ascorbate) that does not penetrate to the interior of the vesicles, the authors then reduced the nitroxide groups of the spin-labeled phospholipid molecules located in the outer layer of the vesicle, resulting in a substantial diminution in the ESR absorption signal because the reduced nitroxide moiety is no longer paramagnetic (there is no unpaired electron). Beginning at this stage, the ESR peak height was monitored as a function of time, and observed to decrease with a half-life of about 6.5 hr at 30°C. A series of control experiments demonstrated that the cause of this decrease was a slow "flip-flop" of spin-labeled

* *For more details, see R. K. Leute, E. F. Ullman, A. Goldstein, and L. A. Herzenberg,* Nature New Biol. 236, *93 (1972).*

phospholipid from the *inner* layer of the vesicle (where the nitroxide group still had an unpaired electron and thus an ESR signal) to the *outer* layer (where the nitroxide group was promptly reduced by ascorbate in the surrounding solution. One possible mechanism for the "flip-flop" is shown at the right in Fig. 14-32, in which there is a simultaneous movement of one phospholipid molecule from the outer to the inner layer while the other moves from the inner to the outer layer of the vesicle. This experiment represents the first direct determination of phospholipid flip-flop, and it is of further interest to note that the rate at which a phospholipid molecule flips back and forth between the inner and outer layers of the vesicle is large enough to account for the (separately observed) chloride transport across similar vesicles, suggesting that the flip-flop may be the mechanism for passive transport of at least some ions across some biological membranes.

Spin-labeled phosphatidylcholine:

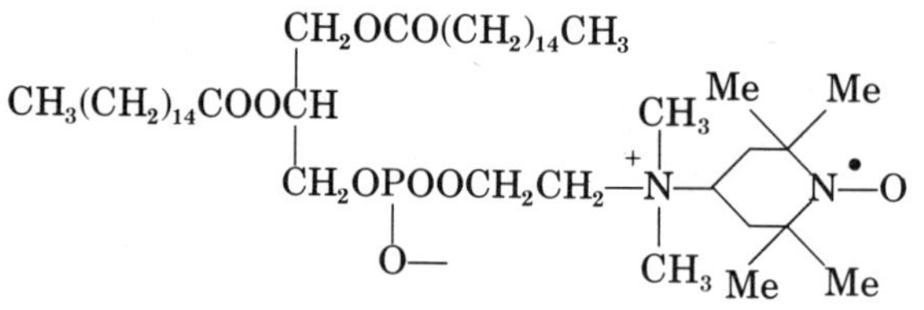

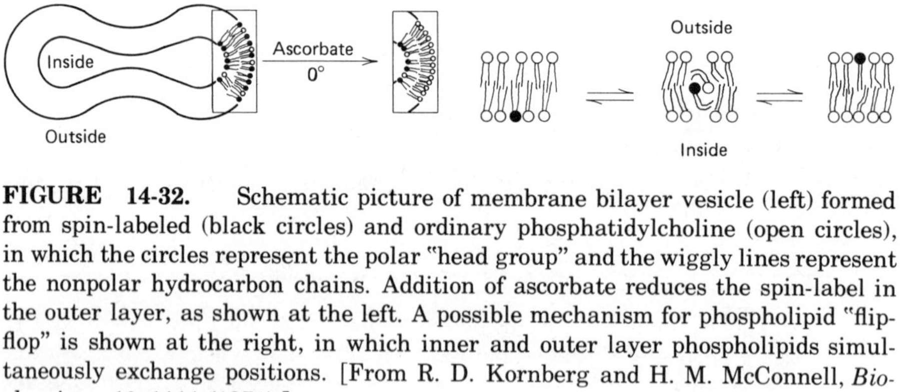

FIGURE 14-32. Schematic picture of membrane bilayer vesicle (left) formed from spin-labeled (black circles) and ordinary phosphatidylcholine (open circles), in which the circles represent the polar "head group" and the wiggly lines represent the nonpolar hydrocarbon chains. Addition of ascorbate reduces the spin-label in the outer layer, as shown at the left. A possible mechanism for phospholipid "flip-flop" is shown at the right, in which inner and outer layer phospholipids simultaneously exchange positions. [From R. D. Kornberg and H. M. McConnell, *Biochemistry 10,* 1111 (1971).]

14.E. ELECTRONIC CIRCUITS AS SPRING MODELS: CAPACITANCE, RESISTANCE, INDUCTANCE, RESONANCE, AND RELAXATION

Historically, the behavior of electronic circuits was understood very quickly, once it was appreciated that the flow of electrical current is mathematically very similar to the flow of water. Electrical *voltage* is analogous to mechanical *pressure.* The accumulation of charge on the two plates of a capacitor results in a voltage, V, proportional to the amount of charge, q, just as the accumulation of water behind a dam leads to a hydrostatic pressure proportional to the amount of water behind the dam. Since the area (for example) of both capacitors and reservoirs can vary, the proportionality constant (known as $(1/C)$, where C is called *capacitance* in the electrical case) is a property of the particular capacitor (reservoir):

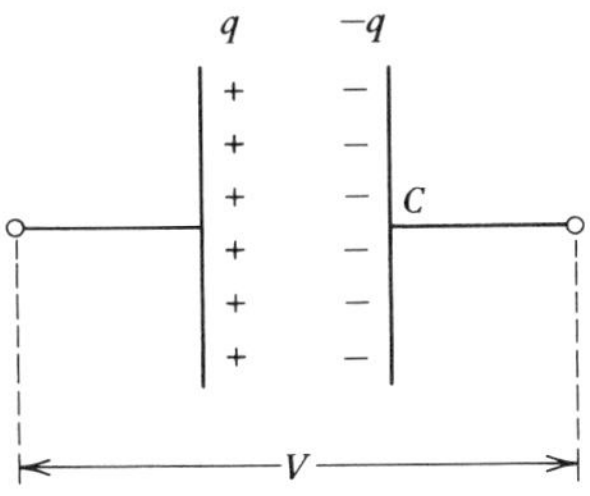

$$V = q/C \tag{14-19}$$

Similarly, just as flow of water through a pipe is proportional to the applied hydrostatic pressure and inversely proportional to the resistance of the pipe, flow of electrical *current*, $I = dq/dt$, is proportional to the applied voltage, and inversely proportional to the *resistance* of the circuit, in the relation usually written in the form (Ohm's law):

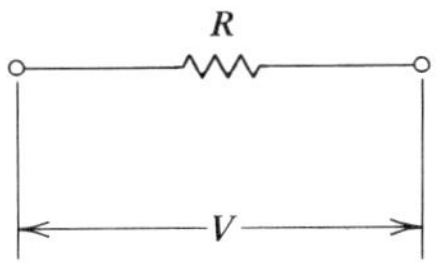

$$V = RI = R(dq/dt) \tag{14-20}$$

Examination of Equations 14-19 and 14-20 shows that they are mathematically very similar to our previous equation for the motion of a weight on a damped spring, Eq. 13-15, provided that we identify spring position x with charge q, velocity dx/dt with electrical current dq/dt, frictional coefficient f with resistance R, and spring constant k with reciprocal capacitance $(1/C)$. Surprisingly, there exists a third circuit element, *inductance* L, which is analogous to mass in the mechanical case (see Table 14-4). An inductor is a coil that resists *changes* in current with a proportionality constant, L:

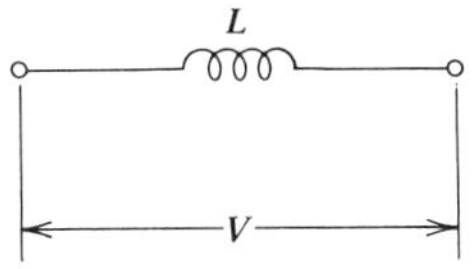

$$V = L(dI/dt) = L(d^2q/dt^2) \tag{14-21}$$

The equation for a circuit containing all three elements subject to a time-varying voltage, $V(t)$, is thus entirely analogous to the equation for a damped weight on a spring driven by a time-varying force, $F(t)$:

Table 14-4 Comparison of properties of mechanical and electrical oscillators

General Characteristic	Weight on a Spring	Electrical Circuit
Independent variable	time, t	time, t
Dependent variable	position, x	charge, q
Inertia	mass, m	inductance, L
Resistance	friction coefficient, f	resistance, R
Stiffness	spring constant, k	$(\text{capacitance})^{-1}$, $(1/C)$
Resonant frequency	$\omega_0^2 = k/m$	$\omega_0^2 = 1/LC$
Time constant for return to equilibrium after driving force is removed (RC circuit)	$\tau = f/k$	$\tau = RC$
Time constant for LR or LRC circuit[a]	$\tau = 2\,m/f$	$\tau = 2L/R$
Time-varying driver	$F = F_0 \cos(\omega t)$	$V = V_0 \cos(\omega t)$

[a] *See section 14.F.*

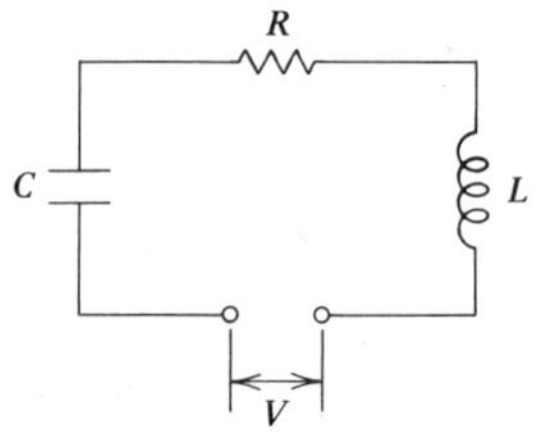

$$L(d^2q/dt^2) + R(dq/dt) + (1/C)q = V(t) \quad (14\text{-}22)$$

$$m(d^2x/dt^2) + f(dx/dt) + kx = F(t) \quad (13\text{-}15)$$

Since the mechanical and electrical differential equations are of the same form, we expect to observe oscillations in electrical charge just as we saw oscillations in mechanical spring position, on application of a steady sinusoidal driving voltage. By analogy to Eq. 13-20, we can write the solution at once for the electrical oscillation:

$$q = \frac{V_0}{[(1/C) - \omega^2 L] + iR\omega} \quad (14\text{-}23)$$

or

$$q = \frac{V_0}{L(\omega_0^2 - \omega^2) + iR\omega} \qquad \omega_0^2 = (1/LC) \quad (14\text{-}24)$$

where complex notation is used for compact display. As with the mechanical case, it is possible to separate the charge response into components that vary either in-phase (dispersion) or 90°-out-of-phase (absorption) with regard to the driving sinusoidal voltage. Electrical circuits thus provide a convenient *model* for all the phenomena discussed in this chapter, as well as an inexpensive lecture or laboratory demonstration of absorption and dispersion line shape. We press this analogy further in Chapter 16.A.2.

Zero-mass Limit

A particularly interesting limit for Equations 13-15 and 14-22 occurs when the mass (inductance) goes to zero. Rewriting those equations in the form

$$\tau \frac{dx}{dt} + x = \frac{F_0}{k} \cos(\omega t); \; \tau = \frac{f}{k} \tag{13-25}$$

$$\tau \frac{dq}{dt} + q = CV_0 \cos(\omega t); \; \tau = RC \tag{14-26}$$

the steady-state solutions can be written (see Table 13-4):

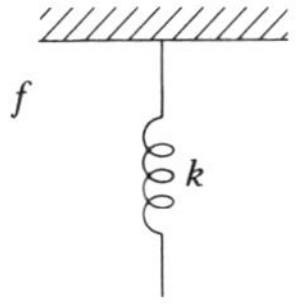

$$x = x' \cos(\omega t) + x'' \sin(\omega t) \tag{14-27a}$$

$$x' = \frac{F_0}{k} \frac{1}{1 + \omega^2 \tau^2} \tag{14-27b}$$

$$x'' = \frac{F_0}{k} \frac{\omega\tau}{1 + \omega^2 \tau^2} \tag{14-27c}$$

$\tau = \frac{f}{k}$ (Massless weight on a spring)

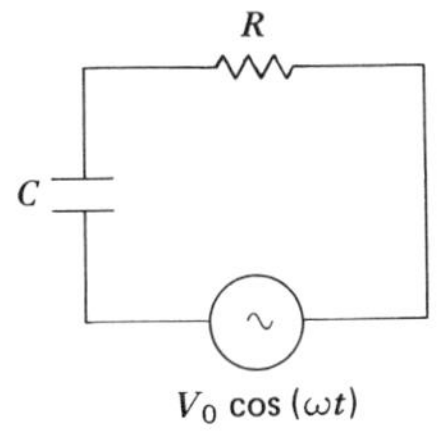

$$q = q' \cos \omega t + q'' \sin \omega t \tag{14-28a}$$

$$q' = CV_0 \frac{1}{1 + \omega^2 \tau^2} \tag{14-28b}$$

$$q'' = CV_0 \frac{\omega\tau}{1 + \omega^2 \tau^2} \tag{14-28c}$$

$\tau = RC$ (Inductorless RC circuit)

Equation 14-28 is of immediate use in Chapter 14.F (dielectric relaxation), and Eq. 14-27 in Chapter 14.G (ultrasonic relaxation).

As a final point on terminology, "relaxation" phenomena (as the phrase is currently used in the literature) are of two types: (1) experiments involving the *steady-state* response of a sinusoidally *driven, massless* weight on a spring (e.g., dielectric and ultrasonic relaxation), and (2) experiments involving the *transient* response of an *undriven* weight of *finite* (e.g., nuclear magnetic relaxation; fluorescence depolarization; gamma-ray correlations) or *zero* mass (e.g., transient chemical reaction kinetics) on a spring. Dielectric and ultrasonic relaxation are treated in the following two sections of this chapter; transient phenomena are discussed in Chapter 16.

14.F. DIELECTRIC RELAXATION: ZERO MASS ON A SPRING. ROTATIONAL DIFFUSION OF MACROMOLECULES.

In this section, we begin by considering a very simple circuit consisting of a capacitor across which a static voltage is applied. When the capacitor is empty (left-most plot of Fig. 14-33), the capacitance of the capacitor is denoted as C_0. However, when the space between the capacitor plates is filled with polar molecules (i.e., with electric dipoles), application of a voltage to the capacitor builds up charge on the plates as before, but the charged plates produce an electric field that generates a force tending to align the dipoles along a direction connecting the two plates (middle plot of Fig. 14-33). [Because random thermal motion tends to make the molecules tumble about in random directions, the alignment produced by even very large applied voltages is far from perfect (compare middle and right-most plots of Fig. 14-33).] The important point is that because the polar molecules are partially lined up between the plates, part of the charge built up on the capacitor is effectively neutralized by the aligned dipoles, so that *additional* charge can be built up on the plates to give the same *net* charge as obtained for the empty capacitor (compare plate charge in left-most and middle pictures of Fig. 14-33). One way of describing this situation is to say that the effective *capacitance* of the capacitor increases when polar molecules are

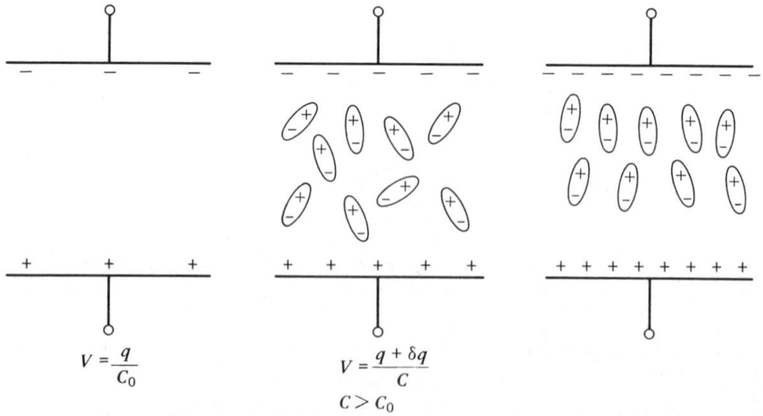

FIGURE 14-33. Three schematic drawings of the charge on the same capacitor when empty (left), and when filled with polar molecules (middle and right). The applied voltage is the same in the left and middle plots, and larger in the right-most plot. Because the polar molecules tend to line up between the plates when the voltage is applied, an additional charge δq may be built up on the same capacitor when filled with polar molecules, since some of the charge on the capacitor is effectively neutralized by the aligned dipoles (see text). Dielectric constant is defined by Eq. 14-29.

placed between the plates: *dielectric constant,* ϵ', is the proportionality constant of this relation:

$$\epsilon' = C/C_0 \qquad (14\text{-}29)$$

in which C and C_0 are the capacitances of the capacitor in the presence and absence of polar filling material.

The increased charge on the filled (compared to the empty) capacitor is related to the square of the dipole moments of the individual polar molecules

$$\epsilon' - 1 = \frac{4\pi N p_0^2}{3kT} \qquad (14\text{-}30)^*$$

in which p_0 is the electric dipole moment of a given polar molecule, N is the number of molecules per unit volume, k is the gas constant per molecule, and T is absolute temperature. The first result from our treatment is thus that the *electric dipole moment* of a polar molecule may be determined from the dielectric constant (i.e., ratio of filled to empty capacitance) of a capacitor filled with the polar molecules.

Equation 14-30 describes the dielectric constant for an applied *static* voltage. If we now apply a *sinusoidally time-varying* voltage, $V_0 \cos(\omega t)$, the dipoles will be forced to line up with their positive ends pointed first toward one capacitor plate and then toward the other, as the voltage across the capacitor changes sign with time. There are two important aspects to this new situation. First, it is necessary to overcome *rotational frictional resistance* in turning the molecules around, so that an alternative mathematical description to a circuit consisting of a filled capacitor and an applied voltage is a circuit consisting of a fixed-capacitance capacitor and a resistor (see Fig. 14-34). Second, we already know the behavior of a circuit consisting of a capacitor and resistor and applied voltage (see preceding section of this chapter), which may be described by in-phase and 90°-out-of-phase charge, as shown in Eq. 14-28. The time constant, $\tau = RC$, for this "equivalent circuit" description of the actual filled capacitor and voltage, is a measure of the time it takes for a given average polar molecule to reorient in direction (via rotational diffusion) of the order of one radian in angle. Experimentally, the in-phase and 90°-out-of-phase components of ϵ, namely ϵ' ("dielectric constant") and ϵ'' ("dielectric loss") in

$$\epsilon = \epsilon' \cos(\omega t) + \epsilon'' \sin(\omega t) \qquad (14\text{-}31)$$

* *To be more accurate, the* relative *dielectric constant of Eq. 14-29 is unitless for cgs calculations and is taken equal to unity in a vacuum. The* absolute *value of the dielectric constant for free space is* 8.84×10^{-14} *F/cm. See any text on electricity and magnetism for more details.*

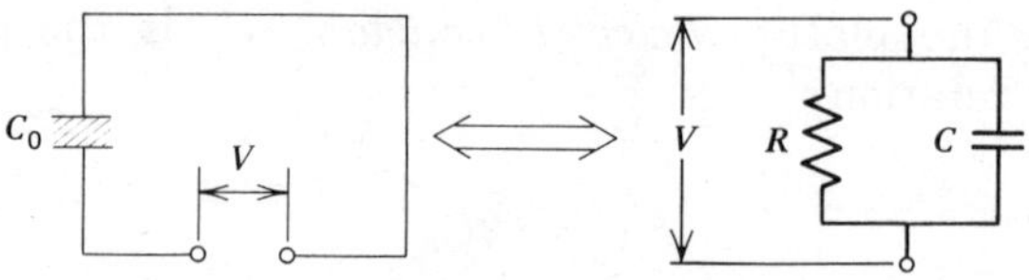

FIGURE 14-34. Electric circuit for a voltage applied to a filled capacitor (left) whose capacitance varies with the frequency of the applied time-varying voltage, V; and equivalent circuit (right) whose capacitance and resistance satisfy the relation, $C\,R = \tau$, in which τ is a time constant approximately equal to the time required for an average dipolar molecule in the filled capacitor to reorient diffusionally by of the order of one radian in angle. If the applied oscillating electric field (corresponding to the applied oscillating voltage) is written as, $E_0 \cos(\omega t)$, then the total electric field in the filled capacitor can be expressed as, $\epsilon\, E_0$, in which $\epsilon = \epsilon' \cos(\omega t) + \epsilon'' \sin(\omega t)$ (Eq. 14-31). ϵ' and ϵ'' are determined by measuring the apparent capacitance, C, and apparent resistance, R, of the circuit (right-hand diagram), according to:
ϵ' = dielectric constant = C/C_0,
ϵ'' = dielectric loss = $1/(R\, C_o\, \omega)$,
where C_0 is the capacitance of the empty capacitor, and ω is the (angular) frequency of the sinusoidally oscillating applied voltage.

[corresponding to q' and q'' in Eq. 14-28a according to Equations 14-28b,c (see Problems)] are determined from measurement of capacitance and conductivity (see Fig. 14-34).

The expressions for ϵ' and ϵ'' of Eq. 14-31 are usually expressed in the form

$$\epsilon' = \epsilon_\infty + \frac{\epsilon_0 - \epsilon_\infty}{1 + \omega^2\tau^2} \qquad (14\text{-}32a)$$

and

$$\epsilon'' = \frac{(\epsilon_0 - \epsilon_\infty)\,\omega\tau}{1 + \omega^2\tau^2} \qquad (14\text{-}32b)$$

in which ϵ_0 and ϵ_∞ are defined graphically in Fig. 14-35, which represents the behavior of a solution consisting of polar macromolecules dissolved in water. Using an argument again based on classical rotational diffusion (see Chapter 21.B) it can be shown that the "relaxation time," or "time constant," $\tau = RC$ for the equivalent circuit that led to Eq. 14-32 is given by

$$\tau = \frac{4\pi\eta a^3}{kT} \qquad (14\text{-}33)$$

in which η is solution viscosity, a is the radius of the polar molecule of interest, k is gas constant per molecule, and T is absolute temperature. For

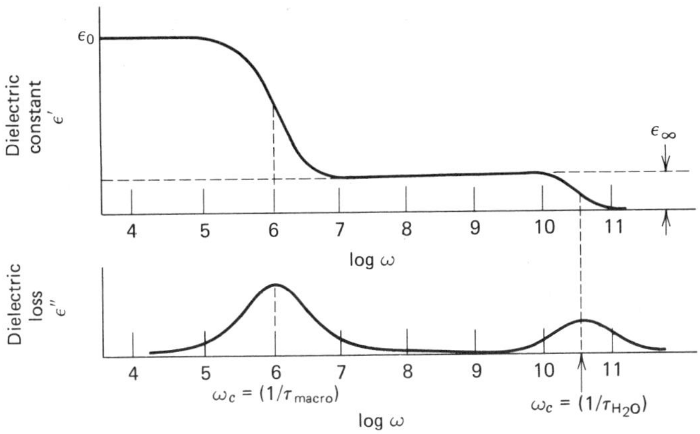

FIGURE 14-35. Frequency dependence of dielectric constant ϵ' (upper plot) and dielectric loss ϵ'' (lower plot), with frequency plotted on a log scale. The left-hand portions of each curve are plotted using Eq. 14-32, and correspond to experiments involving a macromolecule dissolved in water; the "critical" frequency ω_c corresponding to the half-way point in ϵ' versus ω (and the maximum value of ϵ'' versus ω) is the reciprocal of the dielectric "relaxation time," τ_{macro}, for the macromolecule to reorient rotationally of the order of 1 radian in angular distance. The (unobserved) critical frequency and dielectric relaxation time for the solvent, water, are manifested in similar behavior at higher frequency (since water molecules can reorient much faster than macromolecules) at the right-hand portion of the plots. Experiments are usually conducted at frequencies below about 200 MHz.

water itself, $\eta = 0.01$ in cgs units, and we may take $a = 2 \times 10^{-8}$ cm, so that Eq. 14-33 predicts a relaxation time for reorientation of water molecules of about $\tau = 2.4 \times 10^{-11}$ sec. This relaxation time would produce a step in the plot of dielectric constant, ϵ', versus frequency at a frequency of about $\nu = (1/2\pi\tau) = 6.6 \times 10^9$ Hz, which is well above the experimentally observed frequencies that can conveniently be transmitted through wires. Because the experimental frequency range thus falls short of the cut-off in ϵ' versus ν, there will always be an apparent constant contribution, ϵ_∞, to the observed dielectric constant, ϵ', as included explicitly in Eq. 14-32. On the other hand, for a macromolecule, such as chymotrypsin with a molecular radius of about 20 Å, we expect $\tau = 2.4 \times 10^{-8}$ sec, corresponding to a frequency of about 6.6 MHz, which is readily observed electronically, so that we would expect to be able to follow the entire critical frequency range in the vicinity of $\nu = (1/2\pi\tau)$ as shown in Fig. 14-35.

Dielectric relaxation time, τ, is a convenient measure of molecular size, as indicated by Eq. 14-33, and is readily determined experimentally from plots of the type shown in Fig. 14-35. Further information can be obtained from such plots, as seen in Fig. 14-36, which shows the theoretical behavior of ϵ' versus frequency for macromolecules of increasingly elongated shape.

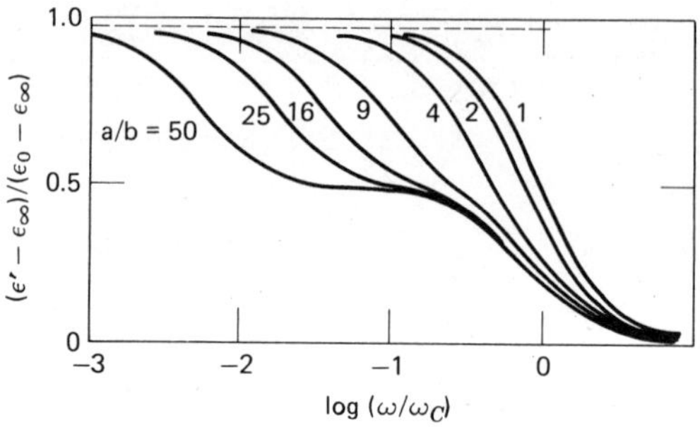

FIGURE 14-36. Theoretical plots of the frequency-dependence of dielectric constant, ϵ', for prolate ellipsoids of various axial ratio, a/b. (a/b is the ratio of the long to the short molecular axis as shown in Fig. 7-22.) The smooth curve obtained for a sphere $[(a/b) = 1]$ begins to break into two "steps" as the axial ratio is increased. The shape of the frequency-dependence of ϵ' thus provides an indication of the asymmetry in macromolecular shape. (From J. L. Oncley, in *Proteins, Amino Acids, and Peptides,* ed. E. J. Cohn and J. T. Edsall, Reinhold, New York, 1943, p. 543.)

EXAMPLE *Macromolecular Size and Shape from Dielectric Relaxation*

Table 14-5 is a collection of dielectric relaxation times and axial ratios derived from fitting experimental plots of ϵ' versus frequency to curves of the type shown in Fig. 14-36.

Table 14-5 Dielectric relaxation time(s) and axial ratios for various proteins, determined by fitting experimental data to plots of the type shown in Fig. 14-36.

Protein	M.W. ($\times 10^3$)	$\tau \times 10^8$ (sec)	a/b
Ovalbumin	45.0	18; 4.7	5
Horse serum albumin	70.0	36; 7.5	6
Horse carboxyhemoglobin	67.0	8.4	1.6
Horse serum pseudoglobulin	142.0	250; 28	9
Insulin	40.0	1.6	
β-Lactoglobulin	40.0	15; 5.1	4
Myoglobin	17.0	2.9	

From J. E. Oncley, in Proteins, Amino Acids, and Peptides, *ed. E. J. Cohn and J. T. Edsall, Reinhold, New York (1943), p. 543.*

EXAMPLE *Cole-Cole Plot for Myoglobin*

Close examination of Fig. 14-36 shows that there would be some difficulty in determining small deviations from spherical shape $[(a/b) = 5$ or less] from a direct fit of experimental data to a theoretical plot of ϵ' versus log(frequency).

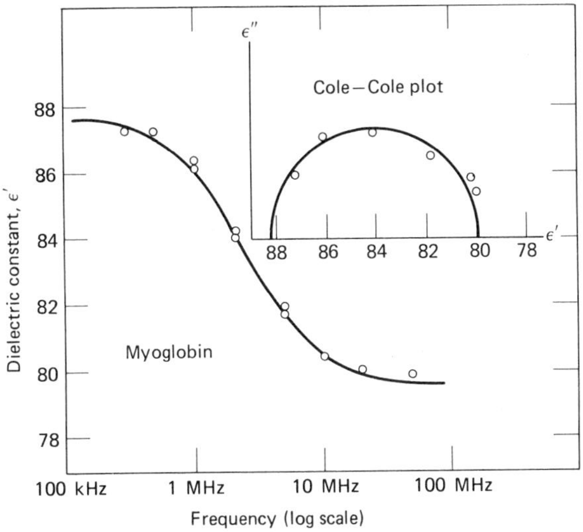

FIGURE 14-37. Dielectric dispersion (i.e., plot of dielectric constant ϵ' versus frequency) for a myoglobin solution. The solid line is calculated using Eq. 14-32. The drawing at the upper right corner is a Cole-Cole plot (see text). (From S. Takashima, in *Physical Principles and Techniques of Protein Chemistry, Part A,* ed. S. J. Leach, Academic Press, New York, 1969, p. 313.)

A quick calculation (see Problems) shows that one of the properties of the quantities ϵ'' and ϵ' is that

$$[\epsilon' - (1/2)(\epsilon_0 + \epsilon_\infty)]^2 + \epsilon''^2 = [(1/2)(\epsilon_0 - \epsilon_\infty)]^2 \tag{14-34}$$

which simply states that a plot of ϵ'' versus ϵ' will produce a circle of radius $(1/2)(\epsilon_0 - \epsilon_\infty)$, with center on the x axis at $\epsilon'' = (1/2)(\epsilon_0 + \epsilon_\infty)$, *provided* that the system is characterized by a single dielectric relaxation time, τ. If the system is characterized by two or more different relaxation times, the plot will still generally approximate a circle, but the center will be displaced below the ϵ'' axis. This so-called Cole-Cole plot turns out to be a rather sensitive and useful test for the presence of multiple relaxation times, such as might be expected for molecules of nonspherical shape. An example of its use is shown in Fig. 14-37 for dielectric relaxation of an aqueous solution of myoglobin. The accurate fit of the direct plot of ϵ' versus log(frequency) to a theoretical equation (Eq. 14-32) having a single relaxation time, τ, is confirmed by the Cole-Cole plot that shows a semicircle centered on the x axis. This graph provides perhaps the most direct evidence that myoglobin in solution has very nearly spherical shape.

EXAMPLE *Dispersion-versus-Absorption Plots in Spectroscopy*

Equation 14-34 is a special case (zero-mass limit) of a general plot of the in-phase component versus the 90°-out-of-phase component of the steady-state

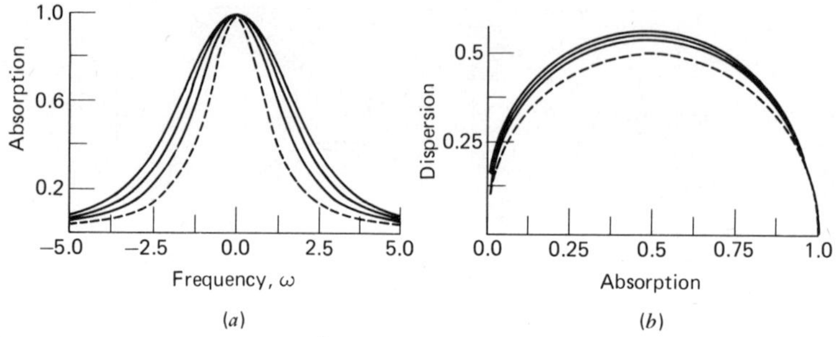

FIGURE 14-38. Theoretical plots of normalized absorption versus frequency (*a*) and dispersion versus absorption (*b*), for a single Lorentzian line shape (Eq. 13-36, dotted curves in the present figure), and for a superposition of Lorentzian lines whose resonant frequencies (absorption "peak" positions) are described by a Gaussian distribution (solid curves). In both (*a*) and (*b*), the outermost curves correspond to the broadest distribution in ω_0. Frequency in (*a*) is in units of $(1/\tau)$, where $(1/\tau)$ is the half-width of any single Lorentzian absorption line at half its maximum peak height. See text for discussion of the value of such plots. (A. G. Marshall, unpublished results.)

displacement of a sinusoidally driven, damped weight on a spring. In fact (see Problems) a semicircle will result from the general plot of x' versus x'' (Eq. 13-36) for *any* Lorentzian spectral line shape. Although such data reduction has been applied only recently to spectral line shape (nuclear magnetic resonance), it appears that these plots may provide a useful way for "diagnosing" the origin of line-broadening in simple spectra. For example, we have already noted that when the line shape results from a superposition of many Lorentzians of different *line width,* the dispersion-absorption plot of experimental data will form a curve displaced *below* the "reference" semicircle expected for a single Lorentzian line shape; however, when the line shape results from a superposition of many Lorentzians of different *resonant frequency* (Fig. 14-38), the dispersion-absorption data form a curve that is displaced *above* the "reference" semicircle.

14.G. ULTRASONIC ABSORPTION AND VELOCITY DISPERSION: ZERO MASS ON A SPRING AGAIN—A NEW TOOL FOR MEDICAL DIAGNOSIS AND TREATMENT

As noted in Chapter 13.A, sound waves constitute a *longitudinal* disturbance, consisting of alternating compressions and rarefactions of the medium *along* the direction of propagation. It is readily shown (see Feynman, Vol. I., pp. 47-4 to 47-7) that such a system satisfies a wave equation, and the analogy may be pressed further to encompass such phenomena as absorption of sound energy as a function of frequency, reflection and refraction at the interface between two layers of medium which differ in the speed at which sound is transmitted (different "refractive index" for sound), and variation of sound wave velocity with frequency. In this section, we try to outline the mechanism and applications for these phenomena.

In contrast to electromagnetic waves, which travel through space even in a *vacuum,* sound waves are transmitted by *molecules.* To achieve the compression and rarefaction required to propagate the sound wave, the molecules in a liquid (the case of major interest here) must be made to slide past one another ("shear"). A liquid will resist such a shear force, according to the magnitude of the applied shear *and* according to the suddenness with which the change is carried out. This situation is closely modeled by our usual driven, damped, weight on a spring: the spring also resists extension according to the magnitude of the required displacement (i.e., the spring constant, $k \neq 0$), and the spring is surrounded by a viscous medium that resists positional changes according to the speed at which they occur (i.e., friction constant, $f \neq 0$). But just as a spring won't remain extended without a weight suspended from one end, the liquid will not permanently withstand a shear stress. Thus, the mathematical description of the response of a liquid to a steady applied sound wave driving force reduces to the problem of a driven, damped, *massless* weight on a spring (Eq. 13-25). We have already obtained the steady-state displacement, x, for such a case:

$$x = x' \cos(\omega t) + x'' \sin(\omega t) \tag{14-27a}$$

with

$$x' = \frac{F_0}{k} \frac{1}{1 + \omega^2 \tau^2} \tag{14-27b}$$

and

$$x'' = \frac{F_0}{k} \frac{\omega \tau}{1 + \omega^2 \tau^2} \tag{14-27c}$$

Equations 14-27b and 14-27c resemble the Lorentz line shape (Eq. 13-36) that has become familiar from our prior discussion of dispersion and absorption of ultraviolet-visible, infrared, microwave, and radio-frequency electromagnetic radiation, except that the "resonance" is now centered at zero driving frequency rather than at what used to be the "natural" frequency of the spring with finite mass. We now try to connect Eq. 14-27 to the frequency-dependence of sound velocity and absorption in matter.

First, a sound wave may be thought of as a periodic variation in pressure (or density) along the direction of propagation. By analogy to our previous treatment of optical absorption and refractive index (Chapter 14.A), we thus write the instantaneous pressure as the real part of the complex expression:

$$\text{Pressure} = \text{Re}(P) = \text{Re}\left[P_0 \exp\left[i\omega\left(t - \frac{y}{v}\right)\right]\right] \tag{14-35}$$

where t is time, y is the distance from the point in question to the source of the sound wave, and v is the ("phase") velocity (i.e., rate of movement of a wave "crest") of the wave. In order to include the effect of energy absorption (compare our previous extension from Eq. 14-3 to Eq. 14-6), we write v as a complex quantity

$$(1/v) = (1/v_{\text{real}}) - i(\alpha/\omega) \tag{14-36}$$

in which α is called the "amplitude absorption coefficient." Substituting Eq. 14-36 into Eq. 14-35, and recalling that the energy per unit volume of the wave will be proportional to the square of the (pressure) amplitude, we find that

$$\text{Intensity} = I = \text{constant} \cdot P \cdot P^*$$

or

$$\boxed{I = I_0 \exp[-2\alpha y]} \tag{14-37}$$

Equation 14-37 simply states that the intensity of a sound wave decreases exponentially (with exponential constant, 2α) as the wave passes through matter, because of the energy dissipated in forcing molecules to slide past one another to achieve the periodic density variations that constitute the wave.

Ultrasonic absorption data are seldom expressed directly as α-values. A common display is based on the "absorption per unit wavelength"

$$I = I_0 \exp[-2\alpha y] = I_0 \exp[-2\alpha\lambda(y/\lambda)]$$

or

$$I = I_0 \exp[-\mu(y/\lambda)]; \; \mu = 2\alpha\lambda \tag{14-38}$$

μ is related to the decrease in sound wave energy when the wave crest progresses one wavelength along the direction of propagation. Since

$$\lambda = 2\pi v_{\text{re}}/\omega,$$
$$\text{we obtain } \mu = 4\pi v_{\text{re}}\alpha/\omega \tag{14-39}$$

We shall shortly find that the sound velocity, v_{re}, varies only slightly (usually less than 1%—see Fig. 14-39) over a wide range of frequency, so that μ may be written

$$\mu \cong \text{constant} \cdot (\alpha/\omega) \tag{14-40}$$

Ultrasonic absorption is usually expressed either as (α/ω^2) or as $\mu = \text{constant} \cdot (\alpha/\omega)$, as shown in Fig. 14-39.

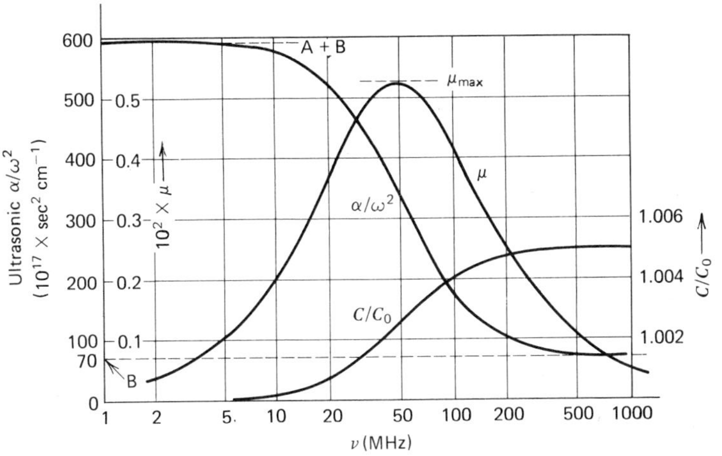

FIGURE 14-39. Plots of ultrasonic α/ω^2, absorption per unit wavelength (μ), and relative sound velocity in the medium (c/c_0), versus logarithm of applied sound wave frequency, assuming the existence of a single type of frictional ("relaxation") process whose critical frequency, $\nu_c = 50$ MHz. The limiting speed of sound (at zero frequency) is taken as $c_0 = 1.2 \times 10^5$ cm/sec. See text for further discussion. (After J. Lamb, in *Physical Acoustics,* ed. W. P. Mason, Vol. II-Part A, Academic Press New York, 1965, p. 208.)

Under the approximation that led to Eq. 14-40, it can be shown (see References) that sound velocity, v_{re}, and energy absorption per unit wavelength, μ, are directly related to our expressions for the in-phase (x') and 90°-out-of-phase (x'') components of the steady-state displacement of a sinusoidally driven, damped, massless weight on a spring (Equations 14-27b and 14-27c):

$$\alpha = C_0 \omega x'' = C_1 \frac{\omega^2 \tau}{1 + \omega^2 \tau^2} \tag{14-41}$$

$$\mu = C_2(\alpha/\omega) = C_3 \frac{\omega \tau}{1 + \omega^2 \tau^2} \tag{14-42}$$

$$\alpha/\omega^2 = C_1 \frac{\tau}{1 + \omega^2 \tau^2} \tag{14-43}$$

$$v_{re} = C_4(1 - C_5 \cdot x') = C_4\left(1 - C_6 \frac{1}{1 + \omega^2 \tau^2}\right) \tag{14-44}$$

($C_0, C_1, C_2, C_3, C_4, C_5, C_6$ are constants)

As shown from the plots of (α/ω^2), μ, and v_{re} in Fig. 14-39, the "relaxation time," τ, of Equations 14-41 to 14-44 is readily found as the reciprocal of the "critical" frequency at which (α/ω^2) and v_{re} are at the halfway point between their extreme values and μ is a maximum (see Problems). The critical frequency can be thought of as a measure of the maximum rate at which the substance can change its density (or equivalently, its volume). Now, for many chemical reactions (such as the helix-coil transition of the following example), there is a change in volume on going from reactants to products. We would thus expect that power will be absorbed by the helix-coil system as the passing sound wave drives the chemical reaction alternately forward and backward in response to the periodic pressure variation from the sound wave. However, when the sound wave frequency is higher than the rate constant for the reaction, the chemical system can no longer "follow" the rapid pressure variations, and so relatively little power is absorbed from the high-frequency sound waves. Thus, it is possible to determine the rate constant(s) for very fast reactions by finding the corresponding critical frequency (frequencies) from one of the plots of Fig. 14-39. Finally (as for the dielectric relaxation results of the preceding section of this chapter), the critical frequencies for changes in density of the solvent (about 0.5×10^{12} sec^{-1} for water) are so high that they are not usually reached experimentally; thus, the experimental plots of, for example, α/ω^2 versus ω consist of a sum of two terms of the form of Eq. 14-43, where $\tau_2 << 1/\omega$, so that

$$\frac{\alpha}{\omega^2} = C_1 \frac{\tau_1}{1+\omega^2\tau_1^2} + C_2 \frac{\tau_2}{1+\omega^2\tau_2^2} \qquad (14\text{-}45)$$

$$\boxed{\lim_{\tau_2<<1/\omega} \left(\frac{\alpha}{\omega^2}\right) = C_1 \frac{\tau_1}{1+\omega^2\tau_1^2} + C_2\tau_2 = \frac{A}{1+\omega^2\tau_1^2} + B} \qquad (14\text{-}46)$$

EXAMPLE *Ultrasonic Absorption for Extracting Rate Constants for Fast Chemical Reactions: Helix-coil Transition of Poly* L*-glutamic acid*

Figure 14-40*a* shows the optical absorbance versus pH for a solution of poly (L-glutamic acid), PGA. This curve provides a measure of the fraction of the PGA polymer segments that are in the helix (as opposed to the random-coil) configuration, as shown by the dashed line in the figure. From these data, it is clear that at pH 5.1, PGA exists in approximately 50% helix and 50% random-coil forms. Ultrasonic absorption data as a function of frequency for PGA at pH 5.1 are plotted in Fig. 14-40*b*, and may be fitted by Eq. 14-46 to give a single relaxation time of about 10^{-6} sec. Although we will defer a detailed discussion of the extraction of chemical reaction rate constants from such relaxation times until Chapter 16.A.1, it can be noted that the present data provide the value, $8 \pm 5 \times 10^7$ sec^{-1}, for the rate constant, k_f, for helical growth, as represented by

$$\left\{\begin{matrix} hcc \\ \text{or} \\ cch \end{matrix}\right\} \underset{k_r}{\overset{k_f}{\rightleftharpoons}} \left\{\begin{matrix} hhc \\ \text{or} \\ chh \end{matrix}\right\} \qquad (14\text{-}47)$$

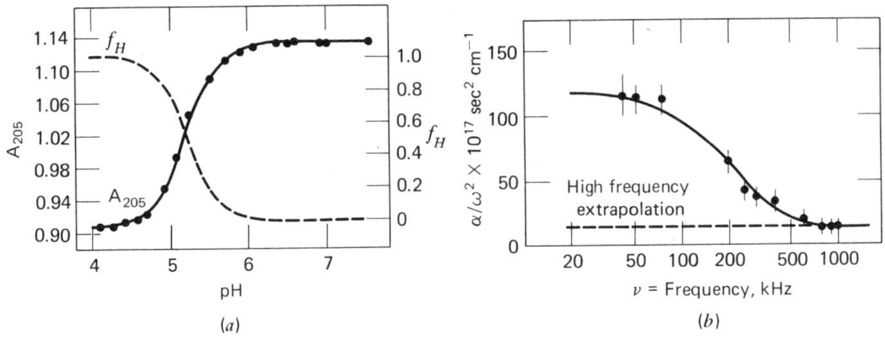

FIGURE 14-40. (*a*) pH dependence of the optical absorbance at 205 nm (solid curve) and the fraction of helicity (dashed curve) for 3.9×10^{-3} mole-residues per liter of PGA at 37°C. (*b*) Typical ultrasonic α/ω^2 versus frequency curve for 2.7×10^{-3} mole-residues per liter PGA at 37°C, pH 5.11, and 0.03M NaCl. The solid line is that obtained from Eq. 14-46. [From A. D. Barksdale and J. E. Stuehr, *J. Amer. Chem. Soc. 94,* 3334 (1972).]

in which "*h*" denotes a polymer segment in the helical (i.e., hydrogen-bonded) conformation, while "*c*" denotes a random-coil segment.

There have been relatively few definitive ultrasonic examples for conformational changes of macromolecules in solution, because for most other macromolecules that have been studied, other (usually less interesting) processes also contribute to the ultrasonic absorption, including changes in the nature of attachment of solvent molecules to the macromolecule (remember that any process that generates a change in volume will contribute to ultrasonic absorption).

The best examples of the use of ultrasonic radiation in biology are based on irradiation of bulk tissue. To appreciate these techniques, it is necessary to have at hand a few typical ultrasonic properties of various tissues, listed in Table 14-6.

Table 14-6 Ultrasonic properties of some common materials.

Tissue	Ultrasonic Propagation Velocity (m sec^{-1})	Ultrasonic Absorption Coefficient at 1 MHz (dB cm^{-1})
Air	330	12
Blood	1570	0.2
Fat	1450	0.6
Muscle (average)	1590	2.3
Water (distilled)	1480	0.002
Skull-bone	4080	13
Human soft tissue (average)	1540	0.8

From P. N. T. Wells, Ultrasonics in Clinical Diagnosis, *Churchill Livingstone, Edinburgh (1972), p. 5.*

EXAMPLE *Ultrasonic Microscopy*

The resolving power of any microscope is of the order of the wavelength of the irradiating waves. Thus since wavelength λ, frequency ν, and wave velocity c are related by $c = \lambda\nu$, it is clear that best resolution is obtained at the highest available frequency. Because different biological tissues exhibit different ultrasonic *absorption,* it is possible to design an ultrasonic microscope in which contrast is based on differences in ultrasonic absorption by different parts of the specimen. The source of ultrasonic irradiation is a "piezoelectric" crystal, which simply changes its mechanical dimensions when subjected to an electrical potential. Since ionic crystals consist of separated charges, it seems reasonable that application of an external electric field should lead to deformation of the crystal, providing a "transducer" that converts electrical oscillations into mechanical oscillations which are then transmitted through the sample (see Fig. 14-41).

Detection of the transmitted waves is tricky, since it is impractical to construct a multi-channel array of tiny piezoelectric crystal detectors—instead, it is common to allow the transmitted radiation (after passing through the microscope) to impinge on a plastic mirror, whose deformations by the ultrasonic waves can then be monitored optically with high spatial resolution. For a 2 GHz sound wave, the sound wavelength is about 7500 Å in typical biological tissue, which compares favorably with the wavelength of optical electromagnetic radiation. A comparison of the performance of an ultrasonic and an optical microscope is shown in Fig. 14-42 for a sample of fresh onion skin. The principal difference in the two images is that the cell nuclei are seen much more clearly (black dots in lower photograph) in the ultrasonic image. The technique is relatively new, and it seems reasonable to expect that there will be many situations in which it may be preferable to optical microscopy, since staining of the specimen is not necessary.

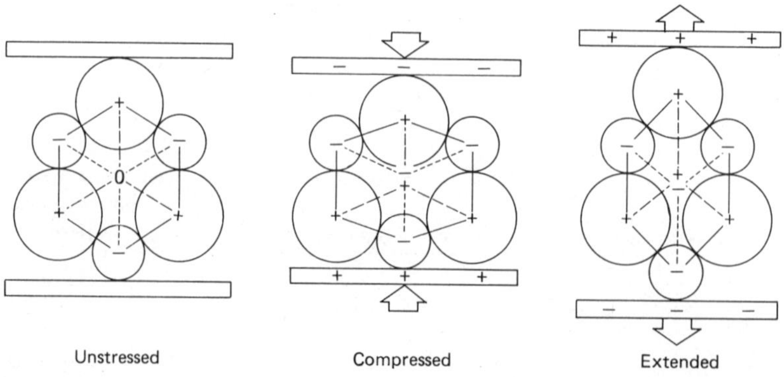

FIGURE 14-41. Diagrams illustrating the interaction between force and electric charge distribution in a piezoelectric transducer. Quartz (silicon dioxide) is represented here. The arrows indicate the directions of the applied stresses, and the resultant surface charges are indicated at the electrodes. The converse effect leads to deformation of the transducer in response to an applied voltage.

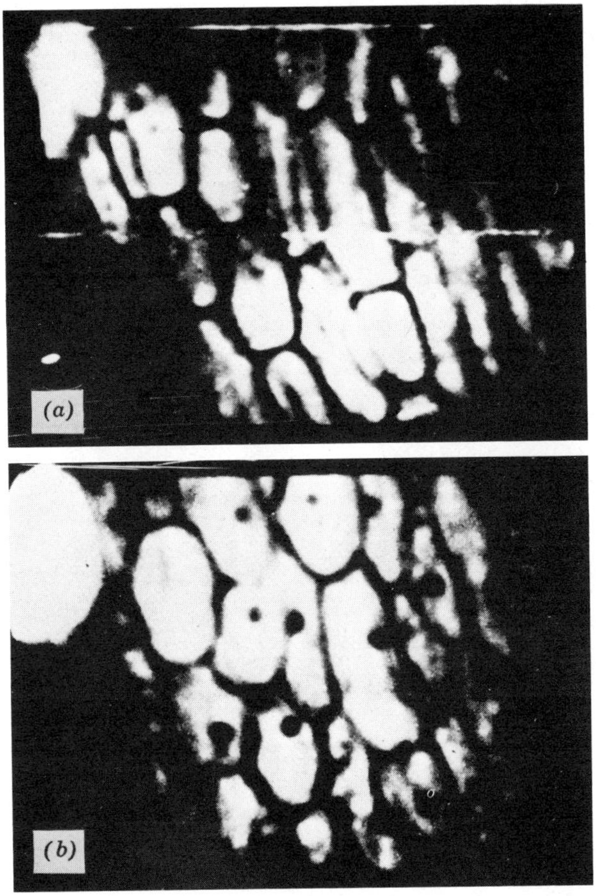

FIGURE 14-42. Micrographs of fresh onion skin, based on optical absorption (above) and ultrasonic absorption (below). Acoustic image formed at 220 MHz irradiating frequency. Field of view is 750 μm horizontally in both cases. Note the much higher contrast for cell nuclei in the acoustic image. [From L. W. Kessler, in *Ultrasonics International 1973: Conference Proceedings,* IPC Science and Technology Press Ltd, Guildford, England (1973), p. 175.]

By geometrical constructions exactly analogous to the case of refraction and reflection of electromagnetic radiation at an interface between two layers of matter that differ in refractive index (i.e., in speed of light in the medium), it can be shown that the relative intensity of ultrasonic radiation reflected at a boundary between two layers of different densities, ρ_1 and ρ_2, and different ultrasonic velocity in the media, c_1 and c_2, is given by:

$$\frac{\text{Reflected intensity (normal to boundary)}}{\text{Incident intensity}} = \left[\frac{\rho_2 c_2 - \rho_1 c_1}{\rho_2 c_2 + \rho_1 c_1}\right]^2 \qquad (14\text{-}48)$$

(a)

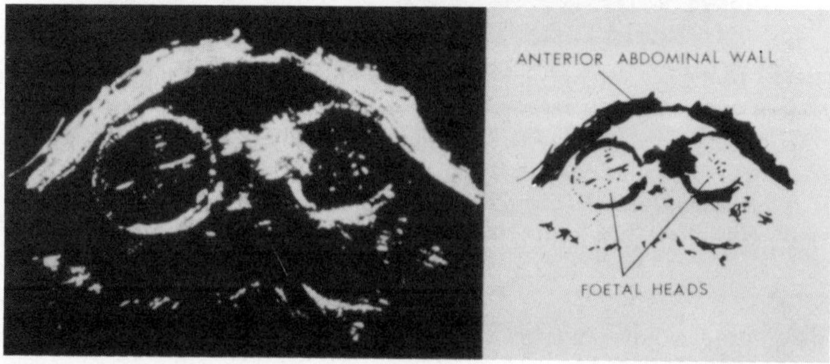

(b)

FIGURE 14-43. Diagnostic ultrasonic tomographs. Above: Scan showing normal right breast, and cyst in left breast, made using a water-immersion scanner operating at 2 MHz. The artifacts that appear on both sides of the scan (but more markedly at the left) are due to reflections at the water surface. Below: view of twin heads in utero at 25 weeks gestation (transverse abdominal section at 2.5 MHz). [Above photo from P. N. T. Wells and K. T. Evans, *Ultrasonics 6* 220 (1969); Below photo from P. N. T. Wells, *Ultrasonics in Chemical Diagnosis,* Churchill Livingstone, Edinburgh (1972), p. 83.]

Examination of Table 14-6 quickly leads to the conclusion that very high relative reflection is expected at interfaces between air and tissue, or between tissue and bone, and even between water (as in a fluid-filled cyst) and tissue.

EXAMPLE *Ultrasonic Tomography*

Sonar, echo-location by bats and porpoises, and ultrasonic tomography are all based on sending a brief burst of ultrasonic sound outward, and then recording the length of time (which is a measure of distance from the source to the object and back) required for an ultrasonic echo to reflect from an object and return to the ultrasonic transducer that serves as both transmitter and receiver. With sufficiently high frequency (and correspondingly small wavelength), it is possible to resolve relatively fine detail (porpoises transmit at frequencies up to 150 kHz and can resolve objects as small as 0.2 mm even while wearing blinder cups over the eyes). The echo pattern following a single pulse gives a one-dimensional image of the tissue boundaries located at varying distance from the transmitter. By recording several such one-dimensional patterns as the transmitter is moved linearly to several locations along the surface of the object (see Chapter 22.C.1 for a similar device based on X-ray scattering), and then coding the response to the scanner for a TV screen, where echo intensity is converted to brightness of the TV image corresponding to that echo location, it is possible to obtain a cross-sectional image of a biological object. "Tomography" comes from Greek roots meaning "cut" (i.e., the cross section), and "write" (the display).

Ultrasonic tomography in medical diagnosis is most useful in visualizing soft tissue (liver, kidney, brain, etc.), where X-ray image contrast is poor. A dramatic example of the value of the technique is seen in Fig. 14-43a showing the image obtained in scanning human breast tissue. [Since air and skin differ widely but water and skin differ only slightly in the product of density and sound velocity (see Eq. 14-48), it is necessary to immerse the subject in water to minimize reflection of ultrasonic waves immediately after they reach the subject.] Although clinical palpation of the breast makes *detection* of lesions relatively easy, it is not so easy to decide whether the lesion is a benign tumor, a malignant tumor, or a cyst. Since these three types of tissue all differ in density·(sound velocity) compared to each other, and compared to normal breast tissue, it is often possible to distinguish those possibilities from examination of an ultrasonic tomographic two-dimensional cross-sectional image (Fig. 14-44).

A second area in which ultrasonic images are useful is in monitoring of the disposition and health of a fetus while it is still in the womb. An obvious application is the detection of multiple births (see Fig. 14-43b) – such pictures are especially useful near the time of delivery to locate the infant's position.

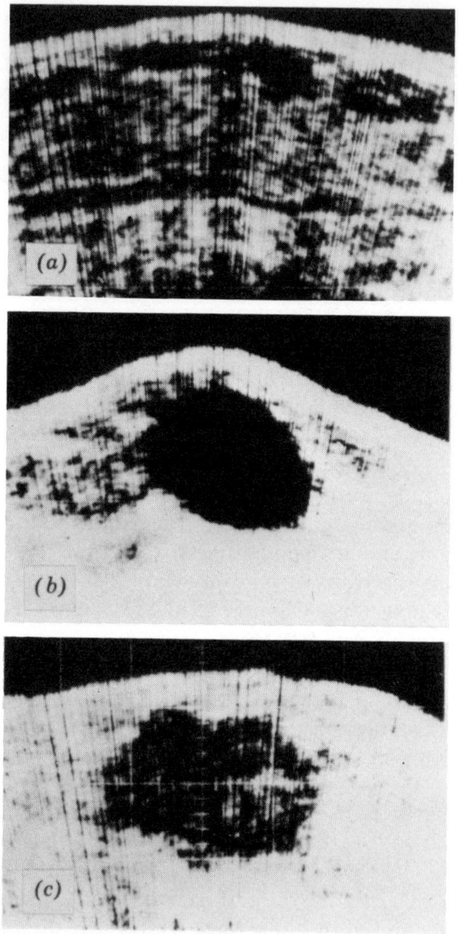

FIGURE 14-44. Differential diagnosis using ultrasonic tomography. (*a*) normal human breast cross section; (*b*) cyst in breast; (*c*) breast fibroadenoma (tumor). All "echo-grams" made using 5 MHz ultrasonic radiation. [T. Wagai, in "Ultrasonics in Medicine," ed. M. de Vlieger, D. N. White, and V. R. McReady, Excerpta Medica–American Elsevier Publishing Company, Inc., N. Y. (1974), p. 188.

EXAMPLE *Miscellaneous Applications of Ultrasound in Biology*

If the object viewed by ultrasound reflection is moving toward or away from the transmitter, there will be a Doppler shift, $\Delta\nu$, in the apparent frequency, ν, of the received wave

$$\Delta\nu = \nu(1 \pm (v/c)) \tag{14-49}$$

in which v and c are the speed of the object and of the transmitted wave. This Doppler effect has been used to monitor blood flow and the actual motion of the mitral valve in the living heart in diagnosis of heart function.

Because of the high absorption of ultrasound energy by bone and fibrous or nerve tissue relative to the surrounding tissue, high-intensity ultrasonic transmitters have been used as surgical knives in delicate operations in which mechanical manipulation of ordinary cutting tools is difficult (middle ear, eye, and pituitary operations, for example). Moreover, because of the relatively low absorption by blood vessels, ultrasonic waves have relatively little destructive effect on blood vessels, greatly reducing post-surgical scarring and complication in many cases.

The most familiar application of ultrasonic radiation is in cleaning and sterilization of glassware and tools. The applications derive from "cavitation" (sudden formation and destruction of "bubble-like" cavities) as the solvent molecules are alternately torn apart from each other and then pushed back together by the ultrasonic pressure waves. The effect is largest at lower frequencies (20 to 40 kHz), and is successful in breaking down bacteria and in dislodging solid material from surfaces (particularly when the surface would be hard to reach otherwise).

PROBLEMS

1. Consider a plane-polarized monochromatic light wave propagating along the y axis:

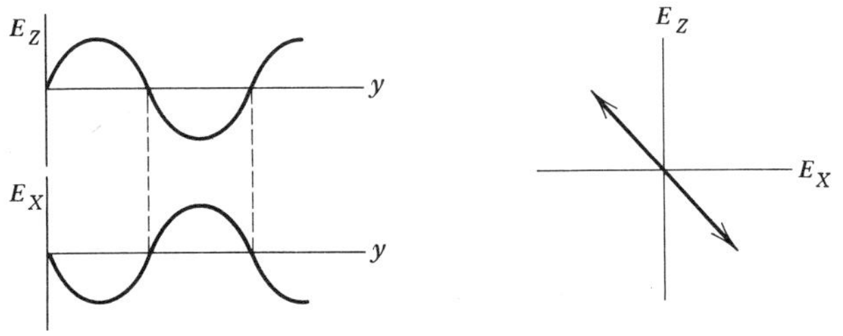

Now suppose that this wave passes through a birefringent material whose thickness is such that the E_x component is slowed down by exactly 90° (1/4 wavelength) compared to the E_z component. Show that the emergent beam will be circularly polarized. Next, suppose that E_x is slowed down by 180° (1/2 wavelength). Show that the emergent beam is now plane-polarized, but in a plane rotated 90° from the initial plane of polarization. This problem suggests that one way to observe birefringence is to detect the light transmitted through crossed polarizers (i.e., two filters, each of which passes light polarized in just one direction, with one of the directions fixed at 90° relative to the other), with the birefringent material located between the polarizers. If the material is *not* birefringent then no light will be transmitted; if the material is birefringent, some of the plane-polarized intensity transmitted through the first polarizer will be effectively rotated before reaching the second polarizer, and some light will get through the second polarizer.

2. Account for the following seeming paradox. No light is transmitted through crossed polarizers. However, if we put a third polarizer between the original two, with the third polarizer transmission direction at, say, 30° to the first polarizer, some light now comes through the second polarizer. It would appear that we have created light without using a source. Explain.

3. Circular birefringence and circular dichroism are most conveniently measured experimentally in terms of the "optical rotation" and "ellipticity" defined by the following diagram of electric field versus time for emergent light resulting from the passage of plane-polarized incident light through an optically active medium:

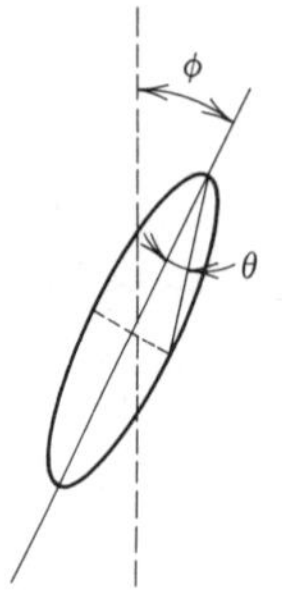

ϕ = optical rotation
θ = ellipticity

(a) Show that the ellipticity is given by

$$\theta = \tan^{-1}\left[\frac{(\exp[-2.303I_R/2] - \exp[-2.303I_L/2])}{(\exp[-2.303I_R/2] + \exp[-2.303I_L/2])}\right]$$

in which I_L and I_R are the relative intensities of left and right circularly-polarized transmitted light.

(b) Because the C.D. is much smaller than the absorbance, it is permissible to expand the exponentials (and then the arctangent) in the above expression, keeping only the first two terms. With that approximation, obtain Eq. 14-12a; that is, show that circular dichroism is proportional to ellipticity.

4. To avoid masking the spectral absorption peaks of interest by the huge peak from solvent H_2O, it is common to conduct proton nuclear magnetic resonance experiments in D_2O as the solvent. However, there are two complications introduced by this substitution. First, the pH meter reading is different in D_2O than in H_2O by about 0.41 pH unit:

$$\text{pH}_{\text{meter reading in H}_2\text{O}} = \text{pH}_{\text{in H}_2\text{O}} = -\log_{10}[\text{H}^+]$$

but

$$\text{pH}_{\text{meter reading in D}_2\text{O}} = \text{pD}_{\text{in D}_2\text{O}} - 0.41 = -\log[\text{D}^+] - 0.41$$

Second, the dissociation constants for deuterated acids in D_2O are smaller than for protonated acids in H_2O:

$$pK_D \cong 1.02\, pK_H + 0.42$$

where $pK = -\log_{10} K$, and K is the acid dissociation constant in D_2O (K_D) or H_2O (K_H).

(a) Suppose the pH of an H_2O solution is 7.5. What should the pH meter reading in a D_2O solution be, in order that the *concentration* of D^+ in the D_2O solution is the same as the *concentration* of H^+ in the H_2O solution?

(b) This is of course usually *not* the desired situation. Typically, we desire that a given buffer or weak acid have the same *degree of ionization*, $[A^-]/([A^-] + [DA])$, in D_2O as in H_2O. Obtain an expression for the pH meter reading in D_2O that will produce the same degree of dissociation in D_2O as in H_2O (i.e., $([A^-]/[DA])$ in $D_2O = ([A^-]/[HA])$ in H_2O), for a weak acid with $pK_H = 6.5$. Work out the actual pH meter reading in D_2O for a 50% degree of dissociation: $[A^-]/[DA] = 1.0$. Based on these calculations, how important is it to readjust the pH of a solution in going from H_2O to D_2O?

5. In the presence of a large static magnetic field, H_0, in the z direction, a group of magnetic nuclei (or electrons) will acquire a macroscopic magnetic moment, M_0 along the z axis. If a small additional magnetic field, $H_1\cos(\omega t)\vec{i} - H_1\sin(\omega t)\vec{j}$, rotating at frequency, ω, about the z axis is now applied, a straightforward geometrical construction predicts that there will be magnetic moment components, u and v, that are rotating about z either exactly in-phase or 90°-out-of-phase with this driving magnetic field, according to the equations

$$du/dt = (\omega_0 - \omega)v - (u/T_2)$$

and

$$dv/dt = -(\omega_0 - \omega)u - (v/T_2) + \gamma H_1 M_0$$

where T_2 is a constant, and H_1 is small in magnitude.

For steady-state conditions, solve these equations to obtain expressions for u and v. (Power absorption in magnetic resonance is proportional to v.)

6. Nuclear magnetic resonance is particularly useful for extracting chemical "exchange" rates from steady-state absorption spectra. Expressing the equations of the previous problem ("Bloch equations") in complex form

$$M = u + iv = \text{complex magnetic moment}$$
$$\hat{\omega}_0 = \omega_0 - i/T_2 = \text{complex frequency}$$
$$dM/dt + i(\hat{\omega}_0 - \omega)M = i\gamma H_1 M_0$$

Suppose that the nucleus can reside at either of two sites of different resonant frequency, ω_A and ω_B. If there is no jumping between A and B sites, the Bloch equations for M_A and M_B are simply

$$dM_A/dt + i(\hat{\omega}_A - \omega)M_A = i\gamma H_1 M_0^A$$

and

$$dM_B/dt + i(\hat{\omega}_B - \omega)M_B = i\gamma H_1 M_0^B$$

If we now permit nuclei to jump back and forth between A and B sites,

$$A \underset{k_{BA}}{\overset{k_{AB}}{\rightleftharpoons}} B$$

then the fractions of nuclei at the two sites will be

$$f_A = \frac{k_{BA}}{k_{AB} + k_{BA}} \qquad f_B = \frac{k_{AB}}{k_{AB} + k_{BA}}$$

and the total magnetic moment in the z direction will be broken down according to:

$$M_A^0 = f_A M_0 \qquad M_B^0 = f_B M_0$$

Finally, since the magnetization transferred from sites A to sites B is proportional to the number of nuclei moving from sites A to sites B, we can write:

$$dM_A/dt + i(\hat{\omega}_A - \omega)M_A + k_{AB}M_A - k_{BA}M_B = if_A\, \gamma H_1 M_0$$
$$dM_B/dt + i(\hat{\omega}_B - \omega)M_B + k_{BA}M_B - k_{AB}M_A = if_B\, \gamma H_1 M_0$$

(a) Use these last two equations to calculate the steady-state total magnetic moment, $M = M_A + M_B$.
(b) Now simplify your expression for M by assuming that both types of sites are equally populated (i.e., $k_{AB} = k_{BA} = k$), and by assuming that the "natural" line width at each site in the absence of exchange is zero: $(1/T_{2_A}) = (1/T_{2_B}) = 0$.
(c) Finally, decompose your simplified expression for M into its real and imaginary parts (i.e., find u and v such that $M = u + iv$), and show that for sufficiently fast exchange rates, $k >> (\omega_A - \omega_B)$, the magnetic resonance steady-state power absorption (proportional to v) will consist of a single averaged resonance. This calculation provides a means for obtaining chemical reaction rate constants from NMR spectra (see Figures 14-25 to 14-27 for examples).

7. A common application for the calculations of the preceding problem is in connection with the rapid binding and dissociation of some small molecule, I, to a macromolecule, E. When the chemical exchange is very fast,

the NMR spectrum of a given proton of I will consist of a single peak, whose resonant frequency is a weighted average of the resonant frequencies, ω_I and $\omega_{\epsilon I}$, of free (I) and bound (EI) species:

$$E + I \underset{}{\overset{k_I}{\rightleftharpoons}} EI$$

$$K_I = \frac{[E][I]}{[EI]}$$

$$\omega_{\text{obs}} = f_I \omega_I + f_{EI} \omega_{EI}$$

Show that when $[I]_0 >> [E]_0$ (i.e., small molecules in great excess), a plot of $[I]_0$ versus $(1/\delta)$ will have slope $= [E]_0 \Delta$ and y intercept $= -K_I$, where

$$\delta = \omega_{\text{obs}} - \omega_I \quad \text{and} \quad \Delta = \omega_{EI} - \omega_I$$

This plot thus provides for simple graphical determination of K_I and Δ (see Figure 14-26 and 14-27) from experimental measurement of NMR peak position. Similar data reduction can be used to extract K_I from optical absorbance data.

8. From the complex version of the equation for an RC circuit (massless weight on a spring),

$$\tau(dq/dt) + q = C\ V_0 \exp[i\omega t]$$

show that a steady-state (complex) solution is given by $(q' - iq'')\exp[i\omega t]$, where q' and q'' are real. [As usual, the real part of this solution

$$\text{Re}(q) = q' \cos(\omega t) + q'' \sin(\omega t)$$

is the same as the real solution of the real equation,

$$\tau \frac{dq}{dt} + q = CV_0 \cos(\omega t).]$$

9. (a) Show that for the dielectric constant (ϵ') and dielectric loss (ϵ'') of Eqs. 14-32, the following equation (Eq. 14-34) holds:

$$[\epsilon' - 1/2(\epsilon_0 + \epsilon_\infty)]^2 + (\epsilon'')^2 = (1/2(\epsilon_0 - \epsilon_\infty))^2 \qquad (14\text{-}34)$$

(b) Show that a plot of ϵ'' versus ϵ' will be a circle of radius, $(1/2)(\epsilon_0 - \epsilon_\infty)$, centered at $\epsilon' = (1/2)(\epsilon_0 + \epsilon_\infty)$ on the ϵ' axis (abscissa). (See Fig. 14-37 for an example.)

10. Show that a plot of the Lorentz line shape for dispersion, $\dfrac{\omega\tau^2}{1 + \omega^2\tau^2}$, versus the Lorentz line shape for absorption, $\frac{\tau}{1+\omega^2\tau^2}$, will also give a circle, centered at $(\tau/2)$ on the absorption axis, with radius $(\tau/2)$. (See Fig. 14-38 for an example.)

11. Ultrasonic and dielectric relaxation parameters show a frequency-dependence that may be related to any of the following three expressions:

	dielectric parameter	ultrasonic parameter
$A = \frac{1}{1+\omega^2\tau^2}$	ϵ'	α/ω^2; sound velocity
$B = \frac{\omega\tau}{1+\omega^2\tau^2}$	ϵ''	μ
$C = \frac{\omega^2\tau}{1+\omega^2\tau^2}$	conductivity	α

Devise a means for extracting the relaxation time, τ, from a plot of A (or B or C) versus $\log(\tau)$. [Plots of A or B versus $\log(\omega)$ are most commonly used.]

REFERENCES

Driven, Damped Weight on a Spring: Dispersion and Absorption

R. P. Feynman, R. B. Leighton, and M. Sands, *The Feynman Lectures of Physics,* Addison-Wesley, Reading, Mass. (1963), Chapters 21,23.

Ultraviolet-Visible Spectra

J. W. Donovan, in *Physical Principles and Techniques of Protein Chemistry,* Vol. A, ed. S. J. Leach, Academic Press, New York (1969), pp. 102–170.

Infra-red Spectra

F. S. Parker, *Applications of Infrared Spectroscopy in Biochemistry, Biology, and Medicine,* Plenum Press, New York (1971).

Circular Dichroism and Optical Rotation

K. Imahori and N. A. Nicola, in *Physical Principles and Techniques of Protein Chemistry,* part C, ed. S. J. Leach, Academic Press, New York (1973), pp. 358–445; D. W. Sears and S. Beychok, *ibid.,* pp. 446–594.

Magnetic Resonance

T. L. James, *Nuclear Magnetic Resonance in Biochemistry,* Academic Press, New York (1975).

Spin Labeling: Theory and Applications, ed. L. J. Berliner, Academic Press, New York (1976).

Electronic Circuits as Spring Models

R. P. Feynman, R. B. Leighton, and M. Sands, *The Feynman Lectures on Physics,* Addison-Wesley, Reading, Mass. (1963), Chapter 23.

Dielectric Relaxation

S. Takashima, in *Physical Principles and Techniques of Protein Chemistry,* part A, ed. S. J. Leach, Academic Press, New York (1969), pp. 291–334.

Ultrasonic Absorption and Dispersion

M. J. Blandamer, *Introduction to Chemical Ultrasonics,* Academic Press, New York (1973).

Ultrasonics in Clinical Diagnosis, ed. P.N.T. Wells, Churchill Livingstone (Edinburgh & London, 1972).

M. Eigen and L. de Maeyer, "Relaxation Methods," in *Techniques of Organic Chemistry,* Vol. VIII, ed. S. L. Friess, E. S. Lewis and A. Weissberger, Interscience, New York (1963), pp. 952–964.

CHAPTER 15 Scattering Phenomena: Steady-state Response of a Driven, Undamped Weight on a Spring

In Chapter 13, we computed the steady-state displacement for a sinusoidally driven, damped weight on a spring. It was noted that the general result assumed especially simple form in either of two limits: (1), when the driving frequency was close to the natural frequency, ω_0, of the undamped weight on a spring; and (2), when the driving frequency was sufficiently far from resonance that damping could be neglected. Examples of phenomena (spectroscopy; dielectric and ultrasonic relaxation) that can be described by the Lorentz line shape (Eq. 13-36) from limit (1) have been discussed in Chapter 14. In this chapter, we treat the "scattering" phenomena (light-, X-ray, and electron-scattering) that can be described by the formulas (Eq. 13-27) from limit (2). Although the derivations and results from Chapter 14 were often rendered more compact by use of *complex* notation, we can treat all the phenomena in this chapter by using *real* notation exclusively.

We begin (as in Chapter 14) by supposing that electrons in matter can be treated as if they are bound to their respective atoms by springs, provided that we introduce only small displacements in the electron positions. Since an electron of charge e, subjected to an electric field, E, will experience a (driving) force, eE, the sinusoidally time-varying electric field component of a monochromatic electromagnetic wave (e.g., a light wave of one color) will drive an electron on a spring according to

$$\boxed{m(d^2x/dt^2) = -kx + eE_0 \cos(\omega t)} \tag{15-1}$$

Net force = Spring restoring force + Driving force from sinusoidally time-varying electric field from monochromatic electromagnetic wave

in which x is the electron displacement from its equilibrium position, k is the force constant of the spring to which the electron is attached, m and e are the mass and charge of the electron, and we suppose that the driving frequency is sufficiently far from the "natural" frequency, $\omega_0 = \sqrt{(k/m)}$, that we may neglect any damping forces (see Chapter 13.B.1). We have already shown (Eqs. 13-16a and 13-27) that the steady-state displacement, x, for Eq. 15-1 is completely in-phase (i.e., is described by $x = x' \cos(\omega t)$) with the driving electric field, according to

$$x = \frac{E_0 e}{m} \frac{1}{(\omega_0^2 - \omega^2)} \cos(\omega t) \qquad \omega_0 = \sqrt{\frac{k}{m}} \tag{15-2}$$

To proceed further, one must recognize that the electron is in fact bound to an atom of net charge equal to one positive electronic charge, *e,* and that positive charge is so massive compared to the electron that the positive charge can be taken as stationary (this assumption was called the central-field approximation when the reader encountered it in elementary chemistry treatments of the hydrogen atom). In other words, there is now an electric *dipole moment,* charge·displacement = *ex,* whose magnitude oscillates sinusoidally at the frequency of the driving electric field. But Maxwell's equations predict that an oscillating electric dipole produces an oscillating electric field that creates an oscillating magnetic field at right angles: these two fields now constitute an electromagnetic wave that propagates away from the oscillating driven electron source. The *amplitude* of the electric field of this re-radiated, or *scattered* wave is proportional to the second derivative of dipole *ex* with respect to time:

$$E_{\text{scatt}} = (1/c^2) \frac{d^2(ex)}{dt^2} \frac{\sin\theta}{r} \tag{15-3}$$

in which c is the speed of light, r is the distance between the oscillating electron and the point in space at which the electric field amplitude is measured, and $\sin\theta$ represents the projection of the scattered amplitude along the observer's line of sight (see Fig. 15-1). Substituting for *ex* from Eq. 15-2 into Eq. 15-3,

$$E_{\text{scatt}} = -E_0 \frac{e^2}{mc^2} \frac{\sin\theta}{r} \left[\frac{1}{\left(\frac{\omega_0}{\omega}\right)^2 - 1} \right] \cos(\omega t) \tag{15-4}$$

Finally, since the *intensity* of electromagnetic radiation is proportional to the *square* of electric field *amplitude* (Chapter 13.A),

$$\frac{I_{\text{scatt}}}{I_0} = \frac{\text{intensity of re-radiated (scattered) radiation}}{\text{intensity of incident (driving) radiation}} = \frac{E_{\text{scatt}}^2}{E_{\text{driving}}^2} \tag{15-5}$$

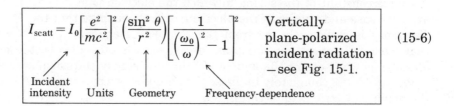

or

$$I_{\text{scatt}} = I_0 \left[\frac{e^2}{mc^2}\right]^2 \left(\frac{\sin^2\theta}{r^2}\right) \left[\frac{1}{\left(\frac{\omega_0}{\omega}\right)^2 - 1}\right]^2 \tag{15-6}$$

Vertically plane-polarized incident radiation – see Fig. 15-1.

(I_0: Incident intensity; $\left[\frac{e^2}{mc^2}\right]^2$: Units; $\left(\frac{\sin^2\theta}{r^2}\right)$: Geometry; $\left[\frac{1}{(\omega_0/\omega)^2 - 1}\right]^2$: Frequency-dependence)

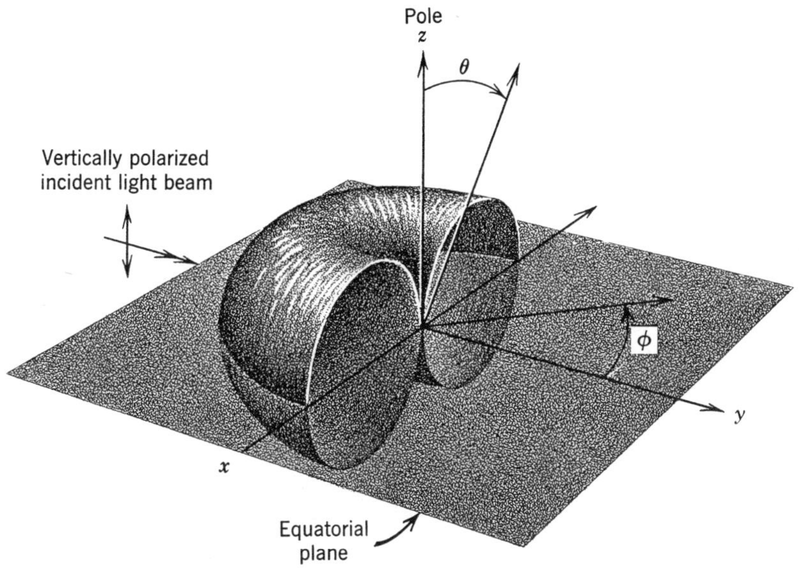

FIGURE 15-1. Diagram of the relative magnitude of scattered amplitude from incident vertically (plane-) polarized radiation along the y direction. The driven electron is located at the origin. Length of radius vector from that origin indicates relative scattered amplitude. (Front half of distribution is omitted for clarity in presentation; it is identical to that shown and joins smoothly to it.) (From D. F. Eggers et al., *Physical Chemistry,* John Wiley & Sons, New York, 1964, p. 496.)

in which it is recalled that the magnitude of the driving force $= eE_0 \cos \omega t$ from the incident radiation. Equation 15-6 forms the basis for our treatments of light scattering, X-ray and neutron diffraction and scattering, and electron microscopy, and thus deserves detailed scrutiny.

Equation 15-6 may be factored into four terms as shown, in which each term carries direct intuitive meaning. Since I_{scatt} is proportional to the incident intensity, I_0, the system is said to be "linear," which will turn out to be useful in relating the present steady-state response to the "transient" response to be discussed in Chapter 16.B. The $(e^2/mc^2)^2$ factor simply accounts for the correct units, for our purposes. The geometry factor includes the familiar inverse-square dependence of the intensity of electromagnetic radiation on distance between the radiation source (the oscillating electron) and the observer. Figure 15-1 shows the significance of the $\sin^2\theta$ factor. Since electromagnetic radiation is a transverse wave, propagation must be in a direction perpendicular to the direction of oscillation of the dipole producing the electric field of the radiation; but since the motion of an electron driven by incident radiation plane-polarized in the y-z plane has no component along the x direction, there can be no radiation scattered ("re-radiated") along the z direction $[\sin(0) = 0]$. On the other hand, radiation scattered by the same system will be of maximal intensity in any

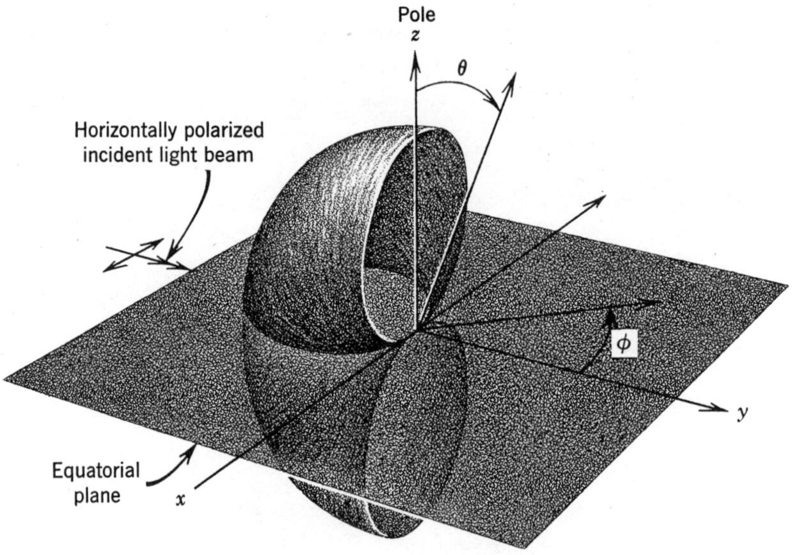

FIGURE 15-2. Diagram of relative scattered amplitude from an electron at the origin, driven by horizontally (plane-) polarized radiation incident along the y direction. Length of radius vector from the origin indicates scattered relative amplitude. (Front half of distribution omitted as in Fig. 15-1.) (From D. F. Eggers et al., *Physical Chemistry,* John Wiley & Sons, New York, 1964, p. 496.)

direction in the x-y plane $[\sin(90°) = 1]$, since those directions are all perpendicular to the direction of oscillation of the electric dipole scattering source.

Clearly, if we had chosen radiation initially plane-polarized along the x direction, a similar pattern would result, with the exception that the scattered intensity would now vary as $\cos^2\phi$ in the equatorial plane (Fig. 15-2):

For the more general and usual case that the incident radiation is *unpolarized,* we may regard unpolarized radiation as being composed of two mutually perpendicular linearly polarized components. The intensity of the scattered radiation is then the average of the intensity from each of the two perpendicular components. For example, in the equatorial (xy) plane, the angular dependence of scattered intensity is either $\sin^2(90°) = 1$, or $\cos^2\phi$ for the two individual components, and the intensity for unpolarized radiation (sum of the two components) is $(1 + \cos^2\phi)/2$ in the equatorial plane. If the more convenient angle, θ_s, is defined as the angle between the observer's line of sight and the direction of propagation of the original incident beam (see Fig. 15-3), then the overall angular factor becomes $(1 + \cos^2\theta_s)/2$:

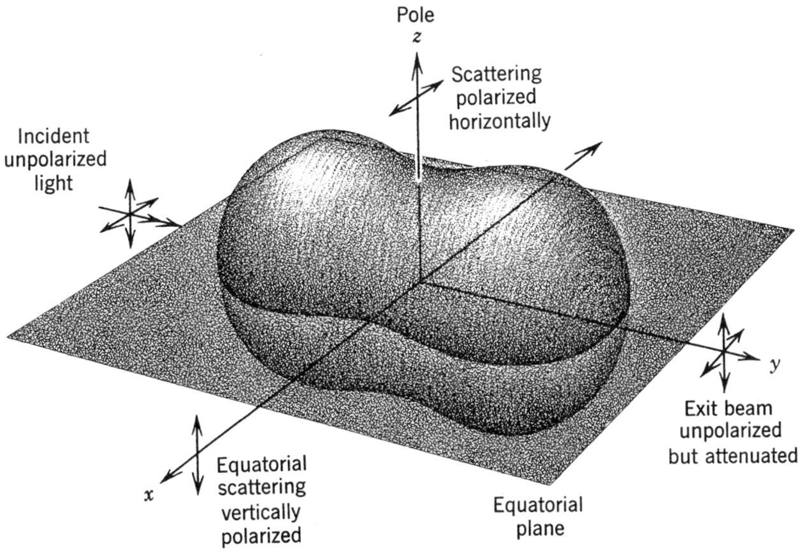

FIGURE 15-3. Diagram of relative scattered amplitude from an electron at the origin, driven by unpolarized radiation incident along the y direction. (From D. F. Eggers et al., *Physical Chemistry,* John Wiley & Sons, New York, 1964, p. 497.)

Unpolarized incident radiation—see Fig. 15-3.	$$I_{\text{scatt}} = I_0\left[\frac{e^2}{mc^2}\right]^2\left(\frac{1+\cos^2\theta_s}{r^2}\right)\left[\frac{1}{\left(\frac{\omega_0}{\omega}\right)^2 - 1}\right]^2$$	(15-7)

The angular dependence of scattered intensity in Equations 15-6 and 15-7 offers two major qualitative features. First, the scattering at right angles to the incident direction of propagation is always *plane*-polarized, even if the incident light is *un*polarized (the reader can note that the apparent brightness of a clear sky through polarized sunglasses changes as the sunglasses are rotated). Second, the scattering process is experimentally manifest in two ways, either by observation of the reduced intensity along the original direction of propagation (as monitored in "turbidity" measurements for light-scattering—see below), or by observation of light (re-) radiated from the scattering center at various angles to the incident direction of propagation (as measured in various diffraction experiments with X rays and neutrons or as the basis for operation of the electron microscope).

The frequency-dependence of the scattered radiation, the final factor in Equations 15-6 and 15-7, is most easily identified in two simple limits (Eqs. 13-30, 13-31):

Rayleigh Scattering ($\omega << \omega_0$)

$$I_{\text{scatt}}\begin{pmatrix}\text{unpolarized}\\ \text{incident}\\ \text{radiation}\end{pmatrix} = I_0\left[\frac{e^2}{mc^2}\right]^2\left[\frac{1+\cos^2\theta_s}{r^2}\right]\left(\frac{\omega}{\omega_0}\right)^4 \tag{15-8}$$

Thomson Scattering ($\omega >> \omega_0$)

$$I_{\text{scatt}}\begin{pmatrix}\text{unpolarized}\\ \text{incident}\\ \text{radiation}\end{pmatrix} = I_0\left[\frac{e^2}{mc^2}\right]^2\left[\frac{1+\cos^2\theta_s}{r^2}\right] \tag{15-9}$$

Because the "natural" frequency for electrons on springs in matter corresponds to the ultraviolet point in the spectrum, Rayleigh scattering is expected for light of visible and lower frequency, while Thomson scattering is expected for X-ray and higher frequencies associated with the wave properties of particles such as neutrons and electrons. Furthermore, it may be noted that while X rays and electrons are scattered *independent* of their energy (frequency) (i.e., all objects look "gray" to X rays and to the electron microscope), visible light is scattered according to the *fourth power* of its frequency. Thus, blue light is scattered much more intensely than red light, which explains why the sky appears blue when viewed at angles well away from the sun, and why the setting sun is red viewed at angles close to the sun (since blue light is preferentially removed from the direction of incidence of the sunlight). (Clouds are another story and are discussed later.)

To account for the radiation scattered from a *macroscopic* sample, it is necessary to consider the combined scattering from two different electrons separated in space, as shown in Fig. 15-4, using a coherent source (for ease in illustration) so that each electron is driven with the same phase initially. Each driven electron will scatter independently, and the amplitudes of the two scattered waves will add as shown in the figure for an observer positioned either at large scattering angle (top diagram of Fig. 15-4) or small angle (bottom diagram). It is qualitatively clear from the figure that when the scattering angle is large (and/or the driving radiation wavelength is small compared to the distance between the two electrons), then there will be some degree of *destructive* interference between the two scattered waves. On the other hand, for small scattering angle (and/or wavelength of the driving radiation large compared to the distance between the two driven electrons), there will be *coherent* addition of the amplitudes of the two scattered waves. One can thus project that the combined scattered waves will yield an intensity at a given observation location that varies between zero and four times the intensity scattered from either electron alone.

A number of new conclusions follow directly from Fig. 15-4. First, if the two electrons are very close together (as for example, two electrons on the

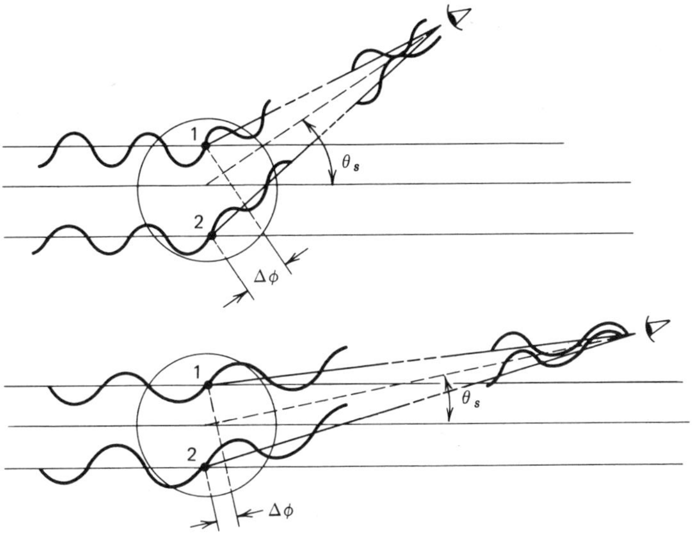

FIGURE 15-4. Schematic diagram of the addition of scattered electric field amplitudes of radiation scattered simultaneously from two electrons (1 and 2 in the figure) separated in space. The driving radiation is assumed coherent (i.e., both electrons are driven with the same phase) in order to show directly the phase difference, $\Delta\phi$, which results for two waves scattered toward the same observation point. Upper diagram is for large scattering angle, θ_s; lower diagram for small θ_s. (See text for interpretation.)

same *atom*) compared to the wavelength of the driving radiation ($\Delta\phi << \lambda$), then the addition of scattered electric fields will be coherent no matter what the scattering angle,

$$E_{\text{scatt}}(\text{combined}) = N\, E_{\text{scatt}}(\text{from any one driven electron}) \qquad (15\text{-}10)$$

and the scattered combined *intensity* will be proportional to the *square*, N^2, of the number of scattering centers:

$$I_{\text{scatt}}(\text{combined}) = N^2\, I_{\text{scatt}}(\text{from any one driven electron}) \qquad (15\text{-}11)$$

Equation 15-11 explains why *heavy atoms* (by which we mean a scattering center with many electrons close together) scatter visible light, X rays, and electrons so intensely, and are thus often used as "markers" in X-ray diffraction and electron microscopy applications. On the other hand, if the distance between the two driven electrons *varies randomly* (as when the two driven electrons are bound to different molecules in a gas or liquid), then the composite scattered amplitude at a given location will be proportional to the *square root* of the number of such electrons, by an argument essentially identical to that used in the random walk calculation. Since intensity varies as the square of amplitude, we thus obtain:

$$I_{\text{scatt}}(\text{composite}) = N\, I_{\text{scatt}}(\text{from any one driven electron}) \qquad (15\text{-}12)$$

Equation 15-11 will apply to X-ray diffraction and electron scattering from solid samples, while Eq. 15-12 will apply to many light-scattering and low-angle X-ray scattering situations.

15.A. SMALL OBJECTS, BIG WAVES, LENSES DON'T HELP: RAYLEIGH LIGHT SCATTERING. TURBIDITY, RADIUS OF GYRATION, ZIMM PLOT: SIZE, SHAPE, AND MOLECULAR WEIGHT OF MACROMOLECULES IN SOLUTION

Relation between Light Scattering and Molecular Weight of a Macromolecule

In this section, we shall be concerned with the scattering of *visible light* by solutions containing dissolved macromolecules (over and above the small scattered intensity from the pure solvent itself). To show why solutions of large molecular weight solutes scatter light so much more effectively than solutions of small molecular weight species, it is necessary to recast the problem in slightly different variables (as usual).

It is convenient to express the magnitude of the electric dipole moment, ex, that results from application of an electric field, E, in terms of the molecular *polarizability*, α, which can be thought of as a measure of how much force is required to pull the electron away from its nucleus:

$$ex = \alpha E = \alpha E_0 \cos(\omega t) \qquad (15\text{-}13)$$

in which E in Eq. 15-13 is the driving electric field from the incident light. Substituting for ex from Eq. 15-13 into Eq. 15-3, we obtain the scattered steady-state electric field amplitude from the driven electron

$$\lim_{\omega << \omega_0} E_{\text{scatt}}\begin{pmatrix}\text{vertically}\\ \text{polarized}\\ \text{incident}\\ \text{light}\end{pmatrix} = -(\omega^2/c^2)\, \alpha\, E_0((\sin\theta)/r)\cos(\omega t) \qquad (15\text{-}14)$$

in which all symbols other than α have their usual meaning. By the same arguments that led from Eq. 15-4 to Eq. 15-8, we then obtain the scattered intensity per driven electron, I_{scatt}, for unpolarized incident light, as

$$\lim_{\omega << \omega_0} I_{\text{scatt}}\begin{pmatrix}\text{unpolarized}\\ \text{incident}\\ \text{light}\end{pmatrix} = I_0\left[\frac{\omega^4}{c^4}\right]\alpha^2\left[\frac{1+\cos^2\theta_s}{2\,r^2}\right] \qquad (15\text{-}15)$$

The remainder of the calculation is to express I_{scatt} and α in experimentally relevant and measurable quantities. It is now necessary to bring forth two more of Maxwell's relations without proof (Equations 15-3 and 15-16 are really the only formulas we have had to quote without explanation):

$$\epsilon' = n^2 = 1 + 4\pi N\alpha \qquad \text{for gases} \tag{15-16}$$

in which ϵ' is dielectric constant, n is refractive index, and N is the number of scattering centers per unit volume. Thus, if there were only one driven electron per molecule, we would have

$$\alpha = \frac{n^2 - 1}{4\pi N} \qquad \text{for a gas molecule} \tag{15-17a}$$

or, analogously,

$$\alpha = \frac{n^2 - n_0^2}{4\,\pi\,N} \qquad \text{for a molecule in solution} \tag{15-17b}$$

in which n_0 is the refractive index of pure solvent, and n the refractive index of the solution. Equation 15-17b may be rewritten as

$$\alpha = \frac{(n + n_0)(n - n_0)\,M}{4\,\pi\,[c]\,N_0} \tag{15-18}$$

in which

$$N\ (\text{particles/volume}) = \frac{[c]\ (\text{grams/volume})\ N_0\ (\text{particles/mole})}{M\ (\text{grams/mole})} \tag{15-19}$$

For a *dilute* solution

$$n + n_0 \cong 2n_0 \tag{15-20}$$

$$\begin{aligned}(n - n_0)/[c] \equiv \psi &= \text{Refractive index increment} \\ &= \text{Change in refractive index with concentration of solute}\end{aligned} \tag{15-21}$$

Combining Equations 15-18 to 15-21,

$$\alpha = \frac{n_0\,\psi\,M}{2\,\pi\,N_0} \tag{15-22}$$

which may be substituted in Eq. 15-15 to give

$$I_{\text{scatt}}^{\text{unpol}} = I_0\,\frac{\omega^4}{c^4}\,\frac{1 + \cos^2\theta_s}{r^2}\,\frac{n_0^2\,\psi^2\,M^2}{8\pi^2\,N_0^2}\ \text{per particle} \tag{15-23}$$

and multiplying the scattered intensity per particle times the number of particles per unit volume (Eq. 15-19), we obtain the scattered intensity per unit volume of solution:

$$I_{\text{scatt}}^{\text{unpol}} = I_0 \frac{\omega^4}{c^4} \frac{1+\cos^2\theta_s}{r^2} \frac{n_0^2\,\psi^2\,[c]\,M}{8\pi^2 N_0} \text{ per unit volume} \tag{15-24}$$

Finally, from Eq. 13-4, $\lambda\nu = c$, and Eq. 13-7, $\omega = 2\pi\nu$

$$\omega = 2\pi c/\lambda \tag{15-25}$$

which, when substituted for ω in Eq. 15-24, gives the result in the desired form

$$I_{\text{scatt}}\begin{pmatrix}\text{unpolarized}\\ \text{incident}\\ \text{light}\end{pmatrix} = I_0 \frac{2\pi^2\,n_0^2\,\psi^2}{\lambda^4 N_0} \frac{1+\cos^2\theta_s}{r^2}[c]\,M \quad \text{per unit volume} \tag{15-26}$$

showing that light scattering is directly proportional to solute concentration and solute molecular weight. As a minor additional point, Eq. 15-26 was derived under the (tacit) assumption of one driven electron per molecule. If there are f driven electrons per molecule, the scattered amplitudes from all of them will add coherently (Eq. 15-11), and the scattered intensity will be proportional to f^2. It is usual to avoid this issue by incorporating f into the polarizability, α, since the constants in the second factor of Eq. 15-26 are typically determined experimentally as a single parameter, K:

$$K = \frac{2\pi^2\,n_0^2\,\phi^2}{\lambda^4 N_0} \tag{15-27}$$

Since it is relatively easy to measure intensity *ratios,* and rather difficult to measure *absolute* intensities, Eq. 15-26 is customarily expressed in terms of a "*Rayleigh* (intensity) *ratio,*" $R(\theta_s)$,

$$R(\theta_s) = \frac{I_{\text{scatt}}\begin{pmatrix}\text{unpolarized}\\ \text{incident}\\ \text{light}\end{pmatrix}}{I_0} \frac{r^2}{1+\cos^2\theta_s} = K[c]M \tag{15-28}$$

Equation 15-28 shows that light-scattering intensity ratios may be used to determine molecular weight of a macromolecule in solution. When the solution contains a mixture of macromolecules of different molecular weights, M_i, the total scattering [as represented in the Rayleigh ratio, $R(\theta_s)$] will be composed of a sum of the scattered intensities from molecules of a given molecular weight M_i, according to the respective weight/volume concentrations, $[c]_i$, of each molecular-weight species:

$$R(\theta_s) = \sum_i R(\theta_s)_i = \sum_i K[c]_i M_i = \frac{K\sum_i [c]_i M_i}{\sum_i [c]_i} [c] = K[c]M_w \qquad (15\text{-}29)$$

in which

$$M_w = \frac{\sum_i [c]_i M_i}{\sum_i [c_i]} = \frac{\sum_i N_i M_i^2}{\sum_i N_i M_i} \qquad (15\text{-}29b)$$

and

$$\sum_i [c_i] = [c] \qquad (15\text{-}29c)$$

where N_i = number of moles of component M_i

Thus, if the heterogeneous solution is (wrongly) assumed to consist of a weight/volume concentration, $[c]$, of a single species of (average) molecular weight, M_w, then M_w is related to the individual M_i by Eq. 15-29b. Equation 15-29b shows why it is necessary to keep light-scattering solutions scrupulously clean of dust particles, since even a very low concentration of high molecular-weight dust particles will contribute substantially to the apparent weight-average molecular weight, M_w, of a macromolecular solute. (See Problems for an example.)

One of the simplest experimental approaches to light-scattering is to measure the reduced intensity of an incident beam along the incident direction. Although the *measurement* thus resembles the measurement of light *absorption,* the *origin* of the phenomenon is quite different. In light *absorption,* radiation enters the sample but some of it never gets out; in light *scattering,* essentially all of the incident radiation is either transmitted or re-radiated away in various directions, but since the re-radiated light is generally in directions *other* than "forward," there is an apparent diminution in the intensity along the incident direction on passing through the sample. From the important feature of Eq. 15-28 that the scattered intensity is proportional to the incident intensity, we can write the infinitesimal loss in "forward"-direction intensity, $-dI$ (the minus sign denoting an intensity *loss*), as being proportional to the infinitesimal thickness of the sample solution, dy, and to the intensity entering the infinitesimal region, I, and to the *turbidity*, τ, (analogous to the "absorbance" in absorption spectroscopy—compare Eq. 14-4 ff.), representing the scattered intensity per unit sample thickness per unit incident intensity:

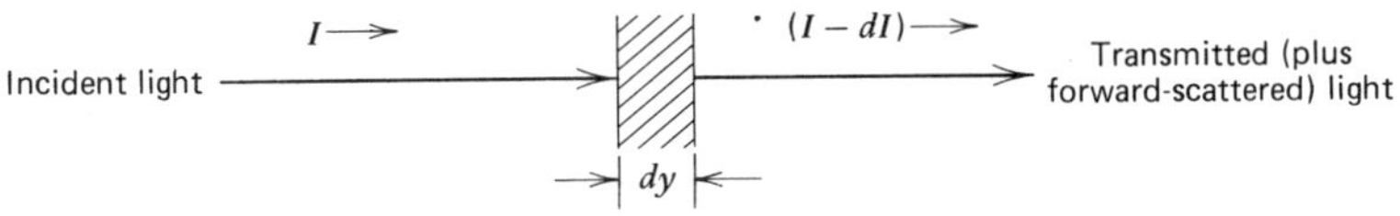

$$-dI = I\,\tau\,dy \qquad (15\text{-}30)$$

Integrating Eq. 15-30 to extend the argument to a sample of finite thickness,

$$I = I_0 \exp\,[-\tau y] \tag{15-31}$$

in which I is the (transmitted intensity) along the direction of incident light of intensity I_0, after passing through a sample of thickness, y, and turbidity, τ. It is left as an exercise to show (see Problems) that the turbidity in this simple experiment is directly related to the Rayleigh ratio

$$\tau = \frac{16\pi}{3} R(90°) \tag{15-32}$$

so that measurement of *turbidity* (i.e., apparent "absorbance"—see Eq. 14-5) provides a direct access to *molecular weight* of a macromolecule in solution. In the following discussion we show that measurement of scattered intensity as a function of *angle* away from the incident direction, θ_s, offers additional information about macromolecular *shape* and *size* in solution.

Relation between Light Scattering and Macromolecular Size and Shape

Figure 15-5 shows that the electric field amplitudes of the two light waves scattered from two driven electrons of the same macromolecule will in general be out of phase at the position of a remote observer for two reasons: first, since the two electrons are in general located at different distances from the radiation *source*, their driven oscillations will be out of phase; and second, because the two electrons are in general at different distances from

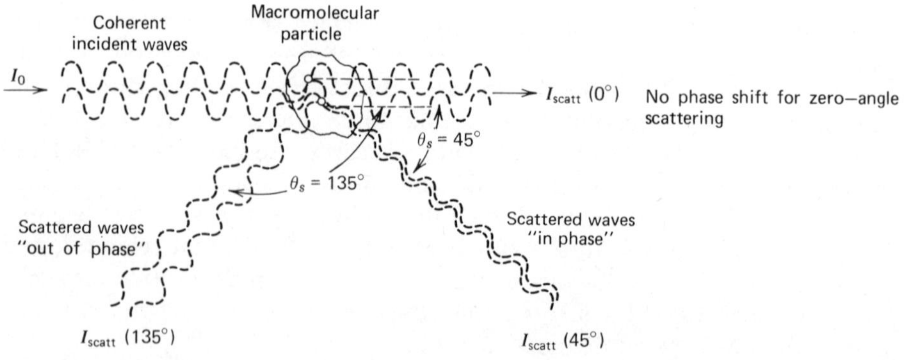

FIGURE 15-5. Phase relationship between incident and scattered light waves from two driven electrons located at different places on the same macromolecule. (Incident light waves are chosen as coherent to make it easier to see phase shifts.) For the particular location angle (between the incident direction and the inter-electron direction) and viewing angles (θ_s) shown, the scattered waves happen to differ in phase angle by 0° at $\theta_s = 45°$ and by 180° at $\theta_s = 135°$. For a simplified formal treatment, see Eq. 15-33 ff.

the *observer,* there will be a further difference in phase of the two scattered waves as perceived by an observer at a given scattering angle, θ_s, measured away from the direction of incident light.

The electric field, E, of light scattered toward an observer positioned at arbitrary scattering angle θ_s, from two spatially separated driven electrons on the same macromolecule (Fig. 15-5) can thus be written:

$$E = E_s \cos(\omega t + \phi_1) + E_s \cos(\omega t + \phi_2) \tag{15-33}$$

in which ϕ_1 and ϕ_2 represent the respective phase differences (for light from electron #1 and #2) between the driver oscillation and the oscillation at the observer's position, and both electrons are driven with the same initial amplitude, E_s. Using the trigonometric identity

$$\cos A + \cos B = 2 \cos((A - B)/2) \cos((A + B)/2) \tag{15-34}$$

we obtain

$$E = \underbrace{2E_s \cos((\phi_1 - \phi_2)/2)}_{\text{Amplitude}} \underbrace{\cos((2\omega t + \phi_1 + \phi_2)/2)}_{\text{Time-dependence}} \tag{15-35}$$

Equation 15-35 shows that the combined amplitude of the two scattered waves varies between zero and twice the amplitude, E_s, for either individual scatterer; when $(\phi_1 - \phi_2) = 0$, there is no phase difference between the two scattered waves at the observer's position, and the amplitude is $2E_s$, but when $(\phi_1 - \phi_2) = 180°$, the two scattered waves cancel completely to give a combined amplitude of zero. The combined scattered *intensity* thus varies between zero and $4I_s$, where I_s is the scattered intensity from either individual electron alone, as claimed in our earlier qualitative discussion.

Equation 15-35 represents the combined electric field from two driven electrons with a single orientation. For macromolecules in solution, the inter-electron vectors are oriented in all directions, so it is necessary to take an average of Eq. 15-35 with the two electrons located anywhere (on opposite sides) of an infinitesimally thin spherical shell. Completion of this calculation requires explicit expression of the path difference, Δ, leading to a phase difference, $2\pi\Delta/\lambda$, at the observer, for each possible orientation of the inter-electron vector: the exercise is a straightforward but tedious (see Tanford, pp. 298–302) application of solid trigonometry and is omitted here. The result is most usefully expressed in the following form:

$$\frac{\text{amplitude of } E_{\text{scatt}} \begin{pmatrix}\text{from two driven electrons located anywhere} \\ \text{on a thin spherical shell of radius, } r_{ij}\end{pmatrix}}{\text{amplitude of } E_{\text{scatt}} \begin{pmatrix}\text{from two driven electrons located close} \\ \text{together compared to } \lambda\end{pmatrix}} = \frac{\sin \mu r_{ij}}{\mu r_{ij}} \tag{15-36a}$$

where μ is defined as

$$\mu = (4\pi/\lambda) \sin (\theta_s/2) \tag{15-36b}$$

and r_{ij} is the distance between the two electrons. For an actual macromolecule with many driven electrons, it would be necessary to sum the scattered electric field amplitudes over all possible pairs of driven electrons. For example, for a spherical macromolecule in which the scattering centers are assumed to be uniformly distributed throughout the volume of the sphere, one need merely add up the scattering from spherical shells (Eq. 15-36) from radius zero to the radius of the sphere:

$$\frac{\text{amplitude of } E_{\text{scatt}} \begin{pmatrix}\text{from a macromolecular solid sphere} \\ \text{of diameter, } D\end{pmatrix}}{\text{amplitude of } E_{\text{scatt}} \begin{pmatrix}\text{Rayleigh scattering from the same} \\ \text{number of driven electrons located} \\ \text{close together compared to } \lambda\end{pmatrix}}$$

$$= \frac{\displaystyle\int_0^{(D/2)} \frac{\sin (\mu r)}{\mu r} 4\pi r^2 \, dr}{\displaystyle\int_0^{(D/2)} 4\pi r^2 \, dr}$$

$$\uparrow [\sin (\mu r)/\mu r \rightarrow 1 \text{ as } r \rightarrow 0]$$

$$= \frac{3}{a^3} [\sin (a) - a \cos (a)]; \; a = (\mu D/2) \tag{15-37}$$

Finally, since intensity varies as the square of amplitude, it is usual to define an "internal interference" factor, $P(\theta_s)$, as the ratio of scattered *intensity* when internal (intramolecular) interference is taken into account to the *intensity* computed in the absence of internal interference:

$$\boxed{P(\theta_s) = \frac{I_{\text{scatt}} \text{ (internal interference included)}}{I_{\text{scatt}} \text{ (internal interference ignored)}}} \tag{15-38}$$

Using Eq. 15-28 the Rayleigh ratio, $R(\theta_s)$ may now be written

$$\boxed{R(\theta_s) = K[c]MP(\theta_s)} \tag{15-39}$$

Scattering functions, $P(\theta_s)$ for macromolecules of various shapes are plotted in Fig. 15-6, and listed in Table 15-1.

Table 15-1 shows that $P(\theta_s)$ takes on a wide variety of functional forms, depending on the macromolecular shape. It would be desirable to express the scattering factor more simply, in terms of some shape parameter that could readily be calculated for a given molecular shape. Consider the scat-

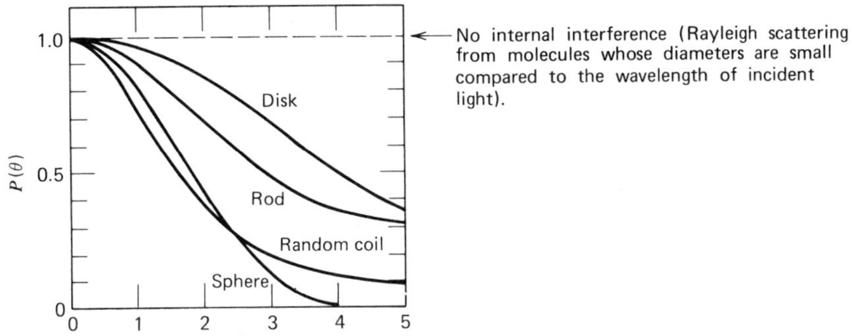

FIGURE 15-6. Scattering function, $P(\theta_s)$ for macromolecules of various shapes. The abscissa represents either $(\mu D/2)$ [where $\mu = (4\pi/\lambda)\sin(\theta_s/2)$ and D is the diameter of the sphere, the length of the rod, or the diameter of the disk], or $\sqrt{\mu^2 D^2/6}$, where D^2 is the root-mean-square end-to-end distance for the random coil. The abscissa thus represents a combination of scattering angle and molecular size: for a macromolecule of a given shape and size, $P(\theta_s)$ decreases as the scattering angle increases; similarly, for a given scattering angle, $P(\theta_s)$ decreases as macromolecular size increases. For macromolecules whose size is less than about 50 Angstrom (upper dotted line), the scattering factor is essentially unity for any scattering angle, using incident visible light (see text). [From C. Sadron and J. Pouyet, *Proc. 4th Intern. Congr. Biochem.*, Vol. 9, Vienna, 1958 (Pergamon Press, Oxford) p. 52.]

tering function, $P(\theta_s)$ for a solid sphere for angles approaching $\theta_s = 0$. In this limit, $a \ll 1$ in Eq. 15-37 and we may approximate $\sin a$ and $\cos a$ by the first few terms of their power series (see Appendix):

$$\sin a = a - \frac{a^3}{3!} + \frac{a^5}{5!} + \cdots \tag{15-40a}$$

$$\cos a = 1 - \frac{a^2}{2!} + \frac{a^4}{4!} + \cdots \tag{15-40b}$$

Table 15-1 Scattering Function, $P(\theta_s)$ for Various Macromolecular Structures

Structure	$P(\theta_s)$	Dimensions	Radius of Gyration, R_G
Sphere	$[(3/a^3)[\sin(a) - a\cos(a)]]^2;\ a = \frac{\mu D}{2}$	D = diameter of sphere	$(3/20)^{1/2}D$
Thin rod	$\frac{1}{a}\int_0^{2a} \frac{\sin x}{x}\,dx - \left(\frac{\sin a}{a}\right)^2;\ a = \frac{\mu D}{2}$	D = length of the rod	$D/\sqrt{12}$
Random coil	$\frac{2}{a^2}[\exp(-a) + a - 1];\ a = \frac{\mu^2 <D^2>}{6}$	$<D^2>$ = root-mean-square end-to-end distance	$\sqrt{<D^2>/6}$
Thin disk	$\frac{2}{a^2}[1 - (1/a)J_1(2a)];\ a = (\mu D/2)$	D = diameter of disk J_1 = Bessel function order 1	$(D/2\sqrt{2})$

so that $P(\theta_s)$ for a sphere becomes

$$P(\theta_s) \cong \left(1 - \frac{a^2}{10}\right)^2 \cong 1 - \frac{a^2}{5} = 1 - \frac{4\pi^2 D^2}{5\lambda^2} \sin^2(\theta_s/2) \quad \text{for a sphere of diameter, } D \qquad (15\text{-}41)$$

Defining a "radius of gyration," R_G, by Eq. 15-42:

$$\boxed{R_g^2 = \frac{\sum_i m_i r_i^2}{\sum_i m_i}} \qquad (15\text{-}42)$$

in which m_i represents the molecular mass located at distance, r_i, from the center of mass of the macromolecule, it is possible to compute the radius of gyration of a sphere according to

$$R_G^2(\text{sphere}) = \frac{\int_0^{(D/2)} r^2 4\pi r^2 dr}{\int_0^{(D/2)} 4\pi r^2 dr} = \frac{(r^5/5)\Big|_0^{(D/2)}}{(r^3/3)\Big|_0^{(D/2)}} = (3D^2/20) \qquad (15\text{-}43)$$

where D is the diameter of the sphere, and the amount of mass located between r and $r + dr$ from the center of the sphere is proportional to $4\pi r^2\, dr$.

Substituting Eq. 14-43 into Eq. 15-41

$$\boxed{P(\theta_s) \cong 1 - \frac{\mu^2 R_G^2}{3}, \text{ where } \mu = (4\pi/\lambda) \sin(\theta_s/2)} \qquad (15\text{-}44)$$

The value of Eq. 15-44 is that although it was derived for a *spherical* macromolecule, the equation is valid for macromolecules of *any* shape. Since calculation of R_G is straightforward for a given molecular shape (see Problems), Eq. 15-44 provides a direct means for determining the *size* of a macromolecule of *stipulated shape*, from measurement of the Rayleigh ratio, Eq. 15-39, in the limit of either low scattering angle, θ_s, or small macromolecule size, D, such that ($\mu D \ll 1$), where μ is defined by Eq. 15-36b. Less sensitively (note the similarity in the curves of Fig. 15-6 for molecules of *different* shape), it is possible to establish the *shape* of a macromolecule of a *stipulated size*. A quick calculation, based on the above Equations 15-36b and 15-44 shows that unless the macromolecular radius of gyration, R_G, is greater than about one-fiftieth the wavelength of visible light in aqueous solution ($R_g > 80$ Å), the scattering function $P(\theta_s)$ is essentially unity for any scattering angle. Since $R_g = 80$ Å corresponds to a molecular weight of about 400,000 for a spherical protein, it is clear that visible light scattering intensity ratio measurements can provide size and shape information about only the very largest natural and synthetic macromolecules. In the next section of this chapter, we will see that the much shorter incident radiation wavelength of X rays makes low-angle X-ray scattering much more suitable for characterization of macromolecular shape for typical proteins and enzymes.

Relation between Light Scattering and Solute Concentration: Zimm Plots

Equation 15-28 relating Rayleigh scattered intensity to macromolecular weight and concentration was based on Eq. 15-12, in which the total scattered intensity is equal to the scattered intensity from one source times the number of sources. Equation 15-12, in turn, was based on the assumption that the angular orientation and distance between two scattering sources (in this context, two different macromolecules in solution) varies randomly. However, for real solutions, it is clear that the inter-molecular distance and orientation are not free to vary randomly: (1) the macromolecules themselves occupy a fixed volume, so that the closest distance of approach between two macromolecular centers cannot be closer than a macromolecular diameter, and (2) there may be net attractive or repulsive forces between macromolecule and solvent, or solvent and solvent molecules, tending to favor nonrandom intermolecular distance and angle variation. These correlations in relative positions of solute molecules are manifested in the presence of an additional term for Eq. 15-28:

$$R(\theta_s) = K[c]\left[\frac{1}{\dfrac{1}{M} + 2B[c]}\right] \tag{15-45}$$

in which B is a molecular "interaction constant" known as the "second virial coefficient," whose origin was discussed in Chapter 2.B in the osmotic pressure section. Combining Eq. 15-45, which accounts for the nonideal nature of any real solution, with Eq. 15-39, which accounts for the partly destructive interference between light waves scattered from different parts of the same molecule, we obtain a result that provides three distinct pieces of information about macromolecules in solution:

$$\boxed{\frac{K[c]}{R(\theta_s)} = \frac{1}{P(\theta_s)}\left(\frac{1}{M} + 2B[c]\right)} \tag{15-46}$$

First, since $P(\theta_s) \rightarrow 1$ as $\theta_s \rightarrow 0$, Eq. 15-46 shows that the *macromolecular weight, M,* may be determined by extrapolating the Rayleigh ratio, $R(\theta_s)$, to the limit of zero concentration and zero scattering angle. Second, from experimental Rayleigh ratios at various scattering angles (but still

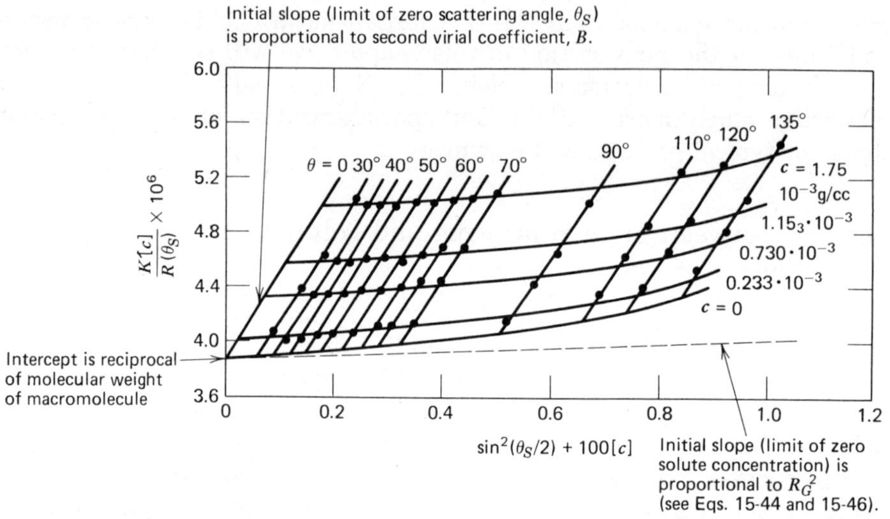

FIGURE 15-7. Zimm plot (see Equations 15-44 and 15-46 for rationale) for a synthetic large polymer in dilute salt solution. Note molecular weight (from intercept) = 260,000 Dalton. [From G. Ehrlich and P. Doty, *J. Amer. Chem. Soc.* *76,* 3764 (1954).]

extrapolated to zero macromolecular concentration), $P(\theta_s)$ may be determined, and information about *macromolecular size and shape* obtained as discussed for Fig. 15-6 and Table 15-1. Finally, from Rayleigh ratio determinations at various concentrations (but this time all extrapolated to zero scattering angle, θ_s), the interaction constant, *B*, may be found, yielding information about the *association* of macromolecules with each other and with the solvent (see Chapter 2.B, osmotic pressure). An example of a graphical procedure (proposed by Zimm) for conducting such an analysis is shown in Fig. 15-7, for a large synthetic polymer.

15.B. BIG OBJECTS, SMALL WAVES, NO LENSES: X-RAY SCATTERING; DETAILED MOLECULAR SHAPE

As pointed out in the previous section, scattering of visible light provides a means for determining molecular weights of typical biological macromolecules, but the size and shape of the molecule can be obtained only for very large macromolecules (M_w = 1,000,000 or larger) for which the particle dimensions (say 100 Å diameter for a sphere) become larger than about 1/50 the wavelength of the incident radiation (say 4000 Å). For such large macromolecules, it is uncommon to find molecules of $M_w = 10^6$ that are monodisperse (i.e., all molecules in the mixture have the same molecular weight) in solution; moreover, molecules as large as 100 Å diameter may be visible directly in the electron microscope (see next Section). One is thus led to seek shape information from scattering measurements using radiation of shorter wavelength. However, as we found in Chapter 13.B, our scattering treatment breaks down when the frequency of the radiation approaches the "natural" frequency of the electron-on-a-spring. Thus, since proteins and nucleotides begin to absorb radiation at wavelengths below about 3000 Å (and since even water and air begin to absorb below about 2000 Å), we must go even higher in frequency to find suitable radiation. However, there is then an absence of radiation *sources* until the X-ray region ($\lambda = 1.54$ Å for the typical copper Kα line) is reached.

Although the formal scattering treatment of the previous section applies with the same *theoretical* validity to X-ray scattering, a number of *practical* differences can be noted. First, examination of Fig. 15-6 (and Eq. 15-44) shows that even for relatively small macromolecules (say, lyzozyme, a small enzyme, $M_w = 14{,}100$), the scattered X-ray intensity falls essentially to zero at only a few degrees of arc away from the incident direction. For larger macromolecules, the scattering essentially disappears for scattering angles greater than a fraction of one degree. Thus, experimental measurement of scattered intensity must be carried out at *very small scattering angle.* Fortunately, X rays can be collimated to within 0.001 rad, so low-angle scattering measurements are feasible. Second, the 2000-fold decrease in wavelength of incident radiation in going from visible to X ray makes it possible to determine the sizes and shapes of most enzymes, proteins, and nucleic acids; moreover, since dust particles are so large compared to X-ray wavelength, the scattering from dust particles decreases to zero by such small scattering angle that dust provides essentially no contribution to the observed low-angle X-ray scattering, thus making solution preparation simpler and less critical.

With regard to treatment of data, an additional simplification (compared to light scattering) follows from the very small scattering angle

$$\sin(\theta_s/2) \cong \theta_s/2,\ \theta_s << 1 \text{ radian} \tag{15-47}$$

so that Eq. 15-44 becomes

$$P(\theta_s) \cong 1 - \frac{16\pi^2 R_G^2 (\theta_s/2)^2}{3\lambda^2},\ \text{for } \theta_s << 1 \tag{15-48}$$

To provide a more direct experimental determination of R_G, Guinier pointed out that since

$$\exp[-(16\pi^2 R_G^2(\theta_s/2)^2/3\lambda^2] = 1 - \frac{16\pi^2 R_G^2(\theta_s/2)^2}{3\lambda^2} + \frac{128\pi^4 R_G^4(\theta_s/2)^4}{9\lambda^4} + \ldots \tag{15-49}$$

and since the first two terms of Eq. 15-49 are the same as for Eq. 15-48 (and even the third term of Eq. 15-49 is the same as Eq. 15-48 for the case of a sphere), it seems reasonable to obtain the radius of gyration, R_G, from the slope, $[-16\pi^2 R_G^2/3\ \lambda^2]$, of a plot of $\log_e\left(\frac{I_{\text{scatt}}}{I_0}\right)$ versus $(\theta_s/2)^2$ (Guinier plot).

Table 15-2 shows some illustrative molecular weights and radii of gyration determined by either light- or low-angle X-ray-scattering from macromolecules in solution.

Table 15-2 Low-angle X-ray and Light-scattering ([a]) Determinations of Molecular Weight and Radius of Gyration for Representative Macromolecules.

Substance	State	R_G	Molecular Weight
Lysozyme	solution	14.3 ± 0.3 Å	14,100
	crystal	13.8[b]	
α-chymotrypsin	solution	18.0	24,500
	crystal	16.0[b]	
α-lactoglobulin A	solution	21.6 ± 0.4	36,600 (monomer)
Myoglobin	solution	16.3 ± 0.5	17,000
	crystal	15.5 ± 0.5[b]	
Myosin	solution	468[a]	493,000
DNA	solution	1170[a]	4,000,000
Bushy stunt virus	solution	120	10,600,000
Tobacco mosaic virus	solution	924[a]	39,000,000

[b] *R_G-values for crystals are based on structures determined by X-ray diffraction* (*see Chapter 22.A*).

The table shows that scattering data can be used to find molecular weights and sizes of molecules ranging over three to four orders of magnitude. It may be noted that while the crystal and solution structures of, for example, myglobin are the same within experimental error, there is a marked difference between the crystal and solution structures of the enzyme, α-chymotrypsin (the differences are even more pronounced when X-ray crystallographic data are used in attempts to fit X-ray scattering results in a Guinier plot). It has been shown that the axial ratio for chymotrypsin changes from 1.3 to 2.0 on going from crystal to solution phase, presumably because the enzyme molecule unfolds partly to a more asymmetric shape. In the following examples, we see how the information from X-ray low-angle scattering data can be used to reach important qualitative conclusions about the structure of macromolecules in solution.

EXAMPLE *Assembly of Protein Subunits: α-lactoglobulin A*

α-lactoglobulin A is a protein consisting of subunits of molecular weight 36,000, whose subunits combine to form a tetrameric species having a radius of gyration, $R_G = 34.4 \pm 0.4$Å. From the known radius (see Table 15-2) of

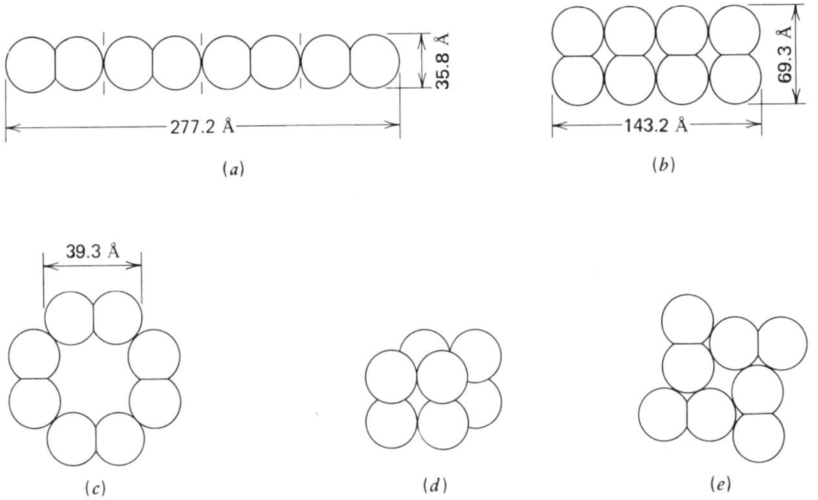

Calculated Radii of Gyration (R_G) for Various Tetrameric Models

Model	R_G(Å)
Sphere	27.7
(*a*)	81.0
(*b*)	47.1
(*c*)	48.7
(*d*)	35.4
(*e*)	40.5
Experimental value	34.4 ± 0.4

FIGURE 15-8. Various models for the assembly of α-lactoglobulin A tetramer from four monomeric units that are themselves formed from the attachment of two spherical portions. Comparison of the calculated and experimental X-ray low-angle scattering values of the radius of gyration, R_G, permits immediate selection of assembly *d* as the most likely [From J. Witz, S. Timasheff, and V. Luzzati *J. Amer. Chem. Soc. 86,* 168 (1964).]

gyration of the 36,600 species, and from the fact that the 36,600 species can itself be cleaved to form two spherical units of molecular weight about 18,000 each, we are in a position to evaluate the feasibility of several possible modes of assembly of those subunits to form the tetrameric entity (see Fig. 15-8). It is immediately easy to rule out the various extended structures, with the direct conclusion that a cubic arrangement of the eight (18,000 molecular weight) spheres is the most likely structure. This example shows how just a single piece of data (in this case, R_G) can vastly narrow down the number of possible macromolecular structures.

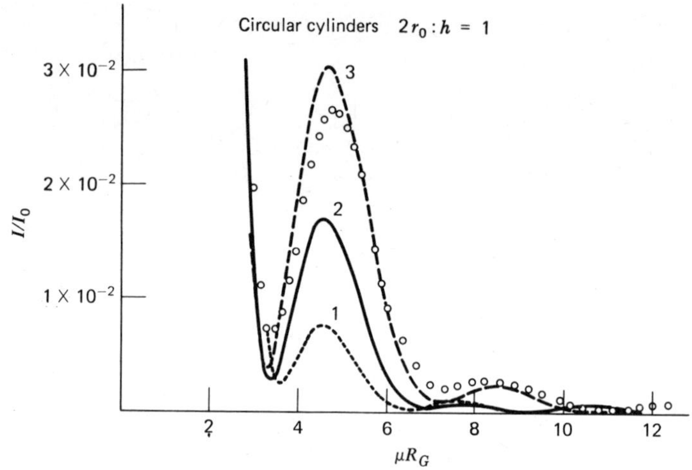

FIGURE 15-9. Comparison of the experimental scattering curve (○) for hemocyanin in water with theoretical curves for different circular cylinders, the ratio of diameter to length of which is 1:1. Curve 1 (dotted line) corresponds to a solid cylinder, while curve 2 (solid line) corresponds to a hollow cylinder with a ratio of inner radius to outer radius r_i/r_0 of 0.3 and curve 3 (dashed line) to a hollow cylinder with a ratio of r_i/r_0 of 0.5. $\mu = (4\pi\theta_s/\lambda)$, R_G = radius of gyration. [From I. Pilz, O. Kratky, and I. Moring-Claesson, *Z. Naturforsch. 25b,* 600 (1970).]

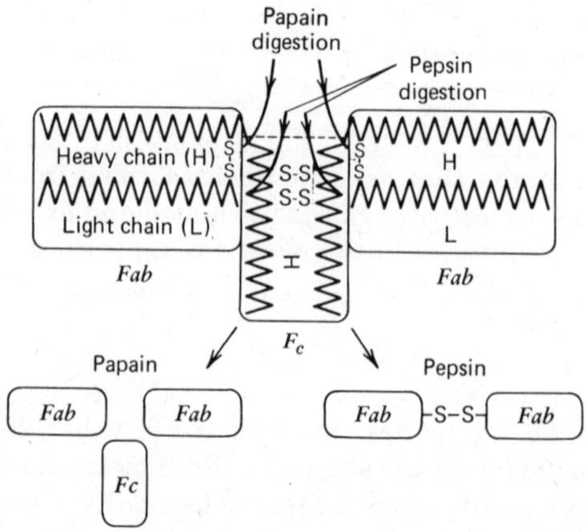

FIGURE 15-10. Schematic diagram of the structure and cleavage into different fragments by the enzymes papain and pepsin of the IgG-immunoglobulin molecule. The two specific binding sites for antigens are expected to be located at the opposite ends of the molecule in the "Fab" regions. [From I. Pilz, *Allg. Prakt. Chem. 21,* 21 (1970).]

EXAMPLE *Finding Holes in Macromolecules: Hemocyanin*

Hemocyanin, the nonheme copper-containing blue respiratory pigment found in snails, has been shown from X-ray scattering to have a molecular weight of 9×10^6 Dalton and a radius of gyration of $R_G = 164$ Å. Comparison of the experimental plot of scattered intensity as a function of scattering angle with theoretical plots for molecules of various shapes (see following example), shows that hemocyanin behaves in solution as a circular cylinder with about the same height as diameter. However, an immediate difficulty is that from the radius of gyration and the ratio of diameter to height, the volume of the corresponding solid cylinder can be calculated as 3.66×10^7 Å^3, which is more than twice as large as the volume determined by other measurements, 1.74×10^7 Å^3. This anomaly can be resolved, only if it is supposed that there is a hollow space in the interior of the cylinder. Figure 15-9 then shows that the best fit to experimental scattered intensity data as a function of scattering angle is obtained for a cylinder with a height of 360 Å, and inner and outer radii, r_i and r_0, of 74 Å and 164 Å.

EXAMPLE *Detailed Macromolecular Shape from X-ray Low-angle Scattering: IgG Immunoglobulins*

Following the availability of the extremely homogeneous immunoglobulins from patients with multiple myeloma, the amino acid sequence and location of disulfide bonds in the IgG-immunoglobulin molecule were established in 1969, as shown in Fig. 15-10.

From the papain digestion, the dimensions of the individual *Fab* unit could be determined from its radius of gyration and the shape of the Guinier plot. Subsequent X-ray scattering from the $F(ab')_2$ dimer from pepsin digestion gave a radius of gyration, $R_G = 53$ Å, corresponding nicely with a proposed "hinged" connection between the two *Fab* portions:

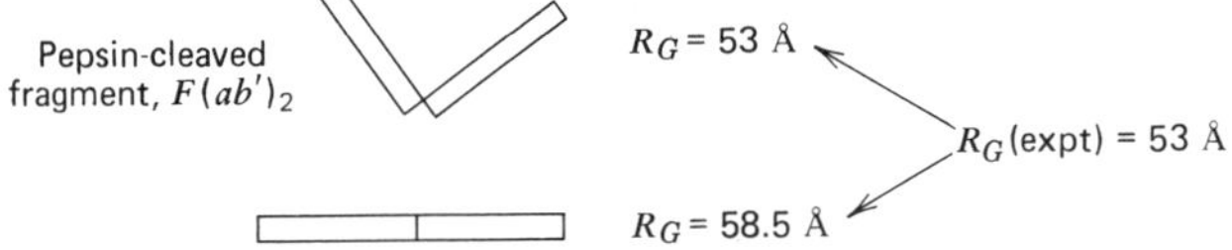

Finally, a number of possible structures for the fully intact IgG immunoglobulin were then used to generate theoretical plots of $\log(I_{\text{scatt}})$ versus $(\theta_s)^2$, using the known dimensions of the *Fab* and *Fc* components, and the resultant curves compared to the experimental scattering shown in Fig. 15-11. The best agreement was for structures in which the ends of the various fragments overlapped, as shown in hypothetical structure #9 in the figure. It is interesting to note that *subsequent* crystal X-ray diffraction data gave a structure that confirmed the solution structure deduced from these scattering measurements, in contrast to many sophisticated biophysical studies that have confirmed the X-ray crystal structures of various molecules *after* the crystal structure was known!

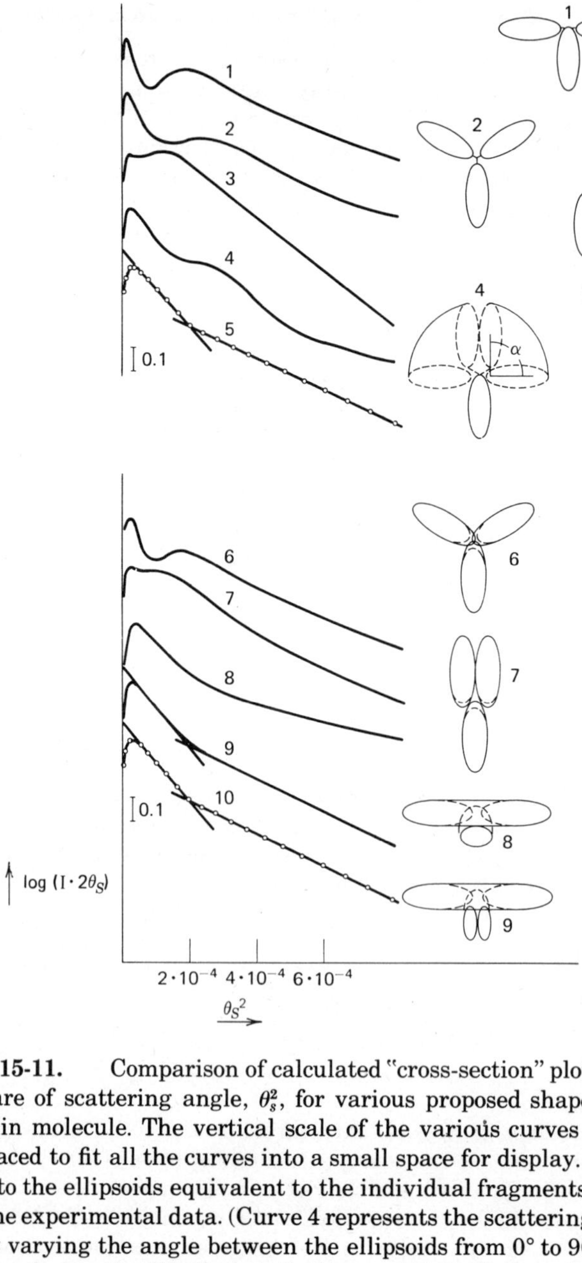

FIGURE 15-11. Comparison of calculated "cross-section" plots of $\log[I_{\text{scatt}} \cdot Z\theta_s]$ versus square of scattering angle, θ_s^2, for various proposed shapes of the IgG immunoglobulin molecule. The vertical scale of the various curves has been successively displaced to fit all the curves into a small space for display. The dashed lines correspond to the ellipsoids equivalent to the individual fragments. Curves 5 and 10 represent the experimental data. (Curve 4 represents the scattering curve for shapes obtained by varying the angle between the ellipsoids from 0° to 90°.) [From I. Pilz, G. Puchwein, O. Kratky, M. Herbst, O. Haager, W. E. Gall, and G. M. Edelmann, *Biochemistry 9,* 211 (1970).]

15.C. BIG OBJECTS, SMALL WAVES, LENSES AVAILABLE: ELECTRON SCATTERING AND THE VARIOUS ELECTRON MICROSCOPES

In the previous two sections, we have presented the basis for experiments designed to learn about molecular size and shape for molecules that are continually moving about *in solution,* and we found we could obtain the most detailed information when the wavelength of the incident radiation was of the order of the particle dimension. One might next inquire as to the best means for study of *solid* objects, using the same kinds of incident radiation. For the case of a *crystalline* solid, with near-perfect regular order in the arrangement of molecules, it is possible to present a relatively simple brief explanation of the nature of scattering of radiation (typically X-rays) whose wavelength is of the order of the spacing between adjacent molecules in the array (see Chapter 22.A). For the more usual situation of interest, in which we are interested in visualizing a noncrystalline object, such as a biological cell or organelle or macromolecule, the requisite mathematical treatment is based on the same principles of adding up the scattered electric field amplitudes from all the driven electrons in the object, but a detailed analysis would take us far afield into the subject of geometrical optics. For this discussion, we shall therefore satisfy ourselves with the principal result from geometrical optics as related to microscopes, namely, that the "resolving power" of a microscope (defined as the minimum distance, d, between two points in the specimen that can just be seen as distinct) is given by

$$d = \frac{0.61\ \lambda}{n\ \sin(\alpha)} \qquad (15\text{-}50)$$

in which λ is the wavelength of the radiation used, n is the index of refraction of the medium between the object and the lens, and α is the angle between the axis of the instrument and the outermost rays leaving the object that will still pass through the microscope objective lens.

Equation 15-50 indicates that, as for study of molecules in solution, the finest detail resolvable for solid objects is of the order of the wavelength of the incident radiation. It is also clear why oil is often placed between the cover slip and objective lens of a light microscope, since the oil acts to increase both n and α, and thus increases resolving power. Equation 15-50 shows that resolving power is proportional to wavelength: the best light microscope resolving power should thus be obtained using ultraviolet light (λ = 2000 Å). The main problems in attempting to improve the resolving power by using radiation of shorter wavelength are, first, that water and air absorb strongly in the wavelength region immediately below 2000 Å, and, second, the next logical available radiation sources (X-ray at about 1 Å and gamma rays at wavelengths down to many times shorter than 1 Å) is that there are no lenses capable of focusing X-rays or gamma rays, so there is no way to build a microscope using those sources. The reader should recall that the basic reason that lenses work at optical wavelengths is that it is possible to find a substance whose refractive index is appreciably greater than 1.0—as we saw in Chapter 14.A, a refractive index greater than 1.0 is directly associated with an absorption of radiation. Thus, since X-rays and gamma rays are negligibly absorbed in passing through most small objects, there is no substance that can be used to make a lens, and hence no microscope.

We see in Section 5 that electromagnetic radiation displays a *particle* ("quantum") behavior, most directly manifested in spectroscopy. De Broglie proposed in 1927 that particles could similarly display a *wave* character, according to

$$\lambda = (h/mv) \tag{15-51}$$

in which λ is the wavelength associated with a particle of mass m moving at velocity v, and h is Planck's constant $= 6.62 \times 10^{-27}$ erg sec. Equation 15-51 suggests that "radiation" of very small wavelength should be associated with rapidly moving particles of small mass. Specifically, electrons are a logical choice for the "light source" for a microscope based on Eq. 15-51, since electrons are charged and can thus be accelerated to very high velocity (and thus very small associated wavelength), and the motion of the electrons is easily controlled by application of electric and magnetic fields so that "lenses" can be constructed. It is left as an exercise for the reader to show that the effective wavelength, λ, of an electron accelerated by a voltage, V volts, is given by

$$\lambda = \frac{12.3}{\sqrt{V}}\ \text{Å} \tag{15-52}$$

Because electron microscopes are typically operated at voltages ranging from about 40keV to about 3 MeV, the corresponding radiation wavelength varies from about 0.06 Å to less than 0.01 Å. Light and electron microscope design is shown schematically in Fig. 15-12.

Unfortunately, while microscopes using *visible* light can be made to approach rather closely the theoretical resolving power of Eq. 15-50, the *electron* microscope typically shows actual "resolution" (minimal spatial separation at which images of two point objects are distinct) of real objects (as opposed to "resolving power" for hypothetical objects) about 1000 times larger than the inherent electron wavelength (see Table 15-3). In the first place, it is theoretically impossible to construct a net divergent cylindrically symmetric electron lens, so that there is no means for correcting spherical aberration in the lenses. Since spherical aberration is most pronounced for rays farthest off-axis along the optical path, this (severe) constraint requires

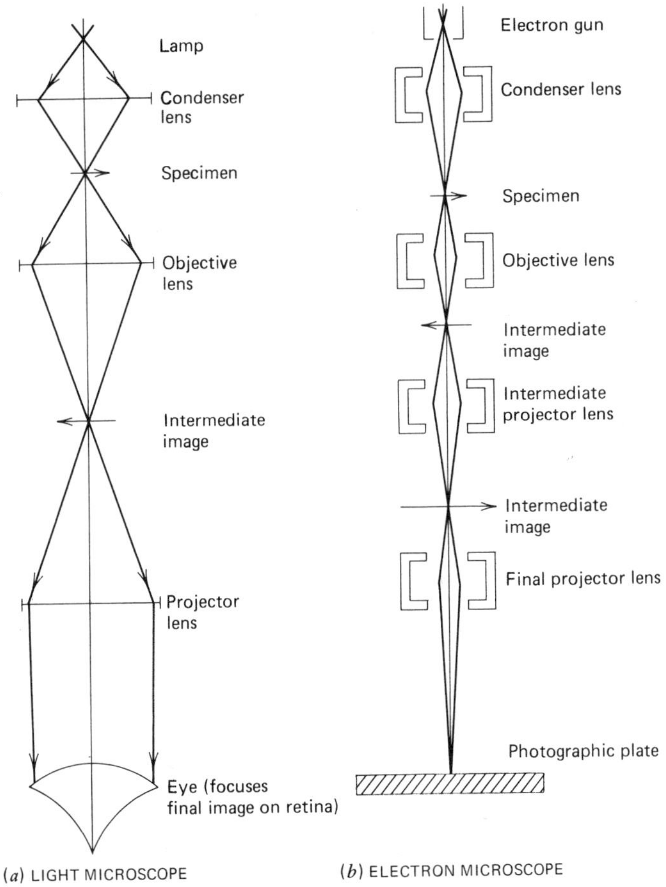

FIGURE 15-12. Schematic comparison of the optical components of (*a*) the compound light microscope and (*b*) the transmission electron microscope. Each instrument consists of a source (lamp, electron gun), condenser lens, imaging lenses (shown here as a two-lens system in the light microscope and a three-lens system in the electron microscope); and a detector (eye, photographic plate). Although the construction of the two microscopes is similar, the basis for operation is completely different, as shown in Table 15-3. (Diagram taken from E. M. Slayter, in *Physical Principles and Techniques of Protein Chemistry,* Part A, ed. S. J. Leach, Academic Press, New York, 1969, p. 5.)

that the aperture of the electron lens be stopped down to a very small value of α in Eq. 15-50, limiting practical resolution to about 2 Å even for electrons of wavelength 0.02 Å.

Since the "color" of a moving electron (i.e., its associated wavelength) is related to the accelerating voltage by Eq. 15-52, and since *chromatic aberration* (focussing of different colors at different places) in the electron microscope cannot be corrected using lenses, the only way to eliminate chromatic aberration is by monochromating the source. The accuracy to within which a typical 100 keV accelerating voltage can be regulated thus limits practical electron microscope resolution to a value not better than about 2 Å. One advantage of the limited aperture dictated by the need to avoid spherical aberration is that the *depth of focus* (i.e., the thickness of the specimen within which sharp focus can be realized) for the electron microscope is much greater than for the light microscope—the electron microscope is thus much better suited for study of *surfaces* (as opposed to thin sections) than the light microscope at the same magnification (see "scanning" electron microscope below).

Table 15-3 Comparison of Properties of Light Microscope and Electron Microscope (See Text for Discussion)

Light Microscope	Property	Electron Microscope
2000–6000 Å	Wavelength, λ, of radiation used	40 keV − 3 MeV electrons, corresponding to $\lambda = 0.06 - 0.01$ Å
$\frac{0.61\ \lambda}{n\ \sin(\alpha)} \cong 2000$ Å	Theoretical limit to resolving power	$(0.61\ \lambda/\alpha) \cong 2$ Å (see spherical aberration)
Correctable using net divergent lenses	Spherical aberration	Net divergent lenses do not exist; correctable only by limiting aperture angle, α, to about 0.01 rad, so that $\sin(\alpha) \cong \alpha$ in Eq. 15-50.
Correctable using compound lenses	Chromatic aberration	Limited by constancy in accelerating voltage and in current to deflection lenses (about 1 part in 10^5), thus limiting resolution to no better than about 2 Å.
1	Relative depth of focus at a given magnification	500 (improvement due to narrower aperture relative to light microscope)
Different *absorption* by different parts of specimen; or different *path length* (and thus different *phase*) for light passing through different parts of specimen	Origin of contrast in the image	Different *scattered intensity* for electrons impinging on different parts of specimen.
Use of dyes to increase the *absorption* by the specimen relative to background (positive dyes) or background relative to specimen (negative dyes), for absorption microscope; dark-field technique for phase-contrast microscope	Means for enhancing image contrast	Use of positive (or negative) stains to increase electron *scattering* from specimen relative to background (or vice versa); shadowcasting to highlight surface relief; metal coating to make surface visible; spherical aberration in lens or reduced aperture to improve contrast; dark-field techniques.

At this stage, it would appear that even with the limitations imposed by spherical and chromatic aberration, the practical resolution of an electron microscope ought to be sufficient to distinguish individual chemical bonds and atoms in molecules. The reason that electron microscopes seldom show detail resolved to better than about 50 Å is poor image *contrast*. Contrast is defined as the difference in apparent brightness of the specimen versus the background in the image:

$$\% \text{ contrast} = \frac{I_{\text{background}} - I_{\text{specimen}}}{I_{\text{background}}} \times 100 \tag{15-53}$$

In the light microscope, it is often possible (particularly with use of dyes) to approach 100% contrast (white background, black object), since there may be very great differences in the *absorption* of light by the specimen compared to the background. However, in the electron microscope, contrast arises when electrons are scattered beyond the reach of the lens (see Fig. 15-13) by the specimen. The trouble is that since the relative intensity of electron scattering is directly related to the electron density of the atoms in the specimen, and since the molecules of biological interest consist primarily of atoms of low atomic number (carbon, oxygen, nitrogen), the scattering from the specimen is bound to be similar to the scattering from the (carbon) film on which the specimen rests (the thin 20 Å-thick carbon film is supported by a wire grid to hold its shape). A quick calculation shows that a 70 Å thick layer of carbon (specimen) is required at an electron accelerating voltage of 100 keV to produce a 5% change in intensity at the image plane (i.e., 5% contrast). The contrast problem in electron microscopy may be compared to the difficulty in conducting light microscopy using a deeply colored stained-glass slide!

Before examining actual electron micrographs, it is important to judge correctly the relation between relative sizes of objects and their respective mass (or molecular weight). For spherical objects of molecular weight, M,

$$M = (4/3)\pi r^3 \, \rho \, N_0 \tag{15-54}$$

in which r is the molecular radius, ρ is the density (g cm^{-3}) of the molecule, and N_0 is Avogadro's number. In other words, the apparent diameter, D, of a macromolecule in an electron micrograph varies as the *cube root* of the molecular weight:

$$D = (6M/\pi\rho N_0)^{1/3} \text{ cm} \tag{15-55}$$

so that two macromolecules must differ in molecular weight by a factor of *eight* so that their apparent sizes differ by a factor of *two* in a micrograph. Therefore, electron micrographs are a poor criterion for *homogeneity* in molecular weight for macromolecules, since it is difficult to distinguish

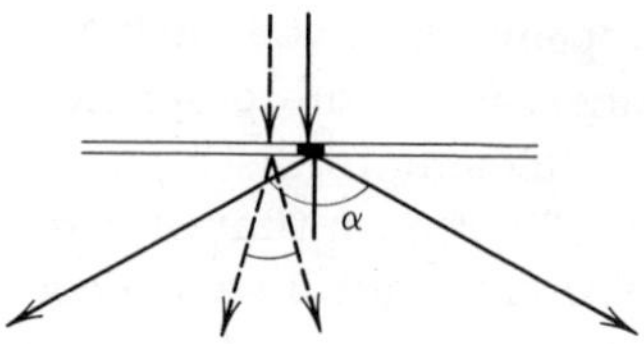

(a)

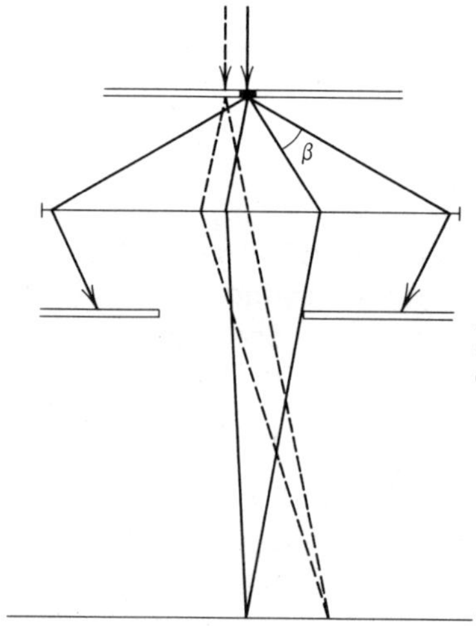

(b)

FIGURE 15-13. Contrast in the electron microscope due to differential scattering by specimen and background. Only those electrons transmitted or scattered into the aperture of the objective lens can contribute to image intensity. If the specimen is thicker or more electron-dense than the supportive film, then electrons emerging from the specimen will, on the average, be scattered to wider angles than electrons emerging from the background film, by the same arguments used to account for the angular-dependence of light scattering earlier in this Chapter. (*a*) Unapertured lens: since essentially all scattered electrons fall within the lens aperture, there is no way to distinguish specimen from background, and contrast is zero. (*b*) By using a limited aperture, electrons scattered to wide angles (i.e., from the specimen) are preferentially prevented from reaching the image plane, so that the intensity of the image of the specimen is reduced (by the electrons scattered over angular region β in the figure) while the intensity of the image of the background is essentially unaffected, leading to increased contrast. The same sort of improvement in contrast occurs because of spherical aberration in the lens – in this case, only electrons passing near the transmission axis are properly focussed, while electrons (mostly from the specimen) that pass through the outer region of the lens are poorly focussed and again there is reduced intensity at the point where the specimen would appear in the image plane. (Diagram from: E. M. Slayter, in *Physical Principles and Techniques of Protein Chemistry,* Part A, ed. S. J. Leach, Academic Press, N. Y., 1969, p. 8.)

even between two macromolecules differing by a factor of 2 in molecular weight. As a calibration, it is easy to show from Eq. 15-55 that a spherical macromolecule of molecular weight, $M = 100{,}000$, with a density of 1.3 g/cm^3 (typical density for anhydrous proteins), will have a diameter of about 60 Å.

Additional operating features of the electron microscope are (1) the electrostatic and magnetic lenses (similar to the "focus" and "convergence" controls on a color television set) provide continuously variable focal length, so that *magnification is continuously controllable,* (2) since electrons are scattered by air, the operating pressure along the optical path must be $\leq 10^{-3}$ mm Hg, (3) the "slide" of an optical microscope is replaced by a thin carbon film supported by a copper screen mesh, and (4) the detection is by means of a fluorescent screen (as in a television) or with photographic film.

EXAMPLE *Positive Staining—Visualization of RNA Synthesis*

Many of the stages of nucleic acid replication, transcription, and protein synthesis have recently been demonstrated graphically from images from positively stained samples (i.e., specimen stained with heavy metal atoms more than background). In the striking picture shown in Fig. 15-14, it is possible to distinguish the extrachromosomal nucleolar DNA gene, on which some 80 to 100 RNA polymerase molecules are simultaneously transcribing each gene to form strands of a protein-RNA precursor molecule that will later be cleaved to form ribosomal RNA. Successive stages of synthesis are clearly evident as a graded series of lengths for the product molecules. The identification of various components as DNA, RNA, and protein is based on use of stains that are selective for those components.

EXAMPLE *Positive Staining—Biological Membranes*

Figure 15-15 shows one of the clearest electron micrographs of a cross section of a cell plasma membrane, taken from intestinal epithelial microvilli "brush border" cells. The surface proteins of each membrane are stained with uranyl acetate and lead citrate (i.e., heavy metal attachment to enhance scattering). We refer to this structure in our discussion of protein movement in membranes in the next chapter.

EXAMPLE *Negative Staining—Assembly of Enzyme Subunits*

Some of the most spectacular electron micrographs of individual macromolecules have been obtained using negative staining (or more accurately, negative contrast), in which a heavy-metal-containing stain is chosen to bind preferentially to the background film, leaving the molecule of interest unstained. Because electrons passing through the specimen are thus focussed relatively completely on the image plane, the photographic film will be *darkest* where the specimen image falls, and the photographic print will thus be *lightest* at the location of the specimen.

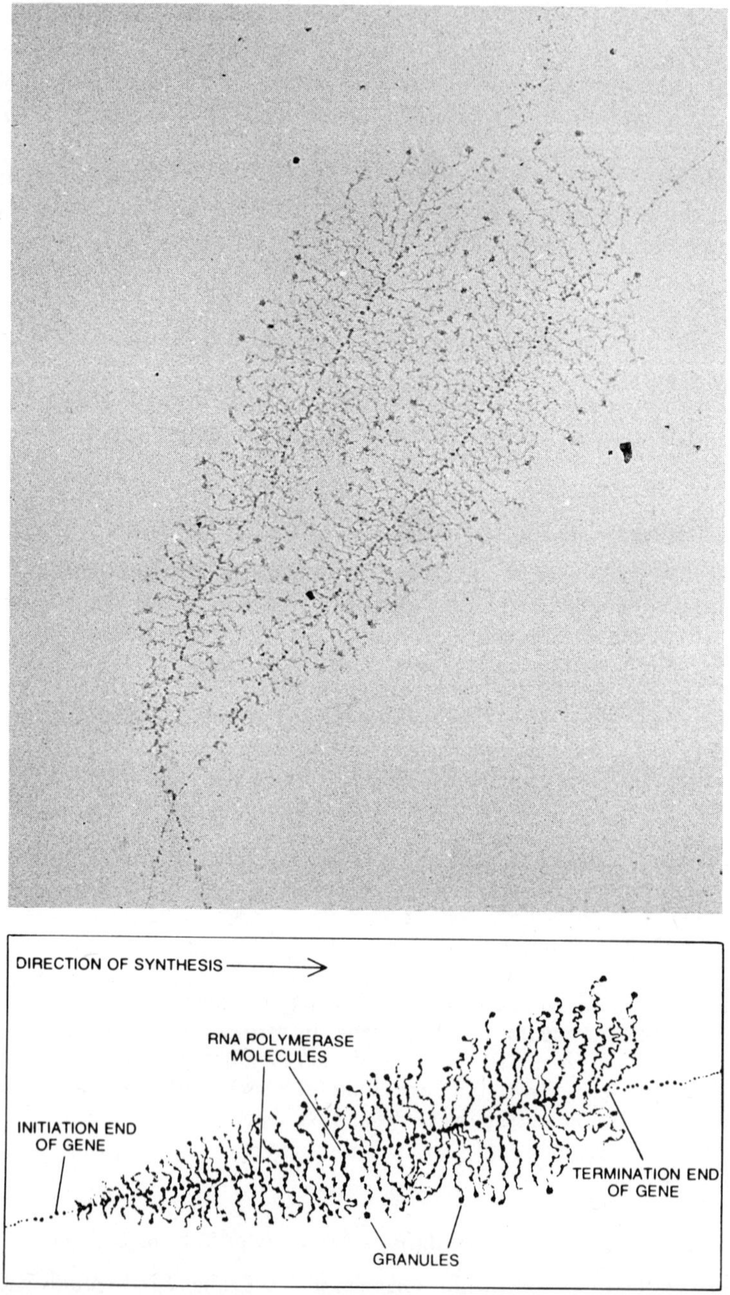

FIGURE 15-14. 48,000-fold magnification of RNA polymerase molecules actively engaged in transcription from DNA to ribosomal RNA precursor molecules. The polymerase molecules are visible as dense granules attached to the DNA gene (long thin central fibril), with each polymerase moving from left to right as transcription proceeds. The synthesized RNA molecules appear as individual strands of the fringe-like structure growing out from the gene. [Courtesy O. L. Miller, Jr., from "The Visualization of Genes in Action" by O. L. Miller, Jr. Copyright © March 1973 by Scientific American, Inc. All rights reserved.]

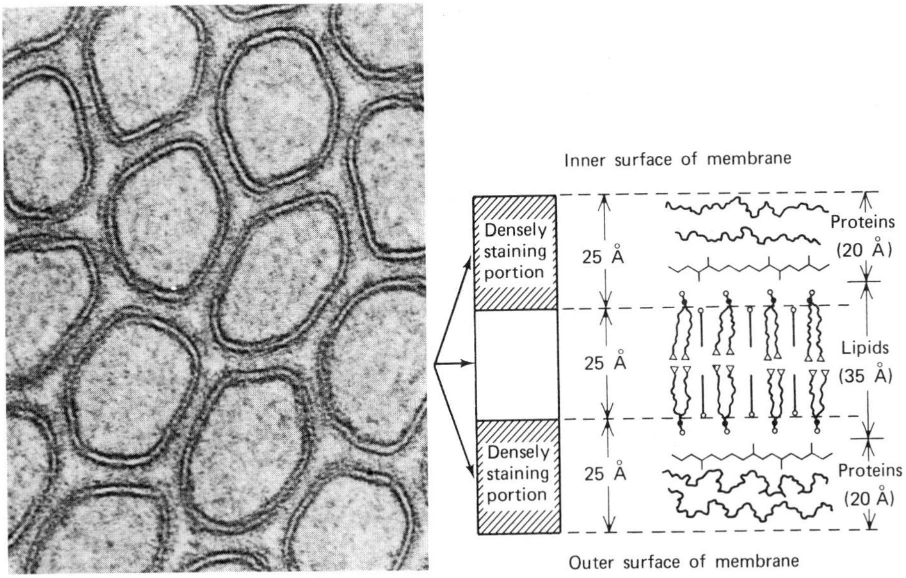

FIGURE 15-15. "Trilaminar" structure of a cell plasma membrane. The proteins on the inner and outer surface of each membrane are positively stained with heavy metal salts. Magnification = 230,000X in electron micrograph. The photograph is taken from a section cut parallel to the main inside surface (from which many small hollow cell projections called microvilli stick out) of cat intestine. (Micrograph from D. W. Fawcett, *An Atlas of Fine Structure: The Cell,* W. B. Saunders Co., Philadelphia, 1966, p. 342; schematic diagram from M. K. Jain, *The Bimolecular Lipid Membrane: A System,* Van Nostrand Reinhold Co, New York 1972, p. 11.)

Assembly of enzymes with roughly spherical subunits is shown in Fig. 15-16. Interpretation of the proposed polymeric structure is made more convincing by the observation of the molecule from several angles.

Assembly of an enzyme composed of nonspherical subunits is shown in Fig. 15-17 for the enzyme, aspartate transcarbamylase (ATCase). The arms of the regulatory subunit are clearly visible in the top view of the enzyme in the electron micrographs (negatively contrasted with phosphotungstate), and the "sandwich" structure shows clearly in the side view. Further corroboration is provided by the triangular shape (with no projecting arms) shown by the catalytic subunit alone. The subunit organization shown in Fig. 15-17 has recently been confirmed by X-ray diffraction studies on the crystalline enzyme.

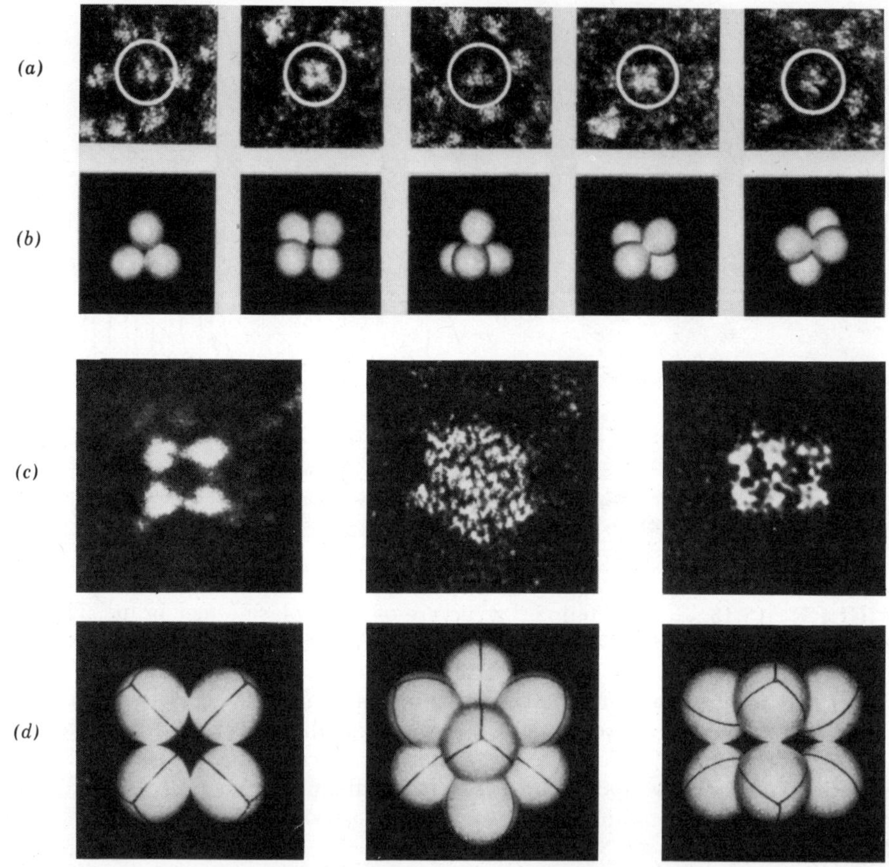

FIGURE 15-16. Electron micrograph images and interpretative models of the enzymes, pyruvate dehydrogenase (rows A and B) and dihydrolipoyl transacetylase (rows C and D). Row A shows individual images (700,000X magnified), with the tetrahedral model viewed from the corresponding orientation in row B. Row C shows various orientations of the transacetylase, with corresponding views of the model in Row D, at 350,000X magnification. [From L. J. Reed and R. M. Oliver, *Brookhaven Symp. Biol. 21,* 397 (1968), as reprinted in L. J. Reed and D. J. Cox, in *The Enzymes,* ed. P. D. Boyer, Vol. 1, 3rd ed., Academic Press, New York, 1970, p. 218.]

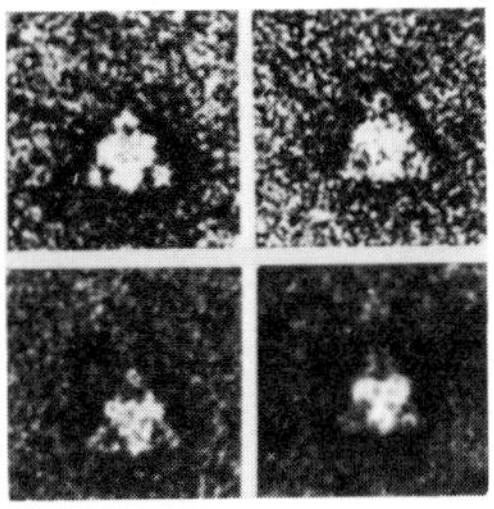

Top view, holoenzyme

Side view of holoenzyme

Top view of catalytic subunit

Model for the arrangement of the catalytic subunits and regulatory subunits of ATCase. The regulatory unit consists of a two-armed moiety which has one arm bound to the upper trimer of catalytic units and the other arm bound to the lower trimer of catalytic units as shown.

FIGURE 15-17. Negative-contrast electron micrographs (upper 3 photos) and hypothetical model for assembly of enzyme subunits. The top right photo shows the trimer that forms the catalytic subunit; in the native enzyme, two such trimers are stacked on top of each other in eclipsed fashion, and held together by three sets of regulatory subunits. Leftmost micrograph is at 400,000X, other two photos at 360,000X magnification. [Micrographs from K. E. Richards and R. C. Williams, *Biochemistry 11,* 3393 (1972); Model from J. A. Cohlberg, V. P. Pigiet, Jr., aad H. K. Schachman, *ibid.,* 3407.]

Shadow-Casting and Freeze-Fracture Electron Microscopy

One of the older techniques in electron microscopy, shadow-casting, has recently been revived with great interest in connection with freeze-fracture studies of the interior of biological membranes. The basic idea in shadow-casting is again dependent on the strong electron-scattering property of heavy metal atoms. Heavy metal atoms are directed at an angle toward the specimen, with the result that metal accumulates to form a layer some 20 Å thick as shown in Fig. 15-18. When the metal-coated specimen is now examined in a top view in a transmission electron microscope, any projections from the surface are readily apparent from the additional accumulation of metal on the object, as well as an absence of metal in the uncoated ("shadowed") region immediately next to the object. Thus, contrast is greatly enhanced, and it is possible to detect surface relief of about 20 Å in

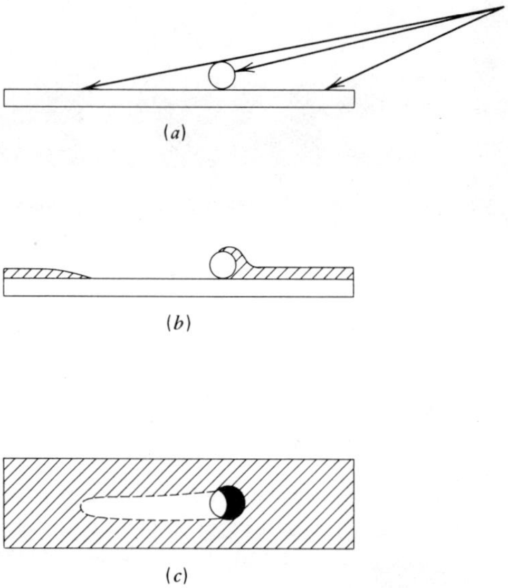

FIGURE 15-18. Schematic description of shadow-casting. (*a*) Heavy atom trajectories, impinging on the specimen to be shadowed. (*b*) metal accumulation on completion of shadowing. (*c*) relative electron density for specimen viewed at right angles to (*a*) and (*b*) (From E. M. Slayter, in *Physical Principles and Techniques of Protein Chemistry,* Part A, ed. S. J. Leach Academic Press, New York 1969, p. 23.)

height, so that protein molecules of typical size (say, larger than 50,000 molecular weight) are often visualized.

Although shadowing has to a large extent been replaced by staining methods for electron microscopy of macromolecules, it is still extremely useful in conjunction with freeze-fracture methods. The freeze-fracture technique is based on the principle that when a biological specimen is quickly frozen and then made to fracture by a suitable hard blow, the fracture lines will follow a path of least resistance as the solid breaks. Because the forces holding the outer and inner halves of a biological membrane together are apparently weaker than the forces holding the (crystallized) solvent together, or that hold the solvent to the membrane surface, breakage often occurs right down the middle of the membrane, exposing the interior surface, as shown in Fig. 15-19. The pattern by which protein molecules are arranged laterally on the membrane interior is readily visualized by shadowing the freeze-fractured surface. By such means, for example, it has been found that membrane proteins at cell junctions, as between two liver cells, show a characteristic closely packed hexagonal arrangement ("nexus") that appears to be instrumental in facilitating transport of various molecules from one cell to another.

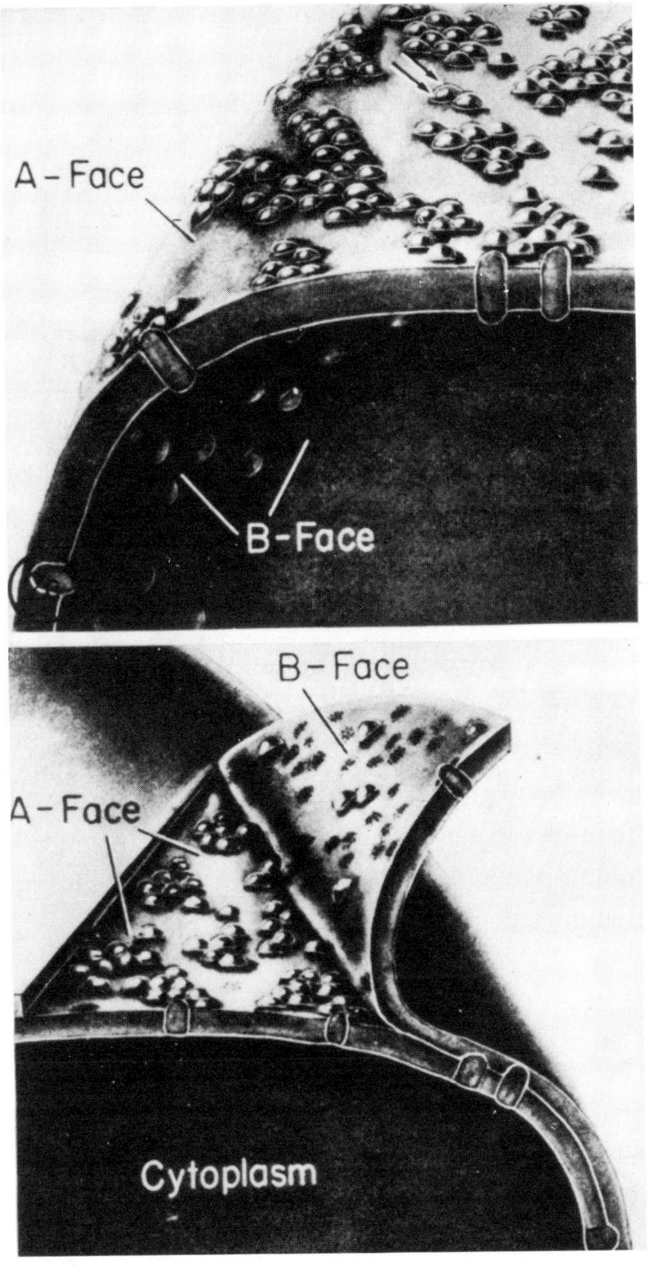

FIGURE 15-19. Schematic representation of the result of freeze-fracturing a plasma membrane. The surface shows bumps and holes where a given protein molecule has been pulled off one of the cleaved layers. [From N. S. McNutt and R. S. Weinstein, *Progr. Biophys. & Mol. Biol. 26,* 57 (1973).]

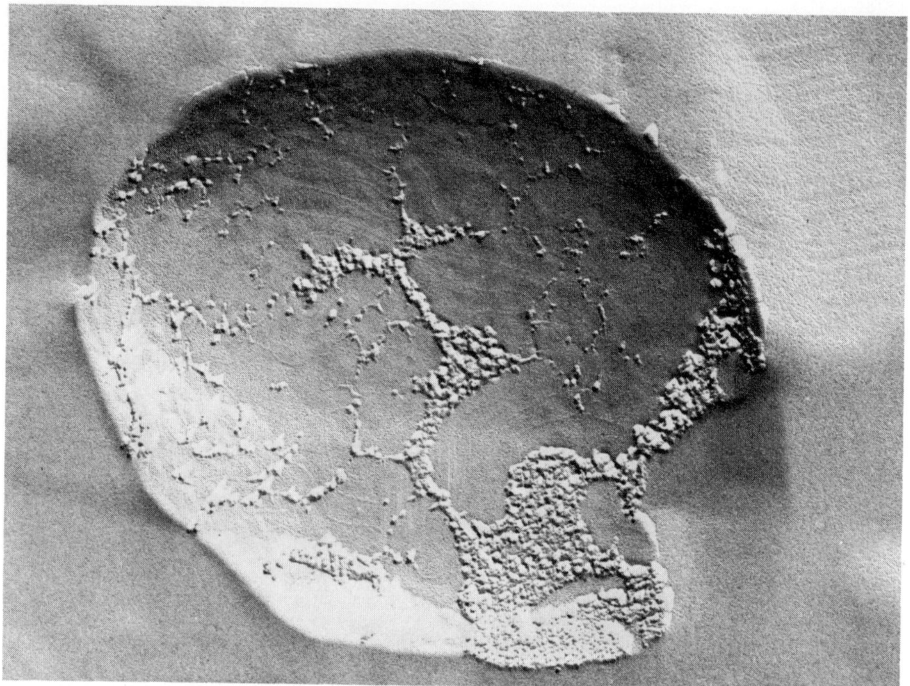

(a)

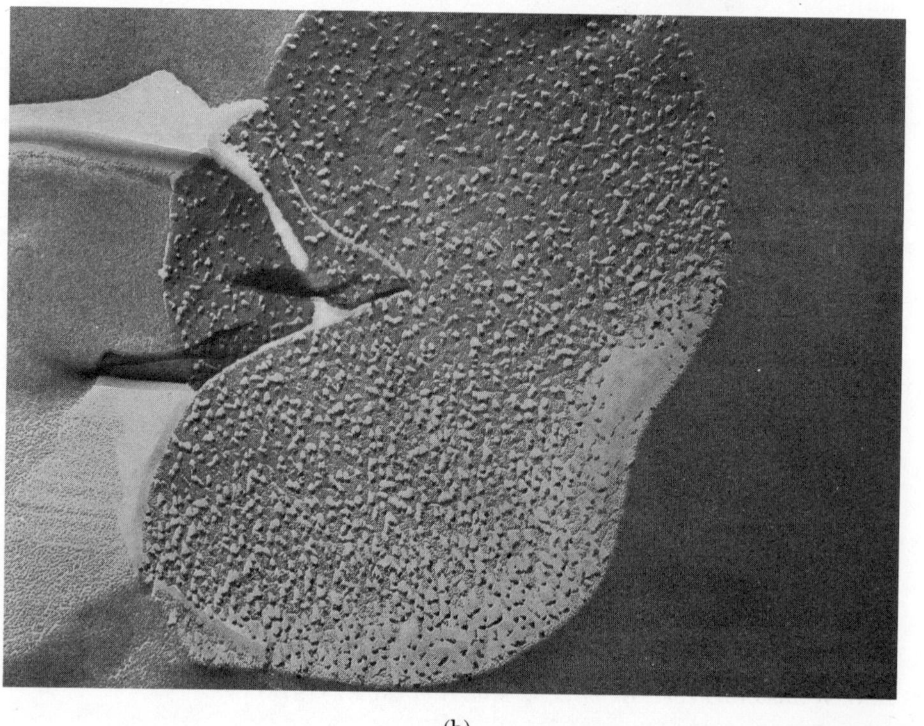

(b)

EXAMPLE *Freeze-fracture Electron Microscopy of Rhodopsin in Synthetic Bilayer Membranes*

When light strikes a retinal rod cell, a visual response is thought to be initiated by the effect of light on the protein, rhodopsin, located in the 1000-odd flattened disks stacked on top of each other to form the outer segment of the rod cell. Since the disks are not structurally connected to the rod cell wall, from which the action spike must be initiated and sent to reach the optic nerve and then the brain, it is thought that some chemical transmitter (probably Ca^{++}) is somehow released from the disks on arrival of light, and the Ca^{++} then diffuses to the outer rod wall and starts the nerve impulse. Some of the most direct molecular evidence about the early events in this scheme is provided by the freeze-fracture appearance of a dispersion of rhodopsin protein in a synthetic phospholipid bilayer membrane, as shown in Fig. 15-20.

Dark-field and Scanning Electron Microscopy

Figure 15-21 shows two schemes for increasing contrast in electron microscopy, by moving the optical detection system off-axis from the observing electron beam. When back-scattered electrons are observed, the method is called "scanning" electron microscopy, because measurements are usually conducted by scanning a pencil-beam of electrons over a surface, and sending the appropriate response to a television camera whose "raster" (scanning from line to line on the television screen) is controlled by the position of the scanning electron beam, providing a continuous television display of the scattered electron pattern from the specimen. For the "dark-field" arrangement, the optical detector is placed as in the ordinary transmission electron microscope (Fig. 15-13), but since the detector is far enough off-axis to miss the irradiating beam and most of the electrons scattered from the background, contrast is improved by greatly reducing the background intensity.

81.1

FIGURE 15-20. Freeze-fractured, shadowed faces of the interior of a synthetic phospholipid bilayer membranes into which rhodopsin protein has been incorporated. (*a*) and (*b*) represent electron micrographs showing the distribution of rhodopsin for dark-adapted (*a*) and light-exposed (*b*) situations. Note that on exposure to light, the rhodopsin spreads out within the membrane from an initially highly aggregated form to a uniformly dispersed form. It is thought that this model system may show that the first functional result of the arrival of light is to cause rhodopsin to spread out in such a way as to cause release of Ca^{++}, perhaps by changing the conformation of an appropriate nearby protein associated with Ca^{++} transport. (*a*) ×70,000; (*b*) ×80,000. Synthetic phospholipid used was 1,2-di(*trans*)-9-octadecenoyl)phosphatidyl choline. [From Y. S. Chen and W. L. Hubbell, *Exp. Eye Res. 17,* 517 (1973).]

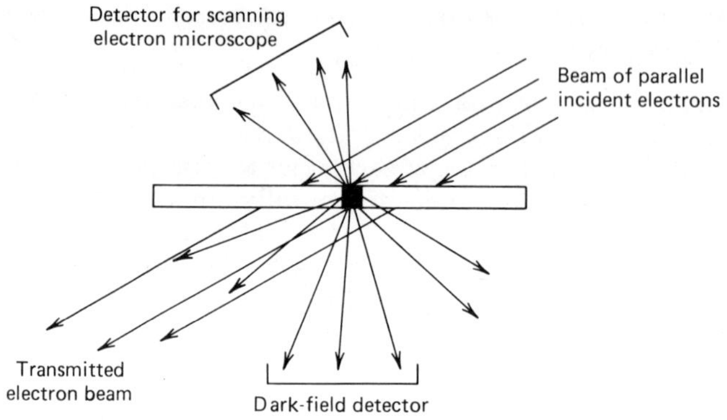

FIGURE 15-21. Schematic diagram of detection of back-scattered electrons (scanning electron microscope) or electrons scattered through the film (dark-field detector) as a means for improving image contrast by greatly reducing the intensity of the background.

EXAMPLE *Dark-field Electron Microscopy at Ultra-high Resolution*

Because of the very high contrast possible from the dark-field method, it becomes possible to approach the resolution limit of the electron optics, about 2 Å. In the striking picture of Fig. 15-22c, *single atoms* in a small molecular ring are resolved.

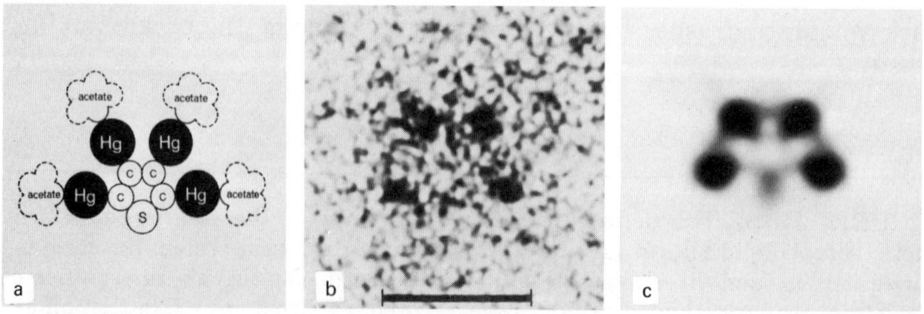

FIGURE 15-22. Dark-field electron micrographs. (*a* to *c*) a small heterocyclic sulfur molecule, labeled with mercuric acetate. The mercury and sulfur are clearly visible (the bar represents 10 Å), (*b*) shows a single micrograph and (*c*) is the image obtained by averaging 64 separate micrographs—since the background "noise" is random, its intensity is reduced relative to specimen image on averaging. The sulfur atom in (*c*) is the smallest atom so far for which an electron micrograph image has been obtained. [From F. P. Ottensmyer, *Canadian Research & Development 6,* 37 (1973).]

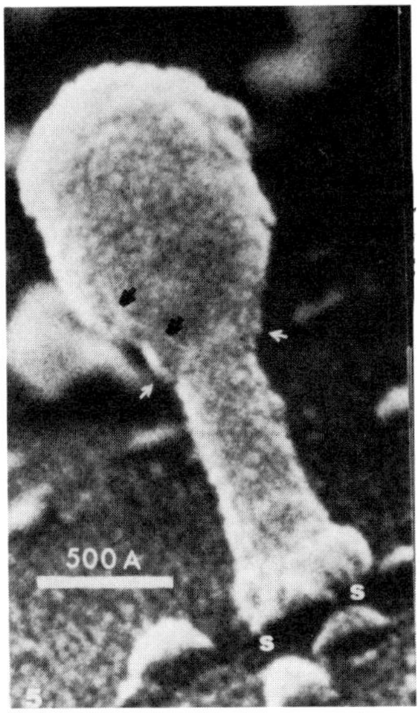

FIGURE 15-23. Scanning electron micrograph of untriggered T4 coliphage. The head, collar, (white arrows), tail, and end plate are visible. There also appear to be small projections extending from the end plate, which may be fibers (see Chapter 21.C.1. for more structural details). [From A. N. Broers, B. J. Panes, J. F. Gennaro, Jr., *Science 189,* 637 (1975).]

EXAMPLE *Scanning Electron Microscope (S.E.M.): Image of Individual Bacteriophage*

Resolution for scanning electron microscopes (S.E.M.) has recently improved to a stage permitting high-resolution visualization of individual bacteriophages 3C (staphylococcal phage) and T4 (coli phage), an example of which is shown in Fig. 15-23. In this and other micrographs, it is possible to see the "head," "tail," "end plate," "tail collar," and small additional small projections extending from the end plate. When the T4 is viewed from the tail end, the sixfold symmetry of the end plate is clearly evident. (See Chapter 21.C.1 for studies of the assembly of the T4.)

PROBLEMS

1. For a given solvent, refractive index increment, light wavelength, and scattering angle, the observed Rayleigh (intensity) ratio is of the form

$$R(\theta_s) = K'[c]M$$

in which c is the weight/volume concentration of solute, M is the solute molecular weight, and K' is a constant. Use this equation to show that the value of M determined for a mixture of solutes of different molecular weight will give a "weight-average molecular weight," M_w, according to:

$$M_w = \frac{\sum_i [c_i] M_i}{\sum_i [c_i]} = \frac{\sum_i N_i M_i^2}{\sum_i N_i M_i} \tag{15-29b}$$

Hint: consider a solution containing arbitrary amounts of three solutes of different molecular weight—compare to Eq. 2-41 ff.

2. Because molecular weights obtained from light-scattering represent a weight-average molecular weight (see preceding problem), it is experimentally necessary to exclude even trace amounts of high molecular weight impurities if accurate molecular weights are to be obtained.
   (a) Show that contamination of an enzyme sample with 5% by weight of trimer will produce an apparent (weight-average) molecular weight that is 10% larger than the actual monomer molecular weight. This calculation shows how light scattering can be used to study oligomerization of macromolecules.
   (b) Find the molecular weight determined from light scattering for a sample of molecular weight, 34,000, contaminated by 1% of an impurity of molecular weight, 1,000,000 Dalton.

3. In the text, it was shown that the intensity of light scattered from electrons uniformly distributed within a solid sphere of diameter, D, is reduced by the factor on the left-hand side of the equation shown below, in comparison to the intensity that would be observed if all the electrons in the sphere were concentrated at its center. Show that this factor is given by the right-hand side of the equation shown below.

$$\frac{\int_0^{D/2} \frac{\sin(\mu r)}{\mu r} 4\pi r^2 dr}{\int_0^{D/2} 4\pi r^2 dr} = \frac{3}{a^3}[\sin(a) - a\cos(a)]; \quad a = \mu D/2$$

$$\mu = \frac{4\pi}{\lambda} \sin(\theta_s/2)$$

4. By integrating (summing) the scattered light intensity Eq. 15-26 over all angles
   (a) Show that

$$I_{\text{scattered}}^{\text{total}} = \int_{\theta_s=0}^{\pi} \int_{\phi=0}^{2\pi} I_{\text{scatt}}(\theta_s) \sin(\theta_s)\, d\theta_s d\phi = \left(\frac{16\pi}{3}\right) I_{\text{scatt}}(90°)$$

   (b) Then, by considering a sample sufficiently dilute that $\exp[-\tau] \cong 1-\tau$, where τ is "turbidity" (see Eq. 15-31), show that

$$\tau = \text{turbidity} = (16\pi/3)\, R(90°)$$

   in which $R(90°)$ is the Rayleigh ratio (Eq. 15-28) for 90° scattering angle.

5. Compute the apparent diameter for each of the following macromolecules: (Assume each macromolecule is a solid sphere of density, 1.3 g/cm^3.) (a) $M = 25{,}000$ Dalton: (b) $M = 75{,}000$; (c) $M = 500{,}000$; (d) $M = 10{,}000{,}000$.

6. Tobacco mosaic virus is a long rod-shaped macromolecule, whose experimental radius of gyration (from light-scattering measurements) is 924 Å. Derive a relation between the length of a long rod and its radius of gyration (use Eq. 15-42, as in Eq. 15-43). Then calculate the length of a tobacco mosaic virus particle from the experimental radius of gyration, and compare to the observed length of 3200 Å from electron microscopy.

7. In this problem various estimates of the size of a given macromolecule are calculated, based on an assumed shape. For a macromolecule of molecular weight, 500,000,

(a) Calculate the radius of the molecule, if the molecule is assumed to be a solid sphere of unit density

(b) Calculate the radius of gyration (i.e., the apparent size of the molecule as determined from light-scattering experiments), if the molecular shape is

(1) a random coil (assume length/link = 4.6 Å; molec wt/link = 50 Dalton)

(2) a long rod of diameter, 25 Å

(3) a long rod of diameter, 15 Å

(4) a long rod of diameter, 10 Å

(c) Does it matter whether the long rod is assumed to have square or circular cross section in these calculations?

8. When the coenzyme NAD binds to yeast glyceraldehyde 3-phosphate dehydrogenase (GPDH), there is a change in enzyme volume on binding, and that change can be monitored quantitatively from low-angle X-ray scattering measurements. For various bound fractions of NAD, the following data were obtained [H. Durchschlag, G. Puchwein, O. Kratky, I. Schuster, and K. Kirschner, *FEBS Lett. 4,* 75 (1969)]:

Fraction of NAD Bound to GPDH	Volume (Å^3)	Degree of Volume Contraction
0 (apoenzyme)	2.64×10^5	0%
0.23	2.58×10^5	
0.46	2.53×10^5	
0.72	2.49×10^5	
0.99 (holoenzyme)	2.45×10^5	100%

A Koshland sequential allosteric mechanism (see Fig. 11-12) predicts a linear relationship between fraction NAD bound and degree of volume contraction, while a Monod model predicts a nonlinear curve. Plot the data tabulated above to establish whether this allosteric enzyme follows the Koshland mechanism or not.

9. From the deBroglie relation, $\lambda = h/mv$, and the relation, $eV = mv^2/2$, in which V is accelerating voltage, e is electronic charge, m is electronic mass, h is Planck's constant, v is electronic velocity, and λ is the wavelength associated with the moving electron, show that:

$$\lambda = (12.3/\sqrt{V}), \text{ in Å}$$

Find the wavelength, in Å, associated with an electron accelerated by a voltage of: (a) 15,000 volt, (b) 40,000 volt, (c) 1,000,000 volt.

REFERENCES

Light Scattering

D. F. Eggers, Jr., N. W. Gregory, G. D. Halsey, Jr., and B. S. Rabinovitch, *Physical Chemistry,* John Wiley & Sons, New York (1964), Chapter 15.

Small-angle X-ray Scattering

I. Pilz, in *Physical Principles and Techniques of Protein Chemistry,* part C, ed. S. J. Leach, Academic Press, New York (1973), pp. 141–245.

Electron Microscopy

E. M. Slayter, in *Physical Principles and Techniques of Protein Chemistry,* part A, ed. S. J. Leach, Academic Press, New York (1969), pp. 2–58.

L. J. Reed and D. J. Cox, *The Enzymes,* Vol. 1, 3rd ed., ed. P. D. Boyer, Academic Press, New York (1970), pp. 213–240.

CHAPTER 16 Transient Phenomena: Initial Response of a Suddenly Displaced Weight on a Spring

We have already given abundant examples of the wide range of information derived from experiments in which one observes the in-phase and/or 90°-out-of-phase components of the *steady-state* response to a *sinusoidal constant-amplitude small driving force.* For that particular experiment, the response amplitude always was proportional to the driving amplitude (i.e., the system was *linear*), and the response was always sinusoidally time-varying at the *same* frequency as the driver. In this section, we obtain the same information (i.e., the "natural" frequency and damping constant of the spring) from a different type of experiment, in which the weight on a spring is subject to a sudden displacement (as from a sharp blow), and the spring response is monitored in the *absence of any driving force.* Under these conditions, we find that the spring oscillates near its *natural frequency,* ω_0, with an amplitude that decreases *exponentially* with time, where the time constant for the exponential decay is called a "relaxation time" (since the system is "relaxing" back toward its equilibrium position), and is proportional to the reciprocal of the friction (damping) coefficient, f.

It is reasonable to wonder why anyone would want to conduct one of these new "transient" experiments, if the same information could be obtained using the steady-state methods we have spent so much time analyzing in the previous two chapters. The answer is occasionally *experimental,* in that the available apparatus may lend itself more easily to one experiment than the other; the *interpretation* may be simpler or more general, as with chemical relaxation compared to steady-state kinetics; or the *convenience* may be better for the transient experiment, as we shall learn in Chapter 20, where a given spectrum is produced up to 10,000 times faster with transient than with steady-state techniques.

Displaced Weight on a Spring; No Driving Force

All the applications in this chapter can be accommodated to the solution of the problem of a weight on a spring, where the weight has been pulled to a nonequilibrium position and then released, with no further external driving force present:

$$m(d^2x/dt^2) + f(dx/dt) + kx = 0 \qquad (13\text{-}37)$$

16.A. DAMPED SPRING WITH ZERO MASS: RATE OF RETURN TO EQUILIBRIUM

An important special case is (again) the limit of zero mass, for which the basic differential Eq. 13-37 reduces to

$$f(dx/dt) + kx = 0 \quad (13\text{-}43)$$

whose solution is readily verified to be

$$x = x_0 \exp[-t/\tau];\ \tau = (f/k) \quad (13\text{-}46)$$

We begin this section with applications based on the simpler zero-mass limit (principally chemical fast-reaction transient kinetics), and then progress to applications of the solution to the more general Eq. 13-37 (optical and magnetic depolarization). [The general solution of Eq. 13-37 is the same as the solution of Eq. 13-43, except that the spring oscillates near its natural frequency, $\omega_0 = \sqrt{(k/m)}$ as the weight relaxes exponentially back to its equilibrium position (see Chapter 16.B).]

16.A.1. Fast-reaction Transient Chemical Kinetics

Slow chemical reactions may obviously be observed simply by mixing the reactants and monitoring the concentration of reactant(s) or product(s) as the reaction proceeds. However, as seen from Problem 16-1 and Table 13-2, most chemical reactions of interest proceed at rates which are much faster than the time it usually takes (say, a few seconds) to mix the reactants. We have already seen in Section 3 that steady-state methods provide a powerful tool for study of fast reactions involving intermediates such as enzyme-substrate complexes. Unfortunately (see Chapter 10.B.2), steady-state Michaelis-Menten enzyme kinetics, for example, is completely insensitive to the presence of multiple intermediates, where the intermediates differ either chemically (e.g., different degree of protonation, different chemical bonding) or physically (different enzyme and/or substrate configuration). Moreover, we quickly learned in Section 3 that even very *simple* steady-state schemes of connected chemical reactions can lead to very *complicated* algebraic manipulations. The principal advantages of observing chemical reactions following some sudden displacement from equilibrium rather than by steady-state methods are: (1) the technique gives access to *fast* chemical reaction rates, up to 10^6 sec^{-1} first-order rate constant; (2) it is often possible to detect *intermediates* between reactant and product, which greatly narrows down the possible mechanisms; and (3) provided that the initial displacement from equilibrium is sudden and small, the algebra describing the adjustment of any species concentration back to equilibrium will always follow a *simple exponential time-behavior* as shown in Eq. 13-46, and worked out explicitly for a general example below.

The Chemical Relaxation Experiment

The procedure in any pulsed chemical relaxation experiment is always the same:

(a) Let the system come to equilibrium.

(b) Perturb the system *suddenly* (i.e., rapidly compared to the rates of any chemical reactions of interest) and *slightly* (i.e., such that any induced *changes* in concentration are small compared to the *total* concentration of any species).

(c) Determine the exponential decay time constant(s) ("relaxation" time(s)) that characterize the adjustment of the concentration of one species (say, one reactant) to the new equilibrium concentration.

(d) Repeat the previous steps for several concentrations of each component, and, from the concentration-dependence of each "relaxation time," deduce the mechanism for the chemical reaction system.

Since chemical equilibrium constants generally vary with pressure, temperature, electric field, and concentration (in the sense that changing one concentration will lead to changes in all the other concentrations of species in that equilibrium), one can imagine producing the required small, sudden change in equilibrium by application of a short, strong pulse in pressure ("P-jump"), temperature ("T-jump"), electric field ("E-jump") or concentration ("stopped-flow" or "concentration-jump"). Before proceeding to applications of these various experiments, it is desirable to work out a single example of the algebra required to relate the "relaxation time" to rate constants and concentrations for a particular chemical kinetic scheme. The scheme chosen for explicit analysis is the simplest case that exhibits all the elements involved in more complicated schemes and that (unlike our steady-state problems, which required separate solution for each variation in the reaction scheme) makes it possible to predict *without further calculation* the results expected for a variety of *other* kinetic schemes.

The Simplest General Example of Chemical Relaxation: $A + B \rightleftharpoons C$

Consider the system of chemical reactions initially at equilibrium:

$$A + B \underset{k_{-1}}{\overset{k_1}{\rightleftharpoons}} C, \quad K_{eq} = \frac{k_1}{k_{-1}} = \frac{[C]_0}{[A]_0[B]_0} \tag{16-1}$$

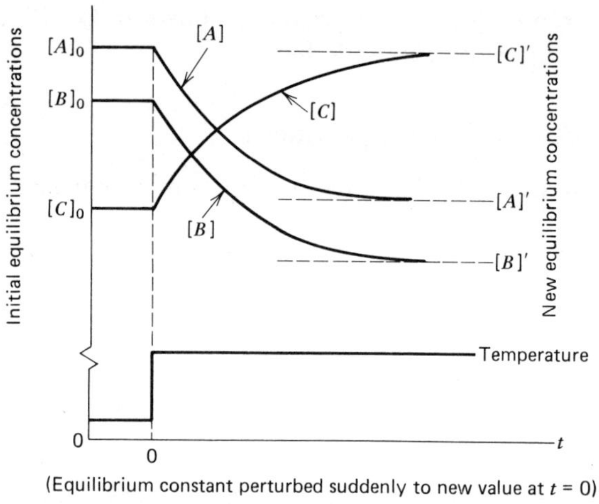

FIGURE 16-1. Reactant (A and B) and Product (C) concentrations as a function of time, following a sudden, small change in equilibrium constant for the system, $A + B \rightleftharpoons C$. The broken y axis is intended to indicate that the *new* equilibrium concentrations differ from the *initial* equilibrium concentrations by an amount which is small compared to the *total* values, $[A]_0$, $[B]_0$, and $[C]_0$.

in which the *initial* equilibrium concentrations are those shown in Fig. 16-1. Now suppose that this system is subjected to, say, a small, sudden temperature change to a *new* constant temperature with corresponding *new* equilibrium constant, K'_{eq}:

$$K'_{eq} = \frac{k'_1}{k'_{-1}} = \frac{[C]'}{[A]'[B]'} \tag{16-2}$$

in which $[A]'$, $[B]'$, and $[C]'$ are the new equilibrium concentrations appropriate to the new temperature. We are interested in what can be learned from analyzing the rate of "relaxation" (readjustment) toward the new equilibrium concentrations. Since

$$d[A]/dt = d[B]/dt = -d[C]/dt \tag{16-3}$$

it suffices to consider the time-behavior of just one of the component concentrations, say, $[A]$:

$$d[A]/dt = -\, k'_1[A][B] + k'_{-1}[C] \tag{16-4}$$

Next, since we are concerned with small *changes* in concentration, rather than with the *total* concentration of any one species, we define the

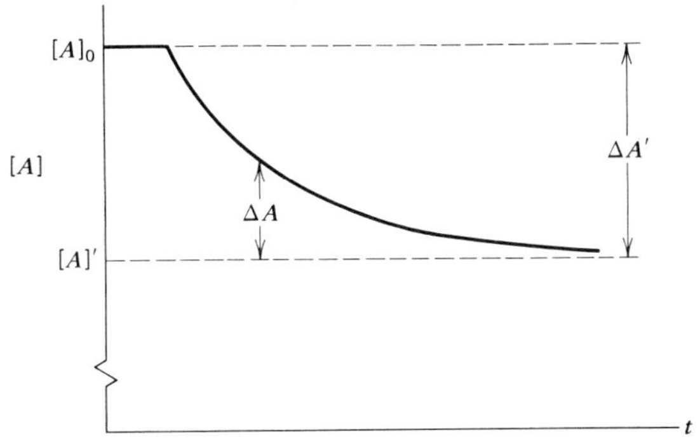

FIGURE 16-2. Relation between *total* concentrations, $[A]$, $[A]_0$, and $[A]'$, and concentration *changes*, ΔA and $\Delta A'$ for reactant, A in Eq. 16-5a. Similar relations for ΔB and ΔC are defined in Eq. 16-5b,c.

instantaneous concentration differences, ΔA, ΔB, and ΔC according to (see Fig. 16-2):

$$\Delta A = [A] - [A]' \text{ or } [A] = [A]' + \Delta A \tag{16-5a}$$

$$\Delta B = [B] - [B]' \text{ or } [B] = [B]' + \Delta B \tag{16-5b}$$

and

$$\Delta C = [C] - [C]' \text{ or } [C] = [C]' + \Delta C \tag{16-5c}$$

Substituting for $[A]$, $[B]$, and $[C]$ from Eq. 16-5 into Eq. 16-4,

$$\frac{d([A]' + \Delta A)}{dt} = \frac{d[A]'}{dt} + \frac{d\Delta A}{dt} = \frac{d\Delta A}{dt}$$

↑ (since $[A]'$ is constant with time)

$$= -k_1'([A]' + \Delta A)([B]' + \Delta B) + k_{-1}'([C]' + \Delta C) \tag{16-6}$$

Furthermore, since one C molecule is produced for every A and B that react,

$$\Delta A = \Delta B = -\Delta C \tag{16-7}$$

Equation 16-6 may be simplified to give

$$\frac{d\Delta A}{dt} = -k_1'([A]' + \Delta A)([B]' + \Delta A) + k_{-1}'([C]' - \Delta A)$$

$$= -k_1'[A]'[B]' + k_{-1}'[C]' - \Delta A(k_1'([B]' + [A]' + \Delta A) + k_{-1}') \tag{16-8}$$

Finally, since the forward and reverse rates must be equal at equilibrium,

$$\text{forward rate} = \text{reverse rate}$$
$$k_1'[A]'[B]' = k_{-1}'[C]' \tag{16-9}$$

and since we are considering a concentration change that is small compared to total concentration,

$$\Delta A \ll [A]', [B]' \tag{16-10}$$

Equations 16-9 and 16-10 may be applied to Eq. 16-8 to give

$$\frac{d\Delta A}{dt} = -\Delta A(k_1'[B]' + k_1'[A]' + k_{-1}') = -\Delta A/\tau \tag{16-11}$$

in which

$$\boxed{(1/\tau) = k_1'[B]' + k_1'[A]' + k_{-1}' = \text{Chemical "relaxation" time}} \tag{16-12}$$

Equation 16-11 is of the form, $dy/dx = -ky$, which we solved in Chapter 9:

$$\boxed{\log_e (\Delta A)_{\text{at time, } t} = \log_e (\Delta A)_{\text{at time, zero}} - \frac{t}{\tau}} \tag{16-13}$$

Equation 16-13 predicts that a plot of $\log_e (\Delta A)$ versus time following the sudden, small change in equilibrium constant should give a straight line whose y intercept is $\log_e (\Delta A')$, and whose slope is $-(1/\tau)$, where τ is defined by Eq. 16-12 (see Fig. 16-3a). Lastly, if such an experiment is repeated for several different concentrations of reactants, then a plot of $(1/\tau)$ versus $([A]' + [B]')$ will give a straight line whose slope is k_1' and whose y intercept is k_{-1}' (see Fig. 16-3b). In other words, we have found a way to extract the *rate constants,* k_{-1}' and k_1' for this particular kinetic scheme.*

The reader should not be distracted by the large number of steps in the derivation leading from Eq. 16-4 to the plots in Fig. 16-3; the important result (which turns out to be general for a wide variety of reaction schemes) is that when the sudden shift in equilibrium is *small* and *sudden,* the concentrations of all participants to the equilibrium readjust *exponentially* to their new equilibrium values, and the time constant for the exponential curve is related in a simple way to the rate constants for that particular reaction mechanism.

* *For a temperature-jump experiment such as that treated in Fig. 16-1 ff., the rate constants,* k_1' *and* k_{-1}', *which are determined from the experiment, correspond to the* new *temperature. Thus, if it is desired to find the rate constants at 25°C, for example, one might first cool the mixture to an initial equilibrium temperature of, say, 20°C, and then perform a 5°C temperature jump (say, by suddenly discharging a capacitor enclosing the mixture) and monitor* $[A]$ *as a function of time.*

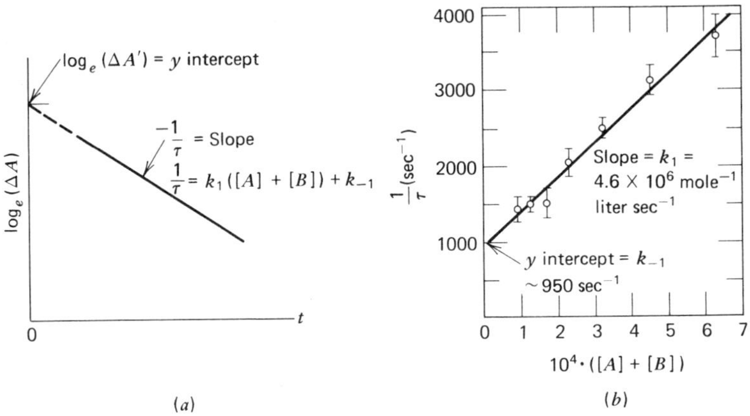

FIGURE 16-3. Schematic plot of logarithm of concentration change versus time for the equilibrium

$$A + B \underset{k_{-1}}{\overset{k_1}{\rightleftharpoons}} C$$

immediately following a sudden small change in equilibrium constant (as from a temperature jump) at time zero. (*b*) Experimental data for a series of experiments of the type shown in (*a*) taken at different concentrations of A and B. A plot of $(1/\tau)$ versus $([A] + [B])$ gives a straight line whose y intercept and slope are k_{-1} and k_1. In this particular case, B is the enzyme, lysozyme, and A is the inhibitor, di-N-acetylglucosamine (see steady-state NMR example of Fig. 14-26 for other experiments on similar systems). The linearity of the plot shown in (*b*) confirms that binding of this inhibitor is indeed a one-step process. [Lysozyme data from D. M. Chipman and P. R. Schimmel, *J. Biol. Chem. 243,* 3771 (1968).]

Interpretation of Chemical Relaxation Lifetime, τ

The preceding derivation of a data reduction yielding a chemical relaxation time, τ, which in turn gave a means for determining the rate constants for a particular simple kinetic scheme, may appear to be a specialized calculation, particularly because of the several tricks and approximations involved. On the contrary, the result provides all the features required to predict what would happen with similar experiments on a wide variety of possible kinetic schemes. To see why this should be true, it is necessary to digress very briefly back to elementary first- and second-order chemical reaction kinetics.

First-order Lifetime. Early in our discussion of steady-state kinetics in Chapter 9 we found that increase or decrease in concentration of a species, $[A]$, subject to a *first-order* rate process

$$d[A]/dt = \pm k[A]$$

$$A \xrightarrow{k} \qquad \text{(9-1)}$$

or

$$\xrightarrow{k} A$$

could be characterized by a *first-order lifetime,* τ_A, such that

$$[A] = [A]_0 \exp[\pm t/\tau_A] \tag{9-9}$$

where

$$\boxed{(1/\tau_A) = k} = \text{The first-order rate constant for increase or disappearance of } A \tag{9-10}$$

For a first-order disappearance, τ_A can be thought of as the time it takes for the concentration of species A to decrease to $(1/e)$th of its initial value, $[A]_0$.

Second-order Lifetime. For a second-order chemical reaction, such as

$$A + A \xrightarrow{k} \text{products} \tag{16-14}$$

the time-course of the reaction is no longer exponential, since

$$d[A]/dt = -k[A]^2, \text{ or } d[A]/[A]^2 = -kdt \tag{16-15}$$

which is readily integrated to give

$$(1/[A]) = (1/[A]_0) + kt \tag{16-16}$$

which is clearly not an exponential process. However, if we were unaware that the reaction was really second-order, and treated it as a first-order process,

$$d[A]/dt = -(k[A])[A] = -(1/\tau_A)[A] \tag{16-17}$$

we would have effectively defined a "pseudo-first-order lifetime," τ_A, such that

$$(1/\tau_A) = k[A] \tag{16-18}$$

Now in *ordinary* kinetics, where concentration changes are large and perhaps even comparable to the initial concentrations of reactants, the error of the definitions, Equations 16-17 and 16-18, would quickly become apparent from plotting some experimental concentration data as a function of time (see Eq. 16-16). *However,* the remarkable feature of the *pulsed* kinetic experiment described earlier in this section is that because the observed concentration changes are so *small,* it is approximately correct to write Equations 16-17 and 16-18 in describing the readjustment to equilibrium for a second-order chemical reaction process. For the more general case,

$$A + B \xrightarrow{k} \text{products} \tag{16-19}$$

we have

$$d[A]/dt = -k[B][A] = -(1/\tau_A)[A] \tag{16-20}$$

in which

$$\boxed{(1/\tau_A) = k[B]} \tag{16-21}$$

Similarly, from the rate of disappearance of B, we find that

$$\boxed{(1/\tau_B) = k[A]} \tag{16-22}$$

In summary of Equations 16-14 to 16-22, a *first-order* chemical rate process will contribute a factor, k, to the *overall* reciprocal lifetime, $(1/\tau)$ for (exponential) readjustment of the concentrations of all the chemical components, and a *second-order* reaction of A with B will contribute a factor, $k[B]$, where $k[B]$ would be called the lifetime for (exponential) change in concentration of A:

$$\log_e(\Delta i)_{\text{at time, } t} = \log_e(\Delta i)' - (1/\tau)t \tag{16-13}$$

in which

$$(1/\tau) = \sum_i (1/\tau_i) \tag{16-23}$$

where

$$(1/\tau_i) = k \quad \text{for each first-order reaction of component, } i \tag{16-24a}$$

and

$$(1/\tau_i) = k[j] \quad \text{for each second-order reaction of component } i \text{ with component } j \tag{16-24b}$$

and the sum in Eq. 16-23 is over all components in the equilibrium.

For example, we have already shown that the concentration of either

A or B or C in the series of reactions, $A + B \underset{k_{-1}}{\overset{k_1}{\rightleftharpoons}} C$, will readjust to a new equilibrium following a small sudden shift in equilibrium constant, according to an exponential time-course with overall time-constant, τ:

$$(1/\tau) = k_1[A] + k_1[B] + k_{-1} \tag{16-12}$$

It is now apparent that the terms $k_1[A]$, $k_1[B]$, and k_{-1} represent the respective contributions to the total reciprocal lifetime, $(1/\tau)$, arising from the (sec- (second-order) lifetime for disappearance of B, $(1/\tau_B) = k_1[A]$; the (second-order) lifetime for disappearance of A, $(1/\tau_A) = k_1[B]$; and the (first-order) lifetime for disappearance of C, $(1/\tau_C) = k_{-1}$. In other words, we should now be able to predict the form of the overall "chemical relaxation" lifetime for many kinds of simple equilibria collected in Table 16-1.

Clearly, an equilibrium consisting of two first-order processes (first entry in Table 16-1) would be expected to have an *overall* reciprocal lifetime that is the sum of the two *individual* first-order reciprocal lifetimes for disappearance of A or B (see top entry in Table 16-1). As another example, the situation of $A + B \underset{k_{-1}}{\overset{k_1}{\rightleftharpoons}} C$, in which the concentration of B is held constant (as by buffering, for example—see Table 16-1), will exhibit an overall reciprocal lifetime consisting of contributions from individual lifetimes for disappearance of A, $(1/\tau_A) = k_1[B]$, and of C, $(1/\tau_C) = k_{-1}$, but not from B, since its concentration is constant and the "lifetime" for change in B concentration is effectively infinite so that the contribution from $(1/\tau_B) = 0$. A number of other useful results are also shown in Table 16-1, and the general pattern should now be apparent, at least for simple equilibria.

EXAMPLE *Rate Constants for Enzyme-substrate Equilibria*

We have seen in Chapter 10 that steady-state enzyme kinetics generally provides a rate constant for rate of appearance of product, and a pseudo-dissociation constant, K_A, which is a measure of the strength of binding of substrate to enzyme. From transient chemical relaxation following a sudden small pulse that perturbs an existing equilibrium, we now gain access to *individual rate constants* for the enzyme-substrate interaction, rather than just an inseparable combination of rate constants, $K_A = (k_{-1} + k_2)/k_1$, as with steady-state experiments. The range of rate constants for binding and dissociation of substrate with enzyme is shown in Table 13-2. For the physiological (i.e., the "natural") substrate, the second-order rate constant for binding is usually in the range, 10^7 to 10^8 M^{-1} sec^{-1}, slightly below the value expected for a diffusion-limited combination (see Problems) of reactants. However, when artificial substrates differing only slightly in bonding or configuration from the physiological substrate are used, the "on" rate constant for binding can be considerably smaller, suggesting that the "fit" between enzyme active site

Table 16-1 Relation between Chemical Relaxation Time, τ, and Rate Constants and Concentrations for Various Kinetic Schemes *

Reaction Scheme	Reciprocal Chemical Relaxation Time, $(1/\tau)$
$A \underset{k_{-1}}{\overset{k_1}{\rightleftharpoons}} B$	$(1/\tau) = k_1 + k_{-1}$
$A + B \underset{k_{-1}}{\overset{k_1}{\rightleftharpoons}} C$	$(1/\tau) = k_1[A] + k_1[B] + k_{-1}$
$A + B \underset{k_{-1}}{\overset{k_1}{\rightleftharpoons}} C + D$	$(1/\tau) = k_1[A] + k_1[B] + k_{-1}[C] + k_{-1}[D]$
$A + C \underset{k_{-1}}{\overset{k_1}{\rightleftharpoons}} B + C$ (C = catalyst)	$(1/\tau) = k_1[C] + k_{-1}[C]$
$A + B \underset{k_{-1}}{\overset{k_1}{\rightleftharpoons}} C$ ($[B]$ held constant)	$(1/\tau) = k_1[B] + k_{-1}$
$A + B + C \underset{k_{-1}}{\overset{k_1}{\rightleftharpoons}} D$	$(1/\tau) = k_1[A][B] + k_1[A][C] + k_1[B][C] + k_{-1}$
$A + A \underset{k_{-1}}{\overset{k_1}{\rightleftharpoons}} A_2$	$(1/\tau) = 4k_1[A] + k_{-1}$
$A + B + B \underset{k_{-1}}{\overset{k_1}{\rightleftharpoons}} AB_2$	$(1/\tau) = 4k_1[A][B] + k_1[B]^2 + k_{-1}$
$E + A + B$ $k_1 \diagup\!\!\diagup k_{-1}$ $\quad k_2 \diagdown\!\!\diagdown k_{-2}$ (Exchange equilibrium) $EA + B \underset{k_{-3}}{\overset{k_3}{\rightleftharpoons}} EB + A$	$(1/\tau) = \left(k_3 + \dfrac{k_{-1}k_2}{k_1[A] + k_2[B]}\right)\left([EA] + [B]\right) + \left(k_{-3} + \dfrac{k_{-2}k_1}{k_1[A] + k_2[B]}\right)\left([EB] + [A]\right)$
$A + B \underset{k_{-1}}{\overset{k_1}{\rightleftharpoons}} AB \underset{k_{-2}}{\overset{k_2}{\rightleftharpoons}} C$	$[(1/\tau_1) + (1/\tau_2)] = k_1[A] + k_1[B] + k_{-1} + k_2 + k_{-2}$ $(1/\tau_1) \times (1/\tau_2) = k_1(A] + [B])(k_2 + k_{-2}) + k_{-1}k_{-2}$ [τ_1 is the relaxation time for (exponential) readjustment of either $[A]$ or $[B]$ to its new equilibrium value; τ_2 is the relaxation time for readjustment of $[AB]$ to its new value.]
$E + S \underset{k_{-1}}{\overset{k_1}{\rightleftharpoons}} X_1 \underset{k_{-2}}{\overset{k_2}{\rightleftharpoons}} X_2 \underset{k_{-3}}{\overset{k_3}{\rightleftharpoons}} E + P$ (Michaelis-Menten mechanism with two intermediates)	There are three chemical relaxation times, of which the slowest, τ_3, takes the form $\dfrac{1}{\tau_3} = \dfrac{k_2}{1 + \dfrac{k_1k_3[S] + k_{-1}k_{-3}[E] + k_{-1}k_{-3}[P] + k_{-1}k_3}{k_1k_3[E] + k_1k_{-3}[E][[E] + [S] + [P]]}} + \dfrac{k_{-2}}{1 + \dfrac{k_1k_3[S] + k_{-1}k_{-3}[P] + k_1k_3[E] + k_{-1}k_3}{k_{-1}k_{-3}[E] + k_1k_{-3}[E][[E] + [S] + [P]]}}$

* *The second scheme is treated fully in the text; the remaining top four schemes give $(1/\tau)$ whose form can be deduced from the text example without further calculation. The second scheme from the bottom is included in the Problems and shows the general method for treating a scheme of arbitrary complexity. It should be noted that all the rate constants for any of the mechanisms shown above may be determined experimentally by plotting $(1/\tau)$ versus concentration of the various species involved (see text for examples).*

and substrate is critical to the binding process. The "off" rate constant for first-order dissociation of the enzyme-substrate complex is a measure of the strength of the enzyme-substrate bond. Finally, from analysis of the pH-dependence of the "on" rate constant for binding, one may learn about the pK_a of the ionizable groups on the enzyme that determine the efficiency of the binding process. Interestingly, these pK_as may differ substantially from those observed for the same amino acid in a small polypeptide, indicating that the tertiary structure of the enzyme folds the polypeptide chains so as to juxtapose two or more charged links in the chain. The interested reader is referred to the References for numerous examples of all these effects.

EXAMPLE *Chemical Relaxation as an Indicator of Macromolecular Conformational Change*

In chemical relaxation experiments, concentration changes are usually detected from change in optical absorbance, pH, or conductivity of the solution. However, change in optical absorption may result from conformational change near the absorbing chemical group, so that it is possible to study mechanisms including steps for which there is no change in covalent bonding, such as the conformational change in an enzyme following binding of substrate:

$$E + S \rightleftharpoons X_1 \rightleftharpoons X_2 \tag{16-25}$$

There is a well-defined chemical relaxation time constant associated with the second step of Eq. 16-25 whose value ranges from about 10^{-1} to 10^{-4} sec for a number of observed conformational changes in enzymes. A convincing example of such a process is the relaxation time for conformational isomerization of ribonuclease. The conformational change rate constants change with pH, consistent with the mechanism

$$E + H^+ \overset{K_A}{\rightleftharpoons} EH \underset{k_{-1}}{\overset{k_1}{\rightleftharpoons}} (EH)' \tag{16-26}$$

in which EH and $(EH)'$ represent different enzyme conformations. The relaxation time for this scheme (where the rate constants for the first step are assumed to be much faster than those for the second step), is given by

$$(1/\tau) = k_{-1} + k_1\left(1 + \frac{K_A}{[\mathrm{H}^+]}\right) \tag{16-27}$$

Figure 16-4 shows the experimental and theoretical pH-dependence of $(1/\tau)$, and the agreement is clearly excellent. Combined with other independent studies, these results have led to a proposed mechanism in which the conformational change involves the opening and closing of a groove in the enzyme structure associated with the active site for binding of substrate.

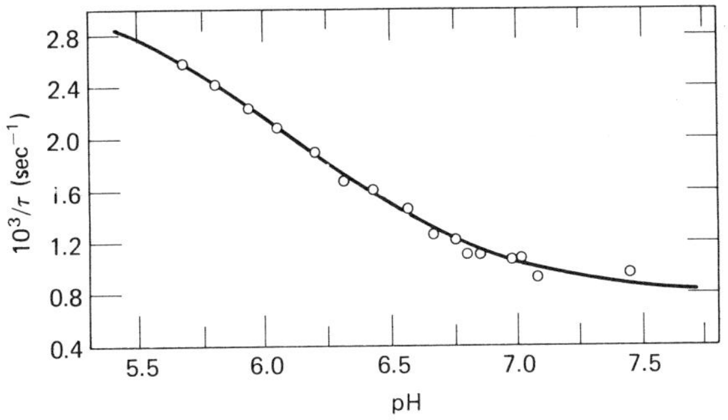

FIGURE 16-4. pH-dependence of the reciprocal chemical relaxation time, $(1/\tau)$, characterizing the conformational isomerization of ribonuclease enzyme at 25°C. The solid line is calculated according to the theoretical model, Eq. 16-26, using Eq. 16-27, with $k_1 = 2468$ sec^{-1}, $k_{-1} = 780$ sec^{-1}, and $pK_a = 6.1$. [From T. C. French and G. G. Hammes, *J. Amer. Chem. Soc. 87,* 4669 (1965).]

EXAMPLE *Multi-step Mechanisms: Cooperative Effects in Glyceraldehyde 3-phosphate Dehydrogenase-NAD*

When many connected steps appear in a chemical reaction mechanism, the individual steps may be resolved, provided that their associated relaxation times, τ, are sufficiently different. A good example (with important implications) is the binding of NAD to yeast glyceraldehyde 3-phosphate dehydrogenase (GPDH), in which three characteristic relaxation times are observed (see Fig. 16-5). There is first a rapid $[(1/\tau_1) = 7{,}000 \text{ sec}^{-1}]$ spectral change associated with binding of NAD to the R-form of the enzyme (see below), then a slower change $[(1/\tau_2) = 690 \text{ sec}^{-1}]$ associated with binding of NAD to

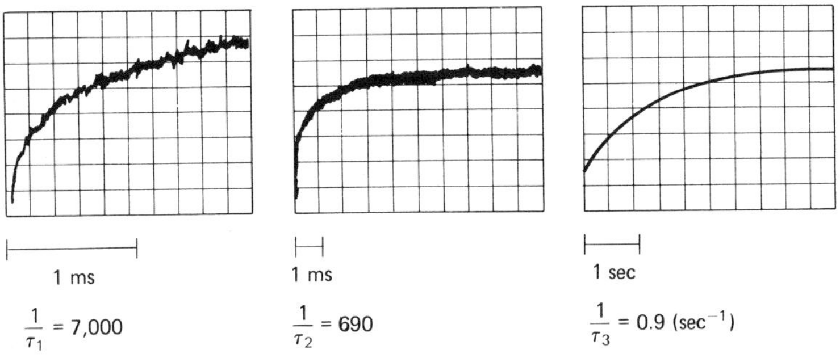

FIGURE 16-5. Changes in optical absorbance (365 nm) as a function of time for a system containing [NAD] = 6×10^{-4} M; [GPDH] = 1.4×10^{-4} M tetramer; pyrophosphate pH 8.5, 0.05 M; [EDTA] = 5×10^{-3} M, following a temperature jump from 40° to 45°C. There are three well-defined relaxation steps that can be correlated to the behavior expected from a Monod-Wyman-Changeux allosteric model (see text). [From K. Kirschner, M. Eigen, R. Bittman, and B. Voight, *Proc. Natl. Acad. Sci. U S A. 56,* 1661 (1966).

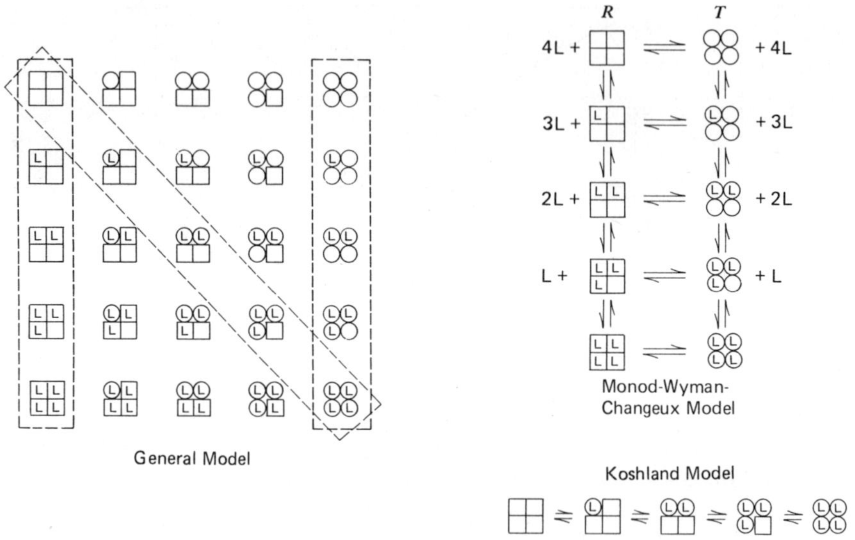

FIGURE 16-6. Review (see Chapter 11.B.2) of allosteric models for cooperative substrate binding to a tetrameric enzyme. The Monod-Wyman-Changeux and Koshland models (right) are seen as special limiting simplifications of the general model (left) in which ligand (L) binds to enzyme subunits that can exist in two conformations (circles and squares).

the T-form of the enzyme, and finally a much slower change [$(1/\tau) = 0.9$ sec^{-1}] associated with a conformational change between R and T forms of GPDH enzyme. (See Fig. 16-6 and accompanying discussion.)

The two simplest mechanistic models for cooperative enzyme-substrate binding behavior were discussed in Chapter 11.B.2, and their principal features are collected in Fig. 16-6.

Detailed analysis of the Koshland model predicts that there should be one second-order relaxation process and four first-order processes, while the Monod-Wyman-Changeux model should produce two second-order and one first-order processes. The transient chemical relaxation method is particularly useful in proving the existence of a *minimum* number of steps, so that observation of more than three characteristic relaxation times would suffice to eliminate the Monod model, for example. For the NAD-GPDH system, three relaxation times are observed (Fig. 16-5) and their respective concentration-dependences are all consistent with a Monod model. The NAD-GPDH system is one of the most convincing examples of where it has been possible to decide between the Monod and Koshland models for allosteric substrate binding to a multimeric enzyme. The key property of this system that made the choice possible was the large difference between the various chemical relaxation times. In oxygen-binding to hemoglobin, for example, cooperative binding is also observed, but transient kinetics has not been particularly helpful because the relevant characteristic relaxation times are too similar in magnitude to be resolved as distinct linear segments in a semi-log plot such as Fig. 16-3a.

EXAMPLE *Electric Birefringence and Dichroism: Tobacco Mosaic Virus (TMV)*

In Chapter 14.F, it was shown that the "friction coefficient" associated with the response of a collection of electric dipole moments to a steady-state oscillatory applied electric field reflects the rate at which the dipoles can reorient rotationally in solution. When that reorientation occurs by many small angular random steps, the process can be treated by a rotational random walk in angle, with an associated rotational diffusion constant, D_{rot}, which varies inversely with molecular volume (big molecules rotate more slowly). In this section, we simply add that very similar information can be obtained by applying a sudden small electric field pulse to a solution containing electric dipoles (actually, the dipole moments need not even be permanent—they can result from "polarization" of charge toward the ends of a molecule once the applied electric field is turned on), and then monitoring the degree of alignment of the molecules as a function of time, just as we monitored the concentration change as a function of time following a shift-in-equilibrium pulse in the chemical relaxation examples. In the steady-state electric field sinusoidal perturbation experiment (Chapter 14.F), the electrical response (capacitance; conductivity) is measured, but in the transient electric field pulse perturbation experiment, it is easier to monitor the optical birefringence or dichroism following application or removal of the electric field pulse, as shown in Fig. 16-7*a*. An experimental example of the decay of optical birefringence as a

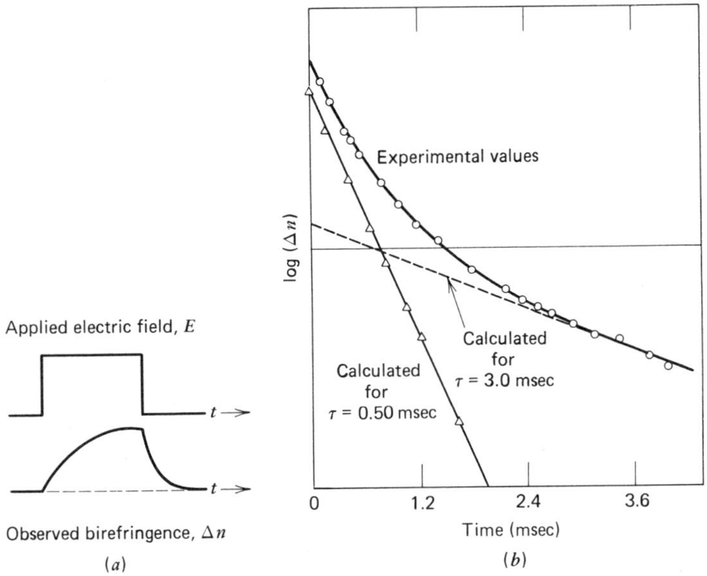

FIGURE 16-7. Theoretical (left) and experimental (right) behavior of optical birefringence, $\Delta n = n_{\parallel} - n_{\perp}$, as a function of time following application or removal of a constant applied electric field, E. The experimental response reflects two separate relaxation processes (see text) corresponding to rotational diffusion of tobacco mosaic virus monomer and end-to-end dimer. [Experimental data from C. T. Okonski and A. J. Haltner, *J. Amer. Chem. Soc. 78*, 3604 (1956).]

function of time following removal of an applied constant electric field is shown in Fig. 16-7*b*, for a solution containing tobacco mosaic virus (TMV). The experimental curve consists of a composite of two characteristic rotational rates [corresponding to the rotational relaxation times (see Chapter 16.B and 21.B) for the two distinct straight-line segments of the semi-log plot]. From this and a control experiment with a solution containing purely monomeric TMV, the results of Fig. 16-7*b* are interpreted to correspond to rotational diffusion of TMV monomer ($\tau_{rot} \cong 1/D_{rot} = 0.5$ msec) and a slower rotational diffusion of an end-to-end TMV dimer ($\tau_{rot} = 3.0$ msec). From the *magnitude* of the observed birefringence (as opposed to its *time-dependence*), it was concluded that TMV (Holmes Rib Grass strain) possesses a huge electric dipole moment of more than 10,000 Debye units. These experiments provide a particularly direct measure of rotational diffusion constants, and thus of molecular size and shape.

EXAMPLE *Ultrasonic Relaxation and Ion-paring of Salts in Water*

One of the earliest observed phenomena in the development of sonar in the 1940s was that sea water showed a much larger ultrasonic absorption than distilled water. Although the principal salt in sea water is NaCl, it was ultimately shown that the principal source of ultrasonic absorption in sea water was from magnesium sulfate. As noted in Chapter 14.G, the critical frequency,

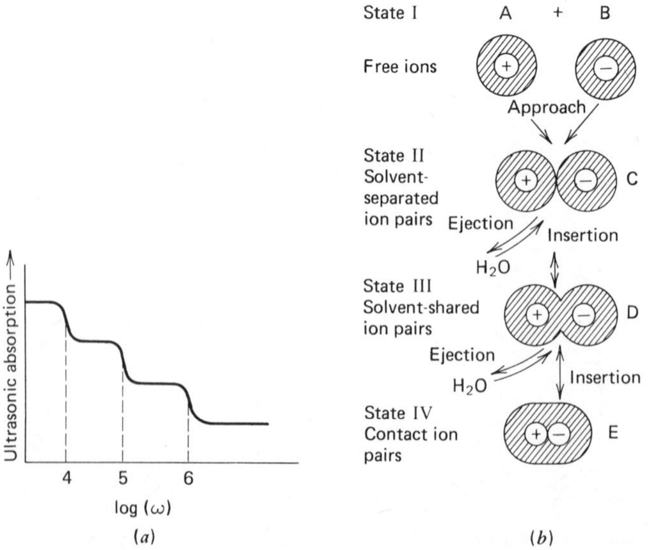

FIGURE 16-8. Ultrasonic absorption as a function of frequency (left) for a typical 2:2 electrolyte, such as $MgSO_4$ in water, and a proposed mechanism (right) that accounts for the number and concentration-dependence of the three characteristic ultrasonic relaxation times corresponding to the three steps in the ultrasonic absorption frequency-dependence. [Based on M. Eigen, *Z. Phys. Chem. Frankfurt 1*, 176 (1954).]

ω_c, in an ultrasonic absorption frequency-dependence defines a relaxation time, $\tau_c = (1/\omega_c)$, which characterizes the process whose change in volume makes the ultrasonic absorption possible. If the process is a chemical reaction with which a net change in volume is associated, then the *ultrasonic* relaxation time can be interpreted in terms of the *chemical* relaxation time for the reaction mechanism of interest. Figure 16-8*a* shows a schematic diagram of the ultrasonic absorption as a function of frequency for a solution containing a "2:2" electrolyte, such as magnesium sulfate. Three distinct steps are visible in the plot, corresponding to at least three distinct chemical equilibria. The generally accepted mechanism to which these steps correspond has been deduced by Eigen from the concentration-dependencies of the three relaxation times:

$$A + B \rightleftharpoons C \rightleftharpoons D \rightleftharpoons E \tag{16-28}$$

as shown in more detail in Fig. 16-8*b*. One of the more interesting conclusions that has followed from these sorts of experiments is that water itself can exist in equilibrium between relatively ordered and relatively disordered regions (the equilibrium is too fast, $\tau = 10^{-12}$ sec, to follow by ultrasonic absorption frequency-dependence, but can be slowed down by adding glycerol to the water and then extrapolating the results back to zero glycerol concentration), and the addition of well-known protein denaturants, such as urea or guanidinium hydrochloride, acts to "break up" the structure of the "ordered" water. Thus, it may be that the primary effect of these denaturants is to change the structure of the *solvent* which in turn changes the structure of the macromolecules. Experiments are currently under way to attempt to sort out such alternatives.

16.A.2. Cybernetics: Black-box Models for Physiology

In Chapter 12, the close analogy between the enzyme-substrate model and the drug-receptor model was used to transfer the mathematical Michaelis-Menten treatment to a physiological problem. In this section, we provide one example of how the same mathematics used to describe chemical concentration changes in a test tube can be applied to analysis of respiratory changes in concentration of CO_2 *in vivo*. The physiological mathematics is called "control theory," and is based on localizing physiology functions in various *boxes* (just as chemical reactions are thought of as arising from the presence of various *equilibria*), and then describing the relation between what goes into the box and what comes out (i.e., the rate constants and reaction "order" in the chemical case) in terms of appropriate differential equations. Because the corresponding differential equations in the physiological case are almost always too difficult to solve exactly, it has become fashionable to build an "equivalent electronic circuit" (Chapter 14.E), just as we did for dielectric relaxation (Chapter 14.F), in which the input and output voltages are related to the physiological inputs and outputs (metabolite concentrations; blood pressure, etc.) according to various resistances and capacitances that are related to the properties of the physiological "box" of interest. The model is judged by whether its outputs correlate with physiological outputs over a wide range of prescribed inputs, and models for a number of physiological systems provide rather good fits to experimental data, including models for the respiratory (see below) and cardiovascular systems, stretch reflex in muscle, regulation of blood glucose, regulation of body temperature, and numerous nervous functions.

EXAMPLE *Black Box Models for Respiration*

In order to understand the models for respiratory function, it is first necessary to be aware of the nature of biological control of breathing rate. The effects of the three most important factors in human breathing rate ("ventilation") are shown in Fig. 16-9*a*. It might be supposed that since oxygen supply to the blood and from there to the tissues is the principal function of breathing, breathing rate ought to be controlled directly by oxygen concentration in the blood. To the contrary, Fig. 16.9*a* shows that oxygen concentration in the blood has almost no effect on breathing, except at very low oxygen levels. The most

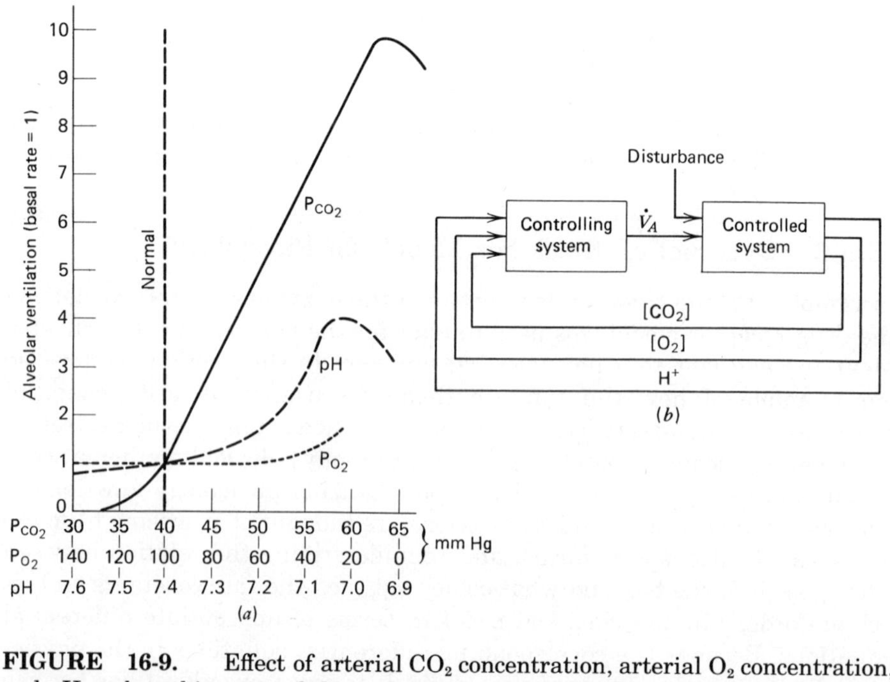

FIGURE 16-9. Effect of arterial CO_2 concentration, arterial O_2 concentration, and pH on breathing rate (left), and representation of the respiratory system as a control system (right) subject to three sensed and controlled parameters. $\dot{V}_A$ is the alveolar ventilation rate (alveoli are the tiny air-filled sacs through which gas molecules diffuse into the blood vessels of the lung). (Ventilation data from A. C. Guyton, *Textbook of Medical Physiology,* 2nd ed., W. B. Saunders Company, Philadelphia, 1961, p. 556.)

important control factor for breathing rate is CO_2 concentration in the blood: when breathing is too slow, CO_2 concentration rises, and the system is stimulated to increase the breathing rate to bring the CO_2 concentration back to normal; similarly, if one deliberately hyperventilates (breathes faster than necessary), CO_2 concentration drops, and the (involuntary) impulse to breathe decreases. These facts account, for example, for the need to force oneself to breathe on reaching high altitude, since CO_2 partial pressure in the atmosphere decreases, and the respiratory control mechanism "thinks" that the body is breathing too fast and deliberately slows down breathing, even though the oxygen pressure is also lower and breathing really ought to be speeded up! (The respiratory control center ultimately readjusts properly after a day or so acclimatization to high altitude.)

The effect of pH on breathing rate is basically due to CO_2 concentration again, since the concentrations of H^+ and CO_2 are related by the reactions

$$CO_2 + H_2O \rightleftharpoons H_2CO_3 \rightleftharpoons \overset{\displaystyle H^+ \atop \displaystyle +}{HCO_3^-} \rightleftharpoons H^+ + HCO_3^- \tag{16-29}$$

Thus when arterial pH is artificially decreased, for example, $[H^+]$ is higher and the equilibria in Eq. 16-29 are all shifted to the left to compensate for the $[H^+]$ increase; this results in an increase in CO_2 concentration that in turn increases breathing rate as we have already noted.

The three factors, $[CO_2]$, $[H^+]$, and $[O_2]$, that control breathing rate may be used to construct a control system model for respiration that in the simplest case might look like Fig. 16-9*b*. Simplifying the model even further, we will neglect the effects of $[H^+]$ and $[O_2]$, and consider only the exchange system for CO_2 as it proceeds from the lungs to the arteries to the tissues to the veins back to the lungs as shown in Fig. 16-10. The system shown in Fig. 16-10 may be described by five equations (two differential equations and three algebraic equations) in five unknowns: $[CO_2]_L$, the CO_2 concentration in the lungs; $[CO_2]_T$, the CO_2 concentration in the tissues; q_1, the rate at which CO_2 leaves

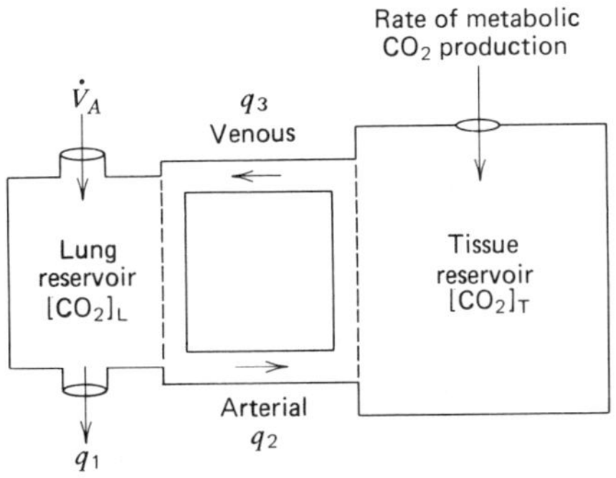

FIGURE 16-10. Schematic diagram of CO_2 exchange between lung, blood, and tissue. The various terms are defined in the text. [After F. S. Grodins, J. S. Gray, K. R. Schroeder, A. L. Norins, and R. W. Jones, *J. Appl. Physiol.* 7, 283 (1954).]

the lung via expired air; q_2, the rate at which CO_2 leaves the lung via arterial blood; and q_3, the rate at which CO_2 enters the lung via venous blood. For example,

$$\frac{d[CO_2]_T}{dt} = \frac{q_2 - q_3 + (\text{rate of metabolic } CO_2 \text{ production})}{(\text{Volume of tissue } CO_2 \text{ reservoir})} \tag{16-30a}$$

A second equation is

$$\frac{d[CO_2]_L}{dt} = \frac{\dot{V}_A F^I_{CO_2} + q_3 - q_1 - q_2}{(\text{Volume of the lung reservoir})} \tag{16-30b}$$

in which $F^I_{CO_2}$ is just the fraction of CO_2 in the inspired ("breathed-in") air. Finally, there are three more equilibrium relations:

$$[CO_2]_L = (q_1/\dot{V}_A) \tag{16-30c}$$

which expresses the equality of alveolar and expired CO_2 concentrations,

$$(q_2/(\text{cardiac output})) = P(\text{slope})[CO_2]_L + \text{intercept} \tag{16-30d}$$

which expresses the equilibrium between arterial and alveolar CO_2 concentrations via an assumed linear curve for absorption of CO_2 as a function of external barometric pressure, P, where the slope and intercept of the curve appear in Eq. 16-30d and finally

$$[CO_2]_T = (q_3/(\text{cardiac output})) \tag{16-30e}$$

expressing the equality of tissue and venous CO_2 concentrations.

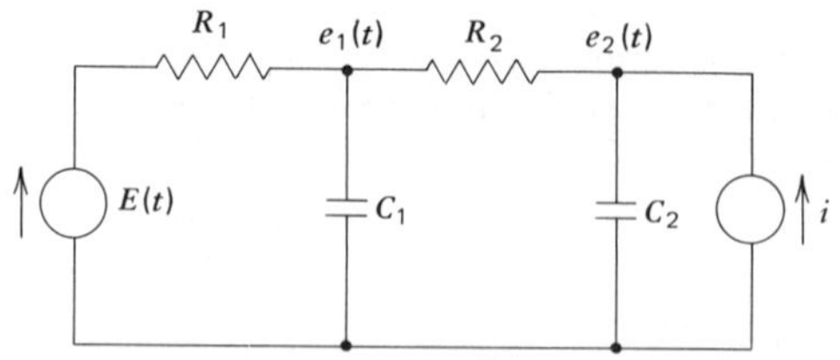

FIGURE 16-11. Electronic circuit analog of isolated controlled system (see Fig. 16-10) portion of respiratory system, in which control based only on CO_2 concentration is assumed. The behavior of this circuit will exactly mimic the behavior of the model that led to Eq. 16-30, with the identifications: $R_1 = \dot{V}_A$ = alveolar ventilation; R_2 = cardiac output; C_1 = volume of lung reservoir; C_2 = volume of tissue reservoir; i = rate of metabolic CO_2 production; $e_1(t) = [CO_2]_L = CO_2$ concentration in the lung; $e_2(t) = [CO_2]_T = CO_2$ concentration in the tissues; and $E(t) = F_{CO_2}{}^I(t)$ = step-function externally applied sudden change in fraction of CO_2 in inspired air. (From F. S. Grodins, *Control Theory and Biological Systems,* Columbia University Press, New York, 1963, p. 131.)

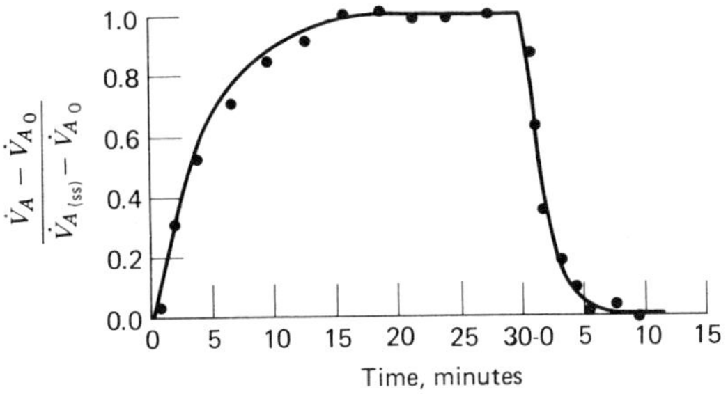

FIGURE 16-12. Experimental (black dots) and theoretical (smooth curve) transient ventilation responses to sudden application (left) and sudden subsequent removal (right) of a 5% CO_2 level in inspired air. (Ventilation is expressed relative to an average value and normalized versus a steady-state level.) [From F. S. Grodins, J. S. Gray, K. R. Schroeder, A. L. Norins, and R. W. Jones, *J. Appl. Physiol.* 7, 283 (1954).]

By choosing $[CO_2]_L$ and $[CO_2]_T$ as the system outputs, and $F^I_{CO_2}$ as the sudden small change applied to the system (i.e., we will suddenly increase the concentration of CO_2 in inspired air), so that all the remaining quantities are time-invariant constants, the five Eqs. 16-30 may be combined to give a differential equation in one variable, subject to the applied force, $F_{CO_2}^I$. Although in this particular case, it would be possible to solve the equation, provided that the perturbation, $F_{CO_2}^I$, is small, it is simpler to construct an equivalent electronic analog circuit (Fig. 16-11), whose components can be identified with the various variables of our equations, and whose response can therefore be monitored to give the same behavior as the equations, with the advantage that we never need to solve the equations!

Evidence for the accuracy of the fit to experimental data resulting from a model somewhat more sophisticated than that of Fig. 16-11 is shown in Fig. 16-12. Since the various normal concentrations of CO_2, cardiac output, metabolic rate, and so on, are fixed in advance, the only adjustable parameter is the volume of the tissue reservoir, which gave a best fit at a value of 30 liters. The mathematical models are of course designed to give fit to experimental data using the *smallest* possible number of parameters. Thus, if the biological system exhibits any redundancy in its control mechanisms, those extra control loops will not appear in the mathematical model. For this reason, it is not always possible to correlate particular "black boxes" in physiological models with actual organs. Furthermore, it is obviously very artificial to isolate a given system, such as the respiratory system, from the rest of the body; for example, $[H^+]$ concentration in the blood is also controlled (with a much slower time constant) by preferential extrusion of NH_4^+ ions by the kidney. With these and other limitations, it still seems remarkable that such complex physiological functions can be described so accurately by such relatively simple electrical circuits based on physiological inputs and outputs.

16.B. DAMPED SPRING WITH FINITE MASS: RINGING A BELL

16.B.1. Fluorescence Depolarization: Site-directed Macromolecular Probes

The problem discussed in Chapter 13.C was the (transient) behavior of a finite weight on a spring initially displaced from equilibrium and then let go with no further driving force. In contrast to the *steady-state* experiment, in which the response eventually settles down to an oscillation at the frequency of the *driver,* the *transient* displacement oscillates at a frequency near the *natural* frequency of the weight on a spring itself (i.e., a tuning fork vibrates at its natural frequency when struck with a sudden sharp blow). In the transient fluorescence depolarization experiment (see Figures 16-13, 16-14) polarized incident light (generally in the ultraviolet range) is directed at the sample near a resonant frequency for absorption. In contrast to the usual light-scattering experiment, however, the radiation that we might expect to be broadcast by the driven absorption dipole does not appear immediately; some of the energy of the driven dipole is lost very quickly and subsequent emission of radiation is thus at a lower frequency (than the incident light) since some of the energy is lost before the radiation could escape from the molecule. The times associated with each of these processes are shown in Fig. 16-13: initial absorption of the incident ray occurs in about 10^{-15} sec; it then takes about 10^{-12} sec for rearrangement of the internal energy of the molecule to leave a dipole still oscillating at a lower frequency from the (delayed) influence of the original incident driving electric field; if the incident light is now removed, radiation con-

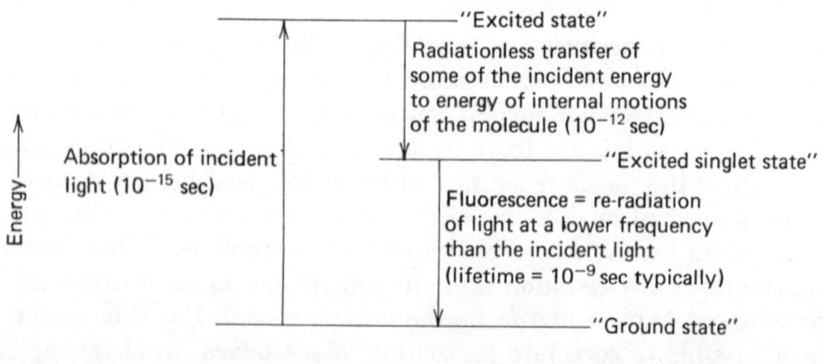

FIGURE 16-13. Schematic diagram of the absorption and delayed re-emission of light by a fluorescent molecule. Ground and excited state are quantum mechanical terms discussed in Section 5. The *frequency* of oscillation of the second "emission" dipole is lower than that of the original driven "absorption" dipole, and the *direction* along which the oscillation of the electron on a spring occurs may also be different for the emission and absorption dipoles, even if the molecule does not rotate at all during the time between absorption and re-emission (see text).

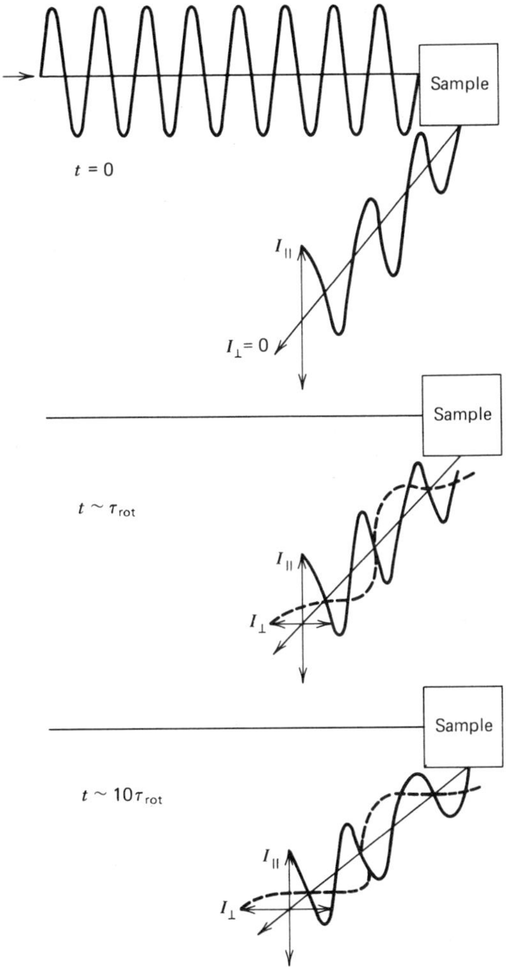

FIGURE 16-14. Relative intensities of vertically polarized light, $I_{\parallel}$, and horizontally polarized light, $I_{\perp}$, emitted by fluorescence at right angles to the direction of propagation of vertically polarized incident light that is turned off at time zero. (Emission and absorption dipole oscillation directions are taken as the same at time zero.) Fluorescence is initially plane-polarized vertically, but as the molecules rotate, the emission dipole oscillation direction also rotates, and a horizontally polarized fluorescence component appears (middle diagram). After the molecules have been allowed to rotate for a time long enough such that their directions are essentially random compared to the initial direction at time zero, the fluorescent intensities of vertically and horizontally plane-polarized light are essentially equal (bottom diagram). (In this figure, it is assumed that molecular rotation is fast compared to the lifetime for decay of overall fluorescent intensity, $\tau_{rot} << \tau$, so that the overall fluorescent intensity does not decrease appreciably during these measurements.)

tinues to be broadcast from the low-frequency dipole for a lifetime of 0.1–200 $\times$ 10^{-9} sec depending on the particular molecule involved.

Suppose for simplicity that the direction along which the driven electron-on-a-spring oscillates in the molecule is the same for the initially driven "absorption" dipole and the second driven "emission" dipole, for a molecule that is at rest. Now suppose that *vertically* polarized incident light is directed at the static molecules and that we subsequently monitor the intensity of fluorescent light that is re-emitted (after turning off the incident light beam) either vertically polarized with intensity, $I_{\|}$, or horizontally polarized with intensity, $I_{\perp}$ (see Fig. 16-14). Since the direction of oscillation of the emitting and absorbing dipoles are the same, we would expect that the intensity of fluorescent light emerging along a direction of propagation perpendicular to the direction of propagation of the incident light should be completely plane-polarized in the *vertical* direction (compare with Fig. 15-1 for the same argument used earlier for light scattering). Next, suppose that we now allow the molecule to rotate somewhat during the time (about 10^{-9} sec) between absorption and emission; clearly, the emitted light intensity along a direction perpendicular to the incident beam will have some component plane-polarized in a *horizontal* direction. Finally, if the molecules are allowed to rotate randomly for a very long time, the direction of oscillation of the emission dipole will be completely random, and we expect to find equal fluorescent intensity observed in both vertical and horizontal planes. If we define an "*anisotropy*" of fluorescence observed along the direction shown in Fig. 16-14 as

$$\boxed{\text{Fluorescence anisotropy} = A(t) = \frac{I_{\|}(t) - I_{\perp}(t)}{I_{\|}(t) + 2I_{\perp}(t)}} \qquad (16\text{-}31)$$

then while the *overall* fluorescent intensity $[I_{\|}(t) + 2I_{\perp}(t)]$ will decrease exponentially in time according to a time constant

$$\tau = \text{lifetime of excited singlet state} \qquad (16\text{-}32)$$

the *anisotropy*, $A(t)$, will decrease exponentially with a time constant, τ_{rot}, which represents the time it takes for a molecule to rotate diffusionally (i.e., by many small angular steps) an angular distance of the order of one radian (see Chapter 21.B):

$$A(t) = A_0 \exp[-t/\tau_{\text{rot}}] \qquad (16\text{-}33)$$

EXAMPLE *Transient Fluorescent Anisotropy as a Probe of Protein Mobility in Membranes*

The molecule 5-dimethylamino-1-naphthalene sulfonyl chloride ("dansyl" chloride)

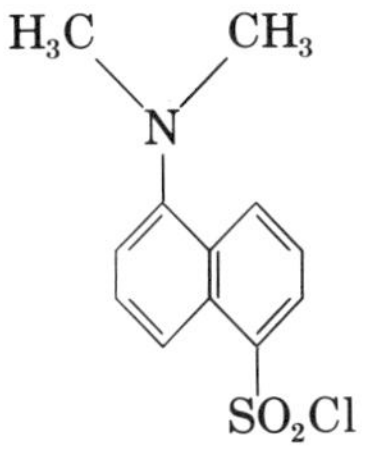

Dansyl chloride

becomes highly fluorescent on binding (covalently) to proteins, and shows an excited singlet state lifetime of about 12 nanoseconds. Since rotational correlation times, τ_{rot}, for proteins are of this order of magnitude, the fluorescence from a dansyl-labeled protein will "depolarize" (i.e., $A(t)$ will decrease appreciably) before all the fluorescent overall intensity has disappeared. The time-dependence of $A(t)$ in Eq. 16-33 can thus be used to determine the rotational correlation time of a dansyl-labeled protein.

Figure 16-15*a* shows a plot of overall fluorescent intensity, $[I_{\parallel}(t) + 2I_{\perp}(t)]$, difference in fluorescent intensity between vertically and horizontally polarized emitted light, $[I_{\parallel}(t) - I_{\perp}(t)]$, and anisotropy, $A(t)$, as functions of time following the removal of a pulse of incident light, for dansyl-labeled protein-containing fragments of excitable biological membranes. The anisotropy appears to decrease according to two characteristic τ_{rot} values, one of which is very much longer (corresponding to much slower rotational motion) than the other:

$$A(t) = f_A \exp[-t/\tau_{rot}^A] + f_B \exp[-t/\tau_{rot}^B] \tag{16-34}$$

$$= f_A \exp[-t/\tau_{rot}^A] + f_B \quad \text{for} \quad \tau_{rot}^B >> \tau_{rot}^A \tag{16-34a}$$

in which τ_{rot}^A and τ_{rot}^B are the rotational correlation times for the two types of rotational motion, and f_A and f_B represent the relative fractions of the observed anisotropy attributable to those respective motions. It is clear that one of the motions is much slower than the other, so that the anisotropy appears to level off to a constant value, as shown in Eq. 16-34a. It seems most likely that the two types of motion correspond to a local rapid rotation due to flexibility at the site of attachment of the dansyl label to the protein ($\tau_{rot}^A = 3 \times 10^{-9}$ sec), and the much slower rotational motion of the dansyl-labeled protein as a whole ($\tau_{rot}^B > 700 \times 10^{-9}$ sec). When the dansyl-labeled proteins are released from the membrane by extraction with Triton detergent (Fig. 16-15*b*), the anisotropy again shows two types of rotational motion, but the slow motion of the protein is now fast enough to cause an observable decrease in the anisotropy ($\tau_{rot}^B = 45 \times 10^{-9}$ sec, corresponding to rotational diffusion of a protein of molecular weight = 75,000 or so). These results are of particular interest when compared with ESR spin-label results (Fig. 14-30). The ESR results show that much of the *membrane itself* is highly "fluid" (i.e., shows appreciable rotational mobility), while the fluorescence results show that the *membrane proteins* appear to be highly immobilized in the membrane. A resolution of this apparent anomaly may be provided by recent results with ESR spin-

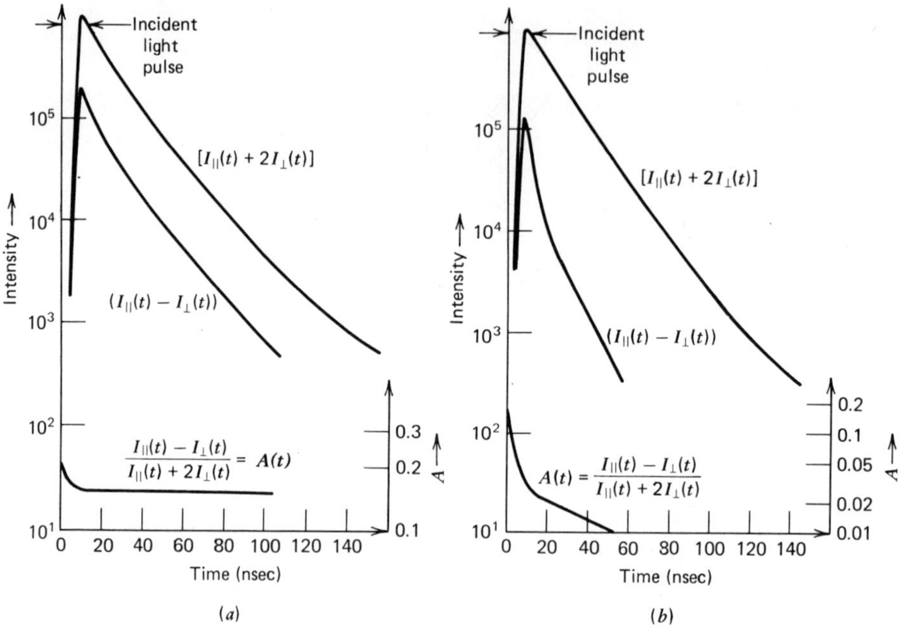

FIGURE 16-15. Plots of various combinations of vertically polarized $[I_{||}(t)]$ and horizontally polarized $[I^{\perp}(t)]$ fluorescent intensity following application of a very short (about 5×10^{-9} sec) pulse of incident light. (*a*) Fluorescence from dansyl-labeled proteins in fragments of biological membranes from the electric organ of the electric eel, *Electrophorus electricus;* (*b*) fluorescence of free dansyl-labeled membrane proteins in solution. In both (*a*) and (*b*) the decay of the anisotropy, $A(t)$, shows components from a fast rotational motion ($\tau_{\text{rot}}^{A} = 3$ nsec) corresponding to local flexibility at the site of attachment of the label to the protein, and a slow rotational motion [$\tau_{\text{rot}}^{\text{B}} > 700$ nsec in (*a*) and $\tau_{\text{rot}}^{\text{B}} = 45$ nsec in (*b*)] corresponding to rotational diffusion of the protein molecule as a whole. The very slow protein rotation in (*a*) shows that this dansyl-labeled membrane-bound protein is very highly immobilized in the membrane, even though the membrane itself (see Fig. 14-30) may be highly fluid in the membrane interior. [From P. Wahl, M. Kasai, and J.-P. Changeux, *Eur. J. Biochem. 18,* 332 (1971).]

labeled phospholipids, in which O. H. Griffith showed that the lipids in immediate vicinity of the protein appear to be much more highly immobilized than those farther away. A modern biological membrane model may therefore consist of bulky slowly rotating proteins floating around in a relatively fluid membrane interior.

EXAMPLE *Steady-State Fluorescence Polarization: Perrin Plot*

Although the newer fluorescence polarization experiments have been performed by monitoring the *transient* anisotropy following a *pulse* excitation, an older and alternative method is to monitor the *steady-state* fluorescence observed during *continuous* irradiation of the sample with incident light. Defining a "polarization" as the steady-state analog of anisotropy:

$$\text{Polarization} = P = \frac{I_{\|} - I_{\perp}}{I_{\|} + I_{\perp}} \tag{16-34}$$

in which $I_{\|}$ and $I_{\perp}$ are now the (time-independent) steady-state values for fluorescent intensity polarized vertically and horizontally, it is clear that the polarization, P, will be a maximum, P_0, when rotational motion is very slow compared to the average lifetime spent by a particular molecule in the fluorescent singlet excited state (i.e., by the time the molecule rotates appreciably, fluorescent emission has already occurred, with polarization preserved). On the other hand, if the molecule rotates rapidly compared to the fluorescent lifetime, $\tau_{rot} << \tau$, then the observed steady-state fluorescence should be completely depolarized (i.e., by the time the fluorescence emission occurs, the direction of oscillation of the emission dipole is random). The qualitative behavior predicted by Eq. 16-35 (derived by Perrin in 1926 and extended to cover anisotropic rotational diffusion of nonspherical molecules by Weber in 1952) should thus be evident (we will not take the time to derive this equation):

$$\frac{[(1/P) \pm (1/3)]}{[(1/P_0) \pm (1/3)]} = 1 + 3(\tau/\tau_{rot}) \tag{16-35}$$

or since rotational correlation time, τ_{rot}, is related to rotational diffusion constant, D_{rot}, according to:

$$\tau_{rot} = (1/2D_{rot}) = \frac{4\pi\eta r^3}{kT} = \frac{3\eta\, V}{RT} \qquad \text{(Compare Eq. 14-33)} \tag{16-36}$$

in which η is solution viscosity, k and R are gas constant per molecule and per mole, r is molecular radius for a spherical molecule, and $V = \frac{4}{3}\pi r^3 N_0$ is molar volume for those molecules, Eq. 16-35 can be rewritten as the "Perrin Equation"

$$\frac{[(1/P) \pm (1/3)]}{[(1/P_0) \pm (1/3)]} = 1 + \frac{R\tau}{V}\,(T/\eta) \tag{16-37}$$

In the limit of very low temperature or very large viscosity [i.e. $(T/\eta) \rightarrow 0$], molecular rotation will be slowed down to the limit, $\tau_{rot} >> \tau$, so that the polarization, P, will assume its maximum value, P_0. More importantly, the *slope* of a plot of $[(1/P) \pm (1/3)]/[(1/P_0) \pm (1/3)]$ versus (T/η) will be $R\tau/V$, allowing for determination of effective molecular volume, V, which in turn allows for calculation of rotational correlation time, τ_{rot}, from Eq. 16-36. Since τ_{rot} as determined from this procedure will be larger for nonspherical molecules than would be predicted from the macromolecular weight used to compute τ_{rot} for a sphere of the same molecular weight, the rotational correlation time may be used as an estimate of macromolecular *shape*. From Chapter 7.C.2, the reader may recall that the frictional ratio, f/f_0 from diffusion measurements also increases with increasing molecular asymmetry in shape. However, the frictional ratio will increase for *any* deviation from compact spherical shape (including random coil), but the rotational correlation time will increase only if the molecule becomes asymmetric but still remains *rigid*. Thus, measurement of τ_{rot} can distinguish not only between spherical and nonspherical molecules, but between *rigid* and *flexible* molecules.

Figure 16-16 shows a Perrin plot for human macroglobulin (MW = 900,000), and also for its subunits obtained by inter-chain disulfide bond cleavage (MW = 180,000), and for the fragments obtained by trypsin digestion (MW = 50,000). Maximum polarization, P_0, was determined directly by measurement of $I_{\parallel}$ and $I_{\perp}$ for proteins dissolved in 90% glycerol at 8°C $[(T/\eta) \times 10^{-4} = 0.005]$ as shown by the data points at the far left of Fig. 16-16. P_0 was also determined by extrapolation of the data obtained for the various proteins in water at various temperatures (data points at right of Fig. 16-16). Rotational correlation time was computed from Eq. 16-36, using the P value obtained for the proteins in water at room temperature (left-most of the right-hand data points of Fig. 16-16). The rotational correlation times were as follows: MW = 900,000, τ_{rot} = 80 nsec; MW = 180,000, τ_{rot} = 69 nsec; MW = 50,000, τ_{rot} = 58 nsec. In other words, even though trypsin cleaves the native macroglobulin to fragments almost 20 times smaller, there is only a 25% reduction in rotational correlation time (rather than the $100 \times [900{,}000/50{,}000](1/3) = 260\%$ reduction that would be expected if both molecules were rigid. A similar argument

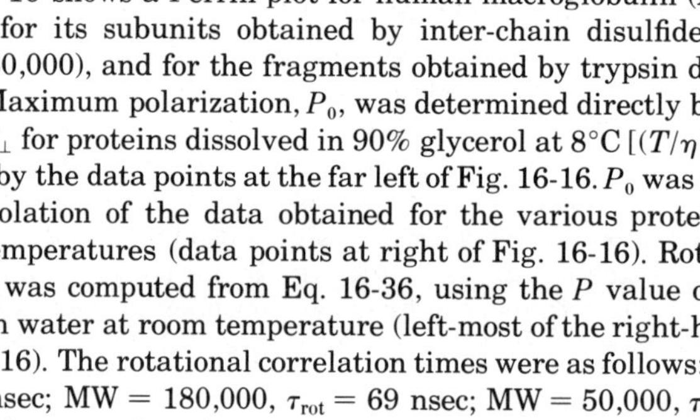

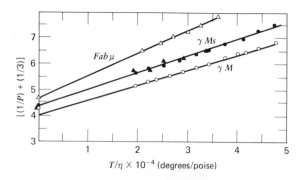

FIGURE 16-16. Perrin plot based on steady-state fluorescence polarization, P, as a function of (T/η), where T is absolute temperature and η is solution viscosity. γM: human macroglobulin (MW = 900,000); γM_s: macroglobulin that has been reduced with dithiothreitol to break all the inter-chain disulfide bonds (MW = 180,000); $Fab\mu$: macroglobulin that has been subjected to trypsin cleavage. In all cases, macroglobulin was first labeled with dansyl chloride to produce a covalently-fluorescent-labeled protein. Rotational correlation times are calculated from the data at $(T/\eta) = 2 \times 10^{-4}$ deg poise^{-1}, using Eq. 16-37 and a lifetime of τ = 12 nsec for the dansyl group. The lowest curve thus corresponds to the macromolecule having the largest τ_{rot}. [From H. Metzger, R. L. Perlman, and H. Edelhoch, *J. Biol. Chem. 241*, 1741 (1966).]

obtains for the 180,000 MW subunit. The conclusion from these τ_{rot} values is that the native MW = 900,000 macroglobulin is a highly flexible molecule.

16.B.2. Gamma-ray Directional Correlations: Same Experiment for Opaque Media.

The gamma-ray directional correlation experiment is conceptually and instrumentally similar to the pulsed-excitation transient fluorescence experiment. In both cases, the first electromagnetic wave is used to "polarize" (i.e., in some way "line up") the sample, and subsequent emission of a second wave occurs in a preferred direction until the sample molecules have had time to rotate, after which the initial "alignment" is destroyed and emission is isotropic (i.e., equal intensity in all directions). The main new features of the gamma-ray case are (a) both the "aligning" and the "observed" electromagnetic waves come from the *sample* (whereas in fluorescence, the initial plane-polarized wave is an *externally* delivered beam), and (b) the gamma rays arise from oscillating *nuclei* on springs rather than oscillating *electrons* on springs, and the energies of the waves are therefore in a different (gamma ray rather than visible light) frequency range than for fluorescence.

The basis and layout for the gamma-ray directional correlation experiment are indicated in Fig. 16-17. Certain nuclei (including ^{62}Zn and ^{111}Cd–

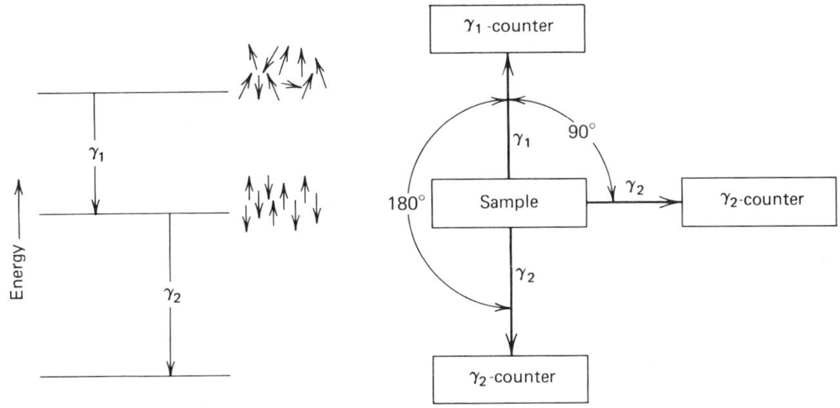

FIGURE 16-17. Origin of anisotropy in direction of gamma-ray emission from initially unaligned nuclear magnetic dipoles. Although initial magnetic dipole directions are random (upper left), restricting the *detection* of those rays to a single direction selects nuclei that must have been aligned according to that emission direction (middle left). Gamma rays *subsequently* emitted (i.e., γ_2) from those aligned nuclei will be observed preferentially at 180° rather than 90° to the direction between the sample and the γ_1-counter (right). As radioactively labeled molecules rotate, the alignment of the nuclear magnetic dipoles is destroyed, and the observed anisotropy in γ_2 emission [Eq. 16-38] decreases. For most cases, the energies of γ_1 and γ_2 are sufficiently different that they are readily distinguishable by the detectors (see Fig. 16-18).

see Fig. 16-18) emit two gamma rays spontaneously in succession (just as fluorescent molecules emit one light ray spontaneously from the excited singlet state). If the emitting (magnetic) dipoles are randomly oriented (as for a collection of randomly oriented ^{111}Cd-labeled macromolecules in solution, then both gamma rays will be emitted with the same intensity in all directions. However, if we record only those initial gamma rays, γ_1, which are emitted in a *particular* direction, then those γ_1 must have originated from nuclei whose magnetic dipoles were *aligned* according to that direction, as shown in the middle left portion of Fig. 16-17. Finally, if we now observe a γ_2 ray that is emitted *by the same nucleus,* we are much more likely to observe it coming out in certain directions (say at 180° with respect to the direction of the first gamma ray) than in others (say at 90° with respect to the first gamma ray). The γ-ray emission pattern for aligned emitting nuclei is similar to light-scattering (Fig. 15-1). If the detectors are thus coded to respond only to γ_1 for direction 0°, and only for γ_2 at 90° or 180°, and if γ_2 events are recorded only if they follow (but not if they precede) γ_1 events, then we are reasonably sure that the γ_1, γ_2 successive rays came from the same nucleus, and we expect to find an anisotropy, $A(t)$, between γ_2 rays arriving at 90° and 180° to the direction of the γ_1-detector. If the molecules are able to rotate during the time between emission of the aligning ray, γ_1, and the observed ray, γ_2, then the anisotropy will be partially lost, and we expect $A(t)$ to decrease with time. As with fluorescence depolarization, the decrease in $A(t)$ with time is exponential in many cases.

$$A(t) = \frac{[(\#\ \gamma_2\text{-counts at } 180°) - (\#\ \gamma_2\text{-counts at } 90°)]}{(\#\ \gamma_2\text{-counts at } 90°)} \qquad (16\text{-}38)$$

A final operational distinction between fluorescence depolarization and gamma-ray depolarization is necessary for interpreting experimental results. In the fluorescence decay, *faster* molecular rotation leads to *faster* disappearance of the fluorescence anisotropy; in gamma-ray studies, *faster* molecular rotation usually leads to *slower* exponential decay of anisotropy. The explanation for the gamma-ray behavior is simple, but postponed until Chapter 21B where it is discussed in common with a variety of other similar phenomena. We are now ready to consider some experimental results.

EXAMPLE *Gamma-ray Directional Correlations for ^{111}Cd-labeled Carbonic Anhydrase*

There is an isotope of cadmium called ^{111m}Cd (m is for "metastable") that follows the desired gamma-ray cascade of Fig. 16-18, with an intermediate state halflife of 85 nsec sufficiently long that molecular rotational diffusion of proteins can markedly affect the anisotropy of the gamma-ray emission before the intensity of the γ_2-emission has dropped too far (compare to similar

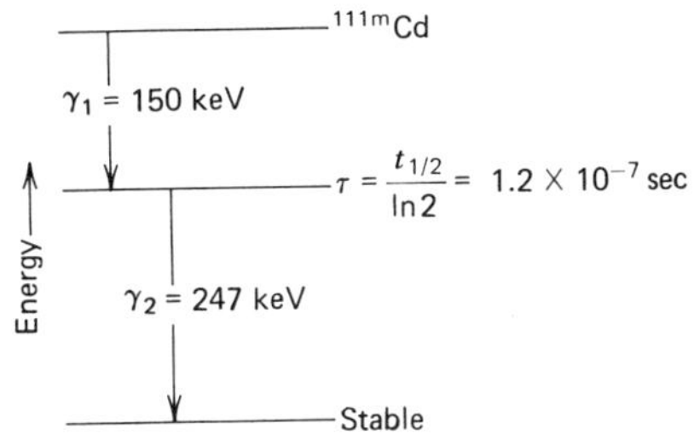

FIGURE 16-18. The gamma-ray cascade for successive emission of two gamma-rays from the metastable ^{111m}Cd nucleus. The energies of γ_1 and γ_2 are independent of the chemical environment of the cadmium nucleus, while the anisotropy of the intensity of the 247 keV gamma-rays as a function of direction is sensitive to molecular rotational motion (see text).

discussion for the 12 nsec fluorescent lifetime for the dansyl fluorescent label). When $^{111m}Cd^{++}$ is free in solution, rotational diffusion of the solvated radioactive metal ion is fast, and the anisotropy decreases only slightly with time (Fig. 16-19*a*). When apo-carbonic anhydrase is added to the solution (apo-carbonic anhydrase is the native enzyme with the active-site Zn^{++} removed), the $^{111m}Cd^{++}$ binds to the active site at the location ordinarily occupied by Zn^{++}; since rotational diffusion of the ^{111m}Cd-labeled protein is now much slower, the gamma-ray anisotropy decreases much more rapidly with time (Fig. 16-19*c*). To show that the change introduced by the enzyme is not simply due to an increase in solution viscosity (which would also cause slower rotational diffusion of $^{111m}Cd^{++}$ and thus a more rapid decay of gamma-ray anisotropy), a control experiment (Fig. 16-19*b*) was introduced, in which the solution contained $^{111m}Cd^{++}$ and *native* carbonic anhydrase (i.e., carbonic anhydrase with a Zn^{++} atom already at the active site, preventing any binding of Cd^{++}); the control experiment shows that the increase in solution viscosity does indeed affect the anisotropy, but to a much lesser degree than the effect of Cd^{++}-binding to the enzyme active site.

Since the anisotropy-versus-time plots such as those shown in Fig. 16-19 are interpreted in terms of a single number, the rotational correlation time, τ_{rot},

$$\tau_{rot} = (1/6D_{rot}) \qquad (16\text{-}39)^*$$

where D_{rot} is rotational diffusion constant, it might seem redundant to display the whole curve just to determine one number. Another display of

* *For the gamma-ray and magnetic resonance experiments (Chapter 21.B) the relation between τ_{rot} and D_{rot} is given by Eq. 16-39. For the dielectric relaxation and fluorescence polarization experiments, D_{rot} has the same meaning, but τ_{rot} may be proportional to D_{rot}^{-1} by a factor of either (1/2) or (1/6), depending on definition or (in the dielectric case) the nature of the induced dipole moment.*

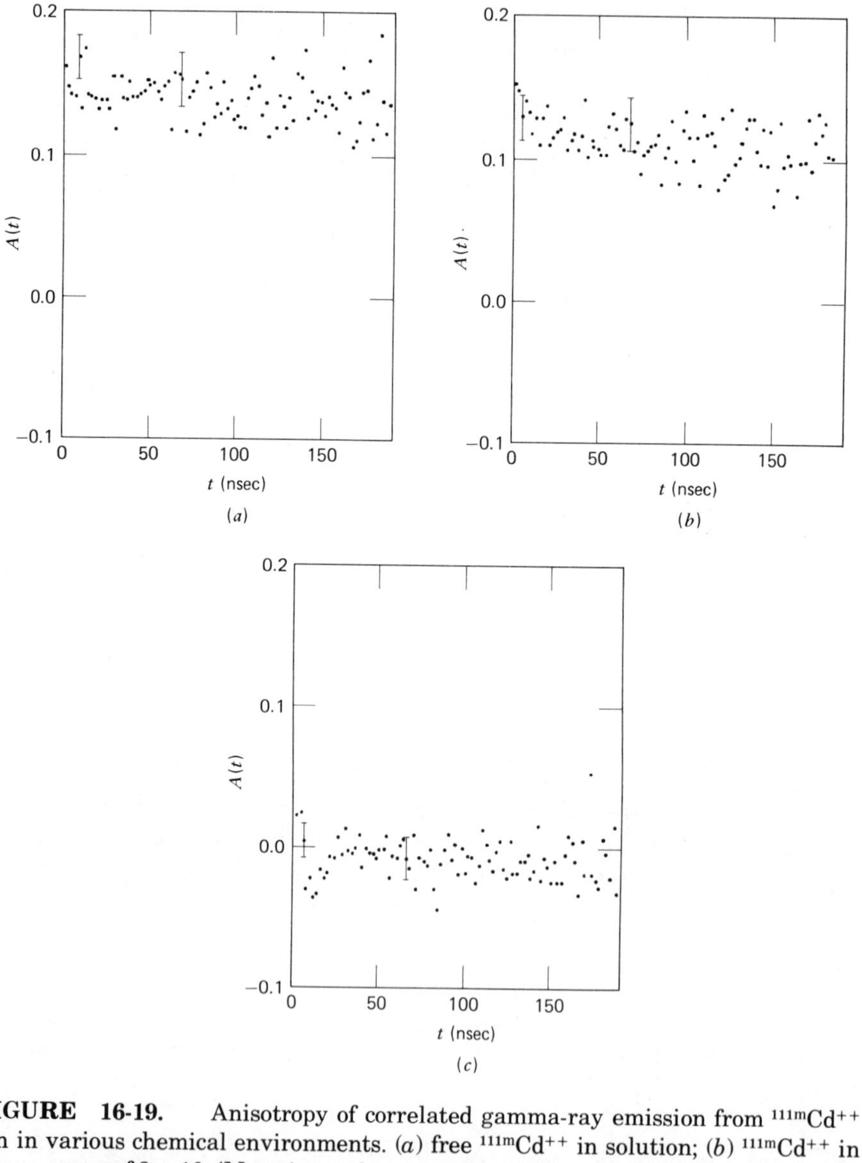

FIGURE 16-19. Anisotropy of correlated gamma-ray emission from $^{111m}Cd^{++}$ ion in various chemical environments. (*a*) free $^{111m}Cd^{++}$ in solution; (*b*) $^{111m}Cd^{++}$ in the presence of 3×10^{-4}M native carbonic anhydrase (i.e., enzyme with no available Cd^{++}-binding sites), and (*c*) $^{111m}Cd^{++}$ in the presence of 2.5×10^{-4}M apo-carbonic anhydrase (i.e., enzyme with one strong Cd^{++}-binding site available). All solutions buffered to pH 6.1 by 0.1 M phosphate, 0.5M NaCl. Note that the anisotropy decays much faster (plot *c*) when the radioactive $^{111m}Cd^{++}$ is bound to an enzyme than when the $^{111m}Cd^{++}$ is free in solution (plot *a*), because of the large difference in rotational diffusion rates in the two cases (see text). [From C. F. Meares, R. G. Bryant, J. D. Baldeschwieler, and D. A. Shirley, *Proc. Natl. Acad. Sci. U. S. A. 64,* 1155 (1969).]

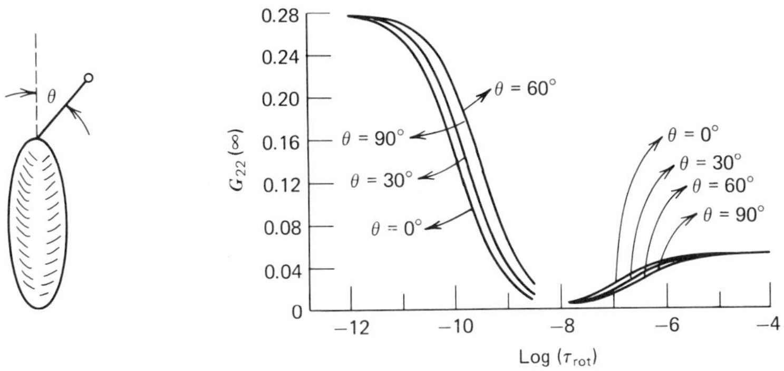

FIGURE 16-20. Plots of integral anisotropy, $G_{22}(\infty)$ versus rotational correlation time, τ_{rot} (log scale), for a ^{111}Cd nucleus, ○, attached at an angle, θ, with respect to the main symmetry axis of a prolate molecular ellipsoid (see Fig. 7-22). An even wider variation in the position of the plot is found when $G_{22}(\infty)$ is plotted versus τ_{rot} for a ^{111}Cd nucleus attached to a spherical molecule, but allowed to rotate at various rates at the site of attachment. [From A. G. Marshall, L. G. Werbelow, and C. F. Meares, *J. Chem. Phys. 57,* 364 (1972).]

data more directly useful is shown in Fig. 16-20, where the vertical axis represents a quantity proportional to the *area* under a plot of $A(t)$ versus time, weighted according to the lifetime for disappearance of γ_2-radiation:

$$G_{22}(\infty) = (\text{constant}) \int_0^\infty A(t) \exp(-t/\tau)\, dt \qquad (16\text{-}40)$$

The importance of the graph in Fig. 16-20 is that one need simply carry out the measurement of $A(t)$, integrated over a time long compared to τ, and then read off the rotational correlation time, τ_{rot}, directly from the graph of $G_{22}(\infty)$ versus τ_{rot}. Moreover, as seen from the additional curves in Fig. 16-20, the shape of the plot of $G_{22}(\infty)$ versus τ_{rot} (where τ_{rot} is just a measure of molecular size) can depend strongly on the *angle* at which the label is attached with respect to the symmetry axes of the overall molecular shape. Similar curves are obtained for labels attached with varying degrees of internal *flexibility* at the site of attachment. The reader may recall that ESR spin labels and fluorescence polarization also provide access to information about molecular shape and flexibility; the unique feature of the gamma-ray experiment is that it supplies the same sort of information at concentrations potentially as low as $10^{-12}M$ and in *opaque media* (in particular, *in vivo* measurements). Gamma-ray directional correlations have in fact already been used to study the binding of $^{111}In^{+++}$ ions to serum proteins in live mice, and it appears that this very new technique will be of unique advantage for such systems.

16.B.3. Magnetic Relaxation: the Spectroscopic Molecular Yardstick

The motion of a magnetic moment, M, in a static applied magnetic field, H_0, was introduced in treatment of the *steady-state* magnetic response to an

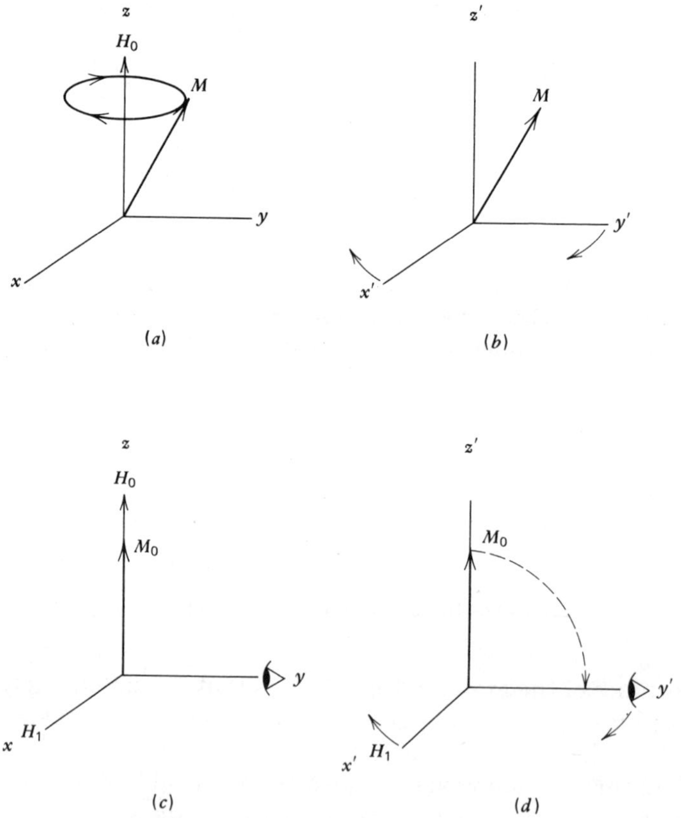

FIGURE 16-21. Behavior of a magnetic moment under the influence of applied magnetic field(s) in a laboratory (i.e., fixed) coordinate frame [(a) and (c)] and in a frame that rotates at the Larmor ("natural") precession frequency ω_0 of a magnetic moment in an applied static magnetic field H_0 [(b) and (d)]. In the lab frame (a), the magnetic moment, M, rotates while the coordinate x-y frame is fixed; in the rotating frame (b), the magnetic moment is fixed and thus behaves as if there were no applied H_0 field at all. Then, beginning from the equilibrium situation (c) with equilibrium magnetization M_0 aligned along the static field H_0, a rotating field H_1 is applied. Finally (d), in the rotating frame, H_0 is effectively absent and H_1 is fixed in magnitude and direction, so that M_0 in the rotating frame precesses about H_1 (i.e., about the x' axis) for as long as H_1 is left "on." If H_1 is left "on" just long enough for M_0 to rotate from the z' direction to the y' direction in the rotating frame, then M_y can be observed by a detector located along the (rotating) y' axis, with the results shown in Fig. 16-22.

oscillatory driving magnetic field in Chapter 14.D; the magnetic moment precesses about the applied static field at a precession frequency

$$\omega_0 = \gamma H_0 \tag{14-15}$$

For discussion of the *transient* magnetic response to a magnetic field pulse, it is particularly convenient to change to a coordinate frame that itself rotates at ω_0, so that the magnetic moment becomes stationary in the new "rotating frame" (see Fig. 16-21*a*,*b*). Since the magnetic moment is now fixed in direction (i.e., it does not precess), it is as if no applied static H_0 field were present—this is the simplification we desire. If we now imagine the application of a second small magnetic field, H_1, which *rotates* at the same rate and in the same sense as the precessing magnetic moment in the *lab* frame (Fig. 16-21*c*), then when we go to the rotating frame, H_1 will become a small *static* field (Fig. 16-21*d*). Now since H_0 is effectively zero, and H_1 is effectively static in the rotating frame, the magnetic moment M_0 in the rotating frame will simply precess about H_1 (in the y'-z' plane) at a frequency

$$\omega_1 = \gamma H_1 \tag{16-42}$$

Thus, by applying the H_1 magnetic field for a time, $T(90°) = (1.57/\omega_1)$ sec, the magnetization that began at equilibrium with magnitude, M_0, directed along the z axis, will have precessed exactly 90° to lie along the positive y' axis. Finally, if we position a detector along the (rotating) y' axis (this is readily accomplished electronically, not mechanically), we can observe the (transient) magnetic moment magnitude along the rotating y' axis as a function of time after removal of the H_1 pulse (see Fig. 16-22). The only purpose of the H_1 pulse is to rotate M_0 into the x'-y' plane. The result of this experiment and its applications is now discussed.

Following application and removal of the rotating H_1-field that is used to prepare the system with magnetization M_0 aligned along the (rotating) y' axis at time zero, the y' magnetization decreases to zero while the z' magnetization recovers back to its equilibrium value according to the exponential relations

$$\boxed{M_y' = M_0 \exp(-t/T_2)} \tag{16-43a}$$

$$\boxed{M_z' = M_0(1 - \exp[-t/T_1])} \tag{16-43b}$$

Since (see Chapter 13) T_2 in the transient experiment is related to absorption-mode line width in the steady-state experiment, $\Delta\omega$, according to (Problem 13-5a).

$$\boxed{(1/T_2) = (\Delta\omega/2) = \pi\Delta\nu} \tag{16-44}$$

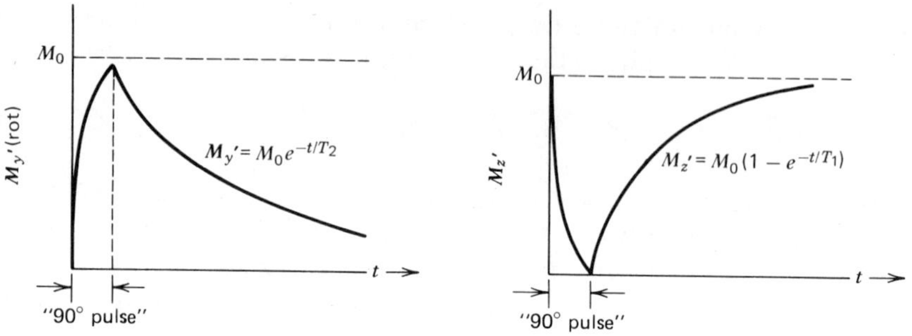

FIGURE 16-22. Behavior of the y' and z' components of magnetization, following application of a "90°-pulse" in which a small rotating H_1 field is used to rotate the equilibrium magnetization M_0 from the z' to the y' direction in a rotating frame (see Fig. 16-21). Once the magnetization is rotated to a new (fixed) direction along the rotating y' axis, the M_y' magnetization decays to zero exponentially with time constant, T_2, while the M_z' magnetization recovers back to its equilibrium value exponentially with time constant T_1. For small molecules, T_1 and T_2 are usually equal, while for macromolecules T_1 may be much larger than T_2. The relation between these "magnetic relaxation times" and molecular rotation is discussed in the text.

and since we have already argued that magnetic resonance steady-state line width is proportional to rotational correlation time, $\tau_{rot} = (1/6D_{rot})$, it is clear that T_2 is directly related to the rate of molecular rotational motion: longer T_2 corresponds to faster rotation of a molecule in solution.

EXAMPLE *Quantitation of Membrane Internal Fluidity from Carbon-13 T_1 Values*

T_2 determination is based most simply on direct monitoring of M_y' in the rotating frame as M_y' decays exponentially to zero. T_1 may be determined similarly by waiting for various lengths of time after the initial 90° pulse (to allow M_z' to recover back to part of its ultimate equilibrium value M_0) and then applying another 90° pulse to rotate the M_z' signal into the rotating y' axis for detection: from the M_z' values corresponding to various waiting times, the M_z' recovery curve can be reconstructed point-by-point, and T_1 determined. For uninteresting practical reasons, T_1 measurements are more easily carried out and interpreted than T_2 measurements for small molecules; and for molecules of molecular weight less than about 10,000, $T_1 = T_2$ in proton magnetic resonance, so T_1 carries the same information about molecular motion as would be obtained from steady-state line width measurements. The advantage of the transient experiment is that it can be conducted about 1000 to 10,000 times faster than the corresponding steady-state experiment (see Chapter 20), and is thus of greatly enhanced practical value.

We have already seen that it is often possible to resolve nuclear magnetic resonance signals for individual carbons in a small molecule (MW < 1000

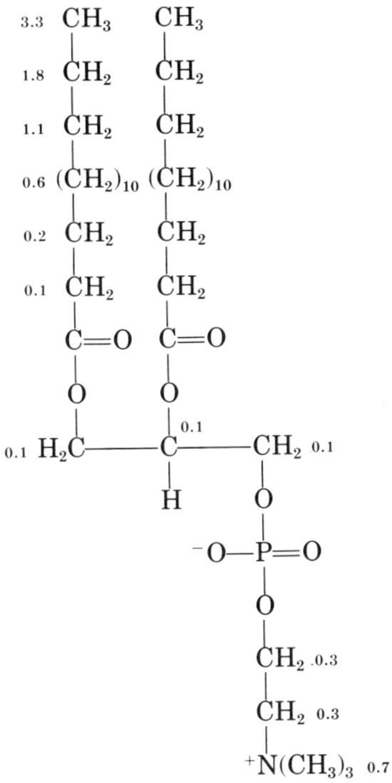

FIGURE 16-23. Carbon-13 NMR T_1 values for the various resolved carbon-13 NMR signals of individual carbons in a phospholipid molecule. Dipalmitoyl lecithin solutions were sonicated to produce vesicles with a bilayer structure similar to that of natural biological membranes. The T_1 values increase progressively in going from the carbons near the surface to the carbons furthest in the interior of the bilayer (see text). [From A. G. Lee, N. J. M. Birdsall, and J. C. Metcalfe, *Chem. Brit. 9,* 116 (1973).]

or so) (see Fig. 14-22), and it is also possible to determine individual T_1 values for each of those carbons in a carbon-13 NMR spectrum. A series of T_1 values for various resolved carbons of a phospholipid in a membrane-type bilayer structure is shown in Fig. 16-23. It is apparent that there is a 30-fold increase in T_1 (and a corresponding increase in rotational flexibility) as one proceeds from the polar "head" group of the glycerol backbone of the terminal methyl group located in the interior of the bilayer. These measurements provide some of the most direct quantitative data on the greatly increased fluidity in the interior of a phospholipid bilayer compared to the relatively restricted rotational motion near the outer surface of the model membrane. Similar experiments have recently confirmed these results, using NMR of deuterium in deuterated phospholipid molecules. Future extensions applied to the mixed-composition lipids that characterize biological membranes should prove most interesting in establishing the extent to which lipids of a particular type tend to aggregate together (or not) in actual membranes.

EXAMPLE *Proton NMR Relaxation Times and the Spectroscopic Molecular Ruler*

We have already discussed the "chemical amplifier" advantage of fast chemical exchange in connection with NMR steady-state measurement of resonant *frequency*. The argument was based on the idea that a magnetic nucleus could jump back and forth between two chemically different sites with different magnetic resonant frequencies; if the jumping rate was fast compared to the difference in resonant frequency between the two sites, a single average signal would be observed at a resonant frequency that was an average of the frequencies from the two sites, weighted according to how much relative time was spent by the nucleus at each site. A similar argument can be applied to the effect of fast exchange on the rate of *exponential decay* of M_y' or *exponential growth* of M_z following a 90° pulse. The M_y' signal will decay at the T_2 time constant appropriate to the first site for as long as it remains there; then on jumping to the other site, the M_y' signal will decay at a rate appropriate to the T_2 of the other site, and so on. As long as the jumping rate, k, is fast compared to the difference in relaxation rates at the two sites

$$k \gg |(1/T_2^A) - (1/T_2^B)| \qquad (\text{"Fast exchange"}) \tag{16-45}$$

then the observed M_y' decay will follow a single (average) exponential decay:

$$M_y' = M_0 \exp\,[-t/T_{2_{\text{ave}}})] \tag{16-46a}$$

in which

$$(1/T_2)_{\text{ave}} = f_A(1/T_2)_A + f_B(1/T_2)_B \tag{16-46b}$$

similarly

$$(1/T_1)_{\text{ave}} = f_A(1/T_1)_A + f_B(1/T_1)_B \tag{16-46c}$$

in which f_A and f_B again represent the relative fractions of nuclei at sites A and B. As with the fast-exchange applications to NMR resonant frequency in Chapter 14.D, the exchange effect can be used to measure the relaxation times, T_{1B} and T_{2B} on a dilute macromolecule, by observing small changes in the average relaxation times T_1 and T_2 for a (much more concentrated and thus stronger NMR signal) small molecule that can bind reversibly to the macromolecule. Provided (as in the following example) that $(1/T_2)_B \gg (1/T_2)_A$, it is thereby possible to measure the "bound" relaxation time for a macromolecule whose concentration is much smaller than could be detected directly.

The "yardstick" information available from T_1 and T_2 follows from relations of the type

$$(1/T_1) = \text{constant}\ (1/r^6) \tag{16-47}$$

in which r is the distance between the magnetic dipole (i.e., the nucleus) of interest and the (much stronger) magnetic dipole of a nearby unpaired elec-

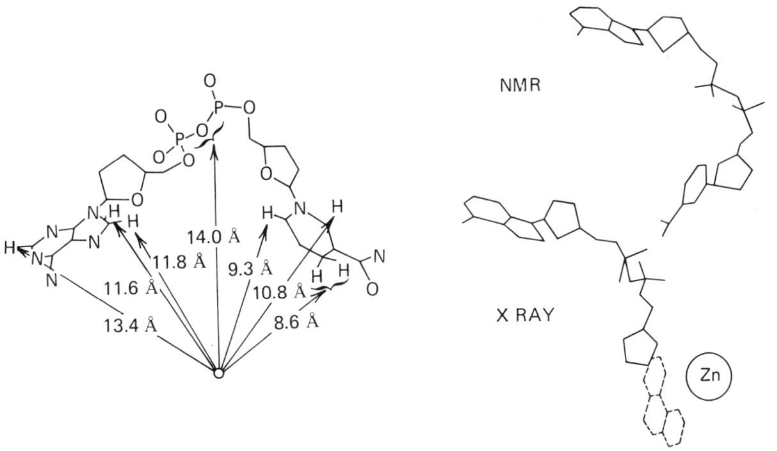

FIGURE 16-24. Configuration of NADH molecule when bound to yeast alcohol dehydrogenase that had been spin-labeled with a paramagnetic nitroxide derivative covalently attached to cysteine-43 of the enzyme polypeptide chain. Left: distances from the nitroxide to the respective protons of NADH, determined from T_1 values of the protons using Eq. 16-47. Right: comparison of the NADH conformation in spin-labeled yeast alcohol dehydrogenase (from NMR T_1 data) with that of ADP-ribose bound to liver alcohol dehydrogenase (from X-ray diffraction data). [NMR data from D. L. Sloan and A. S. Mildvan, *Biochemistry 13,* 1711 (1974): X-ray data from C. I. Branden, H. Eklund, B. Nordstrom, T. Boiwe, G. Soderlund, E. Zeppezauer, I. Ohlsson, and A. Akeson, *Proc. Natl. Acad. Sci. U. S. A. 70,* 2439 (1973).]

tron. Because many separate nuclear resonances can be resolved for different protons, say, on the same molecule, it is thus possible to measure the distances between *each* of those protons and an unpaired electron located, for example, at a particular amino-acid residue of an enzyme. Figure 16-24 shows the results of a series of such determinations from "bound" T_1 and T_2 values for the small molecule, NADH, while bound to alcohol dehydrogenase enzyme. These measurements were made possible by the fast exchange between free and bound NADH, so that the actual enzyme concentration was more than 10 times smaller than that of the observed NADH molecule. Because of the sixth-power dependence of r in Eq. 16-47, it is possible to obtain very precise distance determinations even from rather poor-precision T_1 values. The similarity in the conformation (in solution) of the NADH molecule bound to alcohol dehydrogenase and the conformation of the similar ADP-ribose molecule bound to a dehydrogenase from a different source using X-ray diffraction (on the crystalline enzyme) is striking. Figure 16-24 offers some of the most direct evidence for a similarity between crystal and solution structure of an enzyme, and shows the sort of direct, precise geometrical information obtainable from NMR T_1 and T_2 values.

PROBLEMS

1. In Chapter 16.A.1, frequent recourse is made to the relation between mean lifetime, τ, of a chemical species, and the rate constant, k, for first-order concentration change of that species. The average lifetime of all the molecules of reactant A present at time zero may be expressed as the average over all possible lifetimes of individual A molecules, each weighted by the number of molecules with that particular lifetime. For a simple first-order decay process, this average is given by:

$$\tau = \frac{\int_{[A]_{t=0}}^{0} -t d[A]}{[A]_0} \qquad \text{(a)}$$

in which $[A]_0$ is the concentration of A at time zero. Substitute the concentration of A at time $t, [A(t)]$

$$[A(t)] = [A]_0 \exp[-kt] \qquad \text{(b)}$$

into the rate of disappearance of $[A]$:

$$-d[A] = k\ [A(t)]\ dt \qquad \text{(c)}$$

and then substitute Eq. (c) into (a) to show that $\tau = (1/k)$, as claimed in Eq. 9-10.

Hint:

$$\int_0^\infty x \exp\ [-ax]\ dx = (1/a^2)$$

2. In this exercise, you will derive the Smoluchowski equation that predicts the diffusion-limited rate constant for fast chemical reactions. From the basic law of diffusion (see Eq. 6-51),

$$\text{Flow} = -4\pi r^2\ D\ (dn/dr) \qquad \text{(a)}$$

in which Flow represents the number of particles diffusing to and colliding with a given particle, D is diffusion constant, n is the number of particles per cc, and r is the radius of the spherical surface around the particle, you may integrate to find the number of particles located at arbitrary distance r from the surface of the given particle (where other particles are being continually removed by chemical reaction):

$$n = n_0 - (\text{Flow})/4\pi r D \qquad \text{(b)}$$

In the diffusion-limited reaction where the concentration of particles at the reacting surface, n, is zero, where n_0 is the concentration of particles in bulk solution (i.e., far away from the given particle)

$$\text{Flow} = 8\pi D r_0 n_0 \tag{c}$$

in which the distance of closest approach between two spheres of radius r_0 is $r = 2r_0$. Next, since the central particle itself is not at rest but also diffuses continuously, the number of collisions is actually twice as large as predicted by Eq. (c):

$$\text{Flow} = 16\pi D r_0 n_0 \tag{d}$$

The rate of disappearance of primary particles is thus

$$dn_0/dt = -16\pi D r_0 n_0^2 \tag{e}$$

By comparing Eq. (e) to the usual second order chemical reaction equation

$$d[C]/dt = -k[C]^2 \tag{f}$$

in which $[C]$ is molar concentration, show that

$$k = 16 \times 10^{-3}\pi D r_0 N_0 \tag{g}$$

where N_0 is Avogadro's number. Finally, from the Stokes equation for the diffusion coefficient of a large spherical molecule

$$D = RT/6\pi r\eta N_0 \tag{7-13, 7-55}$$

where η is solution viscosity $\cong 0.00801$ at 30°C, show that

$$\boxed{k = 8 \times 10^{-3}\, RT/3\eta = 8.4 \times 10^9\ \text{M}^{-1}\ \text{sec}^{-1} \text{ for water at } 30°\text{C}} \tag{h}$$

This startling result says that for diffusion limited reactions, the chemical reaction rate is independent of molecular size! Although larger molecules move more slowly by translational diffusion, their cross-sectional area is also larger, and the two effects cancel out to give a reaction rate independent of size.

3. In this problem, you will show how *two* distinct chemical relaxation times result from a kinetic scheme having *two* equilibria. From this calculation, it should seem reasonable that a scheme with N equilibria will in general exhibit N distinct chemical relaxation times [which may or may not be sufficiently different in magnitude to be determined by resolving N separate linear segments in a plot of log (concentration change) versus time for actual experimental data]. Begin by considering the scheme

$$E + S \underset{k_{-1}}{\overset{k_1}{\rightleftharpoons}} X_1 \underset{k_{-2}}{\overset{k_2}{\rightleftharpoons}} X_2 \tag{a}$$

which might correspond to substrate binding to an enzyme, followed by a conformational change in the enzyme:substrate complex, as discussed in the GPDH example (Figures 16-5 and 16-6). Defining concentration differences, ΔS, ΔX_1, and ΔX_2 between instantaneous concentration ($[S]$, $[E]$, $[X_1]$, $[X_2]$) and final (new) equilibrium concentration ($[S]'$, $[E]'$, $[X_1]'$, $[X_2]'$):

$$\begin{aligned} \Delta S &= [S] - [S]' = [E] - [E]' \\ \Delta X_1 &= [X_1] - [X_1]' \\ \Delta X_2 &= [X_2] - [X_2]' \end{aligned} \quad \text{(b)}$$

(a) show that the rate of change of ΔS is given by

$$d(\Delta S)/dt = -k_1'(\Delta S + [E]')(\Delta S + [S]') + k_{-1}'(\Delta X_1 + [X_1]') \quad \text{(c)}$$

(b) Assuming that $\Delta S \ll [S]'$, show that Eq. (c) simplifies to

$$d(\Delta S)/dt = -\Delta S(k_1'([E]' + [S]')) + \Delta X_1(k_{-1}') \quad \text{(d)}$$

Similarly, show that

$$d(\Delta X_1)/dt = \Delta S(k_1'([E]' + [S]')) - \Delta X_1(k_{-1}' + k_2) + \Delta X_2(k_{-2}') \quad \text{(e)}$$

and

$$d(\Delta X_2)/dt = \Delta X_1(k_2') - \Delta X_2(k_{-2}') \quad \text{(f)}$$

Equations (d) to (f) represent three differential equations in three unknowns: ΔS, ΔX_1, and ΔX_2. However, from the conservation equation

$$\Delta S + \Delta X_1 + \Delta X_2 = 0 \quad \text{(g)}$$

we know that

$$d(\Delta S)/dt + d(\Delta X_1)/dt + d(\Delta X_2)/dt = 0 \quad \text{(h)}$$

so that there are really only *two* independent equations in *two* unknowns. By standard methods (not given here), the remaining two *differential* equations can be reduced to two coupled *algebraic* equations, whose solution is given by the roots of the determinant:

$$\begin{vmatrix} a_{11} + (1/\tau) & a_{12} \\ a_{21} & a_{22} + (1/\tau) \end{vmatrix} = 0 = \left(a_{11} + \frac{1}{\tau}\right)\left(a_{22} + \frac{1}{\tau}\right) - a_{21}a_{12} \quad \text{(i)}$$

(c) which you can readily solve to obtain the two values of $(1/\tau)$, where

$$\begin{aligned} a_{11} &= -k_1'([S]' + [E]') \\ a_{12} &= k_{-1}' \\ a_{21} &= k_1'([S]' + [E]') - k_{-2}' \\ a_{22} &= -(k_{-1}' + k_2' + k_{-2}') \end{aligned} \quad \text{(j)}$$

(d) Finally, you should verify that the two roots, $(1/\tau)$, of Eq. (i) satisfy the properties shown in Table 16-1 for this example. For more complicated examples with multiple equilibria, the problem still reduces to solving a determinant of the type shown in Eq. (i), except that there are, in general, N roots for a system with N equilibria. You have thus solved examples of the most typical and most general problems in chemical relaxation transient kinetics by deriving Equations 16-12, 16-13, and Eq. (i) above.

REFERENCES

Fast-reaction Transient Chemical Kinetics

B. H. Havsteen, in *Physical Principles and Techniques of Protein Chemistry,* Part A, ed. S. J. Leach, Academic Press, New York (1969), pp. 245–290.

G. G. Hammes and C.-W. Wu, "Kinetics of Allosteric Enzymes," *Ann. Rev. Biophys. Bioeng. 3,* 1–34 (1974).

Fluorescence Depolarization

R. Rigler and M. Ehrenberg, "Fluorescence Relaxation Spectroscopy in the Analysis of Macromolecular Structure and Motion," *Quart. Rev. Biophys. 9,* 1–20 (1976).

R. F. Chen, H. Edelhoch, and R. F. Steiner, in *Physical Principles and Techniques of Protein Chemistry,* Part A, ed. S. J. Leach, Academic Press, New York (1969), pp. 171–244.

A. J. Pesce, C.-G. Rosen, and T. L. Pasby, *Fluorescence Spectroscopy: An Introduction for Biology and Medicine,* Marcel Dekker, New York (1971).

Magnetic Relaxation

B. D. Sykes and M. D. Scott, "Nuclear Magnetic Resonance Studies of the Dynamic Aspects of Molecular Structure and Interaction in Biological Systems," *Ann. Rev. Biophys. Bioeng. 1,* 27–50 (1972).

T. L. James, *Nuclear Magnetic Resonance in Biochemistry,* Academic Press, New York (1975), Chapters 2, 6, and 8.

Black-Box Models for Physiology

F. S. Grodins, *Control Theory and Biological Systems,* Columbia University Press, New York (1963).

CHAPTER 17 Coupled Springs

In Chapters 14-16, we have treated particular electronic, nuclear, and atomic motions separately, as if each of them was independent of all the others. However, much additional information can be gained from the couplings (usually weak) between different kinds of molecular "springs," including chemical bond strengths, bond lengths, bond angles, and the chemical environment near a particular functional group. In this section, we show how these "couplings" are typically manifested as a "splitting" of the usual single natural frequency of the original spring into two frequencies of maximum steady-state absorption—the *strength* of the coupling then turns out to be related to the *magnitude* of the splitting, and is thus very easily determined from steady-state spectroscopic experiments. We next discuss the result of connecting several springs (e.g., chemical bonds) directly together to produce a composite motion that can be broken down into various "normal" vibrations that become the new natural frequencies (with corresponding spectral absorption "peaks") for the system. Finally, the special coupling between driven electronic motion (i.e., light scattering) and vibrations of the atoms to which those driven electrons are attached, known as the Raman effect, is shown to behave like ordinary "amplitude" modulation, producing a "center band" where the original electronic steady-state response would have occurred, plus "sidebands" spaced on either side of the center band on a frequency scale. Applications of these phenomena are indicated and further discussed in Chapter 18.A, where the actual data reduction for extracting the "coupling constants" from absorption spectra is described.

17.A. COUPLING CONSTANTS AND SPECTRAL SPLITTINGS

Consider the system of Fig. 17-1, consisting of two identical masses, m, mounted on two identical springs (force constant = k_0), where the two springs are also coupled by a third spring (force constant = k). We will first examine the net force, $m(d^2y/dt^2)$, on the left-hand mass:

$$m(d^2y/dt^2) = -k_0 y - ky + kx = -k_0 y - k(y - x) \qquad (17\text{-}1)$$

In other words, the force (left-hand side of Eq. 17-1) on the left-hand mass is the sum of the restoring force from its own spring, $-k_0y$, and the additional force that depends on where the other mass is located, according to

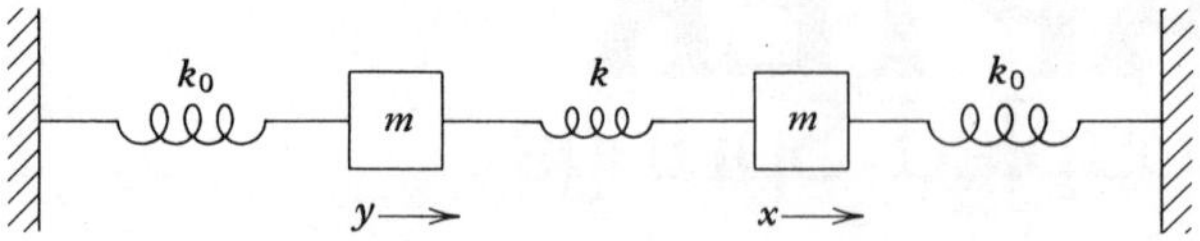

FIGURE 17-1. Schematic diagram of two identical masses mounted on two identical springs (k_0), coupled by a third spring (k). The coordinate for position of the left-hand mass is y, and for the right-hand mass is x. Each mass is subject to a force proportional (with proportionality constant, k_0) to the displacement for its own spring, plus a force proportional to the difference in displacement $(y - x)$ between the two masses due to the presence of the coupling spring (proportionality constant $= k$), as shown by the equations following this figure. (x and y are taken as zero when the masses have their equilibrium displacements.)

the difference is displacement of the two masses relative to each other (e.g., if both masses are moved the same distance to the right, $(y - x) = 0$, the coupling spring is not stretched at all, and the force arising from the presence of the coupling spring is zero). A similar equation can be written for the net force on the right-hand mass in Fig. 17-1:

$$m(d^2x/dt^2) = -k_0 x - k(x - y) \tag{17-2}$$

Next, we will try to find the "natural" frequency(ies) of this system by making the simplest possible assumption that for a given "natural" oscillation ("normal mode"), both masses oscillate at the *same* frequency (we shall return to comment on the significance of this assumption in the next section). In other words, we suppose that the displacements, x and y, have the same frequency of oscillation

$$x = A \cos(\omega t) \tag{17-3a}$$

$$y = B \cos(\omega t) \tag{17-3b}$$

in our usual notation, where ω is the same in both equations, but the amplitudes, A and B, of the motions of the two masses may be different. To solve for ω, substitute for x and y of Equations 17-3a and 17-3b in Equations 17-1 and 17-2 to obtain

$$\left[\omega^2 - \frac{k_0}{m} - \frac{k}{m}\right] A = -\left(\frac{k}{m}\right) B \tag{17-4a}$$

and

$$\left[\omega^2 - \frac{k_0}{m} - \frac{k}{m}\right] B = -\left(\frac{k}{m}\right) A \tag{17-4b}$$

Multiplying Equations 17-4a and 17-4b together, we find

$$\left[\omega^2 - \frac{k_0}{m} - \frac{k}{m}\right]^2 AB = \left(\frac{k}{m}\right)^2 AB \tag{17-5}$$

Setting aside the trivial solution that A and B are zero, we may cancel the AB factor on each side of Eq. 17-5, leaving the solutions

$$\omega = \sqrt{(k_0/m)} \tag{17-6a}$$

and — Two normal mode frequencies

$$\omega = \sqrt{[(k_0/m) + (2k/m)]} \tag{17-6b}$$

The final result, Equations 17-6a and 17-6b, of the treatment of the motion of two coupled springs is thus that the masses can now exhibit *two* natural frequencies, which are separated (in frequency) by an amount *directly related to the strength of the spring that couples the motions* of the two isolated springs. The qualitative nature of this wholly classical mechanical result is preserved in the quantum mechanical result describing the energy levels and spectral splittings resulting from the coupling of, for example, the magnetic precession of one nucleus with that of another. In Chapter 18.A, we shall show how the magnitude of this "scalar coupling" example in magnetic resonance is readily determined from spectral splittings between resolved multiple resonances for a single magnetic nucleus, and how the resulting "coupling constant" provides a direct measure of dihedral angle (i.e., the angle by which the substituents at one end of a carbon-carbon bond are rotated with respect to those at the other end).

17.B. DIRECTLY CONNECTED SPRINGS: NORMAL MODES

A particularly useful model for the intramolecular motions within a molecule is that each of the atoms in the molecule is connected by a spring to other atoms, such that the strength of the spring associated with a given chemical bond is directly related to the strength of that bond (i.e., to the force it would take to break that bond). There are two ways to describe the motions of all the atoms in a molecule. The hard way is to keep track of the positions (i.e., the x, y, and z coordinate of each atom) of all the N atoms in the molecule requiring $(3N)$ total position variables. The easier way is to notice that when the molecule as a whole translates, all the atoms must move in the same direction at once. Thus, if we separate the translations (in any of three independent directions) of the molecule as a whole from the remaining motions, only $(3N - 3)$ variables are needed. Similarly, by separating out the three independent rotational motions of the molecule as a whole, we are left with only $(3N - 6)$ variables associated with vibrational stretching and bending motions of the atoms with respect to each other. It is possible to describe all those stretching and bending motions as a collection of "natural" vibrational frequencies called "normal modes" in much the same way that we went about finding the natural vibrational frequencies of the two masses in the previous coupled spring example. For any *one* normal mode, each of the atoms in the molecule oscillates at the *same* natural frequency, but generally in *different* directions (see below).

Figure 17-2 shows the $[3(4) - 6] = 6$ vibrational normal modes for the carbonate ion, $CO_3^=$. The important features of these modes are, first, that any given intramolecular motion—no matter how complicated—can be expressed as the sum of motions at the normal mode "natural" frequencies, with appropriate amplitudes for each of the normal mode vibrations; and, second, that the steady-state response of the molecule to an applied continuous sine-wave driving force will consist of absorption (and dispersion) peaks (Fig. 13-5), centered at each of the "natural" or "normal" mode frequencies (see Chapter 20.B.3 and 14.A for examples).

The most useful feature of normal modes becomes apparent when we consider the vibrations of molecules containing a large number of atoms. In these cases, it is often possible to treat any one chemical functional group, such as a carbonyl or methylene group, as if it were bound to a very massive

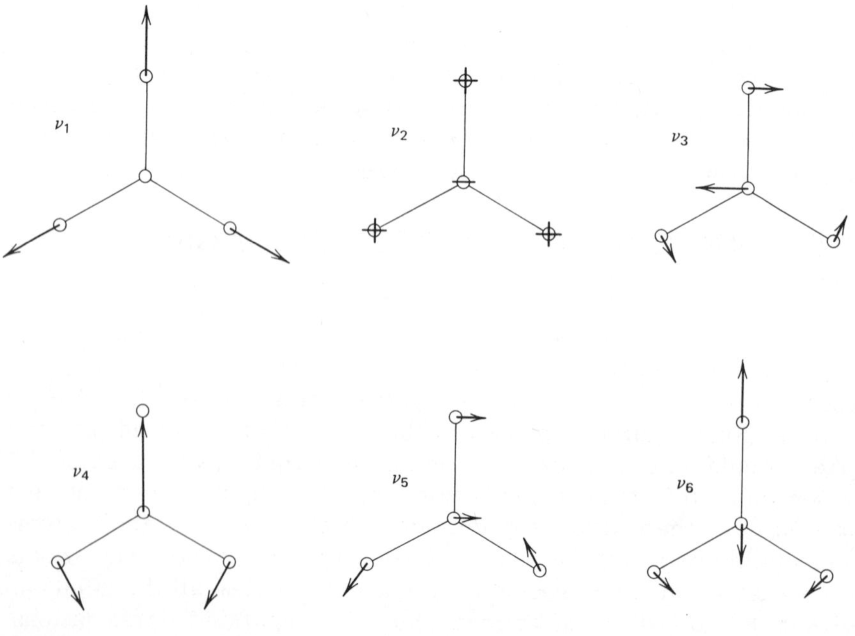

FIGURE 17-2. Directions for oscillating atom displacements in each of the six normal modes of vibration of the carbonate ion, $CO_3^=$ (From F. A. Cotton, *Chemical Applications of Group Theory,* Wiley-Interscience, New York, 1963, p. 247.)

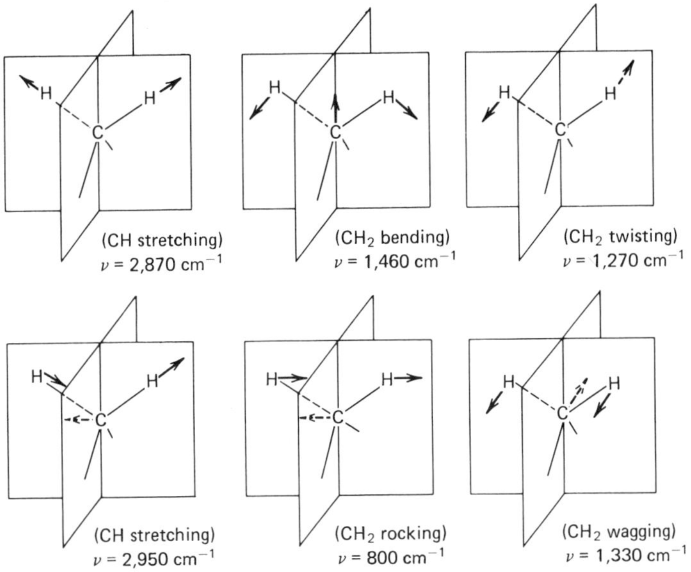

FIGURE 17-3. Directions of motion of oscillating atoms for each of various normal modes of vibration associated with an isolated methylene group. The normal mode frequencies shown here are nearly the same, irrespective of the nearby molecular structure, and thus serve to help identify this chemical group from spectral absorption peaks at these frequencies. (From G. M. Barrow, *Introduction to Molecular Spectroscopy,* McGraw-Hill, New York, 1962, p. 200.)

molecular frame. The reader can appreciate the validity of this view by solving the coupled spring problem of Fig. 17-1 in the limit that one of the masses becomes much larger than the other (see Problems). The effect is that there are normal modes that can now be associated with a given *chemical group,* almost independently of the molecular structure near that group. The modes associated with an isolated methylene group are shown in Fig. 17-3, and there are several "natural" frequencies sufficiently well-separated in frequency so that they could be resolved by an infrared (steady-state) spectrometer (see Chapter 20). This is the formal basis for the "finger-print" value of infrared spectroscopy in identification of chemical functional groups in molecules—the frequency of the electric field (of an electromagnetic wave) required to drive the positively charged nuclei back and forth against their respective inter-atomic springs happens to fall in the infrared frequency range.

17.C. AMPLITUDE MODULATION: RAMAN SPECTROSCOPY

There is a particular special type of light scattering in which the motion of the driven *electron* on a spring is affected by the simultaneous vibrational

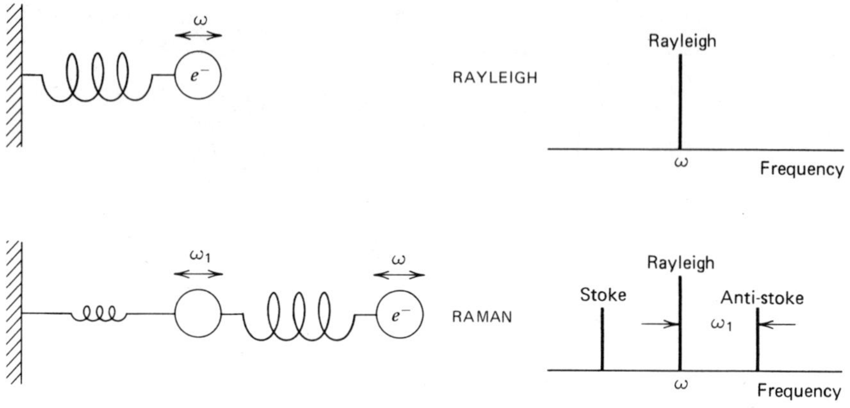

FIGURE 17-4. Schematic diagram of the origin of the Raman effect. The "natural" frequency of a driven electron on a stationary atom is typically in the ultraviolet, and leads to light-scattering with maximum intensity at the "natural" frequency, ω, as shown in the upper right plot. Vibration (at frequency, ω_1) of the atom to which the electron is attached acts to sinusoidally modulate the *amplitude* of oscillation of the driven electron and this amplitude modulation is manifested in the appearance of two new scattered (Raman) components, located at frequencies that are spaced either higher (Anti-Stoke) or lower (Stoke) by $\pm\omega_1$, respectively. Since $\omega_1 << \omega$, the atomic vibration "spring" is shown as much weaker than the electron vibration "spring" in the left-hand diagrams.

oscillations (at much lower frequency) of the normal modes of motion of *atoms* to which those electrons are attached. The "coupling" of the two motions is however not direct, as in the examples given in sections 17.A and 17.B, and requires a somewhat different explanation (see Fig. 17-4).

In our original light-scattering treatment, it proved convenient to define a "polarizability" as the linear proportionality between the magnitude of the driving electric field, $E = E_0 \cos(\omega t)$, and the dipole moment response, ex:

$$ex = \alpha E = \alpha E_0 \cos(\omega t) \tag{15-13}$$

From this equation, we went on to show that the radiation emitted from the driven dipole occurred at just *one* frequency, ω, which was the *same* as the frequency of the driver. However, from our previous discussion of dichroism we have learned to expect that the properties of a molecule are in general different when the molecule is viewed from different directions. In particular, we expect that the induced dipole moment ex for oscillation in the x direction might be different from the induced moment, ey, for motion in the y direction, or ez for motion in the z direction. We would thus predict that the respective polarizabilities (which determine the *amplitude* of the driven *electron* motion along a particular axis) α_x, α_y, and α_z might be different for nonspherical molecules. Furthermore, since the *atoms* of the molecule are continually vibrating, and thus changing the way in which electrons are distributed in the molecule, we expect that the polarizability will change as the atoms move away from their equilibrium positions. If the change in polarizability with distance is small, then we may approximate α as a function of distance by just the first two terms of a Taylor series expansion about the equilibrium atomic positions, r_e (see Appendix):

$$\alpha = \alpha_{r=r_e} + \left[\frac{d\alpha}{d(r - r_e)}\right]_{r=r_e} \cdot (r - r_e) + \cdots, \tag{17-7}$$

in which r is the (nonequilibrium) position of the vibrating atoms, and $(r - r_e)$ represents the distance away from the equilibrium position. But because the atoms are vibrating at, say, frequency ω_1, the distance away from equilibrium, $(r - r_e)$, is also oscillating at frequency ω_1:

$$(r - r_e) = A \cos(\omega_1 t) \tag{17-8}$$

in which A is the amplitude of the atomic oscillation. Substituting Eq. 17-8 into Eq. 17-7, and then Eq. 17-7 into Eq. 15-13, we obtain

$$ex = \alpha_{r=r_e} E_0 \cos(\omega t) + E_0 A (d\alpha/d(r - r_e))_{r=r_e} \cos(\omega t) \cos(\omega_1 t) \tag{17-9}$$

which may be rewritten using a trigonometric identity for the product of cosines (see Appendix)

$$\cos(a)\cos(b) = (1/2)[\cos(a-b) + \cos(a+b)] \tag{17-10}$$

to give the desired result

$$ex = \alpha_{r=r_e} E_0 \cos(\omega t) + (A/2)E_0[d\alpha/d(r - r_e]_{r=r_e} \\ [\cos(\omega - \omega_1)t + \cos(\omega + \omega_1)t] \tag{17-11}$$

The net effect of the molecular vibration is thus to change the steady-state Rayleigh light scattering response from radiation at just one frequency, ω, to radiation at three frequencies: ω, the usual Rayleigh scattering "centerband," and $\omega \pm \omega_1$, radiation at two new "sideband" frequencies located at a spacing of one *vibrational* frequency on either side of the Rayleigh centerband. These new scattered components are called "Raman" components, and it is now clear that they result simply from the sinusoidal "modulation" (change) of the *amplitude* of the Rayleigh response α. The situation seen here is in fact formally identical to the use of voices or music to modulate the amplitude of a radio-frequency "carrier" wave in an AM radio broadcast—the AM receiver then detects the modulated signals and reconstructs the original "audio" signal used to produce them by means of an electronic process that basically converts Eq. 17-9 to Eq. 17-11.

The reader may recall that the presence of normal mode vibrations of the atoms in molecules may be detected directly by irradiation with electromagnetic waves in the *infrared* range (see Chapter 14.A) rather than in this indirect way by detection of scattered radiation in the *visible-ultraviolet* range. It is logical to wonder why it might be desirable to carry out both experiments. There are two main reasons. First, while Rayleigh scattering is observed because the displaced electron on a spring has an *electric dipole moment* (Eq. 15-13), Eq. 17-11 shows that the appearance of the Raman scattering "peaks" requires a *change in polarizability* as the atoms vibrate at that normal mode frequency, ω_1 (i.e., $d\alpha/d(r - r_e) \neq 0$). Furthermore, quantum mechanical "selection rules" (Chapter 19.8.4) predict that absorption (in the infrared) at a given normal mode frequency can occur only when there is a *change in dipole moment* as a result of the vibration. Thus, vibration ν_1 for $CO_3^=$ ion (Fig. 17-2) is infrared "inactive" (i.e., not observable) because there is no change in dipole moment for that vibration. (For the same reason, the O_2 "stretch" frequencies in Table 14-1 were determined from Raman scattering in the ultraviolet visible range, rather than from direct absorption in the infrared.) Since different normal modes tend to be observed by the two techniques (Raman scattering and direct infrared absorption), both measurements are desirable for a complete description of molecular vibrations. Second, direct infrared observation of molecules in solution is made difficult by the very strong infrared absorption by water itself, while this problem in Raman scattering is not significant. In fact (see next Example), Raman "spectroscopy" is one of the few techniques available for equally detailed scrutiny of molecules in both the liquid and solid phases.

EXAMPLE *Comparison of Protein Structure in Crystals and Solution by Raman Scattering*

Figure 17-5 shows Raman spectra obtained for ribonuclease enzyme in solution (top spectrum) and in polycrystalline (powder) form (bottom spectrum). The principal feature of interest is that the polypeptide "backbone" vibrations labeled as the "amide III" region show very good agreement in both peak positions and peak widths between crystal and solution forms, suggesting that the basic backbone conformation of RNase A is the same in both phases. This is an extremely important conclusion, because while the X-ray diffraction determinations of macromolecular structure can be very precise, there is always the question of the validity of comparing the structure of the crystalline molecule with that in solution. Similar conclusions do not apply universally—there are several proteins whose crystal and solution structure differ: for example, a small difference in the "amide III" backbone region Raman spectrum is seen between the solution and crystal forms of another enzyme, carboxypeptidase A. Returning to the spectra in Fig. 17-5, there are differ-

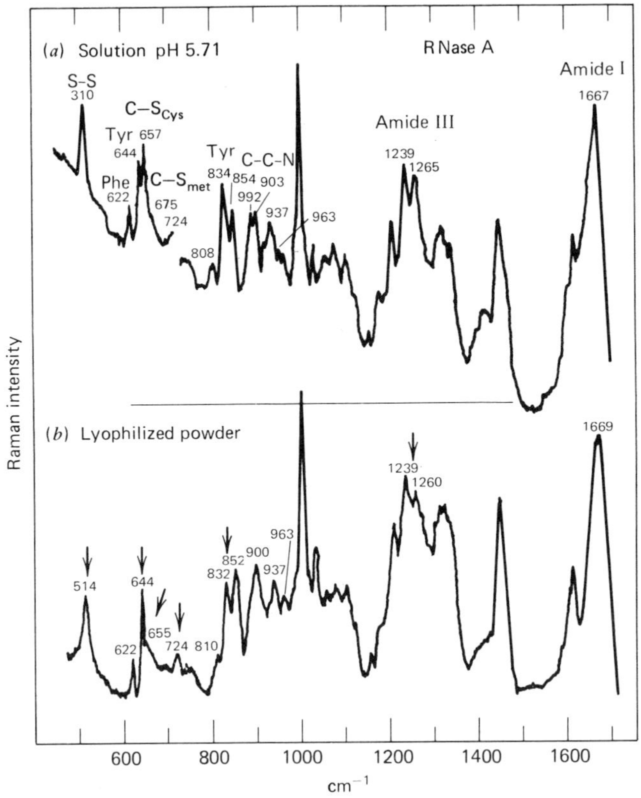

FIGURE 17-5. Raman spectra for the enzyme, ribonuclease A in solution (*top*) and in polycrystalline form (*bottom*). Differences in these spectra are interpreted (see text) to mean that the backbone polypeptide chain conformation is basically preserved on going from crystal to solution, while the disulfide linkages and tyrosine environment change. [From N. Yu and B. H. Jo, *J. Amer. Chem. Soc. 95*, 5033 (1973).]

ences in the 500–700 cm^{-1} region between solution and crystalline form that are associated with changes in the S-S and C–S stretching modes and with the tyrosine ring vibrations, suggesting that while the backbone configuration is preserved, there appear to be changes in the geometry of the disulfide linkages and in the environment of "buried" tyrosines on going from crystal to solution phase. These measurements are becoming increasingly popular, following the availability of laser light sources that provide very high-intensity excitation to enhance the (weak) Raman scattering signals.

PROBLEM

1. (a) For the system of two *nonequal* masses, mounted on identical springs (force constant = k_0), and coupled by a third spring (force constant =

k), calculate the possible "natural" ("normal mode") frequencies for the system.

(b) Show that in the limit that the coupling becomes infinitely weak ($k \to 0$) your result reduces to two natural motions at the "natural" frequencies of the two isolated springs.

(c) More interestingly, show that in the limit that one of the masses (say, m_2) is much heavier than the other, the motion of the lighter mass occurs at a frequency determined by k_0, k, and m_1. In other words, this calculation shows why it is a good approximation, in many cases, to treat a molecular functional group (such as a carbonyl group) as an isolated weight-on-a-spring bound to an infinitely massive framework.

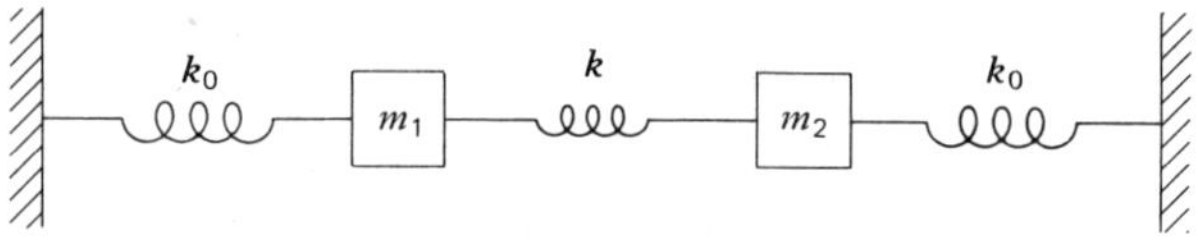

REFERENCES

Raman Spectroscopy

B. Moore, *Chemical and Biochemical Applications of Lasers,* Academic Press, New York (1974).

T. G. Spiro and B. P. Gaber, "Laser Raman Scattering as a Probe of Protein Structure," *Ann. Rev. Biochem. 46,* 553 (1977).

B. G. Frushour and J. L. Koenig, *Advances in Infrared and Raman Spectroscopy,* ed. R. J. H. Clark and R. E. Hester, Heydon & Son, New York (1975), Vol. 1, pp. 35–97.

5

QUANTUM MECHANICS: WHEN IS IT REALLY NECESSARY?

> There was a time when the newspapers said that only twelve men understood the theory of relativity. I do not believe there was ever such a time. There might have been a time when only one man did, because he was the only guy who caught on, before he wrote his paper. But after people read the paper, a lot of people understood the theory of relativity in some way or other, certainly more than twelve. . . . On the other hand, I think I can safely say that nobody understands quantum mechanics.—*Richard Feynman*

In Section 4, we found that many spectral properties ordinarily introduced by quantum mechanics could be understood in terms of simple classical models based on a weight on a spring. The classical model fails, however, in a number of areas important to biophysical applications (and in other areas of interest to different audiences, such as physicists). For example, a classical electron on a spring oscillates at only *one* natural frequency, while the actual electron on a spring in a simple hydrogen atom vibrates at an *infinite* number of "natural" frequencies—quantum mechanics provides the means for predicting the number and magnitudes of those frequencies (Chapter 18). Once it is established that particles may possess only certain, discrete (as opposed to continuous), "quantized" energies, there immediately arises the statistical problem of deciding how many particles possess each particular energy, given that the total number of particles and the total energy of the collection of particles is constant. The resulting Boltzmann distribution (Chapter 19) then predicts a variety of molecular properties such as average electric dipole moment that are unobtainable from classical mechanics, and lead in a natural way to the understanding of lasers and other nonequilibrium situations. Finally, the effect of applied or naturally existing fluctuating electric or magnetic fields on transitions between energy levels (related to spectral absorption intensities and associated chromophore and "reporter group" experiments) follow from the treatment of small sudden (or sinusoidal or random) perturbations on the static energy levels obtained in Chapter 18.

CHAPTER 18 Generalized Geometry: Existence and Positions of Spectral Power Absorption "Lines:" Quantum Mechanical Weight on a Spring

Quantum mechanics is most often introduced using the language of *calculus*. The usual procedure is to propose a suitable (Hamiltonian) differential equation and then show that solutions are possible only for certain discrete ("quantized," often integer) values of some parameter. Many readers of this book may have already solved the "particle-in-a-box" and the "central-force" (i.e., hydrogen atom) problems in this way in elementary chemistry or physics courses. The difficulty with this approach, apart from the often arduous complexities in the mathematical manipulations, is immediately obvious from Table 18-1, which shows that the main content of quantum mechanics is based on *operators,* not *functions*. By learning just a little about operators, we can solve the quantum mechanical weight-on-a-spring problem without using differential equations at all. Then by taking advantage of the one-to-one correspondence between the properties of linear operators and matrices, we use matrices to solve a number of simple and important "spin" problems for which there is no classical analogy, and for which there are many biophysical applications.

As discussed below, two of the postulates on which quantum mechanics is based are (1) the "state" of a particle (from which we can predict the probability of obtaining a given result on experimental measurement of energy, position, momentum, etc.) is described by some *function,* and (2) all the physical "observables" (energy, position, momentum, etc.) are represented by *operators*. Manipulation of these quantities becomes easier once one appreciates that they can be thought of as a generalization of ordinary geometry, as seen in Table 18-2.

The main idea of Table 18-2 is that it is useful to imagine a "space" consisting of *functions* (rather than *vectors*), in which any particular function ψ may be expressed in terms of its components along some basis-function "axes," ϕ_1 to ϕ_n, just as an ordinary vector is expressed in terms of its components along the x, y, and z axes. The procedures involved in quan-

Table 18-1 Names of mathematical recipes for obtaining various types of mathematical objects from other types.*

	Object #1	Object #2	Recipe for Getting from #1 to #2	Example
CLASSICAL MECHANICS	Number	Number	Function	$f(t) = \exp[-kt]$
QUANTUM MECHANICS	Function	Number	Linear functional	$\int_a^b f(t) =$ number; Dirac delta-"function"
	2 functions	Number	Bilinear functional	Moment of inertia; energy of interaction between two dipoles
	Function (vector)	Function (vector)	Linear operator (matrix)	Quantum mechanical energy, Hamiltonian, momentum, position, angular momentum.
	Operator	Operator	Liouville operator ("super-operator")	$L(B) = AB - BA \equiv [A,B]$; A and B are operators

* *Classical mechanics is largely concerned with the calculus of* functions, *while quantum mechanical problems usually involve the other recipes listed in the table. Since there is a one-to-one correspondence between linear operators and matrices, actual quantum mechanical calculations are most readily conducted with matrices (see text).*

Table 18-2 Side-by-Side Comparison of the Properties of Ordinary Vectors in Three-dimensional Space and the Properties of Functions in an n-Dimensional Function "Space."*

Ordinary Geometry		Generalized Geometry (Quantum Mechanics)
Any *vector* can be expressed by its components along three axis vectors of unit length. 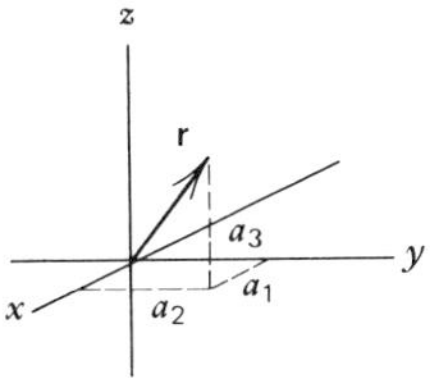 $r = a_1\mathbf{i} + a_2\mathbf{j} + a_3\mathbf{k} = \begin{pmatrix} a_1 \\ a_2 \\ a_3 \end{pmatrix}$ = "column vector" x, y, and z components of r are a_1, a_2, and a_3.	COMPLETENESS (Equations 18-1, 18-2) $\mathbf{i}$ = vector of unit length along x axis $\mathbf{j}$ = vector of unit length along y axis $\mathbf{k}$ = vector of unit length along z axis	Any *function*, ψ, can be expressed in terms of its components for a set of "basis" functions, ϕ_i, of unit "length" (see below). (ψ may be complex—see below.) $\psi = a_1\phi_1 + a_2\phi_2 + \cdots + a_n\phi_n$ $\psi = \sum_i a_i\phi_i$ $= \begin{pmatrix} a_1 \\ a_2 \\ \cdot \\ \cdot \\ \cdot \\ a_n \end{pmatrix}$
$\|r\|^2 = \mathbf{r}\cdot\mathbf{r} = (a_1\ a_2\ a_3)\begin{pmatrix} a_1 \\ a_2 \\ a_3 \end{pmatrix}$ $= a_1^2 + a_2^2 + a_3^2$ Length of vector, $\mathbf{r}$, is $\|r\| = \sqrt{\mathbf{r}\cdot\mathbf{r}}$	LENGTH (Equations 18-3 to 18-5)	$< \psi,\psi > = (a_1^* \ a_2^* \ \cdots \ a_n^*)\begin{pmatrix} a_1 \\ a_2 \\ \cdot \\ \cdot \\ \cdot \\ a_n \end{pmatrix}$ $= \sum_{i=1}^{n} a_i^* a_i$ $= \|a_1\|^2 + \|a_2\|^2 + \cdots + \|a_n\|^2$ "Norm" of a function is $\sqrt{<\psi,\psi>}$
Unit vector has $\mathbf{r}\cdot\mathbf{r} = 1$	UNIT LENGTH	"Normalized" function has $<\psi,\psi> = 1$
Projection of vector, $\mathbf{r}$, onto vector, $\mathbf{s}$, is called "scalar product of r and s," $\mathbf{r}\cdot\mathbf{s}$ $\mathbf{r}\cdot\mathbf{s} = (a_1\ a_2\ a_3)\begin{pmatrix} b_1 \\ b_2 \\ b_3 \end{pmatrix}$ $= a_1b_1 + a_2b_2 + a_3b_3$ where $\mathbf{s} = \begin{pmatrix} b_1 \\ b_2 \\ b_3 \end{pmatrix}$	PROJECTION (SCALAR PRODUCT) (Equations 18-3 to 18-5)	Projection of one function, ψ, onto another function, Φ, is called "scalar product" or "inner product" of ψ and Φ, $<\psi,\Phi>$ $<\psi,\Phi> = (a_1^* \ a_2^* \ \cdots \ a_n^*)\begin{pmatrix} b_1 \\ b_2 \\ \cdot \\ \cdot \\ \cdot \\ b_n \end{pmatrix}$ $= a_1^*b_1 + a_2^*b_2 + \cdots + a_n^*b_n$ where $\Phi = \begin{pmatrix} b_1 \\ b_2 \\ \cdot \\ \cdot \\ \cdot \\ b_n \end{pmatrix}$

Table 18-2 Continued

Ordinary Geometry		Generalized Geometry (Quantum Mechanics)
Vector **r** is said to be *perpendicular* to vector **s** when their scalar product is zero (i.e., **r** has zero projection along **s**). $\mathbf{r}\cdot\mathbf{s} = 0 \Leftrightarrow \mathbf{r} \perp \mathbf{s}$ 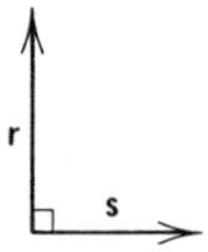	PERPENDICULARITY	Function ψ is said to be *orthogonal* to function Φ when their inner product (scalar product) is zero. $< \psi,\Phi > = 0 \Leftrightarrow \psi$ and Φ are orthogonal
Change vector **r** into vector **s**. $\underset{\approx}{\mathbf{T}}\mathbf{r} = \mathbf{s}$ $\begin{pmatrix} T_{11} & T_{12} & T_{13} \\ T_{21} & T_{22} & T_{23} \\ T_{31} & T_{32} & T_{33} \end{pmatrix}\begin{pmatrix} a_1 \\ a_2 \\ a_3 \end{pmatrix} = \begin{pmatrix} b_1 \\ b_2 \\ b_3 \end{pmatrix}$ $b_1 = T_{11}a_1 + T_{12}a_2 + T_{13}a_3$ $b_2 = T_{21}a_1 + T_{22}a_2 + T_{23}a_3$ $b_3 = T_{31}a_1 + T_{32}a_2 + T_{33}a_3$ $b_i = \sum_{j=1}^{3} T_{ij}a_j$ $\underset{\approx}{\mathrm{T}}$ is a linear transformation T_{ij} is an element (ith row and jth column) of the matrix representation of $\underset{\approx}{\mathrm{T}}$.	TRANSFORMATIONS (Equations 18-6 to 18-8)	Change function ψ into function Φ. $\underset{\approx}{\mathrm{T}}\psi = \Phi$ $\begin{pmatrix} T_{11} & T_{12} & \cdots & T_{1n} \\ T_{21} & T_{22} & \cdots & T_{2n} \\ \cdot & \cdot & & \cdot \\ \cdot & & \cdot & \cdot \\ \cdot & & \cdot & \cdot \\ T_{n1} & T_{n2} & \cdots & T_{nn} \end{pmatrix}\begin{pmatrix} a_1 \\ a_2 \\ \\ \\ \\ a_n \end{pmatrix} = \begin{pmatrix} b_1 \\ b_2 \\ \cdot \\ \cdot \\ \cdot \\ b_n \end{pmatrix}$ $b_1 = T_{11}a_1 + T_{12}a_2 + \cdots + T_{1n}a_n$ $b_2 = T_{21}a_1 + T_{22}a_2 + \cdots + T_{2n}a_n$ $\cdot$ $\cdot$ $\cdot$ $b_n = T_{n1}a_1 + T_{n2}a_2 + \cdots + T_{nn}a_n$ $b_i = \sum_{j=1}^{n} T_{ij}a_j$ $\underset{\approx}{\mathrm{T}}$ is a linear operator T_{ij} is an element of the matrix representation of $\underset{\approx}{\mathrm{T}}$.
Change the *direction* of the vector **r**, but leave the *length* the same [e.g., rotate vector **r**]. $\underset{\approx}{\mathbf{U}}\mathbf{r} = \mathbf{s}$ such that $\lvert\mathbf{s}\rvert = \lvert U\mathbf{r}\rvert = \lvert\mathbf{r}\rvert$ 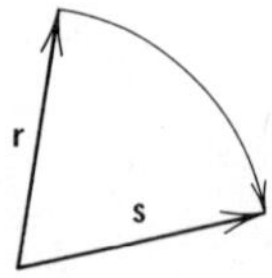	UNITARY TRANSFORMATION	Change function ψ into function Φ while preserving the *norm* of the function. $\underset{\approx}{\mathrm{U}}\psi = \Phi$ If $< \Phi,\Phi > = < U\psi, U\psi > = < \psi,\psi >$, then U is a "unitary" operator.

Table 18-2 Continued

Ordinary Geometry		Generalized Geometry (Quantum Mechanics)
Change the *length* of vector **r**, but not its *direction*. $\underset{\approx}{H}\mathbf{r} = h\mathbf{r}$, h is a real number $\begin{pmatrix} H_{11} & H_{12} & H_{13} \\ H_{21} & H_{22} & H_{23} \\ H_{31} & H_{32} & H_{33} \end{pmatrix}\begin{pmatrix} a_1 \\ a_2 \\ a_3 \end{pmatrix} = h\begin{pmatrix} a_1 \\ a_2 \\ a_3 \end{pmatrix}$ 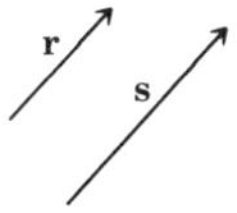	HERMITIAN TRANSFORMATION	Change the *norm* of function ψ, but keep the relative magnitudes of its components along its basis functions the same. $\underset{\approx}{H}\psi = h\psi$, h is a real number, and ψ is said to be an "*eigenfunction*" of the operator H. The operator H is said to be *Hermitian* when (as in quantum mechanics) the "*eigenvalue*" h is real. $\begin{pmatrix} H_{11} & H_{12} & \cdots & H_{1n} \\ H_{21} & H_{22} & \cdots & H_{2n} \\ \vdots & & \ddots & \vdots \\ H_{n1} & H_{n2} & \cdots & H_{nn} \end{pmatrix}\begin{pmatrix} a_1 \\ a_2 \\ \vdots \\ a_n \end{pmatrix} = h\begin{pmatrix} a_1 \\ a_2 \\ \vdots \\ a_n \end{pmatrix}$

The eigenvalues, h, of the Hermitian operator, H, are calculated by solving the "secular" determinant

$$\begin{vmatrix} (H_{11} - h) & H_{12} & \cdots & H_{1n} \\ H_{21} & (H_{22} - h) & & H_{2n} \\ \vdots & & \ddots & \vdots \\ H_{n1} & & & (H_{nn} - h) \end{vmatrix} = 0$$

* *The language of quantum mechanics is thus drawn directly from a generalized geometry in which functions are treated like vectors, with suitable definitions of length, perpendicularity, projection, and the like. Because geometry is more easily visualized than algebra, this format renders quantum mechanical calculations particularly simple and direct. Vector and matrix notation used in this table is explained in the text.*

tum mechanical calculations (e.g., normalization, test for orthogonality, effect of operators on wave functions, solution of eigenvalue problems) can then be understood by analogy to various well-known geometrical concepts such as length, angle, perpendicularity, rotation, and stretching. Because geometry is simpler than algebra (compare a drawing of an ellipse to the formula for an ellipse!), most quantum mechanical calculations are conducted by use of the vector and matrix notation listed in Table 18-2, rather than by solving the relevant differential equations (if in fact a differential equation can even be defined—see "spin" problems below). The rules for vector and matrix multiplication are illustrated by the examples in Table 18-2.

Matrix Manipulations

An n-dimensional vector may be expressed as a column of numbers, in which the numbers represent the magnitude of the vector component ("projection") along each of the defined "basis vectors" or basis functions.

In writing Eq. 18-1, the reader should remember that those components would be different if we had chosen a different basis set [e.g., a unit vector along the x axis would be described as (1,0,0) in the basis **(i,j,k)**, but would be described as (0,1,0) in the basis **(j,i,k)** in an example from ordinary geometry]. In quantum mechanics, the usual problem is first to determine the *eigenfunctions* (Eq. 18.14 ff.) for a particular (Hamiltonian–Table 18-3) operator in terms of some *convenient* set of basis functions, and then to express any given *state* (function) of the actual particle(s) in terms of its components along each of the (new) basis set of *eigenfunctions*.

$$\psi = a_1\phi_1 + a_2\phi_2 + \cdots + a_n\phi_n \tag{18-1}$$

$$\psi = \sum_{i=1}^{n} a_i\phi_i \tag{18-1a}$$

$$\psi = \begin{pmatrix} a_1 \\ a_2 \\ \cdot \\ \cdot \\ \cdot \\ a_n \end{pmatrix} \tag{18-2}$$

in which it is understood that these are the components of ψ for the *particular* basis set, $\phi_1, \phi_2, \cdots \phi_n$

Equation 18-1 is sometimes described as the expression of ψ as a "linear combination" of the basis functions ϕ_i.

In order to evaluate the *scalar product* of two vectors, we first construct a row vector consisting of the complex conjugate (i.e., replace $p \pm iq$ by $p \mp iq$) of the components of the left-hand vector of the scalar product, then multiply the first element of the row vector by the first element of the column vector corresponding to the right-hand function in the scalar product, then multiply the second element of the row vector times the second element of the column vector, and so forth, and finally add all the resulting numbers together, as shown in Equations 18-3 to 18-5a. All of this may sound complicated, but is really a very rapid process when stated in equations rather than words:

$$\psi = \begin{pmatrix} a_1 \\ a_2 \\ \cdot \\ \cdot \\ \cdot \\ a_n \end{pmatrix}; \quad \phi = \begin{pmatrix} b_1 \\ b_2 \\ \cdot \\ \cdot \\ \cdot \\ b_n \end{pmatrix} \tag{18-3}$$

ERRATA

Marshall **Physical Chemistry for the Life Sciences**

Page 569. The shaded parts of the area labeled 18-6 through 18-8 made the text illegible. This should read $\mathbf{T}\psi = \phi$ **T** is an "*operator*" (18-6)

$$\begin{pmatrix} T_{11} & T_{12} & \cdots & T_{1n} \\ T_{21} & T_{22} & \cdots & T_{2n} \\ \cdot & & \cdot & \cdot \\ \cdot & & \cdot & \cdot \\ \cdot & & \cdot & \cdot \\ T_{n1} & T_{n2} & \cdots & T_{nn} \end{pmatrix} \begin{pmatrix} a_1 \\ a_2 \\ \cdot \\ \cdot \\ \cdot \\ a_n \end{pmatrix} = \begin{pmatrix} b_1 \\ b_2 \\ \cdot \\ \cdot \\ \cdot \\ b_n \end{pmatrix} \tag{18-7}$$

$$\begin{aligned} b_1 &= T_{11}a_1 + T_{12}a_2 + \cdots + T_{1n}a_n \\ b_2 &= T_{21}a_1 + T_{22}a_2 + \cdots + T_{2n}a_n \\ &\cdot \\ &\cdot \\ &\cdot \\ b_n &= T_{n1}a_1 + T_{n2}a_2 + \cdots + T_{nn}a_n \end{aligned} \tag{18-8}$$

Page 570. The shaded parts of the area labeled 18-11 made the type illegible. This should read

$$\begin{pmatrix} A_{11} & A_{12} & \cdots & A_{1n} \\ A_{21} & A_{22} & \cdots & A_{2n} \\ \cdot & \cdot & & \\ \cdot & & \cdot & \\ \cdot & & \cdot & \\ A_{n1} & A_{n2} & \cdots & A_{nn} \end{pmatrix} \begin{pmatrix} B_{11} & B_{12} & \cdots & B_{1n} \\ B_{21} & B_{22} & \cdots & B_{2n} \\ \cdot & & \cdot & \\ \cdot & & \cdot & \\ \cdot & & \cdot & \\ B_{n1} & B_{n2} & \cdots & B_{nn} \end{pmatrix} = \begin{pmatrix} C_{11} & C_{12} & \cdots & C_{1n} \\ C_{21} & C_{22} & \cdots & C_{2n} \\ \cdot & & \cdot & \cdot \\ \cdot & & \cdot & \cdot \\ \cdot & & \cdot & \cdot \\ C_{n1} & C_{n2} & \cdots & C_{nn} \end{pmatrix} \tag{18-11}$$

NOTE In general, $AB \neq BA$ (see below).

Page 607. The following pyridine ring structure should be substituted for the one shown in the *Example* on this page.

Pyridine,

Page 679. The right-hand notation below the drawing of Figure 20-10 should read

$(\omega - \omega_0)$, $\sec^{-1}$

Page 690. An asterisk should precede the italicized material below Table 20-2.

$$< \psi,\phi > = (a_1^* a_2^* \cdots a_n^*) \begin{pmatrix} b_1 \\ b_2 \\ \cdot \\ \cdot \\ \cdot \\ b_n \end{pmatrix} \tag{18-4}$$

$$= a_1^* b_1 + a_2^* b_2 + \cdots + a_n^* b_n \tag{18-5}$$

$$= \sum_{i=1}^{n} a_i^* b_i \tag{18-5a}$$

It is possible to represent the transformation of one vector into another vector (also known as "operating" on one function to change it into another function) as the multiplication of the matrix corresponding to that transformation times a column vector to yield the resulting column vector. The rule for this matrix "operation" is simply that we obtain, say, the second element of the new vector by taking the sum of the element-by-element product of the second *row* of the matrix times the original *column* vector as shown in the following equations:

$$\mathbf{T}\psi = \phi \qquad \mathbf{T} \text{ is an "}\textit{operator}\text{"} \tag{18-6}$$

$$\begin{pmatrix} T_{11} & T_{12} \cdots & T_{1n} \\ T_{21} & T_{22} \cdots & T_{2n} \\ \cdot & \cdot & \cdot \\ \cdot & \cdot & \cdot \\ \cdot & & \cdot \\ T_{n1} & T_{n2} \cdots & T_{nn} \end{pmatrix} \begin{pmatrix} a_1 \\ a_2 \\ \cdot \\ \cdot \\ \cdot \\ a_n \end{pmatrix} = \begin{pmatrix} b_1 \\ b_2 \\ \cdot \\ \cdot \\ \cdot \\ b_n \end{pmatrix} \tag{18-7}$$

$$\begin{aligned} b_1 &= T_{11}a_1 + T_{12}a_2 + \cdots + T_{1n}a_n \\ b_2 &= T_{21}a_1 + T_{22}a_2 + \cdots + T_{2n}a_n \\ &\cdot \\ &\cdot \\ &\cdot \\ b_n &= T_{n1}a_1 + T_{n2}a_2 + \cdots + T_{nn}a_n \end{aligned} \tag{18-8}$$

or more compactly

$$b_i = \sum_{j=1}^{n} T_{ij} a_j \tag{18-8a}$$

In quantum mechanics it is often desirable to apply two successive transformations, say, transformation B, followed by transformation A (this process may, for example, be used to find the result of measuring the physical variable corresponding to B, followed by measurement of the physical variable corresponding to A—see below):

Define the product of two operators, AB, as

$$\mathrm{AB}\psi = \mathrm{A}(\mathrm{B}\psi) \tag{18-9}$$

in other words, we operate on function ψ with operator B, and then operate on the resulting column vector using operator A. In this situation, we could alternatively operate on vector ψ using the single operator, C, where a given element of matrix C (say C_{21}) is obtained by multiplying the second *row* of matrix A element-by-element by the first *column* of matrix B.

$$\mathrm{AB} = \mathrm{C} \tag{18-10}$$

$$\begin{pmatrix} A_{11} & A_{12} & \cdots & A_{1n} \\ A_{21} & A_{22} & \cdots & A_{2n} \\ \cdot & & \cdot & \\ \cdot & & \cdot & \\ \cdot & & \cdot & \\ A_{n1} & A_{n2} & \cdots & A_{nn} \end{pmatrix} \begin{pmatrix} B_{11} & B_{12} & \cdots & B_{1n} \\ B_{21} & B_{22} & \cdots & B_{2n} \\ \cdot & & \cdot & \\ \cdot & & \cdot & \\ \cdot & & \cdot & \\ B_{n1} & B_{n2} & \cdots & B_{nn} \end{pmatrix} = \begin{pmatrix} C_{11} & C_{12} & \cdots & C_{1n} \\ C_{21} & C_{22} & \cdots & C_{2n} \\ \cdot & & \cdot & \cdot \\ \cdot & & \cdot & \cdot \\ \cdot & & \cdot & \cdot \\ C_{n1} & C_{n2} & \cdots & C_{nn} \end{pmatrix} \tag{18-11}$$

NOTE In general, $AB \neq BA$ (see below).

Given a particular basis set, $\phi_1, \phi_2, \cdots \phi_n$, we can immediately express the *components*, $a_1, a_2, \cdots a_n$, of an arbitrary *function*, ψ, by projection of ψ along each of the vectors (functions) of the basis set. For example, the second component of ψ may be found from the projection

$$< \phi_2,\psi > = (0100 \cdots 0)\begin{pmatrix} a_1 \\ a_2 \\ a_3 \\ \cdot \\ \cdot \\ \cdot \\ a_n \end{pmatrix} = a_2 \tag{18-12}$$

Similarly, a given *element* of an arbitrary transformation *matrix* may be found by operating with the matrix on a unit-length basis vector, and then taking the scalar product of the resultant vector with another unit-length basis vector, as illustrated below:

$$< \phi_3,A\phi_2 > = (001)\begin{pmatrix} A_{11} & A_{12} & A_{13} \\ A_{21} & A_{22} & A_{23} \\ A_{31} & A_{32} & A_{33} \end{pmatrix}\begin{pmatrix} 0 \\ 1 \\ 0 \end{pmatrix} = (0\ 0\ 1)\begin{pmatrix} A_{12} \\ A_{22} \\ A_{32} \end{pmatrix} = A_{32} \tag{18-13}$$

Finally, once a basis has been chosen, and the components of ψ and the elements of all desired operators have been found, the quantum mechanical problem of interest is to find those particular, special, "eigen" functions for which operation with a given matrix produces just the same function back again differing only by a constant real number factor:

If $H\psi = \lambda\psi$, λ is a real number, then ψ is an eigenfunction, and λ is an eigenvalue associated with that eigenfunction. (18-14)

$$\begin{pmatrix} H_{11} & H_{12} & \cdots & H_{1n} \\ H_{21} & H_{22} & \cdots & H_{2n} \\ \cdot & & \cdot & \cdot \\ \cdot & & \cdot & \cdot \\ \cdot & & \cdot & \cdot \\ H_{n1} & H_{n2} & \cdots & H_{nn} \end{pmatrix} \begin{pmatrix} a_1 \\ a_2 \\ \cdot \\ \cdot \\ \cdot \\ a_n \end{pmatrix} = \lambda \begin{pmatrix} a_1 \\ a_2 \\ \cdot \\ \cdot \\ \cdot \\ a_n \end{pmatrix} \tag{18-15}$$

Careful examination of Eq. 18-15 shows that the eigenvalue problem now reduces simply to solving a set of n equations in n unknowns (refer to almost any elementary text on linear algebra or matrix algebra), and the n eigenvalues, $\lambda_1, \lambda_2, \cdots \lambda_n$, that satisfy Eq. 18-15 are obtained as the roots of the polynomial from the determinant of the matrix

$$\det \begin{pmatrix} H_{11} - \lambda & H_{12} & \cdots H_{1n} \\ H_{21} & H_{22} - \lambda & \cdots H_{2n} \\ \cdot & \cdot & \cdot \\ \cdot & \cdot & \cdot \\ \cdot & \cdot & \cdot \\ H_{n1} & H_{n2} & \cdots H_{nn} - \lambda \end{pmatrix} = 0 \tag{18-16}$$

while the eigenfunctions, $\psi_1, \psi_2, \cdots \psi_n$, corresponding to eigenvalues, $h_1, h_2, \cdots h_n$, are found (see Problems for examples) as

$$\begin{aligned} \psi_1 &= a_{11}\phi_1 + a_{12}\phi_2 + \cdots + a_{1n}\phi_n \\ &\cdot \\ &\cdot \\ &\cdot \\ \psi_n &= a_{n1}\phi_1 + a_{n2}\phi_2 + \cdots + a_{nn}\phi_n \end{aligned} \tag{18-17}$$

where

$$a_{ij} = < \phi_j, \psi_i > \tag{18-18}$$

The great advantage of the matrix approach to quantum mechanics is that it is possible to obtain the quantities of primary physical interest, namely the *eigenvalues*, from a given basis set and Equations 18-13 and

18-16, without ever knowing the *eigenfunctions*, $\psi_1, \psi_2, \cdots \psi_n$, of Eq. 18-17. In contrast, the differential equation approach to quantum mechanical problems requires that the eigenfunctions be obtained first, and then used to obtain the eigenvalues—since the eigenfunctions may be algebraically complicated (as with problems as simple as the weight on a spring) or even inexpressible in classical language (as with eigenfunctions of "spin" operators), the matrix approach is generally shorter, simpler, and often the only possible method.

We shall end this introductory section by indicating the correspondence between physical quantities (and physical measurements) and the generalized geometry of functions that we have created and codified.

Physical Reality and Matrix Manipulations: Postulates of Quantum Mechanics

Here we will dwell neither on the *justification* for the postulates (postulates are in any case meant to be believed rather than understood, where the belief can later be tested by comparison with experiment) nor on the philosophical implications of a theory based, in part, on the idea that the behavior of the universe is governed by probability (and hence uncertainty in any given instance) rather than by certainty based on the prior behavior of the system. We simply lay down the correspondences between actual physical measurements and the associated quantum mechanical matrix procedures that can be used to predict the results of those measurements, and we then proceed directly to the solution of two of the handful of problems for which a solution exists: the quantum mechanical weight on a spring, and the energy spectra for some simple "spin" problems.

Physical Identification # 1: *Possible* results of a measurement of a physical observable. Corresponding to each *physical observable* (see Table 18-3), there is defined an *operator*. The result of any *single* measurement of that observable must be one of the *eigenvalues* of the corresponding operator. Since those eigenvalues are real (in the mathematical sense) for actual physical measurements, the associated operators are "Hermitian," which means that the matrix corresponding to such an operator has the property that when we transpose the rows and columns, the matrix elements differ only by a complex conjugate from the original matrix (this property clearly reduces the number of matrix elements to be evaluated by a factor of two, since we can get the other half by Eq. 18-19):

$$\begin{pmatrix} H_{11} & H_{12} & \cdots & H_{1n} \\ H_{21} & H_{22} & \cdots & H_{2n} \\ \vdots & & \ddots & \vdots \\ H_{n1} & H_{n2} & \cdots & H_{nn} \end{pmatrix} \underset{\substack{\text{(i.e., } H \text{ is} \\ \text{"Hermitian")}}}{\overset{\substack{\mathbf{H}\text{ has real} \\ \text{eigenvalues}}}{=}} \begin{pmatrix} H_{11} & H_{12} & \cdots & H_{1n} \\ H_{12}^* & H_{22} & \cdots & H_{2n} \\ \vdots & & \ddots & \vdots \\ H_{1n}^* & H_{2n}^* & \cdots & H_{nn} \end{pmatrix} \qquad \underset{\text{If } \mathbf{H} \text{ is Hermitian}}{\overset{H_{ij} = H_{ji}^*}{\uparrow}} \tag{18-19}$$

Table 18-3 List of physical observables and their corresponding quantum mechanical operators.*

Physical Observable	Quantum Mechanical Operator
x = distance component along x axis	x (i.e., multiplication by x)
p_x = momentum component along x axis $= mv_x$	$p_x = -i\hbar(\partial/\partial x)$
$\mathbf{r}$ = vector distance	$\mathbf{r} = x\,\mathbf{i} + y\,\mathbf{j} + z\,\mathbf{k}$ (i.e., $\mathbf{r}$ is a *vector* whose *components* are *operators*)
$\mathbf{p}$ = vector momentum	$\mathbf{p} = (-i\hbar)(\partial/\partial x)\mathbf{i} + (-i\hbar)(\partial/\partial y)\mathbf{j} + (-i\hbar)(\partial/\partial z)\mathbf{k}$
t = time	t (i.e., multiplication by t)
$(p_x^2/2m) + V(x)$ = Hamiltonian (i.e., kinetic plus potential energy) for one particle in one dimension	$(p_x^2/2m) + V(x) = -(\hbar^2/2m)(d^2/dx^2) + V(x)$ where $\hbar = h/2\pi$, h is Planck's constant $= 1.055 \times 10^{-34}$ J sec
$(\mathbf{p}\cdot\mathbf{p}/2m) + V(x,y,z)$ = Hamiltonian for one particle in three dimensions	$(\mathbf{p}\cdot\mathbf{p}/2m) + V(x,y,z) = -(\hbar^2/2m)\nabla^2 + V(x,y,z)$ where $\nabla^2 = (\partial^2/\partial x^2) + (\partial^2/\partial y^2) + (\partial^2/\partial z^2)$
$\mathbf{L} = \mathbf{r}\times\mathbf{p}$ = classical ("orbital") angular momentum	$\mathbf{L} = \mathbf{r}\times\mathbf{p} = L_x\mathbf{i} + L_y\mathbf{j} + L_z\mathbf{k}$ where $L_x = (-i\hbar)(y(\partial/\partial z) - z(\partial/\partial y))$, $L_y = (-i\hbar)(z(\partial/\partial x) - x(\partial/\partial z))$, $L_z = (-i\hbar)(x(\partial/\partial y) - y(\partial/\partial x))$ from which it can be shown (see text) that $[L_x,L_y] = L_xL_y - L_yL_x = i\hbar\,L_z$, $[L_y,L_z] = L_yL_z - L_zL_y = i\hbar\,L_x$, $[L_z,L_x] = L_zL_x - L_xL_z = i\hbar\,L_y$
$\mathbf{I}$ = quantum mechanical "spin" angular momentum for which there is no classical analog	$\mathbf{I} = I_x\,\mathbf{i} + I_y\,\mathbf{j} + I_z\,\mathbf{k}$ where the components of I satisfy the relations $[I_x,I_y] = I_xI_y - I_yI_x = i\hbar\,I_z$, $[I_y,I_z] = I_yI_z - I_zI_y = i\hbar\,I_x$, $[I_z,I_x] = I_zI_x - I_xI_z = i\hbar\,I_y$

* **r, p, L,** *and* **I** *are "vector operators"; that is, each is treated as an ordinary three-dimensional* vector *whose* components *are* operators. *Explanation of observed measurements requires the postulation of an additional (purely quantum mechanical) angular momentum,* **I,** *in addition to the angular momentum,* **L,** *obtained by correspondence from classical mechanics. In other words, actual particles exhibit an angular momentum,* $\mathbf{J} = \mathbf{L} + \mathbf{I}$. *The importance of the operators in this Table is that the result of any* single *measurement of any of the observables must be an* eigenvalue *of the corresponding operator (see Physical Identification #1).*

Physical Identification #2: Probability of a *particular* result from measurement of a physical observable. The condition (nature, state) of a particle is described by a function, Ψ. With respect to measurement of a given observable, Ψ may be expressed in terms of its components, $c_1, c_2, \cdots c_n$, along a basis set of eigenfunctions of the operator corresponding to that observable.* The advantage of this expansion,

$$\Psi = c_1\psi_1 + c_2\psi_2 + \cdots + c_n\psi_n \tag{18-20}$$

* *One of the properties of eigenfunctions of a given operator is that each eigenfunction is "orthogonal" ("perpendicular") to other eigenfunctions with different eigenvalues, so that the eigenfunctions form a convenient set of "mutually perpendicular axes" along which to express an arbitrary ("state") function:* $\langle \psi_i,\psi_j \rangle = 0$.

where

ψ_i is an eigenfunction of the desired operator

is that we can now predict that the probability of obtaining the "ith" eigenvalue of the given operator in a *single measurement* on this particle is given by

$$|< \Psi, \psi_i >|^2 = |c_i|^2 \tag{18-21}$$

in which it is assumed that the eigenfunctions have been "normalized" for "length":

$$< \psi_i, \psi_i > = 1 \tag{18-22}$$

Equation 18-21 gives rise to the term "wave function" as a description of the quantum mechanical "state" function, since $|c_i|^2$ is analogous to the *intensity* of a wave (see Chapter 13), and c_i may be thought of as a "probability amplitude" whose square gives the probability of finding that particle in state ψ_i in any particular measurement.

Physical Identification #3: *Average* result from many measurements of a physical observable. Now that we know the probability of obtaining any particular eigenvalue, a_i, of operator, A, we are in a position to compute the average value of many measurements of observable A:

$$\boxed{\text{Average value of many measurements of } A = < \Psi, A\Psi >} \tag{18-23}$$

$$\begin{aligned}
&= < c_1\psi_1 + c_2\psi_2 + \cdots + c_n\psi_n, A(c_1\psi_1 + c_2\psi_2 + \cdots + c_n\psi_n) > \\
&= < c_1\psi_1 + c_2\psi_2 + \cdots + c_n\psi_n, Ac_1\psi_1 + Ac_2\psi_2 + \cdots + Ac_n\psi_n > \\
&= < c_1\psi_1 + c_2\psi_2 + \cdots + c_n\psi_n, a_1c_1\psi_1 + a_2c_2\psi_2 + \cdots + a_nc_n\psi_n > \\
&= < c_1\psi_1, a_1c_1\psi_1 > + < c_1\psi_1, a_2c_2\psi_2 > + \cdots + < c_n\psi_n, a_nc_n\psi_n >
\end{aligned}$$

But

$$< \psi_i, \psi_i > = 1 \quad \text{and} \quad < \psi_i, \psi_j > = 0\,* \tag{18-22}$$

so

$$\boxed{< \Psi, A\Psi > = |c_1|^2 a_1 + |c_2|^2 a_2 + \cdots + |c_n|^2 a_n} \tag{18-24}$$

Equation 18-23 is just a shorthand way of writing Eq. 18-24. Equation 18-24 should be compared to the dice example leading to Eq. 6-38, which provides a non-quantum-mechanical example of exactly the same sort of calculation of average value of a result that can take on only certain discrete (as opposed to continuous) values. In other words, if $|c_i|^2$ is the probability of obtaining result a_i, then Eq. 18-24 must be the recipe for computing the average value of many measurements of A. Identification #3 thus follows directly as a consequence of Eq. 18-21 of Identification #2.

Identifications #1, #2, and #3 give the prescriptions for finding out the *possible results* of a measurement of a given observable (Identification #1), the probability of obtaining any *particular* one of those possible results (Identification #2), and the *average* result of many such measurements (Identification #3). All of these prescriptions are based on a particle that "stands still," so that the result of a later measurement would be the same as an earlier one (more precisely, the *probability* of obtaining a given result is constant in time). Since actual particles are continually changing their positions and other properties, the next problem is to predict how the results of measurements *change with time*. It is simplest to state the recipe in terms of the time-dependence of the "state" function, Ψ, from which we can use Identifications #1, #2, and #3 to find the results of measurements at later times.

Physical Identification #4: Time-dependence of the results of measurements of physical observables. The "equation of motion" (i.e., the equation that predicts the time-dependence) for the state function, Ψ, is given by

$$\boxed{ih\ \partial\Phi(t)/\partial t = \mathscr{H}\Phi(t)} = -\frac{\hbar^2}{2m}\nabla^2\Phi(t) + V(x,y,z,t)\Phi(t) \begin{pmatrix}\text{Schrödinger}\\ \text{Equation}\end{pmatrix} \tag{18-25}$$

for a single particle. We shall most often be concerned with "conservative" systems for which the potential energy, V, does not depend (explicitly) on time. In this case, it is possible to express the solution as the product of two functions:

$$\Phi = \Psi\phi \tag{18-26}$$

where ϕ depends only on time and Ψ is independent of time.

Substituting Eq. 18-26 into Eq. 18-25 and dividing by $\Psi\phi$

$$\frac{1}{\Psi}\left(-\frac{\hbar^2}{2m}\nabla^2\Psi + V\Psi\right) = i\hbar\frac{1}{\phi}\frac{\partial\phi}{\partial t} \tag{18-27}$$

Since the left-hand side of Eq. 18-27 is a function of *coordinates* only, and the right-hand side is a function of *time* only, both sides must be equal to the same constant quantity that is a function of *neither* coordinates nor time. Denoting this constant as E, Eq. 18-27 may be written as two separate equations:

$$\boxed{\mathscr{H}\Psi = E\Psi} \tag{18-28}$$

and

$$\frac{d\phi}{dt} = -\frac{iE}{\hbar}\phi \tag{18-29}$$

The first equation (18-28) is called the time-independent Schrodinger equation, and the second equation (18-29) is readily integrated to give

$$\phi(t) = \exp(-(i/\hbar)Et) \tag{18-30}$$

or

$$\boxed{\Phi(t) = e^{-(i/\hbar)Et}\Phi(0)} \tag{18-31}$$

Equation 18-31 thus provides the desired recipe for obtaining the state function at time t, $\Phi(t)$, given the state function at time zero, $\Phi(0)$.

Physical Identification #5: The final and least intuitive identification concerns the *precision* of *simultaneous* measurements of *two* observables. We have already established that measurement of any *single* observable will in general produce a range of (eigen-)values, with an *average* result

$$\text{Average value of } A = <\Psi, A\Psi> \tag{18-23}$$

and a root-mean-square "width," ΔA, given by (compare to Eq. 6-44)

$$\begin{aligned}(\Delta m)^2 = <(m - <m>)^2> &= <(m^2 - 2m<m> + <m^2>)> \\ &= <m^2> - 2<m><m> + <m>^2 \\ &= <m^2> - <m>^2\end{aligned} \tag{6-44}$$

$$\boxed{(\Delta A)^2 = <\Psi, A^2\Psi> - (<\Psi, A\Psi>)^2} = \text{mean-square deviation in measured } A \text{ values} \tag{18-32}$$

as shown schematically in Fig. 18-1.

There will be a distribution similar to that in Fig. 18-1 for the results of measurement of *any* particular observable, in particular observable B, with its associated average value, $<\Psi, B\Psi>$, and imprecision, ΔB. Now it might be expected (and is often the case) that we should be able to make *independent* simultaneous measurement of both B and A, and that there should be no special relation between the precisions of the two measurements. However, a puzzling but inescapable principle of quantum mechanics is that the very act of measurement changes the state of a particle, so that we sometimes obtain a different result in measurement of B followed by measurement of A, rather than measurement of A and then B. For ex-

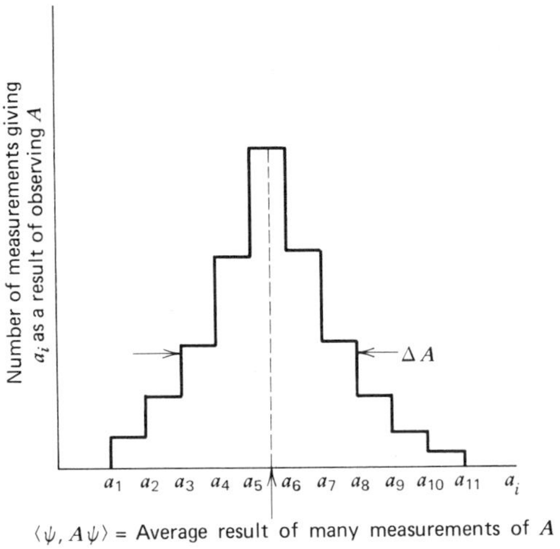

FIGURE 18-1. Schematic diagram of the relative number of measurements of operator A giving a particular result, a_i, as a function of a_i. The average value and width of this distribution of a_i values are indicated in the figure. The predicted result of a measurement of A might thus be expressed

$$\text{Predicted } A \text{ value} = <\Psi, A\Psi> \pm \Delta A.$$

ample, this is mathematically apparent from the successive operation by B and then A compared to operation by A and then B, where A and B are the operators corresponding to the particular physical observables, position and momentum.

Measure (linear) momentum, then position:

$$xp_x f(x) = x(p_x f(x)) = x\left(-i\hbar\frac{df(x)}{dx}\right) = -i\hbar\, x\, f'(x) \tag{18-33}$$

Measure position, then momentum:

$$p_x\, x\, f(x) = p_x(xf(x)) = -i\hbar\frac{d(xf(x))}{dx} = -i\hbar f(x) - i\hbar\, x\, f'(x) \tag{18-34}$$

The difference between these two operations may be expressed as

$$xp_x f(x) - p_x\, x\, f(x) = (xp_x - p_x x)f(x) = i\hbar f(x) \tag{18-35}$$

or more simply

$$(xp_x - p_x x) = \boxed{[x, p_x] = i\hbar} \tag{18-36}$$

in which the brackets in Eq. 18-36 are called a "commutator," and one says that "the commutator of x and p_x is $i\hbar$." It is left as an exercise (see Problems) for the reader to verify the commutators of the "orbital" *angular* momentum operators in Table 18-3.

It can be shown from Eq. 18-36 that when two operators do not commute, there is a relation between the precisions within which the two corresponding variables may be measured:

$$\Delta x \, \Delta p_x \sim \hbar \tag{18-37}$$

Eq. 18-37 is called the "Heisenberg uncertainty principle for position and momentum"; there are similar uncertainty principles for other pairs of noncommuting (i.e., $[A,B] \neq 0$) operators. [There is another ("time-energy") uncertainty principle, but it can be understood in classical terms and will be discussed in Chapter 20.B.]

The Identifications and terminology of the preceding pages of this Chapter provide the basic tools with which we may successfully analyze two important types of problems whose solution requires quantum mechanics (see Section 18.A.).

Existence and Positions of Spectral Lines: the Quantum-mechanical Weight on a Spring

The usual starting point in any quantum mechanical calculation is the computation of the total *energy* of the system, in terms of molecular or particle *properties* (velocity, size and stiffness of the container, rotational speed, etc.). Since the total energy usually takes on only certain discrete (eigen-) values, we can determine the *differences* between the various energy-values ("energy-levels") by irradiating the system with (and detecting the absorption of) electromagnetic energy whose frequency, ν, corresponds to the difference between two energy levels, E_i and E_j:

$$h\nu = E_i - E_j \tag{18-38}$$

Then by fitting the observed pattern of energy spacings to the predicted pattern of energy levels from our energy-eigenvalue calculation, we can compute the molecular parameter of interest (say, the strength of a particular chemical bond). To show the style of the general approach, we calculate the possible energy-eigenvalues of a quantum mechanical weight on a spring, using the language introduced earlier in this chapter.

The classical *kinetic* energy of a particle of mass, m, and velocity, v, moving along the x axis is given by

$$\underset{\text{(classical)}}{\text{Kinetic energy}} = (mv^2/2) = \frac{(mv)^2}{2m} = \frac{p_x^2}{2m} \tag{18-39}$$

By the correspondences defined in Table 18-3, the quantum mechanical operator for kinetic energy must have the form

$$\underset{\text{(quantum)}}{\text{Kinetic energy}} = (p_x^2/2m) = -(\hbar^2/2m)(d^2/dx^2) \tag{18-40}$$

The *potential* energy, $V(x)$, for a particle held by a spring of force constant, k, can be deduced from the force law for a spring

$$\text{Force} = -kx = -\frac{dV(x)}{dx} \tag{18-41}$$

or

$$V(x) = (1/2)kx^2 + \text{constant} = (1/2)kx^2 \tag{18-42}$$

in which the constant of integration has been set equal to zero when $x = 0$. The total energy (i.e., Hamiltonian) operator may now be written as the sum of kinetic and potential energy:

$$\boxed{\mathscr{H} = p_x^2/2m + kx^2/2} \tag{18-43}$$

At this stage, further bookkeeping can be vastly simplified by expressing the total energy in units of $\hbar(k/m)^{1/2}$, where linear momentum, P, is now in units of $(\hbar)^{1/2}(mk)^{1/4}$, and linear position, X, is now in units of $(\hbar)^{1/2}(mk)^{-1/4}$, as shown in Equations 18-45, 18-46a, and 18-46b.

$$\boxed{\mathscr{H} = (1/2)(P^2 + X^2)} \tag{18-45}$$

where

$$P = \hbar^{-1/2}(mk)^{-1/4}\,p_x = -i\hbar^{1/2}(mk)^{-1/4}\,\frac{\partial}{\partial x} = -i\,\frac{\partial}{\partial X} \tag{18-46a}$$

and

$$X = \hbar^{-1/2}(mk)^{1/4}\,x \tag{18-46b}$$

The reader should verify that P and X satisfy the relation

$$\boxed{PX - XP = -i} \tag{18-47}$$

The eigenvalue equation for determination of energy-values (Equations 18-28 and 18-45) becomes

$$(P^2 + X^2)\Psi = 2E\Psi \tag{18-48}$$

We now resort to a trick based on Eq. 18-47:

$$\begin{aligned}(P \pm iX)(P \mp iX) \pm 1 &= P^2 \pm i(XP - PX) + X^2 \pm 1 \\ &= P^2 \pm i(i) + X^2 \pm 1 = P^2 + X^2\end{aligned} \tag{18-49}$$

From Eq. 18-49, we can then deduce that

$$\begin{aligned}(P \mp iX)(P^2 + X^2) &= (P \mp iX)[(P \pm iX)(P \mp iX) \pm 1] \\ &= (P \mp iX)(P \pm iX)(P \mp iX) \pm (P \mp iX) \\ &= [(P \mp iX)(P \pm iX) \pm 1](P \mp iX) \\ &= [P^2 \mp i(XP - PX) + X^2 \pm 1](P \mp iX)\end{aligned}$$

or

$$(P \mp iX)(P^2 + X^2) = (P^2 + X^2 \pm 2)(P \mp iX) \tag{18-50}$$

We will now examine Eq. 18-50 for the $(P + iX)$ case:

$$(P + iX)(P^2 + X^2)\Psi = (P^2 + X^2 - 2)(P + iX)\Psi \tag{18-51}$$

But from Eq. 18-48, we can rewrite the left-hand side as:

$$2E\,(P + iX)\Psi = (P^2 + X^2)(P + iX)\Psi - 2(P + iX)\Psi$$

or just

$$(P^2 + X^2)[(P + iX)\Psi] = 2(E + 1)[(P + iX)\Psi] \tag{18-52}$$

In other words, if Ψ is an eigenfunction of $(P^2 + X^2)$ with (energy) eigenvalue, E, then $(P + iX)\Psi$ is an eigenfunction of $(P^2 + X^2)$ with (energy) eigenvalue $(E + 1)$. $(P + iX)$ is sometimes called a "raising operator" since it acts to raise the energy to the system by one unit. We can construct an energy "ladder" of possible quantum energies by applying the operator $(P + iX)^n$, which will increase the energy by n units, and the ladder extends *upward* without limit (see Fig. 18-2).

Similarly, one can show that the "lowering operator," $(P - iX)$, acts to *decrease* the energy by one unit, so that we can start on any "rung" of the energy "ladder" and move either up or down one energy unit ("rung") at a time:

$$(P^2 + X^2)[(P \pm iX)\Psi] = 2(E \pm 1)[(P \pm iX)\Psi] \tag{18-53}$$

At first, it might seem that the ladder could also extend infinitely downward, but we have already postulated that the eigenvalues of quantum mechani-

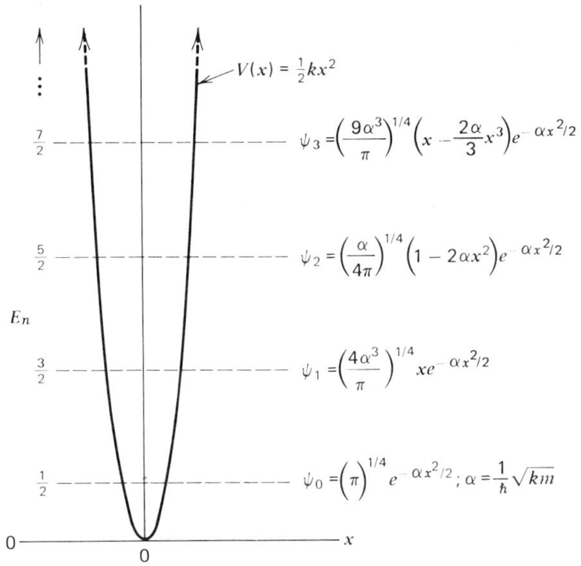

FIGURE 18-2. Energy eigenvalues, E_n, and corresponding energy eigenfunctions, ψ_n, for the quantum mechanical mass (m) on a spring of force constant, k. The energy eigenvalues are expressed in units of $\hbar(k/m)^{1/2}$ for convenience, and form a "ladder" beginning at $E_0 = 1/2$, with equally spaced "rungs" of one unit each extending upward indefinitely.

cal operators that correspond to physical observables (such as x and p_x) are mathematically real (as opposed to complex or imaginary), and the square of any real number must be greater than zero, so the permitted eigenvalues for $(P^2 + X^2)$ must be greater than zero and the ladder must stop either at or above zero energy (see Fig. 18-2).

Having discovered the *spacing* between the energy "rungs," we need only locate the *absolute* energy of any *one* "level" to know the energies of *all* the possible eigenvalues of the Hamiltonian, $(P^2 + X^2)$. Since the effect of the "lowering" operator on the eigenfunction Ψ_0 with the smallest possible eigenvalue must be zero (i.e., we cannot extend the ladder to negative energy),

$$(P - iX)\Psi_0 = (1/i)\left(\frac{d}{dX} + X\right)\Psi_0 = 0 \tag{18-54}$$

It is readily verified that the solution of this equation is

$$\boxed{\Psi_0 = \text{constant}\cdot\exp[-X^2/2]} = \text{"ground" state "wave" function} \tag{18-55}$$

Furthermore,

$$(1/2)(P^2 + X^2)\Psi_0 = E_0\Psi_0 = (1/2)\Psi_0 \text{ in units of } \hbar(k/m)^{1/2}$$

or

$$\boxed{E_0 = (1/2)\hbar(k/m)^{1/2}} \text{ in conventional units}$$

Although we do not need to know them to find their corresponding energy eigenvalues, we could now obtain the energy *eigenfunctions* for the harmonic oscillator by using the "raising" operator as often as necessary:

$$\boxed{\Psi_n = (P + iX)^n\Psi_0 = \text{constant}\cdot\left(\frac{d}{dX} - X\right)^n \Psi_0} \tag{18-56}$$

The first few eigenfunctions are shown explicitly in Fig. 18-2.

For our purposes, the important features of the quantum mechanical treatment of a weight on a spring are (1) there is an *infinite number of "natural" frequencies,* $\nu_{ij} = (E_j - E_i)/h$ near which we expect to find power absorption on application of an electromagnetic field (infrared radiation, for example), rather than the *single* "natural" frequency for a classical mechanical weight on a spring;* (2) using the nonzero commutation relation between x and px (Eq. 18-44), we were able to construct "raising" and "lowering" operators that changed the system energy by one unit up or down, resulting in a "ladder" of equally spaced energy "rungs," and (3) we were able to deduce the energy *spacings* without knowing *any* of the eigenfunctions explicitly. Although the many vibrational "natural" frequencies predicted by the quantum treatment can actually be observed for small molecules in the gas phase, it is usually not possible to resolve the many frequencies for each of the individual "springs" in a large molecule in a liquid, so quantum mechanics offers little more than our classical treatment (Chapter 13) as far as qualitative uses for ultraviolet-visible and infrared spectroscopy. However, the real importance of the preceding exercise is that it is precisely analogous to the treatment of "spin" angular momentum in quantum mechanics, for which there is no classical analog, and for which (in particular) we cannot write an explicit eigenfunction. Because of feature (3) above, we will again be able to deduce energy "level" spacings without ever knowing the eigenfunction explicitly, and the exer-

* *In our later discussion of "selection rules" (Chapter 19.B.4), we will find that only certain of the infinite number of possible transitions are allowed, namely those for which $(j - i) = +1$ ("absorption") or $(j - i) = -1$ ("emission"). Thus, for the simple undamped, quantum mechanical weight on a spring, we would find power absorption at just* one *frequency, $\nu = \hbar(k/m)^{1/2}$. For actual molecules, the potential energy includes nonzero terms proportional to x^3, x^4, and so on, whose effect is to alter the spacings between adjacent energy levels so that $(E_j - E_i)$ is no longer independent of i and j. Thus, for real molecules, even with the "selection rule" that $(j - i) = \pm 1$, there will be an* infinite *number of* different *absorption frequencies for electromagnetic radiation incident on a charged quantum mechanical weight on a spring.*

cise should appear a little less strange for having seen the weight-on-a-spring example for which there *is* a classical analog. In contrast again to the quantum mechanical harmonic oscillator, the quantum mechanical treatment of "spin" is of direct and immediate use in accounting for the observed magnetic resonance absorption from nuclei, for which it *is* possible to resolve the many new "natural" frequencies predicted from the quantum calculation.

18.A. SPIN PROBLEMS: THE SIMPLEST QUANTUM MECHANICAL CALCULATIONS

The energy associated with the interaction of a magnetic moment, μ, with a *static* magnetic field, H_0, is given by $-\mu \cdot H_0$, or just $\mu_z H_0$ if H_0 is taken along the negative z direction (so that we can forget about the minus sign from here on). Since classical magnetic moments arise from spinning charges, it should seem reasonable that a magnetic moment should be proportional (the constant of proportionality is called the "magnetogyric ratio," γ) to the classical angular momentum of the spinning charged particle, **L**:

$$\boldsymbol{\mu}_L = \gamma_L \mathbf{L}; \quad \mathbf{L} = L_x \mathbf{i} + L_y \mathbf{j} + L_z \mathbf{k} \tag{18-57}$$

[We shall later suppose that nonclassical angular momentum (see below) shows the same property for particles that exhibit nonclassical "spin" angular momentum, **I**:]

$$\boldsymbol{\mu}_I = \gamma_I \mathbf{I}; \quad \mathbf{I} = I_x \mathbf{i} + I_y \mathbf{j} + I_z \mathbf{k} \tag{18-58}$$

The *energy* associated with the interaction of a magnetic moment with a static magnetic field may thus be expressed in the form

$$\boxed{\text{Hamiltonian} = \mathcal{H}_0 = \gamma_L H_0 L_z} \tag{18-59}$$

for a particle possessing angular momentum component, L_z, along the z axis.

From Eq. 18-59, it is clear that it is necessary to find the eigenvalues for the operator, L_z, in order to predict the pattern of energy-levels expected in magnetic resonance experiments. Modeling an approach after our earlier successful solution of the weight-on-a-spring problem, we begin by examining the commutators for various pairs of angular momentum operators, whose properties are derived by correspondence from classical mechanics (see Table 18-3 and Problems):

$[L_x, L_y] = i\,L_z$		(18-60a)
$[L_y, L_z] = i\,L_x$	angular momentum in units of $\hbar$	(18-60b)
$[L_z, L_x] = i\,Ly$		(18-60c)

If new operators, L_+ and L_-, are now defined as

$$L_+ = L_x + i\,Ly \tag{18-61a}$$

$$L_- = L_x - i\,Ly \tag{18-62a}$$

then Equations 18-60b, 18-60c, and 18-61a may be combined to yield

$$\begin{aligned} L_zL_+ - L_+L_z &= L_zL_x + iL_zL_y - L_xL_z - iL_yL_z \\ &= (L_zL_x - L_xL_z) - i(L_yL_z - L_zL_y) \\ &= [L_z,L_x] - i[L_y,L_z] \\ &= iL_y - i(i)L_x = L_x + iL_y \\ L_zL_+ - L_+L_z &= L_+ \end{aligned} \tag{18-63}$$

Rearranging Eq. 18-63

$$L_zL_+ - L_+ = (L_z - 1)L_+ = L_+L_z \tag{18-63a}$$

Applying both sides of Eq. 18-63a to ψ_n, which is an eigenfunction of L_z with associated eigenvalue, μ_n, we conclude that

$$(L_z - 1)L_+\psi_n = L_+(L_z\psi_n) = \mu_n(L_+\psi_n)$$

or

$$L_z(L_+\psi_n) = (\mu_n + 1)(L_+\psi_n) \tag{18-64}$$

In other words, if ψ_n is an eigenfunction of L_z with eigenvalue μ_n, then $(L_+\psi_n)$ is also an eigenfunction of L_z with eigenvalue $(\mu_n + 1)$, and the effect of the "raising" operator L_+ has been to *increase* the eigenvalue of L_z by *one unit.*

As might be anticipated from our accumulated experience, L_- turns out to be a "lowering" operator that *decreases* the L_z eigenvalue by one unit. We have thus again managed to find a "ladder" of eigenvalues of L_z that so far appear to extend indefinitely upward and downward (see Fig. 18-3). It is possible to show (see Problems) that there is an upper limit, $\mu_{\max}$, and a lower limit, $\mu_{\min}$, to the length of the "ladder" of Fig. 18-3, and that

$$\mu_{\max} = -\mu_{\min} \tag{18-65}$$

	⋮
$L_+^3\,\psi$	$\mu+3$
$L_+^2\,\psi$	$\mu+2$
$L_+\,\psi$	$\mu+1$
ψ	μ
$L_-\,\psi$	$\mu-1$
$L_-^2\,\psi$	$\mu-2$
	⋮

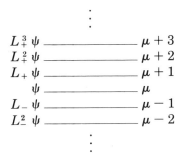

FIGURE 18-3. Eigenfunctions (left) and corresponding eigenvalues (right) for the z component of *angular momentum,* L_z, following application of "raising" or "lowering" operators L_+ and L_-. Since energy for interaction of a magnetic moment with a fixed magnetic field is proportional to L_z, the *energy*-level diagram for this system will also consist of a "ladder" of equally spaced *energy*-(eigen)values.

Now if the rungs of the ladder are one unit apart *and* the middle of the ladder is at zero (Eq. 18-65) whether there is a rung there or not, then it is clear that z-component angular momentum eigenvalues can take on only *integral* or *half-integral* values (see Fig. 18-4). It is then possible to show (see Problems) that the μ values for "orbital" angular momentum z component, L_z, must take on only *integral* values.

Scrutiny of Equations 18-60 to 18-69 shows that the only angular momentum properties required to construct the eigenvalue ladder were the commutation relations, Eq. 18-60. Actual particles often behave as if they possessed an additional angular momentum (beyond that deduced by correspondence from classical mechanics) called "spin" angular momentum, **I**, whose associated quantum mechanical operator therefore satisfies the same commutation relations:

$$\mathbf{I} = I_x\,\mathbf{i} + I_y\,\mathbf{j} + I_z\,\mathbf{k} \qquad \text{a vector whose components are operators} \qquad (18\text{-}66)$$

where

$$[I_x,I_y] = i\,I_z \qquad (18\text{-}67a)$$

$$[I_y,I_z] = i\,I_x \qquad \text{angular momentum in units of } \hbar \qquad (18\text{-}67b)$$

$$[I_z,I_x] = i\,I_y \qquad (18\text{-}67c)$$

From the commutators, Eq. 18-67, we can immediately conclude that "spin" angular momentum z component eigenvalues can be only integral or half-integral multiples of $\hbar$. It turns out that both kinds of particles have been observed, as indicated by the partial list in Fig. 18-4, and there are some

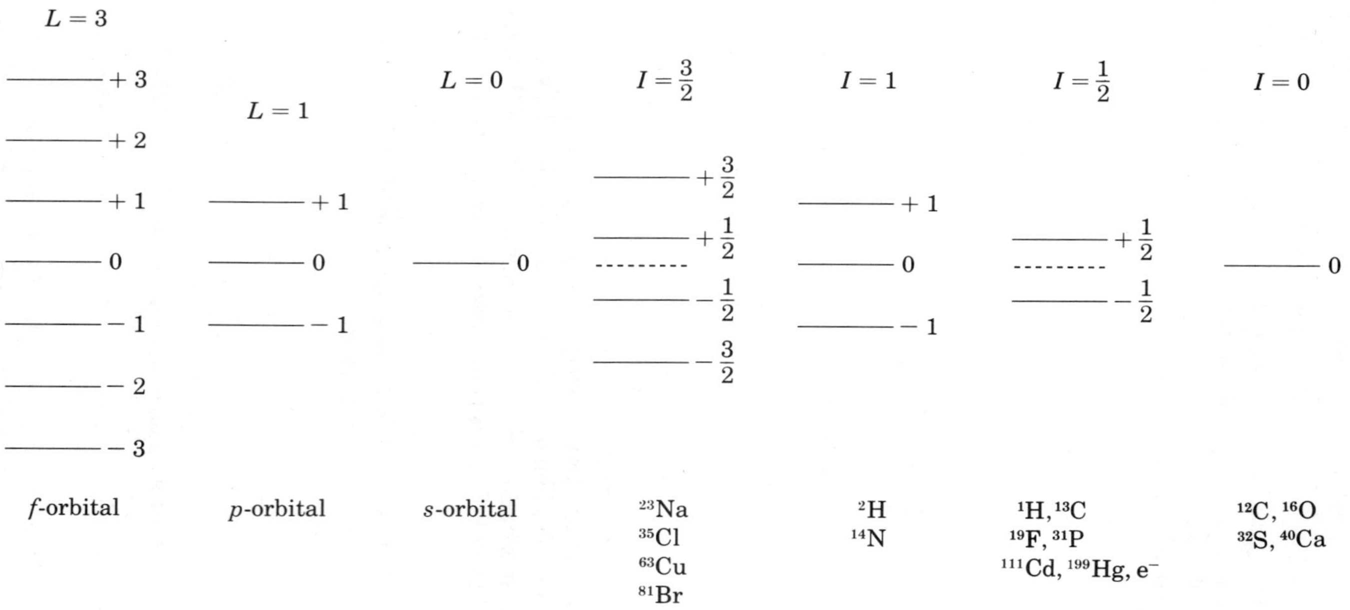

FIGURE 18-4. Schematic diagram of z-component angular momentum operator eigenvalues corresponding to the stated maximum z component of "orbital" (L) or "spin" (I) angular momentum shown at the top of each "ladder." From Eq. 18-59, the magnetic *energy* of a given particle is proportional to its *z-component angular momentum (eigen-)value* and the applied external static *magnetic field strength*. Thus, in the absence of an applied magnetic field, the magnetic energy for each of the possible z-component angular momentum eigenfunctions for a given particle is the same. On application of an external static magnetic field, the (magnetic) energy-level diagram for any of the listed particles is as shown here, where the energy *scale* for a given particle is determined by its γ value in Eq. 18-59. The familiar "orbitals" of the hydrogen atom (see $L = 3, 1, 0$ in the figure) correspond to the orbital angular momentum z-component eigenvalues listed; in addition, the "spin" angular momentum z-component eigenvalues corresponding to some interesting nuclear isotopes are also listed. Actual problems involving "spin one-half" nuclei are treated in this section.

particles (such as carbon-12 nuclei) that exhibit no "spin" at all. "*Spin*" of a nucleus is an intrinsic nuclear property, as are mass, charge, parity, and so on, and is the same for all chemically identical nuclei (e.g., all carbon-13 nuclei have maximum I_z eigenvalue $= \hbar(1/2)$), but the *z component* of "spin" can take on any of the values shown in Fig. 18-4. The simplest and most widespread applications for "spin" operators are connected with magnetic resonance studies of spin one-half nuclei, principally ^1H and ^{13}C, and the remainder of this section will deal with particles of spin one-half.

As a final important point of nomenclature, **L** and **I** are *vectors* (whose components are operators), but it is common practice to say that a particle has orbital angular momentum L or "spin" I (L and I are integers or half-integers), by which is meant that the *maximum* eigenvalue of L_z or I_z is L or I (in units of $\hbar$).

One Nucleus of Spin One-Half in a Static Magnetic Field

From Fig. 18-4, the only possible z-component "spin" angular momentum eigenvalues for a spin one-half particle are $+(1/2)$ and $-(1/2)$, corresponding to respective I_z eigenfunctions that we shall denote as α and β:

$$I_z\,\alpha = \frac{1}{2}\alpha \tag{18-68a}$$

$$I_z\,\beta = -\frac{1}{2}\beta \tag{18-68b}$$

For simplicity, we shall suppose that α and β are normalized to unit "length"; in addition, it can be shown that the eigenfunctions corresponding to two different eigenvalues of a Hermitian operator (such as I_z) are orthogonal ("perpendicular") (see References). In other words

$$< \alpha,\alpha > = < \beta,\beta > = 1 \tag{18-69a}$$

$$< \alpha,\beta > = < \beta,\alpha > = 0 \tag{18-69b}$$

We can now compute all the matrix elements of the operator, $\mathscr{H}_0 = \gamma H_0 I_z$ (Eq. 18-59) required to determine the energy eigenvalues from Eq. 18-16, for a single particle of spin one-half in a static magnetic field H_0, using the basis set,

$$\phi_1 = \alpha \tag{18-70a}$$
$$\phi_2 = \beta \tag{18-70b}$$

$$\mathscr{H}_{11} = <\phi_1, \mathscr{H}\phi_1> = <\alpha, \mathscr{H}\alpha> = \gamma H_0 <\alpha, I_z\alpha> = \gamma H_0 \left\langle \alpha, \frac{1}{2}\alpha \right\rangle$$

$$= \frac{\gamma H_0}{2} <\alpha,\alpha> \overset{\text{Eq. 18-69a}}{\downarrow}{=} \frac{\gamma H_0}{2} \qquad \text{(18-71a)}$$

or in matrix notation

$$= \gamma H_0 (1 \quad 0) \begin{pmatrix} \frac{1}{2} & 0 \\ 0 & -\frac{1}{2} \end{pmatrix} \begin{pmatrix} 1 \\ 0 \end{pmatrix} = \gamma H_0 (1 \quad 0) \begin{pmatrix} \frac{1}{2} \\ 0 \end{pmatrix} = \frac{\gamma H_0}{2} \qquad \text{(18-71b)}$$

$$\mathscr{H}_{12} = <\phi_1, \mathscr{H}\phi_2> = <\alpha, \mathscr{H}\beta> = \gamma H_0 <\alpha, I_z\beta> = \gamma H_0 \left\langle \alpha, -\frac{1}{2}\beta \right\rangle$$

$$= -\frac{\gamma H_0}{2} <\alpha,\beta> \overset{\text{Eq. 18-69b}}{\downarrow}{=} 0 \qquad \text{(18-71c)}$$

or in matrix notation

$$= \gamma H_0 (1 \quad 0) \begin{pmatrix} \frac{1}{2} & 0 \\ 0 & -\frac{1}{2} \end{pmatrix} \begin{pmatrix} 0 \\ 1 \end{pmatrix} = \gamma H_0 (1 \quad 0) \begin{pmatrix} 0 \\ -\frac{1}{2} \end{pmatrix}$$

$$= \gamma H_0 (0) = (0) \qquad \text{(18-71d)}$$

Similarly,

$$\mathscr{H}_{21} = 0 \qquad \text{(18-71e)}$$

$$\mathscr{H}_{22} = -\frac{\gamma H_0}{2} \qquad \text{(18-71f)}$$

The energy-eigenvalues of $\mathscr{H}_0$ are now found by solving the determinant, Eq. 18-16:

$$\det \begin{pmatrix} \mathscr{H}_{11} - \lambda & \mathscr{H}_{12} \\ \mathscr{H}_{21} & \mathscr{H}_{22} - \lambda \end{pmatrix} = \begin{vmatrix} \frac{\gamma H_0}{2} - \lambda & 0 \\ 0 & -\frac{\gamma H_0}{2} - \lambda \end{vmatrix} = 0 \qquad \text{(18-72)}$$

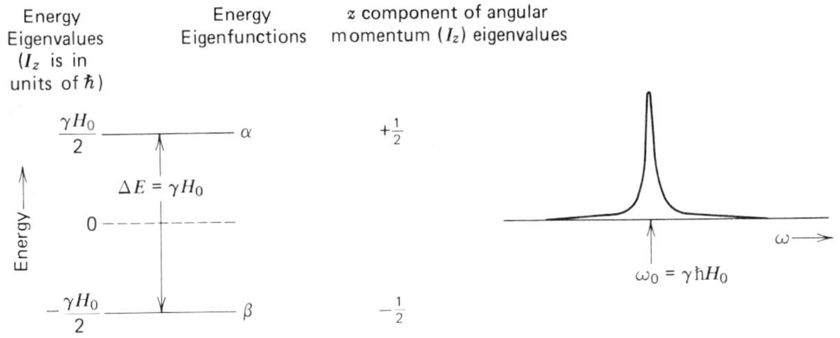

FIGURE 18-5. Schematic energy-level diagram and predicted absorption spectrum for a system of isolated spin one-half particles in a static magnetic field H_0. Absorption line *shape* is discussed in Chapter 13; absorption *intensity* is discussed in Chapter 19.B.4; here we are only interested in the spectral line *position*.

whose roots are simply

$$\lambda_1 = \frac{\gamma H_0}{2} \tag{18-73a}$$

and

$$\lambda_2 = -\frac{\gamma H_0}{2} \tag{18-73b}$$

The energy eigenvalues and eigenfunctions (Eq. 18-17) are shown in Fig. 18-5, and it is seen that there are two energy-levels, and the spacing between the levels is proportional to the strength of the applied static magnetic field, H_0.

Two Chemically Different Isolated Nuclei of Spin One-Half

For a sample containing *two* types of nuclei, each of spin one-half, but with different magnetogyric ratios, $\gamma_1 \neq \gamma_2$, it is intuitively obvious that as long as the two types of nuclei do not interact with each other, we expect to find an absorption spectrum consisting of *two* separate peaks located at $\omega_1 = \gamma_1 \hbar H_0$ and $\omega_2 = \gamma_2 \hbar H_0$, based on the example from Fig. 18-5. (In other words, we expect to find two resonant frequencies from a collection of two kinds of tuning forks, as seen in Fig. 18-6.) To simplify the bookkeeping for this and ensuing problems, it is useful to introduce a Hamiltonian (*total* kinetic and potential energy) that is the sum of the (*individual*) Hamiltonians we would write for either type of nucleus by itself:

$$\mathscr{H} = \mathscr{H}_1 + \mathscr{H}_2$$
$$\mathscr{H} = \gamma_1 H_0 I_{z_1} + \gamma_2 H_0 I_{z_2} \qquad (I_z \text{ is in units of } h) \tag{18-74}$$

in which the operator I_{z_1} operates only on functions for particles with $\gamma = \gamma_1$ and I_{z_2} operates only on particles with $\gamma = \gamma_2$. Since we must specify states for *both* types of nuclei to predict the results of any given measurement, it is useful to define a basis set, ϕ_i, in terms of "product" functions of eigenfunctions of I_{z_1} for γ_1-nuclei and eigenfunctions of I_{z_2} for γ_2-nuclei as shown below.

$$\phi_1 = \alpha_1\alpha_2 = \begin{pmatrix} 1 \\ 0 \\ 0 \\ 0 \end{pmatrix} \tag{18-75a}$$

$$\phi_2 = \alpha_1\beta_2 = \begin{pmatrix} 0 \\ 1 \\ 0 \\ 0 \end{pmatrix} \tag{18-75b}$$

$$\phi_3 = \beta_1\alpha_2 = \begin{pmatrix} 0 \\ 0 \\ 1 \\ 0 \end{pmatrix} \tag{18-75c}$$

$$\phi_4 = \beta_1\beta_2 = \begin{pmatrix} 0 \\ 0 \\ 0 \\ 1 \end{pmatrix} \tag{18-75d}$$

From what we already know about I_z-operators and eigenfunctions,

$$\begin{aligned} I_{z_1}\alpha_1\alpha_2 &= \frac{1}{2}\,\alpha_1\alpha_2 & \qquad I_{z_2}\alpha_1\alpha_2 &= \frac{1}{2}\,\alpha_1\alpha_2 \\ I_{z_1}\alpha_1\beta_2 &= \frac{1}{2}\,\alpha_1\alpha_2 & \qquad I_{z_2}\alpha_1\beta_2 &= -\frac{1}{2}\,\alpha_1\beta_2 \\ I_{z_1}\beta_1\alpha_2 &= -\frac{1}{2}\,\beta_1\alpha_2 & \qquad I_{z_2}\beta_1\alpha_2 &= \frac{1}{2}\,\beta_1\alpha_2 \\ I_{z_1}\beta_1\beta_2 &= -\frac{1}{2}\,\beta_1\beta_2 & \qquad I_{z_2}\beta_1\beta_2 &= -\frac{1}{2}\,\beta_1\beta_2 \end{aligned} \tag{18-76}$$

we can immediately construct the matrices for operators I_{z_1}, I_{z_2}, and $\mathscr{H}$ in this ϕ-basis:

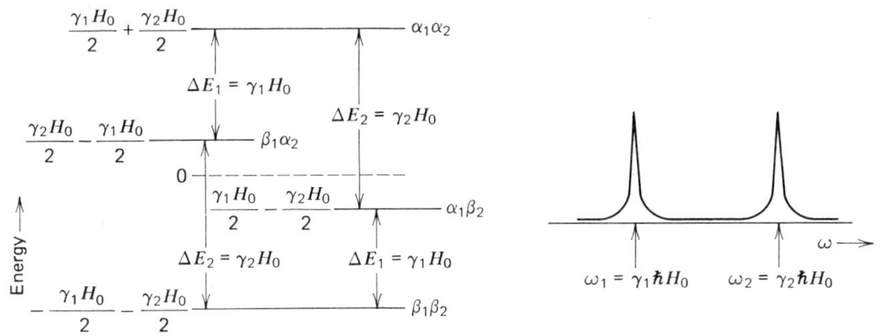

FIGURE 18-6. Schematic energy-level diagram and predicted absorption spectrum for a collection of two types of isolated (i.e., uncoupled) nuclei of spin one-half in a static magnetic field H_0. The two types of nuclei differ in their γ values, and thus in their energy eigenvalues and consequently in the separation between energy levels that determine their absorption frequencies.

$$I_{z_1} = \begin{pmatrix} \frac{1}{2} & 0 & 0 & 0 \\ 0 & \frac{1}{2} & 0 & 0 \\ 0 & 0 & -\frac{1}{2} & 0 \\ 0 & 0 & 0 & -\frac{1}{2} \end{pmatrix} \qquad I_{z_2} = \begin{pmatrix} \frac{1}{2} & 0 & 0 & 0 \\ 0 & -\frac{1}{2} & 0 & 0 \\ 0 & 0 & \frac{1}{2} & 0 \\ 0 & 0 & 0 & -\frac{1}{2} \end{pmatrix} \tag{18-77}$$

$$\det\begin{pmatrix} \left(\frac{\gamma_1 H_0}{2} + \frac{\gamma_2 H_0}{2} - \lambda\right) & 0 & 0 & 0 \\ 0 & \left(\frac{\gamma_1 H_0}{2} - \frac{\gamma_2 H_0}{2} - \lambda\right) & 0 & 0 \\ 0 & 0 & \left(-\frac{\gamma_1 H_0}{2} + \frac{\gamma_2 H_0}{2} - \lambda\right) & 0 \\ 0 & 0 & 0 & \left(-\frac{\gamma_1 H_0}{2} - \frac{\gamma_2 H_0}{2} - \lambda\right) \end{pmatrix} = 0$$

$$\begin{aligned} \lambda_1 &= \frac{\gamma_1 H_0}{2} + \frac{\gamma_2 H_0}{2} \\ \lambda_2 &= \frac{\gamma_1 H_0}{2} - \frac{\gamma_2 H_0}{2} \\ \lambda_3 &= -\frac{\gamma_1 H_0}{2} + \frac{\gamma_2 H_0}{2} \\ \lambda_4 &= -\frac{\gamma_1 H_0}{2} - \frac{\gamma_2 H_0}{2} \end{aligned} \tag{18-78}$$

The energy-levels and spectrum corresponding to Eq. 18-78 are shown in Fig. 18-6.

Two Chemically Different Coupled Nuclei of Spin One-Half: The "AX" Spectrum

When two chemically different* nuclei, such as two protons, are located on the same molecule, their behavior is no longer independent of each other. The result of the interaction between the two nuclei may be described classically by analogy to the coupled-spring model of Chapter 17.A, so that the resonant frequency associated with a given proton becomes "split" into two or more resonant frequencies arising from "coupling" to the other proton. Quantum mechanics is required to calculate the *strength* of the coupling (and thus the *magnitude* of the splitting). In classical mechanics, no two particles may occupy the same space at the same time, yet a quantum mechanical electron (see any elementary treatment of the particle in a box or the hydrogen atom) has a finite probability of being found *at* the nucleus itself, and is thus affected in a nonclassical way (known as the Fermi "contact" interaction) by the nuclear magnetic moment. Since the electron may be found at some later time at a point far removed from the nucleus, this "contact" effect spreads out along any nearby chemical bonds and is "transmitted" along any intramolecular path with electron density along the way (i.e., *through chemical bonds,* as opposed to *through space*). Ultimately, the affected electron(s) reach a second magnetic nucleus that (as in the present case) may be chemically different from the first nucleus; then from a similar "contact" effect, the second nucleus is perturbed by an interaction that originally arose because of the presence of the first magnetic nucleus. To a first approximation (see Problems), this "contact" phenomenon can be described by a "scalar" coupling between the two nuclear spins in a very simple way:

$$\mathscr{H} = \gamma_1 H_0 I_{z_1} + \gamma_2 H_0 I_{z_2} + J I_{z_1} I_{z_2} \qquad (\text{units of } \hbar \text{ for } I_z) \qquad (18\text{-}79)$$

in which J (which has units of $\sec^{-1}$) is a measure of the strength of the nuclear "scalar" or "spin-spin" coupling. We now calculate the energy levels and energy level differences ("transition frequencies") for the Hamiltonian of Eq. 18-79, and then indicate the extremely valuable uses for the experimentally measurable coupling constant, J.

The matrix representation of the operator product, $I_{z_1} I_{z_2}$, is obtained at once from the rule for matrix multiplication (Eq. 18-11) and the already-

* *"Chemically different" here means two nuclei whose γ values are not identical, as for two different nuclear isotopes (^{13}C and ^{1}H, or ^{35}Cl and ^{37}Cl) or two nuclei of the same isotope with different chemical bonding (^{1}H in a CH_3 group and ^{1}H in a CH_2 group).*

established form of the matrices for the individual operators I_{z_1} and I_{z_2} (Eq. 18-77):

$$I_{z_1}I_{z_2} = \begin{pmatrix} \frac{1}{2} & 0 & 0 & 0 \\ 0 & \frac{1}{2} & 0 & 0 \\ 0 & 0 & -\frac{1}{2} & 0 \\ 0 & 0 & 0 & -\frac{1}{2} \end{pmatrix} \begin{pmatrix} \frac{1}{2} & 0 & 0 & 0 \\ 0 & -\frac{1}{2} & 0 & 0 \\ 0 & 0 & \frac{1}{2} & 0 \\ 0 & 0 & 0 & -\frac{1}{2} \end{pmatrix} = \begin{pmatrix} \frac{1}{4} & 0 & 0 & 0 \\ 0 & -\frac{1}{4} & 0 & 0 \\ 0 & 0 & -\frac{1}{4} & 0 \\ 0 & 0 & 0 & \frac{1}{4} \end{pmatrix} \quad (18\text{-}80)$$

Substituting Eq. 18-80 into the left-hand Eq. 18-78, we solve for the energy eigenvalues of the Hamiltonian of Eq. 18-79:

$$\det \begin{pmatrix} \left(\frac{\gamma_1 H_0}{2} + \frac{\gamma_2 H_0}{2} + \frac{J}{4} - \lambda\right) & 0 & 0 & 0 \\ 0 & \left(\frac{\gamma_1 H_0}{2} - \frac{\gamma_2 H_0}{2} - \frac{J}{4} - \lambda\right) & 0 & 0 \\ 0 & 0 & \left(-\frac{\gamma_1 H_0}{2} + \frac{\gamma_2 H_0}{2} - \frac{J}{4} - \lambda\right) & 0 \\ 0 & 0 & 0 & \left(-\frac{\gamma_1 H_0}{2} - \frac{\gamma_2 H_0}{2} + \frac{J}{4} - \lambda\right) \end{pmatrix} = 0 ;$$

$$\lambda_1 = \frac{\gamma_1 H_0}{2} + \frac{\gamma_2 H_0}{2} + \frac{J}{4}$$

$$\lambda_2 = \frac{\gamma_1 H_0}{2} - \frac{\gamma_2 H_0}{2} - \frac{J}{4}$$

$$\lambda_3 = -\frac{\gamma_1 H_0}{2} + \frac{\gamma_2 H_0}{2} - \frac{J}{4}$$

$$\lambda_4 = -\frac{\gamma_1 H_0}{2} - \frac{\gamma_2 H_0}{2} + \frac{J}{4}$$

The Hamiltonian (Eq. 18-79) that leads to the "AX" spectrum of Fig. 18-7 is a good approximation when the difference in "natural" frequency between the two nuclei is much larger than the strength of the coupling between them (see Problems):

$$\omega_1 - \omega_2 >> J \quad (18\text{-}81)$$

Since the resonant frequency ω is proportional to the applied magnetic field strength H_0, the frequency difference in Eq. 18-81 may be made larger

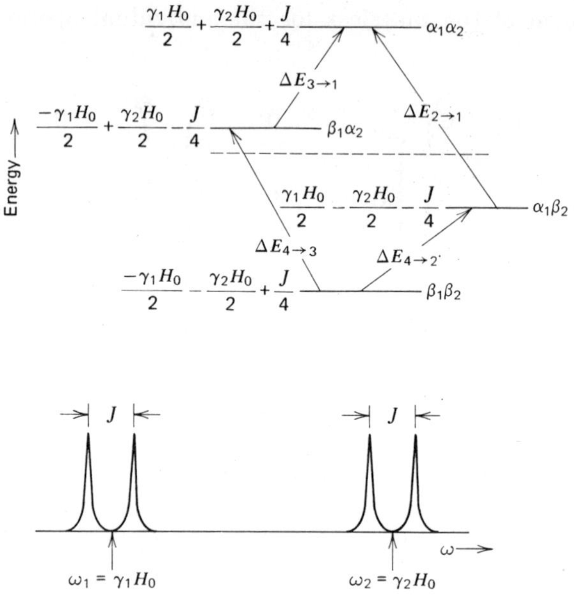

FIGURE 18-7. Schematic energy-level diagram and predicted absorption spectrum for the magnetic resonance of two chemically different coupled nuclear spins one-half in a static magnetic field H_0. The coupling constant, J, is easily measured as the splitting of either resonance. As in Fig. 18-6, the $\Delta E_{4\to1}$ transition is "forbidden" (see Chapter 19.B.4).

simply by increasing the magnetic field strength, so that the situation in Fig. 18-7 can often be closely approached in actual experiments by using very large fields (50 kGauss or more).

It is possible to extend the treatment leading to Fig. 18-7 to situations involving the coupling of more than two nuclei without any more mathematics, using a qualitative argument. The argument is that a given spin one-half nucleus may have a magnetic moment whose measurable direction points either *along* (↑) or *opposed* (↓) to the direction of the applied magnetic field (corresponding to the two states α and β with different energies as shown in Fig. 18-5). If this first nucleus is scalar coupled to a second nucleus, then the second nucleus will be subjected to an applied field that is slightly greater or slightly less (by an amount, $J/2\gamma\hbar$) than H_0, according to whether the direction of the first spin was "up" or "down." If there are now *two* spins of the first type, then there are *four* possible net orientations of their spin-directions and two of those orientations produce the same net magnetic moment and hence the same effect on the nucleus of the second type (see Fig. 18-8). The resulting spectral patterns expected for splitting of a —C—H proton magnetic resonance signal by either another —C—H, a —CH_2, or a —CH_3 group are then evident in Fig. 18-8, and conclude our treatment of the determination of coupling constants from observed nuclear magnetic resonance spectra.

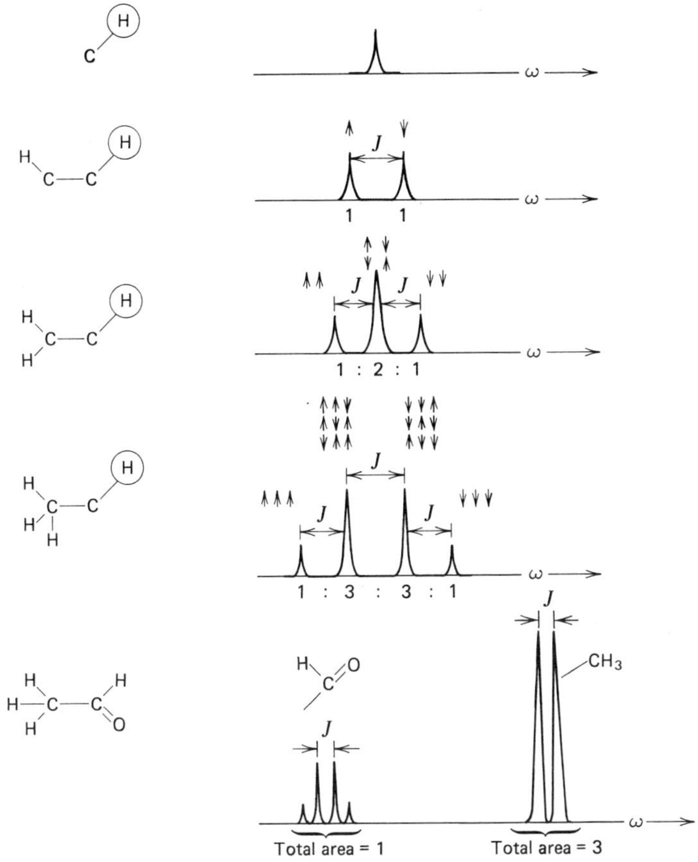

FIGURE 18-8. Qualitative method for predicting the splitting patterns in NMR spectra due to coupling between two chemically different types of protons on the same molecule. Each of the spectra at the right depicts the NMR spectrum due to absorption by the circled proton at the left. The circled proton is subjected to a magnetic field that is the sum of the applied magnetic field (top graph) plus or minus a small field according to whether the spin of the coupled nucleus (nuclei) is "up" or "down." Since, for example, the net magnetic moment for two antiparallel spins is the same for both possible configurations, ⇅ and ⇵, both configurations will yield the same magnetic field at the circled nucleus, and the magnetic resonance corresponding to those configurations will be twice as intense as for the single configurations ↑↑ or ↓↓ (see middle plot). The proton NMR spectrum for the acetaldehyde molecule is shown in the bottom trace: the methyl resonance (relative intensity = 3 for 3 identical protons) is split into two lines corresponding to the two possible orientations of the aldehyde proton spin while the aldehyde proton resonance (relative intensity = 1 for one proton) is split into four lines according to the four possible net magnetic moments associated with the eight possible configurations of the three methyl proton spins (see preceding graph). This figure contains enough information to predict the proton NMR spectrum for any molecule (at sufficiently large H_0), given the resonant frequencies for each unperturbed proton and the coupling constants between all types of protons in the molecule.

The great interest in magnetic scalar coupling constants, J, is that there is a well-defined empirical relation (with theoretical justification beyond the scope of this discussion) between J values and the dihedral angle between the C—H bond directions of protons of adjacent carbon atoms (see Fig. 18-9). Since we have already argued that such scalar couplings are transmitted along chemical bonds, it should seem reasonable that the efficiency of the transmission (and hence the magnitude of J) should depend on the degree of overlap between the (nonspherical and therefore directional) electronic orbitals associated with intervening chemical bonds. The unique value of proton nuclear magnetic resonance (NMR) in determination of the structure and stereochemistry of molecules in solution is based on the variation in proton *resonant frequency* ("chemical shift") according to the *substituents* on the carbon to which that proton is bonded, and the *J value* variation according to the *dihedral angle* between adjacent carbons in the chain.

EXAMPLE Some of the most definitive NMR evidence for preferred conformations of biologically interesting molecules has been obtained for carbohydrate derivatives. The illustrative example in Fig. 18-10 shows the proton NMR spectra for acetylated cellulose (a glucose polysaccharide with β-glycosidic links between glucose units) and acetylated amylose (a glucose polysaccharide with α-glycosidic links). From Fig. 18-9 a scalar coupling constant

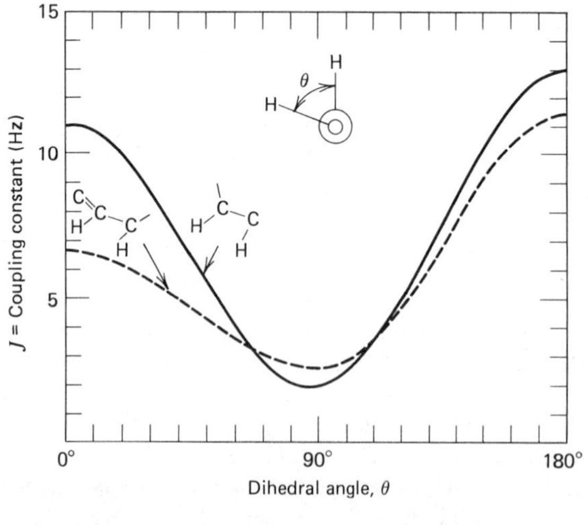

FIGURE 18-9. Proton-proton scalar magnetic coupling constant, J, as a function of dihedral angle between the respective C—H bonds on adjacent carbons in a molecule. This graph is sometimes called a "Karplus" curve, after the individual who offered the first theoretical explanation of its general form. Although the curve is double-valued (for each measured J there are in general *two* possible dihedral angles), the curve is nevertheless useful in establishing the stereochemistry about various bonds in molecules in solution. (From E. D. Pearce, *High-Resolution NMR*, Reed Press, New York, 1967, p. 104.)

of about 8 Hz is predicted for *axial* protons, as is observed for all five ring protons of the substituted cellulose. The smaller coupling constant of about 3 Hz for coupling between proton 1 and proton 2 of the substituted amylose, on the other hand, indicates an α-glycosidic linkage in that polymer.

EXAMPLE While there are many more examples of the use of NMR J couplings in determination of the conformation of *isolated* biochemicals in solution [most notably perhaps the evidence that the reduced form of pyridine dinucleotide, NADH, exists in a *folded* conformation, based on the chemical nonequivalence of the two (geminal) C-4 protons of the dihydropyridine ring], a recent and important example of a conformational *change* of substrate on binding to enzyme has been reported (see Fig. 18-11).

As we have seen in Fig. 18-10, an α-linked glucose in its usual "chair" conformation shows a coupling constant of $J \sim 3$ Hz for the anomeric proton. If,

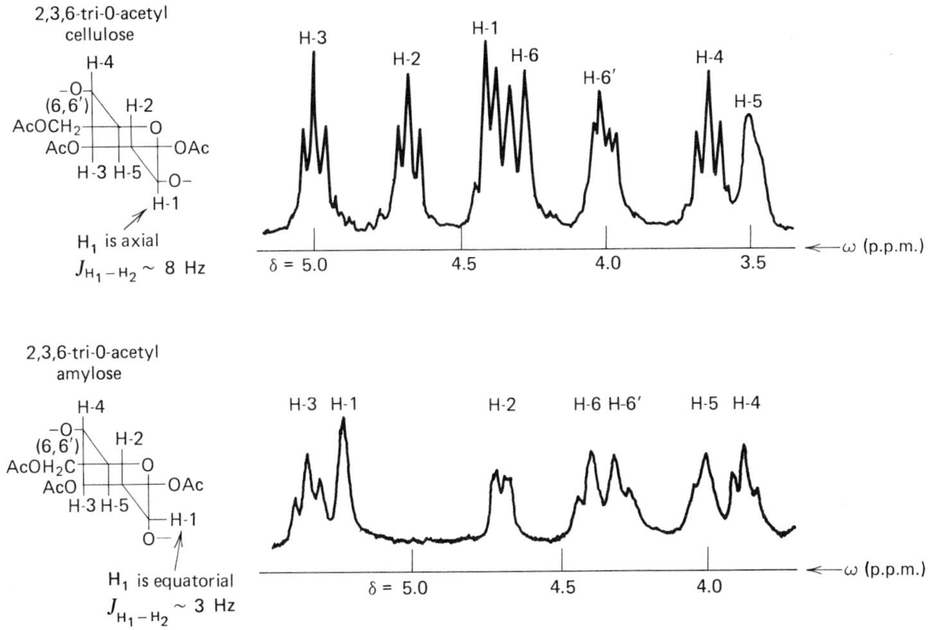

FIGURE 18-10. Conformations and proton NMR spectra for acetylated cellulose (*top*) and acetylated amylose (*bottom*), which differ structurally in whether the anomeric proton, H-1, is axial (cellulose) or equatorial (amylose). The larger (8 Hz) splitting is seen for H-1 of the cellulose derivative, while a smaller (3 Hz) splitting is seen for H-1 of the amylose derivative, confirming the stereochemistry at the anomeric linkage. Because sugar stereochemistry is rather difficult to establish by conventional chemical reactivity methods, NMR is the method of choice for finding the structure of newly isolated sugar oligomers, such as various antigenic saccharides on cell surfaces. The frequency scale for these power absorption spectra is listed in "parts-per-million": thus, for a magnetic field at which protons resonate near 100 MHz, 1 p.p.m. = 100 Hz. [From H. Friebolin, G. Keilich, and E. Siefert, *Angew. Chemie 81,* 791 (1969).]

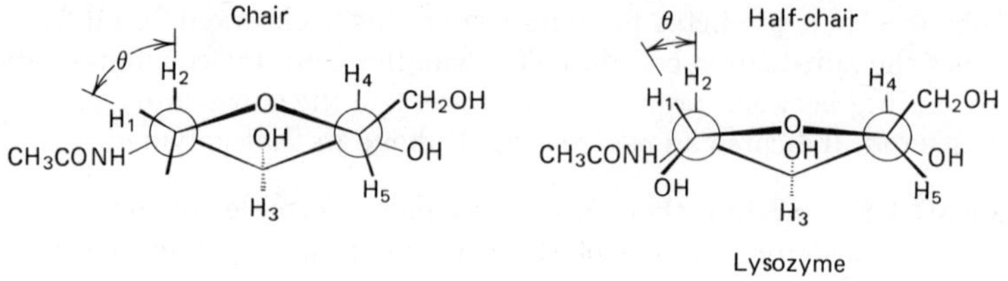

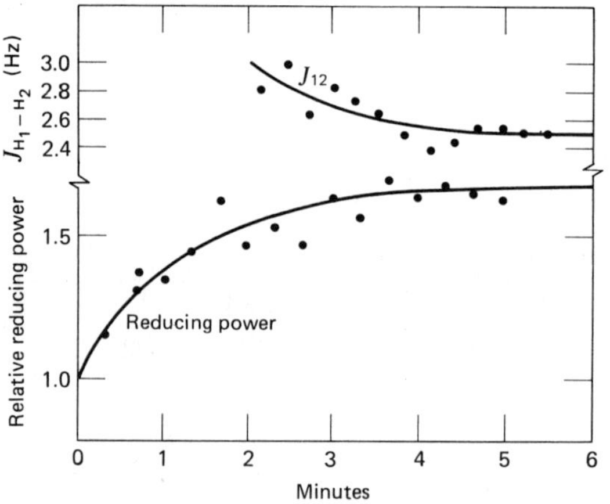

FIGURE 18-11. Distortion of substrate on binding to an enzyme. Undistorted (*top left*, "chair" conformation) and distorted (*top right*, "half-chair" conformation) of the 2-acetamido-2-deoxy-D-glucopyranose ring of the tetrameric (N-acetylglucosamine-N-acetylmuramic acid)$_2$, or (NAG-NAM)$_2$ substrate of the enzyme, lysozyme. The NMR J-coupling between protons 1 and 2 of the terminal reducing end of the substrate (*lower right*) and the reducing power of the mixture (*lower left*) of substrate and enzyme are monitored as a function of time. There is clearly an increase in J on binding, and extrapolation back to time zero (the instant of mixing of substrate and enzyme) confirms that the terminal sugar ring changes from a "chair" ($\theta \sim 60°$) to a half-chair conformation ($\theta \sim 0°$) on binding to lysozyme. This result shows that a major aspect of the lysozyme mechanism is to strain the substrate molecule to facilitate breaking of a sugar-sugar bond link. [From S. L. Patt, D. Dolphin, and B. D. Sykes, *Ann. New York Acad. Sci. 222,* 211 (1973).]

however, the "chair" conformation were distorted into a "half-chair" (Fig. 18-11), then the dihedral angle that determines J between protons 1 and 2 of the sugar ring would decrease, and J would increase. Sykes et al observed just that behavior for the binding of a tetrameric oligosaccharide, (NAG-NAM)$_2$ to the enzyme, lysozyme, which in nature breaks down (NAG-NAM)$_n$ polymers in cell walls. Because (NAG-NAM)$_2$ is a substrate and is quickly broken into smaller pieces that no longer bind as well to lysozyme (and which in any case are no longer distorted in shape on binding), the larger J value expected immediately on binding of substrate to enzyme quickly reverts back to the 3 Hz value as the reaction proceeds. It was thus necessary to monitor the coupling constant as a function of time, and then extrapolate back to the instant of mixing (time zero) to calculate the J value for the substrate-enzyme complex. The extrapolation was made possible by independently monitoring the course of the reaction by measuring the reducing power (a new reducing-end of sugar is exposed every time a reaction occurs) of the saccharides as a function of time after mixing the substrate and enzyme. This experiment is significant in two ways. First, it provides the most definitive evidence for a substrate conformational change on binding to enzyme. Second, the conformational change observed in solution by NMR turns out to be the same as that observed in the crystal by X-ray diffraction studies, so that it is highly likely that other mechanistic conclusions based on X-ray studies of lysozyme will be relevant to the actual solution behavior.

18.B. MOLECULAR ORBITAL THEORY AND DRUG ACTIVITY

It is presumed that the reader has already encountered the hydrogen-atom electronic wave functions ("atomic orbitals") whose corresponding electron densities and electron energies are shown in Fig. 18-12. Although these functions are derived for the two-particle electron-proton Coulomb potential, they furnish a good qualitative description of the behavior of the many electrons in larger atoms, provided that we agree to file at most two electrons in any one "atomic orbital." The reader may also have encountered (usually in an organic chemistry course) a "*hybrid* orbital" description, in which particular *combinations* of the atomic orbitals of a given atom ("sp^3," "sp^2," "sp") are formed from the hydrogenlike orbitals. In either case, we might expect that if *complex atoms* can be described by combinations of the atomic orbitals for a *hydrogen atom*, then *molecules* might be describable from combinations of the *atomic orbitals* of the constituent *atoms*. This hypothesis forms the basis for "*molecular orbital theory*," which in turn provides much of the theory of chemical bonding and molecular properties. Examples in this section are based on the particularly simplified "Huckel" molecular orbital theory; the corresponding calculations are relatively brief, and the qualitative agreement with experiment is surprisingly good.

Molecular orbital theory has proved to be of greatest value in describing molecules in which there is multiple bonding (double bonds, triple bonds) between carbon atoms, and particularly for "conjugated" molecules characterized by alternating single and double bonds: —C=C—C=C—C=, and so on. Double bonds in carbon atoms can be described in terms of two types of bonds, as shown in Fig. 18-13. The σ bond is formed by overlap of one p orbital from each of the two carbons, where the p-orbital main axis lies *along* the axis of the σ bond. A second type of bond, the π bond, is formed by overlap of one p orbital from each of the two carbons, where the p-orbital main axis is *perpendicular to* the axis of the π bond. In our simplified version of molecular orbital theory, we ignore all atomic orbitals except those involved in π-type bonds, in the same spirit that the properties of the periodic

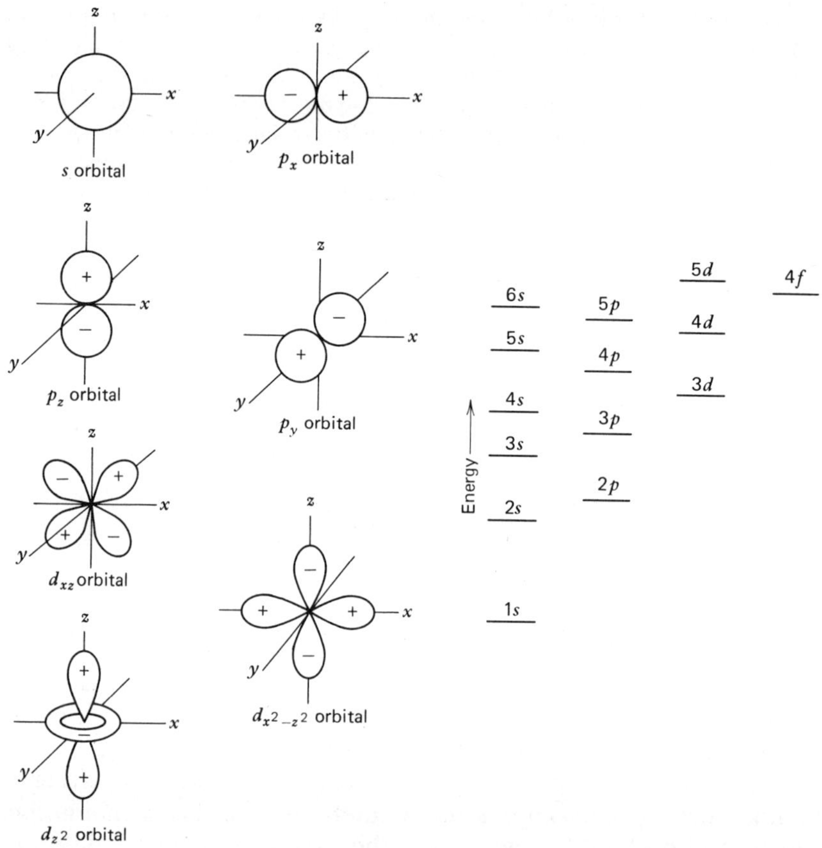

FIGURE 18-12. Wave-functions (left) and corresponding energy-levels (right) for the hydrogen-atom. The energy levels represent the order in which the hydrogen-like orbitals would be filled by successive addition of electrons in complex atoms. The orbital shapes are defined by a surface of constant electron density (just as a topographical outline is defined by a surface of constant altitude). (Orbitals drawn from L. B. Kier, *Molecular Orbital Theory in Drug Research,* Academic Press, New York, 1971, p. 27; energy levels from J. C. Davis, Jr., *Advanced Physical Chemistry,* Ronald Press, New York, 1965, p. 242.)

table are deduced by ignoring all but the "valence" orbitals in a hydrogen-like description of complex atoms. (Mathematically, we would say that σ- and π-molecular orbitals are "orthogonal.")

For a conjugated carbon chain, we expect that each carbon will contribute one p electron for π-type bonding, so that we may write a molecular wave function as the weighted sum ("linear combination") of one $2p$-atomic orbital from each of the carbons in the chain:

$$\psi = c_1\phi_1 + c_2\phi_2 + \cdots + c_n\phi_n \tag{18-82}$$

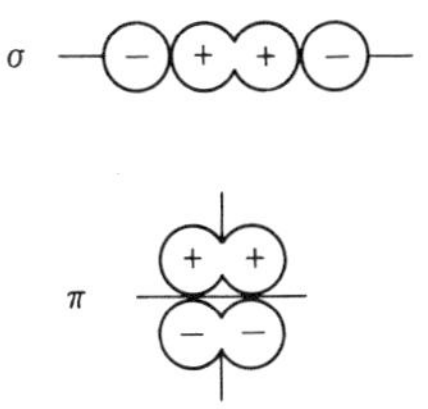

FIGURE 18-13. Molecular orbital description of a carbon-carbon double-bond. In the σ-bond, the bond is formed by overlap of two p-orbitals whose main axes are along the bond direction; in the π-bond, the overlap is between two p-orbitals whose main axes are perpendicular to the bond direction. Simplified molecular orbital theory is restricted to properties of π-bonds in multiply bonded carbon atoms.

where ϕ_i is the $2p$-atomic orbital wave function for the ith carbon, and ψ is the molecular orbital, where we usually "normalize" each of the possible ψ to unit "norm" (unit "length"):

$$|c_1|^2 + |c_2|^2 + \cdots + |c_n|^2 = 1 \qquad (18\text{-}83)$$

We are interested (as usual) to solve first of all for the energy levels of the system, which we obtain from the determinant obtained from the Hamiltonian, $\mathscr{H}$.

$$\mathscr{H}\psi = E\psi \qquad (18\text{-}28)$$

$$\det\begin{pmatrix} \mathscr{H}_{11} - \lambda & \mathscr{H}_{12} & \cdots & \mathscr{H}_{1n} \\ \mathscr{H}_{21} & \mathscr{H}_{22} - \lambda & \cdots & \mathscr{H}_{2n} \\ \vdots & & \ddots & \vdots \\ \mathscr{H}_{n1} & \cdots & & \mathscr{H}_{nn} - \lambda \end{pmatrix} = 0 \qquad (18\text{-}84)$$

Energy-Levels

As the reader may recall, the solution of an $(n \times n)$ determinant involves solution of an nth-degree polynomial. Because we are usually interested in molecules with more than $n = 2$ atoms (!), it would be desirable to seek some simplification in Eq. 18-84 before proceeding. Fortunately, an immediate major simplification follows from an examination of the physical meaning of the various $\mathscr{H}_{ij}$ matrix elements. For example, we have already argued (Eq. 18-23 ff.) that the average energy of a system in state, ψ, is $< \psi, \mathscr{H}\psi >$ (Eq. 18-23), so it is logical to interpret the matrix element, $\mathscr{H}_{ii}$,

$$\mathscr{H}_{ii} = < \phi_i, \mathscr{H}\phi_i > \qquad (18\text{-}85)$$

as the Coulomb energy that binds an electron to a carbon $2p$-atomic orbital of ith carbon in the molecule. Since the π electrons in a conjugated molecule come from identical carbon cores, it should seem reasonable that the Coulomb binding energy, α, should be approximately the same for each of the carbons in the conjugated chain:

$$\boxed{\mathscr{H}_{11} = \mathscr{H}_{22} = \cdot \cdot \cdot = \mathscr{H}_{nn} = \alpha} \tag{18-86}$$

By similar logic, the matrix element, $\mathscr{H}_{ij}$, may be identified as the interaction energy between a $2p$-atomic orbital on the ith carbon with a $2p$-atomic orbital on the jth carbon of the chain. Again, we might expect the interaction between, say, carbons 2 and 3 to be the same as the interaction between carbons 7 and 8, and the common value is denoted as β, and sometimes (confusingly) called the "resonance" energy:

$$\boxed{\mathscr{H}_{12} = \mathscr{H}_{23} = \cdot \cdot \cdot = \mathscr{H}_{(n-1)n} = \beta} \tag{18-87}$$

Finally, since a given atomic orbital has virtually no electron density extending more than one bond length away from the given atom, we can safely assume that the energy of interaction between two $2p$-electrons in the carbon chain is zero when the carbons are separated by more than one bond:

$$\boxed{\mathscr{H}_{ij} = 0 \text{ unless } i \text{ and } j \text{ differ by } \pm 1} \tag{18-88}$$

EXAMPLE *Ethylene, $H_2C{=}CH_2$*

For the two-carbon ethylene molecule, there is one double bond, and one $2p$-carbon electron from each of the two carbons available for π-bonding. The form of the molecular orbital(s), ψ, may thus be written

$$\psi = c_1\phi_1 + c_2\phi_2 \tag{18-89}$$

and the energy-levels obtained by solving the determinant, Eq. 18-90, using the simplifications of Equations 18-86 to 18-88:

$$\begin{vmatrix} \alpha - \lambda & \beta \\ \beta & \alpha - \lambda \end{vmatrix} = 0 \tag{18-90}$$

Although we could solve the determinant, Eq. 18-90, directly, it is useful to introduce a manipulation that will be valuable in succeeding (more complicated) cases: namely, to divide each term by β and set

$$X = \left(\frac{\alpha - \lambda}{\beta}\right) \tag{18-91}$$

so that Eq. 18-90 becomes

$$\begin{vmatrix} X & 1 \\ 1 & X \end{vmatrix} = 0 = X^2 - 1 \tag{18-92}$$

whose solution is

$$X = \pm 1 \qquad \begin{matrix} X_1 = -1 \\ X_2 = 1 \end{matrix} \tag{18-93}$$

or simply,

$$\lambda_1 = \alpha + \beta \tag{18-94a}$$

$$\lambda_2 = \alpha - \beta \tag{18-94b}$$

To find the corresponding molecular orbital (MO) wave function, ψ, we simply write down the eigenvalue equation for energy, using Equations 18-89 and 18-93, and solve for the coefficients, c_1 and c_2:

$$\mathscr{H}\psi_1 = \lambda_1\psi_1 \tag{18-95}$$

$$\begin{pmatrix} \alpha & \beta \\ \beta & \alpha \end{pmatrix}\begin{pmatrix} c_1 \\ c_2 \end{pmatrix} = (\alpha + \beta)\begin{pmatrix} c_1 \\ c_2 \end{pmatrix} \tag{18-96}$$

or

$$\begin{pmatrix} \alpha c_1 + \beta c_2 \\ \beta c_1 + \alpha c_2 \end{pmatrix} = \begin{pmatrix} (\alpha + \beta)c_1 \\ (\alpha + \beta)c_2 \end{pmatrix} \tag{18-96a}$$

Equation 18-96a says that two vectors are equal. But in either ordinary 3-space or in function-space, two vectors are equal if and only if their components along each basis vector (in this case ϕ_1 and ϕ_2 basis vectors) are equal:

$$\alpha c_1 + \beta c_2 = \alpha c_1 + \beta c_1 \tag{18-97a}$$

and

$$\beta c_1 + \alpha c_2 = \alpha c_2 + \beta c_2 \tag{18-97b}$$

whose solution is clearly

$$c_1 = c_2 \tag{18-98}$$

Similarly, from the other energy eigenvalue, $\mathscr{H}\psi = \lambda_2\psi$, the other molecular orbital can be shown to have components

$$c_1 = -c_2 \tag{18-99}$$

Finally, since ψ is to be normalized to unit "length," we must have

$$< \psi_1,\psi_1 > = (c_1^* \; c_1^*)\begin{pmatrix} c_1 \\ c_1 \end{pmatrix} = |c_1|^2 + |c_1|^2 = 1; \, c_1 = c_2 = \frac{1}{\sqrt{2}} \tag{18-100a}$$

and

$$<\psi_2,\psi_2> = (c_1^* \;\; -c_1^*)\begin{pmatrix} c_1 \\ -c_1 \end{pmatrix} = |c_1|^2 + |c_1|^2 = 1;\; c_1 = -c_2 = \frac{1}{\sqrt{2}} \qquad \text{(18-100b)}$$

From Eq. 18-100, we can now write the two molecular orbitals for ethylene as

$$\psi_1 = \frac{1}{\sqrt{2}}(\phi_1 + \phi_2) \qquad \text{(18-101a)}$$

and

$$\psi_2 = \frac{1}{\sqrt{2}}(\phi_1 - \phi_2) \qquad \text{(18-101b)}$$

The energy eigenvalues and eigenfunctions for the π-electrons of ethylene are shown in Fig. 18-14. Since α and β are *negative* numbers (relative to the energy of a π-electron that has been taken an infinite distance away from the molecule), the two electrons in the lowest occupied molecular orbital are "held" in the "bond" by a net attractive energy. The contour for the lower-energy *MO* shows that the electron density is mostly localized in the space between the two carbons, again consistent with our usual concept of a chemical bond. For the higher-energy *MO,* on the other hand, the electron density for the corresponding *MO* (if there were an electron actually in that energy-level) shows that an electron in this "anti-bonding" molecular orbital would be localized in regions outside the space between the carbons, and would thus provide no basis for a chemical bond. (Electrons are placed pairwise in *molecular* orbitals, just as in the more familiar hydrogenlike *atomic* orbitals.)

This example shows how the molecular orbitals, from which we can obtain electron density around each of the carbon atoms, produce a bonding energy and also predict that the π electron will most often be found in the region between the two bonded carbon atoms, in accordance with intuitive expectations. From here on, we will justify *MO* calculations, *not* on the basis that they correctly predict electronic energies or electron densities in actual molecules (although that is sometimes the case), but on the basis that certain *MO*-calculated properties can be *correlated* highly with interesting *biological functions* of the molecules involved. With a few more examples of how the calculations are carried out, the reader will be in a position to conduct a wide

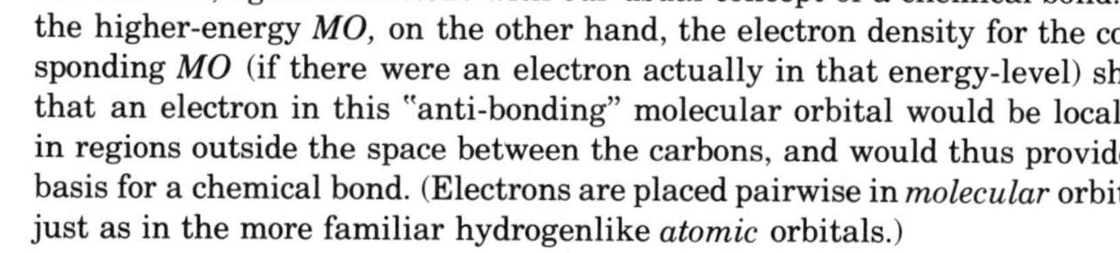

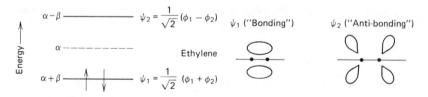

FIGURE 18-14. Schematic diagram of energy levels (left) and electron-density contours (right) for the two π-electron molecular orbitals of ethylene. The arrows in the energy-level picture represent two electrons (of opposite spin) occupying the lowest energy MO. See text for discussion.

range of *MO* calculations him(her)self, and some representative correlations of *MO* properties with drug action are presented.

EXAMPLE *Butadiene, $H_2C{=}CH{-}CH{=}CH_2$*

Proceeding in exactly the same way as for ethylene, we express the molecular orbital(s) as *l*inear *c*ombinations of *a*tomic *o*rbitals ("LCAO," in the "*MO*" jargon):

$$\psi = c_1\phi_1 + c_2\phi_2 + c_3\phi_3 + c_4\phi_4 \tag{18-102}$$

and immediately write the determinant from which energies are computed, using Equations 18-86 to 18-88, as

$$\begin{vmatrix} \alpha-\lambda & \beta & 0 & 0 \\ \beta & \alpha-\lambda & \beta & 0 \\ 0 & \beta & \alpha-\lambda & \beta \\ 0 & 0 & \beta & \alpha-\lambda \end{vmatrix} = 0 \tag{18-103}$$

which under the definition, $X = \dfrac{\alpha-\lambda}{\beta}$ (Eq. 18-91), and dividing each term by β as before, yields

$$\begin{vmatrix} X & 1 & 0 & 0 \\ 1 & X & 0 & 0 \\ 0 & 1 & X & 1 \\ 0 & 0 & 1 & X \end{vmatrix} = 0 \tag{18-104}$$

Equation 18-104 is solved by expanding the determinant according to the co-factors of, for example, the top row (see any reference book on determinants or linear algebra):

$$\begin{vmatrix} X & 1 & 0 & 0 \\ 1 & X & 1 & 0 \\ 0 & 1 & X & 1 \\ 0 & 0 & 1 & X \end{vmatrix} = X\begin{vmatrix} X & 1 & 0 \\ 1 & X & 1 \\ 0 & 1 & X \end{vmatrix} - 1\begin{vmatrix} 1 & 1 & 0 \\ 0 & X & 1 \\ 0 & 1 & X \end{vmatrix} + 0\begin{vmatrix} 1 & X & 0 \\ 0 & 1 & 1 \\ 0 & 0 & X \end{vmatrix} - 0\begin{vmatrix} 1 & X & 1 \\ 0 & 1 & X \\ 0 & 0 & 1 \end{vmatrix}$$

$$= X\begin{vmatrix} X & 1 & 0 \\ 1 & X & 1 \\ 0 & 1 & X \end{vmatrix} - \begin{vmatrix} 1 & 1 & 0 \\ 0 & X & 1 \\ 0 & 1 & X \end{vmatrix} = X^2\begin{vmatrix} X & 1 \\ 1 & X \end{vmatrix} - X\begin{vmatrix} 1 & 1 \\ 0 & X \end{vmatrix} - 1\begin{vmatrix} X & 1 \\ 1 & X \end{vmatrix}$$

$$= X^2(X^2-1) - X(X-0) - (X^2-1)$$

$$= X^4 - 3X^2 + 1 = 0; \quad X = \pm 1.62, \pm 0.62 \tag{18-105}$$

$$\lambda = \alpha \pm 1.62\beta, \alpha \pm 0.62\beta$$

As with ethylene, the butadiene *MO* wave functions, ψ, are found by applying the Hamiltonian operator to the *MO* of energy, λ_1, then to the *MO* of energy, λ_2, and so on. For example,

$$\mathcal{H}\psi_1 = \lambda_1\psi_1 \tag{18-106}$$

$$\begin{pmatrix} \alpha & \beta & 0 & 0 \\ \beta & \alpha & \beta & 0 \\ 0 & \beta & \alpha & \beta \\ 0 & 0 & \beta & \alpha \end{pmatrix}\begin{pmatrix} c_1 \\ c_2 \\ c_3 \\ c_4 \end{pmatrix} = (\alpha + 1.62\beta)\begin{pmatrix} c_1 \\ c_2 \\ c_3 \\ c_4 \end{pmatrix} \tag{18-107}$$

$$\left.\begin{aligned} \alpha c_1 + \beta c_2 &= (\alpha + 1.62\beta)c_1 \\ \beta c_1 + \alpha c_2 + \beta c_3 &= (\alpha + 1.62\beta)c_2 \\ \beta c_2 + \alpha c_3 + \beta c_4 &= (\alpha + 1.62\beta)c_3 \\ \beta c_3 + \alpha c_4 &= (\alpha + 1.62\beta)c_4 \\ \text{and} \qquad & \\ |c_1|^2 + |c_2|^2 + |c_3|^2 + |c_4|^2 &= 1 \end{aligned}\right\} \tag{18-108}$$

It is left as an exercise to compute the four unknowns, c_1, c_2, c_3, and c_4 from Eqs. 18-108. The remaining three molecular orbitals are obtained similarly, with the results shown in Fig. 18-15.

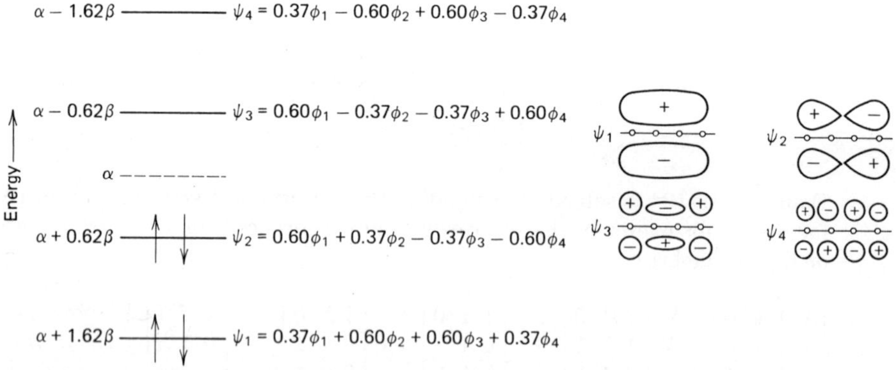

FIGURE 18-15. Schematic diagram of energy levels (left), energy eigenfunctions (middle), and electron-density contours (right) for the four π-electron molecular orbitals of butadiene. The arrows represent occupation of the two lowest MO's by the four π-electrons.

EXAMPLE *Benzene,* 

It is left to the reader (see Problems) to show that the determinant from which the energy-eigenvalues for the π-electrons of benzene are obtained has the form

$$\begin{vmatrix} X & 1 & 0 & 0 & 0 & 1 \\ 1 & X & 1 & 0 & 0 & 0 \\ 0 & 1 & X & 1 & 0 & 0 \\ 0 & 0 & 1 & X & 1 & 0 \\ 0 & 0 & 0 & 1 & X & 1 \\ 1 & 0 & 0 & 0 & 1 & X \end{vmatrix} = 0 = X^6 - 6X^4 + 9X^2 - 4 \qquad (18\text{-}109)$$

whose solutions are: $X = 2,1,1,-1,-1,-2$, resulting in the energy-levels and molecular orbitals shown in Fig. 18-16. (18-110)

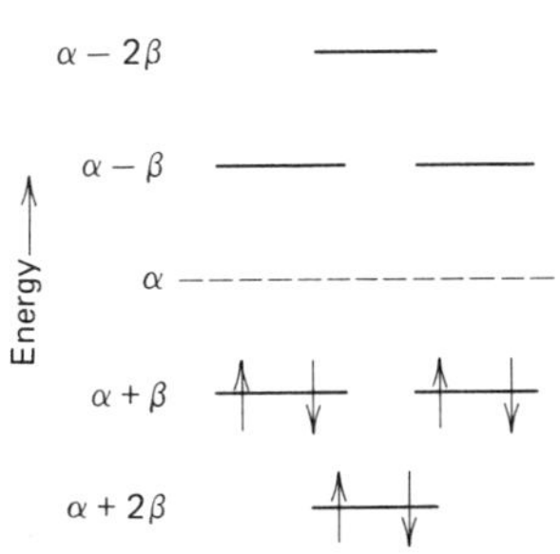

$$\psi_6 = \left(\frac{1}{\sqrt{6}}\right)(\phi_1 - \phi_2 + \phi_3 - \phi_4 + \phi_5 - \phi_6)$$

$$\psi_5 = (1/\sqrt{12})(2\phi_1 - \phi_2 - \phi_3 + 2\phi_4 - \phi_5 - \phi_6)$$

$$\psi_4 = \left(\frac{1}{2}\right)(\phi_2 - \phi_3 + \phi_5 - \phi_6)$$

(18-111)

$$\psi_3 = \left(\frac{1}{2}\right)(\phi_2 + \phi_3 - \phi_5 - \phi_6)$$

$$\psi_2 = (1/\sqrt{12})(2\phi_1 + \phi_2 - \phi_3 - 2\phi_4 - \phi_5 + \phi_6)$$

$$\psi_1 = \left(\frac{1}{\sqrt{6}}\right)(\phi_1 + \phi_2 + \phi_3 + \phi_4 + \phi_5 + \phi_6)$$

FIGURE 18-16. Schematic diagram of energy levels (left) and energy eigenfunctions (right) for the six π-electron molecular orbitals in benzene. The arrows denote occupation of the three lowest MO's by the six π-electrons. Note that both of the molecular orbitals ψ_2 and ψ_3 have the same energy (also that ψ_4 and ψ_5 have the same energy); in this situation, one says that the energy levels for ψ_2 and ψ_3 are "degenerate." There are still only two electrons in each molecular orbital, even though two of the MO's happen to have the same energy.

EXAMPLE *Pyridine,*

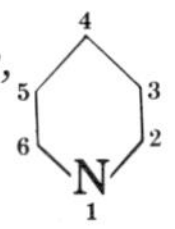

When atoms other than carbon contribute π electrons to the molecular bonds, we can no longer claim that α (or β) is the same for each of the atoms involved. A correction whose form preserves the simplicity of the *MO* analysis we are following is to suppose that the values of α and β for any *new* atom may be obtained from the values for a *carbon* atom according to the linear relations

$$\alpha_{\text{(hetero atom)}} = \alpha_{\text{(carbon)}} + h_{\text{(hetero atom)}}\,\beta_{\text{(carbon-carbon)}} \qquad (18\text{-}112)$$

and

$$\beta_{\text{(carbon-hetero)}} = k_{\text{(carbon-hetero)}}\,\beta_{\text{(carbon-carbon)}} \qquad (18\text{-}113)$$

Relatively self-consistent values of h and k have been determined empirically for a variety of hetero-atoms, as shown from the h and k values of Table 18-4. The values in Table 18-4 are not unique: other self-consistent sets (most notably that of Pullman in B. Pullman, *Quantum Biochemistry,* Wiley-Inter-

Table 18-4 *MO* Parameters for a Simplified Description of Hetero (i.e., Noncarbon) Atoms (see Equations 18-112 and 18-113).*

h	*k*
$\dot{N} = 0.5$	$C—\dot{N} = 1.0$
$\ddot{N} = 1.5$	$C—\ddot{N} = 0.8$
$\dot{O} = 1.0$	$C—\dot{O} = 1.0$
$\ddot{O} = 2.0$	$C—\ddot{O} = 0.8$
$\dot{N}^{\oplus} = 2.0$	$C—\dot{N}^{\oplus} = 1.0$
$\ddot{C}H_3 = 2.0^c$	$C—\ddot{C}H_3 = 0.7^c$

* [*From A. Streitweiser,* Molecular Orbital Theory for Organic Chemists, *Wiley, N. Y., 1961, p. 135.*]

science, N.Y. 1963) are sometimes more useful for certain series of homologous compounds.

Using the $\dot{N}$ values for h and k in Table 18-4, the required determinant for calculation of energy eigenvalues of pyridine may be written immediately in the form

$$\begin{vmatrix} X+0.5 & 1 & 0 & 0 & 0 & 1 \\ 1 & X & 1 & 0 & 0 & 0 \\ 0 & 1 & X & 1 & 0 & 0 \\ 0 & 0 & 1 & X & 1 & 0 \\ 0 & 0 & 0 & 1 & X & 1 \\ 1 & 0 & 0 & 0 & 1 & X \end{vmatrix} = 0 \qquad (18\text{-}114)$$

with the resultant energy-levels and wave functions for occupied levels given in Fig. 18-17.

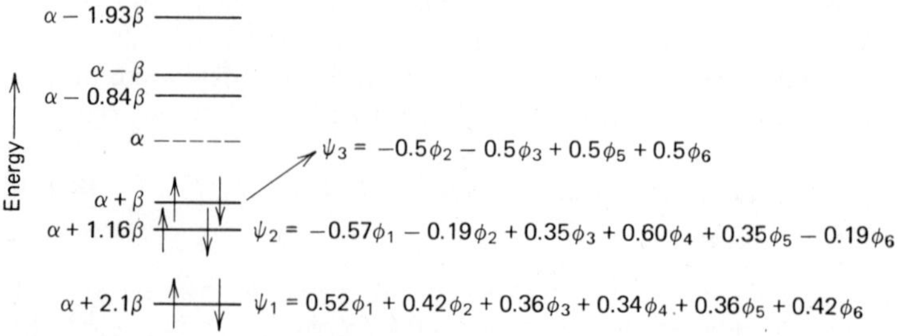

(18-115)

FIGURE 18-17. Schematic diagram of energy-levels (left) and energy eigenfunctions of the occupied molecular orbitals (right) for the six π-electrons of pyridine. Note that the energy-level degeneracy observed for benzene is removed by the introduction of a hetero-atom in the ring.

Application: Correlation of Hallucinogenic Activity with Molecular Orbital Energy

The basis for binding of one molecule to another is sometimes attributed to a process known as "charge transfer," in which it is visualized that a portion of the electron originally located on the "donor" molecule is now associated with the other "acceptor" molecule: $D + A \rightleftharpoons DA$. (Charge-transfer interactions provide the basis for xerography, for example.) In general, one might then expect that an ideal *donor* molecule might have a relatively high-lying "highest-occupied molecular orbital" (HOMO), since relatively little energy would then be required to remove part or all of that electron for transfer to the acceptor molecule. Of the many examples in which donor properties with respect to charge-transfer have been correlated with HOMO energy-levels calculated from MO theory, perhaps the most biologically interesting is that shown in Table 18-5. The table shows a marked correlation between HOMO energy for the π electrons of the drug and its hallucinogenic potency, as determined by the minimum effective dose (see Chapter 12B) relative to mescaline.

Table 18-5 Highest Occupied Molecular Orbital (HOMO) Energy Versus Hallucinogenic Potency (Ratio of Effective Dose of Mescaline to Effective Dose of Stated Drug) for Several Known Hallucinogenic Drugs. (Structures of TMA and TMA-2 are listed below.)*

Compound	Biological Activity	E_{HOMO}
LSD	3700	0.2180
Psilocin	31	0.4603
6-Hydroxydiethyltryptamine	25	0.4700
TMA-2	17	0.4810
TMA	2.2	0.5357
Mescaline	1	0.5357

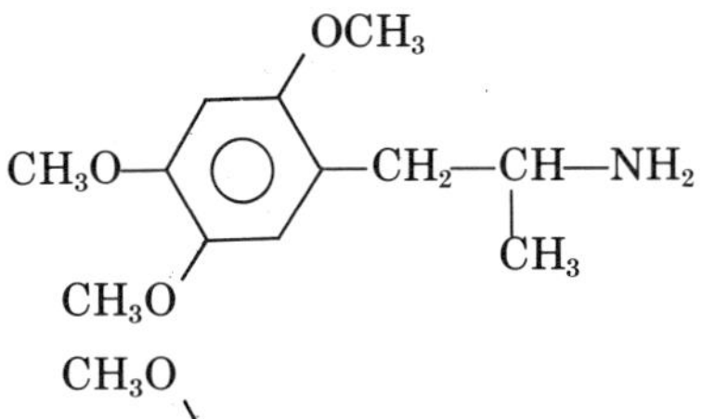

TMA-2

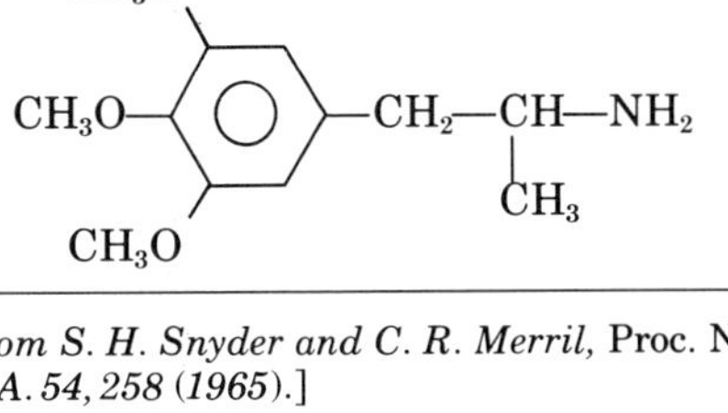

TMA

*[*From S. H. Snyder and C. R. Merril,* Proc. Natl. Acad. Sci. *U.S.A. 54, 258 (1965).*]

In spite of the striking correlation between the HOMO molecular orbital parameter and drug activity in this case, it is important to note that *correlations* do *not* prove or even necessarily indicate *cause and effect.* For example, there are hundreds of molecules with HOMO energy as small or smaller than that of LSD, but which exhibit no hallucinogenic potential whatever. In other words, MO theory can provide correlations that may suggest an aspect of a possible mechanism for a process, but the theory has not been especially useful in predicting which *new* drugs will be more or less effective than known ones.

Electron Density on Various Atoms from Molecular Orbital Theory

We have already calculated the *energy levels* and corresponding *energy* (*MO*) *eigenfunctions* for π electrons in several conjugated organic molecules. From the expression of each *MO* for a given molecule as a linear combination of atomic orbitals (LCAO), we can immediately determine the probability of finding a π electron localized on any single atom, from the square of the coefficient that describes the projection of the *MO* along that particular atomic orbital. For example, in butadiene, there are two π electrons in each of the occupied molecular orbitals. The electron density at atom 1, for example is thus given by:

$$\text{Electron density at atom } 1 = q_1 = 2(0.37)^2 + 2(0.60)^2 = 1 \qquad \text{(18-116a)}$$

in which $(0.37)^2$ is recognized as the probability of finding a π electron in *MO* ψ_1 localized on atom #1, and $(0.60)^2$ is the probability of finding a π electron in *MO* ψ_2 localized on atom #1. Similarly, the reader should verify that

$$q_4 = 2(0.37)^2 + 2(-0.60)^2 = 1 \qquad \text{(18-116b)}$$

and also that

$$q_2 = q_3 = 1 \qquad \text{(18-116c)}$$

A more interesting example is provided by pyridine (Fig. 18-17), for which the reader should verify from the *MO*'s in Fig. 18-17 that the electron densities on the various ring atoms are as shown below:

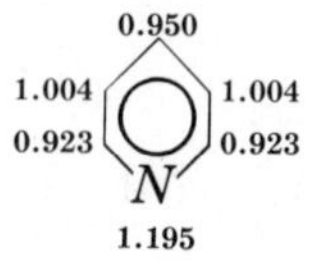

(18-117)

Armed with the parameters of Table 18-4, the reader could in principle compute the energy levels, energy eigenfunctions (*MO*'s), and then the electron densities on each atom for a very extensive number of conjugated organic molecules. Again, we concentrate on correlation of electron density with biological function, rather than on comparison of calculated and experimental electron densities on actual molecules.

EXAMPLE *Electron Density and Carcinogenicity for Aromatic Polycyclic Hydrocarbons*

It might be hoped that some clues as to the causative agent or triggering mechanism for cancer might be found by locating the critical features of molecules with known (often very high) carcinogenic (cancer-forming) potential. Among the anthracene derivatives shown in Table 18-6, there is a strong correlation between the electron density (computed by *MO* theory) in the "*K*-region" of the aromatic ring and carcinogenic activity. More elaborate correlations involving more compounds suggest that the carcinogenic activity seems to be removed somewhat by high electron density in a second "L-region." These results would appear to suggest that carcinogenic activity is correlated with covalent bond formation in the "K-region," while covalent bond formation in the "*L*-region" may be an alternate noncarcinogenic reaction. This area is under active and continuing investigation.

1,2-Benzanthracene

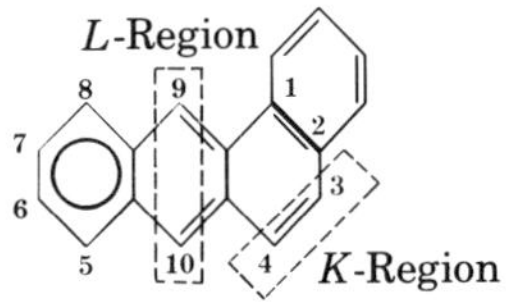

Table 18-6 Relationship between the Total Electron Density in the *K*-Region and Carcinogenic Activity of Some Anthracene Derivatives. Electron density was calculated by *MO* theory, in much the same way as the pyridine result of Eq. 18-117.*

Compound	Total Charge	Carcinogenic Activity
Anthracene 	1.259	−
1,2-Benzanthracene	1.283	+
5-Methyl-1,2-benzanthracene	1.296	++
10-Methyl-1,2-benzanthracene	1.306	+++
5,9,10-Trimethyl-1,2-benzanthracene	1.332	++++

* [*From N. J. Doorenbos, "Physical Properties and Biological Activity," in A. Burger, ed.,* Medicinal Chemistry, *2nd ed., Interscience, N. Y. 1960, pp. 46–71.*

18.C. BIOLOGICAL IRON: MOSSBAUER SPECTROSCOPY

In the preceding sections, we have examined the energy levels that result from the force binding electrons to molecules and the force acting on a magnetic moment in a static magnetic field. Since atomic nuclei are themselves composed of a collection of protons and neutrons, one might expect that the forces between those nucleons might also lead to a set of energy levels between which we might expect to induce transitions by irradiating the nucleus at a suitable frequency. Although such *nuclear* energy levels indeed exist, and while absorption and emission of radiation at frequencies corresponding to the nuclear energy-level differences are indeed observed, present theoretical calculations do not satisfactorily account for the positions of the energy levels of nuclei. The problem is that since all the nucleons are essentially equally massive, it is *not* a good approximation to treat a particular nucleon as if its behavior were governed by interaction with an average force field due to all the other nucleons. [The reason that such an approximation succeeds so well for behavior of *electrons* in atoms is that it *is* a good approximation to treat the nucleus as much more massive than the electron(s), so that the force acting on the electron can be regarded as arising from a "central force field" (i.e., from a nuclear charge located as a point source at the origin), leading to the hydrogen atom quantum mechanical treatment found in most elementary chemistry texts]. Mossbauer spectroscopy simply describes an experiment in which the absorption of electromagnetic radiation in the "gamma-ray" region (ca. 10 keV) by a suitable nucleus is monitored as a function of the frequency of the incident radiation. Absorption is observed at frequencies corresponding to differences in the nuclear energy levels of the target nucleus; since those levels reflect the environment (electronic oxidation state, symmetry of surroundings, nature of ligands) very close to the nucleus, Mossbauer methods provide a very specific probe of just one type of atom (usually ^{57}Fe) in a complex molecule.

Although theory does not really predict *absolute* nuclear energy levels with good generality, it is possible to understand (at least qualitatively) small *relative* nuclear energy-level differences for a given nucleus in various different environments. For example, if the nuclear charge distribution is nonspherical, and if there is an electric field gradient at the nucleus (as from a chemical bond, for example), then there will be a different energy for the nucleus for different relative orientations of the nucleus with respect to its surroundings (see Fig. 18-18).

Just as the energy of interaction of a charge q with an electric field E is of the form, $q \cdot E \cdot x$, where x is the distance at which q is located, the energy of interaction of a nonspherical charge distribution with an electric field gradient, dE/dz, can be written in the form $q \cdot x \cdot z \cdot (dE/dz)$. By matrix techniques beyond the scope of this text, the energy eigenvalues (energy levels) resulting from this potential energy term in the nuclear Hamiltonian can

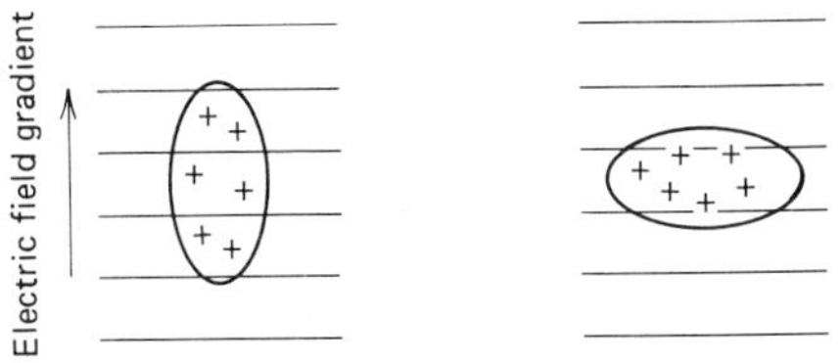

FIGURE 18-18. Schematic diagram of two relative orientations of a nonspherical nucleus in an electric field gradient. The energy of interaction between the nuclear charge distribution and the electric field gradient is different for the two orientations.

be shown to give a "quadrupole splitting" illustrated for the (most common Mossbauer) case of ^{57}Fe in Fig. 18-19. For, say, an iron atom in a spherically symmetric environment, the electric field gradient at the iron nucleus is zero and there is no quadrupole splitting; for more distorted electronic distributions, the splitting will be finite. Thus the quadrupole *splitting* reflects the *symmetry* of the distribution of electrons around the Mossbauer nucleus. In addition, since nuclei interact with electrons on the same atom, there will be small changes in nuclear energy ("isomer shift," in Mossbauer spectroscopists' terminology) according to changes in the number and distribution of electrons about that nucleus. Thus, the *position* (energy) of a Mossbauer absorption peak gives information about *bonding* and *oxidation state* for that atom.

A final minor aspect of Mossbauer spectra concerns the typical reported frequency scale of the spectrum. In the most familiar types of spectroscopy (ultraviolet, visible, and infrared absorption), the radiation source is simply a heated "black-body," which emits radiation over a relatively broad range.

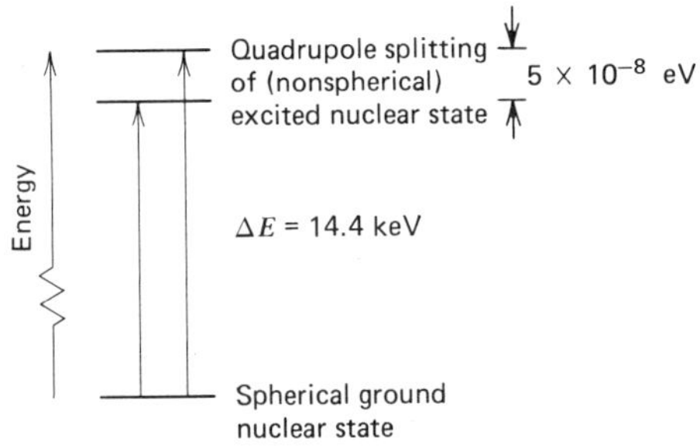

FIGURE 18-19. Schematic diagram of the nuclear energy levels for ^{57}Fe. Because of the small (ca. 5×10^{-8} eV) quadrupole splitting of the excited nuclear state, radiation can be absorbed by ^{57}Fe nuclei at two distinct energies that are both very close to 14.4 keV. (The "natural" linewidth for ^{57}Fe absorption is only about 5×10^{-9} eV, so the quadrupole splitting is about ten linewidths in the spectrum, leading to two well-resolved absorption peaks—see Examples and discussion of natural linewidth in Chapters 19.B and 20.)

For spectroscopy of nuclear energy levels in the Mossbauer experiment, the spectral radiation source is a sample of excited (say, ^{57}Fe) nuclei, resulting from energy decay from a conveniently long-lived higher-excited species (^{57}Co, with half-life of 270 days, in the case of ^{57}Fe Mossbauer spectroscopy). This radiation source is extremely monochromatic, and in order to vary the frequency of the source, the source is physically moved either toward or away from the absorber, so that the radiation frequency "seen" by the absorber is Doppler-shifted to either higher or lower frequency, respectively (just as a train whistle changes from higher to lower pitch according to whether the train is approaching or moving away from the hearer). Since Mossbauer absorption signals range over such a relatively small frequency span, it is possible to cover the necessary spectral range by moving the source toward or away from the absorber at a rate of a few mm/sec. Thus (see Examples), Mossbauer absorption peak positions and splittings are usually reported in units of mm/sec, where a velocity of 1 mm/sec corresponds to an energy shift of about 5×10^{-8} eV.

EXAMPLE *Mossbauer Spectra of Fe-containing Proteins*

Because the Mossbauer effect is a relatively recent discovery (1958), it is not easy to find examples of definitive conclusions about oxidation states or molecular symmetry that were *first* determined by Mossbauer methods. However, because the Mossbauer energy shifts and splittings reflect the environment about just *one* atom (usually ^{57}Fe), the technique offers unusually *specific* information about, say, Fe in biological macromolecules, compared to, for example, optical absorption spectra, which reflect properties of a complex or molecule as a whole.

One definitive recent Mossbauer example is shown in Fig. 18-20, illustrating the Mossbauer gamma-ray absorption spectra for the non-heme iron protein, hemerythrin, which is the protein responsible for oxygen transport in certain marine worms.

Although the deoxyhemerythrin (deoxy-Hr) shows just a single isomer shift appropriate (by comparison to model compounds) to iron in a +2 oxidation state, the oxyhemerythrin (oxy-Hr) exhibits two distinct quadrupole splittings, showing that the two Fe atoms in Oxy-Hr must be nonidentical. The isomer shift for oxy-Hr is substantially different from that of deoxy-Hr, and is interpreted as corresponding to Fe in the +3 oxidation state for both Fe atoms in oxy-Hr. It is of course *possible* that the two Fe atoms in deoxy-Hr might also have different environment but happen to have the same isomer shift and quadrupole splitting; in any case, one can *definitely* conclude that there are two types of Fe in oxy-Hr.

In general, since ^{57}Fe is only about 2.2% abundant, it is necessary to have relatively large amounts of sample for Mossbauer measurements (ca. 100 mg of Fe-containing protein). Also, the sample must be either crystalline or extremely viscous; otherwise the line width becomes large compared to the small energy shifts that one wants to observe [this effect is associated with the re-

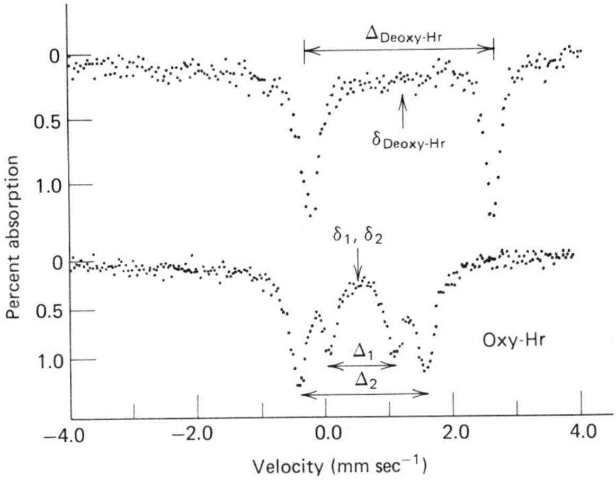

FIGURE 18-20. Mossbauer absorption spectra of oxyhemerythrin (oxy-Hr, bottom) and deoxy-hemerythrin (deoxy-Hr, top). Isomer shifts (δ) and quadrupole splittings (Δ) are shown. For oxy-Hr, two quadrupole splittings are clearly evident, indicating that the two iron atoms in oxy-Hr are nonidentical (see text). Both Fe atoms in deoxy-Hr exhibit the same δ and Δ consistent with a similar environment for both Fe atoms in deoxy-Hr. Isomer shifts suggest that both Fe atoms in deoxy-Hr have +2 oxidation state and both Fe atoms in oxy-Hr have +3 oxidation state. [From K. Garbett, C. E. Johnson, I. M. Klotz, M. Y. Okamura, and R. J. P. Williams, *Arch. Biochem. Biophys. 142,* 574 (1971).]

coil momentum associated with absorption of a gamma-ray by the nucleus—in a solid, the recoil is taken up by the crystal lattice as a whole, rather than by just one atom or molecule, and the recoil energy (which broadens the frequency range over which absorption occurs) becomes vanishingly small so that the observed absorption line is very narrow.] Future applications for the technique will require improved theoretical description of the behavior of electrons in metal complexes.

PROBLEMS

1. (a) Show that the following three basis vectors form an orthonormal set (i.e., that they are mutually "perpendicular" and each have unit "length" (see Table 18-2 and Eqs. 18-3 to 18-5a):

$$\begin{pmatrix}1\\0\\0\end{pmatrix}, \begin{pmatrix}0\\1\\0\end{pmatrix} \text{and} \begin{pmatrix}0\\0\\1\end{pmatrix}$$

(b) Show that the following three basis vectors form a different orthonormal set. This example corresponds to the construction of p_x, p_z, and p_y "orbitals" from combinations of the original p_{+1}, p_0, and p_{-1} functions obtained by solving the angular part of the hydrogen atom Hamiltonian (see any elementary chemistry textbook).

$$\begin{pmatrix} 1/\sqrt{2} \\ 0 \\ 1/\sqrt{2} \end{pmatrix}, \begin{pmatrix} 0 \\ 1 \\ 0 \end{pmatrix}, \begin{pmatrix} -1/\sqrt{2} \\ 0 \\ 1/\sqrt{2} \end{pmatrix}$$

2. From the definition of the components, L_x, L_y, and L_z of the quantum mechanical angular momentum operator L obtained by correspondence from classical mechanics (see Table 18-3), verify the "commutation" relations:

$$[L_x, L_y] = L_x L_y - L_y L_x = i\hbar\, L_z$$
$$[L_y, L_z] = L_y L_z - L_z L_y = i\hbar\, L_x$$

and

$$[L_z, L_x] = L_z L_x - L_x L_z = i\hbar\, L_y$$

3. For a particle of spin one-half (see Fig. 18-4), there are two possible eigenvalues ($\pm\hbar/2$) of the operator for the z component of "spin" angular momentum, I_z. Thus, if the I_z-eigenfunctions corresponding to those eigenvalues are denoted as unit column vectors, $\begin{pmatrix} 1 \\ 0 \end{pmatrix}$ and $\begin{pmatrix} 0 \\ 1 \end{pmatrix}$, then the matrix representation of the operator I_z must be of the form

$$I_z = \begin{pmatrix} \frac{\hbar}{2} & 0 \\ 0 & \frac{\hbar}{2} \end{pmatrix}$$

so that

$$I_z\alpha = \frac{\hbar}{2}\begin{pmatrix} 1 & 0 \\ 0 & -1 \end{pmatrix}\begin{pmatrix} 1 \\ 0 \end{pmatrix} = \frac{\hbar}{2}\begin{pmatrix} 1 \\ 0 \end{pmatrix} = \frac{\hbar}{2}\alpha$$

and

$$I_z\beta = \frac{\hbar}{2}\begin{pmatrix} 1 & 0 \\ 0 & -1 \end{pmatrix}\begin{pmatrix} 0 \\ 1 \end{pmatrix} = -\frac{\hbar}{2}\begin{pmatrix} 0 \\ 1 \end{pmatrix} = -\frac{\hbar}{2}\beta$$

From the properties of the "raising" and "lowering" operators, I_+ and I_-, namely,

$$I_+\alpha = 0$$
$$I_+\beta = \alpha$$
$$I_-\alpha = \beta$$
$$I_-\beta = 0$$

we can immediately construct the matrix representation for I_+ and I_- (see Eq. 18-13):

$$I_+ = \begin{pmatrix} 0 & 1 \\ 0 & 0 \end{pmatrix} \qquad I_- = \begin{pmatrix} 0 & 0 \\ 1 & 0 \end{pmatrix}$$

from which you should next construct the matrices for operators I_x and I_y:

$$I_x = \frac{\hbar}{2}(I_+ + I_-) = \frac{\hbar}{2}\begin{pmatrix} 0 & 1 \\ 1 & 0 \end{pmatrix}; \qquad I_y = \frac{\hbar}{2i}(I_+ - I_-) = \frac{\hbar}{2}\begin{pmatrix} 0 & -i \\ i & 0 \end{pmatrix}$$

(a) Using the rule for matrix multiplication (Eq. 18-11), show that the matrices representing I_x, I_y, and I_z for a spin one-half particle satisfy the established commutation relations for components of angular momentum:

$$[I_x, I_y] = i\hbar I_z; \quad [I_y, I_z] = i\hbar I_x; \quad [I_z, I_x] = i\hbar I_y$$

(b) Similarly, it may be shown that the matrices representing the operators I_x, I_y, and I_z for a particle of "spin" = 1 have the form:

$$I_z = \hbar\begin{pmatrix} 1 & 0 & 0 \\ 0 & 0 & 0 \\ 0 & 0 & -1 \end{pmatrix}; \quad I_x = \frac{\hbar}{\sqrt{2}}\begin{pmatrix} 0 & 1 & 0 \\ 1 & 0 & 1 \\ 0 & 1 & 0 \end{pmatrix}; \quad I_y = \frac{i\,\hbar}{\sqrt{2}}\begin{pmatrix} 0 & -1 & 0 \\ 1 & 0 & -1 \\ 0 & 1 & 0 \end{pmatrix}$$

Show that these three matrices satisfy the commutation rules for operator components of vector angular momentum.

4. (a) Using the usual relationship between Cartesian and angular coordinates, $x = r\sin(\theta)\cos(\phi)$, $y = r\sin(\theta)\sin(\phi)$, and $z = r\cos(\theta)$, show that the operator, L_z, obtained in Cartesian form from classical mechanics (see Table 18-3) can be expressed in angular coordinates much more simply as:

$$L_z = -i\hbar(x(\partial/\partial y) - y(\partial/\partial x)) = -i\hbar(\partial/\partial\phi)$$

(b) Now solve the eigenvalue equation for L_z in angular coordinates

$$L_z\,\Phi = n\,\hbar\,\Phi$$

where n takes on integral or half-integral values, to obtain an expression for the eigenfunction, Φ. Now show that if the eigenfunction, Φ, is to be single-valued (i.e., $\Phi(0) = \Phi(2\pi)$), then the eigenvalues of L_z are restricted to *integral* multiples of $\hbar$. (Note that for "spin" angular momentum, the I_z-eigenvalues can be either integral or half-integral multiples of $\hbar$.)

5. (a) Since most organic molecules of interest contain more than two types of protons within four chemical bonds of each other, it is desirable to extend the two-spin treatment of Chapter 18.A to cases involving three types of coupled magnetic nuclei, each of spin one-half. Beginning from the basic "first-order" Hamiltonian,

$$\mathscr{H} = \gamma_A H_0 I_{z_A} + \gamma_B H_0 I_{z_B} + \gamma_C H_0 I_{z_C} + \frac{J_{AB}}{\hbar} I_{z_A} I_{z_B} + \frac{J_{BC}}{\hbar} I_{z_B} I_{z_C} + \frac{J_{AC}}{\hbar} I_{z_A} I_{z_C}; \quad J \text{ in sec}^{-1}$$

and the basis set of products of eigenfunctions of I_{z_A}, I_{z_B}, and I_{z_C} shown below for the three nuclei (A, B, and C)

$$\begin{aligned} \psi_1 &= \alpha_A \alpha_B \alpha_C & \psi_5 &= \beta_A \alpha_B \beta_C \\ \psi_2 &= \alpha_A \alpha_B \beta_C & \psi_6 &= \beta_A \beta_B \alpha_C \\ \psi_3 &= \alpha_A \beta_B \alpha_C & \psi_7 &= \alpha_A \beta_B \beta_C \\ \psi_4 &= \beta_A \alpha_B \alpha_C & \psi_8 &= \beta_A \beta_B \beta_C \end{aligned}$$

Compute the energy eigenvalues (there are eight of them) for this system of nuclei in the presence of an applied external magnetic field, H_0.

(b) When the differences in resonant frequency between nuclei A, B, and C are large compared to the coupling constants, J_{AB}, J_{BC}, and J_{AC}, the only allowed transitions between energy levels are those for which just one of the three spins changes its I_z-eigenvalue. Using this "selection rule" (see Chapter 19.B.4), find the energies for the 12 allowed transitions. Then, assuming that $|J_{AB}| > |J_{BC}| > |J_{AC}|$, show that the energy absorption spectrum for this system can be analyzed *graphically*, by considering each nuclear resonance to be "split" by coupling to one of the other nuclei, and then "split" again by coupling to the remaining nucleus, as shown schematically below. Thus, while the calculations required to reach this diagram may have been tedious, you are now in a position to analyze any proton (or carbon or fluorine or phosphorus) NMR spectrum, to first order, by working backward from the scheme shown below to obtain the resonant frequencies ("chemical shifts") and coupling constants (J) for an NMR spectrum without doing any quantum mechanical calculations at all!

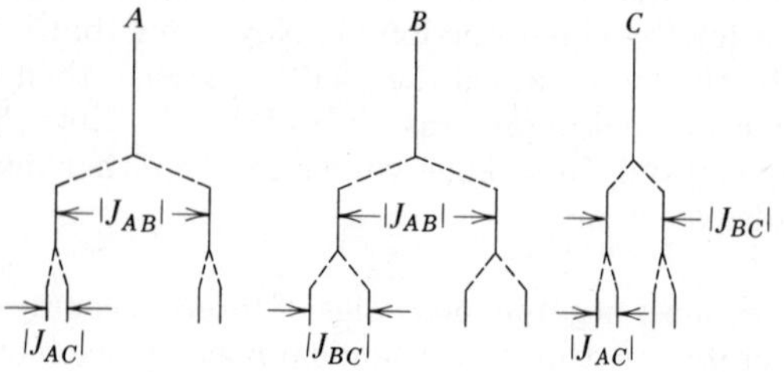

6. Solve the set of equations, Eq. 18-108, to obtain the molecular orbital eigenfunctions for the butadiene molecule.

7. (a) Show that the determinant from which the energy-eigenvalues for the pi-electrons of benzene are obtained has the form

$$\begin{vmatrix} X & 1 & 0 & 0 & 0 & 1 \\ 1 & X & 1 & 0 & 0 & 0 \\ 0 & 1 & X & 1 & 0 & 0 \\ 0 & 0 & 1 & X & 1 & 0 \\ 0 & 0 & 0 & 1 & X & 1 \\ 1 & 0 & 0 & 0 & 1 & X \end{vmatrix} = 0 = X^6 - 6X^4 + 9X^2 - 4$$

(b) Show that the solutions of that determinant are $X = 2,1,1,-1,-1,-2$, and use that information to obtain the energy levels and molecular orbitals for benzene.

8. (a) From the energy eigenvalues for the pi-electrons of pyridine (Fig. 18-17), show that the molecular orbitals corresponding to the three lowest energy eigenvalues are those listed in Eq. 18-115.

(b) Use those three eigenfunctions to compute the electron densities on each of the six atoms of the pyridine ring (answer shown below).

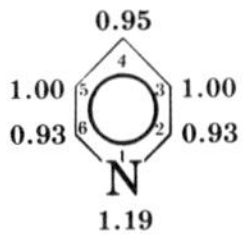

REFERENCES

Spin Problems (Magnetic Resonance Spectra)

A. Carrington and A. D. McLachlan, *Introduction to Magnetic Resonance,* Harper & Row, New York (1967).

E. D. Becker, *High Resolution NMR: Theory and Chemical Applications,* Academic Press, New York (1969).

J. A. Pople, W. G. Schneider and H. J. Bernstein, *High-Resolution Nuclear Magnetic Resonance,* McGraw-Hill, New York (1959). Examples quite dated, but well-organized calculations of NMR spectra in Chapter 6.

Molecular Orbital Theory

L. B. Kier, *Molecular Orbital Theory in Drug Research,* Academic Press, New York (1971).

Basic Quantum Mechanics

L. Pauling and E. B. Wilson, Jr., *Introduction to Quantum Mechanics,* McGraw-Hill, New York (1935). Still an excellent text.

J. C. Davis, Jr., *Advanced Physical Chemistry,* Ronald Press, New York (1965). Excellent treatment.

CHAPTER 19 Putting the Marbles in the Right Bags: Boltzmann Distribution

In the preceding chapter, we showed that the energy of a particle is "quantized" (i.e., the energy may take on only certain discrete values) for the examples of a weight on a spring, a group of magnetic nuclei of spin one-half, and π electrons in conjugated organic systems. For these typical examples, we have so far considered the behavior of just *one* particle or molecule at a time. To understand the behavior of a *large number* of such particles (chemistry, after all, takes place on a scale of 10^{20} or more particles), we are immediately faced with the following question: given the usual situation of a *fixed total number of particles, N,* and a *fixed total energy, E,* for all the particles, how can we predict the relative number of particles, N_i, which are described by the energy eigenfunction, ψ_i, whose energy eigenvalue is ϵ_i: in other words, how do we decide how many particles have a particular energy? [For simplicity, we shall suppose that there is only one eigenfunction, ψ_i, associated with a given energy eigenvalue, ϵ_i: namely, that none of the energy levels is "degenerate" (see p. 18-43).]

$$\sum_i N_i = N = \text{fixed} \tag{19-1}$$

$$\sum_i N_i\epsilon_i = E = \text{fixed} \tag{19-2}$$

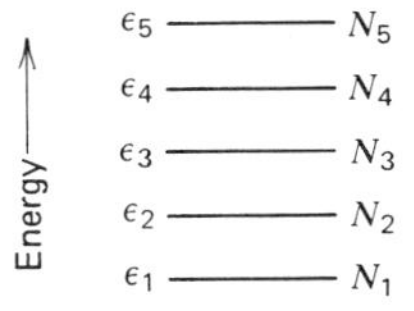

How many particles, N_i, have a given energy, ϵ_i, subject to the constraints that the total number of particles, N, and the total energy, E, are constant?

To understand the answer to this question, it is necessary to return (very briefly) to the bags-and-marbles (coin-tossing, random walk) situation of Chapter 6. In Chapter 6, we considered a problem equivalent to tossing a coin N times and then asking about the probability that m of the N tosses came out "heads." We found that the average number of heads was $(N/2)$, but that the root-mean-square deviation varied as $\sqrt{N}$. A pictorial expression of these results is shown in Fig. 19-1, which shows the relative probability of obtaining a particular fraction of total tosses as heads, versus the relative fraction of total tosses as heads. This time we have normalized that probability such that the most likely result (half the total tosses are heads) has unit probability. Figure 19-1 simply shows that as the number of tosses

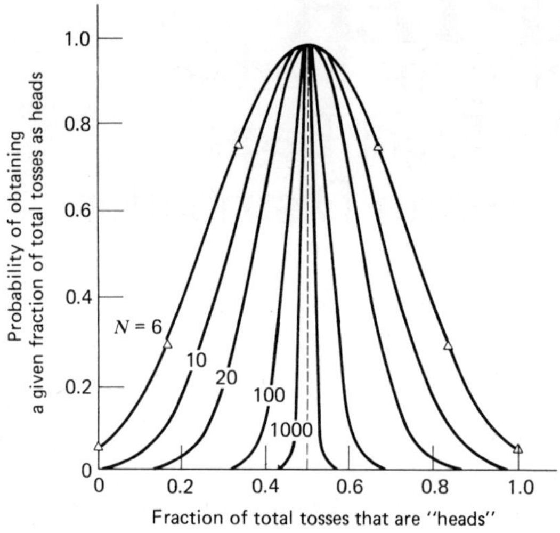

FIGURE 19-1. "Peaking" of the relative probability of obtaining a given fraction as heads in N tosses of a coin. Notice that as N increases, it becomes increasingly likely that a particular experiment will be relatively close to the expected (average) result. (Adapted from L. K. Nash, *Elements of Statistical Thermodynamics,* Addison-Wesley, Reading, Mass., 1968, p. 11.)

increases, it becomes increasingly likely that any *particular* experiment will yield a result close to the *average* result, just as the *relative* precision of a radioactive counting experiment (see Chapter 8.A) increases as the number of observed counts increases.

Figure 19-1 suggests that for very large numbers of tosses, we will be very likely to observe the *most probable* result (i.e., $(N/2)$ "heads"), and that we can safely ignore all but the most probable result with a very high confidence level.

A similar situation exists for the particles-in-energy-levels problem. For example, the (relatively!) simple case of three identical harmonic oscillators (weights on springs) with a specified total energy of three "quanta" of energy, $(1/2)(\hbar)(k/m)^{1/2}$ each (see Chapter 18), is illustrated in Fig. 19-2. For the first "configuration" there is one oscillator with energy, $\epsilon_3 = 3$ units, and two oscillators with energy, $\epsilon_0 = 0$ units. There are three ways to choose which of the three particles will have energy, ϵ_3, two ways to choose which of the remaining two particles will have energy, ϵ_0, and one way of choosing the single remaining oscillator to have energy, ϵ_0; however, it doesn't matter which order we assign the last two oscillators, so there are only $(3 \cdot 2 \cdot 1)/(2 \cdot 1) = 3!/2!$ genuinely distinct arrangements out of the $3! = 6$ arrangements that would be possible if we could tell the oscillators apart from one another. Similarly, the other two "configurations" can be formed by 6 or 1 distinct arrangement, as shown in the figure. In this example, since the number of particles is fixed at $N = 3$, and the total energy of the

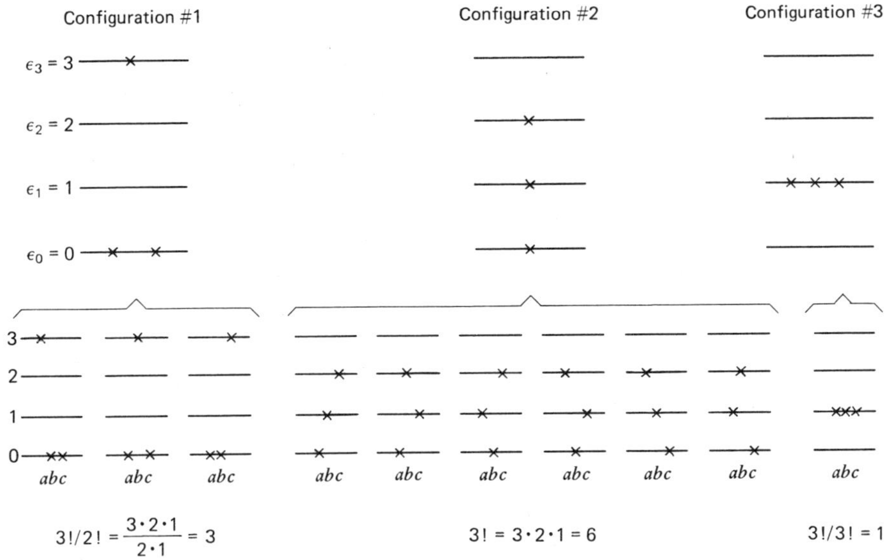

FIGURE 19-2. Configurations (upper diagrams) and number of arrangements leading to those configurations (lower diagrams) for a system of three (*a,b,c*) identical harmonic oscillators ($N = 3$) of fixed total energy $E = 3$ units. A configuration simply specifies how many particles occupy each energy level. In this example, configuration #2 is most probable, since there are more ways of forming it than of forming any of the other possible configurations. When the number of particles becomes very large, the most probable configuration becomes overwhelmingly more important than all of the others combined, and is called the "Boltzmann distribution" (see Eq. 19-3).

three particles is fixed at $E = 3$ units, the most probable configuration (on the grounds that there are more distinct ways of forming it than of forming either of the other two possible configurations) is configuration #2. As with the coin-tossing example just discussed, there is a pronounced "peaking" of the probability distribution near the *most probable* configuration, as the number of particles is increased to very large numbers (say 10^{20} molecules), so that when a large number of molecules is considered, the properties of the system may be determined with extremely high accuracy from the *most probable* distribution of particles among the energy levels. It can be shown (see References) that the most probable distribution of N total particles among various energy levels, $\epsilon_1, \epsilon_2, \ldots$, subject to the constraint that the total energy, E, is constant (Equations 19-1 and 19-2), has the properties:

$$N_i/N_j = \exp[-\beta\epsilon_i]/\exp[-\beta\epsilon_j] = \exp[-\beta(\epsilon_i - \epsilon_j)] \tag{19-3a}$$

so that

$$N_i/N = \frac{\exp[-\beta\epsilon_i]}{\sum_i \exp[-\beta\epsilon_i]} \tag{19-3b}$$

in which N_i and N_j represent the numbers of particles in energy levels ϵ_i and ϵ_j, and β is a constant. By comparing the classical and quantum-mechanical properties of an ideal gas confined in a box (see Problems), it is found that the constant, β, may be identified with $(1/kT)$: this statistical derivation is in fact sometimes used as a *definition* of temperature, as we will shortly find in the laser experiment. Equation 19-3 then becomes the *"Boltzman distribution"*:

$$N_i/N_j = \exp[-(\epsilon_i - \epsilon_j)/kT] \qquad k = \text{Boltzmann constant}, \quad T = \text{Absolute temperature} \tag{19-4a}$$

$$N_i/N = \exp[-\epsilon_i/kT]/\sum_i \exp[-\epsilon_i/kT] \tag{19-4b}$$

$$= (1/q)\exp[-\epsilon_i/kT] \qquad q = \text{"partition function"} \tag{19-4c}$$

Based on the Boltzmann distribution, a temperature is considered "low" when most of the particles have an energy much greater than thermal energy, $\epsilon_i >> kT$, and a "high" temperature corresponds to particles whose energy is much *less* than thermal energy, $\epsilon_i << kT$. At "low" temperature, most of the particles will be in the lowest energy level(s), while at "high" temperature most of the energy levels will be equally populated, as shown schematically in Fig. 19-3. Of the energy-level situations we have examined so far, the "high" temperature condition applies to magnetic resonance experiments, while the "low" temperature condition is valid for infrared and optical-ultraviolet experiments, as shown in Table 19-1. In any case where

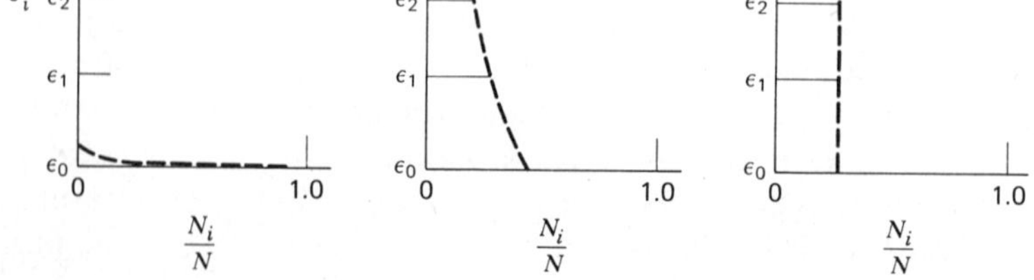

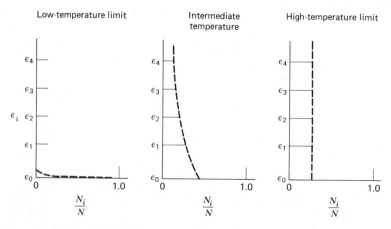

FIGURE 19-3. Relative equilibrium populations (abscissa) of various energy levels (ordinate) for three relative ratios of e_i to thermal energy, kT. Low-temperature: $\epsilon_i >> kT$; intermediate temperature, $\epsilon_i \simeq kT$; high temperature, $\epsilon_i << kT$. Examples of each situation may be found in Table 19-1. Relative populations are obtained from Eq. 19-4a.

Table 19-1 Relative Populations of Two Adjacent Energy Levels, for Various Types of Natural Oscillations (See Eq. 19-4a)

Nature of Oscillation	Typical Natural Frequency of that Oscillation	Energy Difference $(\epsilon_1 - \epsilon_0)$ Associated with that Frequency where $h\nu = (\epsilon_1 - \epsilon_0)$	Relative Populations of Lowermost and Next Highest Energy Levels, N_1/N_0, for Various Temperatures: Room Temp. (295°K)	Liq. N_2 (77°K)	Liq. He (4.2°K)
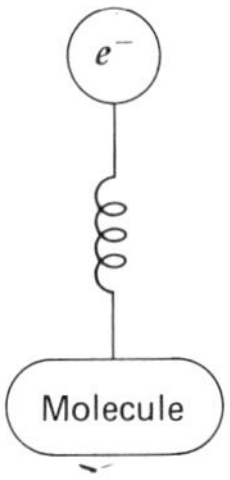	2000Å = 1.5×10^{15} Hz	10^{-11} erg	$10^{-106} \cong 0$	$10^{-404} \cong 0$	$10^{-7440} \cong 0$
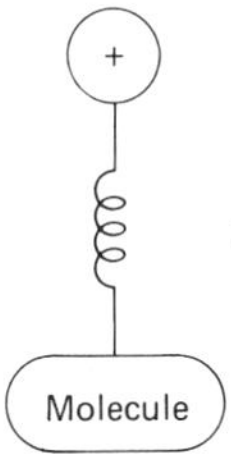	1500 cm^{-1} = 4.5×10^{13} Hz	3×10^{-13} erg	0.00066	$7 \times 10^{-13} \cong 0$	$10^{-233} \cong 0$
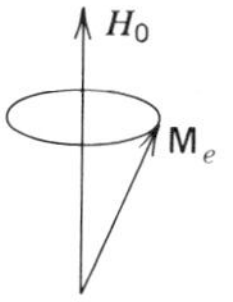	65.8 GHz (electron spin resonance in magnetic field of 23.5 kGauss)	4.4×10^{-16} erg	0.989	0.960	0.471
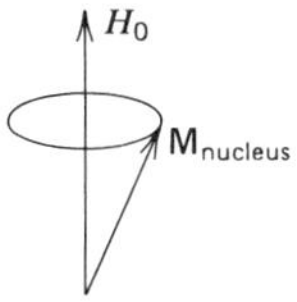	100 MHz (proton magnetic resonance in magnetic field of 23.5 kGauss)	6.6×10^{-19} erg	0.99998	0.99994	0.9989

the system is at thermal equilibrium with its surroundings (i.e., almost all the time, for our purposes), there will be more particles in lower energy levels than in higher energy levels, as seen in Fig. 19-3.

As discussed further in the second half of this chapter, Table 19-1 partly explains why magnetic resonance absorption is so much weaker than infrared and optical absorption. Since the absorption intensity depends on the *difference* in the number of particles in the lower and upper energy levels, proton magnetic resonance absorption is weak (e.g., for every 100,000 protons, there are about 50,001 in the lower energy level and about 49,999 in the upper level, for a population *difference* of only about 0.002% between the two levels), but optical absorption is strong because almost *all* the optical oscillators reside in the lowest energy level.

19.A. LANTHANIDE NMR SHIFT REAGENTS AND MOLECULAR CONFORMATION IN SOLUTION

The mechanism and use of lanthanide NMR "shift" reagents provide a good (and important) example of almost everything we have thus far covered in this Section. Lanthanide ions form 1:1 adducts with a variety of

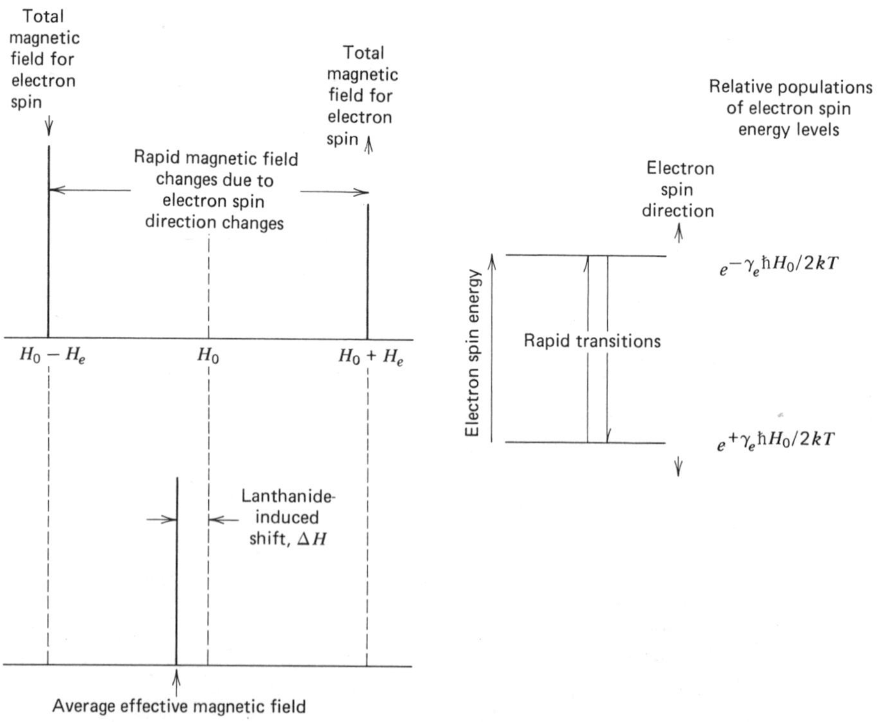

FIGURE 19-4. Highly schematic explanation of the origin of lanthanide-induced nuclear magnetic resonance "shifts." A given proton in the vicinity of a lanthanide ion "sees" one of two possible magnetic fields (upper left), according to whether the lanthanide electron spin is "up" or "down" (right-hand diagram). Since electron spin direction changes rapidly back and forth, the magnetic field at the proton averages to a single effective value (lower trace); since slightly more electrons occupy the lower energy level than the higher one, (see Table 19-1) the average magnetic field will be slightly shifted from the value of the applied field itself, H_0 (see Problem 14-6).

nucleophilic "donor" species, including ROH, RNH_2, and RPO_4H^-. Since the lanthanide ions (except for lanthanum itself) are paramagnetic (i.e., possess an electron magnetic moment), there will be a small magnetic field associated with the lanthanide magnetic moment, and that magnetic field will either add or subtract from the applied magnetic field for each nucleus in the vicinity of the lanthanide ion, according to whether the electronic spin direction is "up" or "down" (compare Fig. 18-8). However, the electron "spin" direction is not constant; we shall soon see in Chapter 19.B that the spin direction changes rapidly back and forth. The small magnetic field associated with that electron spin direction thus also changes sign rapidly, and the effective total magnetic field (i.e., applied magnetic field plus or minus the small magnetic field due to the lanthanide electron spin) at a given nearby proton thus averages to a single effective magnetic field whose value is determined by the relative numbers of electrons in the two possible lanthanide electronic "spin" states. Finally, since there are slightly more electrons in the lower-energy state than in the higher-energy state, the average electron magnetic moment is nonzero, and the effective average magnetic field at a nearby proton nucleus will be slightly shifted (hence "shift reagent") from the magnitude of the applied magnetic field, H_0 (see Fig. 19-4).

The reason that all of this is useful is that the magnitude of the induced "shift" is related to the geometry of the complex between lanthanide and donor, as given by Eq. 19-5. Since lanthanide-donor bond distances are well-known, the relative induced shifts for different nuclei in the same donor molecule may be used to determine the configuration of the donor molecule, as shown in the following example (see Fig. 19-5).

$$\Delta H/H_0 = (\text{constant}) \frac{[3\cos^2\theta - 1]}{r^3} \tag{19-5}$$

in which r is the distance between the lanthanide ion and the nucleus of interest in the donor molecule, and θ is the angle between the lanthanide-donor bond axis and the lanthanide-nuclear vector (see nucleotide example

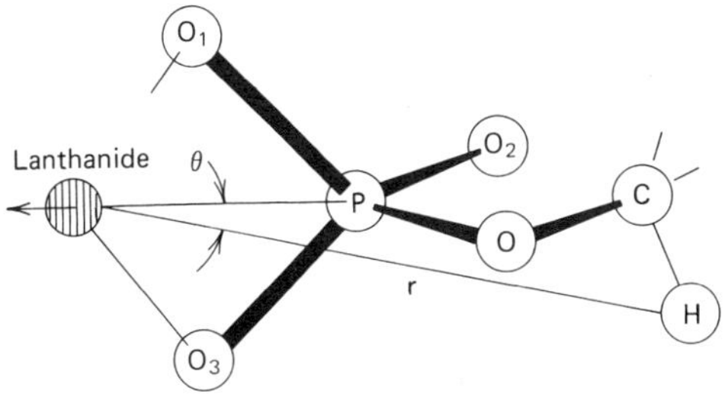

FIGURE 19-5. Illustration of parameters r and θ for the complex of a lanthanide ion with a monophosphate nucleotide. [Adapted from C. D. Barry et al., *Nature* *232*, 237 (1971).]

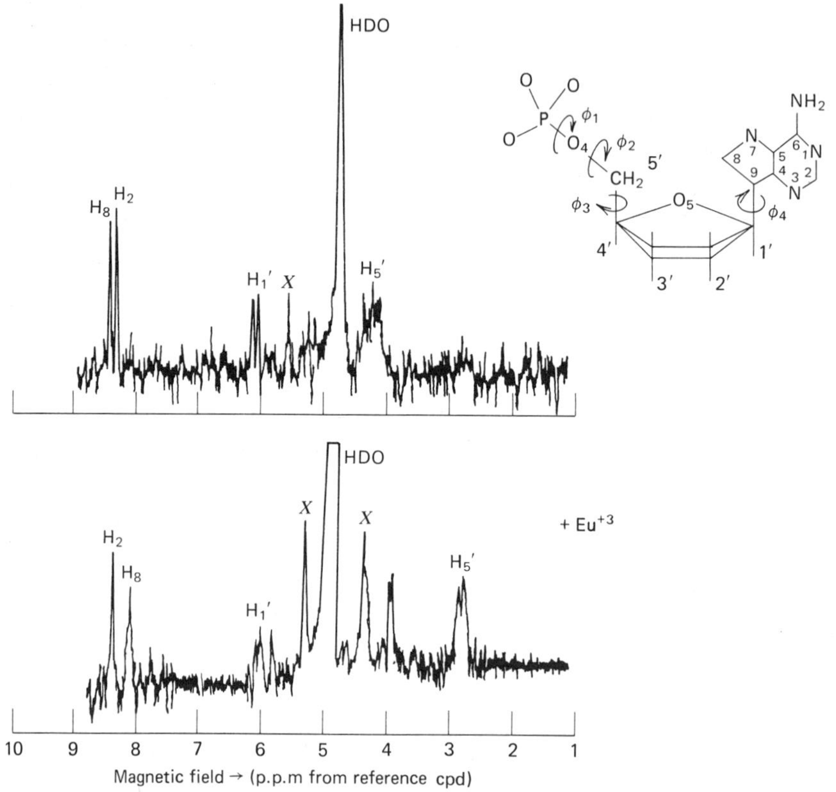

FIGURE 19-6. Proton NMR spectra of AMP (structure shown above) in absence (*top*) and presence (*bottom*) of lanthanide shift reagent, Eu^{+3} ion. Peaks marked "*X*" are artifacts of the instrument and should be ignored. The angular positions, ϕ_1 to ϕ_4 indicate the rotational position about each of the four labile (i.e., nonrigid) bonds. [From C. D. Barry et al., *Nature 232,* 237 (1971).]

below). Since the constant has the same value for different protons in the same donor molecule, *ratios* of induced shifts for different protons in the molecule will reflect differences in θ and r for the two respective protons. By comparing observed induced shift ratios with those calculated for various possible molecular conformations (see following example), it is possible to determine the preferred configuration of a molecule in solution.

Figure 19-6 shows the proton NMR spectrum of adenosine monophosphate (AMP) in the absence (top) and presence (bottom) of Eu^{+3} ion as a shift reagent. It is clear that the various proton NMR signals have been shifted by different amounts (e.g., the 5′-proton closest to the lanthanide ion is shifted most).

The obvious problem in trying to compare observed and calculated shift ratios is that there are four rotationally labile bonds in the AMP molecule, so that one should really consider all possible configurations corresponding to ϕ-values spaced every 4°, say, to make the problem tracta-

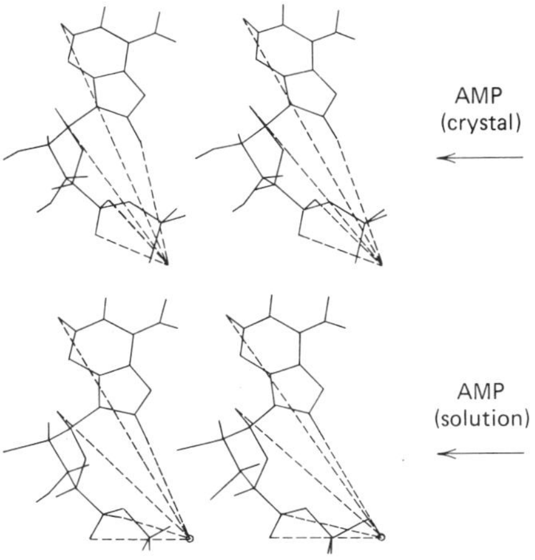

FIGURE 19-7. Stereo views of AMP configuration in crystal (*top*) as determined by X-ray diffraction, and in solution (*bottom*) as determined partly by use of lanthanide shift reagents. [From C. D. Barry, A. C. T. North, J. A. Glasel, R. J. P. Williams, and A. V. Xavier, *Nature 232,* 236–245 (1971).]

ble. With four possible rotationally labile bonds, the number of configurations to be considered is $(360°/4°)^4 = 64 \times 10^6$, a very large number! Fortunately, it is possible to eliminate most of these possibilities by ignoring any configuration that places two atoms closer than (say) a van der Waals radius to each other—this "filtering" procedure leaves some 32000 possible conformations. By next eliminating all configurations inconsistent with the observed lanthanide-induced shift ratios for the various shifted protons, the number of possible configurations is reduced to a mere 195. Finally, by comparing the observed and calculated proton NMR line *widths* in the presence of Gd^{+3} (as in the NAD:YADH example in Fig. 16-24), there remain only 12 very similar possible configurations. The average of these 12 slightly different conformations predicts (Fig. 19-7) a structure very similar to that found in the crystalline form of AMP using X-ray diffraction (see Chapter 22). Some idea as to the significance of this technique may be gained from the fact that the *Nature* publication of this example was one of the most highly quoted scientific papers in the world the year after it was published. The method has since been refined (altering some of the initial conclusions of Fig. 19-7), yet extensions of this method should be useful in establishing the structure *in solution* of a wide range of biologically interesting metabolites, hormones, and drugs.

19.B. TRANSITIONS BETWEEN ENERGY LEVELS

19.B.1 General Formulae

We have already shown in Chapter 18 how to determine the quantum mechanical wave functions and energy levels for some particular time-independent Hamiltonians (i.e., for some particular types of potential energy). However, we have done nothing to indicate how a given particle may *change* its state to another state of different energy (i.e., proceed from one energy level to another). In fact, for a *time-independent* Hamiltonian in the absence of electric and magnetic fields, if we begin at time zero with the system in a given energy eigenstate, Φ_n, then Eq. 18-31 predicts that the state, $\psi(t)$, at any later time will be of the form $\exp[-(i/\hbar)E_n t]\Phi_n$, where E_n is the energy of the eigenstate Φ_n. The probability of finding the system in the original state, Φ_n, after time, t, is just the square of the projection of $\psi_n(t)$, the wave function at time, t,

$$\boxed{\psi_n(t) = e^{-(i/\hbar)E_n t}\,\Phi_n} \tag{19-6}$$

on $\Phi_n = \psi_n(0)$:

$$\begin{aligned} |<\psi_n(t),\psi_n(0)>|^2 &= |<e^{-(i/\hbar)E_n t}\Phi_n, \Phi_n>|^2 \\ &= e^{[-(i/\hbar)E_n t+(i/\hbar)E_n t]}|<\Phi_n,\Phi_n>|^2 \\ |<\psi_n(t),\psi_n(0)>|^2 &= 1 \end{aligned} \tag{19-7}$$

where the initial eigenfunction, Φ_n, is assumed to be normalized. Equation 19-7 verifies that if the system starts out in any given energy level at time zero, it will still be there after an arbitrary interval, t. [We could readily generalize Eq. 19-7 to show that even if the system is initially described by some linear combination (weighted average) of energy eigenstates, the

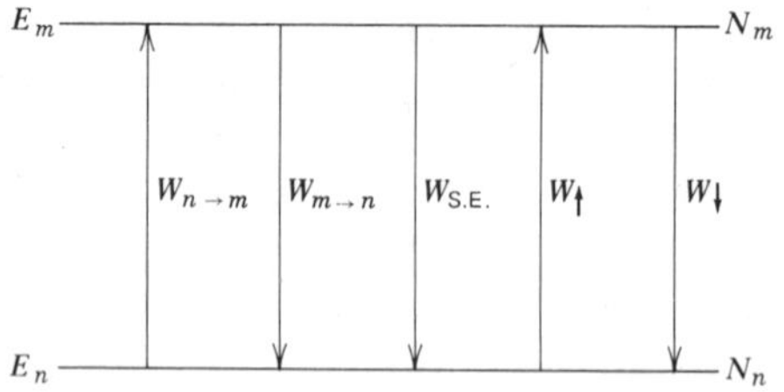

FIGURE 19-8. Schematic diagram of various mechanisms for transitions between the energy eigenstates of a two-state system. $W_{m\to n}$ and $W_{n\to m}$ represent the rate constants for emission and absorption (of a photon) *induced* by an oscillating electric or magnetic field whose frequency, ν, is given by Eq. 19-8: $E_m - E_n = h\nu$. $W_{\text{S.E.}}$ is the rate constant for "spontaneous" emission. $W_\uparrow$ and $W_\downarrow$ are the rate constants for the radiationless absorption or emission of energy, $E_m - E_n$, due to interaction of the system with other particles. It is shown (see below) that $W_{m\to n} = W_{n\to m}$, but $(W_\uparrow / W_\downarrow) = \exp[-(E_m - E_n)/kT$ (see Equations 19-29 and 19-33). Spontaneous emission is important mainly at optical frequencies, where it predicts the maximal rate for fluorescence. Induced absorption and emission rates are calculated in this section. Radiationless absorption and emission rates are most easily calculated for radiofrequency spectroscopy (see Chapters 19.B.3 and 21.B). N_n and N_m are the numbers of particles in energy levels E_n and E_m (see text).

probability of finding the system with any given energy at later time t would be the same as at time zero.]

The mechanisms by which transitions arise are shown schematically in Fig. 19-8. All the mechanisms are in some way related to the presence of perturbing time-varying electric or magnetic fields on the charged particles of which matter is composed. *Induced absorption, induced emission,* and *spontaneous emission* correspond to the taking-in (induced absorption) or release (induced or spontaneous emission) of a *photon* whose energy is related by

$$\boxed{E_m - E_n = h\nu} \qquad \begin{aligned} h &= \text{Planck constant} \\ &= 6.62 \times 10^{27} \text{ erg sec} \end{aligned} \tag{19-8}$$

to the difference in energy between the two energy levels of the system. The remaining *radiationless absorption and emission* correspond to the taking-in or release of an energy amount, $|E_m - E_n|$, where the absorbed or emitted energy is respectively released or absorbed by *other molecules,* so that no photons are involved.

Since the mechanism for *induced* absorption and emission is essentially the same in all forms of spectroscopy, we first treat it in this section. We shortly show that the remaining mechanisms for transitions provide a means for establishing the equilibrium Boltzmann distribution of populations of energy levels. The probability for *spontaneous* emission can be calculated (see Davis reference); we will simply note that it is proportional to $|E_m - E_n|^3$. It thus happens that spontaneous emission is negligible for low-energy spectroscopy (e.g., NMR). Moreover, it is often possible to calculate the rates of radiationless transitions in this energy range (see Chapter 21.B), so we will treat radiofrequency transitions later in this chapter. In the optical region (ultraviolet-visible-infrared), it is extremely difficult to calculate the rates of radiationless transitions, and we will limit our discussion of optical transitions to a qualitative analysis of lasers in the next section of this chapter.

We begin with the *induced* transitions that result from the introduction of a *time-varying* energy [such as the oscillating energy of interaction between the (oscillating) electric field of monochromatic electromagnetic radiation and the (fixed) electric dipole moment of a molecule]. Denoting this time-dependent energy as a perturbation, $\mathscr{H}'(t)$, added to the time-independent kinetic and potential energy, $\mathscr{H}^0(t)$, the time-dependence of the wave function for the system (the "Schrodinger" Eq. 18-25) can be written

$$\boxed{i\hbar\frac{\partial\Psi(t)}{\partial t} = (\mathscr{H}^\circ + \mathscr{H}'(t)\Psi(t)} \tag{19-9}$$

In general, Eq. 19-9 may be solved in the limit that $\mathscr{H}'(t)$ is either *sudden* or *small.* When $\mathscr{H}'(t)$ is *small,* the simplest cases are for perturbations that are *constant* (except for being turned "on" and "off" at particular times) and either *periodic* or *randomly varying* in time. In this section, we examine the effect of the small, periodic energy perturbation resulting from application of electromagnetic radiation to a quantum mechanical system, and the randomly varying case is discussed in Chapter 21 along with other noise phenomena. In the limit that the perturbation, $\mathscr{H}'(t)$ is small, it is possible to express the wave function at later time t in terms of energy eigenfunctions of the unperturbed Hamiltonian, $\mathscr{H}°$,

$$\Psi(t) = c_1(t)\psi_1(t) + c_2(t)\psi_2(t) + \cdots + c_n(t)\psi_n(t)$$

according to

$$\Psi(t) = \sum_{n=1}^{\infty} c_n(t)\psi_n(t) = \sum_{n=1}^{\infty} \underbrace{c_n(t)}\underbrace{e^{-(i/\hbar)E_n t}}\Phi_n \qquad (19\text{-}10)$$

Slowly varying time-dependence due to $\mathscr{H}'(t)$ [arrow to $c_n(t)$]

Rapidly varying time-dependence due to unperturbed Hamiltonian, $\mathscr{H}°$ [arrow to $e^{-(i/\hbar)E_n t}$]

In Eq. 19-10 the problem has been simplified by factoring out the rapidly varying time-dependence (of $\psi_n(t) = e^{(-(i/\hbar)(E_n t)}\Phi_n)$ due to the unperturbed Hamiltonian (Eq. 19-6) from the much slower time-variation due to the (much smaller) $\mathscr{H}'(t)$. Substituting Eq. 19-10 for $\Psi(t)$ into Eq. 19-9, we obtain

$$i\hbar\sum_n \frac{dc_n(t)}{dt}\psi_n(t) + i\hbar\sum_n c_n(t)\frac{\partial}{\partial t}\psi_n(t) = \sum_n c_n(t)\mathscr{H}°\psi_n(t) + \sum_n c_n(t)\mathscr{H}'(t)\psi_n(t) \qquad (19\text{-}11)$$

But since we have the Schrodinger equation (Eq. 18-25) for the wave functions, $\psi_n(t)$, of the unperturbed Hamiltonian, $\mathscr{H}°$

$$i\hbar\frac{\partial}{\partial t}\psi_n(t) = \mathscr{H}°\psi_n(t) \qquad (19\text{-}12)$$

the two middle terms of Eq. 19-11 cancel, leaving

$$i\hbar\sum_n \frac{dc_n(t)}{dt}\psi_n(t) = \sum_n c_n(t)\mathscr{H}'(t)\psi_n(t) \qquad (19\text{-}13)$$

For reasons that presently become apparent, we now take the projection of the vector expressed by each side of Eq. 19-13 along the unperturbed basis function, $\psi_m(t)$,

$$< \psi_m(t), i\hbar \sum_n \frac{dc_n(t)}{dt} \psi_n(t) > = i\hbar \sum_n \frac{dc_n(t)}{dt} < \psi_m(t), \psi_n(t) > = i\hbar \frac{dc_m(t)}{dt} \qquad (19\text{-}14a)$$

where we have used the fact that different eigenfunctions of $\mathscr{H}'$ are orthogonal (compare Eqs. 18-69)

and

$$< \psi_m(t), \sum_n c_n(t) \mathscr{H}'(t) \psi_n(t) > = \sum_n c_n(t) < \psi_m(t), \mathscr{H}'(t) \psi_n(t) >$$
$$= \sum_n c_n(t) \mathscr{H}'(t)_{mn}$$

mth row — nth column of matrix representation of operator $\mathscr{H}'(t)$ in the $\psi_n(t)$ basis set

(19-14b)

Equation 19-14a shows that the effect of this procedure is to pick out the time-dependence of just one of the wave function amplitudes, $dc_m(t)/dt$; equating Equations 19-14a and 19-14b then produces the desired general result

$$\boxed{\frac{dc_m(t)}{dt} = \frac{1}{i\hbar} \sum_n c_n(t) \mathscr{H}'(t)_{mn}} \qquad (19\text{-}15)$$

Equation 19-15 forms a set of n differential equations (one for each choice of m), each of which describes the time behavior of the amplitude for one of the original (unperturbed) energy eigenfunctions. Given a set of initial conditions (e.g., $c_1(0) = 1$ and all other $c_i(0) = 0$ at time zero, so that the system begins in its lowest-energy state), and a particular specified time-dependent perturbation, $\mathscr{H}'(t)$, we can solve for the relative populations, $c_m^*(t)c_m(t)$ of each of the energy levels at time t by solving Eq. 19-15 to obtain each of the $c_m(t)$, as shown by the following calculation of principal interest.

Induced Emission and Absorption: Transitions Induced by an Oscillating Electric Field (i.e., by Electromagnetic Radiation)

In Chapter 14.A, we accounted for the absorption of electromagnetic radiation using a *classical mechanical* model that matter consists of electrons on springs and that radiation consists of an oscillating electric field. In this section we shall use our recently derived expressions for energy eigenfunction amplitudes to compute the "transition probability" for absorption (or emission) of electromagnetic radiation impinging on a *quantum mechanical* system. This calculation forms the basis for understanding of the mecha-

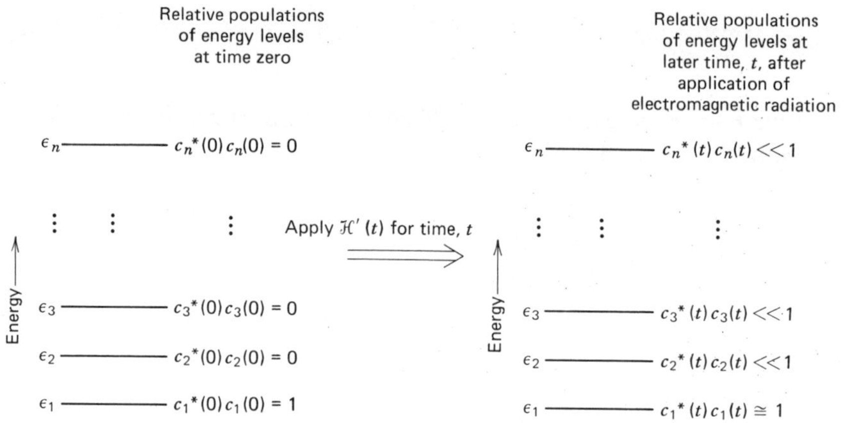

FIGURE 19-9. Schematic diagrams of energy levels and relative populations of those levels before (left) and after (right) application of an electromagnetic field continuously for a time interval, t. The system is taken to be in its ground state with certainty at time zero [$c_1(0) = 1$ and all others $c_m(0) = 0$ at $t = 0$]. The electromagnetic field perturbation is taken as sufficiently small that the ground state population is still essentially the same at time, t, and we calculate the amplitudes, $c_m(t)$, of each of the other energy eigenfunctions to obtain the probability, $c_m^*(t)c_m(t)$, of finding the system in the mth energy level after time, t.

nism of laser operation, as well as "saturation" phenomena that can be used to determine spectral energy-level "lifetimes." We then present a number of biological applications of such phenomena.

For greatest ease in notation, we suppose that a system is initially prepared at time zero in its lowest energy state ("ground" state) with unit probability, as shown in Fig. 19-9. An electromagnetic field is then directed at the sample, producing a time-dependent energy perturbation, $\mathscr{H}'(t)$. Our present task is to calculate the probability of finding that system in any of the *other* energy levels at some later time, t.

Limiting the treatment to plane-polarized monochromatic electromagnetic radiation of frequency, ω, and electric field amplitude, $2E_0$, and neglecting the magnetic component of the radiation (as in Chapter 14.A), the electric field of the radiation,

$$E_x = 2E_0 \cos(\omega t) \tag{19-16a}$$

may be written more conveniently in complex notation as

$$E_x = E_0[\exp(i\omega t) + \exp(-i\omega t)] \tag{19-16b}$$

and the energy of interaction of that radiation with the electric dipole moment μ_x of some molecule or atom may be written

Energy of interaction = (Electric field)·(Electric dipole moment)

or

$$\mathscr{H}'(t) = E_x \cdot \mu_x$$

$$\boxed{\mathscr{H}'(t) = E_0\mu_x[\exp(i\omega t) + \exp(-i\omega t)]} \qquad (19\text{-}17)$$

in which μ_x = (charge)·(x), where x is the distance separating the two charges of the dipole.

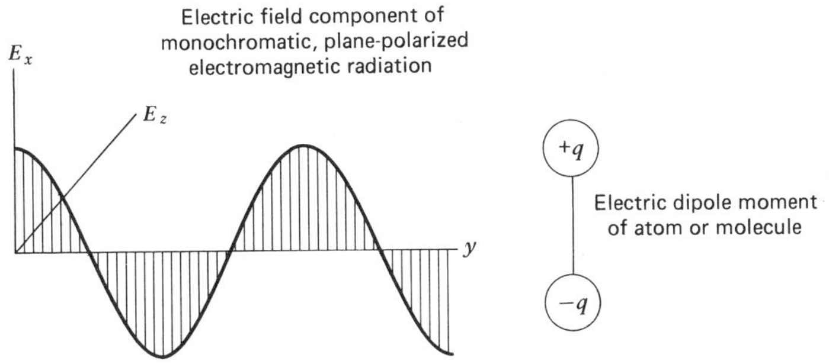

Having obtained an explicit form for the time-varying perturbation energy, $\mathscr{H}'(t)$, all that remains is to substitute Eq. 19-17 into Eq. 19-15 and solve the resulting equations subject to the initial condition

$$\boxed{c_1(0) = 1, c_m(0) = 0, m \neq 1} \qquad (19\text{-}18)$$

Substituting,

$$\frac{dc_m(t)}{dt} = \frac{1}{i\hbar}\sum_n c_n(t) < \psi_m(t), E_0\mu_x(e^{i\omega t} + e^{-i\omega t})\psi_n(t) > \qquad (19\text{-}19a)$$

$$= \frac{E_0}{i\hbar}\sum_n [e^{i\omega t} < \psi_m(t), \mu_x\psi_n(t) > + e^{-i\omega t} < \psi_m(t), \mu_x\psi_n(t) >]$$

But (19-19b)

$$\psi_m(t) = e^{-(i/\hbar)E_m t}\Phi_m \qquad (19\text{-}20a)$$

and

$$\psi_n(t) = e^{-(i/\hbar)E_n t}\Phi_n \quad \text{from Eq. 18-31} \qquad (19\text{-}20b)$$

Substituting Eq. 19-20 into Eq. 19-19b (refer to Eq. 18-24 for this manipulation),

$$\frac{dc_m(t)}{dt} = \frac{E_0}{i\hbar}\left[e^{(i\omega t+(i/\hbar)(E_m-E_n)t)} <\Phi_m, \mu_x\Phi_n> + e^{(-i\omega t+(i/\hbar)(E_m-E_n)t)} <\Phi_m, \mu_x\Phi_n>\right] \tag{19-21}$$

Denoting the mnth matrix element of the operator, μ_x, in the Φ_m basis set as

$$\mu_{x_{mn}} = <\Phi_m, \mu_x\Phi_n> = \text{"transition (electric) dipole moment"} \tag{19-22}$$

(a quantity we need not stop to evaluate now) Eq. 19-21 may be simplified to the form

$$\frac{dc_m(t)}{dt} = \frac{E_0}{i\hbar}(\mu_{x_{mn}})\left[e^{(i/\hbar)(E_m-E_n+h\nu)t} + e^{(i/\hbar)(E_m-E_n-h\nu)t}\right] \tag{19-23}$$

Equation 19-23 gives the *instantaneous* rate of change in the amplitude, $c_m(t)$, of the eigenfunction for the mth energy level; in order to determine the *accumulated* change in $c_m(t)$ after $\mathscr{H}'(t)$ has been applied for time, t, it is necessary to integrate Eq. 19-23 from time zero to time t, corresponding to Fig. 19-9, leaving the desired result

$$c_m(t) = (\mu_x)_{mn}E_0\left\{\underbrace{\frac{1-e^{(i/\hbar)(E_m-E_n+h\nu)t}}{E_m-E_n+h\nu}}_{\substack{\text{Induced } emission^* \\ \text{of radiation when} \\ h\nu \to -(E_m-E_n), \\ \text{where } E_m > E_n}} + \underbrace{\frac{1-e^{(i/\hbar)(E_m-E_n-h\nu)t}}{E_m-E_n-h\nu}}_{\substack{\text{Induced } absorption^* \\ \text{of radiation when} \\ h\nu \to (E_m-E_n), \\ \text{where } E_m > E_n}}\right\} \tag{19-24}$$

There are two very significant features to Eq. 19-24. First, it shows that the rate of transitions [and hence the magnitude of $c_m(t)$] between the mth energy level and the nth energy level will be appreciable, only when one of the denominators of Eq. 19-24 becomes small [the numerators vary between 0 and 2 so it is the denominators that have the greatest effect on $c_m(t)$].* The denominators become small, *either* when the radiation energy is very close to that required to promote the particle *upward* in energy ("*absorption*") from energy level E_n to E_m, *or* when the electromagnetic radiation energy is very close to that released when the particle undergoes

**From Eq. 19-24, it would appear that when the "driving" radiation frequency is exactly "on resonance," namely $h\nu = |E_m - E_n|$, the amplitude of the mth basis set function, $c_m(t)$, becomes infinite. This is analogous to the motion of the classical weight on a spring, whose amplitude also increases without limit when the undamped spring is driven exactly on resonance. As for the classical spring, this anomaly disappears when we introduce a nonzero damping term (in this case, spontaneous emission or radiationless transitions), so that there will always be a finite (i.e., noninfinite) probability of finding the driven electron in any one of the energy levels corresponding to a given basis state function of the unperturbed Hamiltonian (i.e., the energy levels in the absence of any applied fields).*

a transition *downward* in energy ("*emission*") from state E_m to state E_n as written beside Eq. 19-24. The absorption process is doubtless familiar to most readers—the induced emission ("*stimulated emission*") phenomenon is the new aspect. In other words, when there are more particles in the lower energy level E_n than in the upper level E_m, the amplitude of the incident radiation will decrease due to absorption; but when there are more particles in the upper level E_m than in the lower level E_n, the amplitude of the incident radiation will actually *increase* on passing through the sample! Masers and lasers (see next section) are based on this stimulated emission of radiation. The second major feature of Eq. 19-24 is that the coefficients of the first and second terms are the same, which simply indicates that (see next section) the probability of stimulated transitions upward or downward is the same—this property is also essential to laser design. Finally, in order that $c_m(t) \neq 0$ (i.e., that the transition occur with finite probability, or be "allowed"), it is clearly necessary that the matrix element of the electric dipole operator $\mu_{x_{mn}} \neq 0$. Thus, it is always necessary to examine $\mu_{x_{mn}}$ (see Chapter 19.B.4) to determine which of the many possible transitions are actually "allowed"—often the choice is easily stated in a simple "selection rule" (see below).

19.B.2. Population Inversion: Lasers and Their Applications

Lasers and masers (*L*ight or *M*icrowave *A*mplified *S*timulated *E*mission of *R*adiation) rely on the emission of radiation *induced* ("stimulated") by externally applied electromagnetic radiation at the frequency corresponding to the difference in energy between the two energy levels involved (first term of Eq. 19-24). To understand the process, we must first find the relative populations of each of the energy-levels following irradiation of the system. From Eq. 19-24 we immediately obtain the relative populations:

$$c_m^*(t)c_m(t) = E_0^2\,(\mu_{x_{mn}})^2\,\frac{(1 - e^{(i/\hbar)(E_m - E_n - h\nu)t})(1 - e^{(-i/\hbar)(E_m - E_n - h\nu)t})}{(E_m - E_n - h\nu)^2} \qquad (19\text{-}25a)$$

$$c_m^*(t)c_m(t) = 4E_0^2\,(\mu_{x_{mn}})^2\,\frac{\sin^2\left[\left(\frac{1}{2\hbar}\right)(E_m - E_n - h\nu)t\right]}{(E_m - E_n - h\nu)^2} \qquad (19\text{-}25b)$$

where the progression from Eq. 19-25a to 19-25b involves rearrangement using well-known trigonometric identities (see Problems), and where we have limited our attention just to the second term of Eq. 19-24 (i.e., the *absorption* process). Since absorption clearly occurs over a *range* of frequency of the incident radiation, ν, it is necessary to sum the absorption over all incident radiation frequencies to obtain the *total* absorption corresponding to transitions from E_n to E_m:

$$\int c_m^*(t)c_m(t)d\nu = 4E_0^2\,(\mu_{x_{mn}})^2 \int \frac{\sin^2\left[\frac{1}{2\hbar}(E_m - E_n - h\nu)t\right]}{(E_m - E_n - h\nu)^2}\,d\nu \quad (19\text{-}26)$$

in which we suppose that the external radiation source is flat ("white," producing radiation of amplitude which is independent of frequency) for the very narrow frequency range over which the integrand is appreciably larger than zero. Finally, since there will be negligible absorption at frequencies much different than $h\nu = E_m - E_n$, we can extend the limits of the integral to $\pm\infty$, and then apply the definite integral

$$\int_{-\infty}^{\infty} \frac{\sin^2 x}{x^2}\,dx = \pi \quad (19\text{-}27)$$

to simplify Eq. 19-26 to the form

$$\int c_m^*(t)c_m(t)d\nu = \frac{1}{\hbar^2}\,E_0^2\,(\mu_{x_{mn}})^2 t \quad (19\text{-}28)$$

The significance of Eq. 19-28 is that it shows that the probability of a transition from state E_n to state E_m (i.e., absorption of a photon) is proportional to time, so that for a given molecule and given externally applied radiation amplitude, *the transition probability for induced absorption per unit time,* $W_{n\to m}$, *is constant.*

Similarly, by starting from the *first* term of Eq. 19-24 (*induced emission*), it can be shown that the probability for induced emission per unit time, $W_{m\to n}$, is given by the *same* constant:

$$\boxed{W_{n\to m} = W_{m\to n}} = W \quad (19\text{-}29)$$

The immediate question raised by the principal result, Eq. 19-29, is why we ordinarily observe (induced) absorption rather than induced emission on directing incident radiation onto a sample. We now can answer that question.

Consider the two energy levels in question (E_n and E_m, as shown in Fig. 9-8), whose respective populations (i.e., the number of particles in each of the two levels) are N_n and N_m. The number of "upward" stimulated transitions is $W_{n\to m}\,N_n$, and the number of downward stimulated transitions is $W_{m\to n}\,N_m$ per second, so the rate of change of the two energy-level populations N_n and N_m is given by

$$dN_n/dt = W_{m\to n}N_m - W_{n\to m}N_n = W(N_m - N_n) \quad (19\text{-}30a)$$

$$dN_m/dt = W_{n\to m}N_n - W_{m\to n}N_m = W(N_n - N_m) \quad (19\text{-}30b)$$

where Eq. 19-29 has been used to obtain the second equality in each case. Taking the difference between Eqs. 19-30a and 19-30b

$$\frac{dN_n}{dt} - \frac{dN_m}{dt} = \frac{d(N_n - N_m)}{dt} = -2W(N_n - N_m) \tag{19-31}$$

Integrating Eq. 19-31 over the time interval that external radiation has been applied to the system ($t = 0$ to $t = t$), we find that the difference in population $(N_n - N_m)$ between the two affected energy-levels obeys the equation

$$\boxed{(N_n - N_m)_{t=t} = (N_n - N_m)_{t=0}\, e^{-2Wt}} \tag{19-32}$$

We have already indicated (Chapter 19.A) that for a system at thermal equilibrium

$$\frac{N_m}{N_n} \underset{\uparrow}{=} e^{-(E_m - E_n)/kT} \tag{19-4a}$$

Thermal equilibrium

Therefore, since most systems we study are *initially* at thermal equilibrium, $N_n > N_m$, and Eq. 19-32 predicts that radiation will cause more upward than downward transitions, with the net effect that the amplitude of the incident radiation decreases as the radiation goes through the sample (see Fig. 19-10). Equation 19-32 thus accounts for the net *absorption* of radiation

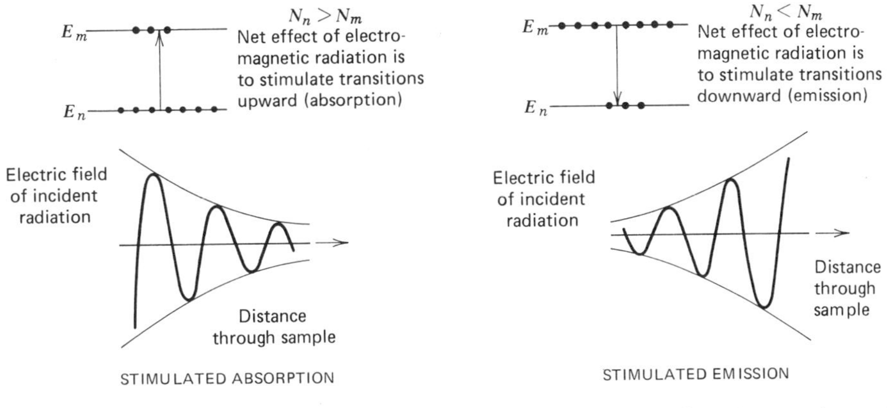

FIGURE 19-10. Pictorial consequences of Eq. 19-32. When (as at thermal equilibrium) there are more particles in the lower energy-level than the higher, incident electromagnetic radiation induces transitions net upward, and the amplitude of the applied radiation decreases with time as the radiation passes through the sample (left). However, when (in a *non*equilibrium initial condition) there are more particles in the upper than the lower energy level, incident radiation induces transitions net downward, and the amplitude of the applied radiation actually increases with time on passing through the sample (right) Equation 19-32 shows that in *both* cases, the applied radiation has an effect that tends to *equalize* the populations of the two energy levels. (In both cases, the frequency of the applied radiation corresponds to the difference in energy between the two levels.)

by matter; even though the quantum-mechanical transition probability (rate constant) for upward and downward transitions is the *same,* there are *more* particles in the *lower*-energy state, so there are more net upward than downward transitions. More generally, Eq. 19-32 predicts that induced absorption or emission will continue until the two energy-level populations are *equalized, no matter which of the two levels is more populated than the other to begin with.* If (by as yet unspecified means) we could arrange that the upper level were more populated than the lower level at time zero, then the incident radiation would stimulate transitions *net downward* (*emission* of radiation by the sample) until the two levels again become equalized (see Fig. 19-10). In this case, the amplitude of the incident radiation would actually *increase* on passing through the sample. Design of lasers and masers is simply based on various tricks for achieving an initial population imbalance between two energy levels with more particles in the higher-energy state, so that subsequent irradiation at a frequency equal to the energy difference between the states leads to stimulated emission, and hence amplification (increase) in the amplitude of the incident radiation.

General Design of Lasers

Figure 19-11 shows the typical energy-level diagram from which most lasers are designed. Since the degree of amplification from the laser (i.e.,

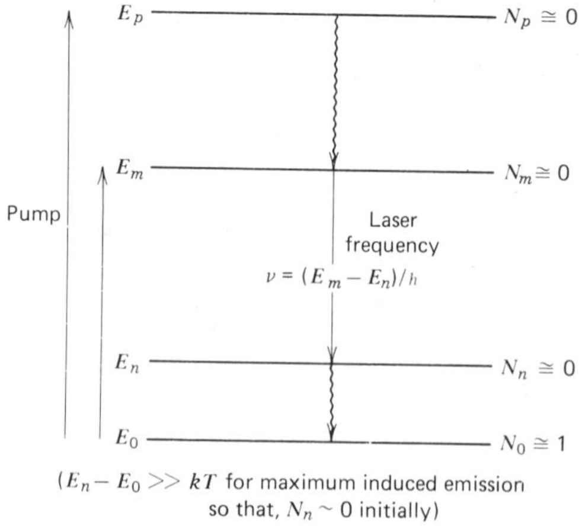

FIGURE 19-11. Energy levels (left) and initial relative populations (right) of a system suitable for use as a laser. By irradiation at frequencies leading to absorption that populates either the E_m or E_p states, followed by (in the latter case) spontaneous transition to the E_m level, it is possible to populate the E_m level much more than the E_n level. Then by irradiation at the frequency, $(E_m - E_n)/h$, induced emission is produced at that frequency. Provided that the particles reaching the E_n state have rapid access to the ground level (lower wavy line), the whole process can be kept going indefinitely.

the amplitude of the induced emission) is clearly largest when the population of the lower energy level of the transition (E_n) is nearly zero initially (see Eq. 19-32), it is desirable to choose an E_n level that is several kT in energy above the ground state (Eq. 19-4a) so that $N_n \ll N_m$ to begin with. Laser operation then begins by using (stimulated) *absorption* to increase the population of the upper state E_m, either directly by irradiation at frequency $(E_m - E_0)/h$ or by irradiation at $(E_p - E_0)/h$, followed by some spontaneous transition down to the desired E_m level. Then by applying radiation at the frequency, $(E_m - E_n)/h$, stimulated emission will occur as the applied radiation begins to equalize the populations of levels E_m and E_n.

The pumping radiation may either be left on continuously while the desired emission is being stimulated ("continuous" lasers, used primarily for the laser "knife" or as the light source in high-resolution spectroscopy), or the pumping process may be used to build up the population of the upper E_m level that is then depleted suddenly by later application of the stimulated emission frequency ("pulsed" lasers, used for study of very fast reactions involving electronically excited molecules).

Finally, since it is desirable to have a very long path length through the sample to realize the largest amplifier "gain" from the stimulated emission process (see Fig. 19-10, right-hand diagram), but undesirable to have a bulky long sample chamber, most lasers are constructed with mirrors at both ends of the sample chamber, so that the irradiating light bounces back and forth many times through the sample before escaping through the (slightly transmitting) mirror at one end of the sample tube. In order for the reflected light not to interfere destructively with other light waves in the sample tube, the distance between the mirrors at the two ends of the chambers is carefully adjusted to a fixed number of wavelengths of the stimulated emission radiation, so that the laser intensity builds up coherently even in the presence of the mirrors. The "pumping" source can then consist simply of a flash tube wrapped around the cylindrical sample chamber (Fig. 19-12). Laser applications are generally based on one of the following properties of laser beams: (1) *high intensity,* (2) *coherence,* (3) *high monochromaticity* (i.e., very little frequency "spread" in the output beam), and (4) very *short duration* (in the case of pulsed lasers). Since lasers are becoming increasingly "tunable" (i.e., variable output frequency), they are becoming the radiation source of choice for optical spectroscopy because of their high *selectivity* (since only a very tiny frequency bandwidth of the spectrum is examined at a time) and high intensity (for study of weak signals). Medical applications are largely based on the large amount of heat produced over a small controllable area using a focussed laser beam. Lasers can be used to provide a short pulse of heat to a chemically reactive mixture to conduct temperature-jump experiments (Chapter 16.A.1) because of the absorption of laser light by the solvent. Very short laser pulses (10^{-12} sec or less) have been used to excite and study superfast reactions in small molecules—these experiments are leading to better understanding of the primary

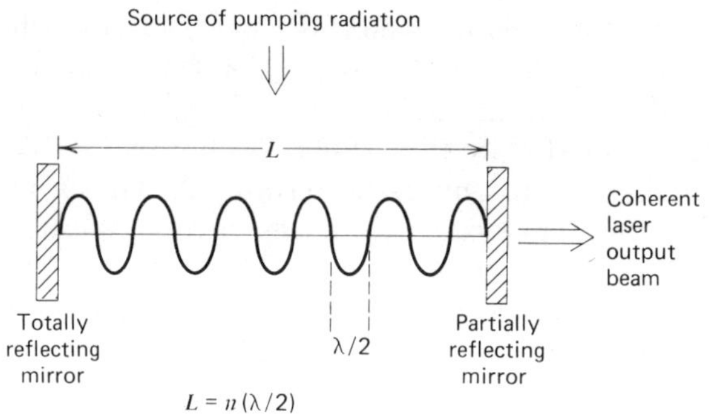

FIGURE 19-12. Highly schematic diagram of a laser. Energy (from a flash lamp wrapped around the sample) is initially fed into the region between the mirrors (the laser "cavity") to populate level E_m of Fig. 19-11. When light is then supplied at the frequency, $(E_m - E_n)/h$ (usually by spontaneous emission from E_m to E_n to begin with), that light then stimulates further transitions, and if the cavity is constructed with the correct length (see figure), the phase of each of those waves is preserved on reflection from either end, and the stimulated emission amplitude grows coherently with each reflection. By making one of the mirrors slightly transmittive, a fraction of this coherent radiation escapes to form the laser output.

events that follow immediately after initial absorption of radiation by matter, as, for example, in primary events of photosynthesis.

EXAMPLE *Bloodless Surgery: The Laser Knife*

Absorption of any radiation results in generation of some heat as the energy is dissipated from the excited molecules to the surroundings. When the radiation is from a highly intense focussed laser source, the heat produced is so substantial that tissue can actually be vaporized near the focussed point. Using such a beam as a knife for surgery has several advantages, principally that the *depth* of cutting can readily be controlled by using a well-focussed beam, so that it is easier to cut through just one layer of tissue at a time for delicate operations; furthermore, since blood is coagulated by heat, any small blood vessels severed by the beam will be sealed shut by coagulation immediately following the cut (see Fig. 19-13 for an impressive example).

The best-known and most significant clinical contribution of the laser is in treatment of retinal detachment, a condition in which the retina of the eye (the layer containing the light-sensitive cells) becomes detached from the choroid layer of cells immediately behind the retina at the back of the eye. Using only a "cyclopegic" drug to dilate the pupil, the operation consists simply of focussing an intense laser beam at the site of detachment: laser-induced coagulation acts to cement the retina back to the choroid layer, preventing further all-out retinal detachment over a wide area.

Lasers should in principle be useful for excision of tumors or in preferential destruction of any tissue whose color is sufficiently different from that of its

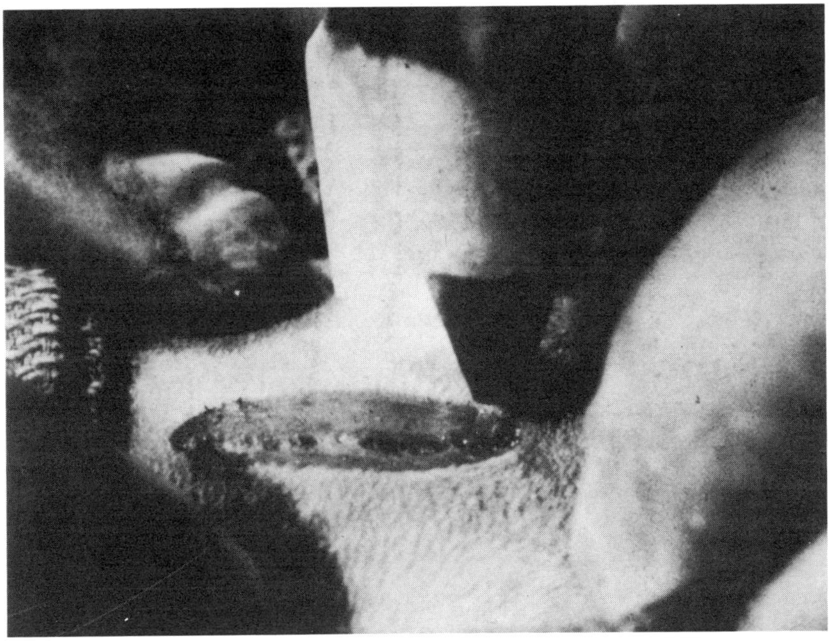

FIGURE 19-13 Bloodless skin incision on a rat, using a continuous CO_2 laser. The cutting power of the focussed beam is about 2500 watts/cm^2! (From American Optical, as printed in L. Goldman and R. J. Rockwell, Jr., *Lasers in Medicine,* Gordon & Breach, New York, 1971.

surroundings. Lasers have in fact been used to lighten "port-wine" birthmarks and tattoos and to treat skin melanoma.

EXAMPLE *Micro-Surgery*

Because of the small size of the focussed laser beam (50 microns), extremely delicate operations should become easier, as for cases of difficult access such as the pituitary gland. It is even possible to selectively destroy certain cell organelles (chromosomes, mitochondria, nucleolus) while leaving the others intact, opening up a wide new range of physiological studies, typified by the example in Fig. 19-14.

EXAMPLE *Laser Light-Scattering*

We will deal with this subject at some length in Chapter 21.C. For now, we simply note that light-scattering *intensity* at a given scattering angle (see Chapter 15.A) depends on the *size* of the object from which the light is scattered. When visible light is used, its wavelength is sufficiently large for the scattering measurement to distinguish between blood cells of different size, and with a suitable flow system, it is possible to scan a cell population according to cell size. It is easy to imagine variants of this technique in which the

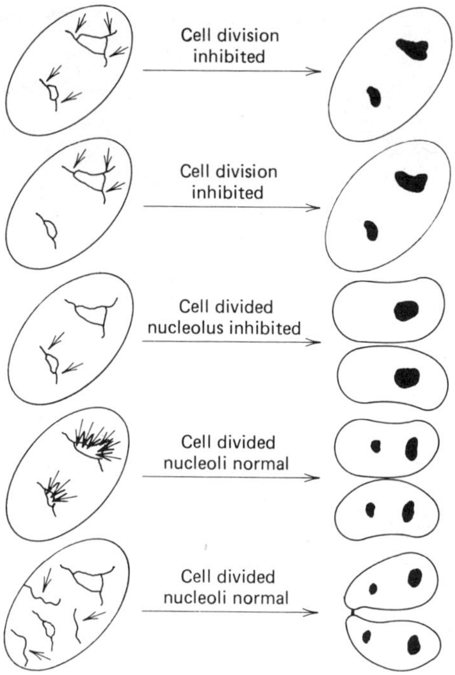

FIGURE 19-14 Effects of argon laser micro-irradiation at various sites of irradiation (lesion diameter approx. 1 micron). From varying effects according to varying irradiation sites, one can begin to deduce which nucleolar-associated chromosome segments are associated with which features of cell division. [From Y. Ohnuki, M. W. Berns, D. E. Rounds, and R. S. Olson, *Exptl. Cell. Res. 71,* 132 (1972).]

cells are stained with light-scattering dyes specific for, say, DNA, so that cells may be scanned and even separated according to a variety of cell sizes, shapes, and contents. Since abnormal cells often have unusual shape and/or size, the technique appears very promising for pre-screening in diagnosis of disease; the data are adequate, for example, to distinguish the full range of cell types (from normal to pre-cancerous abnormal to cancerous) in cells from the female genital tract.

EXAMPLE *Laser Raman Spectroscopy*

Intramolecular vibrations provide an amplitude modulation to the motion of molecular electrons driven by optical radiation, and it is thus effectively possible to obtain the infrared spectrum of a molecule by observing the side-bands from a sample that has been irradiated with optical light (see Chapter 17.C). Since the coupling of the two types of motion (driven electron motion and vibrational motion of nuclei) is weak, the amplitude of the sideband is weak and a very strong optical source is required. Recent Raman experiments are thus being performed increasingly using a laser as the optical source.

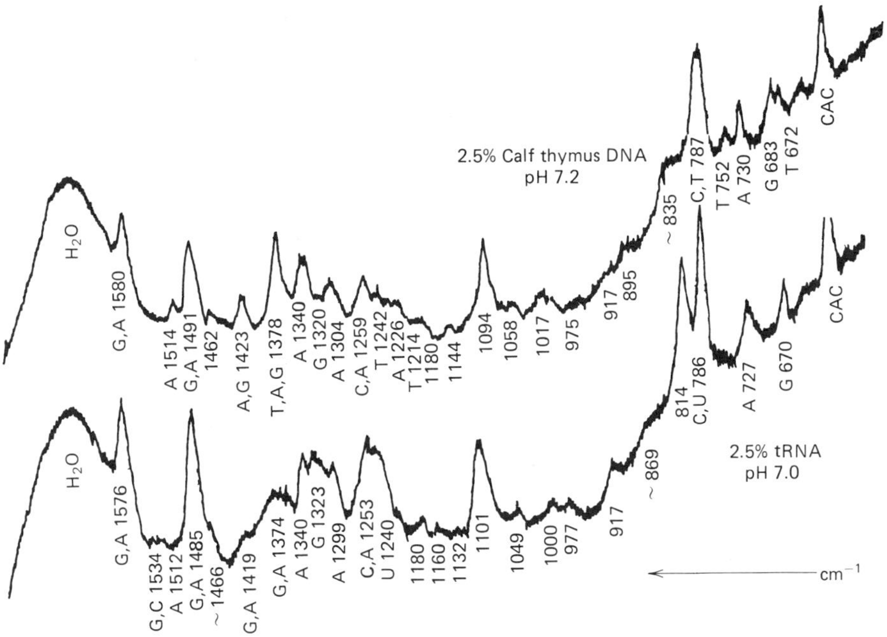

FIGURE 19-15 Laser-induced Raman spectra of calf thymus DNA in 2.5% aqueous solution at pH 7.2 (upper), and yeast transfer RNA in 2.5% aqueous solution at pH 7.0 (lower). Peaks that can be identified as arising from particular nucleotides are labeled accordingly. (Compare this spectrum for detail with the visible absorption spectra of Figure 14-5*b*.) (Adapted from W. L. Peticolas, in *Advances in Raman Spectroscopy,* Vol. 1, ed. J. P. Mathieu, Heydon & Son, Ltd., London, 1973, p. 286.)

Since the laser intensity is concentrated into a frequency bandwidth approximately one-millionth as wide as with a typical lamp source, the power available for the Raman experiment is many orders of magnitude larger when laser sources are used, and the Raman spectra (i.e., the sidebands) are also much more intense. Figure 19-15 shows the Raman spectrum of a typical DNA and a typical RNA system; it is clearly possible to resolve signals from the four bases from one another (compare the much poorer resolution of ordinary visible absorption, Fig. 14-5*b*, for example). By monitoring the change in relative intensities of various bases as functions of various perturbants, it should be possible to make deductions about the solution structure of these molecules.

EXAMPLE *Laser-induced Temperature-Jump Chemical Relaxation Kinetics*

Lasers provide a particularly attractive means for introducing the sudden, small temperature change ("T-jump") that initiates chemical relaxation experiments (see Chapter 16.A.1). The only problem is that the absorption

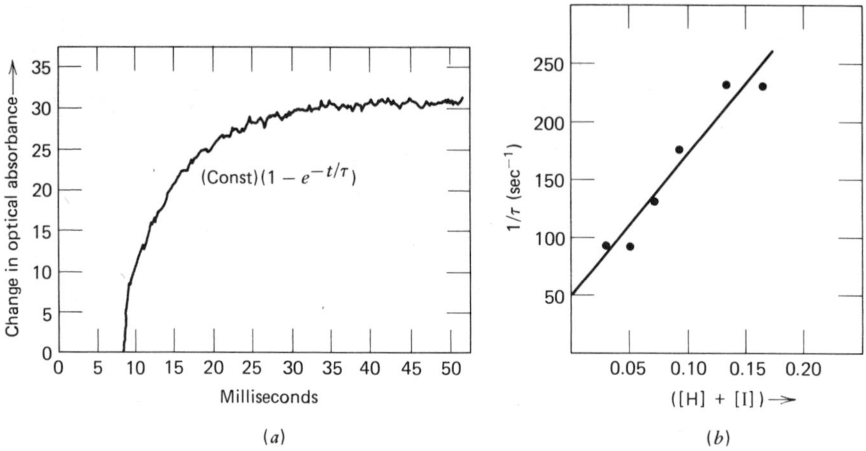

FIGURE 19-16. Laser-induced temperature-jump experiment, for binding of methylisonitrile [I] to the β-chains of hemoglobin [H]. Left: plot of absorbance change versus time following laser heating-pulse, showing exponential change in absorbance. Right: plot of reciprocal exponential relaxation constant, $(1/\tau)$, from plots of type (*a*) taken at several different initial equilibrium concentrations of [H] and [I]. Table 16-1 shows that this reaction is bimolecular: $I + H \rightleftharpoons HI$. [From J. H. Baldo, B. A. Manuck, E. B. Priestley, and B. D. Sykes, *J. Amer. Chem. Soc. 97,* 1684 (1975).]

frequency for water (i.e., the frequency at which irradiation would produce most effective heating of the solution) is appreciably removed from conveniently available high-power laser source frequencies. A recent innovation that solves this problem is illustrated in Fig. 19-16, in which the sample was placed directly inside the laser "cavity," so that an even T-jump of 10°C or larger could be achieved in less than 1 msec. The change in absorbance was monitored as a function of time (Fig. 19-16*a*) for a system consisting of β-chains of hemoglobin [H] and methylisonitrile [I]. The reciprocal exponential time constant for several such curves taken at varying concentrations of [H] and [I] was then plotted (Fig. 19-16*b*) versus ([H] + [I]), and the straight-line behavior indicated a second-order binding process of the type, $I + H \rightleftharpoons HI$ (see Table 16-1). The attractiveness of this method is that it produces heating that is relatively even throughout the sample, in a time which can potentially be made very short (say 10^{-8} sec) using pulsed laser sources. It should thus be possible to study chemical relaxation in solution for processes up to 1000 times faster than presently accessible using conventional T-jump methods (as by discharge of a capacitor).

EXAMPLE *Miscellaneous Laser Applications*

The *coherent* nature of laser radiation allows for the *three*-dimensional imaging of objects with optical resolution, in a process known as *holography*. Since molecules containing different isotopes (^{12}C versus ^{13}C, for example) may absorb light at different frequencies, it has proved possible to *separate isotopes* by irradiating an isotopic mixture at a (laser) frequency at which only one of the isotopes absorbs the light appreciably; the energetically excited isotopic molecule may then decompose or otherwise chemically react to a form readily separable from the other isotopic species. A number of such isotopic separations have been found, though none is yet sufficiently efficient to displace existing methods. Finally, it has been proposed that lasers can be instrumental in achieving controlled thermonuclear fusion, and there is hope that laser-assisted fusion will become feasible in one or two decades.

19.B.3. Saturation Phenomena and Spectral Lifetimes

In the preceding Section, we calculated the effect of monochromatic radiation on the relative populations of a system of *isolated* particles characterized by two energy-levels, E_m and E_n. The principal result of that calculation appeared as Eq. 19-32 showing that externally applied radiation at a frequency of $(E_m - E_n/\hbar)$ causes any initial population difference, $(N_n - N_m)$, to decrease exponentially to *zero* while the irradiation is applied. However, for an actual system in *thermal equilibrium* with its *surroundings,* it is found that in the absence of externally applied radiation, the two populations are not the same, but obey the Boltzmann ratio of Eq. 19-4a. We are therefore led to conclude that there must be some "coupling" or connection between the system and its surroundings that makes "downward" transitions more likely than "upward" transitions by the Boltzmann factor that describes the two populations at *thermal* equilibrium:

$$\frac{N_n}{N_m} = \frac{W\downarrow}{W\uparrow} = e^{-(E_n - E_m)/kT} \quad \begin{array}{l}\text{at thermal equilibrium}\\ \text{(no radiation applied)}\end{array} \qquad (19\text{-}33)$$

Equation 19-33 is conceptually identical to the usual relation between an equilibrium constant and the component rate constants in an ordinary *chemical* equilibrium

$$A \underset{k_{-1}}{\overset{k_1}{\rightleftharpoons}} B; \qquad \frac{[B]}{[A]} = \frac{k_1}{k_{-1}} \quad \text{at chemical equilibrium} \qquad (19\text{-}34)$$

in which $W_\uparrow$ and $W_\downarrow$ represent the "rate constants" for transitions upward or downward (see Fig. 19-8).

In the *absence* of applied radiation, we can write equations for the rate of change of energy-level populations, dN_n/dt and dN_m/dt, where the rate constants $W\uparrow$ and $W\downarrow$ for (radiationless) transitions between levels reflect some sort of interaction between the system and its surroundings (this interaction is discussed more precisely in Chapter 21.A and 21.B):

$$dN_m/dt = -W_\downarrow N_m + W_\uparrow N_n \qquad (19\text{-}35a)$$

$$dN_n/dt = -W_\uparrow N_n + W_\downarrow N_m \tag{19-35b}$$

or just

$$\frac{d(N_n - N_m)}{dt} = (N_m + N_n)(W_\downarrow - W_\uparrow) - (N_n - N_m)(W_\downarrow + W_\uparrow) \tag{19-36}$$

It is convenient to rewrite Eq. 19-36 in the form

$$\boxed{d(N_n - N_m)/dt = \frac{(N_n - N_m)_{eq} - (N_n - N_m)}{T_1}} \tag{19-37}$$

where

$$(N_n - N_m)_{eq} = (N_n + N_m)\left(\frac{W_\downarrow - W_\uparrow}{W_\downarrow + W_\uparrow}\right) \tag{19-37a}$$

$$\boxed{\frac{1}{T_1} = (W_\downarrow + W_\uparrow)} \tag{19-38}$$

For example, starting from a situation in which $N_n = N_m$ at time zero (i.e., both levels equally populated initially), Eq. 19-37 is quickly solved to give

$$(N_n - N_m) = (N_n - N_m)_{eq}\,(1 - e^{-t/T_1}) \tag{19-39}$$

from which it is clear that $(N_n - N_m)_{eq}$ represents the equilibrium population difference between the two levels. Equation 19-39 describes, for example, the exponential increase in magnetic moment for a system of spin one-half nuclei suddenly placed in a constant magnetic field at time zero (see Fig. 19-17).

We may now combine Equations 19-31 and 19-37 to find the behavior of a two-level system subject both to externally applied radiation at fre-

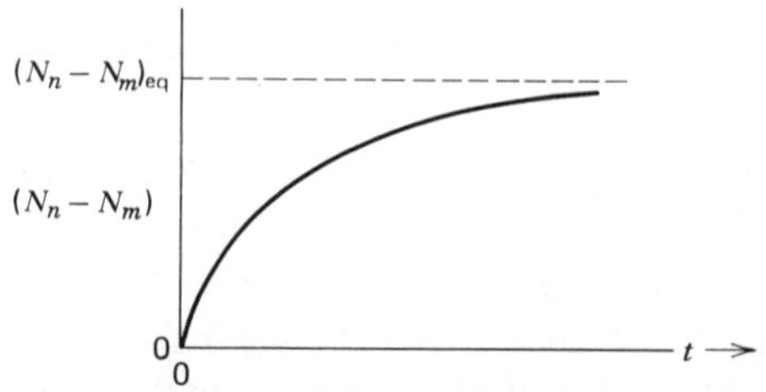

FIGURE 19-17 Plot of population difference versus time, for a two-level system with initial population difference equal to zero at time zero. At infinite time later (i.e., thermal equilibrium) the population difference approaches its thermal equilibrium value $(N_n - N_m)_{eq}$ (see Eq. 19-37a).

quency, $(E_m - E_n)/\hbar$, *and* to the interaction between the system and its surroundings characterized by the parameter, T_1, of Eq. 19-38:

$$\frac{d(N_n - N_m)}{dt} = -2W(N_n - N_m) + \frac{(N_n - N_m)_{\text{eq}} - (N_n - N_m)}{T_1} \qquad (19\text{-}40)$$

whose steady-state solution (i.e., the population difference observed after the applied electromagnetic field has been "on" for a time long compared to T_1) is:

$$\boxed{(N_n - N_m)_{\text{steady-state}} = \frac{(N_n - N_m)_{\text{eq}}}{1 + 2WT_1}} \qquad (19\text{-}41)$$

Equation 19-41 leads to several important conclusions and applications. First, it accounts for the phenomenon of "saturation" in spectroscopy. Based on the radiation-*induced* transition probability for an *isolated* system (Eq. 19-32), there was no steady-state, and the population difference (and hence the observed spectroscopic absorption signal) decreased exponentially to zero while the irradiation was applied. However, because of the countervailing tendency for the lower energy-level to become relatively more populated than the upper due to coupling of the system to the *surroundings*, Eq. 19-41 predicts that there will eventually be a steady-state situation in which the population-*equalizing* effect of the applied radiation exactly cancels the tendency for *increase* in the population *difference* toward a thermal equilibrium value from coupling to the surroundings. So long as $2WT_1 << 1$, this steady-state has the property that $N_n - N_m \cong (N_n - N_m)_{\text{eq}}$, and the applied radiation does not appreciably alter the equilibrium energy-level populations. However, since the rate of (induced) absorption of energy (the ordinate in a typical (power) "absorption" spectrum), dE/dt, is given by

$$\frac{dE}{dt} = (N_n - N_m)(E_m - E_n)W = (N_n - N_m)_{\text{eq}}(E_m - E_n)\left(\frac{W}{1 + 2WT_1}\right) \qquad (19\text{-}42)$$

(based on Eq. 19-31) and since W is proportional to the square of the electric (or magnetic) field amplitude of the applied radiation (Eq. 19-28), it is clear that the power absorption spectrum signal will increase as the applied radiation intensity increases, as long as $2WT_1 << 1$. However, when the applied radiation becomes sufficiently intense that $2WT_1 \cong 1$, then the power absorption begins to level off, and for even more intense radiation levels, the power absorption fails to increase when the radiation intensity is further increased. This last situation is called "saturation": power absorption is determined (in this limit) not by the strength of the irradiating field, but by the rate at which the system can repopulate its lower energy level via coupling to the surroundings.

One immediate consequence of Eq. 19-42 is that it leads to a simple means for determination of the system-surroundings coupling constant, T_1. A plot of power absorption, dE/dt versus square of applied oscillating field strength will level off at high field strength to a value given by $(N_n - N_m)_{eq}(E_m - E_n)/2T_1$, from which T_1 can be evaluated.

Once the parameter T_1 has been determined, its magnitude may be related to either motional or distance information (see Chapter 16.B.3 and Chapter 21.B). Alternatively, it is possible to distinguish between different regions of a macroscopic specimen provided that the T_1 values are different in the two regions (see "zeugmatography" example in Chapter 22.C.2). Finally, it is possible to monitor the rates of various chemical or physical processes when the rate of interest is of the order of $(1/T_1)$, as shown by the following cytochrome-*c* example.

Saturation Transfer As a Measure of Reaction Rates

Suppose a molecule could exist in two equally probable forms, which differ in the energy spacing between the two energy levels, E_m and E_n of the preceding discussion of Chapter 19.B.3, and that the molecule could isomerize from one form to the other as shown schematically below. Beginning with

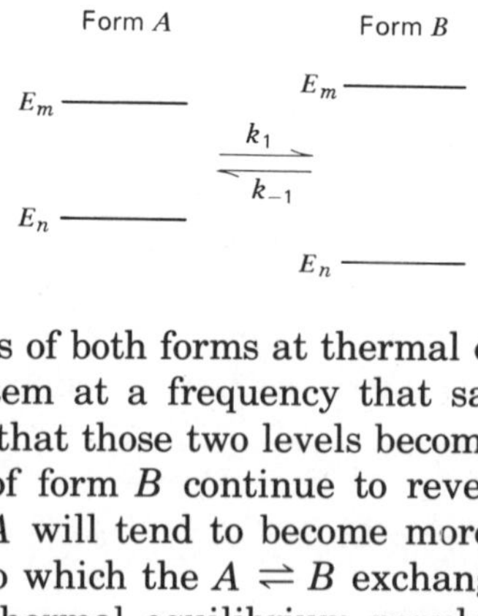

a collection of molecules of both forms at thermal equilibrium, suppose we next irradiate the system at a frequency that saturates the transitions between E_m^B and E_n^B so that those two levels become equally populated. As (saturated) molecules of form *B* continue to revert back to form *A*, the energy levels of form *A* will tend to become more equally populated, according to the extent to which the $A \rightleftharpoons B$ exchange rate is fast compared to the rate at which thermal equilibrium populations are reached [i.e., $(1/T_1^A)$]. Thus, the rate of exchange between forms *A* and *B* can be determined by monitoring the population difference between the two energy levels of form *A* immediately after irradiation of form *B* at a power level which saturates the energy level transitions of form *B*, provided only that k_1, k_{-1} are of the order of $(1/T_1^A)$. The experiment may be appreciated most readily from a recent example.

EXAMPLE *Electron Exchange Between Reduced and Oxidized Forms of Cytochrome c*

Portions of the nuclear magnetic resonance spectrum of oxidized cytochrome *c* are shown in Fig. 19-18*a*. Peak *a* is assigned to the methionine methyl

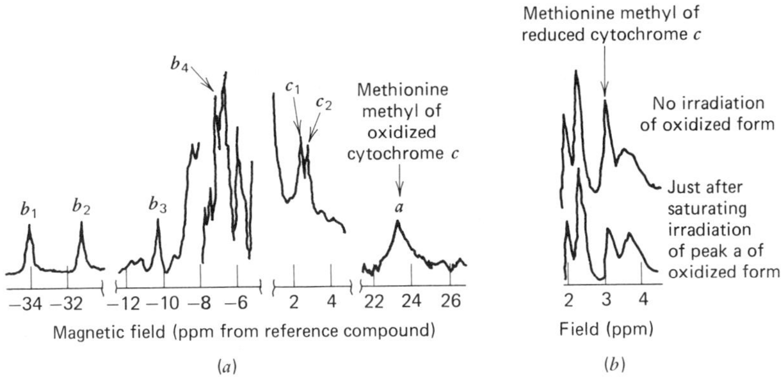

FIGURE 19-18. (*a*) Proton nuclear magnetic resonance spectrum of oxidized cytochrome *c* in aqueous solution (see text for assignment of peaks). (*b*) Partial proton NMR spectrum of a solution of cytochrome *c* containing approximately equal concentrations of oxidized and reduced forms. The intensity of the methionine methyl peak of the *reduced* form is appreciably decreased immediately following saturating irradiation of the methionine methyl of the *oxidized* form, showing that the rate of interconversion between oxidized and reduced forms is about (1/4) second (i.e., T_1 for the reduced form—see text). [From R. K. Gupta and A. G. Redfield, *Science 169,* 1204 (1970).]

protons, peaks b_1 to b_4 to porphyrin ring methyl groups, and peaks c_1 and c_2 to porphyrin side chain methyl groups. The spectrum of the reduced form is similar, except that since the reduced form is diamagnetic (the oxidized form is paramagnetic), the methyl peaks are now observed at different frequencies due to an absence of the paramagnetic-induced frequency shifts discussed in Chapter 19.A. For example, the methionine methyl peak shifts from 23.4 ppm in the oxidized cytochrome to 3.3 ppm in the reduced form, corresponding to the situation diagrammed above. When both oxidized and reduced forms are present, the composite spectrum is simply a superposition of the individual spectra of the two forms, since exchange between the two forms is slow compared to the difference in resonant frequency between the two forms (review this point in Chapter 14.D and Problem 14-6).

If irradiation at high power level is now applied at the frequency of the methionine methyl of the *oxidized* form (23.4 ppm), and the same methionine methyl of the *reduced* form (3.3 ppm) is then observed (see Fig. 19-18*b*), it is seen that the intensity of the methionine methyl of the *reduced* form has decreased intensity, due to a partial equalizing of the populations of the methionine methyl energy levels of the *reduced* form because of exchange of (saturated, and thus equally populated) methionine methyl energy levels from the *oxidized* form. Now if the methionine methyl of the reduced form had a T_1 much shorter than the lifetime for the oxidized $\rightleftharpoons$ reduced exchange process, the intensity of the methionine methyl of the reduced form would be unchanged on irradiation of the oxidized form. Similarly, if the exchange rate were very fast compared to $(1/T_1)$ for the reduced form, the intensity of the methionine methyl of the oxidized form would go to zero on saturation of the oxidized form. Since the observed result in Fig. 19-18*b* is intermediate between these extremes, it is clear that the lifetime for electron transfer between the oxidized and reduced forms of cytochrome *c* is of the order of T_1 for the methionine methyl of the reduced form, namely about one-quarter of a second.

Experiments of this type are presently being attempted on visual and photosynthetic pigments, based on the much shorter saturation parameters (T_1) of transitions between optical (i.e., electronic) energy levels, to detect and quantify the very rapid changes that follow the absorption of a photon by such molecules. These changes occur on a time scale shorter than 10^{-9} sec.

Fluorescent Energy Transfer As a Spectroscopic Ruler

In magnetic resonance, the communication between a given nuclear spin and its surroundings (i.e., the T_1 process) is most commonly attributable to a dipole-dipole interaction between the magnetic dipole of the given nucleus and the dipole moments of surrounding nuclei. As discussed in Chapter 21.B, T_1 in this situation varies as the *sixth power* of the distance between the two interacting nuclear dipoles (see Chapter 16.B.3 for an example). A similar interaction between the two electric (as opposed to magnetic as in the nuclear example) dipoles of electronically excited species has recently been shown to provide a sensitive measure of the distance between two fluorescent moieties on the same molecule. Thus, if a macromolecule were to be labeled with two such moieties, one could in principle measure the distance between the two labeled sites on the macromolecule, based on this fluorescence experiment. A schematic picture of the relevant processes is shown in Fig. 19-19.

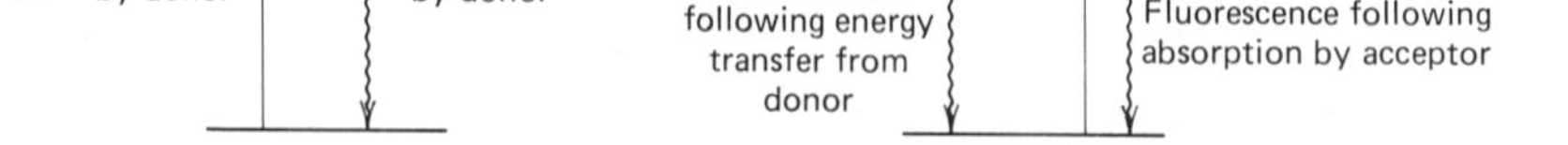

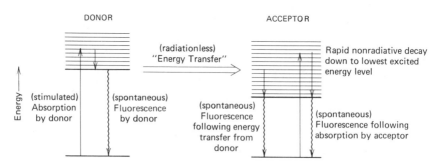

FIGURE 19-19. Schematic energy diagram of the relevant processes involved in fluorescent energy transfer. The acceptor region of the molecule can reach an excited energy state *either* by stimulated absorption of externally applied radiation (solid upward arrow at right), *or* by transfer of energy from a similarly excited nearby donor region of the same molecule ("energy transfer" arrow). Following excitation, there is rapid nonradiative (i.e., no radiation is emitted) decay of the acceptor energy down to a lowest excited level from which *spontaneous* emission occurs. Since the energy transfer process depends very strongly on the distance between the donor and acceptor ends of the molecule, that distance can be deduced by comparing the fluorescence for the acceptor moiety in the presence and absence of a nearby donor group (see text).

Fluorescence itself is an example of *spontaneous emission* (i.e., emission of radiation in the apparent *absence* of any externally applied oscillating fields). While the origin of this phenomenon is (slightly) beyond the scope of this discussion, we need only be aware that the rate of spontaneous emission varies as the cube of the energy-level difference involved, so that this process is clearly much more important at optical frequencies than at microwave or radiofrequencies (moreover, spontaneous emission in magnetic resonance is weaker by an additional factor of about 10^5 because magnetic interactions are so much weaker than electrical ones). Figure 19-19 provides the basis for the experimental application of fluorescent energy transfer as a distance indicator in molecules in solution.

EXAMPLE *Calibration of the Fluorescent Yardstick*

Although the distance-dependence of the energy-transfer process is reasonably well-established, it is still necessary to calibrate the observed fluorescence intensity as a function of known distance between donor and acceptor moieties in a model rigid compound to establish the proportionality constant. Such a compound is shown below, based on donor and acceptor groups separated by a known number of L-proline links—the compound was chosen because the poly-L-proline chain adopts a rigid "type II trans helix" configuration whose end-to-end distance is readily measured from a molecular model.

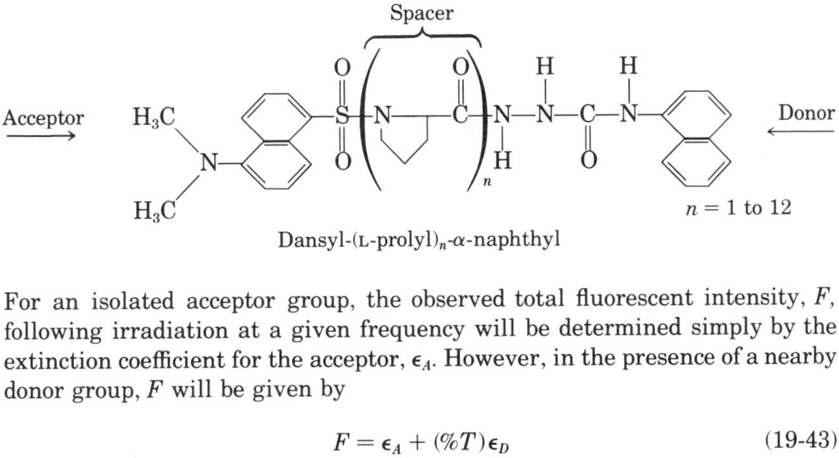

Dansyl-(L-prolyl)$_n$-α-naphthyl

For an isolated acceptor group, the observed total fluorescent intensity, F, following irradiation at a given frequency will be determined simply by the extinction coefficient for the acceptor, ϵ_A. However, in the presence of a nearby donor group, F will be given by

$$F = \epsilon_A + (\%T)\epsilon_D \tag{19-43}$$

in which ϵ_D is the extinction coefficient of the donor group and $(\%T)$ is the percent of the excited donor energy transferred to the acceptor group. [Because the energy of the acceptor fluorescence is so much less (see Fig. 19-19) than that of the donor fluorescence, it is easy to discriminate between the two, so as to detect only the fluorescence from the acceptor group.]

Figure 19-20 shows the experimentally determined $(\%T)$ for various distances of separation between donor and acceptor groups of dansyl-(L-prolyl)$_n$-α-naphthyl, for values of n ranging from 1 to 12. The energy transfer is 50%

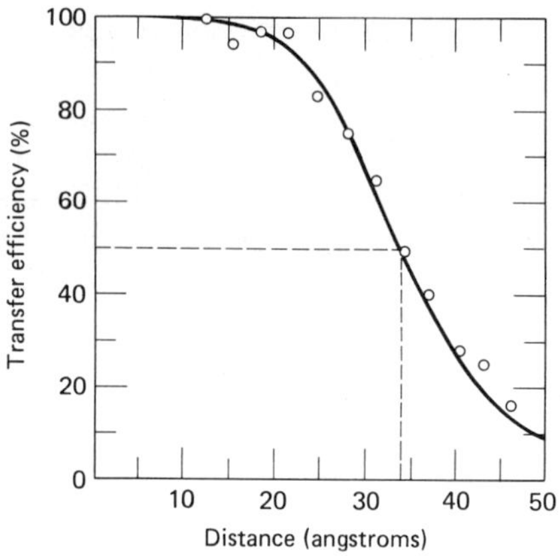

FIGURE 19-20 Calibration of the fluorescent spectroscopic "ruler." Plot of efficiency of fluorescent energy transfer from donor to acceptor (see Fig. 19-19) as a function of the (known) distance between the donor and acceptor groups in the dansyl-(L-prolyl)$_n$ $-\alpha$-naphthyl molecule, with n varying from 1 to 12, corresponding to a donor:acceptor separation between 12 and 46 Å. The solid line is a theoretical fit to the data, based on a choice of $R_0 = 34.6$ Å as the distance at which fluorescent energy transfer is 50% efficient. [From L. Stryer and R. P. Haugland, *Proc. Natl. Acad. Sci.* U. S. A. *58* 719 (1967).]

efficient at a donor:acceptor separation of 34.6 Å, and the experimental data can be fitted to the theoretical expression

$$(\%T) = \frac{(R_0/r)^6}{\left(\frac{R_0}{r}\right)^6 + 1} \tag{19-44}$$

in which r is the donor:acceptor separation distance, and R_0 is the separation distance at which the energy transfer is 50% efficient ($\%T = 0.5$ at $R_0 = 34.6$ Å in this case).

The excellent agreement between experimental fluorescent energy transfer efficiency and a theoretical sixth-power distance dependence on donor:acceptor separation distance indicates that this sort of measurement should be useful in determining the distance between two labeled regions (one labeled with donor and one with acceptor) on a macromolecule in solution, when the distance is in the range of about 25 to 40 Å. By using different donor:acceptor pairs with different R_0 values, it should be possible to construct other "rulers" to measure shorter or longer distances, so that this technique could be used to determine distances in the range of 10 to 60 Å, which correspond to the dimensions of all but the largest proteins.

19.B.4. Intensities of Spectral Absorption Lines

Selection Rules and Relative Line Intensities According to our previously derived Eq. 19-24, a transition (more precisely, an electric dipole transition) between energy state E_n and energy state E_m will occur when the system is irradiated with an electromagnetic field oscillating at the corresponding frequency, $(E_m - E_n)/h$, *provided* that the matrix element of the electric dipole moment operator, Eq. 19-22, between the wave functions ψ_n and ψ_m is *nonzero*. The latter part of the preceding sentence states the "*selection rule*" for transitions between energy levels. Since (see Chapter 19.B.3) the *power absorption* by a quantum mechanical system is proportional to the probability for induced transitions (Eq. 19-42)—provided that irradiation power is kept below the saturation limit ($2WT_1 << 1$)—and since the transition probability is proportional to the square of the matrix element of the dipole operator between ψ_n and ψ_m (Eq. 19-28), the selection rule thus predicts the *presence* (or absence) and *relative intensities* of the possible transitions between the various energy levels resulting from a quantum mechanical calculation. Selection rules are most easily illustrated for the relatively simple energy-level schemes resulting from magnetic resonance phenomena, and examples follow. The main feature of these calculations is that it is now necessary to determine the actual *wave functions*, ψ_i, corresponding to the already computed energy levels, E_i, to evaluate the dipole operator matrix elements in question. As long as we were concerned only with energy *levels*, it was *not* necessary to write down the associated wave functions themselves, but it *is* necessary to know the wave functions to determine which transitions are "allowed" (i.e., those for which $|\mu_{x_{mn}}|^2 \neq 0$) and their relative intensities.

Using a "nuclear tetherball" analogy, we argued in Chapter 14.D that in order to transfer power to a *classical* magnetic moment precessing around a static externally applied magnetic field, it is necessary to apply a force in a direction perpendicular to both the axis and radius of the precession. (In other words, the best way to accelerate a tetherball is to push it around in a direction parallel to the ground.) Using quantum mechanical language, we would say that the driving energy is provided by an applied external *oscillating* magnetic* field, $H_x\cos(\omega t)$, interacting with the x component of the *magnetic* dipole moment of the particle(s), $\mu_x = \gamma\hbar I_x$, where I_x is the x component of the quantum mechanical nuclear angular momentum ("spin") operator:

$$\text{Driving energy} = \gamma\hbar\, H_x\, I_x \cos(\omega t) \tag{19-45}$$

But Eq. 19-45 describes an interaction formally identical to that already treated (Eq. 19-17 ff.) for the effect of an oscillating *electric* field interacting with the *electric* dipole moment of a system of particles, except that the

* *This oscillating field may be thought of as being composed of two counter-rotating fields of equal magnitude. Since only one of the rotating fields will be rotating in the same sense as the precessing moment, it can be shown that the other rotating field may be ignored—a common mathematical trick.*

transition probability for the magnetic case is now proportional to the square of the matrix element, $|I_{x_{mn}}|^2$, instead of the matrix element, $|\mu_{x_{mn}}|^2$. We are now in a position to evaluate some spectral absorption intensities for some simple spin examples.

EXAMPLE *One Type of Nucleus of Spin One-half in a Static Magnetic Field*

For a single type of magnetic nucleus of spin one-half in a static magnetic field, H_0 along the z direction, the energy levels and wave functions are shown in Fig. 18-5. There is only one transition involved in absorption, namely from state β to state α. We need therefore simply evaluate the matrix element (scalar product, projection)

$$\langle \alpha, I_x \beta \rangle$$

to find out if that transition is "allowed." Such calculations are greatly facilitated by expressing the I_x operator in terms of the "raising" and "lowering" operators (Equations 18-61a, 18-61b) discussed in Chapter 18.A:

$$\boxed{I_x = \frac{I_+ + I_-}{2}} \qquad (19\text{-}46)$$

Then, using the simple properties of the raising and lowering operators (Prob. 18-3),

$$I_+ \beta = \alpha \qquad (19\text{-}47a)$$

$$I_+ \alpha = 0 \qquad (19\text{-}47b)$$

$$I_- \beta = 0 \qquad (19\text{-}47c)$$

$$I_- \alpha = \beta \qquad (19\text{-}47d)$$

and the by now familiar orthogonality ("perpendicularity") properties of the basis functions, α and β

$$<\alpha, \beta> = <\beta, \alpha> = 0 \qquad (19\text{-}48a)$$

$$<\alpha, \alpha> = <\beta, \beta> = 1 \qquad (19\text{-}48b)$$

the desired matrix element is immediately obtained:

$$<\alpha, I_x \beta> = \left\langle \alpha, \left(\frac{I_+ + I_-}{2}\right) \beta \right\rangle$$

$$= \frac{1}{2} (<\alpha, I_+ \beta> + <\alpha, I_- \beta>)$$

$$= \frac{1}{2}(<\alpha, \alpha> + <\alpha, 0>)$$

$$<\alpha, I_x\beta> = \frac{1}{2}(1 + 0) = \frac{1}{2} \neq 0 \tag{19-49}$$

↑ This transition is "allowed"

EXAMPLE *Two Chemically Different Isolated Nuclei of Spin One-half*

A more interesting case involves the energy-level scheme shown in Fig. 18-6, for two (uncoupled) nuclei of different resonant frequency. In this case, there are six possible transitions between the four energy levels. The individual calculation of transition probability for each pair of levels is carried out as before, recognizing that two particles are involved, so that the total x component of angular momentum, I_x, is the sum of the x components for the two particles individually, I_{x_1} and I_{x_2}:

$$I_x = I_{x_1} + I_{x_2} \tag{19-50a}$$

$$I_x = \frac{(I_+ + I_-)_1}{2} + \frac{(I_+ + I_-)_2}{2} \tag{19-50b}$$

Since the four wave functions are $\alpha_1\alpha_2$, $\alpha_1\beta_2$, $\beta_1\alpha_2$, and $\beta_1\beta_2$, the probability for a transition between states $\alpha_1\alpha_2$ and $\beta_1\beta_2$, for example, is proportional to

$$|<\alpha_1\alpha_2, I_x\beta_1\beta_2>|^2 = |<\alpha_1\alpha_2, (I_{x_1} + I_{x_2})\beta_1\beta_2>|^2$$

$$= \left|\left\langle\alpha_1\alpha_2, \left(\frac{(I_+ + I_-)_1}{2} + \frac{(I_+ + I_-)_2}{2}\right)\beta_1\beta_2\right\rangle\right|^2$$

$$= \frac{1}{4}(<\alpha_1\alpha_2, I_{+1}\beta_1\beta_2> + <\alpha_1\alpha_2, I_-\beta_1\beta_2>$$
$$+ <\alpha_1\alpha_2, I_{+2}\beta_1\beta_2> + <\alpha_1\alpha_2, I_{-2}\beta_1\beta_2>)^2$$

$$= \frac{1}{4}(<\alpha_1\alpha_2, \alpha_1\beta_2> + <\alpha_1\alpha_2, 0> + <\alpha_1\alpha_2, \beta_1\alpha_2>$$
$$+ <\alpha_1\alpha_2, 0>)^2$$

$$<\alpha_1\alpha_2, I_x\beta_1\beta_2> = \frac{1}{4}(0 + 0 + 0 + 0)^2 = 0 \tag{19-51}$$

↑ This transition is "forbidden" (not "allowed")

In obtaining the final line of Eq. 19-51, the following orthogonality relations have been used, based on the mutual "perpendicularity" of the four eigenfunctions (see note at bottom of page 573):

$$<\alpha_1\alpha_2, \alpha_1\beta_2> = <\alpha_1\alpha_2, \beta_1\alpha_2> = 0 = <\alpha_1\alpha_2, \beta_1\beta_2> \tag{19-52}$$

Similarly, the remaining transition probabilities are proportional (with the same proportionality constant (see Eq. 19-24) to:

$$|<\alpha_1\alpha_2, I_x\alpha_1\beta_2>|^2 = \frac{1}{4} \qquad W_{\alpha_1\beta_2\to\alpha_1\alpha_2}:W_{\beta_1\alpha_2\to\alpha_1\alpha_2}:W_{\beta_1\beta_2\to\alpha_1\beta_2}:W_{\beta_1\beta_2\to\beta_1\alpha_2} = 1{:}1{:}1{:}1$$

$$|<\alpha_1\beta_2, I_x\beta_1\alpha_2>|^2 = 0$$

$$|<\alpha_1\beta_2, I_x\beta_1\beta_2>|^2 = \frac{1}{4} \qquad (19\text{-}53)$$

$$|<\beta_1\alpha_2, I_x\beta_1\beta_2>|^2 = \frac{1}{4}$$

$$|<\alpha_1\alpha_2, I_x\beta_1\alpha_2>|^2 = \frac{1}{4}$$

Note that of the six possible transitions, only four are "allowed" in this "first-order" treatment: the "double-quantum" transition, $\beta_1\beta_2 \rightarrow \alpha_1\alpha_2$, is forbidden, and the $\alpha_1\beta_2 \rightarrow \beta_1\alpha_2$ transition is also forbidden. The resulting allowed transitions and corresponding absorption spectrum are shown in Fig. 18-6. There are *two* equally intense absorption lines, since both $W_{\beta_1\beta_2\to\alpha_1\beta_2}$ and $W_{\beta_1\alpha_2\to\alpha_1\alpha_2}$ occur at the same frequency (energy difference) and both $W_{\beta_1\beta_2\to\beta_1\alpha_2}$ and $W_{\alpha_1\beta_2\to\alpha_1\alpha_2}$ also occur at the same frequency (please see Fig. 18-6). It is left as an exercise (see Problems) for the reader to show that the "*AX*" spectrum resulting from all allowed transitions for the Hamiltonian of Eq. 18-79 gives the four equally intense peaks shown in Fig. 18-7.

From these examples, the reader is now in a position to compute the relative intensities of all transitions involving the energy levels for any number of spin one-half nuclei. This covers a very large class of problems in the description of line positions and intensities for proton or carbon-13 nuclear magnetic resonance spectra (electron spin resonance spectra are similar). There exist generally available computer programs that will fit an observed NMR spectrum to that calculated for a system consisting of a stated number of spin-one-half nuclei all coupled together by terms of the type, $J_{ij}I_{i_z}I_{j_z}$, and such programs have proved very successful in extracting the desired spectral positions and coupling constants from spectra for use in assigning configuration of molecules in solution (Chapter 18.A). See Problems for some informative general cases.

PROBLEMS

1. Obtain the ratio of the population of the first "excited" state to the population of the "ground" state, N_1/N_0, given in Table 19-1, from the energy differences for the various types of spectroscopy listed in that Table, for the three temperatures (295 K, 77 K, and 4.2 K) given in the table.

2. Suppose that the effect of a paramagnetic lanthanide atom coordinated to some organic molecule is to split each proton (or carbon-13) magnetic

resonance signal for the organic molecule into two resonances, separated in frequency by an amount equal to $K[(3 \cos^2(\theta) - 1/r^3)$, where r is the distance from the lanthanide atom to the proton (or ^{13}C) of interest, θ is the angle between the lanthanide-proton (lanthanide-carbon) direction and the symmetry axis for the complex, and K is a constant characteristic of that lanthanide atom. Further suppose that the rapid transitions between the two electron spin states (due to very short electron T_{1_e}—see Chapter 19.B.3) effectively act to cause rapid "chemical" exchange between the two proton (carbon-13) resonant frequencies. Use this information to show that this "exchange" will lead to a coalescence of the two proton peaks into a single peak that is shifted from the "un-split" original frequency by an amount proportional to $[(3 \cos^2(\theta) - 1)/r^3]$, in the limit that $(1/T_{1e}) \gg [K(3 \cos^2\theta - 1)/r^3]$. You may assume that the energy difference between the two electron spin states satisfies a "high-temperature limit"

$$e^{-(\Delta\epsilon/kT)} \cong 1 - (\Delta\epsilon/kT)$$

3. Convert from complex to real notation in the expression derived in the text for the relative population of the mth energy level following application of an oscillating electric field—that is, obtain Eq. 19-25b from 19-25a.

4. Show that four of the six possible "absorption" transitions for the "AX" NMR spectrum of Fig. 18-7 (Eq. 18-79 Hamiltonian) are allowed with equal probability. Since each of the allowed transitions falls at a different frequency, you will thus show that the predicted power absorption spectrum consists of four equally intense lines located as shown in Fig. 18-7.

5. For particles of spin one-half (Eqs. 18-61 ff.) and spin one (Problem 18-3), we have shown that the operators, $I_+ = I_x + iI_y$ and $I_- = I_x - iI_y$, act to raise or lower (respectively) the eigenvalue of I_z. Assuming that this result is true for particles of arbitrary spin (Fig. 18-4), show that induced magnetic resonance transitions can occur only between adjacent "rungs" on the I_z eigenvalue "ladder." An analogous calculation shows that transitions between energy levels for the harmonic oscillator (Fig. 18-2) can also occur only between adjacent rungs of the energy "ladder." These two selection rules are sometimes stated as $\Delta m = \pm 1$ and $\Delta v = \pm 1$, where m (or v) simply denotes the integers (or half-integers) that specify a particular "rung" on the respective "ladder."

6. The eight eigenfunctions for the first-order Hamiltonian for three coupled spin one-half nuclei are:

$$\psi_1 = \alpha_1\alpha_2\alpha_3 \qquad \psi_5 = \beta_1\alpha_2\beta_3$$
$$\psi_2 = \alpha_1\alpha_2\beta_3 \qquad \psi_6 = \beta_1\beta_2\alpha_3$$
$$\psi_3 = \alpha_1\beta_2\alpha_3 \qquad \psi_7 = \alpha_1\beta_2\beta_3$$
$$\psi_4 = \beta_1\alpha_2\alpha_3 \qquad \psi_8 = \beta_1\beta_2\beta_3$$

Using the magnetic selection rule, $< \psi_m, I_x \psi_n > \neq 0$ for allowed transitions, show that 12 of the possible 28 induced absorption transitions are allowed, and give their relative intensities—see Problem 18-5b for the appearance of the spectrum.

Hint: $I_x = I_{x_1} + I_{x_2} + I_{x_3}$, and $I_x = (1/2)(I_+ + I_-)$

REFERENCES

Boltzmann Distribution

L. K. Nash, *Elements of Statistical Thermodynamics,* Addison-Wesley, Reading, Mass. (1968).

J. C. Davis, Jr., *Advanced Physical Chemistry,* Ronald Press, New York (1965).

Transitions Between Energy Levels

J. C. Davis, Jr., *Advanced Physical Chemistry,* Ronald Press, New York (1965).

L. Pauling and E. B. Wilson, Jr., *Introduction to Quantum Mechanics,* McGraw-Hill, New York (1935).

Lasers

A. E. Siegman, *An Introduction to Lasers and Masers,* McGraw-Hill, New York (1971).

L. Goldman and R. J. Rockwell, Jr., *Lasers in Medicine,* Gordon and Breach, New York (1971).

M. W. Berns, *Biological Microirradiation,* Prentice-Hall, Englewood Cliffs, N. J. (1974).

Selection Rules

E. Feenberg and G. E. Pake, *Notes on the Quantum Theory of Angular Momentum,* Stanford University Press, Stanford, Calif. (1959).

J. A. Pople, W. G. Schneider and H. J. Bernstein, *High-Resolution Nuclear Magnetic Resonance,* McGraw-Hill, New York (1959), Chapter 6.

R. Chang, *Basic Principles of Spectroscopy,* McGraw-Hill, New York (1971).

6

HARD PROBLEMS INTO SIMPLE PROBLEMS: TRANSFORMS, A PICTURE BOOK OF APPLICATIONS

In this Section, we shall be concerned principally with applications of *Fourier* transform methods. *Mathematically,* a Fourier transform is simply a weighted sum (or integral) of a discrete (or continuous) set of data points. *Physically,* Fourier transforms arise naturally in several sorts of problems. For example, an ordinary lens acts as a Fourier transformer, and since there are no *physical* lenses available in the X-ray region, a *mathematical* transform is used to convert X-ray diffraction patterns into a three-dimensional image of the object in question, as discussed in Chapter 22.A. The human ear or eye also acts as a Fourier transformer, by analyzing a weighted sum of *coherent* sound or electromagnetic waves into its corresponding pitches or colors (frequencies), and a similar mathematical transform of complicated mixtures of spectral oscillations can be used to analyze for the spectral frequency components of the mixture. These spectroscopic Fourier methods (Chapter 20) have had tremendous impact since their introduction about 1965—the mathematical paper making these methods routinely feasible is the most highly cited mathematics article in the entire mathematics literature. Furthermore, transforms can be used to pick out the frequency component(s) of *randomly* fluctuating electric or magnetic fields that fall at the natural frequency for electronic or nuclear motions, and which can then account for the observed line widths in scattering or resonance (spectroscopy) phenomena (Chapter 21). Finally, it is important to note that the critical features of transforms and their applications can be apprehended without extensive mathematical manipulations (note the virtual lack of equations in Chapter 20, for example), and we shall resort frequently to pictures rather than algebra in introducing these subjects.

CHAPTER 20 Weights on a Balance: Shortcuts to Spectroscopy. Multichannel and Multiplex Methods

To understand why Hadamard and Fourier methods have proved so valuable in spectroscopy, it is first necessary to recognize the *disadvantage* of scanning a narrow observation window across the spectrum in the conventional way. Surprisingly, this same basic disadvantage is inherent to the way we normally weigh objects on an ordinary double-pan balance. We therefore begin by looking for better ways to use the balance, and the improvements are more or less directly applicable to the spectroscopic case.

Weights on a Balance

Consider the problem of determining the weights of three unknown objects, using the schematic balance shown in Fig. 20-1. Conventionally, we would solve the problem by weighing the unknowns one at a time in, say, the left pan, by putting the appropriate (known) weights on the right, as shown schematically below in Fig. 20-1. The obvious advantage of this pro-

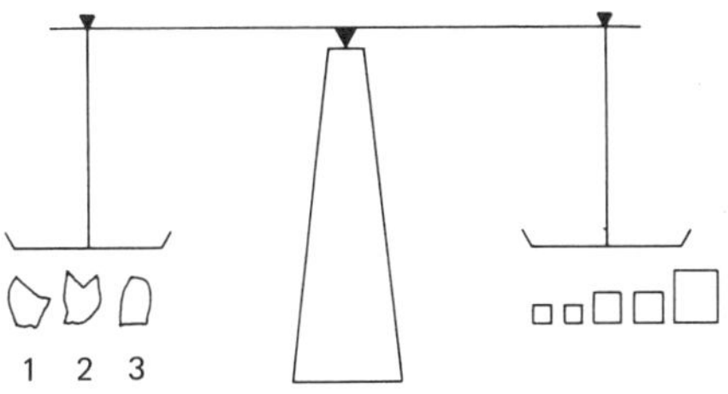

FIGURE 20-1. Schematic diagram of double-pan balance, set of standard weights (right), and three unknown weights (left).

	Unknown		
Measurement	#1	#2	#3
#1	1	0	0
#2	0	1	0
#3	0	0	1
Number of weighings of each unknown =	1	1	1

cedure is that each measurement yields an unknown weight directly (i.e., no data reduction is required). However, the disadvantage (see below) is that each unknown object has been weighed only *once.*

In any experimental measurement characterized by a certain level of imprecision (random noise), it is desirable to repeat the measurement *many* times to obtain a more accurate result. As discussed in Chapter 8.B.4, the signal (in this case, the weight of a given unknown) will accumulate as the number of weighings, N. But, if the noise is random, its magnitude may be treated as a random walk about zero (the average noise level), and the average absolute distance away from zero after N steps of a random walk (more precisely, the root-mean-square distance) is proportional to $(N)^{1/2}$. Thus, the true measure of the precision of the *repeated* measurement, namely, the signal-to-noise ratio, is proportional to $(N/\sqrt{N})$, or just $\sqrt{N}$.

Returning to the balance problem, suppose we now place *two* unknown weights on the left pan of the balance, and then measure the weights of three linearly independent combinations of unknown objects, weighed two at a time:

	Unknown		
Measurement	#1	#2	#3
#1	1	1	0
#2	1	0	1
#3	0	1	1
Number of times each unknown is weighed =	2	2	2

This time, the three desired unknown weights are related to the three observed total weights by three linear algebraic equations, which may then be solved to yield the desired unknown weights. *However,* since each unknown has now been weighed *twice,* the precision (signal-to-noise ratio) for each calculated unknown weight is now better by a factor of $(2/\sqrt{3})$ than for the original conventional experiment (i.e., the signal is 2 times as large and the noise is $\sqrt{3}$ times bigger since three weighings are involved). Furthermore, since the *same* total number of weighings (three) is required, the total *time* required to conduct the new experiment is also the same. For an arbitrary number, N, of unknown weights, in which approximately *half* the weights are put on the balance at once, the general improvement in signal-to-noise ratio for the proposed encoding-decoding scheme is $(N/2)/\sqrt{N} = \sqrt{N}/2$ for an experiment that takes no longer to carry out than N conventional single-object weighings.* Although the details of the spec-

* *Strictly speaking, the precision of the calculated weight will be better, only if the average error in a particular weighing depends only on the* balance, *and is* independent *of the magnitude of the measured weight of the* unknown *(signal). This general condition is called "detector-limited" noise, and is distinguished from "source-limited" noise in which the root-mean-square*

troscopy experiment are obviously different from those of the balance experiment (see below), the preceding encoding-decoding scheme still applies, and forms the basis for *Hadamard transform* spectroscopy (Chapter 20.A).

By logical extension of the preceding argument, one might think of putting *all three* unknown weights on the two sides of the balance in different combinations, while keeping track of the (known) weight required to balance any particular arrangement of unknowns:

Measurement	Unknowns Placed in Left Pan	Unknowns Placed in Right Pan
#1	#1, #2	#3
#2	#1, #3	#2
#3	#2, #3	#1
(Number of weighings of each unknown = 3)		

Again, it is possible to extract the three desired individual unknown weights by straightforward solution of three coupled linear algebraic equations. However, since each unknown has now been weighed *three* times, the signal-to-noise ratio for each calculated unknown weight is improved by a factor of $3/\sqrt{3} = \sqrt{3}$ times compared to conventional one-at-a-time weighing. For an arbitrary number of unknowns, N, it follows that the general improvement in signal-to-noise ratio will be a full factor of $\sqrt{N}$. This improvement also applies to the (experimentally different) *Fourier* transform spectroscopy experiment treated in Chapter 20.B.

Hadamard and Fourier methods, then, provide a means for improving the precision (signal-to-noise ratio) in a weighing experiment by a factor of about $\sqrt{N}/2$ or $\sqrt{N}$, and require the *same* total time for measurements as conventional one-at-a-time weighing. This improvement is known as the *Fellgett* advantage*; all that remains is to show that we ordinarily perform spectroscopy as inefficiently as we ordinarily use a double-pan balance, and then explore the available means for exploiting the potential Fellgett advantage in the situation.

---

noise is proportional *to the square root of the* signal *magnitude. When the noise is source-limited, the above encoding-decoding scheme will* not *improve the signal-to-noise ratio for the calculated weights compared to that obtained in the simpler one-at-a-time weighing procedure. Spectroscopic examples in which noise is detector-limited include (low-energy) infrared, microwave, nuclear magnetic resonance and ion cyclotron resonance experiments. Examples in which noise is source-limited include (high-energy) optical (ultraviolet-visible) and charged-particle (photoelectron, ESCA, electron impact) spectroscopy. Noise becomes source-limited when the photon (or particle) energy becomes large compared to thermal energy (kT), so that it is possible to count individual particles, as in the "shot-noise" example of Chapter 8.B.1. Thus, we expect the encoding-decoding "tricks" of this section to be most useful for relatively low-energy spectroscopy, as demonstrated later.*

* *Again, the Fellgett advantage can be realized, only if the noise is "detector-limited" and not when the noise is "source-limited" (see preceding footnote).*

Direct Multichannel Spectrometers

Figure 20-2 is a highly schematic diagram of a generalized spectrometer. The dispersive element might be a prism or grating (infrared, optical), for example, and the slit might be a band-pass filter for a low-frequency (microwave, radiofrequency nuclear magnetic resonance) case. The slit width is chosen sufficiently narrow that when detector readings are collected from a number of individual slit positions (bottom of Fig. 20-2), there is sufficient resolution (i.e., sufficient number of data points per absorption line) to distinguish spectral features of interest. The most important feature of such a spectrometer is that its detection of an absorption spectrum requires a procedure formally identical to the one-at-a-time method of determining the weights (spectral intensities) of N different unknown objects (spectral slit positions). It would thus be desirable to open up the slit aperture to the full width of the desired spectral window by using N separate single-channel detectors as shown in Fig. 20-3. Since *all* the slit positions are now monitored at once, rather than just one at a time, the "multichannel" spectrometer of Fig. 20-3 offers (in principle) an improvement of the full factor of $\sqrt{N}$ Fellgett advantage in *signal-to-noise ratio,* compared to the spectrum obtained in the same length of time by scanning one slit at a time as in Fig. 20-2. Alternatively, it would be possible to obtain a spectrum having the *same* signal-to-noise ratio in $(1/N)$ the *time* required to scan the N individual slit positions one at a time.

Because of the conceptual *simplicity* of the *multichannel* spectrometer of Fig. 20-3, it is logical to investigate its *feasibility*. It is desirable to be able to resolve spectral detail as narrow as the width of a typical absorption line; therefore, the minimum number of channels that will be required in a

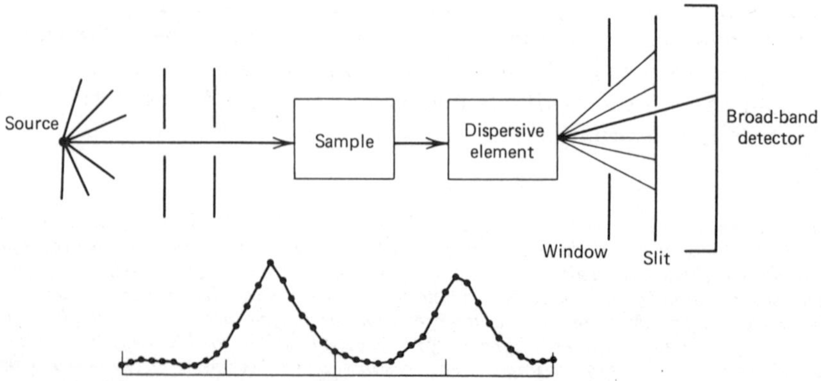

FIGURE 20-2. *Top:* schematic diagram of single-slit scanning *absorption* spectrometer. Single-slit scanning *emission* spectrometer lacks only broad-band source. *Bottom:* detector readings from a number of individual slit positions.

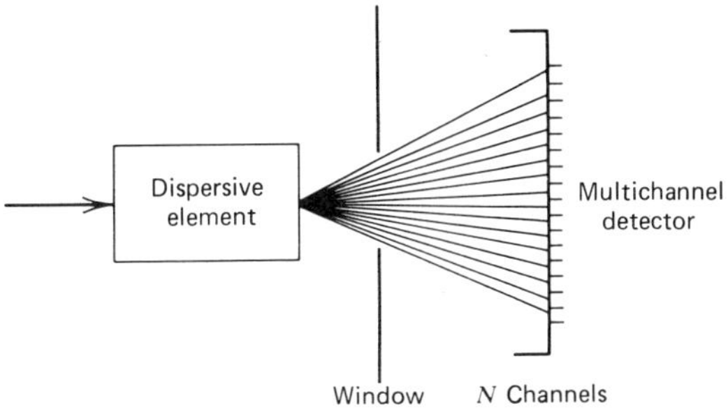

FIGURE 20-3. Schematic diagram of detector section of direct multichannel spectrometer composed of many separate single-channel detectors.

multi-channel spectrometer is simply the width of the entire spectral range of interest, divided by the width of a single spectral line. The resultant necessary number of channels for various forms of spectroscopy is calculated in Table 20-1.

From Table 20-1, it would appear that electronic (ultraviolet-visible) spectroscopy is the least likely candidate for success with a direct multichannel spectrometer, but in fact, multichannel detection of ultraviolet-visible radiation is readily accomplished photographically. The resolution of a fine-grain photographic plate is sufficient to provide for the huge required number of channels, since the desired spectrum may be dispersed over the necessary distance (a few meters) without undue effort.

In ESCA (electron spectroscopy for chemical analysis) and photoelectron spectroscopy, electrons are dislodged from atoms or molecules by X-ray or ultraviolet radiation and the released electrons possess a translational energy (and thus translational velocity) that depends on the energy of the bound state occupied by that electron in the original atom or molecule. By scanning the energy of the observed dislodged electrons, the energies

Table 20-1 Minimum Number of Channels Required for Various Types of Direct Multichannel Spectrometers

Type of Spectroscopy	Energy for Resonant Absorption (Hz)	Typical Spectral Range	Width of 1 Line	Min No. of Channels[a]
Mössbauer	6×10^{18} Hz	10^8 Hz	10^7 Hz	10
ESCA	3.5×10^{17}	10^{17}	10^{14}	1,000
Photoelectron	5×10^{15}	3×10^{15}	10^{12}	3,000
Electronic	1.5×10^{15}	1.2×10^{15}	10^9	1,250,000
Vibrational	2×10^{14}	1.5×10^{14}	3×10^9	50,000
Rotational	4×10^{10}	3×10^{10}	10^4	300,000
^{13}C NMR	8×10^7	2×10^4	0.5	40,000
ICR	2×10^6	2×10^6	10^2	20,000

[a] *Number of channels is obtained by dividing the typical spectral range by the width of one line.*

of the original molecular electronic levels can be determined. By passing the electrons between two charged parallel plates, the dislodged electrons may be dispersed in space according to their velocities, to achieve the arrangement shown in Fig. 20-3. Such a multichannel electron detection scheme has recently become feasible with the advent of the "vidicon" detector, in which an arriving electron strikes a fluorescent screen on the face of a television camera. Because electrons of different velocity can be dispersed to strike different regions of the screen, their arrival will be recorded independently by different elements of the television camera grid. Because of the small required number of detector channels (see Table 20-1), the multichannel Fig. 20-3 spectrometer is thus now feasible for ESCA and photoelectron spectroscopy.

For the lower-energy forms of spectroscopy listed in Table 20-1, direct multichannel methods are less attractive. For microwave (electron spin resonance or "pure rotational" spectroscopy), for example, there is no broad-band radiation *source* available. A black-body radiation source, such as employed for higher radiation energies (xenon or hydrogen discharge for ultraviolet, hot tungsten wire for visible, globar for near (to visible) infrared and infrared, mercury vapor (for far infrared) would have to be operated at an unreasonably high temperature to achieve sufficient radiation flux for use as a radiation source at microwave or lower energies. It would be conceivable to construct an array of individual (narrow-band) microwave transmitters (ca. \$5000 each) as the "broad-band" radiation source, but the number of required channels (Table 20-1) shows that the cost would be excessive (\$$10^9$ . . . !). With respect to multichannel *detectors*, photographic film does not respond to radiation whose wavelength is much longer than about 12,000 Å, so photographic methods are no longer usable. For infrared spectroscopy, the necessary broad-band *source* is available, but it would be necessary to disperse the spectrum over about 100 meters to resolve the desired spectral detail with existing individual (thermopile) detectors whose size is about 1 mm. At a cost of about \$200 per detector, the cost of multichannel infrared spectroscopy again becomes unmanageable. Finally, for radiofrequency (nuclear magnetic resonance, ion cyclotron resonance) spectroscopy, broad-band sources are also available, but the cost of an array of tens of thousands of individual narrow-band mixer-filter detectors (see below) is again unreasonably high.

For infrared, microwave, and radiofrequency spectrometers, the direct multichannel approach is just not feasible, either geometrically or economically. We now consider two recent and valuable indirect approaches: *Hadamard* transform spectroscopy and *Fourier* transform spectroscopy.

20.A. HADAMARD TRANSFORM ENCODING-DECODING ("MULTIPLEX") SPECTROMETERS

Figure 20-4 shows the instrumental modification that allows for the use of the Hadamard scheme: a mask is inserted between the desired spectroscopic

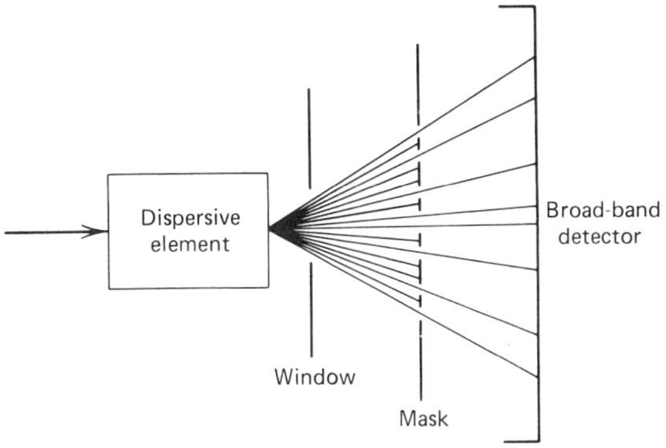

FIGURE 20-4. Disperser, mask and (single, broad-band) detector of Hadamard spectrometer. Approximately half of the possible slit positions are open.

"window" and the original (inexpensive) broad-band detector of Fig. 20-2. The mask is constructed so that its smallest opening is the same as the (narrow) slit width of the conventional spectrometer (Fig. 20-2), but with approximately *half* the total possible slit positions open. The pattern of open and shut slits is a particular random sequence (see below).

Let the spectrum of transmitted intensities (the dots at the bottom of Fig. 20-2) be represented by spectral elements: $x_1, x_2, \ldots, x_N$ (i.e., x_n represents the spectral intensity impinging on the nth slit position). When the mask in Fig. 20-4 is in position, the detector *total* (intensity) *response, y,* consists of a *sum* of all the desired spectral (intensity) elements, each weighted by a factor, a_n, of either zero or one, depending on whether that particular slit was shut or open

$$y = a_1x_1 + a_2x_2 + \cdots + a_Nx_N; \qquad a_n = 0 \text{ or } 1 \tag{20-1}$$

About half the a_n are equal to one and the other half equal to zero

In other words, the detector has provided *one observable* (y) expressed in terms of N *unknowns* (x_1 to x_N) according to the "code" (a_1 to a_N) of Eq. 20-1. This situation is precisely analogous to that of putting half the unknown weights on the left pan of the balance, as already discussed. To recover the desired spectrum of unknown intensities, x_1 to x_N, the next step is to remove the first mask and introduce in its place a *second* mask, again with a (different) random arrangement of open and shut slits with approximately half the slits open, chosen such that this second slit arrangement is linearly *independent* from the first. Proceeding in this way, one readily obtains N observables (the total transmitted intensity through each of N different masks, y_1 to y_N) expressed in terms of N unknowns (the desired spectral intensity elements, x_1 to x_N), according to a "code" in which all the coefficients are either zero or one, and roughly half the coefficients in any one row are zero:

$$\begin{aligned}
y_1 &= a_{11}x_1 + a_{12}x_2 + \cdots + a_{1N}x_N \\
y_2 &= a_{21}x_1 + a_{22}x_2 + \cdots + a_{2N}x_N \\
&\ \ \vdots \\
y_N &= a_{N1}x_1 + a_{N2}x_2 + \cdots + a_{NN}x_N
\end{aligned} \tag{20-2}$$

Referring to Table 20-1, it would obviously be impractical to construct, say, 50,000 different masks to collect infrared spectral intensities in this fashion; moreover, even the largest available digital computers cannot necessarily solve a set of that many linear algebraic equations reliably. The Hadamard "trick" is based on a particularly useful choice of the slit pattern so that both these problems can be solved simultaneously.

A particularly convenient method for providing the N linearly independent slit arrangements is illustrated for the $N = 3$ case in Fig. 20-5. Instead of N separate masks, it is sufficient to construct just a single movable mask consisting of $(2N - 1)$ potentially open slit positions. Then, by translating the mask across the (fixed) spectral window to the first position shown in the figure, we produce the first slit arrangement at upper right of Fig. 20-5; by moving the mask to the second (or third) position, we produce the second (or third) slit arrangement. *Mechanically,* one thus changes from one slit arrangement to the next, simply by translating the single $(2N-1)$-position mask across the window by one slit width per move. *Computationally,* these particular slit arrangements make it possible to obtain the *desired* spectral intensity elements $(x_1, x_2, \cdots, x_N)$ simply by combining the *observed* total experimental intensities passing through the mask $(y_1, y_2, \cdots, y_N)$ according to a similar "code" obtained by substituting -1 for 0 in the original "code" that gave the y_i in terms of the x_i (see Fig. 20-5). In other words, we can recover the x_i from the experimental y_i without ever having to solve the set of N equations in N unknowns in the usual way.

On both computational and mechanical grounds, the Hadamard approach of Figures 20-4 and 20-5 conveniently accomplishes an improvement of a factor of $\sqrt{N}/2$ in signal-to-noise ratio over the conventional one-slit-at-a-time scanning spectrometer of Fig. 20-2, because *half* the N possible slits are open during each measurement, rather than just one, while *both* experiments require the *same* time for obtaining the N separate intensity measurements (the computations are so simple that they can be completed by a small dedicated computer in a time that is short compared to that required for data collection). Alternatively, the Hadamard spectrometer can provide the *same signal-to-noise ratio* in a factor of $(4/N)$ as much time as would be required by the conventional single-slit instrument.

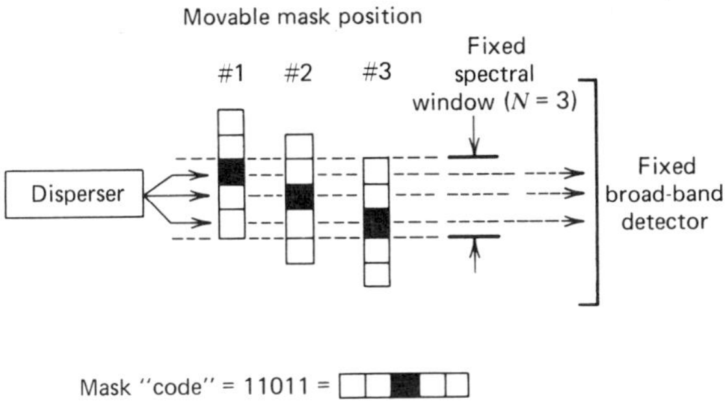

Observed experimental total intensities passing through the masked window

$$y_1 = x_1 + x_2 + 0$$
$$y_2 = x_1 + 0 + x_3$$
$$y_3 = 0 + x_2 + x_3$$

	x_1	x_2	x_3	
$y_1 \rightarrow$	1	1	0	Hadamard "code"
$y_2 \rightarrow$	1	0	1	
$y_3 \rightarrow$	0	1	1	

Calculated intensity for each fixed slit position at the window, summed over all three mask positions

$$2x_1 = y_1 + y_2 - y_3$$
$$2x_2 = y_1 - y_2 + y_3$$
$$2x_3 = -y_1 + y_2 + y_3$$

FIGURE 20-5. Schematic operation of an $N = 3$ channel Hadamard spectrometer. Simply by positioning a $(2N - 1) = 5$-slit mask in three successive locations relative to the fixed spectral window, we can produce the three desired Hadamard-coded combinations of total intensity at the detector. The *calculational* simplicity of the scheme is shown at the lower left of the figure: the experimental intensities (y_1, y_2, y_3) are each given by (different) weighted sums of the desired individual intensities (x_1, x_2, x_3), with weight factors of zero or one (upper right), while the desired individual intensities are readily recovered from the observed intensities by simple addition and subtraction (lower left) using the original Hadamard "code" with 0 replaced by -1 (see text).

The alert reader may have noted that the three slit arrangements shown in Fig. 20-5 differ from each other by "cyclic permutation"; that is, each successive arrangement differs from the preceding one by removing the first slit choice from the left of a given row and placing it at the far right of that row to obtain the next row. This feature is common to all Hadamard experiments, and it turns out to be possible to construct N linearly independent arrangements of slits that exhibit this cyclic permutation property whenever $N = (2^n - 1)$—see Problems. The above Hadamard technique is quite new, and has so far been applied most prominently in infrared spectroscopy, a technique discussed in further instrumental detail below.

Incoherent and Coherent Spectrometers

In order to proceed to Fourier methods, it is important to understand that the spectrometers discussed up to now (Figures 20-2 to 20-5) can operate with an "incoherent" radiation source [i.e., one for which there is no necessary common phase relationship (see below) between the various radiation components issuing from the source]. For such *incoherent* source spectrometers, *Hadamard* coding techniques provide a means for effectively opening up the slit width without sacrificing resolution (Fig. 20-6). There are, however, some major advantages (see below) in using a scanning spectrometer having a "coherent" source: the *coherent* source makes possible another type of encoding-decoding scheme (*Fourier*) for opening the spectral window while preserving resolution (Fig. 20-6). Finally, one reason both Hadamard (incoherent source) and Fourier (coherent source) methods can be applied to infrared spectroscopy is that the Michelson interferometer (a device used in Fourier transform infrared spectrometers) can be thought of as a means for effectively converting incoherent into coherent radiation in the present context (see Chapter 20.B.3).

A coherent radiation source and coherent detector in a spectrometer provide two important practical advantages. First, since the frequency of the coherent radiation source is readily determined to very high accuracy by use of an electronic counter, the line *positions* in a spectrum may be determined very *accurately,* simply by measuring the frequency of the source as it is (slowly) scanned over the spectral window. Second, coherent radiation permits the implementation of electronic filtering techniques that can make the spectrometer *resolution* arbitrarily high. Thus, by making the instrumental spectral line-broadening arbitrarily small, the observed spectral line shape can be made to approximate the characteristic spectrum of the *sample* rather than artifacts of the *instrument.*

The basic operation of a coherent-source spectrometer is shown in Fig. 20-7 for a hypothetical infrared laser source spectrometer for use in vibrational spectroscopy. The radiation issuing from the source consists of

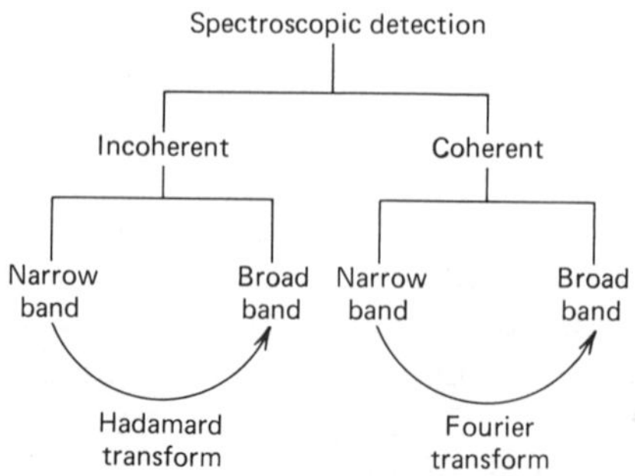

FIGURE 20-6. Spectroscopy classified by radiation source and spectrometer bandwidth.

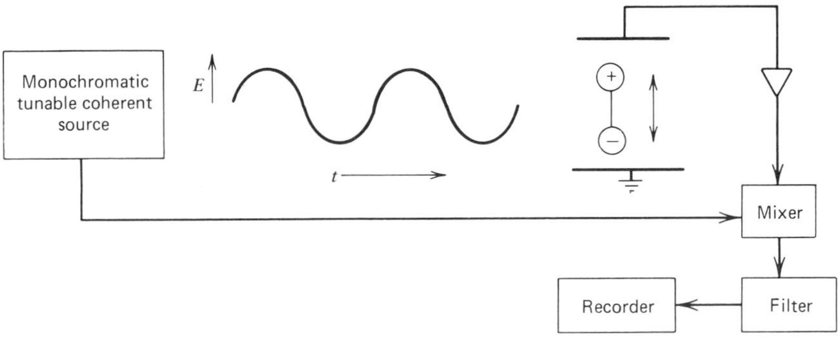

FIGURE 20-7. Hypothetical infrared laser source spectrometer.

a plane-polarized electric field whose magnitude varies sinusoidally with time. Upon encountering an electric dipole (i.e., a polar molecule), the electric field will force the dipole to oscillate at the frequency of the radiation, and the amplitude of that dipolar oscillation will be greatest when the "driving" electric field oscillation frequency is the same as ("in resonance with"—see Chapter 13.B.2) the "natural" vibration frequency of the dipole. If the source radiation is *coherent,* then all dipoles in a given region of space will oscillate together, forming a *macroscopic* oscillating electric dipole in the sample. That macroscopic oscillating dipole then induces an oscillating charge (and thus a corresponding oscillating voltage) on the parallel plates of the capacitor enclosing the sample in Fig. 20-7. That induced voltage may then be amplified and (in the most important step) multiplied (in a "mixer") by the oscillating signal from the source, and the product decomposed *electronically* into the sum and difference of the two sine wave frequencies, just as the product of two sine waves may be decomposed *algebraically* (by a trigonometric identity) into sine waves of the sum and difference frequencies. The low-pass filter rejects the (higher) "sum" frequency and passes only the (lower) "difference" frequency, which is then recorded. The above mixing process effectively extracts a small spectral frequency segment that is centered at the source frequency and whose width is determined by the bandwidth of the low-pass filter.

In more familiar language, this sort of spectrometer provides a slit *position* determined by the (precisely measurable) frequency of the source, and a slit *width* determined by the bandwidth of the electronic low-pass filter and which therefore may be made arbitrarily wide or narrow without any mechanical adjustment of the spectrometer geometry. Spectrometers in which a macroscopic change in a physical property of the sample is induced by radiation from a coherent source, followed by electronic detection of that macroscopic change in the manner described above (e.g., Fig. 20-7), have long been employed in radio-frequency nuclear magnetic resonance (NMR), and ion cyclotron resonance (ICR) spectroscopy, and have recently been introduced in microwave and infrared spectroscopy.

20.B. FOURIER TRANSFORM SPECTROSCOPY

Fourier transform methods at first seem foreign to our intuition, because we are prejudiced by our eyes and ears to analyze our surroundings in the *frequency* domain—we judge light by its *color* and sound by its *pitch*. However, we could equivalently express the light or sound wave as an (electric field or pressure) magnitude that oscillates continuously with *time*. The Fourier transform is simply the mathematical operation that converts the time-domain description (e.g., a sine wave that oscillates indefinitely) to a frequency-domain description (e.g., a sharp spike at the frequency of the sine wave oscillation in this example). A more interesting example is given in Fig. 20-8, which shows a simple d.c. (i.e., zero-frequency) time-domain pulse, which is turned "on" and time zero and "off" at time, T. Intuition would suggest (as in the preceding example) that the frequency representation of such a pulse should consist simply of a signal at *zero* frequency, but the actual frequency representation (upper right of Fig. 20-8) consists of a signal spread over a *range* of frequencies near zero. For a *shorter* pulse (middle of Fig. 20-8), the frequency representation is spread over an even *wider* range, and in the limit that the d.c. pulse is made *infinitely* narrow (bottom, Fig. 20-8), the frequency representation is a completely *flat* spectrum. Thus, a very short, intense electric field pulse effectively produces broad-band electromagnetic radiation over a frequency range proportional to the reciprocal of the duration of the pulse. [The mathematical correspondence between the time-domain (left) and frequency-domain (right) diagrams of Fig. 20-8 is called a Fourier transformation.]

The diagrams of Fig. 20-8 suggest that the *broad-band* frequency *excitation* required for a multichannel or multiplex spectrometer could be

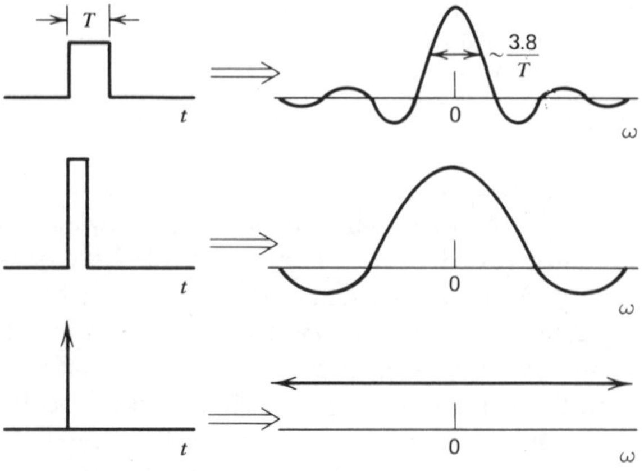

FIGURE 20-8. Time domain (left) and frequency domain (right) representations of dc pulses of three different durations.

$y(t) = \cos(\omega t);\ 0 \leq t \leq T$
$y(t) = 0 \qquad ;\ t < 0 \text{ or } t > T$

$y(t) = e^{-(t/\tau)} \ ;\ t \geq 0$
$y(t) = 0 \qquad ;\ t < 0$

$y(t) = e^{-(t/\tau)}\cos(\omega t);\ 0 \leq t \leq T$
$y(t) = 0 \qquad ;\ t < 0 \text{ or } t > T$

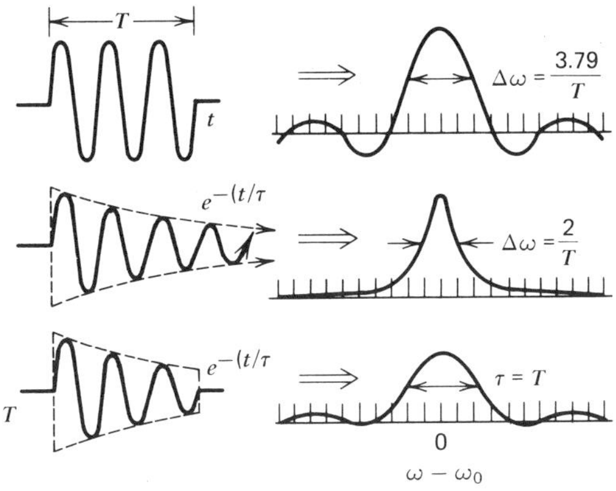

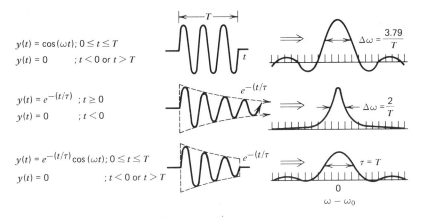

FIGURE 20-9. Frequency representations (right) of three types of time domain (left) spectrometer signals.

generated by use of a sufficiently *short* pulse of electromagnetic radiation. (If the pulse consists of an a.c. rather than a d.c. waveform, then the pictures of Fig. 20-8 still apply, except that the frequency representation is now centered at the a.c. frequency rather than at zero frequency—see top trace of Fig. 20-9.)

For NMR spectroscopy, for example, Table 20-1 indicates that an excitation bandwidth of about 10 kHz is required—Fig. 20-9 (top trace) indicates that such an excitation may be produced simply by applying a radiofrequency pulse whose duration is of the order of 10^{-5} sec. As another example, electron impact spectroscopy is based on the rapid passage of an electron past a molecule. This passing electron produces a very short, sharp pulse of electric field at the molecule, and thus acts as a very broad-band, nearly flat source of irradiation. In that case, the frequency bandwidth is sufficient to excite the same sorts of transitions as are more conventionally studied in photoelectron and ESCA spectroscopy. Having presented a technique for borad-band *excitation,* we now discuss broad-band *detection* of resonant responses at various frequencies for a sample.

When a given single oscillator is subjected to irradiation at its resonant frequency, the amplitude of oscillation will increase (Chapter 13.B). If the irradiating excitation is then removed, the oscillation will persist with an amplitude that decreases (usually exponentially—Chapter 13.C) with time, as illustrated for three convenient limiting situations at the left of Fig. 20-9. For instance, if the oscillatory response is *not* appreciably reduced during the observation time, T (top trace of Fig. 20-9), then the corresponding frequency representation has a functional form resembling the amplitude of (Fraunhofer) diffraction by a slit. If, on the other hand, the oscillation is observed for *several* lifetimes of its decay (middle trace of Fig. 20-9), the spectral frequency representation approaches the familiar Lorentzian line shape encountered in many forms of spectroscopy. Finally, the bottom trace of Fig. 20-9 illustrates an *intermediate* case in which the observation (data acquisition) time T is of the order of the decay lifetime, τ. The irreversible decay of the oscillation may be due to "spontaneous emission" ("radiative damping"—see Chapter 19.B.1) and to interactions of the sample (nucleus, ion, molecule) with its surroundings so as to disrupt the continued regular oscillation of the sample. The interaction with surroundings may be neutral-neutral collisions (microwave, infrared, ultraviolet-visible); ion-molecule collisions (ion cyclotron resonance); rotational diffusion (NMR, ESR); or depletion of the excited species due to chemical reaction.

The multichannel (Fellgett) advantage of the Fourier approach can now be understood. Suppose that the time-domain response from a collection of different driven oscillators, $y(t)$, is sampled at N equally spaced intervals during a total acquisition time, T. Each of these sampled time-domain points, $y(t_n)$, $n = 1$ to N, is then a sum of all the discrete frequency-domain spectral amplitudes, $x(\omega_n)$, according to the "code" shown below:

$$\begin{aligned}
y(t_1) &= a_{11}x(\omega_1) + a_{12}x(\omega_2) + \cdots + a_{1N}x(\omega_N) \\
y(t_2) &= a_{21}x(\omega_1) + a_{22}x(\omega_2) + \cdots + a_{2N}x(\omega_N) \\
&\;\;\vdots \\
y(t_N) &= a_{N1}x(\omega_1) + a_{N2}x(\omega_2) + \cdots + a_{NN}x(\omega_N)
\end{aligned} \tag{20-3}$$

in which

$$a_{nm} = \exp\,[2\pi i m t_n/T] \tag{20-4}$$

or just

$$a_{nm} = \exp\,[2\pi i m n/N] \tag{20-5}$$

Equation 20-3 should be compared to Eq. 20-2: since there are now N independent observed sampled time-domain data points, $y(t_1)$ to $y(t_n)$, each expressed in terms of *all* N discrete frequency-domain amplitudes, $x(\omega_1)$ to $x(\omega_N)$, it is again possible to "decode" the observed data to obtain the desired spectrum by solving the set of linear algebraic Eq. 20-3. Fortunately, the decoding procedure—known as a "discrete" Fourier transformation—may be carried out very rapidly by a digital computer, using a famous algorithm developed in 1965 by Cooley and Tukey. Furthermore, in contrast to the *Hadamard* technique, in which *half* the possible spectrum is detected during any given observation (i.e., *half* the a_{nm} in any one row of Eq. 20-2 are zero), the magnitude of *each* a_{nm} in the *Fourier* experiment of Eq. 20-3 is unity:

$$|a_{nm}| = |\exp\,[2\pi i n m/N]| = 1 \tag{20-6}$$

so that in the Fourier experiment, it is as if *all* the possible slits are open.

By the arguments previously used for the double-pan balance example, it is now clear that detection of the time-domain response, followed by Fourier transformation to obtain the frequency-domain response, provides a frequency spectrum exhibiting *either* signal-to-noise improvement of a factor of $\sqrt{N}$ in the *same* total observation period, *or* a spectrum having the *same* signal-to-noise ratio in a factor of $(1/N)$ as much time as required by a conventional spectrometer that scans the spectrum one slit at a time with a narrow-band detector.*

The final consideration for interpreting frequency spectra obtained by Fourier transformation of a time-domain response is a comparison to the line shape obtained by a conventional slow-sweep spectrometer. Under the very general condition that the system response be linear (i.e., the response be proportional to the amplitude of the irradiating excitation—see Chapter 19.B.3), the slow-sweep and Fourier transform spectra are identical in the limit of long acquisition time (middle trace of Fig. 20-9).

EXAMPLE *Fourier Transform Spectral Line Shape*

In this example, we shall work out the spectral line shape shown at the middle of Fig. 20-9 (the remaining two cases are left as Problems). Consider therefore a spectral time-domain response that consists of an exponentially damped sine wave observed for an infinite length of time:

$$f(t) = e^{-(t/\tau)} \cos(\omega_0 t) \quad 0 \leq t < \infty \tag{20-7}$$

We may define three distinct spectral frequency-domain line shapes, $A(\omega)$, $B(\omega)$, and $C(\omega)$, obtained from that time-domain response:

$$A(\omega) = \frac{1}{\pi} \int_{-\infty}^{\infty} f(t) \cos(\omega t) dt \tag{20-8a}$$

$$B(\omega) = \frac{1}{\pi} \int_{-\infty}^{\infty} f(t) \sin(\omega t) dt \tag{20-8b}$$

$$C(\omega) = [\sqrt{A^2(\omega) + B^2(\omega)}] \tag{20-8c}$$

in which $A(\omega)$ is called the "cosine Fourier transform" of $f(t)$, $B(\omega)$ is called the "sine Fourier transform" of $f(t)$, and $C(\omega)$ is the "magnitude" or "absolute-

* *For the unique case of Fourier transform infrared spectrometers based on the Michaelson interferometer, the spectrum is obtained by discrete Fourier transformation of the (spatially dispersed) sampled interferogram. Since half the spectral intensity is lost at the half-silvered mirror of the interferometer, a Fourier transform infrared spectrometer provides only half the full (factor of N) Fellgett time-saving advantage (see Chapter 20.B.3).*

value" frequency representation of $f(t)$. $(C(\omega))^2$ is called the "power spectrum" of $f(t)$. From the definite integrals (see an integral Table)

$$\int_0^\infty e^{-at} \cos(bt)dt = (a/(a^2 + b^2)) \tag{20-9a}$$

and

$$\int_0^\infty e^{-at} \sin(bt)dt = (b/(a^2 + b^2)) \tag{20-9b}$$

and the trigonometric identities

$$\cos A \cos B = \frac{1}{2}[\cos(A - B) + \cos(A + B)] \tag{20-10a}$$

and

$$\sin A \cos B = \frac{1}{2}[\sin(A - B) + \sin(A + B)] \tag{20-10b}$$

$A(\omega)$ and $B(\omega)$ are readily evaluated by substituting Eq. 20-7 into Eq. 20-8, then using Eq. 20-10 to simplify the integrals to the forms given in Eq. 20-9 (see Problems):

$$A(\omega) = \frac{1}{2\pi}\left\{\frac{\left(\frac{1}{\tau}\right)}{\left(\frac{1}{\tau}\right)^2 + (\omega - \omega_0)^2} + \frac{\left(\frac{1}{\tau}\right)}{\left(\frac{1}{\tau}\right)^2 + (\omega + \omega_0)^2}\right\} \cong \frac{1}{2\pi}\left(\frac{\tau}{1 + (\omega - \omega_0)^2\tau^2}\right) \tag{20-11a}$$

$$B(\omega) = \frac{1}{2\pi}\left\{\frac{(\omega - \omega_0)}{\left(\frac{1}{\tau}\right)^2 + (\omega - \omega_0)^2} + \frac{(\omega + \omega_0)}{\left(\frac{1}{\tau}\right)^2 + (\omega + \omega_0)^2}\right\} \cong \frac{1}{2\pi}\frac{(\omega - \omega_0)\tau^2}{1 + (\omega - \omega_0)^2\tau^2} \tag{20-11b}$$

from which $C(\omega)$ is quickly obtained as

$$C(\omega) \cong \frac{1}{2\pi}\left(\frac{\tau}{\sqrt{1 + (\omega - \omega_0)^2\tau^2}}\right) \tag{20-11c}$$

[Since we are typically interested in the frequency response of the sample *near* resonance ($\omega \cong \omega_0$), we are justified in neglecting the second term in both $A(\omega)$ and $B(\omega)$, as indicated in Equation 20-11a and 20-11b.]

The frequency-domain spectral line shapes for $A(\omega)$, $B(\omega)$, and $C(\omega)$ are shown (bottom diagram) in Fig. 20-10. The reader should now note that $A(\omega)$ and $B(\omega)$ represent the "absorption" and "dispersion" line shapes that we origi-

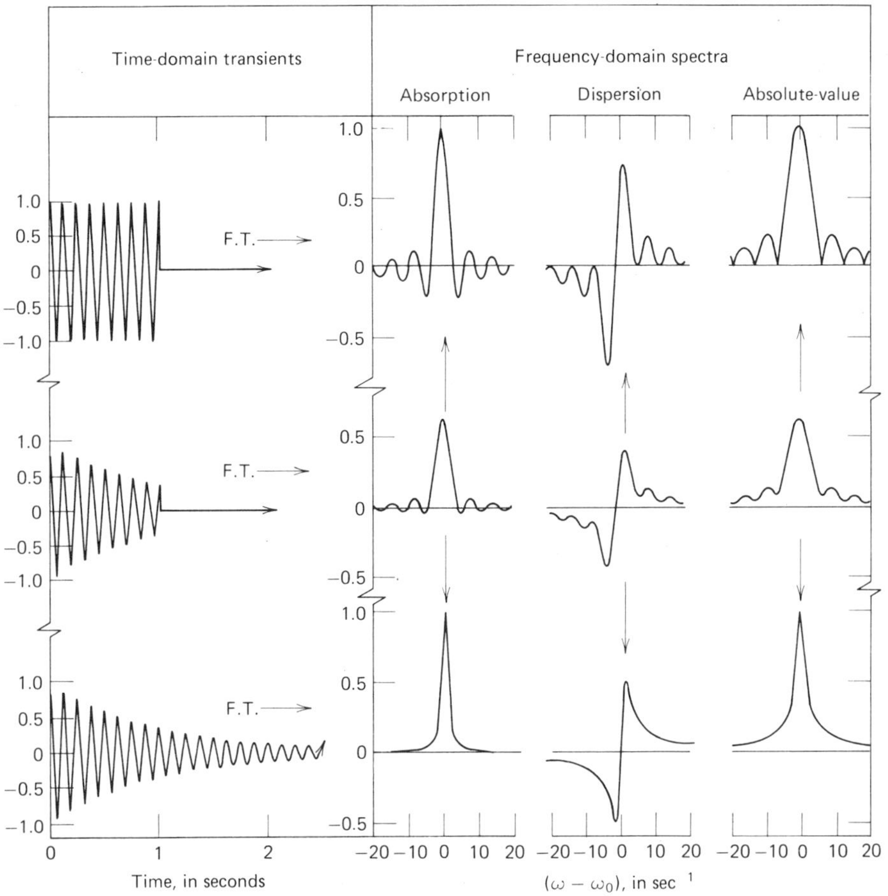

FIGURE 20-10. Time-domain transients and frequency-domain spectra obtained by Fourier transformation of those transients, for three representative observations of the damped oscillatory time-domain response of a single type of oscillator. The oscillator is driven until time zero (excitation not shown), and the excitation is turned off at time zero during the detection thereafter. The time-domain signal is taken to have unit amplitude, so that the frequency-domain vertical scale is in units of $(1/2\pi)$ for each line shape. $A(\omega)$, $B(\omega)$, and $C(\omega)$ of Eq. 20-11 are denoted as "absorption," "dispersion," and "absolute-value" in the figure. Compare with Fig. 20-9 for further discussion.)

nally derived for the *steady-state* response to a *continuous* oscillatory excitation (Chapter 13.B, esp. Fig. 13-5*e*). Thus, we have provided one example of the equivalence of the Fourier transform of the *transient* response to a *pulsed* oscillatory excitation and the *steady-state* response to a *continuous* oscillatory excitation.

The properties of the "Lorentz" line shapes of Eq. 20-11 can readily be computed as follows.

Peak Height. Peak extrema (maxima and minima) are first located by evaluating the slope of each curve and setting the slope equal to zero, and then computing the value of the function at that point(s).

$A(\omega)$ is clearly a maximum at $\omega = \omega_0$, at which the peak height is $\left(\frac{1}{2\pi}\right) \cdot \tau$.

$C(\omega)$ is also maximal at $\omega = \omega_0$, at which the peak height is also $\left(\frac{1}{2\pi}\right) \cdot \tau$.

The extrema of $B(\omega)$ are quickly found to occur for $(\omega - \omega_0) = \pm \frac{1}{\tau}$, at which the values of $B(\omega)$ are $\pm \left(\frac{1}{2\pi}\right)\left(\frac{\tau}{2}\right)$.

Peak Width. A simple definition for peak width is the width between the points of half maximal height [for $A(\omega)$ and $C(\omega)$], or the peak-to-peak frequency difference for $B(\omega)$. By setting $A(\omega)$ [or $C(\omega)$] equal to half its maximal height, and solving for ω, we quickly obtain:

$A(\omega)$ has a peak width of $(2/\tau)$ at half its maximal height.
$C(\omega)$ has a peak width of $(2\sqrt{3}/\tau)$ at half its maximal height.
$B(\omega)$ has a peak-to-peak frequency separation (see peak height calculation) of $(2/\tau)$.

Calculation of the remaining frequency-domain line shapes shown in Fig. 20-10 is similar, and is left as an exercise (see Problems).

Finally, it is clear from Fig. 20-10 that it is desirable to observe a damped time-domain response for a very long time so that the frequency-domain line shape obtained by Fourier transformation is an accurate representation of the steady-state Lorentz line shape. In the Problems, the reader will show that the former line width is within 10% of the latter, when the time-domain transient signal is observed for about three damping constants in time. These calculations provide a relatively complete description of Fourier transform spectral line shapes, and we are now in a position to examine some experimental spectra.

EXAMPLE *Fourier Transform Nuclear Magnetic Resonance (FT-NMR)*

Fourier methods have probably had more impact to date on NMR spectroscopy than any other form of spectroscopy. The experiment is based on *excitation* consisting of a short *pulse* of radiofrequency oscillating magnetic field at or near the "natural" Larmor precession frequencies of the nuclei in the sample, followed by *detection* using the same sort of mixing-filtering scheme as for the hypothetical infrared spectrometer of Fig. 20-7, except that the nuclear precessional motion is *circular* rather than *linear*.

Specifically, a short pulse of radiofrequency magnetic field near the Larmor frequencies of the nuclei of interest (say, carbon-13 nuclei) is applied to the sample, for a time sufficient to cause the equilibrium magnetic moment of the carbon-13 nuclei to rotate 90° to lie in the *x-y* plane where the detector is located (Fig. 16-21). If the exciting pulse is short enough, its frequency repre-

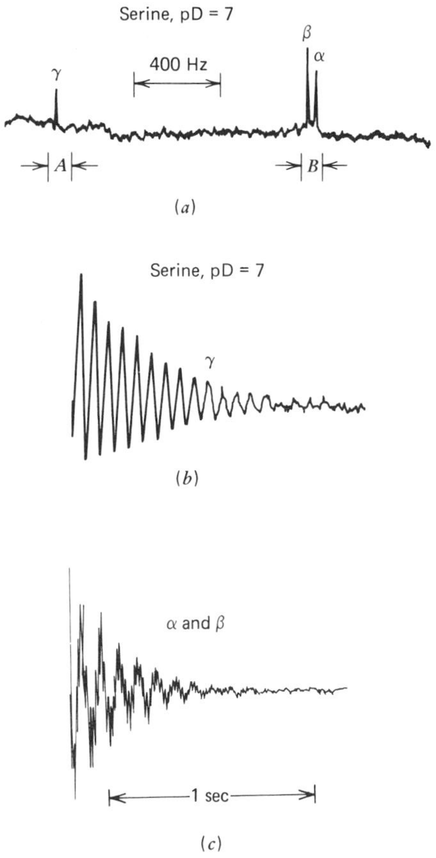

FIGURE 20-11. Frequency-domain (*a*) and time-domain (*b*) carbon-13 NMR signals for the amino acid, serine (0.5 M, 15% enriched in carbon-13, neutral pD). The topmost trace is a conventional steady-state slow-sweep carbon-13 frequency spectrum of serine. Traces *b* and *c* represent time-domain signals obtained following broad-band pulsed excitation of the sample, using a mixer, reference frequency and low-pass filter chosen to extract only frequencies in the region marked "*A*" [for time-domain signal trace (*b*)] or "*B*" [for time-domain signal trace (*c*)] (see text). Fourier transformation of, say, time-domain trace (*c*) would provide the frequency spectrum of the region marked "*B*" in Fig. 20-11*a*, so that two distinct carbon-13 NMR absorption peaks are detected simultaneously in (*c*) in the same time it would take to scan through just one of those peaks using conventional steady-state detection. [Adapted from W. Horsley, H. Sternlicht and J. S. Cohen, *J. Amer. Chem. Soc. 92,* 680 (1970).]

sentation will correspond to a broad-band excitation source with essentially flat power over the spectral region of interest (Fig. 20-9). The pulse is then turned off, and the nuclear magnetic moments precess at their respective Larmor frequencies. Using the mixer: low-pass-filter detector scheme of Fig. 20-7, only the signals from the desired frequency band are recorded and stored in a small computer, and the time-domain signal is then Fourier transformed to yield the spectrum of NMR signals in the frequency band of interest as shown in Fig. 20-11.

Figure 20-11*a* shows the conventional *steady-state* absorption spectrum of the carbon-13 nuclear magnetic resonances from serine at neutral pD. Figure 20-11*b* shows the time-domain carbon-13 NMR signal obtained by *pulsed* excitation followed by *mixer-filter* detection, using a low-pass filter with an upper frequency limit of about 100 Hz. Since the reference signal (Fig. 20-7) frequency was about 14 Hz different from the carbon-13 NMR frequency of interest, the time-domain signal consists of a damped oscillation at about 14 Hz. Similarly, Fig. 20-11*c* shows the time-domain signal resulting from an exciting pulse in the vicinity of the other two carbon-13 resonant frequencies, followed by a similar mixer-filter detection. In Fig. 20-11*c*, the reference frequency was chosen to lie about 10Hz smaller than the "α" resonance and about 68 Hz smaller than the "β" resonance, so that both signals are passed by the filter and the time-domain signal consists of a superposition of two oscillations of about equal amplitude. A Fourier transform of the time-domain data of Fig. 20-11*b* would produce a single absorption signal corresponding to the region shown as "*A*" in Fig. 20-11*a*, while a Fourier transform of the tran-

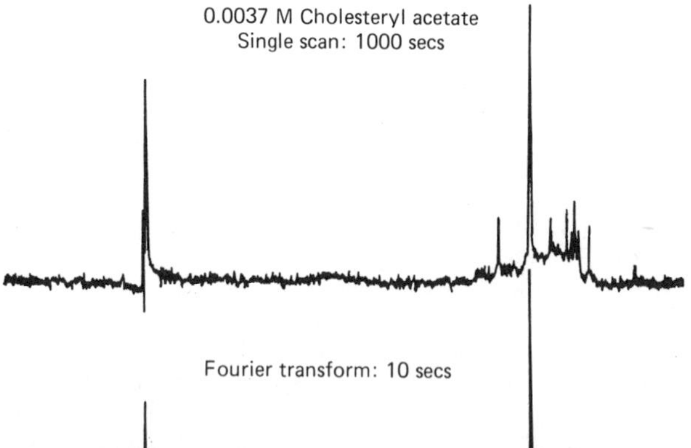

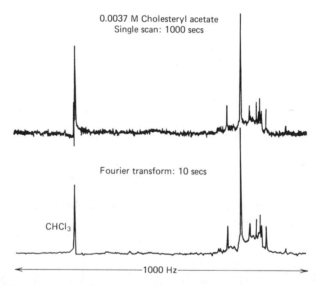

FIGURE 20-12. Time-saving advantage of Fourier transform over conventional NMR detection. *Top:* conventional steady-state slow-sweep proton NMR spectrum of cholesteryl acetate. *Bottom:* FT-NMR proton spectrum obtained in *1/100* of the time required to obtain a conventional spectrum (*top*) having the *same* signal-to-noise ratio. Alternatively, it is possible to obtain an FT-NMR spectrum with much better *signal-to-noise* ratio by accumulating many time-domain transients in the *same* time it would take to obtain a *single* conventional spectrum (see Fig. 8-5). (From H. Hill and R. Freeman, *Introduction to Fourier Transform NMR,* Varian Associates Analytical Instrument Division, Palo Alto, Calif., 1970.)

sient of Fig. 20-11*c* would yield a spectrum showing two absorption peaks corresponding to the region shown as "*B*" in Fig. 20-11*a*.

Figure 20-11*c* illustrates the multi-channel advantage of Fourier transform spectroscopy, since it is clear that *two* distinct absorption signals at different frequencies are both detected at once, rather than just *one* at a time in a conventional slow-scanning spectrometer. A more impressive example is shown in Fig. 20-12, in which the desired spectrum was obtained 100 times faster using Fourier transform of a time-domain signal (Fig. 20-12*b*) than with conventional steady-state slow scanning of one peak at a time (Fig. 20-12*a*). Alternatively, it would be possible to obtain much better signal-to-noise ratio for FT-NMR versus conventional NMR spectra by acquiring many FT-NMR time-domain signals in the time it would take to acquire just one conventional NMR spectrum, as illustrated in Fig. 8-5.

The great *selectivity* provided by carbon-13 NMR (Fig. 20-13) has been appreciated by chemists for 20 years, but the carbon-13 NMR spectra of all but the smallest molecules were essentially inaccessible until the advent of

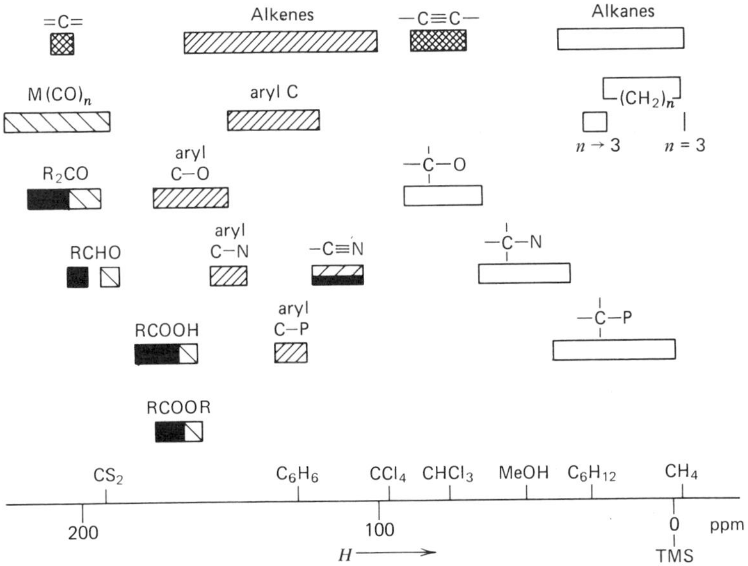

FIGURE 20-13. Range of carbon-13 NMR absorption frequencies in neutral organic compounds. Carbon hybridization denoted as ▭ for sp^3, ▨ for sp^2, ▩ for sp. Carbonyl carbons are distinguished as saturated ■ and conjugated ▧. Infrared absorption for various types of carbon bonding shows a similar dispersion in frequency absorption values, but the infrared absorption of most species in aqueous solution is obscured by the very strong infrared absorption of water itself (see FT-infrared examples). (From J. B. Stothers, *Carbon-13 NMR Spectroscopy,* Academic Press, New York, 1972, p. 13.) Tetramethylsilane (TMS) is simply a reference compound providing a reference frequency from which the other carbon-13 "chemical shifts" are measured.

Fourier methods about 1965. Today hundreds of carbon-13 FT-NMR spectrometers are in routine laboratory use. Biochemical examples of FT-NMR spectra are shown in Chapter 14.D.

EXAMPLE *Fourier Transform Ion Cyclotron Resonance (FT-ICR) Mass Spectroscopy*

The two most chemically useful analytical spectroscopic techniques are probably NMR and mass spectroscopy. Although space does not permit an extended discussion of conventional mass spectrometers, the typical experiment is based on ionizing gaseous neutral molecules directly or indirectly with a beam of accelerated electrons, followed by deflection of ions of a particular charge-to-mass ratio (using electrically charged plates or magnetic fields, much as with the "convergence" or "focus" controls of a color television tube) so that those ions pass through a small slit to an ion detector. Then by varying the electric or magnetic (or both) deflection fields, ions of different charge-to-mass ratio can be observed in turn, essentially in the spirit of the *single-slit* photon spectrometer of Fig. 20-2. (There is a type of mass spectrometer in which the various charge-to-mass ions are simultaneously dispersed in space and recorded on a photographic plate, as in the *multichannel* detector of Fig. 20-3, but then it is necessary to process the photograph before seeing the spectrum, removing much of the time-saving advantage of the Fellgett argument presented earlier). One is therefore led to consider possible *multiplex* mass spectrometric detection schemes. The Hadamard method will not be applicable, since noise is in this case "source-limited" (see p. 664), so that the Fellgett advantage is lost. However, there is one kind of mass spectrometer, the "ion cyclotron resonance" (ICR) spectrometer (for which *coherent* broad-band excitation and detection are possible), which is susceptible to the use of Fourier methods in much the same way they are used in NMR spectroscopy.

Figure 20-14 shows a schematic diagram of an ion cyclotron resonance (ICR) spectrometer. An ion moving in a magnetic field is constrained to a circular path; the frequency of this circular motion is proportional to the charge-to-mass ratio for that ion:

$$\boxed{\nu = (qB/2\pi m)} \tag{20-12}$$

When such ions are irradiated by a circularly polarized (i.e., rotating) electric field whose frequency is close to the ion "cyclotron" frequency of Eq. 20-12, the resulting ion motion becomes spatially coherent (i.e., all the ions move in a group) as the ions absorb energy from the irradiation by increasing the radii of their cyclotron orbits (see Fig. 20-14). Once the ions are all moving essentially together, their composite cyclotron motion will induce a macroscopic voltage in the surrounding plates, and that response may be amplified, mixed, filtered, and recorded as for the hypothetical infrared spectrometer of Fig. 20-7. Except that the "natural" system motion is circular rather than linear, the spectrometers of Figures 20-7 and 20-14 are conceptually very similar. For an applied magnetic field strength of 1 tesla (10 kGauss), ICR frequencies for typical singly charged ions of mass 16 to mass 400 fall in the (radio-

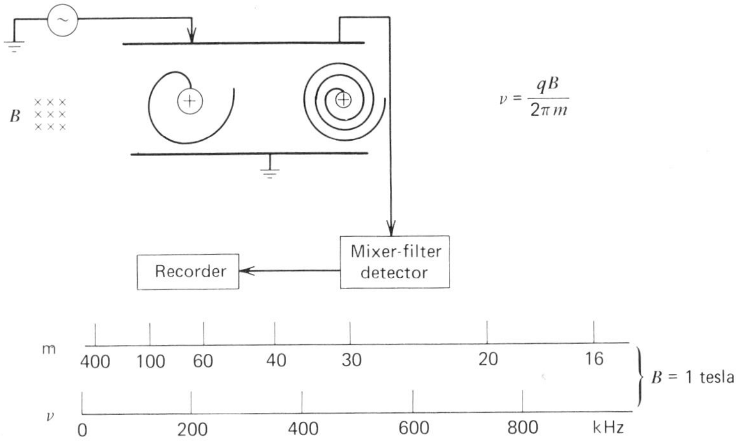

FIGURE 20-14. Schematic diagram of an ion cyclotron resonance (ICR) mass spectrometer. Operation is described in the text.

frequency) range between about 35 kHz and 1 MHz, as shown at the bottom of Fig. 20-14.

The ICR spectrometer thus produces a signal whenever the irradiation frequency matches the ion cyclotron frequency of ions of a given charge-to-mass ratio present in the sample. In other words, the device can function as a *mass* spectrometer to detect ions spanning a range of charge-to-mass ratios, by irradiating the sample with an oscillating electric field whose frequency is scanned slowly over the range required by Eq. 20-12, while monitoring power absorption, in the same spirit as for the conventional single-slit scanning spectrometer of Fig. 20-2.

Figure 20-15*a* shows the ICR time-domain response following *broad-band* excitation of a bandwidth sufficient to excite ion cyclotron resonances for ions in the $(m/e) = 213$ to 215 range. The beat pattern of Fig. 20-15*a* results from the sum of the two ICR oscillating signals from the *smaller* ion ($m/e = 213$) with the *higher* ICR frequency and the *larger* ion ($m/e = 215$) having the *smaller* ICR frequency. (Both ICR frequencies have been subjected to the mixing and filtering procedure that effectively lowers both ICR frequencies down to a convenient frequency range for illustration.) Fourier transformation of the time-domain signal of Fig. 20-15*a* gives the frequency-domain (mass) spectrum of Fig. 20-15*b*. This compound was chosen for illustration, since the two bromine isotopes have approximately equal natural abundance, and thus provide peaks of about equal intensity.

Although Fourier methods in mass spectroscopy are very new (first published paper appeared in 1974), they already promise to reduce the time required to obtain an ICR mass spectrum by a factor of up to 10,000 (compare to a factor of about 1000 time-saving Fellgett advantage from use of Fourier methods in NMR). In addition, the *mass resolution* obtained with *Fourier* methods in ICR is up to 100 times better than for *conventional* ICR detection (Fourier methods

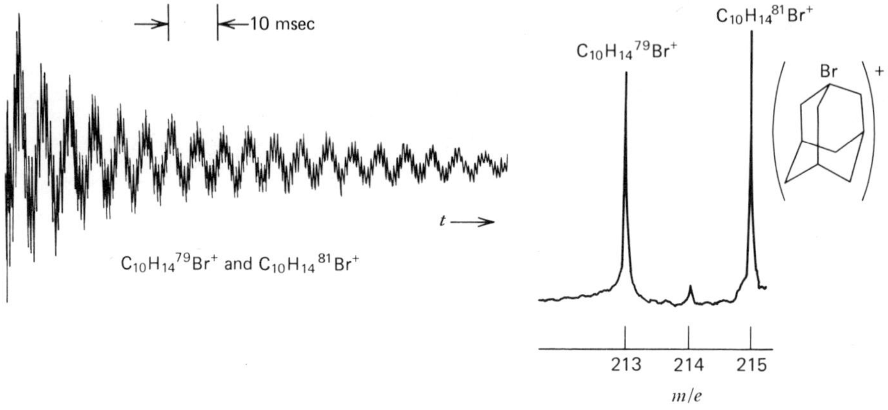

FIGURE 20-15. Time-domain (left) and frequency-domain (right) spectrum obtained by Fourier transformation of the time-domain data, for the ion cyclotron resonance response of a sample of bromoadamantane ions to a broad-band frequency excitation (see text). The small peak at $(m/e) = 214$ is a carbon-13 satellite of the $C_{10}H_{14}{}^{79}Br^+$ species (^{13}C is about 1% abundant, but there are 10 carbons in the molecule, so the species containing a single ^{13}C is 10% abundant). The time-domain trace clearly shows the superposition of two sine waves whose frequencies are related to the natural ICR frequencies of the two major ions present in the sample. The time-domain data were obtained in about 0.1 sec. [From M. B. Comisarow and A. G. Marshall, unpublished data, and M. B. Comisarow and A. G. Marshall, *J. Chem. Phys. 62,* 293 (1975).]

in NMR did not really improve NMR spectral resolution). Figure 20-16 shows why these improvements should be of interest outside the usual scope of ICR mass spectroscopy as a means for study of ion-molecule reactions in the gas phase (an area of limited direct biochemical significance). The mass spectra of morphine and a nucleotide shown in Fig. 20-16 (obtained with a conventional ICR mass spectrometer) point up a potentially major operating advantage for this type of mass spectrometer: namely, it is possible to work at very low sample pressure, so that relatively involatile substances (peptides, nucleotides, carbohydrates, etc.) can be observed. Until now, this operating advantage could not really be exploited, since the mass resolution of conventional ICR spectrometers was so poor (note that the width of a single mass peak is of the order of several mass units in Fig. 20-16) and it took so long (up to an hour) to obtain a spectrum. It seems reasonable to expect that with the much better mass resolution and speed of the FT-ICR spectrometer, it should be possible to follow up on the analytical potential of ICR for mass spectrometric studies of natural products.

EXAMPLE *Fourier Transform Infra-Red Spectroscopy*

The use of Fourier methods in infrared spectroscopy (absorption of radiation at frequencies characteristic of the natural vibrations of the "springs" that connect various nuclei to each other via chemical bonds) involves essentially

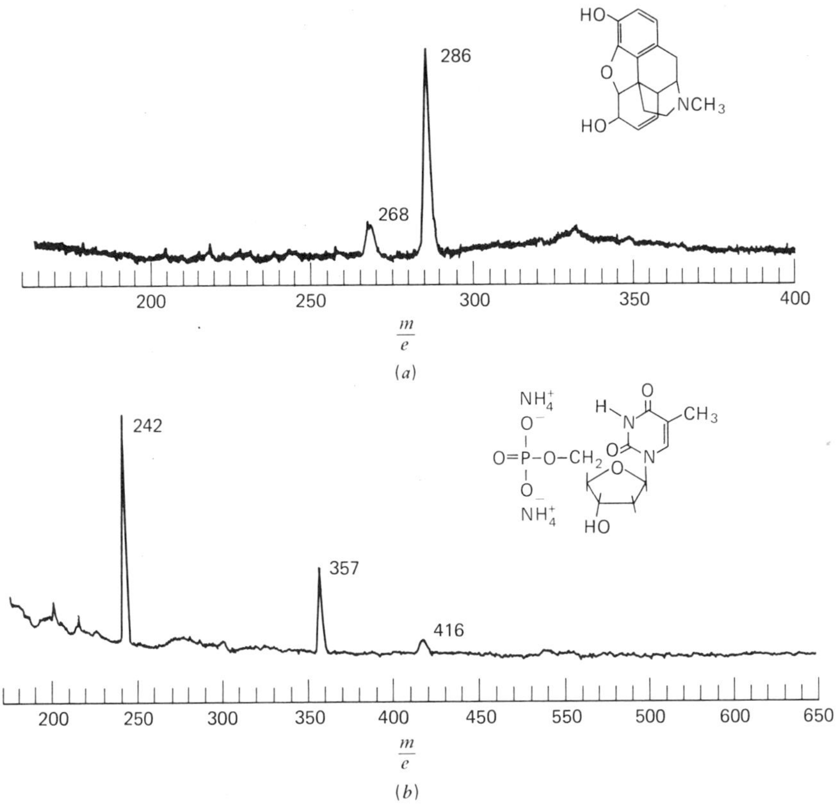

FIGURE 20-16. Conventional ICR mass spectra of morphine (*a*) and a nucleotide (*b*). Sample pressure was in the 10^{-7} torr range (i.e., very low by usual mass spectrometric standards). [From R. T. McIver, Jr., E. B. Ledford, Jr., and J. S. Miller, *Anal. Chem. 47,* 692 (1975).]

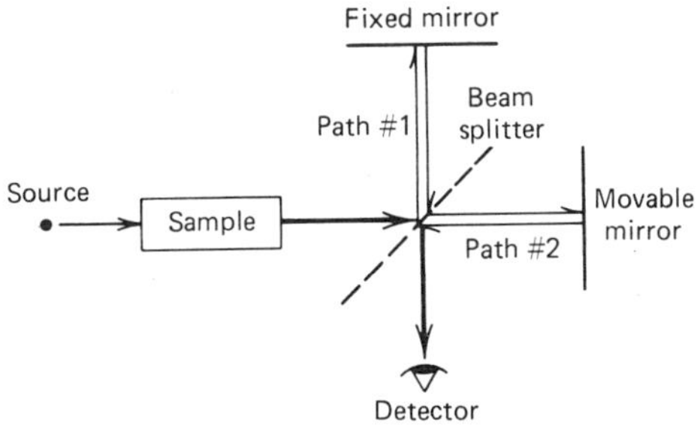

FIGURE 20-17a. Schematic diagram of Michaelson interferometer. The beam splitter divides the incident beam from the sample into two beams of equal amplitude that travel out and back from the two mirrors; the two reflected beams then recombine and are sent to the detector by the beam splitter. (Half the intensity is lost at the beam splitter.) The device thus provides a means for *adding together* two sample beams that have traveled *different path lengths* on their way to the detector.

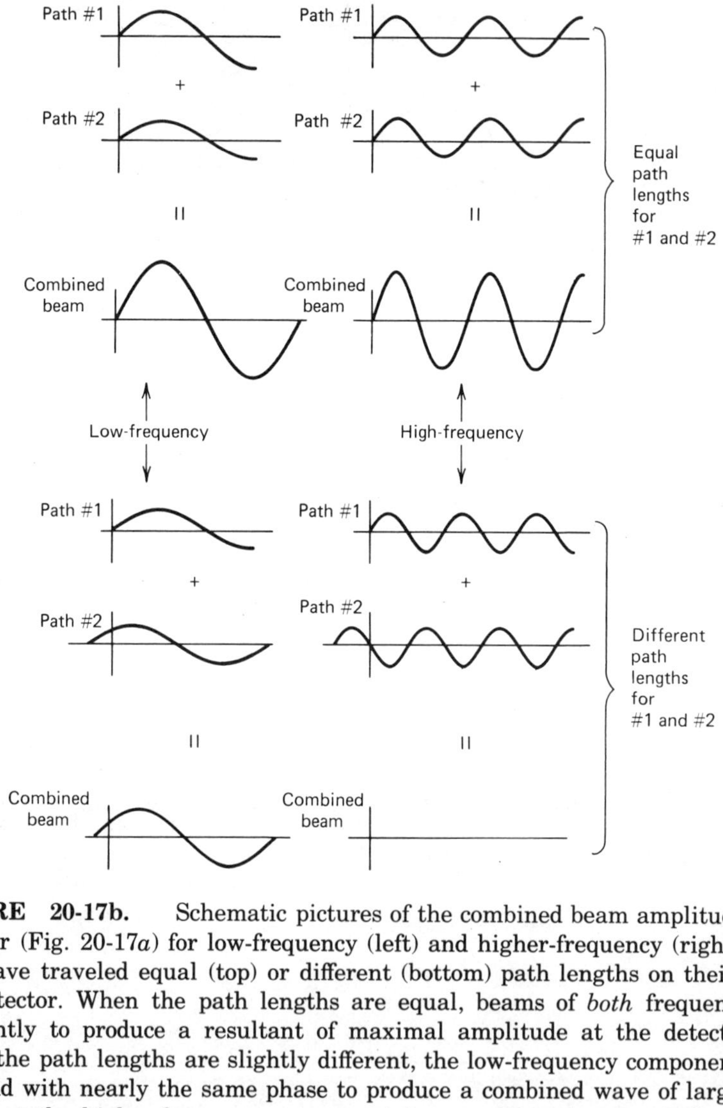

FIGURE 20-17b. Schematic pictures of the combined beam amplitude at the detector (Fig. 20-17*a*) for low-frequency (left) and higher-frequency (right) beams that have traveled equal (top) or different (bottom) path lengths on their way to the detector. When the path lengths are equal, beams of *both* frequencies add coherently to produce a resultant of maximal amplitude at the detector (top). When the path lengths are slightly different, the low-frequency component beams still add with nearly the same phase to produce a combined wave of large amplitude, but the higher-frequency component beams add with appreciably different phase (here shown is 180° phase difference) to produce a resultant combined beam of much reduced amplitude. As the path difference is increased (by moving the movable mirror), even the relatively lower-frequency components add with much different phase, and thus the amplitude of an interferogram (see Fig. 20-18) decreases in general from a maximum value (for equal paths) to smaller and smaller values as the path difference between the two beams increases, just as the amplitude of an NMR or ICR time-domain signal generally decreases with time when the spectrum consists of signals at many different frequencies.

identical data reduction, but a rather different experimental arrangement. It is possible to disperse an infrared spectrum using a device known as the Michaelson *interferometer* (see any elementary physics text for detailed analysis of this device, shown schematically in Fig. 20-17*a*). The interferometer is simply a means for adding the amplitudes of two beams of light that have traveled a *different path length* since leaving the sample. When the beams have traveled the *same* path length (equal distances between the beam splitter and the fixed and movable mirrors of Fig. 20-17), the two beams add to give a maximum amplitude, since electromagnetic waves of any frequency each recombine with the same "phase." However, if the distance between the beam splitter and the fixed mirror is *different* from the distance between the beam splitter and the movable mirror, then the amplitude of the recombined light wave will depend on the *frequency* of the wave (see diagram in Fig. 20-17*b*). As the two light beam paths are made increasingly different, more and more frequency components become out-of-phase in the recombined beam, resulting in an "interferogram" pattern of recombined intensity as a function of movable mirror position (Fig. 20-18). As with the previous NMR and ICR examples, the interferogram represents a sum of the amplitudes of *all* the frequency components of the radiation from the sample, so that the Fellgett advantage appears, and the various frequency components become out-of-phase with increasing *path difference* in the same way that the NMR or ICR time-domain signals became out-of-phase with increasing *time*. The only

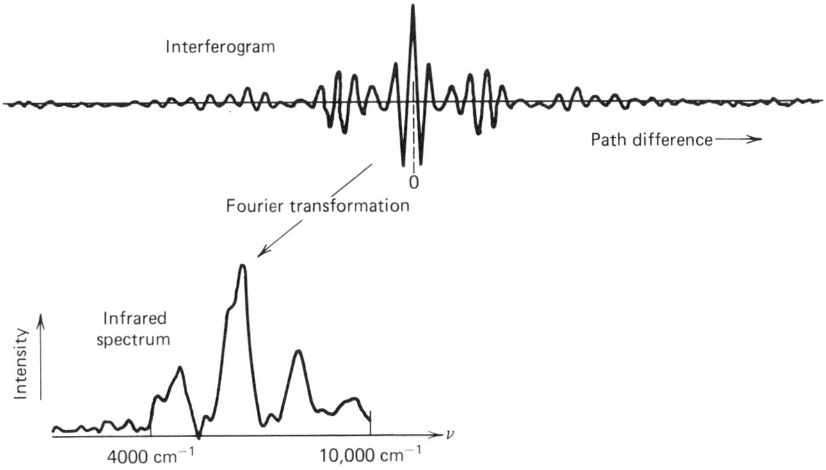

FIGURE 20-18. Infrared interferogram (signal at detector of Fig. 20-17*a* as a function of path difference between the two split beams) and infrared spectrum of the planet Jupiter. Spectrum was obtained by Fourier transformation of the interferogram data. Since this spectrum is taken through the earth's atmosphere, the absorption peaks are primarily due to earth-atmosphere gases, not to specific emission peaks from the source (see further examples). [From P. Pellgett, *J. de Physique,* Colloque C2, Tome *28,* p. C2-165 (1967).]

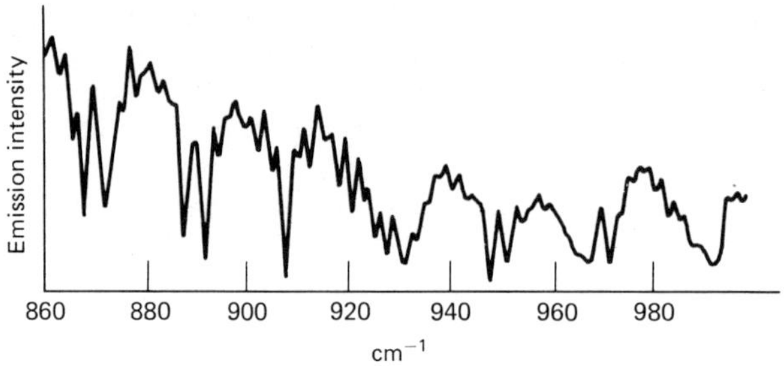

FIGURE 20-19. Far infrared FT-IR emission spectrum of Jupiter, corrected for atmospheric absorption. The dominant absorption features are due to NH_3 in the Jovian atmosphere. [From S. T. Ridgeway and R. W. Capps, *Rev. Sci. Instrum. 45,* 676 (1974).] This FT-IR spectrum was obtained in three hours; a conventional single-slit scanning spectrometer would have required 500 hours to produce a similar spectrum.

difference is that the interferogram is "two-sided" (i.e., symmetric about the mirror position that gives zero path difference) while the NMR and ICR transients are defined only after time zero and are thus "one-sided." Thus, a Fourier transformation of the infrared interferogram should (and does) generate an infrared spectrum of the sample.

Infrared Spectra of Planets and Stars

Figure 20-18 shows the Fourier transform infrared spectrum of the planet Jupiter as seen through the earth's atmosphere (the spectrum thus primarily exhibits the infrared absorption of the earth's atmosphere rather than the spectrum of emission from the planet itself). More can be learned by studying the infrared emission at lower frequencies, where atmospheric water and CO_2 do not remove most of the intensity on its way through the earth's atmosphere, as shown in Fig. 20-19. The spectrum in Fig. 20-19 has been corrected for the (small) absorption by the earth's atmosphere, using recently obtained infrared

Table 20-2 Carbon Isotopic Ratios for Some Representative Star Types *

Star Type	Example	$^{12}C/^{13}C$
Mira	χ-Cyg	30
M Supergiant	α-Ori	5
Carbon star	ω-Ori	12–15

Determined from the infrared spectral intensities from $^{12}C^{16}O$ versus $^{13}C^{16}O$ in the star emission as seen from earth (corrected for atmospheric absorption). [From data of H. L. Johnson and M. E. Mendez, Astronom. J. 75, *785 (1970); this paper presents infrared spectra for 32 stars.]*

data gathered at high altitude (i.e., above most of the atmosphere). The peaks in the Fig. 20-19 spectrum arise primarily from transitions between various excited vibrational states of ammonia in Jupiter's atmosphere. Although limited information about atmospheric composition and temperature (see Chapter 19) for energy-level population as a function of temperature) can be gained from such infrared spectra, the most detailed information concerns isotopic ratios of various nuclei, since the vibrational frequencies of, say, $^{12}C^{16}O$ and $^{13}C^{16}O$ are slightly different, because the masses on the ends of the respective "springs" (chemical bonds) are different. The $^{12}C/^{13}C$ ratios for some representative star types are shown in Table 20-2; such data is useful in confirming theories of stellar evolution pathways.

EXAMPLE *FT-IR Detection of Dilute Atmospheric Pollutants*

Because chemically different small molecules have different effective masses (nuclei) at the ends of their interatomic "springs" (chemical bonds), they are characterized by different vibrational "natural" frequencies, which can be driven by electromagnetic radiation at those frequencies, which lie in the infrared part of the electromagnetic spectrum. However, to best distinguish different small molecules from each other, it is desirable to record the spectrum at very high *resolution* (Table 20-1). Thus, Fourier methods are valuable, since they provide a large effective number of "channels" to give high resolution without sacrificing the high sensitivity (signal-to-noise) required for study of dilute species such as atmospheric pollutants. Figure 20-20 shows how infrared spectra can be used to identify and quantify atmospheric pollutants (approximate concentrations of pollutants shown at right of Fig. 20-20). Most importantly, the infrared detector can be located at a site *remote* from the pollution source (say, a smokestack plume), allowing for easy monitoring, particularly at night.

EXAMPLE *Fourier Transform Infrared Difference Spectra: Hemoglobin*

In the usual optical absorption experiment, the absorption from the *solvent* is eliminated by taking the *difference* between the absorption from the dissolved sample and the absorption from the solvent. However, when the solvent absorption happens to be very strong (as for the absorption of infrared radiation by water), it then becomes necessary to take the difference between two very small numbers (the transmitted intensity through the sample minus that through the solvent), and the effect of a slight amount of stray light in a conventional single-slit instrument renders the comparison essentially useless. The interferometer is, however, much less sensitive to stray light problems, and provides infrared spectra for molecules dissolved in water, as shown in Fig. 20-21. The figure shows that while virtually no useful detail can be seen in the FT-IR spectrum of hemoglobin in aqueous solution, subsequent subtraction of the spectrum of the solvent yields a spectrum in which the typical amide "I" and "II" vibrations at 1657 and 1547 cm^{-1} are clearly evident. It should now be possible to make statements about the secondary structure of

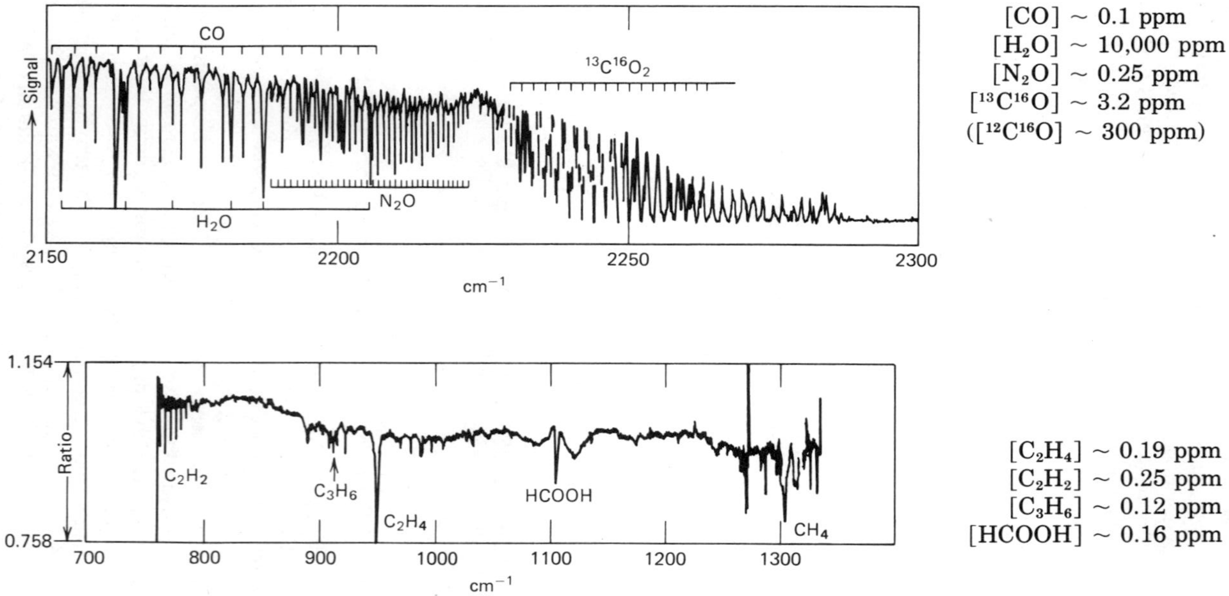

FIGURE 20-20. Fourier transform infrared spectra of air samples containing small amounts of various pollutants. Approximate concentrations of respective pollutants are determined from observed absorption peak heights. In the lower spectrum, small peaks have been blown up by calculating the *ratio* of the observed spectrum, point by point, to the corresponding spectral amplitude for an air sample containing no pollutants. [From P. L. Hanst, A. S. Lefohn, and B. W. Gay, Jr., *Applied Spectrosc. 27*, 188 (1973).]

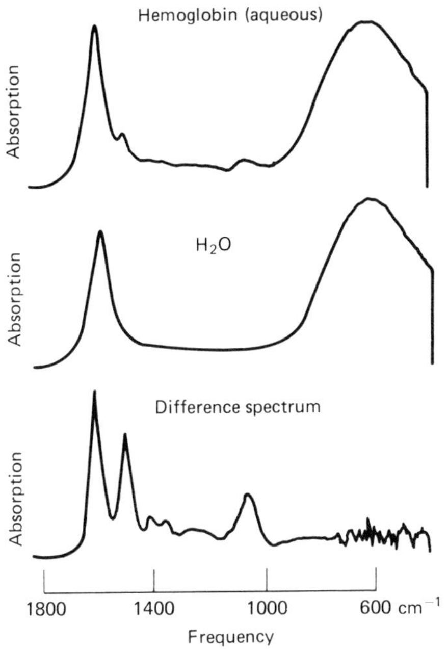

FIGURE 20-21. FT-IR absorption Spectrum. *Top:* aqueous solution of hemoglobin. *Middle:* water alone. *Bottom:* spectrum obtained by point-by-point subtraction of the middle spectrum (water) from the top spectrum (aqueous hemoglobin) so as to eliminate the very strong absorption due to solvent. Note the greatly increased detail in the difference spectrum compared to the top spectrum. [From J. L. Koenig, *Applied Spectrosc. 29,* 293 (1975).]

proteins, based on detailed study of conformationally sensitive infrared absorption frequencies for the protein in water solution.

PROBLEMS

1. Hadamard masks can be constructed for cases when the number of potentially open slits is $(2^n - 1)$, where n is an integer. Figure 20-5 shows how the scheme works for $n = 2$, for a 3-slit system. Given that one of the necessary linear combinations of slit positions for the $n = 7$ case is a cyclic permutation of the arrangement, 1 0 0 1 0 1 1, show that the seven cyclic permutations of that arrangement are linearly independent, and give the prescription (as at lower left of Fig. 20-5) for obtaining the desired intensity passing through a given slit from measurements of the total intensity passing through each of the seven slit arrangements.

2. Calculate the sine and cosine Fourier transforms of the time-domain function

$$f(t) = \cos(\omega_0 t) \qquad 0 \le t \le T, \text{ and } f(t) = 0 \text{ for } t < 0 \text{ or } t > T$$

and combine them to obtain the magnitude spectrum of this time-domain signal. The resulting $A(\omega)$, $B(\omega)$, and $C(\omega)$, using Eqs. 20-8, should produce the top row of curves in Fig. 20-10. These curves describe (among other things) the spectrum from a short burst time-domain response, the diffraction from a slit, and scattering from a point source.

3. Calculate the sine and cosine Fourier transforms of the time-domain function

$$f(t) = \exp(-t/\tau)\cos(\omega_0 t) \qquad 0 \le t < \infty;\ f(t) = 0 \text{ for } t < 0$$

and combine them to obtain the magnitude spectrum of this time-domain signal. The resulting $A(\omega)$, $B(\omega)$, and $C(\omega)$ should produce the bottom row of curves in Fig. 20-10. These curves describe a variety of spectral line shapes observed in magnetic resonance, light scattering, fluorescence, and many other types of spectroscopy.

4. Calculate the sine and cosine Fourier transforms of the time-domain function

$$f(t) = \exp(-t/\tau)\cos(\omega_0 t), \qquad 0 \le t \le T;\ f(t) = 0 \text{ for } t < 0 \text{ or } t > T$$

and combine them to obtain the magnitude spectrum of this time-domain signal. The resulting $A(\omega)$, $B(\omega)$, and $C(\omega)$ should produce the middle row of curves in Fig. 20-10 when $\tau = T$. (The calculations in Problems 2, 3, and 4 are in increasing order of algebraic difficulty.) These curves describe spectral line shapes observed in Fourier transform magnetic resonance and ion cyclotron resonance spectra, and represent general formulae from which the results of problems 2 and 3 represent particular simple limits ($1/\tau \to 0$ or $T \to \infty$).

5. Calculate the full width at half maximum height for $A(\omega)$ and $C(\omega)$, and the peak-to-peak distance for $B(\omega)$, for each of the three representative Fourier transform spectra of Problems 2, 3, and 4. As shown in Fig. 20-10, it should then be clear that the absorption-mode display, $A(\omega)$, always provides the narrowest signal, and is thus preferred when high-resolution spectra are required.

6. Problem 5 also shows that for a given damping constant, τ, the minimum line width (for, say, $A(\omega)$) is obtained in the limit that the experimental data acquisition period T approaches infinity. It would be desirable to know how long T must be made to give a line width within, say, 10% or 1% of the limiting minimal linewidth, since in practice we can record only so much data before our patience or computer storage space is exhausted! Therefore, calculate the full line width at half maximum height for $A(\omega)$ from Problem 4, for the following values of T:

(a) $T = \tau$
(b) $T = 2\tau$
(c) $T = 3\tau$
(d) $T \to \infty$

Then use these four points to plot a graph of line width versus acquisition time, T, to answer the question posed above.

7. In actual Fourier transform spectroscopy experiments, we deal with the Fourier transform of a finite number of (discrete) time-domain data points obtained by sampling the continuous waveforms of Problems 2–4 at regularly spaced times. A general sampling theorem (which should be intuitively obvious) is that it is necessary to sample a wave at least twice per wave cycle to estimate its frequency properly (i.e., we must be able to tell that the wave went up and down during the measuring period in order to know that a cycle occurred!). In other words, we must sample at least as fast as

$$F = 2\nu_{\max}$$

where $\nu_{\max}$ is the largest frequency in the spectrum of interest. Furthermore, the number of points we can store, N, is determined by the size of the available computer memory, so that the longest possible data acquisition time (and therefore the best possible resolution or smallest line width—see Problem 6) is limited by N:

$$FT = N$$

Using these equations, obtain an expression for spectral line width in the simplest case that line width is determined solely by acquisition time (Problems 2 and 5). Use this formula to compute the smallest possible spectral line width for an NMR spectrum whose maximum frequency is 10,000 Hz, 1,000 Hz, or 100 Hz, using a computer with a memory size of of 4096 data words for time-domain data.

REFERENCES

Hadamard and Fourier Transform Spectroscopy

A. G. Marshall and M. B. Comisarow, "Multichannel Methods in Spectroscopy," in *Transformations in Chemistry,* ed. P. R. Griffiths, Plenum, New York (1978).

Fourier Transform Magnetic Resonance

H. Hill and R. Freeman, *Introduction to Fourier Transform NMR,* Varian Associates, Analytical Instrument Division, Palo Alto, Calif. (1970).

T. C. Farrar and E. D. Becker, *Pulse and Fourier Transform NMR,* Academic Press, New York (1971).

Fourier Transform Ion Cyclotron Resonance

M. B. Comisarow and A. G. Marshall, *J. Chem. Phys. 64,* 110 (1976), and references therein.

Fourier Transform Infrared Spectroscopy

J. L. Koenig, *Applied Spectroscopy 29,* 293 (1975). General review.

D. M. Hunten, "Fourier Spectroscopy of Planets," *Science 162,* 313 (1968).

CHAPTER 21 Fourier Analysis of Random Motions: Autocorrelation and Spectral Density

By now the reader has come to appreciate that a spring (whether classical—Chap. 13—or quantum mechanical—Chapter 18) can be made to vibrate by application of some sort of perturbation whose motions contain *frequency components* near the "natural" vibrational frequency of the spring. In Chapters 13 and 18, the perturbation consisted of a *coherent* oscillating electric field, so that all the driven springs in any given region of space oscillate *in phase*. [Although we did not stop to point it out at the time, the *pulsed* electric (or magnetic) field that serves as a broad-band frequency source in FT-NMR spectroscopy (Chapter 20.B.1) is also coherent.] In this section, we wish to determine the frequency components of *incoherent* radiation sources or molecular motions. It will then become evident that a *randomly time-varying* electric (or magnetic) field can have frequency components that also induce transitions between energy levels (quantum mechanical language) or drive electrons on springs near their "natural" frequencies (classical language) even in the *absence* of externally applied radiation, thus explaining the "T_1" process that accounts for the "natural line width" of spectral lines discussed in Chapters 14 and 19. We first present a general formula for computing the frequency components of *any* time-varying random process, and then show how some *particular* types of random motion (e.g., chemical exchange, rotational diffusion, translational diffusion) are related in a natural way to spectral line-broadening in magnetic resonance, fluorescence depolarization, and quasielastic Rayleigh light-scattering experiments.

By a *random* or *stochastic* process, we mean a process in which the amplitude, $f(t)$, of the instantaneous molecular position or direction, or electromagnetic field depends on time in a way not completely definite. For example, the three time-domain traces in Fig. 21-1 might represent the instantaneous positions of each of three different molecules in one-dimensional random walks, as a function of time (see Chapter 6.A). In that example, we found that the *average* position of a molecule away from the origin was zero, but we could calculate a root-mean-*square* (average) position that was directly related to the rate of molecular translational diffusion. Suppose we now record an arbitrary random amplitude, $f(t)$, for a specified time interval, T; it turns out that we can express $f(t)$ in terms of an

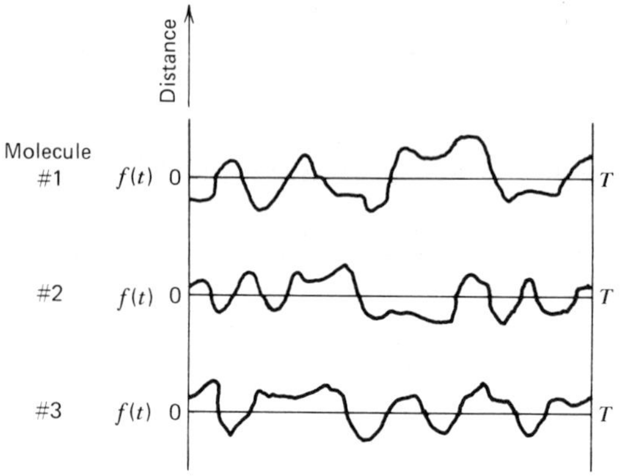

FIGURE 21-1. Hypothetical recordings of position as a function of time for three molecules undergoing one-dimensional translational diffusion. The behavior is different for different molecules, and the frequency components of the motion vary with time (sometimes the oscillations are slow and sometimes fast), so that meaningful frequency analysis of the motion must involve a time-average and average over different particles (ensemble-average).

infinite series of sinusoidal oscillations having component amplitudes, a_n and b_n, at corresponding frequencies, ω_n according to Eq. 21-1 (compare Eq. 13-16a):

$$f(t) = \sum_{n=1}^{\infty} a_n \cos(\omega_n t) + b_n \sin(\omega_n t) \tag{21-1a}$$

where

$$\omega_n = (2\pi n/T) \tag{21-1b}$$

As for the driven (Chapter 13.B) or undriven (Chapter 13.C and 20.B) weight-on-a-spring amplitude, we have broken down the response into components that vary with time as either cosine or sine oscillations, but with one major difference. In the preceding examples, the motion or oscillation was *coherent;* that is, all the particles (or electromagnetic waves) in one region of space were oscillating with the same phase, so we could predict the *group* ("ensemble") behavior simply by knowing the behavior of any *one* particle (or wave). Now, however, the amplitude of the *random* signal varies from time to time and from one particle to another (Fig. 21-1), so that the observed experimental result is based on some sort of *average* (over *time* and over *particles*) of the behavior of individual particles (waves). If we attempt a simple time-average of $f(t)$ in Eq. 21-1, however, averaged over the observation period from time zero to time T, the time-average, $< f(t) >$, goes to zero because the average of either $\cos((2\pi n/T)t)$ or $\sin((2\pi n/T)t)$ goes to zero over that interval:

$$\int_0^T \cos\left(\frac{2\pi nt}{T}\right)dt = \int_0^T \sin\left(\frac{2\pi nt}{T}\right)dt = 0 \qquad (21\text{-}2)$$

a_n and b_n in the present analysis represent the point-by-point amplitudes of the cosine and sine oscillations at a series of equally spaced discrete frequencies (Eq. 21-1), and are the *discrete* version of the corresponding $A(\omega)$ and $B(\omega)$ that we defined for a *continuous* amplitude-versus-frequency representation in Eq. 20-8. The new feature is that while a_n and b_n for any one particle at any *one* time are nonzero, the *time-average* values of a_n and b_n are zero, and we can no longer obtain a_n and b_n directly from observing the *average* behavior of an ensemble of particles.

However, if we conduct an experiment that measures the *square* of $f(t)$, such as a measurement of radiation *intensity* (rather than radiation amplitude), then the time-averaged frequency-components of the resulting spectrum can now be determined, using (see Problems)

$$\int_0^T \cos^2\left(\frac{2\pi nt}{T}\right)dt = \int_0^T \sin^2\left(\frac{2\pi nt}{T}\right) = \frac{1}{2} \qquad (21\text{-}3)$$

so that

$$< (f(t))^2 > = \sum_{n=1}^{\infty}\left(\frac{a_n^2 + b_n^2}{2}\right) = \text{constant for any one particle} \qquad (21\text{-}4)$$

Finally, when we take an additional "ensemble"-average over the various particles (or radiation emitted by various particles), we obtain what is called the "power spectrum" of the random motion, $P(\omega)$:

$$P(\omega_n) = \overline{\left(\frac{a_n^2 + b_n^2}{2}\right)} \qquad (21\text{-}5)$$

so that

$$\overline{<(f(t))^2>} = \sum_{n=1}^{\infty} P(\omega_n) \qquad (21\text{-}6)$$

in which the superscript line denotes an ensemble-average.* $P(\omega_n)$ thus represents the time-averaged, ensemble-averaged frequency component (at frequency ω_n) of the squared amplitude of the random motion of interest.

* *Throughout this discussion, we denote a* time-*average by brackets,* < >, *and an* ensemble-*average by a superscript line.*

[The name, "power" spectrum arises from the example that when $f(t)$ is an electric current flowing through unit resistance, the instantaneous power dissipation is $(f(t))^2$.] In our examples, $P(\omega_n)$ will more often represent an *energy (or intensity) spectrum* for some sort of radiation, since radiation intensity is proportional to the square of the electric (or magnetic) field amplitude (see Chapter 13.A). $P(\omega_n)$ for random incoherent radiation corresponds to $[C(\omega)]^2$ for monochromatic coherent radiation discussed in Eq. 20-8c: it is simply that for incoherent sources, we cannot determine $A(\omega)$ or $B(\omega)$ of Equations 20-8a and 20-8b, so that we must deal solely with *intensity* rather than *amplitude* of the radiation.

From the pictures in Fig. 21-1, it seems clear that we cannot a priori state the value that $f(t)$ will have at a particular time for a particular particle. However, if we know something about the random process (such as the magnitude of the diffusion constant in a translational diffusion situation), then we *can* say that if $f(t)$ has a given value at a particular time for a given particle, then it will have a similar value for a little while afterward (i.e., the value at time, $t + \tau$, will be "correlated" with the value at time, t), and if we wait a sufficiently long time, we will no longer be able to predict the value of $f(t + \tau)$ based on the initial value of $f(t)$ [i.e., $f(t + \tau)$ is "uncorrelated" with $f(t)$ for sufficiently long τ]. Since the length of time it takes for $f(t + \tau)$ to become uncorrelated with $f(t)$ has something to do with the characteristic *rate* (e.g., diffusion constant in the translational diffusion case) of the random process, we might expect to learn about the frequency components of the random process from study of the "correlation function," $G(\tau)$,

$$G(\tau) = \overline{< f(t)\, f(t + \tau) >} \tag{21-7}$$

in which the brackets denote a time-average and the bar denotes an ensemble-average over all particles undergoing the random process. A remarkable property of the correlation function appears when we represent $f(t)$ and $f(t + \tau)$ in terms of their frequency components as defined by Eq. 21-1:

$$G(\tau) = \overline{< f(t) f(t + \tau) >}$$

$$= \overline{\int_0^T \left(\sum_{n=1}^{\infty} a_n \cos\omega_n t + b_n \sin\omega_n t\right)\left(\sum_{m=1}^{\infty} a_m \cos\omega_m (t + \tau) + b_m \sin\omega_m (t + \tau)\right) dt} \tag{21-8}$$

$$= \sum_{n=1}^{\infty} \overline{\left(\frac{a_n^2 + b_n^2}{2}\right)} \cos\omega_n t = \sum_{n=1}^{\infty} P(\omega_n) \cos\omega_n t \tag{21-9}$$

↑ See Problems   ↑ from Eq. 21-5

Replacing the sum in Eq. 21-9 by an integral, we obtain

$$\overline{< f(t)f(t+\tau) >} = G(\tau) = \int_0^\infty P(\omega)\cos\omega\tau\, d\omega \tag{21-10}$$

which is more typically written in the equivalent form *

$$\boxed{P(\omega) = 2/\pi \int_0^\infty G(\tau)\cos(\omega\tau)d\tau} \tag{21-11}$$

Equation 21-11 is a recipe for finding the relative amount of power at any given frequency, resulting from random particle or electromagnetic wave motion with correlation function, $G(\tau)$. Thus, as soon as we work out the form of $G(\tau)$ for a particular type of motion (Chapter 21.A, 21.B, and 21.C), we will immediately be able to compute the power spectrum as a function of frequency from Eq. 21-11. For all the specific random processes that we will consider, the correlation function, $G(\tau)$, decreases exponentially with time

$$G(\tau) = G(0)e^{-\tau/\tau_c} \tag{21-12}$$

in which the "*correlation time*," τ_c, is a measure of the time it takes for $f(t+\tau)$ to become uncorrelated with the earlier value, $f(t)$, and $G(0)$ is the mean-square instantaneous amplitude of the random motion (see below). When the correlation function is of the form in Eq. 21-12, we can compute the power spectrum easily from Eq. 21-11:

$$P(\omega) = \frac{2}{\pi}G(0)\int_0^\infty e^{-\tau/\tau_c}\cos(\omega\tau)dt$$

$$P(\omega) = \frac{2}{\pi}G(0)\left(\frac{\tau_c}{1+\omega^2\tau_c^2}\right) \tag{21-13}$$

The power spectrum ("spectral density"), $P(\omega)$, thus takes the familiar Lorentzian form, and is illustrated in Fig. 21-2 as a function of $\log_{10}(\omega)$. The power spectrum is flat from zero frequency up to frequencies of the order of $(1/\tau_c)$, and falls off rapidly near $(1/\tau_c)$. The "correlation time, τ_c, is thus a measure of the shortest time in which the particle position or direction (or electromagnetic field) mean-square amplitude can change significantly.

Figure 21-2 shows that such a random process effectively acts as broadband *source* of frequencies up to about $(1/\tau_c)$ sec^{-1}. Thus, when the random

* *The connection between Equations 21-10 and 21-11 is a property of Fourier transforms — see any Fourier reference for the mathematical proof.*

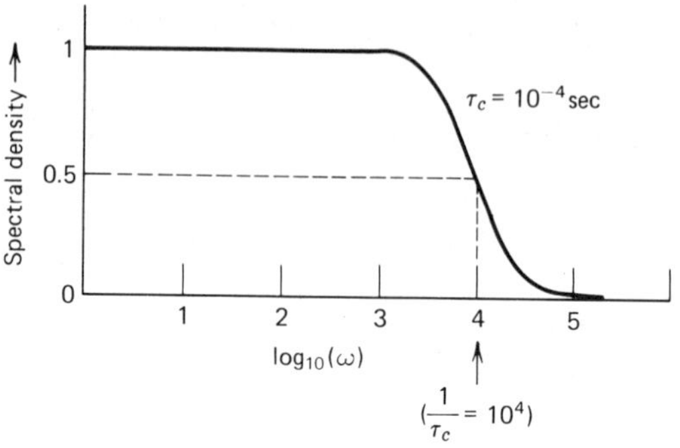

FIGURE 21-2. Power spectrum (spectral density) as a function of frequency (log scale) for a random process described by an exponentially decaying correlation function with correlation time, $\tau_c = 10^{-4}$ sec. [For convenience, $G(0)$ in Eq. 21-12 has been taken as $(\pi/2)$.] The spectrum is "white" (i.e., constant in magnitude) from zero frequency up to frequencies of the order of $(1/\tau_c)$. [The spectral density, $P(\omega)$, of Eq. 21-13 is simply a Lorentzian curve, whose form in this plot is distorted by the log scale for the abscissa.]

motion is associated with an electric (or magnetic) field amplitude change, the random motion can act as a broad-band source of *electromagnetic radiation* over that range of frequencies. If the "natural" frequency of the sample of interest happens to fall at a frequency less than about $(1/\tau_c)$ sec^{-1}, such a random process will drive that system at its natural frequency (classical language) so as to cause transitions between the corresponding energy levels (quantum mechanical language) even in the *absence* of an external radiation source. Since those transitions occur at a rate that defines the "lifetime," T_1, (see Chapter 19.B.3) of the population difference between the two energy levels involved, we can now account for the "natural line width," $(2/T_1)$, of several types of spectral absorption lines, in terms of microscopic motion of the particles in the sample. It remains only to show how the spectral *line width* for a sample (related to the rate of transitions between the two energy levels) is related to the *correlation time* for the particle random motion, and then (see subsequent sections) how that *correlation time* is determined by some particular types of *random motion*.

For a given random process, there is a well-defined mean-square value for the instantaneous amplitude of the motion:

$$\overline{|f(t)|^2} = \text{constant} \tag{21-14}$$

But $\overline{|f(t)|^2}$ in Eq. 21-14 is just the value of the correlation function, $G(\tau)$, for $\tau = 0$. Recalling Eqs. 21-7 and 21-10, we thus reach the immediate important conclusion,

$$\boxed{\overline{|f(t)|^2} = G(0)} = \int_0^\infty P(\omega)d\omega \tag{21-15}$$

or

$$\int_0^\infty P(\omega)d\omega = \begin{pmatrix}\text{area under a plot of the spectral} \\ \text{density function versus frequency}\end{pmatrix} = \text{constant} \qquad (21\text{-}16)$$

independent of the correlation time, τ_c, for that process. The significance of Eq. 21-16 is evident from Fig. 21-3, which shows plots of spectral density versus frequency (log scale) for three different choices of correlation time for a given random process. Since the area under each curve must be the same (Eq. 21-16), while the cut-off frequency $(1/\tau_c)$ varies from plot to plot, the *height* of each curve must also vary as shown in the figure. If we now specify that this random process involves a fluctuating electric (or magnetic) field, then the plots show the relative power spectrum of radiation produced by the process. Finally, if we consider the radiation energy at a particular frequency, ω_0, chosen as the natural frequency for motion of a particular type of molecule, then the rate constant for induced (radiationless) transitions, $(1/T_1)$, will be proportional to the amount of power reaching the sample (Chapter 19.B.1). For example, when the correlation time is very long $((1/\tau_c) << \omega_0)$, very little power reaches the sample and relatively infrequent transitions are induced and $(1/T_1)$ is small; when $(1/\tau_c) \sim \omega_0$, there is maximal spectral density at ω_0, and the rate of induced transitions is also a maximum; lastly, when the correlation time is very short $[(1/\tau_c) >> \omega_0]$, the spectral density at ω_0 is again small and $(1/T_1)$ is small. From the consequences of Eq. 21-16 as shown in Fig. 21-3, it is clear that the rate constant for induced transitions (in the *absence* of externally applied radiation), $(1/T_1)$, must show the dependence on correlation time illustrated in Fig. 21-4, with a maximum spectral line-broadening (minimum T_1) when the random motion has a correlation time of the order of the natural frequency of the sample, $(1/\tau_c) \sim \omega_0$. Figure 21-4 thus accounts for natural *spectral*

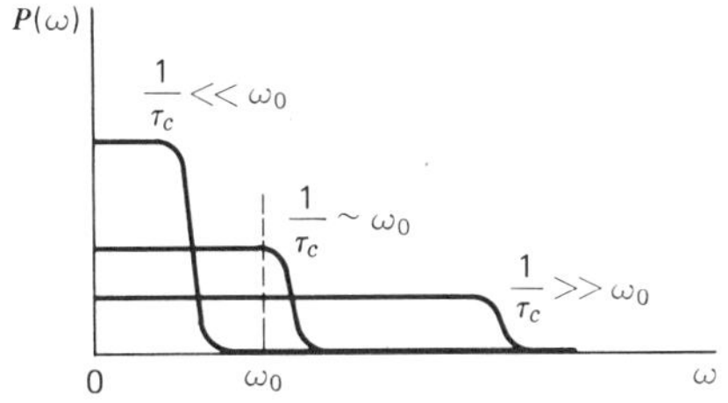

FIGURE 21-3. Spectral density, $P(\omega)$, as a function of frequency (log scale), for random processes characterized by three different correlation times. The quantity of interest (see text) is the spectral density at the fixed frequency, ω_0, of the natural vibration of some oscillator in a sample. For either long (top-most curve) or short (lowermost curve) correlation time, τ_c, the spectral density at ω_0 is small; the spectral density at ω_0 is a maximum when $(1/\tau_c) \sim \omega_0$. These curves account for dependence of the induced transition rate constant $(1/T_1)$ on correlation time shown in Fig. 21-4.

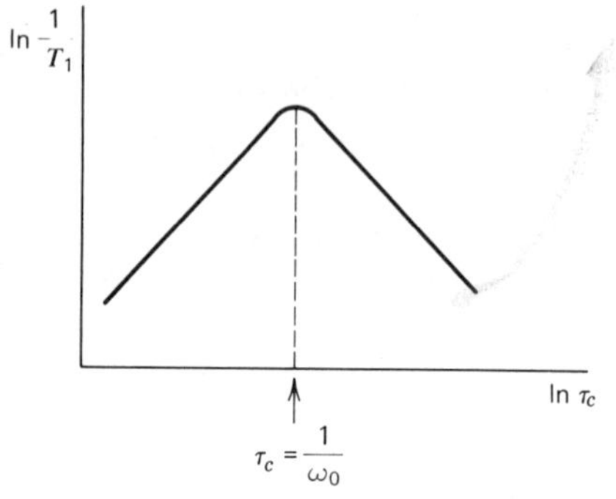

FIGURE 21-4. Plot of transition probability, $(1/T_1)$ (log scale) versus correlation time, τ_c (log scale) for a particular random process. As explained by the plots in Fig. 21-3 (see text), the transition probability is a maximum when $(1/\tau_c) \sim \omega_0$, as seen in the present figure. These transitions occur in the absence of externally applied radiation, and thus define the "natural" spectral line width for the sample in question (see text).

line width of a sample, $\Delta\omega = (2/T_1)$, in terms of the characteristic *correlation time* for change in the mean-square amplitude of a randomly varying local electromagnetic field. In the next few sections, we will show how to calculate correlation functions (and thus correlation times and thus spectral line widths) for some particular types of random molecular motion (chemical exchange, rotational diffusion, and translational diffusion) that produce corresponding randomly fluctuating electromagnetic fields. Finally, we can then use those results in reverse order, to determine characteristic rate constants (e.g., chemical reaction rate constant; rotational or translational diffusion constant) from measured spectral line widths.

21.1. RANDOM JUMPS BETWEEN TWO SITES OF DIFFERENT "NATURAL" FREQUENCY

Perhaps more than for any of the formal descriptions encountered thus far, the intuitive content of the random noise treatment becomes clear only after working through a few examples. Our first example concerns the effect of chemical "exchange" (random jumps) of, say, a magnetic nucleus between two sites at which the magnetic field magnitude is different. If the two sites correspond to a proton located on two different molecules, for example, then there is a well-defined (time-averaged) *exchange rate constant, k,* for the random process. The problem then consists of evaluating the correlation function for the magnetic field strength by taking an ensemble-average over all possible (i.e., two) initial sites at time zero and all possible final sites at later time, τ. From the magnetic field correlation function, we can obtain the power spectrum of the magnetic radiation (Eq. 21-11) and thus calculate the probability for transitions between the two energy states of the proton in a fixed applied magnetic field.

In calculating the average value of any function (in this case magnetic field), we construct a sum of terms of the type, (value of the function) · (probability of finding that value), as discussed on pp. 142 to 143. We must therefore first compute the required probabilities of finding a nucleus at either of the two initial sites, and of finding a nucleus at each of the two final sites at a later time, τ, given that the nucleus was found at a given initial site at time zero. Therefore, let P_A and P_B be the probabilities of finding the proton at sites A or B. From the chemical equilibrium between the two sites

$$A \underset{k}{\overset{k}{\rightleftharpoons}} B \tag{21-17}$$

we can immediately write the equations for rate of change of P_A and P_B with time:

$$\begin{aligned} \frac{dP_A}{dt} &= -kP_A + kP_B \\ \frac{dP_B}{dt} &= -kP_B + kP_A \end{aligned} \tag{21-18}$$

Since we have chosen a simple case for which the forward and reverse rates are equal, there is equal probability of finding a proton at site A or site B initially:

$$P_A(0) = P_B(0) = \frac{1}{2} \tag{21-19}$$

Next, Equations 21-18 are readily solved (Chapter 10.A.3) to give the probability that a proton is located at a stated final site at time, τ, given that it was found at a given initial site at a time zero: P (final site, τ; initial site, 0)

$$P(A,\tau;A,0) = \frac{1}{2}(1 + e^{-2k\tau}) \tag{21-20a}$$

$$P(B,\tau;A,0) = \frac{1}{2}(1 - e^{-2k\tau}) \tag{21-20b}$$

$$P(A,\tau;B,0) = \frac{1}{2}(1 - e^{-2k\tau}) \tag{21-20c}$$

$$P(B,\tau;B,0) = \frac{1}{2}(1 + e^{-2k\tau}) \tag{21-20d}$$

The correlation function for magnetic field fluctuations due to chemical exchange between two sites of different magnetic field may now be constructed using the probabilities of Equations 21-19 and 21-20 to compute the ensemble-average values of the product of the magnetic field at the initial site at time zero, H_{initial}, and the magnetic field at the final site at (later) time τ, H_{final}:

$$G(\tau) = \overline{<H(0)H(\tau)>} = \sum_{\substack{\text{initial}\\ \text{sites}}} \sum_{\substack{\text{final}\\ \text{sites}}} \left(P_{\substack{\text{initial}\\ \text{site}}} H_{\substack{\text{initial}\\ \text{site}}}\right) \left(P(H_{\substack{\text{final}\\ \text{site}}}, \tau; H_{\substack{\text{initial}\\ \text{site}}}, 0) H_{\substack{\text{final}\\ \text{site}}}\right) \tag{21-21}$$

It is convenient to define the magnetic field at site A or site B, H_A or H_B, as

$$H_A = +h \tag{21-22a}$$

$$H_B = -h \tag{21-22b}$$

from which the correlation function of Eq. 21-21 may be evaluated as:

$$\begin{aligned} G(\tau) = \overline{<H(0)H(\tau)>} &= P_A H_A P(H_A,\tau;H_A,0)H_A \\ &+ P_A H_A P(H_B,\tau;H_A,0)H_B \\ &+ P_B H_B P(H_A,\tau;H_B,0)H_A \\ &+ P_B H_B P(H_B,\tau;H_B,0)H_B \end{aligned} \tag{21-23}$$

$$= \frac{1}{4} h^2(1 + e^{-2k\tau}) + \frac{1}{4}(-h^2)(1 - e^{-2k\tau}) + \frac{1}{4}(-h^2)(1 - e^{-2k\tau}) + \frac{1}{4}(h^2)(1 + e^{-2k\tau})$$

$$= h^2 e^{-2k\tau}$$

$$\overline{<H(0)H(\tau)>} = h^2 e^{-\tau/\tau_c} = G(\tau) \tag{21-24a}$$

where

$$\frac{1}{\tau_c} = 2k \tag{21-24b}$$

From the correlation function, Eq. 21-24a, we quickly obtain the spectral density, $P(\omega)$, from Eq. 21-11

$$P(\omega) = \frac{2}{\pi}\int_0^\infty h^2\, e^{-\tau/\tau_c}\cos(\omega\tau)d\tau$$

$$\boxed{P(\omega) = \frac{2}{\pi} h^2 \frac{\tau_c}{1+\omega^2\tau_c^2}, \qquad \frac{1}{\tau_c} = 2k} \tag{21-25}$$

EXAMPLE *Random Magnetic Field Fluctuations and NMR Relaxation Times*

Consider a collection of spin one-half nuclei, subjected to a large uniform magnetic field in the (usual) z direction *and* a small magnetic field that fluctuates (with correlation time, τ_c) between the values $+h$ and $-h$ in each of the three orthogonal directions (x, y, and z). Using methods formally very similar to those we have just examined, it can be shown that the nuclear magnetic lifetime, T_1, is determined by the Fourier transform of the correlation function for magnetic field fluctuations, at frequency $\omega_0 = \gamma H_0$:

$$\left(\frac{1}{T_1}\right) = \pi\gamma^2 P(\omega_0) = 2\gamma^2 h^2 \frac{\tau_c}{1+\omega_0^2\tau_c^2} \tag{21-26}$$

since transition probability is proportional to $(1/T_1)$ [see Eq. 19-38].

There is another characteristic decay constant, T_2, in magnetic resonance, which describes the (exponential) time-decrease of a magnetic moment introduced in the x-y plane (see Equation 16-43a).

Using the methods of time-dependent perturbation theory presented in Chapter 19.B, it can be shown that T_2 is related to the spectral density for the same random process at the frequencies, ω_0 and zero frequency:

$$\left(\frac{1}{T_2}\right) = \frac{\pi}{2}\gamma^2\ [P(0) + P(\omega_0)]$$

$$\left(\frac{1}{T_2}\right) = \gamma^2 h^2 \left[\tau_c + \frac{\tau_c}{1+\omega_0^2\tau_c^2}\right] \tag{21-27}$$

Equation 21-26 simply states that NMR spectral "lifetime," T_1, is determined by the spectral density at the resonant frequency, ω_0; in other words, the randomly fluctuating magnetic field affects the nuclei in proportion to the (mean-square) amplitude of the magnetic field frequency component at the "natural" nuclear magnetic precession frequency. [The reason that $(1/T_2)$ is determined by the spectral density at ω_0 *and* at zero frequency is that the y magnetization can decrease in two ways: by growth of M_z back to its thermal equilibrium value (requiring transitions between the two energy levels and thus determined by spectral density at the corresponding frequency for that transition, ω_0) *and* by the decrease in M_y resulting from the fact that different nuclei may precess at slightly different rates so that their resultant My decreases with time as the nuclei "get out of phase" with each other in their precession. Since the energy of a magnetic moment does not change when its direction in the x-y plane changes, the "natural" frequency for this loss of phase coherence can be thought of as zero, and T_2 is thus determined partly by the spectral density at zero frequency.]

Equations 21-26 and 21-27 are of a form that is general to the effect of a variety of random processes on NMR (and other types of) spectral lifetime and line width. The important features are first, that in the limit of very rapid fluctuations (i.e., very short correlation time, $\omega_0\tau_c << 1$)

$$\frac{1}{T_1} = \frac{1}{T_2} = 2\gamma^2 h^2 \tau_c, \quad \omega_0\tau_c \ll 1 \tag{21-28}$$

However, since the spectral density at zero continues to increase, while the spectral density at ω_0 is negligible for slower correlation times ($\omega_0\tau_c > 1$), as evident from Fig. 21-3, $(1/T_2)$ continues to grow larger as τ_c increases, while $(1/T_1)$ decreases in this region, as shown in Fig. 21-5.

A particularly useful application of the preceding analysis is the extraction of *chemical exchange rates* from measurements of NMR T_1 and T_2 values, for the "fast-exchange limit," $(1/\tau_{ex}) \gg (1/T_2), (1/T_1)$. (We have already discussed an example of chemical exchange rates from NMR relaxation times for the "slow-exchange limit" on pp. 650 to 652.)

The more important aspect of the present discussion is that the basic nature of the argument is preserved, no matter what the origin of the fluctuating

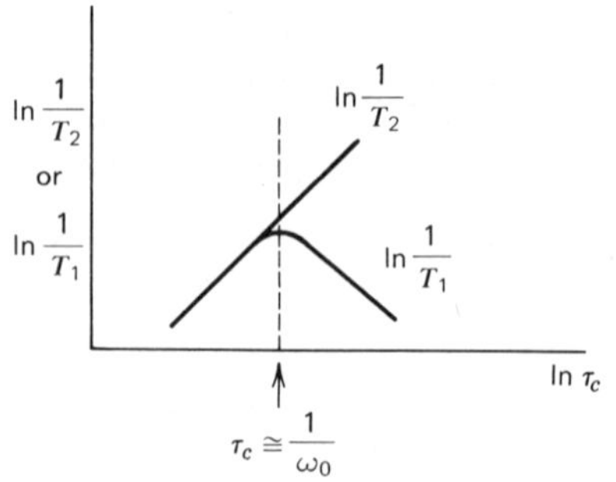

FIGURE 21-5. NMR relaxation rates, $(1/T_1)$ and $(1/T_2)$ (log scale) as a function of correlation time (log scale) for a random process in which the magnetic field fluctuates randomly between the values $+h$ and $-h$ in each of three mutually perpendicular directions. In the "extreme-narrowed" limit, $\omega_0\tau_c << 1$, at the left of the figure, $T_1 = T_2$. For longer correlation times, T_2 becomes shorter while T_1 becomes longer as τ_c increases; thus T_2 is more sensitive to *slow* fluctuations, while T_1 and T_2 are equally sensitive to *rapid* motions. Applications for NMR T_1 and T_2 measurements are based on this general property.

magnetic field. In the next section, we analyze the magnetic field fluctuations that result from *rotation* of a molecule, and the conclusions are formally very similar to those reached in Equations 21-26-28 and Fig. 21-5, except that the correlation time will now be a measure of the length of time it takes for a molecule to reorient approximately a radian in angle. It will thus be possible to deduce the rate of rotation of a molecule in solution from NMR T_1 and T_2 measurements (see Chapter 21.B for an example).

21.B. Random Rotational Motion: Rotational Diffusion Constants from Dielectric Relaxation, Magnetic Resonance, and Fluorescence Depolarization

Suppose that some molecular property (energy, dipole moment, etc.) varies as the *orientation* of the molecule changes. Then random fluctuations in the angle between some well-defined molecular axis and a fixed laboratory axis will lead to a fluctuation in, say, dipole moment, which may have frequency components at the "natural" frequencies for a sample of interest. Thus, random rotational motion can serve as a broad-band radiation source of frequencies up to an upper frequency limit determined by the maximal rate of rotation of the molecule. We will treat this situation first for the case of rotational diffusion about just one axis (compare to one-dimensional translational diffusion in Chapter 6), and then extend the treatment to three-dimensional rotational diffusion of actual macromolecules in solution.

One-dimensional Rotational Diffusion

As in the previous example of random jumps between two types of sites, we begin by computing a correlation function for some molecular property, $f(\phi)$, that varies with the angle ϕ about one laboratory axis, by averaging over all initial *orientations* and final *orientations,* just as we formerly averaged over all initial and final *sites.* Since the number of possible orientations is infinite (i.e., is a *continuous* variable, in contrast to the finite number of *discrete* sites in the preceding example), the sum over orientations becomes an integral over angle (compare to Eq. 21-21)

$$\overline{<f(0)f(\tau)>} = \int_{\phi=0}^{2\pi} f(\phi)P(\phi,\tau;\ \phi_0,0)\ d\phi \int_{\phi_0=0}^{2\pi} f(\phi_0)P(\phi_0)d\phi_0 \qquad (21\text{-}29)$$

in which $P(\phi_0,0)\ d\phi_0$ is just the probability of finding a molecule pointing at an angle between ϕ_0 and $\phi_0 + d\phi_0$ at time zero, and $P(\phi,\tau;\ \phi_0,0)\ d\phi$ is the probability of finding a molecule pointing at an angle between ϕ and $\phi + d\phi$ at time τ, given that it was pointing at an angle between ϕ_0 and $\phi_0 + d\phi_0$ at time zero.

Since in the usual situation, we are equally likely to find a molecule initially pointing in any direction, $P(\phi_0)$ will be independent of ϕ_0, so that

$$\int_{\phi_0=0}^{2\pi} P(\phi_0)d\phi_0 = 1 \tag{21-30a}$$

$$= P(\phi_0) \int_0^{2\pi} d\phi_0 = 2\pi P(\phi_0) \tag{21-30b}$$

or simply

$$P(\phi_0) = \frac{1}{2\pi} \tag{21-31}$$

All that remains is to find $P(\phi,\tau;\ \phi_0,0)$ in the same way we found $P(H_{\text{final site}},\ \tau;\ H_{\text{inital site}},\ 0)$ in the preceding example, namely from the equation that describes the time-variation in the quantity of interest (in this case, molecular orientation angle, ϕ). By analogy to our experience with the equation of motion for *translational* diffusion (see Eq. 6-55)

$$\frac{\partial P(y,t)}{\partial t} = D \frac{\partial^2}{\partial y^2} P(y,t) \tag{21-32}$$

we would expect to be able to describe rotational diffusion (i.e., reorientation by the resultant of many tiny angular jumps clockwise or counterclockwise) by a similar equation

$$\boxed{\frac{\partial P(\phi,t)}{\partial t} = D_{\text{rot}} \frac{\partial^2}{\partial \phi^2} P(\phi,t)} \tag{21-33}$$

in which D_{rot} is called the *rotational diffusion coefficient.* The reader can readily verify that the following equation is a solution of Eq. 21-33: *

$$P(\phi,t;\ \phi_0,0) = \sum_{n=0}^{\infty} e^{-in\phi_0}\, e^{in\phi}\, e^{-n^2 D_{\text{rot}} t} \tag{21-34}$$

We are now in a position to consider some *particular* angle-dependent molecular properties, $f(\phi)$. For example, an electric dipole moment oriented at an angle, ϕ, with respect to some fixed laboratory axis, has a component along that lab axis that varies with angle according to cos (ϕ).† As the molecule (and its associated electric dipole moment) rotates, the electric field compo-

* *In Eq. 21-34 ff., we employ complex notation in the same spirit as in Chapters 13 and 19. The same results could be obtained using cosines rather than exponentials, but the algebraic manipulations are much simpler with complex quantities. Since the final results must always be real, we need only re-define the correlation function (Eq. 21-7) as* $\overline{\langle f(\tau)^* f(0)\rangle} = G(\tau)$, *and the power spectrum as* $P(\omega) = 1/2\pi \int_{-\infty}^{\infty} G(\tau) e^{-i\omega\tau}\, d\tau$.

† *exp* $(i\phi)$ *in complex notation.*

nent along the lab axis changes, so that random rotations of the molecule produce radiation along the lab-fixed axis at frequencies predicted by the spectral density calculated from the correlation function for the random rotational motion. The calculations are simpler than the language:

$$\overline{\frac{\langle f(\tau)^* f(0)\rangle}{\langle (f(0))^2\rangle}} = \left(\frac{1}{2\pi}\right)^2 \int_{\phi_0=0}^{2\pi} \int_{\phi=0}^{2\pi} e^{i\phi_0} e^{-i\phi} \sum_{n=0}^{\infty} e^{-in\phi_0} e^{in\phi}\, e^{-D_{\text{rot}} n^2 t}\, d\phi\, d\phi_0$$

$$= \left(\frac{1}{2\pi}\right)^2 \sum_{n=0}^{\infty} e^{-n^2 D_{\text{rot}} t} \int_{\phi=0}^{2\pi} e^{i(n-1)\phi}\, d\phi \int_{\phi_0=0}^{2\pi} e^{-i(n-1)\phi_0}\, d\phi_0 \qquad \text{(21-35)*}$$

But since

$$\int_{\phi=0}^{2\pi} e^{i(n-1)\phi}\, d\phi = 0 \text{ for } n \neq 1 \qquad \text{(21-36a)}$$

and

$$\int_{\phi=0}^{2\pi} e^{i(n-1)\phi}\, d\phi = \int_{0}^{2\pi} d\phi = 2\pi \text{ for } n = 1 \qquad \text{(21-36b)}$$

Equation 21-35 simplifies to a single term

$$\overline{\frac{\langle f(\tau)^* f(0)\rangle}{\langle (f(0))^2\rangle}} \underset{\substack{\uparrow \\ \text{one-dimension}}}{=} e^{-D_{\text{rot}} t} \qquad \text{(21-37)}$$

Equation 21-37 is of the same functional form as Eq. 21-24 for two-site random jumps, and all the discussion following Eq. 21-24 is thus applicable to the effect of rotational diffusion (about a single axis) on spectral density as a function of frequency.

For other examples of interest (magnetic relaxation, fluorescence depolarization) the molecular parameter of interest varies with angle in a slightly more complicated way, namely as a function of *two* (spherical) angles, θ and ϕ (see Fig. 21-6 for definition of θ and ϕ), according to "spherical harmonic" functional form (Eq. 21-38). It may be noted that the electric dipole moment angle-variation can be thought of as a spherical harmonic of "rank" one ($l = 1$ in Table 21-1; the magnetic and fluorescent parameters turn out to vary as spherical harmonics of "rank" 2 ($l = 2$ in Table 21-1).

* *We have divided the correlation function by the mean-square amplitude of the random variable, so that Eq. 21-35 ff. will apply to* any *randomly-varying angular function whose instantaneous (complex) amplitude is proportional to exp ($i\phi$).*

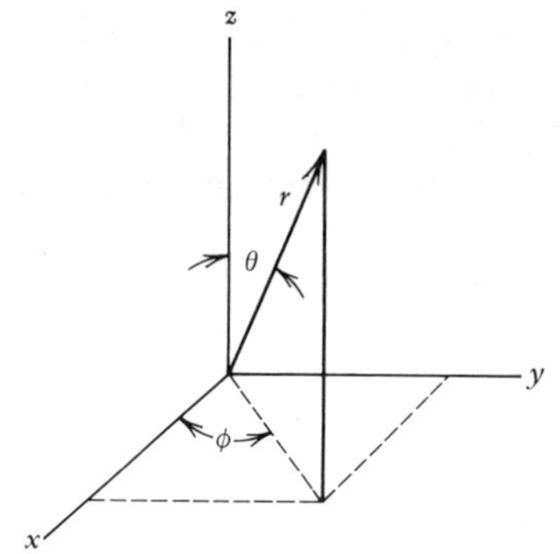

FIGURE 21-6. Definition of spherical coordinates r, θ, and ϕ. Electric dipole component along an arbitrary axis varies as a function of only one angle, while magnetic resonance and fluorescence parameters vary as spherical harmonic functions of both angles θ and ϕ.

Table 21-1 Functional Form of Some of the Spherical Harmonic Angular Functions $Y_{lm}(\theta,\phi)$, for $l = 0$, 1, and 2†

$l = 0$	$m = 0$	$Y_{00}(\theta,\phi) = \sqrt{1/4\pi}$
$l = 1$	$m = 1$	$Y_{11}(\theta,\phi) = -\sqrt{3/8\pi}\,\sin\theta\, e^{i\phi}$
	$m = 0$	$Y_{10}(\theta,\phi) = \sqrt{3/4\pi}\,\cos\theta$
	$m - 1$	$Y_{1-1}(\theta,\phi) = \sqrt{3/8\pi}\,\sin\theta\, e^{-i\phi}$
$l = 2$	$m = 2$	$Y_{22}(\theta,\phi) = \sqrt{15/32\pi}\,\sin^2\theta\, e^{2i\phi}$
	$m = 1$	$Y_{21}(\theta,\phi) = -\sqrt{15/8\pi}\,\sin\theta\cos\theta\, e^{i\phi}$
	$m = 0$	$Y_{20}(\theta,\phi) = \sqrt{5/16\pi}\,(3\cos^2\theta - 1)$
	$m = -1$	$Y_{2-1}(\theta,\phi) = \sqrt{15/8\pi}\,\sin\theta\cos\theta\, e^{-i\phi}$
	$m = -2$	$Y_{2-2}(\theta,\phi) = \sqrt{15/32\pi}\,\sin^2\theta\, e^{-2i\phi}$

$$m = (l), (l-1), (l-2), \cdots, 0, -1, -2, \cdots -(l-1), -l$$

$$\int_{\theta=0}^{\pi}\int_{\phi=0}^{2\pi} Y^*_{lm}(\theta,\phi)Y_{l'm'}(\theta,\phi)\sin\theta\, d\theta\, d\phi = 0 \text{ for } l \neq l' \text{ and/or } m \neq m' \quad \text{(21-38a)}$$

$$= 1 \text{ for } l = l' \text{ and } m = m' \quad \text{(21-36b)}$$

† *Angle-dependent parameters for dielectric relaxation vary as $Y_{10}(\theta,\phi)$, while angle-dependent parameters for magnetic and fluorescent relaxation vary as $Y_{2m}(\theta,\phi)$. In this section, we compute the spectral density obtained from the correlation function for the appropriate $Y_{lm}(\theta,\phi)$ using a diffusional model for the rotational motion of the molecule of interest. (The reader should recognize these functions as the "angular" part of the quantum mechanical solution of the hydrogen atom wave functions derived in most introductory treatments of quantum mechanics.) Some properties of the $Y_{lm}(\theta,\phi)$ are also listed in the table.*

Three-dimensional Rotational Diffusion

Since actual molecules in solution can obviously rotate about more than one axis, we are now led to consideration of three-dimensional *rotational* diffusion: namely, that a molecule may reorient by many small *angular* jumps about each of three mutually perpendicular axes. (In Chapter 6 we treated *translational* diffusion in three dimensions by considering small *linear* jumps along each of three mutually perpendicular directions.) Again, by extending the thinking that led to Eq. 6-55 for translational diffusion

$$\frac{\partial P(x,y,z,t)}{\partial t} = D_{\text{trans}}\ \nabla^2 P(x,y,z,t) = D_{\text{trans}}\left(\frac{\partial^2}{\partial x^2} + \frac{\partial^2}{\partial y^2} + \frac{\partial^2}{\partial z^2}\right) P(x,y,z,t) \qquad (21\text{-}39)$$

we obtain the desired equation for rotational diffusion in three dimensions

$$\frac{\partial P(\theta,\phi,t)}{\partial t} = D_{\text{rot}}\ \nabla^2 P(\theta,\phi,t) \qquad (21\text{-}40)$$

in which $P(\theta,\phi,t)\sin\theta d\theta d\phi$ represents the probability of finding a molecule pointing in a direction between θ and $\theta + d\theta$ and between ϕ and $\phi + d\phi$ at time, t; D_{rot} is the rotational diffusion constant (large D_{rot} corresponds to rapid diffusion); and ∇^2 in Eq. 21-40 is expressed in polar coordinates (Fig. 21-6 with $r = 1$).

As in the one-dimensional case, we suppose that all initial molecular directions are equally likely, so that $P(\theta_0,\phi_0,0)$ is independent of θ and ϕ:

$$\int_{\theta_0=0}^{\pi}\int_{\phi_0=0}^{2\pi} P(\theta_0,\phi_0)\sin\theta_0 d\theta_0 d\phi_0 = 1$$

(i.e., we are certain to find the initial direction somewhere between the limits of the two integrals)

$$= P(\theta_0,\phi_0)\int_0^{\pi}\int_0^{2\pi}\sin\theta_0 d\theta_0 d\phi_0 = 4\pi P(\theta_0,\phi_0) \qquad (21\text{-}41)$$

$$\boxed{P(\theta_0,\phi_0,0) = \frac{1}{4\pi}} \qquad (21\text{-}42)$$

Finally, it may be verified that the solution of the rotational diffusion equation, Eq. 21-40, can be written

$$\boxed{P(\theta,\phi,t;\ \theta_0,\phi_0,0) = \sum_{l=0}^{\infty}\sum_{m=0}^{\infty} Y_{lm}^{*}(\theta_0,\phi_0) Y_{lm}(\theta,\phi)\, e^{-l(l+1)D_{\text{rot}}t}} \qquad (21\text{-}43)$$

in which $P(\theta,\phi,t;\ \theta_0,\phi_0,0)$ represents the probability that a molecule is oriented in a direction near θ and ϕ at time t, given that it was oriented near θ_0 and ϕ_0 at time zero. Using Equations 21-38, 21-42 and 21-43, the reader can now show that the correlation function for any of the spherical harmonic angular functions will be of the form

$$\overline{<Y^*_{lm}(\theta,\phi,\tau)Y_{lm}(\theta_0,\phi_0,0)>}$$

$$= \int_\theta \int_\phi \int_{\theta_0} \int_{\phi_0} \sum_l \sum_m Y^*_{lm}(\theta_0,\phi_0)Y_{lm}(\theta,\phi)\, e^{-l(l+1)D_{\text{rot}}t}\, Y^*_{lm}(\theta,\phi)Y_{lm}(\theta_0,\phi_0)$$

$$\times (\sin\theta d\theta d\phi)(\sin\theta_0 d\theta_0 d\phi_0) \quad (21\text{-}44)$$

$$\boxed{\overline{<Y^*_{lm}(\theta,\phi,\tau)Y_{lm}(\theta_0,\phi_0)>} = c_{lm}\, e^{-l(l+1)D_{\text{rot}}t}} \quad (21\text{-}45\text{a})$$

$$= c_{lm} \exp[-t/\tau_{\text{rot}}]$$

where (21-45b)

$$\boxed{\tau_{\text{rot}} = \frac{1}{l(l+1)D_{\text{rot}}} \begin{cases} \xrightarrow{l=1} 1/2D_{\text{rot}} \text{ for dielectric relaxation} \\ \xrightarrow{l=2} 1/6D_{\text{rot}} \text{ for NMR, ESR, and fluorescence depolarization} \end{cases}} \quad (21\text{-}46)$$

and c_{lm} is a function of l and m, but is independent of time.

While the algebra leading to the three-dimensional rotational diffusion result, Eq. 21-45, is appreciably more involved than for either the two-site jump case of Eq. 21-24 or the one-dimensional rotational diffusion case of Eq. 21-37, the functional form is again a simple exponential decay for the correlation function, leading to the simple spectral density picture of Fig. 21-2.

Finally, it is possible to relate *rotational* diffusion constant to *rotational* friction coefficient [steady-state rotational velocity per unit rotational force (torque)] by an equation essentially the same as that between *translational* diffusion coefficient and *translational* friction coefficient (steady-state translational velocity per unit linear force); refer to Eq. 7-13

$$D_{\text{trans}} = \frac{kT}{f_{\text{trans}}} \quad (7\text{-}13)$$

$$D_{\text{rot}} = \frac{kT}{f_{\text{rot}}} \quad (21\text{-}47)$$

Similarly, just as we were able to relate the *translational* friction coefficient for a spherical macromolecule to molecular size and solution viscosity

$$f_{\text{trans}}(\text{sphere}) = 6\pi\eta R \quad (7\text{-}55)$$

in which R is the radius of the sphere and η is the solution viscosity, an analogous treatment provides a similar relation between *rotational* friction coefficient and molecular size and solution viscosity for spherical macromolecules:

$$f_{\mathrm{rot}}(\text{sphere}) = 8\pi\eta R^3 \tag{21-48}$$

Combining Equations 21-47 and 21-48 produces the desired relation between rotational diffusion constant D_{rot} (obtainable from the various experiments we are about to describe) and molecular size, R, of a spherical macromolecule:

$$\boxed{D_{\mathrm{rot}} = \frac{kT}{8\pi\eta R^3}} \leftarrow \text{(Spherical macromolecule)} \tag{21-49}$$

EXAMPLE *Dielectric Relaxation*

In Chapter 14.F, we discussed experiments (dielectric loss; dielectric constant) that reflect the rate at which a molecule with a permanent electric dipole moment can reorient (i.e., rotate) in an electric field. Since the component of the electric dipole along a fixed laboratory direction (the direction normal to the two parallel capacitor plates in Fig. 14-33) varies with angle as cos (θ), where θ is the angle between the dipole moment vector and a vector normal to the plates, the steady-state spectral line width is obtained from the spectral density calculated from the correlation function of $f(\theta) = \cos(\theta)$. (See Eq. 21-45 for $l = 1$, $m = 0$):

$$\overline{<\cos(\theta,\tau)\cos(\theta_0,0)>} = c_{10}\, e^{-l(l+1)D_{\mathrm{rot}}\tau} \tag{21-50a}$$

$$= c_{10}\, e^{-\tau/\tau_{\mathrm{rot}}} \tag{21-50b}$$

where

$$\boxed{\tau_{rot} = \frac{1}{2D_{\mathrm{rot}}}} \tag{21-51}$$

Thus, for dielectric relaxation studies, the characteristic line width (actually, half the line width at half maximum height) for dielectric constant as a function of frequency of the applied electric oscillating field (Fig. 14-35) is given by

$$\frac{1}{\tau_{rot}} = 2D_{\mathrm{rot}} = \frac{kT}{4\pi\eta R^3} \tag{21-52}$$

as claimed in Chapter 14-F (Eq. 14-33). An intuitive picture of this phenomenon is that the dipoles are continually being aligned by the influence of the applied electric field, but are continually being unaligned by random rotational motion (rotational diffusion) of the molecules themselves. The observed steady-state dielectric constant at a given frequency is a measure of that alignment. However, since we now know the power spectrum for the random rotational diffusion process from the Fourier transform of Eq. 21-50b, we have therefore accounted for the steady-state dielectric constant as a function of frequency of an applied electric field of fixed amplitude (i.e., if the applied field amplitude (and its associated "aligning" tendency) is fixed, then any variation in dielectric constant must result from changes in the power spectrum of the random rotational motion). Since the power spectrum (of the correlation function, Eq. 21-50b) must be of the familiar Lorentzian form of Eq. 21-13, we have thus accounted for the functional frequency-dependence of dielectric loss and dielectric constant previously discussed in Chapter 14 (Eq. 14-32).

Applications for the rotational diffusion constant information obtained from dielectric relaxation measurements have already been discussed (see Chapter 14.F), and the reader is referred to those sections for examples.

EXAMPLE *Magnetic Relaxation*

Associated with any nuclear (or electronic) magnetic moment, there is a small "local" magnetic field, whose strength varies as $(1/r^3)$, where r is the distance between the magnetic moment and the point at which the local field is measured. For a molecule that possesses two, say, magnetic nuclei near each other, each magnetic nucleus produces a "dipolar" magnetic field at the other. When such molecules are now placed under the influence of a large static applied magnetic field, the net magnetic field at a given nucleus will consist of the vector sum of the large applied static field and a small field whose direction fluctuates as the molecule rotates in solution. The mean-square average of that small fluctuating field will thus vary as $(1/r^6)$, where r is now the distance between the two nuclei, and the power spectrum of the fluctuating field will be determined by spectral densities obtained as the Fourier transform of correlation function of the appropriate $Y_{2m}(\theta,\phi)$ factors that describe the angle-dependence of the interaction. Since a system of two spins one-half (see Fig. 18-6) is described by essentially three different energies, whose differences (the "transition" energies) correspond to "natural" frequencies of ω_0, $2\omega_0$, and 0, where $\omega_0 = \gamma\hbar H_0$, we would expect that the NMR relaxation rates, $(1/T_1)$ and $(1/T_2)$, should be directly related to spectral densities at those frequencies. Although a full quantum mechanical time-dependent perturbation treatment is required to provide the correct bookkeeping, the final result is in accord with the above intuitive reasoning, and is qualitatively similar (as promised) to the relaxation rates obtained earlier (Equations 21-26 and 21-27) for the simpler case of two-site exchange:

$$\frac{1}{T_1} = \frac{9}{8}\,[J^{(B)}(\omega_0) + J^{(C)}(2\omega_0)] \qquad (21\text{-}53a)$$

$$\frac{1}{T_2} = \frac{3}{4}\left[\frac{3}{8}J^{(A)}(0) + \frac{15}{4}J^{B}(\omega_0) + \frac{3}{8}J^{(C)}(2\omega_0)\right] \qquad (21\text{-}53b)$$

Due to interaction of two magnetic dipoles for particles of spin one-half

where

$$J(\omega) \equiv \int_{-\infty}^{\infty} \overline{<f(\theta,\phi,\tau)^* f(\theta,\phi,0)>}\, e^{-i\omega\tau} d\tau \qquad (21\text{-}54)$$

and

$$f^{A}(\theta,\phi) = \frac{\gamma^2\hbar^2}{r^3}\sqrt{\frac{16\pi}{5}}\,Y_{20}\,(\theta,\phi) \qquad (21\text{-}55a)$$

$$f^B(\theta,\phi) = \frac{\gamma^2\hbar^2}{r^3}\sqrt{\frac{8\pi}{15}}\,Y_{2,-1}\,(\theta,\phi \qquad (21\text{-}55b)$$

$$f^C(\theta,\phi) = \frac{\gamma^2\hbar^2}{r^3}\sqrt{\frac{32\pi}{15}}\,Y_{2,-2}\,(\theta,\phi) \qquad (21\text{-}55c)$$

where we have again employed complex notation for brevity.

Equations 21-53 and 21-45 may be used, either to compute the internuclear distance r when the rotational correlation time, τ_{rot} of Eqs. 21-46 and 21-49 is known; or to compute the rotational correlation time when the internuclear distance is known. In Chapter 16 (Fig. 16-24), we gave an example of determination of the complete configuration of a cofactor in an enzyme active site, based on measurements of nuclear magnetic T_1 and T_2 and known enzyme size (from which rotational correlation time can be computed from Equations 21-46 and 21-49), where the proton-electron distances were obtained from equations similar to Eq. 21-53.

An example of the utility of the τ_{rot} parameter obtained when the internuclear distance is known is provided by a recent study of binding of various regulators to the enzyme, aspartate transcarbamylase. [The structure (Fig. 15-17) and function (Figures 11-9, 11-16, 16-6) of this enzyme have already been described earlier in the text.] This enzyme is feedback-inhibited by a metabolic product (cytidine triphosphate = CTP) produced several steps after the original reaction. Measurement of T_1 and T_2 for the protons of various molecules that bind to the feedback inhibitor binding site [R. E. London and P. G. Schmidt, *Biochemistry 13,* 1170 (1974)] show that *5-methylcytidine* protons have a rotational correlation time essentially the *same* as that of the protein as a whole (i.e., this molecule is rigidly bound to the feedback inhibitor site), while *thymidine* protons have a rotational correlation time about an order of magnitude *shorter* (faster rotational diffusion) than for the protein as a whole (i.e., thymidine is bound relatively loosely at the same site). Since the methylcytidine is more analogous to the biological inhibitor, CTP, than is thymidine, these results suggest that a "tight fit" of the feedback inhibitor in its binding site is essential for allosteric function of the molecule. Furthermore, we have already pointed out that this particular enzyme can be taken apart into "catalytic" and "regulatory" units, and that enzyme catalysis is essentially just as efficient using the *catalytic* unit alone as for use of the entire assembled enzyme. In the present NMR study, the authors also compared the binding of the two inhibitors to the isolated *regulatory* unit and the whole enzyme and found that again the methylcytosine was firmly bound and the thymidine was loosely bound to the isolated regulatory subunit, just as for the whole enzyme. Thus, it appears that for this enzyme, the catalytic and regulatory functions can be completely separated for study, even though binding of either substrate or regulator affects binding of the other in the assembled whole enzyme.

When a chemical reaction is mediated by a macromolecule, much information about the mechanism can be gained from kinetics studies (see Section 3 and Chapter 16.A.1). When no chemical reaction occurs, other parameters (such as τ_{rot} in the above example) can provide much of the same sort of information.

Another example of rotational correlation time information obtained from magnetic relaxation times has already been described in Fig. 16-23. In that case, a variation in T_1 for different carbons in a phospholipid molecule was interpreted in terms of varying degrees of rotational freedom for different parts of the fatty acid chains of this membrane constituent. Essentially the same arguments are operative in accounting for *electron* (rather than nuclear) spin resonance, examples of which have also been discussed earlier (Chapter 14.D).

EXAMPLE *Fluorescence Depolarization*

The fluorescence depolarization experiment has been described in Chapter 16 (Fig. 16-14). Vertically polarized light is directed at the sample; the light is absorbed, and "fluorescent" light is then emitted by the excited sample within a time interval between zero and several nanoseconds after the incident light is removed. If the molecule rotates during the time interval between absorption and fluorescent emission, then there will be a decrease in the "polarization" (i.e., in the intensity of the vertically polarized fluorescent beam relative to the horizontally polarized fluorescent beam) of the fluorescent emission. Thus, the fluorescent polarization decreases with time after absorption, at a rate that reflects the rate at which the fluorescent molecule reorients rotationally.

The intensity of fluorescence polarized in various planes represents a complicated average over various molecular orientations, both initially at the time of absorption and later at the time of emission. Now the component of incident electric field along the direction of the "absorption dipole" (i.e., the axis of the electron on a spring for the absorption process) varies as $\cos(\theta_{L\to A})$, where $\theta_{L\to A}$ is the angle between the absorption dipole axis and the axis of the plane-polarized incident light. However, the *intensity* of that component varies as the *square* of the electric field amplitude, and thus varies as $\cos^2(\theta_{L\to A})$. Similarly, the fluorescent emission intensity of light plane-polarized in two perpendicular (laboratory-defined) directions can be shown to vary as $\cos^2(\theta_{L\to F})$ and $\sin^2(\theta_{L\to F})\cos^2(\phi_{L\to F})$, where $\theta_{L\to F}$ and $\phi_{L\to F}$ are angles describing the orientation of the emission dipole with respect to the laboratory frame. Thus, even though the relevant electric field *amplitude* components vary as $Y_{10}(\theta,\phi)$, the observed fluorescent *anisotropy* $A(\tau)$,

$$A(\tau) = \frac{I_{||}(\tau) - I_{\perp}(\tau)}{I_{||}(\tau) + 2I_{\perp}(\tau)} \tag{16-31}$$

varies as radiation *intensities,* and the resultant angle-dependence of $A(\tau)$ is described by $Y_{2m}(\theta,\phi)$, just as in the magnetic resonance example. The fluorescent anisotropy (in intensity) thus decreases as the mean-square degree of alignment of the emission dipoles in the excited fluorescent molecule, and is thus proportional to the correlation function for angular position of the emission dipoles:

$$A(\tau) \propto e^{-6D_{\text{rot}}\tau} = e^{-\tau/\tau_{\text{rot}}} \tag{21-56}$$

with

$$\tau_{\text{rot}} = \frac{1}{6D_{\text{rot}}} \quad \begin{bmatrix}\text{Magnetic resonance or}\\ \text{fluorescence depolarization}\end{bmatrix} \tag{21-57}$$

Thus, fluorescence depolarization measurements provide a measure of *rotational diffusion constant* (as do magnetic resonance or dielectric relaxation experiments), with the distinction that the experimentally determined rotational correlation time varies as $(1/6D_{rot})$ in the fluorescence and magnetic resonance cases, but varies as $(1/2D_{rot})$ for the dielectric relaxation case. This difference arises because the relevant correlation times for the two kinds of correlation function derive from experimental parameters that vary as spherical harmonic functions of rank 2 and rank 1.

It has proved possible to attach a number of fluorescent labels covalently to various macromolecules. Since the rotational correlation times that are calculated from the decay of fluorescence polarization following a pulse of excitatory polarized incident light compare closely to those computed from a simple model of the macromolecule as a sphere in a continuous medium (Equations 21-49 and 21-57)

$$\tau_{rot} = \frac{4\pi\eta R^3}{3\,kT} \tag{21-58}$$

as seen from the data in Table 21-2, the fluorescent labels generally appear to be attached rigidly to the macromolecular backbone, and provide a direct measure of the rotational flexibility of the macromolecule itself.

As rotational correlation times are being determined from various types of experiments, the reliability of the values increases. For example, a recent NMR determination of the τ_{rot} for serum albumin gave a value of 4×10^{-8} sec,

Table 21-2 *Rotational Correlation Times,* τ_{rot}, of Proteins, Determined Either from Experimentally Measured Fluorescence Depolarization Decay Rates or from a Theoretical Frictional Model of the Protein as a Sphere in a Continuous Medium.*

Protein	Molecular Weight	τ_{rot} (from fluorescence depolarization)[a]	τ_{rot} (from Eq. 21-58)[b]	$\frac{\tau_{rot}(\text{expt})}{\tau_{rot}(\text{calc})}$
Apomyoglobin	17,000	8.3 nsec	4.4 nsec	1.9
β-lactoglobulin (monomer)	18,400	8.5	4.7	1.8
Trypsin	25,000	12.9	6.4	2.0
Chymotrypsin	25,000	15.1	6.6	2.3
Carbonic anhydrase	30,000	11.2	8.0	1.4
β-lactoglobulin (dimer)	36,000	20.3	9.7	2.1
Apoperoxidase	40,000	25.2	10.5	2.4
Serum albumin	66,000	41.7	17.4	2.4

* The experimentally determined τ_{rot} values are consistently within a factor of two of the theoretical values, and thus reflect the rotational diffusional motion of the protein molecule as a whole. [Adapted from J. Yguerabide, in *Methods in Enzymology,* Vol. 26, part C, ed. C. H. W. Hirs and S. N. Timasheff, Academic Press, N. Y. (1972), p. 528.]
[a] All experimental τ_{rot} are adjusted to give the value that would be observed in water at 25°C.
[b] Calculated τ_{rot} obtained assuming a rigid unhydrated sphere of the molecular weight of the protein, assuming a partial specific volume of 0.73 cm³/g (see Eq. 21-58).

in excellent agreement with the fluorescence depolarization value. While the rotational correlation time, τ_{rot}, is defined as the time constant for exponential decay of the correlation function for rotational diffusion, τ_{rot} may also be thought of as the characteristic time it takes for a typical macromolecule to rotate (diffusionally) through an angle of the order of a radian (see Problems).

Recent fluorescence depolarization studies of an antibody to which a fluorescent hapten was bound have provided some of the most definitive evidence for *flexibility* in an immunoglobulin molecule. Antibodies were grown specific to a dansyl hapten (see Chapter 16.B.1 for dansyl group chemical structure). The dansyl group was then allowed to bind to its specific antibody, and to hapten-binding fragments of the antibody as shown below:

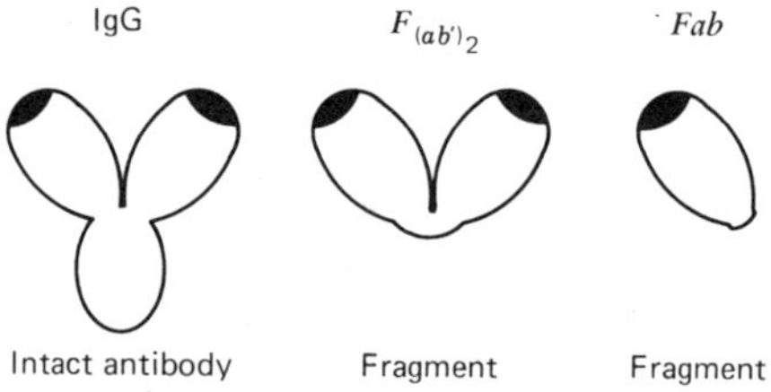

The proposed hapten-binding region is shaded in each sketch. The anisotropy of bound dansyl fluorescence was then monitored as a function of time follow-

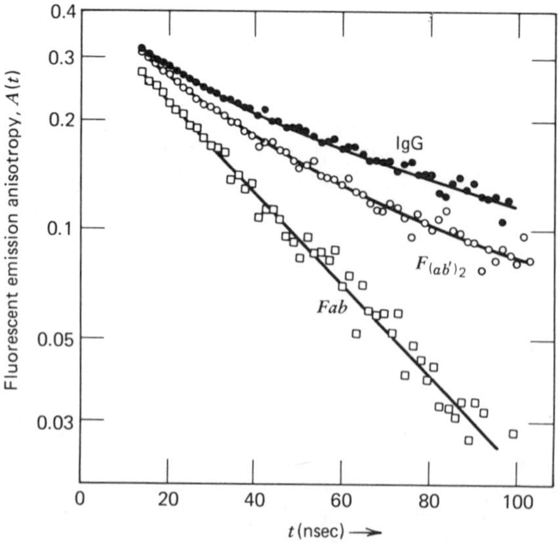

FIGURE 21-7. Fluorescent emission anisotropy, $A(t)$, (log scale) as a function of time, for dansyl-L-lysine hapten bound to antidansyl antibody (IgG) or to the indicated antibody fragments. The solid lines are least-square best fits to a single exponential decay (*Fab* curve) or to a sum of two exponentials [IgG and $F_{(ab')2}$ curves]. See text for interpretation of these results in terms of segmental flexibility of the IgG molecule. [From J. Yguerabide, H. F. Epstein, and L. Stryer, *J. Mol. Biol. 51*, 573 (1970).]

ing a pulse of incident plane-polarized light, and the behavior of the various labeled antibody components is shown in Fig. 21-7. The slope of each curve is a measure of the rotational diffusion constant for the bound dansyl group for that molecule. For the smallest fragment (*Fab*), there is relatively rapid rotational diffusion with $\tau_{rot} = 33$ *n*sec, and the experimental data are well-fitted by a theoretical curve based on rotational diffusion of a spherical *Fab* macromolecule. However, for the dansyl label bound to either the $F(ab')_2$ or IgG macromolecules, the fluorescent anisotropy decay is nonexponential. However, the curves can be fitted by the sum of two exponentials, a rapid initial decay due (presumably) to internal macromolecular flexibility in the region of the "hinge" that joins the two F_{ab} fragments ($\tau_{rot} = 33$ *n*sec), and a slower decay component corresponding to rotation of the intact antibody as a whole (uppermost curve of Fig. 21-7, with $\tau_{rot} = 168$ *n*sec). These experiments thus provide convincing evidence that there is internal segmental flexibility within the antibody molecule. It is possible that this flexible joint may have biological significance in facilitating the formation of the antigen-antibody complex.

21.C. RANDOM TRANSLATIONAL MOTION: TRANSLATIONAL DIFFUSION.

21.C.1 Light Scattering: Translational Diffusion Coefficients for Macromolecules (or Bacteria) in Solution

We have already discussed at some length (Chapter 15.A) the steady-state intensity of light scattered at some angle, θ, with respect to an incident beam. However, from the arguments presented in Fig. 21-8 below, it would seem that the observed scattered light intensity from a uniformly dispersed solution should be zero. We are now in a position to resolve this paradox, by supposing that at any one instant, the *solvent density* in any two small regions (Fig. 21-8*b*) is not identical, so that the cancellation of scattered amplitudes described in the figure is not complete, and some light is scattered to an observer. Similarly, for a solution containing macromolecular solute, any additional light scattering over and above that from the solvent must be due to fluctuations in the *solute concentration*. Since the only way that concentrations can fluctuate is for solute molecules to move (translate) from one place to another, and since the intensity of the observed scattered light depends on the magnitude of those fluctuations, we might hope to learn about the rate of macromolecular *translational diffusion* from measurement of scattered light intensity from a solution.

Intuitively, we expect that the wavelength of the scattered light should be related to the distance between concentration maxima and minima as we scan across different regions of the solution at a given instant. It should thus seem logical to analyze the overall concentration fluctuations in terms of an infinite number of components, in which each component is a concentration "wave" of appropriate amplitude and wavelength—this procedure

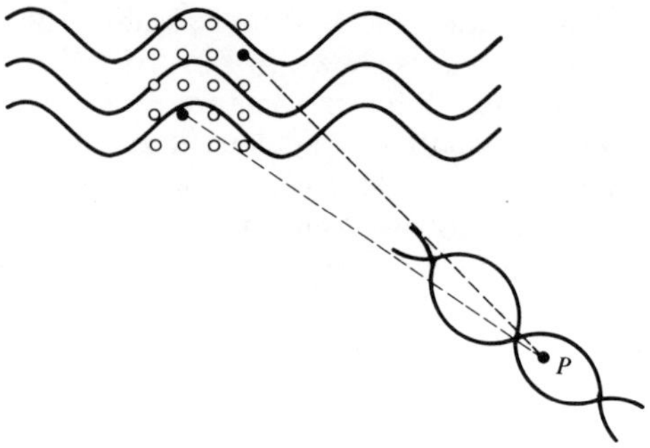

(a)

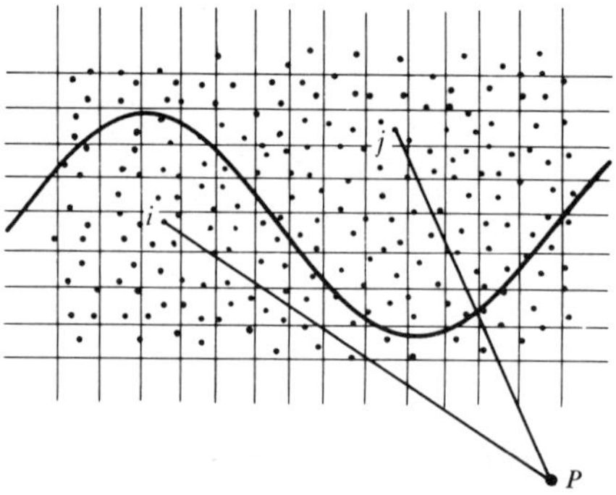

(b)

FIGURE 21-8. Scattering of long-wavelength radiation (light) from a perfect colorless crystal (*a*) and from a pure liquid (*b*). (*a*) If the crystal is large enough and the interparticle spacing small enough, then all the particles in the crystal can be paired off in such a way that for every scatterer, *i*, there will be another scatterer, *j*, so that the two scattered amplitudes will exactly cancel at a given observation location, *P*. Since the argument applies to any initial scatterer, *i*, the total scattered intensity at any angle is zero. (*b*) The pure liquid is divided into many small cubes, each of which contains a (small) number of molecules, and each cube is small compared to the wavelength of the driving radiation. Each cube is considered to be a scattering center. Suppose *i* and *j* are again placed [as in (*a*)] such that their two scattered amplitudes are again out of phase at observation location, *P*. Now if *i* and *j* contained equal numbers of molecules (and thus produced equal scattering amplitude), their combined waves would cancel at *P*, but since fluctuations in solvent density and solute concentration prevent complete cancellation from occurring, some light is scattered. (From K. E. van Holde, *Physical Biochemistry*, Prentice-Hall, Englewood Cliffs, N. J., 1971, p. 186.)

is formally very similar to our previous (Chapter 20.B and 21.A) analysis of an arbitrary coherent or random signal into sinusoidal components at a series of different frequencies. Thus, if we now consider a small sinusoidal variation in the solute concentration (and thus in the refractive index)

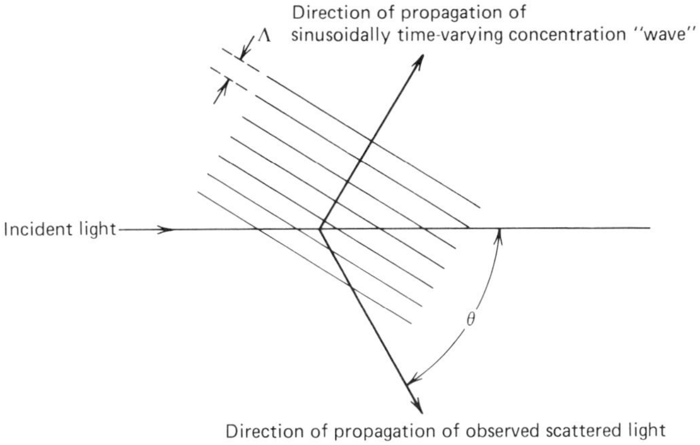

FIGURE 21-9. Schematic illustration of directions of incident and scattered light, for a solution whose concentration at a given point in space varies sinusoidally with time (see text). [After N. C. Ford, Jr., *Chemica Scripta 2,*193 (1972).]

of a solution (see Fig. 21-9), then it is found that light is scattered, provided that the condition

$$\Lambda = \frac{\lambda}{n}\,\frac{1}{2\sin(\theta/2)} \tag{21-59}$$

is met, in which $\left(\frac{\lambda}{n}\right)$ is the wavelength of incident light in the solution having index of refraction, n, θ is the scattering angle, and Λ is the wavelength (Fig. 21-9) of the sinusoidal solute concentration fluctuation (wave) responsible for the scattering. This relationship is very similar to the familiar Bragg equation (see Chapter 22.A) for diffraction of X rays by an atomic lattice (crystal), for essentially the same reasons.

Obviously the solute concentration of any real liquid solution does not vary sinusoidally with time, but rather varies randomly due to Brownian motion according to the translational diffusion discussion of Chapter 6. However, we have already learned that it is possible to analyze any random time-varying process in terms of its mean-square components at various frequencies; thus Eq. 21-59 is all we need to describe the effect, at any one wave length, of a whole spectrum of fluctuation frequencies.

Specifically, let the component of concentration fluctuation at the frequency corresponding to wavelength, Λ, be noted as $\delta c(\Lambda, t)$. If we employ a coordinate frame in which the y direction is along the propagation direction of the concentration "wave" shown in Fig. 21-9, then we can describe the concentration fluctuation by a one-dimensional(translational) diffusion equation:

$$\frac{\partial(\delta c(y,t))}{\partial t} = D_{\text{trans}}\,\frac{\partial^2}{\partial y^2}\delta c(y,t) \tag{21-60}$$

in which it is convenient to express the solution of Eq. 21-60 as the product of time-varying and space-varying factors:

$$\delta c(y,t) = \delta c(\Lambda,t)\,\sin\left(\frac{2\pi y}{\Lambda}\right) \tag{21-61}$$

Substituting Eq. 21-61 into Eq. 21-60 and solving for the time-dependent part,

$$\delta c(\Lambda,t) = \delta c(\Lambda,0)e^{-D_{\text{trans}}(2\pi/\Lambda)^2 t} \tag{21-62}$$

We can now construct a correlation function for *concentration fluctuations*

$$G_{\text{conc}}(\tau) = <\overline{\delta c(\Lambda,\tau)\delta c(\Lambda,0)}> = <\overline{(\delta c(\Lambda,0))^2}>e^{-D_{\text{trans}}(2\pi/\Lambda)^2 t} \tag{21-63}$$

or

$$G(\tau) = G(0)\,e^{-K^2 D_{\text{trans}}\tau}, \qquad K = \frac{4\pi n}{\lambda}\sin\frac{\theta}{2} \tag{21-63a}$$

Finally, by assuming that changes in solution refractive index are proportional to changes in solute concentration, and by recalling that scattered light intensity varies as the *square* of refractive index (see Eq. 15-26), it should seem reasonable that the correlation function for scattered light *intensity* fluctuations should vary as $|G_{\text{conc}}(\tau)|^2 = (\text{constant})\exp[-2K^2 D_{\text{trans}}\tau]$. The observed scattered light intensity (Fig. 21-10) will thus show a frequency-dependence determined by the Fourier transform of that correlation function to give the (increasingly familiar) Lorentzian line shape, with a line width proportional to translational diffusion constant, D_{trans}. Experimentally, early measurements of Rayleigh scattered intensity spectral line width were made by determining the full spectral line shape, point-by-point, while more recent measurements are based on direct calculation of the correlation function from a recording of the time-varying scattered intensity. In the following examples, it will become clear why this new technique has developed into the method of choice for determination of translational diffusion coefficients to high accuracy, and some other applications will suggest themselves.

A final practical note is in order before proceeding to the examples. The calculated expected spectral line width (at half maximum height) due to concentration fluctuations arising from translational diffusion of macromolecular solutes is of the order of 1000 Hz for a macromolecule whose molecular weight is about 50,000, for typical scattering angles. Since the frequency of the incident (visible) light itself is of the order of 5×10^{14} Hz, direct measurement of the scattered spectral line shape would require a resolving power of better than $10^{14}/10^3 = 10^{11}$, or about 10,000 times higher than available from the best gratings or interferometer dispersive devices. Therefore, typical measurements are conducted by "beating" the signal of interest against some other coherent source (i.e., adding the two beams together). If the frequencies of the two signals (sample and reference beams) are ω_1 and ω_2, then their resultant electric field consists of a simple sum

$$E = E_1(\cos\omega_1 t) + E_2 \cos(\omega_2 t) \tag{21-64}$$

Since a photomultiplier tube is a detector for radiation *intensity* (rather than electric field *amplitude*), the detected intensity will be proportional to the square of the electric field of Eq. 21-64:

$$I_{\text{detected}} \propto |E|^2 = |E_1 \cos\omega_1 t + E_2 \cos\omega_2 t|^2 \tag{21-65}$$

Expanding Eq. 21-65, and using a trigonometric identity for $\cos(\omega_1 t)$ $\cos(\omega_2 t)$,

$$I_{\text{detected}} \propto E_1^2 \cos^2(\omega_1 t) + E_2^2 \cos^2(\omega_2 t) + \frac{E_1 E_2}{2} (\cos(\omega_1 - \omega_2)t + \cos(\omega_1 + \omega_2)t) \tag{21-66}$$

However, the anode of the photomultiplier detector will not respond to frequencies greater than about 10^{10} Hz, so the signals at frequencies, ω_1, ω_2, and $(\omega_1 + \omega_2)$ in Eq. 21-66 will be observed as their time averages. Since the time average of $\cos^2(\omega_1)t = (1/2)$ while the time average of $\cos(\omega_1 + \omega_2)t = 0$, the anode output current of the detector will consist of a *constant* component and a component *oscillating* at the "beat" or difference frequency, $(\omega_1 - \omega_2)$.* Since the difference frequency is to be of the order of the scattered spectral line width (a few kHz), and since it is possible to measure such frequencies very accurately (i.e., to within the "natural" line width of the optical laser source itself), it is possible to obtain the scattered light intensity at frequencies extremely close to the frequency of the incident (laser) light itself. Accurate measurement of 1000 Hz line widths requires that the incident light beam be monochromatic to better than 100 Hz; thus, the experiment has only become possible since the advent of highly monochromatic (laser) light sources.

EXAMPLE *Translational Diffusion Coefficients and Molecular Weights*

As just discussed (Eq. 21-63 ff.), the spectrum of Rayleigh scattered light intensity from a macromolecular solute is expected to show a Lorentzian line

* *This process has the same net effect (although via a different mechanism) as the mixer-filter detection of Fig. 20-7.*

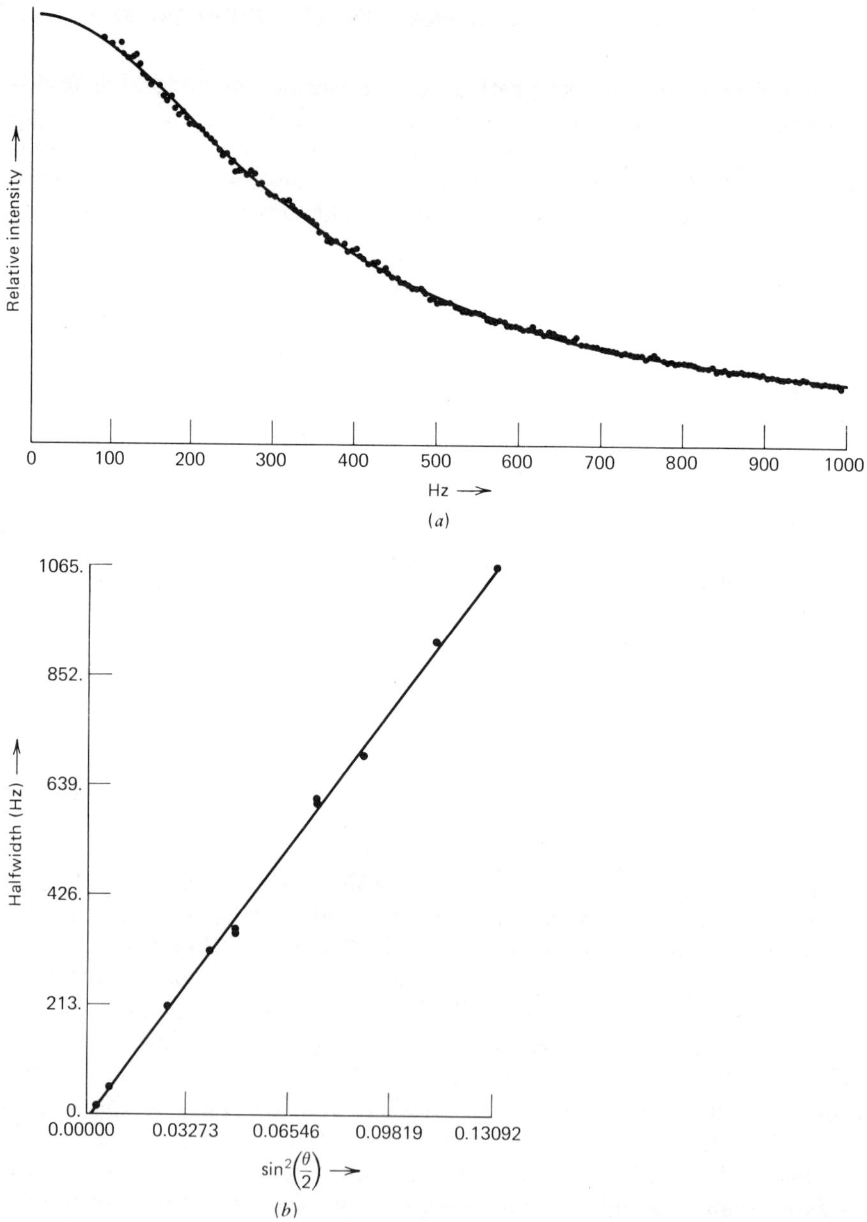

FIGURE 21-10. Determination of translational diffusion constant from Rayleigh scattered laser light. (*a*) relative scattered intensity as a function of frequency, using helium-neon laser source of wavelength, $\lambda = 632.8$ nm, at scattering angle of $\theta = 23°$, for a solution of approximately 1 mM carbon monoxy-hemoglobin. (*b*) Plot of light scattering half-line width at half maximum height (in Hz) as a function of $\sin^2(\theta/2)$. To within experimental error, curve *A* can be fitted by a Lorentzian line shape (solid line), and curve *B* can be fitted by a straight line that passes through the origin. The slope of curve *B* is proportional to the translational diffusion coefficient, D_{trans} (see text). [From D. D. Haas, R. V. Mustacich, B. A. Smith, and B. R. Ware, *Biochem. Biophys. Res. Commun. 59,* 174 (1974).]

shape, with a width at half maximum height that is proportional to K^2D_{trans}, where

$$K = (4\pi n/\lambda) \sin (\theta/2) \quad (21\text{-}67)$$

in which n is the index of refraction of the solution, λ is the wavelength of the incident light, and θ is the angle between the incident and scattered beams. A typical spectrum for carbon monoxy-hemoglobin is shown in Fig. 21-10A, for a scattering angle of $\theta = 23°$. By plotting the half-width at half maximum height of such spectra as a function of $\sin^2 (\theta/2)$, a straight line plot is observed (Fig. 21-10*b*), from which a translational diffusion coefficient is computed to be $D_{20,w} = 6.9 \pm 0.3 \times 10^{-7}$ cm²/sec. The accuracy of the determination is evidently high, judging from the excellent fits of the data to appropriate curves in Figs. 21-10*a* and 21-10*b*, of the order of ±1% in most cases. Some translational diffusion constants determined in this manner are listed in Table 21-3. Advantages of the light-scattering determination include: (a) required concentrations of macromolecular solute are low, typically 1 mg/cc, so that determinations can be made accurately using relatively small samples, (b) accurate temperature control is not required, since macroscopic convection does not affect the spectrum when a suitable optical design is used, and (c) the determination can be made very rapidly, typically in an hour or so after sample preparation.

In Chapter 7, it was noted that sedimentation *rate* experiments provide an attractive rapid determination of molecular weight, provided that the diffusion constant of the solute is known. Combination of sedimentation coefficients from sedimentation rate measurements with diffusion constants from light scattering thus provides a new and useful means for obtaining molecular weights of macromolecules in solution, particularly for larger macromolecules (molecular weight greater than about 1,000,000) for which conventional molecular weight determinations are impractical or inaccurate. Table 21-4 lists some recently determined molecular weights of various viruses in solu-

Table 21-3 Diffusion Constants of Several Natural and Synthetic Macromolecules[a]

Sample	D_{trans} (from light scattering)	D_{trans} (hydrodynamic methods)
Lysozyme	11.5 ± 0.3 × 10^{-7} cm² sec^{-1}	11.6 × 10^{-7} cm² sec^{-1}
Ovalbumin	7.1 ± 0.2	8.3
Bovine serum albumin	6.7 ± 0.1	(no direct comparison available)
Tobacco mosaic virus	0.40 ± 0.02	0.3
DNA (calf thymus)	0.2 ± 0.1	0.13
Polystyrene latex spheres, r = 440 ± 40 Å	0.59 ± 0.02	0.56 ± 0.06[b]
Polystyrene latex spheres, r = 630 ± 30 Å	0.368 ± 0.006	0.38 ± 0.02[b]
Polystyrene latex spheres, r = 1830 ± 30 Å	0.134 ± 0.004	0.134 ± 0.002[b]

[a] *From S. B. Dubin, J. H. Lunacek, and G. B. Benedek,* Proc. Natl. Acad. Sci. U.S.A. 57, *1164 (1967).*
[b] *Calculated directly from known size and* $D = kT/6\pi\eta r$ *(see Equations 7-13 and 7-55).*

Table 21-4 Translational Diffusion Coefficients and Molecular Weights of Viruses [From R. D. Camerini-Otero, P. N. Pusey, D. E. Koppel, D. W. Schaefer, and R. M. Franklin, *Biochemistry 13,* 960 (1974).]

Virus	$D_{20,w}$ ($\times 10^{-7}$ cm^2 sec^{-1})	$s_{20,w}$ ($\times 10^{-13}$ sec)	$\bar{v}_{\text{virus}}$ (cm^3 g^{-1})	Molecular Weight ($\times 10^6$)
R17	1.53 ± 0.02[a]	79 ± 1	0.689 ± 0.012	4.02 ± 0.17[b]
Q	1.42 ± 0.01	88 ± 2	0.668 ± 0.009	4.55 ± 0.16
BSV	1.25 ± 0.01	133 ± 2	0.706 ± 0.002	8.81 ± 0.17
PM2	0.65 ± 0.01	294 ± 3	0.771 ± 0.007	47.9 ± 1.7
T7	0.64 ± 0.01	487 ± 5	0.639 ± 0.006	50.9 ± 1.1

[a] *Obtained from autocorrelation function for Rayleigh scattered light* [*D calculated from slope of a plot of log(autocorrelation of observed scattered intensity) versus time* (*see Eq. 21-63 ff.*).]

[b] *Calculated from Eq. 7-31, using sedimentation coefficient determined from sedimentation rate measurements.*

tion; from these results (review Chapter 7.B and 7.C), the degree of hydration (assuming spherical molecular shape) could be calculated as about 1 cm^3 H_2O per gram of dry virus in most cases.

EXAMPLE *Macromolecular Reaction Rate Constants from Rayleigh Light Scattering*

If the width of a Rayleigh scattered light spectrum depends on macromolecular translational diffusion coefficient, then we would expect that light scattering line width measurements should be useful in monitoring any reaction with which there is associated a significant change in diffusion coefficient. The logical place to look for such a reaction is in the association of two large molecules to form a complex whose molecular weight is thus significantly larger than either of the two reactant macromolecules. A striking example of this idea is provided by the following recent experiment with bacteriophage fragments.

For a bacteriophage T4D particle to be able to "infect" a bacterium (i.e., attach itself to the bacterium surface and inject nucleic acid into the bacterium to induce synthesis of more bacteriophage particles), the "phage" particle must be fully assembled from the three principal components shown below. Since the partly reconstituted "tail-fiberless" particles are *not* infective, it is desirable to learn as much as possible about the assembly process to find out what makes the bacteriophage infective. Conventional kinetic studies of the assembly process rely on a bio-assay for infectivity of the product, and thus reflect the rate of the *overall* process. However, since the tail fibers have much smaller molecular weight than either the heads or tails themselves, light scattering measurements will be relatively insensitive to the presence of tail fibers. Thus, light scattering measurements will reflect primarily the extent of reaction I shown below. Combination of the reaction rate constants for step I (from light scattering) and for the overall steps I and II (from bio-assay of final product) will thus show which of these steps is the critical (i.e., rate-determining) step in the overall process.

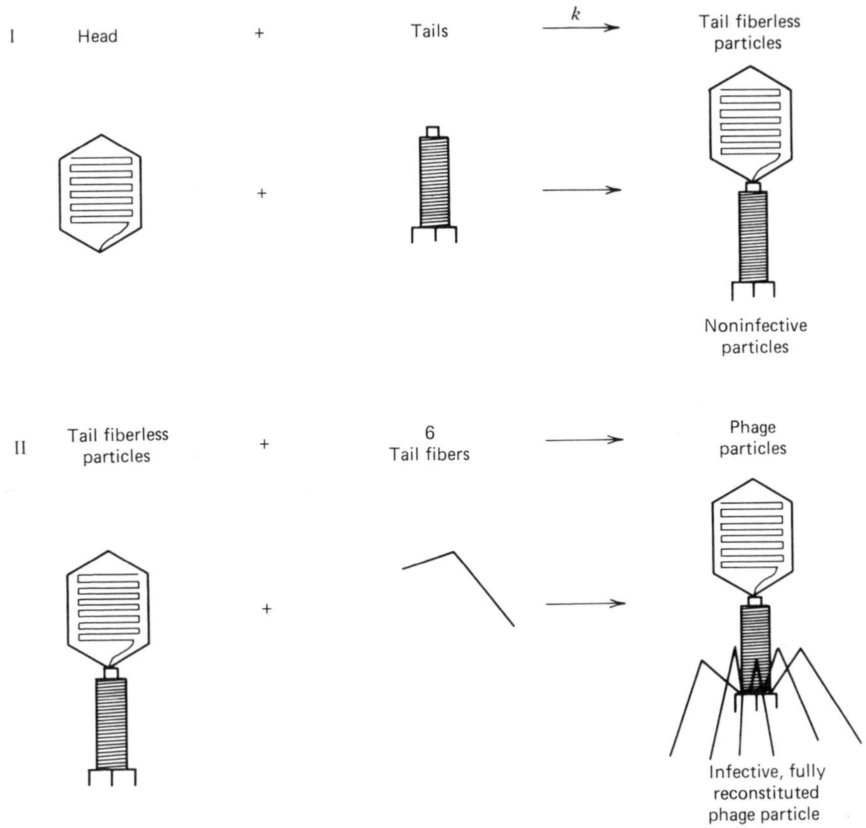

To begin with, it is necessary to determine the diffusion coefficients of the components that contribute most to the observed light scattering: the heads and the tailfiberless particles. These coefficients were computed as for the data in Table 21-4 and had the values, 3.60×10^{-8} and 3.14×10^{-8} cm^2/sec, corresponding to respective molecular weights of 1.76×10^8 and 1.95×10^8. Just as the intensity autocorrelation function for a single type of (spherical) particle can be represented as

$$\frac{I(\tau)}{I(0)} = \exp\left[-2K^2 D_{\text{trans}}\tau\right] \tag{21-68}$$

the autocorrelation for a solution consisting of a mixture of two types of (spherical) macromolecules can be shown to be of the form

$$\frac{I(\tau)}{I(0)} = \left[A_1\, e^{-K^2 D_1 \tau} + A_2\, e^{-K^2 D_2 \tau}\right]^2 \tag{21-69}$$

in which $I(0)$ is the value of $I(\tau)$ at time zero, and D_1 and D_2 are the diffusion coefficients of the two types of macromolecule in the solution. The constants in

Eq. 21-69 are the weight-average concentration fractions of components #1 and #2 at a given time:

$$A_1 = \frac{c_1 M_1}{c_1 M_1 + c_2 M_2}; \quad A_2 = \frac{c_2 M_2}{c_1 M_1 + c_2 M_2} \text{ at time, } t \qquad (21\text{-}70)$$

Now the autocorrelation data can be collected in about 10^{-3} sec (corresponding to Rayleigh scattered light spectral line width of a few hundred Hz), and for appropriate initial concentration of reactant heads and tails, the reaction of interest has a lifetime of the order of 1000 sec. Thus, it is possible to measure the autocorrelation intensity function at several stages of the reaction; then by rearranging Eq. 21-69 to the form

$$[I(\tau)/I(0)]^{1/2} - \exp\,[-K^2 D_2 \tau] = A_1\,(\exp\,[-K^2 D_1 \tau] - \exp\,[-K^2 D_2 \tau]) \qquad (21\text{-}70)$$

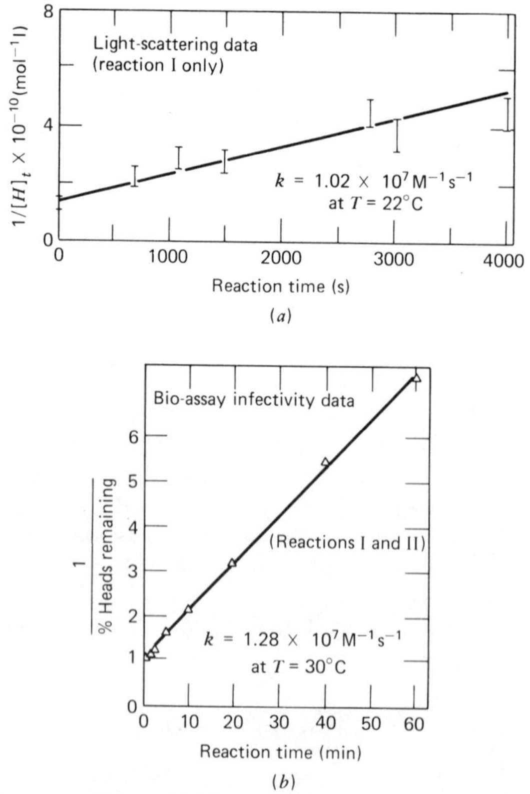

FIGURE 21-11. Determination of second-order reaction rate constant for the assembly of bacteriophage T4D particles from head and tail fragments, using a plot of (1/[H]) versus time, where [H] is the concentration of heads at time t. (a) data from light scattering intensity autocorrelation function measurements (see text). (b) data from bio-assay for infectivity to obtain concentration of final assembled phage. [From J. Aksiyote-Benbasat and V. A. Bloomfield, *J. Mol. Biol.* *95*, 335 (1975).]

and plotting the left-hand side of Eq. 21-70 versus the quantity in parentheses on the right-hand side, the weight-average concentration of reactant can be monitored throughout the course of the reaction. The second-order reaction rate constant k may then be extracted in the usual way (see p. 514), based on the concentration of, say, heads as a function or reaction time, t, as shown in Fig. 21-11a:

$$(1/[H])_t - (1/[H])_0 = kt \tag{21-71}$$

The comparable data for the rate of the overall reaction (judged by bio-assay for concentration of final product) are shown in Fig. 21-11*b*. Within experimental error, *both rates are the same* when corrected for a small difference in temperature between the two experiments, indicating that reaction I (the reaction of heads and tails to form tailfiberless particles) is the rate-limiting step in our oversimplified mechanism. Furthermore, since the calculated rate constant $k = 1 \times 10^7\ M^{-1}\ \text{sec}^{-1}$, is about 500 times slower than would be predicted for a reaction whose rate is diffusion-limited, it seems clear that the molecules must not only come together to react, but must have the correct relative orientation for the correct part of the head to react with the right end of the tail. For example, the head particles are essentially elongated icosahedra, and the tails are essentially rods; if we suppose that only one face of the icosahedron (12 faces total) is reactive and that only one of the two ends of the tail is reactive, then the probability that both particles will be properly oriented on collision is only (1/24). Requiring both molecules to be co-linear would further reduce this factor. This example is the first convincing measurement of a macromolecular rate constant from Rayleigh scattered spectral line width; it is logical to suppose that the technique will be exploited extensively in the next few years to study assembly of proteins, nucleic acids, and other polymers in solution.

EXAMPLE *Rotational Diffusion Constants for Macromolecules in Solution from Depolarized Light Scattering Spectrum*

In the examples given previously, the detector measured *total* scattered light intensity as a function of scattering angle. It is, however, quite feasible to employ a plane-polarized incident light source, and (by using suitably oriented polarizing plates in front of the detector) to detect scattered light that is vertically or horizontally polarized, in much the same scheme as for fluorescence polarization measurements (e.g., by monitoring the light scattered at 90° to the incident beam, in either vertical or horizontal plane-polarization) (see Fig. 16-14). Under these conditions, one might expect that the scattered light intensity spectral line width should depend on *rotational* diffusion (as in the fluorescence case) as well as on *translational* diffusion as just discussed. The only real difference is that the scattered light is re-emitted from the same excited state that was populated by the incident light absorption, whereas in fluorescence the re-emitted light comes from a state of lower energy than the state initially populated by incident light absorption. As for fluorescence, the rotational orientation factor for the scattered light is manifested as a rank 2 spherical harmonic, and since translation and rotation can be assumed to be independent (i.e., uncorrelated), their correlation functions multiply to give a spectral density for scattered (plane-polarized) light intensity from a solution of macromolecular concentration, $[c]$,

Horizontally polarized scattered
light intensity from a vertically
polarized incident light source

$$= \text{"Depolarized" intensity} = I_D \propto [c] \quad (\alpha_{\|} - \alpha_{\perp})^2 \frac{(K^2 D_{\text{trans}} + 6D_{\text{rot}})}{(\omega - \omega_0)^2 + (K^2 D_{\text{trans}} + 6D_{\text{rot}})^2} \tag{21-72}$$

in which $\alpha_{\|}$ and $\alpha_{\perp}$ represent the polarizability of the molecule in two mutually perpendicular directions. The quantity $(\alpha_{\|} - \alpha_{\perp})$ is simply a measure of the optical "anisotropy" of the molecule, namely, the degree to which scattering in different directions is different for different molecular orientations. If the molecule is optically "isotropic" (i.e., $(\alpha_{\|} - \alpha_{\perp}) = 0$), then scattering is independent of the orientation of the molecule and is thus unaffected by molecular rotations. Returning to Eq. 21-72, it is clear that for measurements at very small scattering angle, θ, the importance of the K^2D_{trans} term decreases as $\sin^2(\theta/2)$ and thus becomes negligible for sufficiently small scattering angle:

$$\theta \rightarrow 0^{\lim} I_D \propto [c]\ (\alpha_{\|} - \alpha_{\perp})^2 \quad \frac{6D_{\text{rot}}}{(\omega - \omega_0)^2 + (6D_{\text{rot}})^2} \tag{21-73}$$

Equation 21-73 thus provides a simple means for determination of *rotational* diffusion constant, D_{rot}, from measurement of "depolarized" scattered light intensity. The I_D spectrum consists of a single Lorentizian line whose width is a direct measure of the rotational diffusion constant D_{rot}. The question now becomes: how does the limit given in Eq. 21-73 depend on macromolecular size? For a spherical macromolecular shape, we already know that

$$D_{\text{trans}} = \frac{kT}{6\pi\eta r} \tag{21-74}$$

$$D_{\text{rot}} = \frac{kT}{8\pi\eta r^3} \tag{21-49}$$

so that

$$\frac{D_{\text{rot}}}{D_{\text{trans}}} \cong \frac{1}{r^2} \text{ for a spherical macromolecule of radius, } r \tag{21-75}$$

For visible light, λ is of the order of 10^{-5} cm, so that the parameter, $K = (2\pi/\Lambda)$ (where Λ is given in Eq. 21-59), varies between zero and about 10^{10} for scattering angles between zero and 180°. Thus, for small molecules with $r = 10^{-8}$ cm, $(1/r^2) = 10^{16}$ and we can always neglect the K^2D_{trans} term compared to the $6D_{\text{rot}}$ term in Eq. 21-72. However, for larger molecules such as tobacco mosaic virus with an effective radius of about 3×10^{-5} cm, $(1/r^2) = (1/9) \times 10^{10}$, and K^2D_{trans} will be negligible compared to $6D_{\text{rot}}$ only for very small scattering

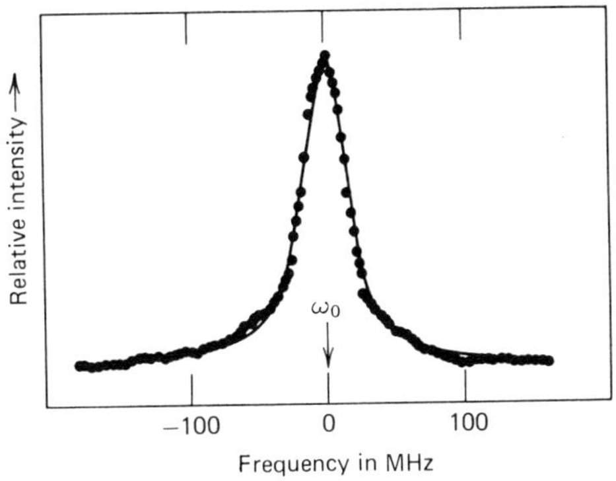

FIGURE 21-12. Depolarized Rayleigh scattered light intensity spectrum from a sample of muscle calcium-binding protein. The smooth curve is a best-fit single Lorentzian line. The rotational diffusion constant for this molecule was determined from the width at half maximum height of this curve (see text). [From D. R. Bauer, S. J. Opella, D. J. Nelson, and R. Pecora, *J. Amer. Chem. Soc. 97,* 2580 (1975).]

angle (say, $\theta < (1/10)$ radian to give $D_{\text{rot}}/K^2D_{\text{trans}} \cong 100$). Thus, for all but the largest macromolecules (molecular weight = 1,000,000 or so), depolarized light intensity spectral line shape can be made to satisfy Eq. 21-73 and may be used to determine the rotational diffusion constant for the macromolecule in solution.

An experimental example is shown in Fig. 21-12 for the depolarized Rayleigh spectrum from muscle calcium-binding protein, an approximately spherical small protein of molecular weight 12,000. The experimental data are fitted very well by a single Lorentzian line shape, from which a rotational correlation time was calculated as $\tau_{\text{rot}} = 12 \pm 1$ nsec. From this result and Eq. 21-58, a molecular radius of about 22 Å was calculated. This radius is about 5 Å larger than the average crystallographic radius of the "dry" protein, suggesting that a significant amount of water is firmly bound to the protein in solution. In contrast to some of the other methods we have discussed for determination of rotational diffusion constant (e.g., fluorescence depolarization, paramagnetic spin-labelling), the light scattering experiment does not require the attachment of a chromophore or other label to the macromolecule. This advantage suggests that this technique may well become the method of choice for determination of *rotational* diffusion constants of macromolecules in solution.

21.C.2. Electrophoretic Light Scattering: Electrophoresis without Boundaries

A charged macromolecule in an electric field, E, will move at a constant average velocity, v_E, given by Eq. 7-3, as discussed in Chapter 7.A:

$$v_E = \mu E \tag{21-76}$$

in which μ is the electrophoretic mobility. If we now choose to observe the light scattered by these molecules in any direction other than perpendicular to their electrophoretic motion, we would expect that the frequency of the scattered light should be Doppler-shifted by an amount proportional to the drift velocity, v_E. When this feature is incorporated into our prior treatment of the Rayleigh spectral line shape, the total scattered intensity at a particular angle is of the form,

$$I = (\text{const})\ [c]\ \frac{K^2 D_{\text{trans}}}{((\omega - \omega_0) \pm K v_E)^2 + (K^2 D_{\text{trans}})^2} \tag{21-77}$$

in which $[c]$ is concentration of solute and the other parameters have their usual meanings. Since the width of the Rayleigh line varies as K^2, it is clear that the narrowest line shape (and thus the best electrophoretic resolution of Rayleigh lines for molecules of different electrophoretic mobility) will be observed at small scattering angle.

EXAMPLE *Rapid Electrophoresis of Human Blood Plasma*

Figure 21-13 shows the Rayleigh spectrum from a sample of human blood plasma exposed to a strong (350 volt/cm) electric field. The fastest-moving component (albumin) shows the greatest Doppler shift and forms the leading (rightmost) peak in the spectrum. The remaining peaks (not yet assigned)

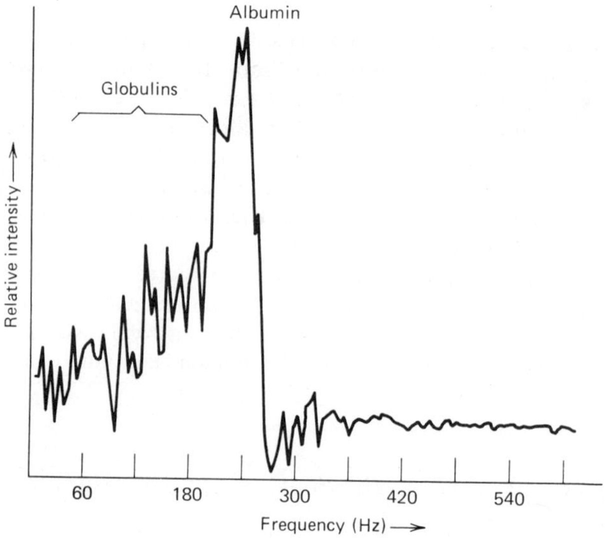

FIGURE 21-13. Electrophoretic light scattering spectrum of human blood plasma at pH 8.4 and ionic strength 0.006. Scattering angle was 7°30′, and electric field strength 350 V/cm. [From B. R. Ware, *Adv. Colloid and Interface Science 4*, 1 (1974).]

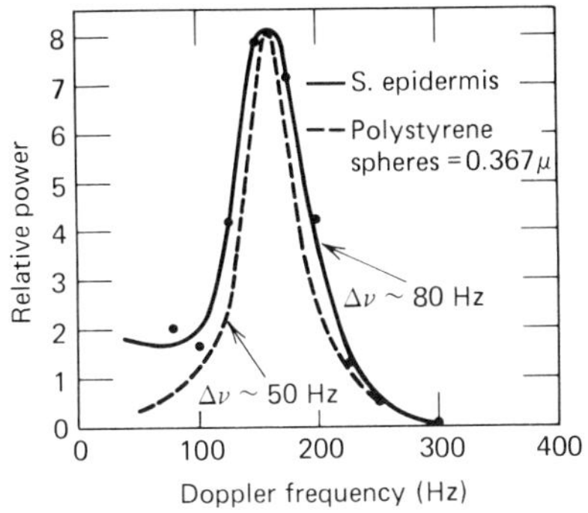

FIGURE 21-14. Electrophoretic light scattering spectra (superimposed from separate samples) for the bacterium, *Staphylococcus epidermidis,* and for polystyrene spheres. Since the bacterium signal is broader, it is deduced that there must be appreciable heterogeneity in bacterial size, shape, and/or surface charge. [From E. E. Uzgiris, *Opt. Commun.* 6, 55 (1972).]

correspond to the various globulins (compare to Fig. 7-4 and Table 7-1 for conventional electrophoresis). An interesting feature (and probable advantage) of the light scattering electrophoresis experiment is that the relative intensity for any species is proportional to its weight concentration multiplied by its molecular weight—thus, the most massive species will be much more prominent than a comparable weight concentration of lower molecular weight species. This accounts for the increased relative height of the globulin peaks in Fig. 21-13 compared to the much more concentrated albumin (consult Table 7-1). An additional advantage is that since no gels are required, the electrophoretic resolution of very large molecules can be achieved (macromolecules much bigger than about 100,000 molecular weight do not easily pass through polyacrylamide electrophoresis gels); recent experiments have demonstrated the electrophoretic motion of whole bacteria (Fig. 21-14). These studies open up a new range of experiments for characterization of surface charge and shape on a quantitative basis. Finally, it should be noted that the illustrated experiments were conducted in solutions with very low salt concentration (Fig. 21-14 data were taken in distilled water solvent), in contrast to conventional electrophoresis experiments that require high salt concentration to achieve good conductivity in the solution. It should thus be possible to perform light scattering electrophoresis separations under conditions that more closely reflect the native environment of many macromolecules. The light scattering electrophoresis experiments can be completed in a few minutes, compared with a few hours for conventional electrophoresis, which should be useful for possible clinical assays.

EXAMPLE *How Do Bacteria Swim?*

The data in Fig. 21-14 show that when all the (charged) particles in a sample move at the same speed, their (uniform) velocity is manifested as a Doppler shift in the Rayleigh spectrum. Therefore, in considering a sample of bacteria that move *on their own* (i.e., in absence of electric field) at uniform speed but random directions, we would expect a Rayleigh spectrum that consists of the superposition of the (constant) Doppler shifts arising from bacteria moving in various directions. (In spectroscopy, this situation is called "inhomogeneous line-broadening.") On the other hand, in the limit that the bacteria move at constant speed, but change direction very frequently, we would expect to be able to predict the Rayleigh line shape as a single Lorentzian with a linewidth determined by the appropriate effective translational diffusion constant. In summary, the light-scattering spectrum from bacteria in solution is expected to show a line width that increases with increasing maximum *swimming speed* ("inhomogeneous broadening") and/or with increasing *distance* between direction changes and more frequent *changes of direction* (Eq. 6-60). Rayleigh spectra from actual bacteria tend to fall between the two extreme types of motion just described.

Figure 21-15*a* shows the intensity autocorrelation functions for three different *E. coli* bacterial strains. The motional behavior of the three bacterial strains is clearly much different, as evidenced by the large difference in the auto-

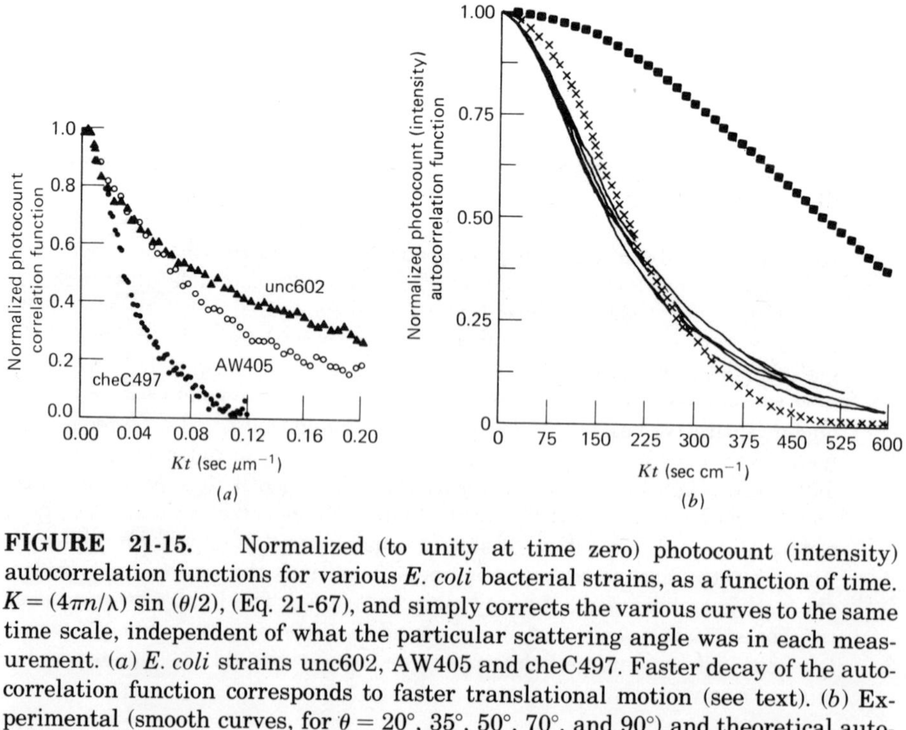

FIGURE 21-15. Normalized (to unity at time zero) photocount (intensity) autocorrelation functions for various *E. coli* bacterial strains, as a function of time. $K = (4\pi n/\lambda)\sin(\theta/2)$, (Eq. 21-67), and simply corrects the various curves to the same time scale, independent of what the particular scattering angle was in each measurement. (*a*) *E. coli* strains unc602, AW405 and cheC497. Faster decay of the autocorrelation function corresponds to faster translational motion (see text). (*b*) Experimental (smooth curves, for $\theta = 20°$, 35°, 50°, 70°, and 90°) and theoretical autocorrelation functions for *E. coli* cheC497, using a model of pure translational motion (top line of squares) or helical motion (crosses). For the helical model, translational velocity is taken as 50 μm sec^{-1} with rotational velocity 60 rad sec^{-1}. [From (*a*) D. W. Schaefer and B. J. Berne, *Biophys. J. 15*, 785 (1975); (*b*) D. W. Schaefer, G. Banks and S. S. Alpert, *Nature 248*, 162 (1974).]

correlation behavior. An example of more detailed information from such studies is shown in Fig. 21-15*b*. The figure shows experimental (smooth curves) and theoretical autocorrelation functions for models of pure translational motion (squares) and helical motion (crosses); clearly the helical motion gives much better fit to the experimental data. This result obviously does not prove that the bacteria swim in helical paths, but it does show that they must change direction frequently.

The interest in bacterial motion is connected with bacterial swimming response to various external agents: thermotaxis (response to heat), chemotaxis (response to chemical gradients), and phototaxis (response to light), since it is conjectured that bacteria must possess a crude form of memory to change direction at a later time, based on what the surroundings were like at an earlier time.

PROBLEMS

1. If the amplitude of a randomly varying quantity can be expressed as

$$f(t) = \sum_{n=1}^{\infty} (a_n \cos(\omega_n t) + b_n \sin(\omega_n t)) \qquad \text{where } \omega_n = (2\pi n/T)$$

show that the time-average value of $(f(t))^2$ is given by $\sum_{n=1}^{\infty} \frac{a_n^2 + b_n^2}{2}$

(In other words, show that Eq. 21-4 follows from Eq. 21-1.)

2. For the same $f(t)$ as in Problem 1, show that the correlation function

$$G(\tau) = \overline{\langle f(t)\, f(t+\tau)\rangle}$$

is given by

$$\sum_{n=1}^{\infty} \overline{\frac{a_n^2 + b_n^2}{2}} \cos(\omega_n t)$$

where the brackets indicate a time average from time zero to time, T. (In other words, show that the correlation function is the Fourier transform of the power spectrum: Eq. 21-9 from Eq. 21-1.)

3. Calculate the power spectrum, $P(\omega) = (2/\pi) \int_0^\infty G(\tau)\, d\tau$, and the area under a plot of $P(\omega)$ versus frequency (between zero and infinite frequency), for a correlation function of the form

$$G(\tau) = G(0) \exp(-\tau/\tau_c)$$

in which $G(0)$ represents the mean-square amplitude of the randomly varying quantity. In this problem, you have shown that the area under the power spectrum is independent of τ_c, which is necessary for understanding of Figures 21-3 and 21-4.

4. In Chapter 21.A, we derived the correlation function and spectral density for a two-site chemical exchange process, in which the two rate constants (Eq. 21-17) were equal. Now calculate the correlation function and power spectrum (i.e., retrace the steps from Eq. 21-18 to Equations 21-24 and 21-25) for a two-site exchange in which the equilibrium populations of the two sites are not equal:

$$A \underset{k_{-1}}{\overset{k_1}{\rightleftharpoons}} B;\ k_1 \neq k_{-1}$$

Hints: See Equations 10-10 and 10-11 for the populations of the two sites as functions of time, and simplify the algebra by letting

$$H_A = \frac{k_1 + k_{-1}}{k_{-1}} h \text{ and } H_B = \frac{k_1 + k_{-1}}{k_1} (-h)$$

5. Calculate the correlation function and spectral density for three-site exchange between three kinetically equivalent sites, A, B, and C, at which the magnetic field values are $+h$, 0, and $-h$:

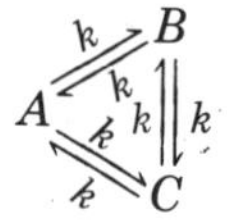

Hint: For the initial condition that $[A] = [A]_0$, $[B] = [C] = 0$, the concentrations at the three sites are:

$$[A] = [A]_0 \left(\frac{1}{3} + \frac{2}{3} e^{-3kt}\right)$$

$$[B] = [A]_0 \left(\frac{1}{3} - \frac{1}{3} e^{-3kt}\right)$$

$$[C] = [A]_0 \left(\frac{1}{3} - \frac{1}{3} e^{-3kt}\right)$$

(These calculations are very similar to those that lead to the correlation function and power spectrum for the rotation of the three protons of a methyl group about the axis that binds the methyl group to a molecule; the result is central to the interpretation of NMR relaxation times in terms of the shape and flexibility of the molecule in question.)

6. It has been shown in the text that rotational diffusion leads to a correlation function for spherical harmonic angular functions of the type

$$\overline{\langle f(\theta,\phi)^* f(\theta_0,\phi_0)\rangle} = \exp[-\tau/\tau_{\text{rot}}]$$

in which τ_{rot} is called the "rotational correlation time." In order to gain a physical picture for τ_{rot}, calculate the average angle by which a molecule has rotated during the time, τ_{rot}, when the physical quantity of interest varies as a spherical harmonic rank one or rank 2.

Hint: let $\theta_0 = 0$, and consider just the $m = 0$ case.

7. Calculate the rotational correlation time (Eq. 21-58) for macromolecules of molecular weight 25,000, 40,000 and 66,000, assuming that viscosity of water = 0.01 poise (cgs units), temperature = 25°C, and partial specific volume of each protein is approximately 0.73 cm^3/g. Your results should compare approximately to the values listed in Table 21-2.

REFERENCES

Fourier Transforms

D. C. Champeney, *Fourier Transforms and Their Physical Applications,* Academic Press, London (1973).

R. Bracewell, *The Fourier Transform and Its Applications,* McGraw-Hill, New York (1965).

Spectral Densities and Correlation Functions

C. P. Slichter, *Principles of Magnetic Resonance,* Harper & Row, N. Y. (1963), Chapter 5. (Ignore quantum mechanical language; the basic argument is classical.) Appendix C (two pages) gives the correlation function derivation for two-site chemical exchange.

Correlation Functions for Random Rotational and Translational Motion

A. Abragam, *The Principles of Nuclear Magnetism,* Clarendon Press, Oxford (1961), pp. 298–302.

Fluorescence Depolarization and Rotational Correlation Times

J. Yguerabide, in *Methods in Enzymology 26C,* 528 (1972).

Laser Light Scattering and Macromolecular Diffusion

N. C. Ford, Jr., "Biochemical Applications of Laser Rayleigh Scattering," *Chemica Scripta 2,* 193–206 (1972). Translational diffusion.

R. Pecora, "Light Scattering Spectra and Dynamic Properties of Macromolecular Solutions," *Disc. Faraday Soc. 49,* 222–228 (1970). Rotational diffusion.

Electrophoretic Light Scattering

B. R. Ware, "Adv. Colloid and Interface," *Science 4,* 1–44 (1974).

CHAPTER 22 Reconstruction of Objects from Images

In previous Chapters (14.G and 15.C), we have described some images based on the interaction of waves (ultrasonic or electromagnetic) with matter. In this section, we examine the means by which structural information may be extracted at about 1 Å resolution from the pattern of very short-wavelength ("X-rays" of about 1 Å wavelength) electromagnetic radiation scattered from ordered arrays (crystals) of macromolecules. Radiation scattered from an ordered array of scattering centers is called "diffraction." The qualitative analysis of X-ray diffraction experiments becomes relatively simple when we combine what we already know about scattering (Chapter 15) and about Fourier transforms (Chapters 20 and 21).

22.A. X-RAY CRYSTALLOGRAPHY: DETERMINATION OF THE CARBON SKELETON OF A MACROMOLECULE

The simplest picture that accounts for the appearance and nature of X-ray diffraction patterns is that proposed by W. L. Bragg as a graduate student in 1912, shown in Fig. 22-1. When X-rays impinge on atoms, radiation (of the same wavelength) is scattered in various directions according to the formal treatment given in Chapter 15. However, when the atoms are located at regularly spaced positions in a crystal, then there will be special angles of incidence for which the X-rays scattered by a whole plane of atoms are in phase, so that the scattered total intensity is a maximum. Figure 22-1 shows that the condition that X-ray waves scattered by atoms in two adjacent positions in such a plane be in phase [so that their amplitudes add to give an intense "spot" in the pattern of scattered (diffracted) radiation] is simply that the difference in path length traveled by the two waves be an integral multiple of an X-ray wavelength, λ:

$$2d \sin \theta = n\lambda \qquad (22\text{-}1)$$

in which d is the spacing between the two layers of atoms and n is an integer. From this picture, it seems plausible that it should be possible to deduce the *spacings between atoms* in a crystal by varying the angle of incidence of the source X-rays on the sample, and observing the angles for which scattered intensity is a maximum (diffraction "spots"). Furthermore, since (see Chapter 15.B) we have already argued that scattered intensity from a given atom increases directly with atomic number, we might also

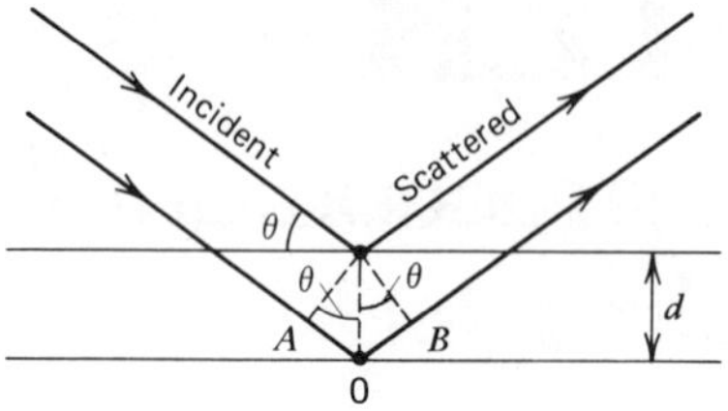

FIGURE 22-1. Bragg model for reflection of X-rays from adjacent planes of atoms in a crystal lattice. The scattered waves from the two atoms in the figure will recombine to give a resultant scattered wave of maximum amplitude when the two scattered waves have the same phase, namely, when their difference in path length, OA + OB = 2d sin θ, is an integral number of X-ray wavelengths, $n\lambda$, with n an integer (Eq. 22-1).

expect to be able to distinguish atoms that are *chemically* different (i.e., different atomic number) in a crystal by comparing relative scattered intensities corresponding to layers of different types of atoms in the crystal. Although the intuitive content of the problem of obtaining the molecular *structure* of a crystalline molecule from the inter-atomic *spacings* derived from the X-ray scattering pattern is clear from the above picture, the actual *mechanics* of the solution are most easily performed using a connection between X-ray diffraction and Fourier theory first appreciated by W. H. Bragg (father of W. L. Bragg) about 1915.

22.A.1. The Diffraction Image: The Lens as a Fourier Transform Device

The Bragg result, Eq. 22-1, indicates that the combined amplitudes of the X-ray waves scattered by various layers of atoms in a crystal form a "diffraction" (intensity) pattern, in which the *diffraction image* intensity peaks are located at positions related to the *spacings* between peaks of electron density (i.e., atom positions) in the *original* object. The *closer* the spacings between atoms in the original crystal, the *farther apart* the intensity peak positions in the diffraction image (i.e., for smaller *d*, it takes a larger θ to satisfy Eq. 22-1). This same sort of behavior is familiar from our previous discussion of time-domain and frequency-domain spectral signals (Chapter 20.B), in which we noted that compression in one domain (shorter time-domain signal) corresponded to expansion in the other domain (broader frequency-domain signal). In fact, the mathematical relation between the original object and its diffraction image is the same Fourier transform that we encountered in the spectral line shape example earlier.

Since the reader has already encountered the formation of *optical* images in elementary physics context, it is simplest to consider the stages by which an *optical* image is formed (Fig. 22-2) in order to understand the

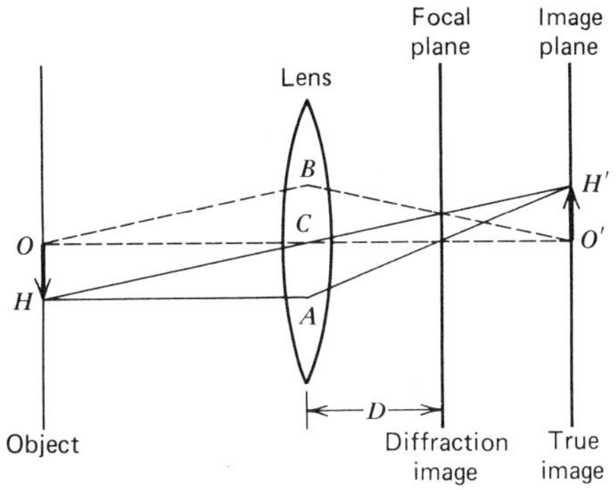

FIGURE 22-2. Images formed behind an optical biconvex lens. The "true" image at the far right is formed by intersection of different rays from the *same part* of the object. The "diffraction" image at the focal plane is formed by intersection of rays traveling in the *same direction* from different parts of the original object. These two kinds of images are related by a Fourier transform (see text). D is the lens focal length. (After M. J. Buerger, *Contemporary Crystallography*, McGraw-Hill, New York, 1970, p. 230.)

problems faced in obtaining an *X-ray* image of an object from an X-ray *diffraction* image. Figure 22-2 shows that there are two types of images formed behind the biconvex lens used in many optics examples: the usual image formed in the "image" plane, and another "diffraction" image formed in the other special "focal" plane. With a suitable geometric analysis (see reference for Fig. 22-2), it may be shown that these two types of image form a Fourier transform pair, just as the time-domain and frequency-domain signals we have analyzed earlier.

To understand the corresponding processes in image reconstruction from X-ray diffraction (where there are no lenses, since the index of refraction of X-rays is essentially unity through virtually any substance), we will separate the biconvex lens into two halves, separated by twice the lens focal length, D, as shown in Fig. 22-3. With the separated lens arrangement, the "true" image is formed at the far right as before, but the diffraction image is now formed *between* the two lenses; thus the first lens forms a diffraction image from the original object rays, while the second lens forms a "true" image from the diffraction image. Phrased a little differently, the "true" image is obtained by diffraction of the diffraction image (mathematically, one would say that the Fourier transform of the Fourier transform of a function yields the original function again). In any case, the principal point of Fig. 22-3 is that experimental diffraction of X-rays by a crystal has the effect of the *first* lens in Fig. 22-3; but since there are no X-ray lenses (Chapter 15.B) to carry out the function of the *second* lens in the optical analog in Fig. 22-3, the second stage in X-ray "true" image formation consists of converting the diffraction image intensity data via *mathematical* Fourier

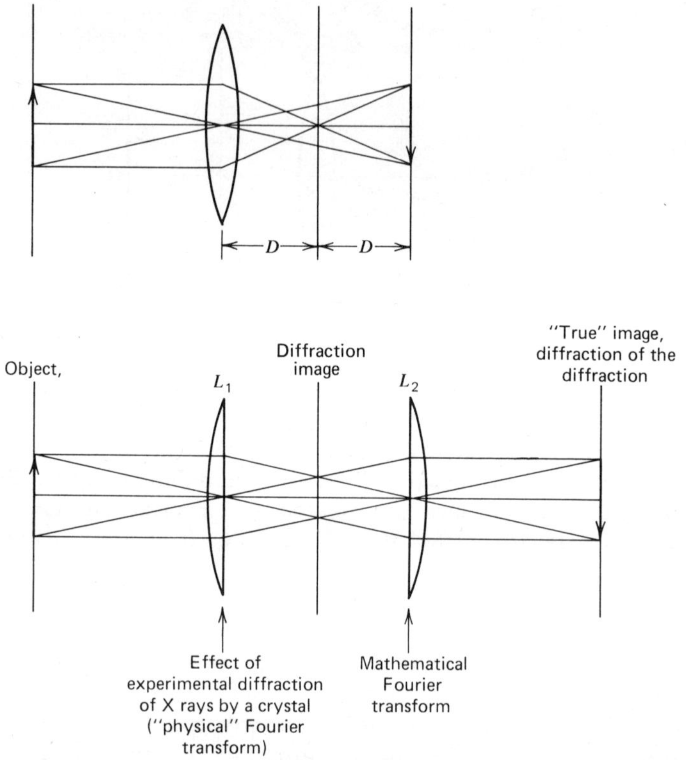

FIGURE 22-3. Separation of optical image formation into two stages: formation of diffraction image from original object rays (lens L_1), followed by formation of "true" image from diffraction image by second lens (L_2). The equivalent stages in formation of a "true" image from an X-ray diffraction experiment are shown at the bottom of the figure and discussed in the text. (After M. J. Buerger, *Contemporary Crystallography,* McGraw-Hill, New York, 1970, p. 240.)

transformation into the true electron density image of the original object molecules in the crystal. Now we know that evaluation of the Fourier transform of a function requires knowledge of the amplitude *and phase* (see following pictorial examples) of the function at each point, but the X-ray detector produces only *intensity* information, so there is no direct experimental means for finding the necessary *phases* of the X-ray waves at various points in the diffraction image. We shall return to this "phase problem" later; for now, we will suppose that the phases can somehow be found, and concentrate on how to form the "true" image from the diffraction image.

22.A.2. (Pictorial) Fourier Synthesis

In Chapter 20.B, we discussed Fourier *analysis,* in which a particular (time-domain) waveform was analyzed into its periodic (frequency-domain) com-

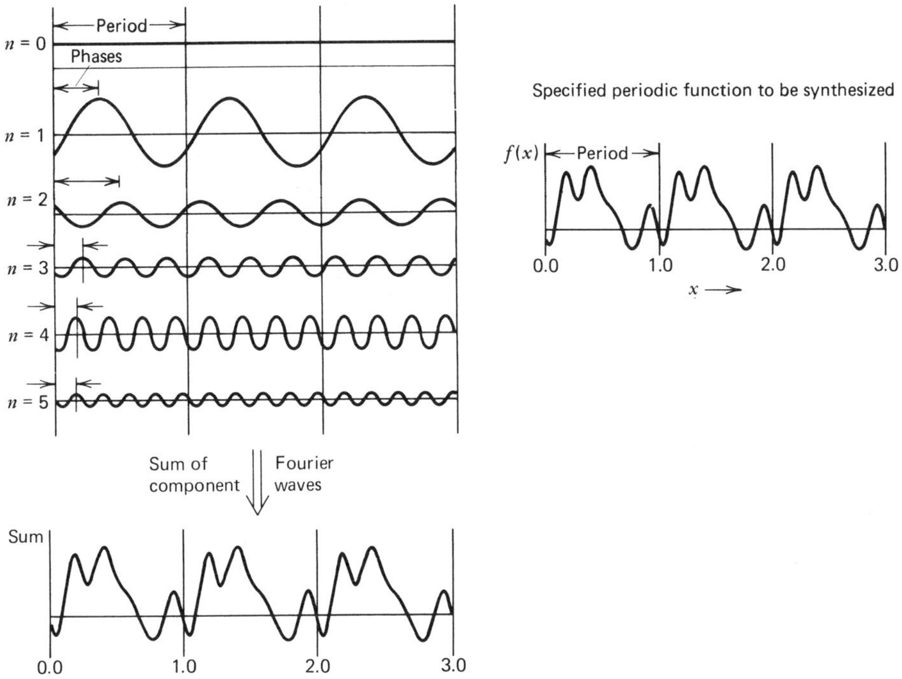

FIGURE 22-4. Fourier synthesis of a given periodic function (upper right), from a sum of the waves whose periods are $(1/n)$ of the period of the original function, $n = 0,1,2, \ldots$ [From J. Waser, *J. Chem. Educ. 45,* 446 (1968).]

ponent amplitudes and phases (the "spectrum" of the signal). In X-ray diffraction, we are interested in the converse problem (Fourier *synthesis*), in which we construct the true image from a sum of the component waves obtained from the X-ray diffraction image. For now, we need only appreciate that because the crystal itself consists of a repeating array of atoms, the diffraction image intensity pattern will also be periodic in space; we therefore begin by considering how an arbitrary periodic function (Fig. 22-4) may be synthesized by knowing the amplitudes and phases of the component sinusoidal waves, each with a period that repeats an integral number of times during the period of the arbitrary periodic function of interest.

X-rays are scattered (diffracted) from real crystals according to the electron density associated with particular atoms. In the next figure (Fig. 22-5), we therefore consider the Fourier synthesis of the electron density in a one-dimensional crystal of a simple (diatomic) molecule, from the amplitudes and phases of the component waves obtained (see below) from the X-ray diffraction image. Figure 22-5 shows a number of features characteristic of X-ray structure determination. First, it requires relatively few Fourier component waves (up to $n = 2$) to locate the approximate location of the *molecules* in the crystal lattice. Second, the addition of more and

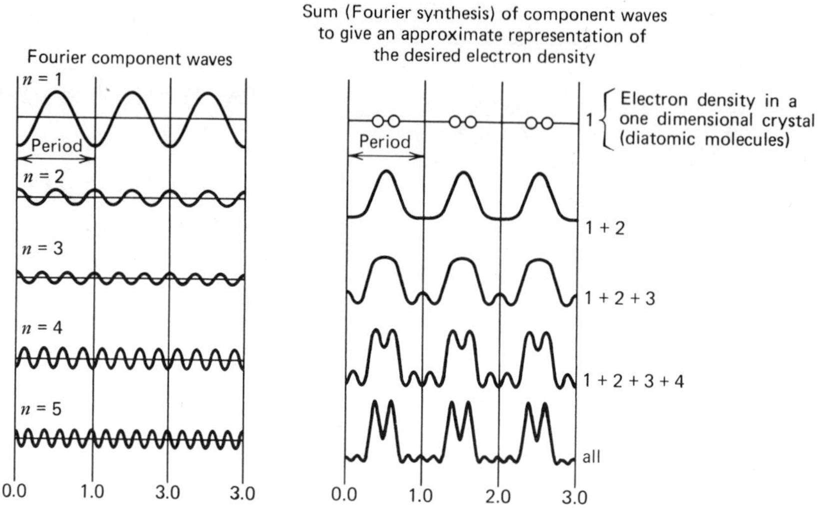

FIGURE 22-5. Fourier synthesis of the electron density in a one-dimensional crystal of diatomic molecules, by adding together the indicated number of Fourier component waves of proper amplitude and phase. See text for discussion. [After J. Waser, *J. Chem. Educ. 45,* 446 (1968).]

more Fourier components (corresponding to inclusion of more and more "intensity spots" in an experimental X-ray diffraction pattern) results in better and better *resolution* – inclusion of waves up to $n = 5$ gives a relatively accurate representation of the location of individual *atoms* in the crystal. [Mathematically, one would say that the Fourier series *converges rapidly* to give the desired (electron density function).] Third, we can neglect the $n = 0$ Fourier component since we are interested only in the positions of atoms (electron density *peaks*) and not in the absolute level of the (constant) *baseline* determined by the $n = 0$ component. Finally, it may be noted that for this particular example, the phases of each of the component waves are either 0° or 180°, but never any other angle. This feature is characteristic of *centrosymmetric* structures, namely ones in which the location of all atoms in a given unit cell is unchanged by reflection of each atom through the center of the unit cell. For now, we will only remark that the centrosymmetric condition provides a great simplification in structure determination, since the phases can take on only two values (see "phase problem" below). Even for crystals that are not centrosymmetric, it is often possible to find particular Bragg reflection planes that are centrosymmetric, to take advantage of this feature.

Fourier synthesis in two (or three) dimensions proceeds along entirely along analogous lines (Fig. 22-6), and the results are most usually expressed as a contour map (Fig. 22-6 for two dimensions) or as a superposition of

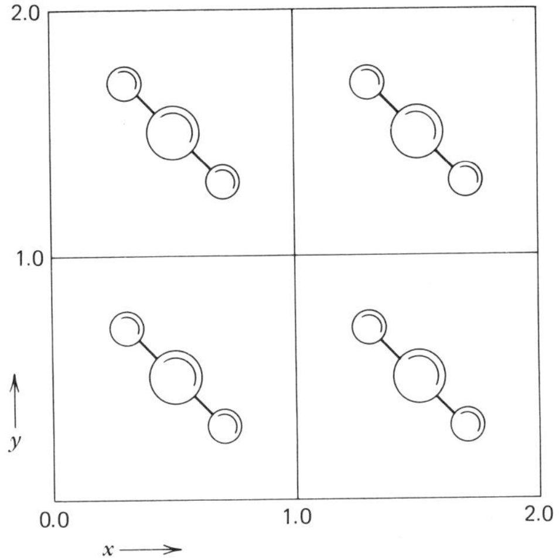

(*a*) Model of the two-dimensional crystal. Four unit cells of a "crystal" are shown, each containing a linear, triatomic molecule AB_2. The next four figures show the representation of the electron density of this "crystal" by the superposition of waves.

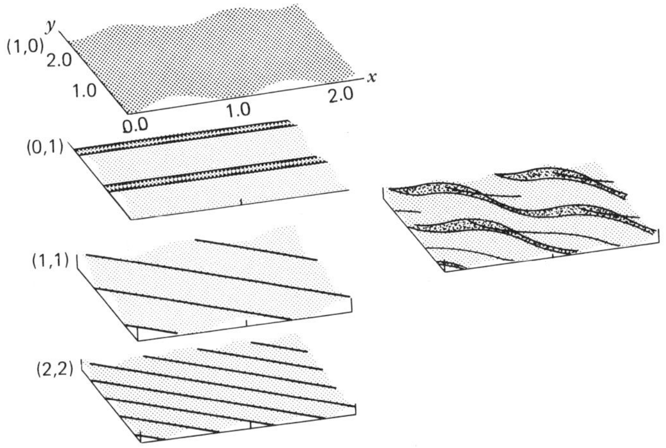

(*b*) The four largest waves. The four waves with the largest amplitudes required to represent the electron density of the "crystal" of Figure 22-6a are shown on the left and their sum on the right. The resulting mountain ranges begin to show the outlines of the molecules. The pairs of integers (1,0), (0,1), (1,1), and (2,2) on the far left characterize the different waves. The first integer is equal to the number of wave crests or troughs within the repeat distance along the x direction. The second integer does the same for the y direction.

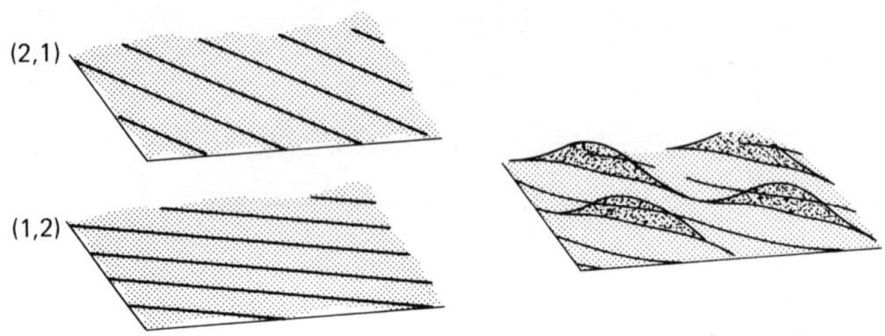

(c) The two waves next in size. Adding the two waves shown on the left improves the definition of the molecular outlines, but no structural details of the molecules are as yet visible. The reason is that up to now all waves of higher frequency than the lowest have been oriented more or less *along* the mountain ranges.

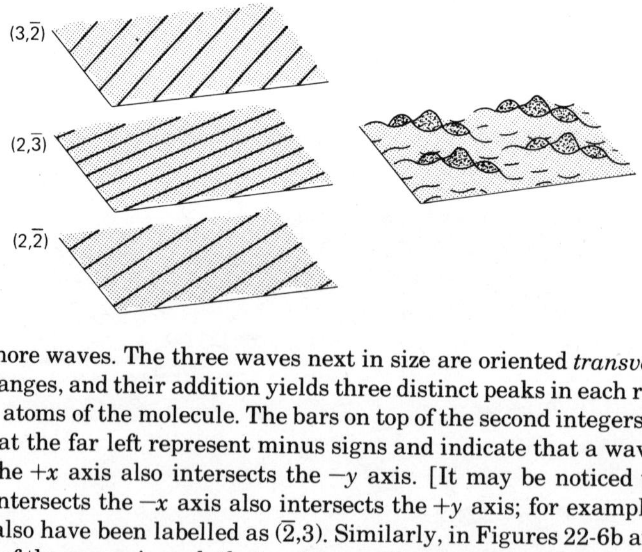

(d) Three more waves. The three waves next in size are oriented *transversely* to the mountain ranges, and their addition yields three distinct peaks in each range, representing the atoms of the molecule. The bars on top of the second integers in the pairs of integers at the far left represent minus signs and indicate that a wave crest that intersects the $+x$ axis also intersects the $-y$ axis. [It may be noticed that a wave crest that intersects the $-x$ axis also intersects the $+y$ axis; for example, the wave $(2,\bar{3})$ could also have been labelled as $(\bar{2},3)$. Similarly, in Figures 22-6b and 22-6c the orientation of the waves is such that a given crest simultaneously cuts across either the $+x$ and the $+y$ axes, or else the $-x$ and the $-y$ axes; these waves could therefore also have been labelled by a pair of negative rather than positive integers. For example, (2,1) and $(\bar{2},\bar{1})$ indicate the same wave.]

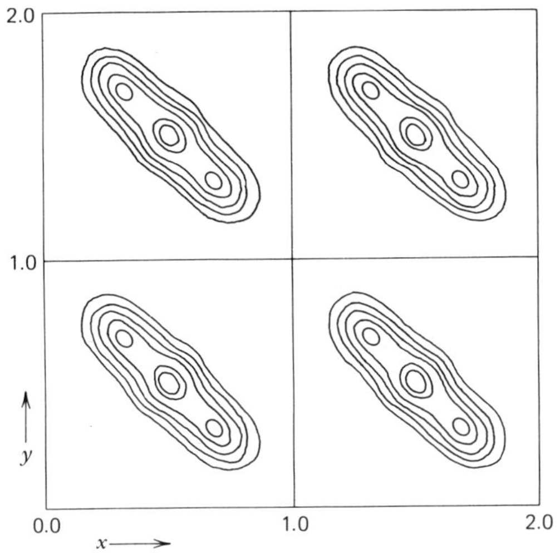

(*e*) Representation by contour mapping. The mountain ranges and peaks of the diagram on the far right of Figure 22-6d are shown by contour lines of constant altitude, altitude being synonymous with electron density here. This is the common way in which electron densities are represented in X-ray diffraction work. The increment in electron density from one contour line to the next is constant.

FIGURE 22-6. Fourier synthesis of the electron density in a two-dimensional crystal of a triatomic molecule. (Subcaptions self-explanatory). [From J. Waser, *J. Chem. Educ. 45,* 446 (1968).]

many layers of contour maps (Fig. 22-7 for three dimensions) of electron density.

22.A.3. Location of Atoms in the Unit Cell: Patterson Synthesis

In the previous section, we discussed how an electron density map of a molecule could be generated as the sum of suitable component amplitudes and phases of the components of the X-ray diffraction image from a crystal. X-ray diffraction images are generated as shown in Fig. 22-8*a,* and yield patterns such as those in Fig. 22-8*b* for a particular orientation of a lysozyme crystal. The *unit cell* dimensions* are determined from the *spacings* between the various adjacent intensity spots in the diffraction image—by rotating the crystal about various axes, one obtains the three unit cell dimensions. The *location* of each atom *within* each unit cell (i.e., the molecular structure) is obtained from analysis of the *relative intensities* of the various spots. Now if the amplitude and phase of the diffracted X-rays has the value, $\sqrt{s}\,|F|e^{i\phi}$ at one of the spots in the diffraction pattern (where s is

* *The unit cell is the smallest repeating three-dimensional unit in the crystal, and may contain one or more molecules.*

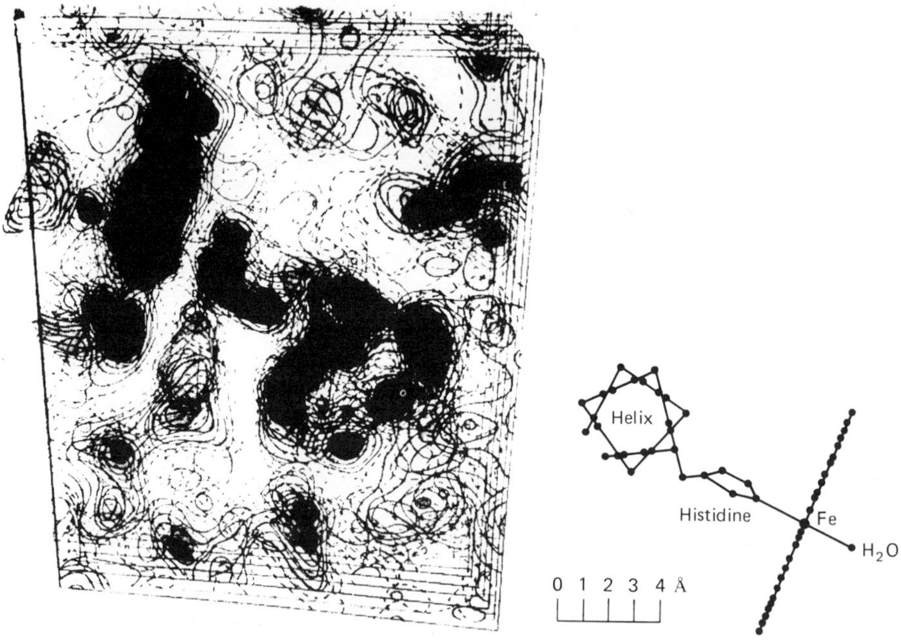

(b)

FIGURE 22-7. Diagram (right) showing how the heme-group, viewed edge-on, is linked to a helical region of the polypeptide chain through a histidine residue in myoglobin. This was derived from the three-dimensional Fourier synthesis at 2 Å resolution shown at the left. [From J. C. Kendrew, R. E. Dickerson, B. E. Strandberg, R. G. Hart, D. R. Davies, D. C. Phillips, and V. C. Shore, *Nature 185,* 422 (1960).]

a "scale" factor—see Fig. 22-9), then the intensity measurement will provide information only about $\sqrt{s}\,|F|e^{i\phi}\sqrt{s}\,|F|e^{-i\phi} = s|F|^2$, so the phase information would appear to be lost, leaving us with no apparent means for conducting the desired Fourier synthesis. However, Patterson showed that it is often possible to recover some information about the positions of atoms in the unit cell by taking a Fourier transform of $|F|^2$, rather than of $|F|e^{i\phi}$. This new function is called the Patterson function, and the result of such manipulations is illustrated pictorially in Fig. 22-9. The Patterson function is related in a simple way to the desired electron-density map, and for unit cells with just a few atoms, it is usually possible to make a good guess as to the relative positions of atoms within the unit cell from looking at the Patterson map (see Problems). Then, by taking the Fourier transform of those estimated electron densities, one obtains diffraction amplitudes, $|F|$, which may be compared with the experimental diffraction intensities, since intensity is proportional to (amplitude)^2. By systematically varying the positions of the atoms in the proposed theoretical structure, one can mini-

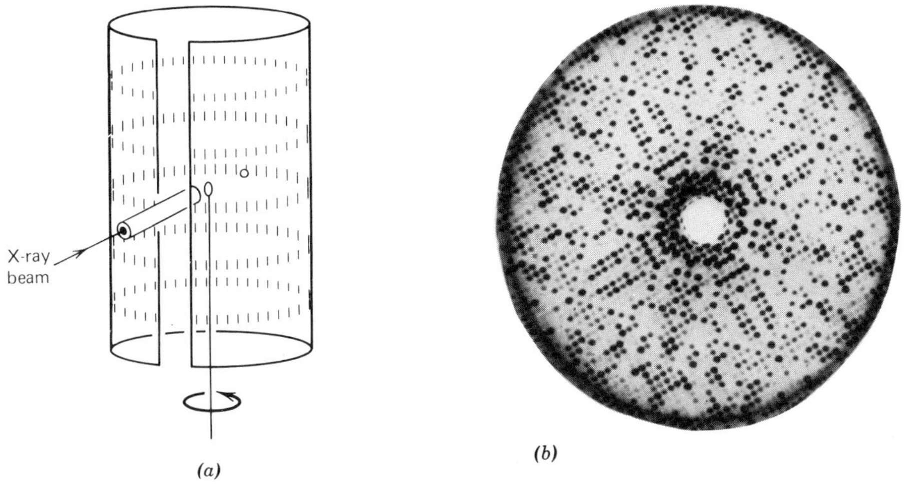

FIGURE 22-8. (*a*) Generation of X-ray diffraction pattern. Monochromatic collimated X-ray beam impinges on crystal; many diffracted beams are scattered at angles (Eq. 22-1) depending on orientation of crystal. To produce all possible reflections, crystal must be rotated about different axes. Images are usually recorded photographically, with center hole cut in film to prevent exposure from incident X-ray beam itself. [From C. W. Bunn, *Crystals, Their Role in Nature and Science*, Academic Press, New York, 1965.] (*b*) X-ray diffraction pattern taken down the (particular) 4-fold axis of a (tetragonal) crystal of hen egg white lysozyme protein. The effects of the 4-fold symmetry axis on the diffraction pattern are clearly evident. *Spacings* between successive spots are related to the distance between successive Bragg planes (Fig. 22-1) while *intensities* of various spots are related to the locations of various atoms in the unit cell (see text). (From R. D. B. Fraser and R. P. MacRae, "X-Ray Methods," in *Physical Principles and Techniques of Protein Chemistry*, Vol. A, ed. S. J. Leach, Academic Press, New York, 1969, p. 68.)

mize the difference between observed and calculated $|F|^2$ to obtain the theoretical structure that best accounts for the observed diffraction pattern. Although this "trial and error" scheme is feasible for small molecules, it seems reasonable that since the number of points in the Patterson image increases as n^2 [with n peaks superimposed at the origin and the remaining $(n^2 - n)$ peaks elsewhere], where n is the number of atoms per unit cell, this method rapidly loses its appeal as the molecule grows larger. For very large molecules, a modified scheme is thus required (next section).

22.A.4. Structure Determination for Large Molecules: Solutions to the Phase Problem

The very strong dependence of Patterson function amplitude on atomic number (Fig. 22-9) suggests that if a large organic molecule (consisting primarily of relatively light atoms of $Z = 1, 6, 7$, and 8) were somehow covalently labeled with a heavy atom, M (say, $Z_M = 25$ or greater), then the

FIGURE 22-9. Images of electron density in a simple three-atom system. Left: electron density obtained by Fourier transform of $\sqrt{s}|F|e^{i\phi}$, where $\sqrt{s}|F|e^{i\phi}$ is the scattered wave amplitude and phase at each point that satisfies Eq. 22-1 in the diffraction image. (Fourier synthesis). Right: Patterson function obtained by Fourier transform of $|F|^2$, where $|F|$ is the scattered wave amplitude. The intensities of the various spots in the X-ray crystal diffraction pattern are each proportional to $s|F|^2$ for that point in the diffraction image; $|F|^2$ is thus readily determined from experimental scattered intensities in the diffraction pattern (Patterson synthesis). The relative amplitudes of the numbers in the *Fourier* synthesized image are related to the relative scattering power of the respective atom (i.e., to Z_A, Z_B, or Z_C, where Z is atomic number—namely, the number of electrons per atom). The relative amplitudes of the numbers in the *Patterson*-synthesized image are proportional to Z_AZ_B, Z_AZ_C, or Z_BZ_C. The positions of the points in the Fourier image are simply the positions of the atoms in the unit cell; the points in the Patterson image are obtained by placing each atom in turn at the origin and marking the positions of the other atoms relative to it. When the number of atoms involved is small, as in the present case, it is easy to deduce the correct atom *positions* (left-hand diagram) that are required to account for the Patterson pattern (right-hand diagram) (see Problems). This principle is the basis for isomorphous replacement methods for obtaining protein structure (see text) since the heavy atom (large Z) is easily located from its large amplitude spot in the Patterson map.

Patterson peaks associated with M—M distances in the crystal should be much stronger than Patterson peaks associated with M-Carbon or Carbon-Carbon peaks (by a factor of $\frac{(25)^2}{(25)(6)} = 4.2$ or $\frac{(25)^2}{(6)^2} = 17.4$ in the present example). This result leads to the two principal means for determination of structures of large (MW up to about 1000) and very large (MW greater than 1000) organic molecules.

Heavy Atom Method (MW 1000 or so) It is convenient in the ensuing discussion to represent the diffraction wave at a given spot in the diffraction pattern as a complex number whose modulus is the diffraction amplitude and whose phase is the phase of the diffraction wave. In this way, we can conduct our discussion graphically by use of Argand diagrams (Fig. 22-10). The phase angle of interest then simply becomes the angle between the diffraction complex "vector" and the real axis in the Argand diagram.

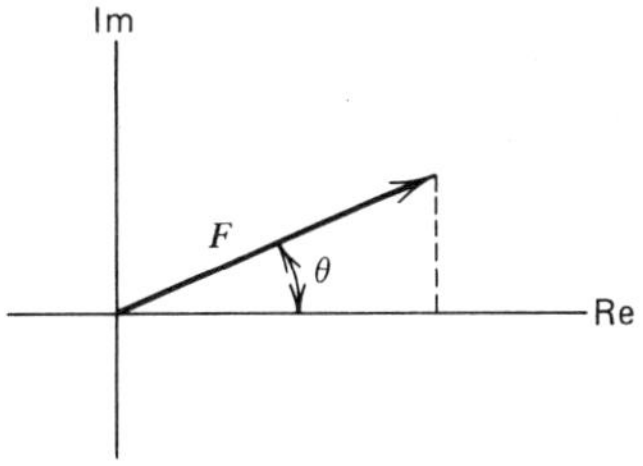

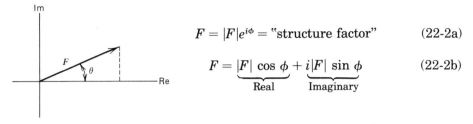

$$F = |F|e^{i\phi} = \text{"structure factor"} \quad (22\text{-}2a)$$

$$F = \underbrace{|F| \cos \phi}_{\text{Real}} + \underbrace{i|F| \sin \phi}_{\text{Imaginary}} \quad (22\text{-}2b)$$

FIGURE 22-10. Representation of a complex number according to its real part (x axis) and its imaginary part (y axis). The complex quantity of interest is the diffraction wave (amplitude = $|F|$ and phase = ϕ) at a given spot in an X-ray diffraction pattern. When there are several atoms in a unit cell, the resultant F is the amplitude and phase of the *vector sum* of the diffraction waves from *all* the atoms in the cell.

When an organic molecule contains a heavy atom, and when the number of other light atoms is still sufficiently small that the "structure factor" (diffraction wave at a given spot in the diffraction pattern) contribution from the heavy atom, F_M, is still large compared to the total scattering from all the light atoms, F_L, then it is possible to pick out the M—M peaks from the M-Carbon or Carbon-Carbon peaks in the Patterson function calculated from the Fourier transform of the $|F|^2$ from the observed diffraction pattern intensities. Using the arguments in Fig. 22-9, it is then usually possible to locate the heavy atom position(s) in the unit cell. Then, by taking a ("reverse") Fourier transform of the heavy atom electron density just proposed, the heavy atom amplitude and phase can be calculated for each spot in the diffraction pattern. The trick that comes next is to notice that if $|F_M| \gg |F_L|$, then the phase, ϕ, for the diffraction wave $F = F_M + F_L$ (where the Argand vector sum is shown in Fig. 22-11), of the molecule as a whole is almost the same as for the heavy atom alone, ϕ_M. Therefore, to a first approximation, we may assign the phase of each diffraction spot as the phase

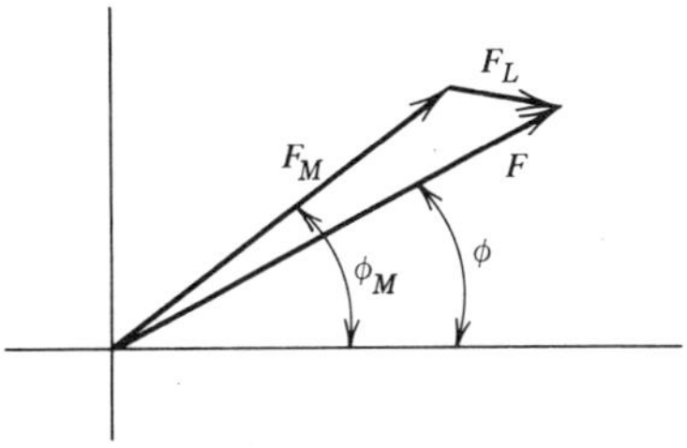

FIGURE 22-11. Graphical illustration of the basis of the heavy-atom phase determination. As long as $|F_M| >> |F_L|$, then ϕ_M is a good approximation to the desired phase, ϕ, of the structure factor, F, of the whole molecule. [ϕ_M is obtained from Fourier transform of the heavy metal atom positions, which are obtained from a Patterson map (see text).]

value calculated from the (known) heavy atom positions. Then, by taking a Fourier transform of the quantities, $|F|e^{i\phi}$, where $|F|$ are derived from the observed diffraction spot intensities and ϕ are calculated from the heavy atom phases, we obtain an approximate electron density map for the whole molecule. Since at least some of the light atoms should now be visible in this electron density map, we can use their approximate positions to compute part of the contribution of the light atoms to F_L. Then by returning to Fig. 22-11, we can produce a better approximation to the phase of F for the whole molecule by including this estimate of F_L in vector addition with F_M to give an improved phase angle for F. Constructing a new electron density map from the Fourier transform of this set of F with improved phases, we should now be able to locate more of the light atoms, and continue the refinement until all the light atoms have been located and the structure is determined.

The most complex structure that has been solved by this heavy atom method is that of vitamin B_{12} by Hodgkin and her collaborators. Vitamin B_{12} contains nearly a hundred atoms (not counting hydrogen), one of which is a cobalt atom of atomic number 27 that functions as the heavy atom. Clearly, the method suffers from an upper limit in molecular weight, because if the molecule is too large (or the heavy atom not heavy enough, or both), then we no longer have the condition, $|F_M| \gg |F_L|$, required for the phase from the heavy metal position alone to be a good approximation to the phase for F for the molecule as a whole.

Isomorphous Replacement Method (MW > 1000) For larger molecules, such as proteins, the previous method breaks down for two reasons. First, it is no longer easy to locate the heavy atom position from a Patterson map calculated from the observed diffraction intensities, because the total scattering from all the light atoms is no longer negligible compared to that from the heavy atom (i.e., even though the scattering from any *one* light atom is much weaker than from the heavy atom, there are so many light atoms that their resultant *total* scattering may be larger than from the single heavy atom). Second, since it is no longer true that $|F_M| \gg |F_L|$ as just indicated, the graphical construction of Fig. 22-11 shows that it is no longer a good approximation to take $\phi \cong \phi_M$. More experiments are therefore required.

Suppose that X-ray diffraction experiments are conducted on a protein before and after introduction of (preferably) a single heavy atom into the protein structure in a way that does not change the conformation of the protein or the unit cell dimensions ("isomorphous" means same shape). The purpose of these measurements is first to locate the heavy atom and then use its location to find the light atoms. Specifically, the structure factors (for each "spot") for the heavy atom derivative, F_M, can be expressed (as before) as the sum of the contributions due to the protein itself, F_P, and due to the heavy atom alone, F_M:

$$F = F_M + F_P \tag{22-3}$$

The only difference between Eq. 22-3 and Fig. 22-11 is that F_M is no longer large compared to F_P, so that it would not be easy to pick out the M—M locations from a Patterson function constructed directly from the $|F|^2$. However, if we *subtract* the protein structure factor from that of the heavy atom derivative to obtain $(F - F_P)$, we would expect to be able to locate the heavy atoms from the Patterson function constructed from these structure factor *differences*, $|F - F_P|^2$. Since we don't know the phases associated with either F or F_P, we can't actually compute the required $(F - F_P)$, but we can do nearly as well by computing $(|F| - |F_P|)^2$, and constructing a Patterson function from it to locate the heavy atom. This solves the first half of the problem outlined in the preceding paragraph.

Once the heavy atoms have been located in the heavy atom derivative, we can take a Fourier transform of those positions and compute the heavy atom contribution (both amplitude and phase) to the structure factor of the heavy atom derivative as shown in Fig. 22-12. As shown in the figure, we now know the amplitudes of F, F_M, and F_P, and we know that the vector sum (in an Argand diagram) of F_M and F_P must give F, and we know the phase angle (direction in an Argand diagram) of F_M. Therefore, considering the protein structure factor first, we know that the vector F_P must terminate at the point, 0, in the figure, with a phase as yet unknown (i.e., between 0° and 360°). Thus the possible orientations of F_P trace out a circle of radius $|F_P|$ centered at 0. Similarly the orientations of the heavy atom derivative structure factor, F, trace out a circle of radius $|F|$ centered at A in the figure. Finally, since F and F_P must add vectorially to give F_M, the only possible phase choices are those measured from points in the Argand diagram at

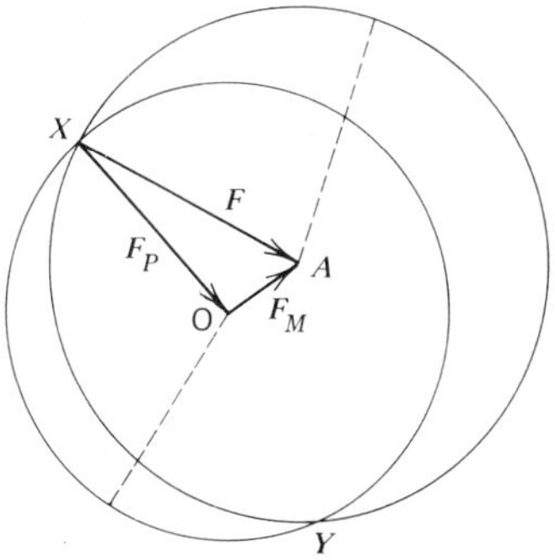

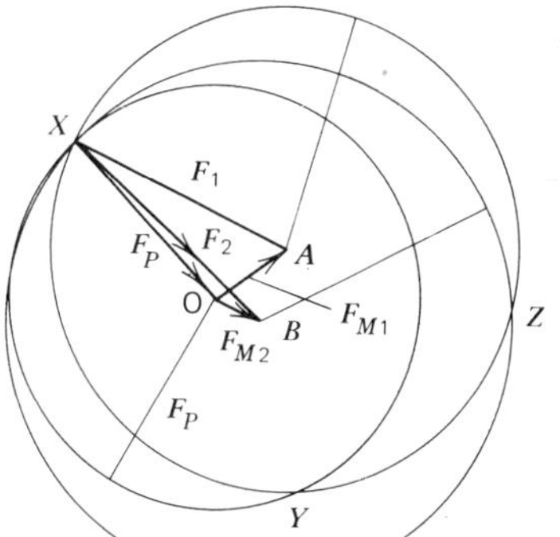

FIGURE 22-12. Phase determination using one (left) or two (right) isomorphous heavy atom derivatives of a protein. Phase is determined to within two values (X and Y at left, or X and Z at right) by use of a single heavy atom derivative, and is determined completely by use of two isomorphous derivatives (right), as explained in text. (After H. R. Wilson, *Diffraction of X-rays by Proteins, Nucleic Acids, and Viruses*. Edward Arnold Publishers, London, 1966, p. 58.)

which the two circles intersect. Thus, the phases in the heavy atom derivative (and in the protein itself) are narrowed down to *two* possible values. By repeating the same experiments and graphical constructions using a *second* heavy atom derivative, it is possible to choose between the two possible phases for F_P and proceed to solve the structure as indicated earlier (i.e., by using those phases to make a Fourier map of electron density, then using the locations of some light atoms to obtain improved phase for the various F_P, then making an improved electron density map, etc., until all atoms are found). This discussion should make it clear why at least *two* different isomorphous (F_P must be the same in both derivatives) heavy atom derivatives of a crystalline macromolecule are required to establish its crystal structure (see Example).

EXAMPLE *Myoglobin*

Myoglobin (as the reader is doubtless aware) is the monomeric unit from which (tetrameric) hemoglobin is constructed. It is a protein of molecular weight 17,000 consisting of 153 amino acids, or about 1260 atoms (not counting hydrogen). [As with most molecules, the scattering from hydrogens is so weak compared to that from carbon, nitrogen and oxygen ($Z = 1$ rather than $Z = 6$, 7, or 8) that it is not possible to locate hydrogen atoms in the electron density map.] The unit cell in a myoglobin crystal contains two protein molecules packed into a monoclinic region whose dimensions are 64.6 × 31.3 × 34.8 Å. In order to determine the structure of myoglobin, Kendrew et al. used the isomorphous replacement method, employing three kinds of heavy atom derivatives in various combinations to generate six types of crystals from which intensities of about 400 reflections were experimentally measured for each type of crystal. At this stage, the structure was resolved down to about 6 Å, which was sufficient to identify the course of the highly folded polypeptide chain to produce a relatively compact protein structure. Further study using data from 9600 reflections has improved the resolution to about 2 Å, which is sufficient to identify and resolve each of the individual amino acids and the disposition of their side chains. Examples of the use of the heavy atom derivatives in establishing the structure factor phases required for construction of the Fourier electron density map of myoglobin are shown in Fig. 22-13, for Kendrew's experiments. The figure (especially diagram *c*) shows why it may be necessary to use more than two heavy atom derivatives to obtain unambiguous phase information. A picture of the three-dimensional structure of myoglobin may now be found in almost any modern biochemistry text, along with discussion of the organization of myoglobin units in the oxygen-binding function of hemoglobin.

A recent list of proteins whose three-dimensional structures were known at atomic resolution (about 2 Å) as of late 1975 are listed in Table 22-1. The increasing rate at which structures are being determined is ascribable primarily to reductions in the cost (and increases in the speed and size) of the digital

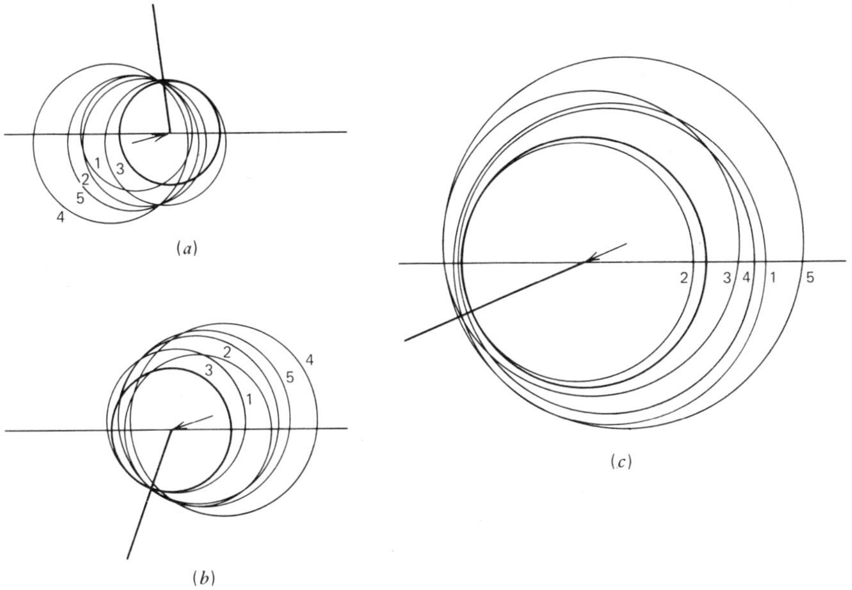

FIGURE 22-13. Phase determination for structure factors corresponding to three different reflections (*a*, *b*, and *c*) for crystals of myoglobin and its heavy-atom derivatives. In each case, the heavy circle represents the amplitude of the reflection from the unsubsituted protein, and the light circles those from the following derivatives:
1. Myoglobin + PCMBS (p-chloro-mercuri-benzenesulfonate)
2. Myoglobin + $HgAm_2$ (mercury diamine)
3. Myoglobin + Au (aurichloride)
4. Myoglobin + PCMBS + $HgAm_2$
5. Myoglobin + PCMBS + Au

The short lines from the centers are the vectors for the heavy atoms; the heavy straight line in each case indicates the phase angle eventually selected. [From G. Bodo, H. M. Dintzis, J. C. Kendrew, and H. W. Wyckoff, *Proc. Roy. Soc.* (A) *253*, 91 (1959).]

computers that reduce the reflection data. Also, the very high intensity X-radiation from synchrotrons has made it possible to obtain data from smaller samples.

EXAMPLE *Nucleotide-Binding Enzymes: Lactate Dehydrogenase (LDHase) and Glyceraldehyde-3-phosphate Dehydrogenase (GPDHase)*

The relation between structure and function in enzymes is particularly well-illustrated by recent structural determinations (using X-ray crystallography) of two nucleotide-binding enzymes, LDHase and GPDHase. Both enzymes are glycolysis dehydrogenases that bind the NAD^+ molecule; both enzymes are tetramers constructed of four chemically equivalent subunits, where each sub-

Table 22-1 Chronology of X-Ray Crystal Structure Determinations for Proteins

Year	Protein	Species	Molecular Weight	Resolution (Å)
1960	Myoglobin	Sperm whale	17,800	2.0, 1.4
1965	Lysozyme	Hen egg white	14,600	2.0
1967	Carboxypeptidase A	Bovine	34,600	2.8, 2.0
	α-Chymotrypsin	Bovine	25,000	2.0
	Ribonuclease A	Bovine	13,700	2.0
	Ribonuclease S	Bovine	13,700	3.5, 2.0
1968	Hemoglobin, oxy	Horse	66,500	2.8, 2.5
	Papain	Papaya	23,000	2.8
1969	γ-Chymotrypsin	Bovine	25,000	2.7
	Hemoglobin	Chironomous	17,000	2.8, 2.5
	Insulin	Porcine	5,800	2.8, 1.5
	Rubredoxin	*Clostridium pasteurianum*	6,000	3.0, 1.5
	Subtilisin BPN′	*Bacillus amyloliquefaciens*	27,500	2.5, 2.0
1970	Chymotrypsinogen	Bovine	25,000	2.5
	Elastase	Porcine	25,900	3.5
	Hemoglobin, deoxy	Horse	66,500	2.8
	Hemoglobin, deoxy	Human	66,500	3.5
	Lactate dehydrogenase	Dogfish	140,000	2.8, 2.0
	Trypsin inhibitor	Bovine pancreas	6,500	2.5, 1.5
1971	Carbonic anhydrase C	Human	30,000	2.0
	Cytochrome b_5	Calf liver	11,000	2.8, 2.0
	Cytochrome *c*, ferri	Horse	12,500	2.8
	Cytochrome *c*, ferro	Tuna	12,500	2.5, 2.0
	Flavodoxin	*Clostridium pasteurianum*	16,000	3.25, 1.9
	Hemoglobin	*Glycera dibranchiata*	18,200	2.5
	Hemoglobin	Lamprey	18,000	2.0
	High potential iron protein	Chromatium	9,650	2.25, 2.0
	Myogen	Carp	11,000	2.0, 1.85
	Nuclease	*Staphylococcus aureus*	16,800	2.0, 1.5
	Subtilisin Novo	*Bacillus subtilis*	27,500	2.8
	Trypsin	Bovine	25,000	2.7
1972	Concanavalin A	Jack bean	108,000	2.0
	Concanavalin A	Jack bean	108,000	2.4
	Cytochrome *c*, ferro	Bonito	12,500	2.3
	Ferredoxin	*Peptococcus aerogenes*	6,000	2.5, 2.0
	Flavodoxin	*Desulfovibrio vulgaris*	16,000	2.5
	Insulin	Porcine	5,800	3.1
	Malate dehydrogenase	Porcine	72,000	3.0, 2.5
	Ribonuclease A	Bovine	13,700	2.5
	Thermolysin	*Bacillus thermoproteolyticus*	34,600	2.3
1973	Alcohol dehydrogenase	Horse liver	80,000	2.9, 2.4
	Bence–Jones protein Mcg	Human	23,000	3.5, 2.3

Table 22-1 Chronology of X-Ray Crystal Structure Determinations for Proteins (Continued)

Year	Protein	Species	Molecular Weight	Resolution (Å)
	α-Chymotrypsin	Bovine	25,000	2.8
	Cytochrome *c*, ferro	Tuna	12,020	2.45, 2.0
	Cytochrome c_2	*Rhodospirillum rubrum*	12,480	2.0
	Cytochrome c_{550}	*Micrococcus denitrificans*	14,000	4.0, 2.5
	Fab fragment	Human	55,000	2.8, 2.0
	Glyceraldehyde-3-phosphate dehydrogenase	Lobster	143,000	3.0, 2.9
	Lactate dehydrogenase, ternary complex	Dogfish	140,000	3.0
	Trypsin-pancreatic trypsin inhibitor	Bovine	30,000	2.8, 1.9
1974	Adenylate kinase	Porcine	22,000	3.0
	Bence–Jones protein REI	Human	11,000	2.8, 2.0
	Fab Fragment, McPC 603	Mouse	55,000	3.0
	Flavodoxin, oxidized	*Clostridium MP*	16,000	1.9
	Insulin	Porcine	5,800	1.8
	Lysozyme	Human	14,500	2.5
	Lysozyme	Bacteriophage T4	18,700	2.5
	Phosphoglycerate kinase	Horse muscle	38,000	3.0
	Phosphoglycerate kinase	Yeast	45,000	3.5
	Phosphoglycerate mutase	Yeast	110,700	3.5
	Prealbumin	Human	54,000	2.5
	Rhodanese	Bovine liver	32,000	3.9, 3.0
	Trypsin-soybean trypsin inhibitor	Porcine, soybean	43,500	2.6
1975	Bacteriochlorophyll protein	*Chlorobium limicola*	127,000	2.8
	Bence–Jones fragment Au	Human	11,000	2.5
	Bence–Jones fragment Rhe	Human	12,500	3.0
	Carbonic Anhydrase B	Human	30,000	2.2
	Carboxypeptidase B	Bovine	34,000	2.8
	Glucagon	Porcine	3,600	3.0
	Hexokinase	Yeast	51,000	2.7
	Lysozyme, triclinic	Hen egg white	14,600	1.5
	Pepsin	Porcine	35,000	2.7
	Phospholipase A_2	Porcine	14,500	3.0
	Protease SGPB	*Streptomyces griseus*	18,400	2.8
	Superoxide dismutase	Bovine	16,000	3.0
	Thioredoxin S_2	*Eschericia coli*	11,700	2.8
	Triose phosphate isomerase	Chicken muscle	52,000	2.5
	Trypsin	Bovine	25,000	1.8

[*From B. W. Matthews, "X-Ray Crystallographic Studies of Proteins," in* Ann. Rev. Phys. Chem. 27, *493–523 (1976)*.]

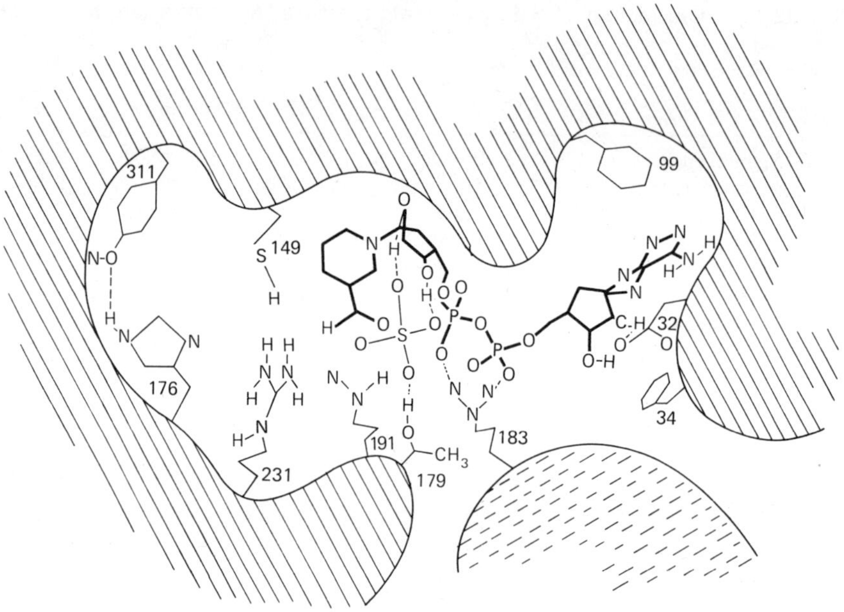

FIGURE 22-14. Schematic diagram of the binding site for NAD^+ in glyceraldehyde-3-phosphate dehydrogenase (GPDHase) enzyme. Positions of amino acids in the polypeptide sequence are indicated by small numbers near the appropriate residues. [From M. Buehner, G. C. Ford, D. Moras, K. W. Olsen, and M. G. Rossmann, *J. Mol. Biol. 90,* 25 (1974).]

unit possesses one catalytic site. The configuration of NAD^+ bound to the GPDHase enzyme is shown schematically below in Fig. 22-14. While it is obvious that one may begin to formulate reaction mechanisms for catalysis using the detailed information in those diagrams, a feature of more general interest is shown schematically in Fig. 22-15. Rossmann and coworkers have shown that the structure of the NAD^+-binding regions in both enzymes is very similar (and a similar conclusion appears to apply to several other nucleotide-binding enzymes whose structures have recently been determined). Furthermore, it is interesting to note that for the *noncooperative* LDHase enzyme, the active sites are well-separated from each other in the tetrameric LDHase molecule (Fig. 22-15), while for the *cooperative* GPDHase enzyme, the active sites place adjacent bound NAD^+ coenzyme molecules in close proximity in the tetrameric enzyme. Thus, it appears that for GPDHase (as for the earlier and better known hemoglobin example determined by Perutz et al.), cooperative ligand binding can be explained in detail from subunit-subunit contacts observed in the X-ray crystal structure.

The preliminary finding that the three-dimensional structure of such enzymes in their nucleotide-binding regions is similar suggests that the structure in this region must have been "frozen" at a very early stage of evolutionary development, from the need to provide a suitably shaped pocket to bind the

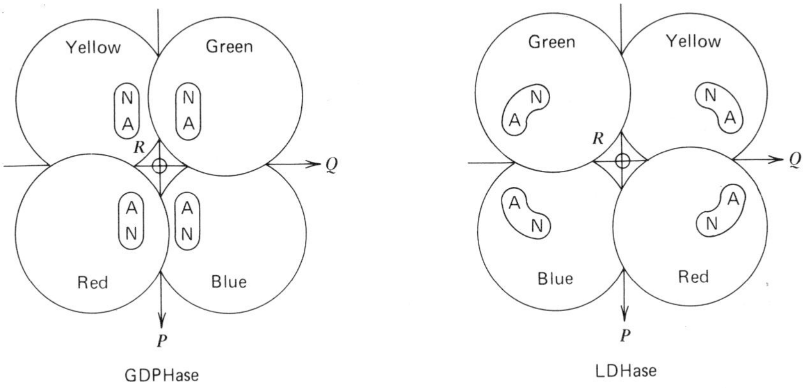

FIGURE 22-15. Diagrammatic comparison of the association of subunits in GPDHase (left) and LDHase (right). The active sites are well-separated in the noncooperative LDHase but are closely juxtaposed in the cooperative GPDHase. GPDHase was the first multimeric enzyme whose X-ray structure was determined. [From M. Buehner, G. C. Ford, D. Moras, K. W. Olsen, and M. G. Rossmann, *J. Mol. Biol. 90,* 25 (1974).]

nucleotide. Since the particular amino acid sequence of the enzyme need not have been "frozen" so early, the amino acid sequences for these enzymes now differ for different enzymes, even though the *shape* of the binding pocket is essentially the same. By comparing the amino acid sequences in this "frozen" structural region for different enzymes, in order to see which amino acid residues have changed, it should be possible for geneticists to determine the evolutionary stage (sequence) at which particular enzymes evolved into distinct types, and which enzymes are most directly related to each other in an evolutionary context. This sort of analysis for other proteins has already been developed back about 1.5 billion years, and application to the nucleotide-binding enzymes promises to extend the analysis back perhaps twice as far.

EXAMPLE *Nucleic Acids*

The most famous example (and deservedly so) of macromolecular structure determined via X-ray crystallography is unquestionably the double-helix DNA structure of Crick and Watson. Unfortunately, the analysis of this structure can be discussed sensibly only from knowledge of the Fourier transform of a helical curve, and the result is mathematically complicated by convenient expression in a Bessel function expansion. The calculation was first carried out by Crick, and led immediately to the correct identification of helical structure in DNA, as had also been proposed by Pauling for polypeptide chains in various proteins (myoglobin chains are approximately 77% in the α-helix configuration). We shall therefore omit any attempt at quantitative appreciation of helical diffraction, and the interested reader is referred to the H. R. Wilson text (see References) for a readable discussion.

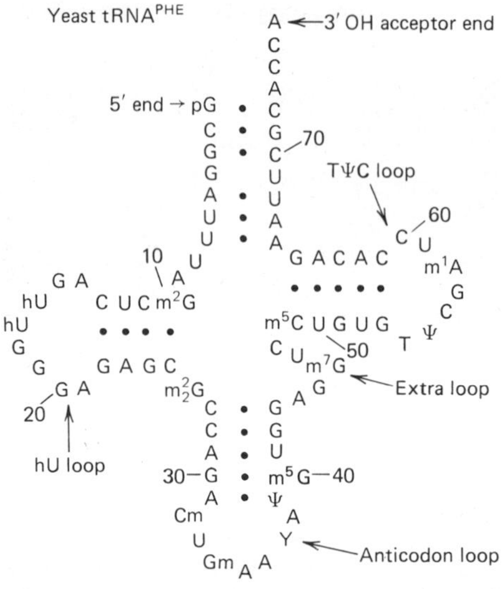

(a)

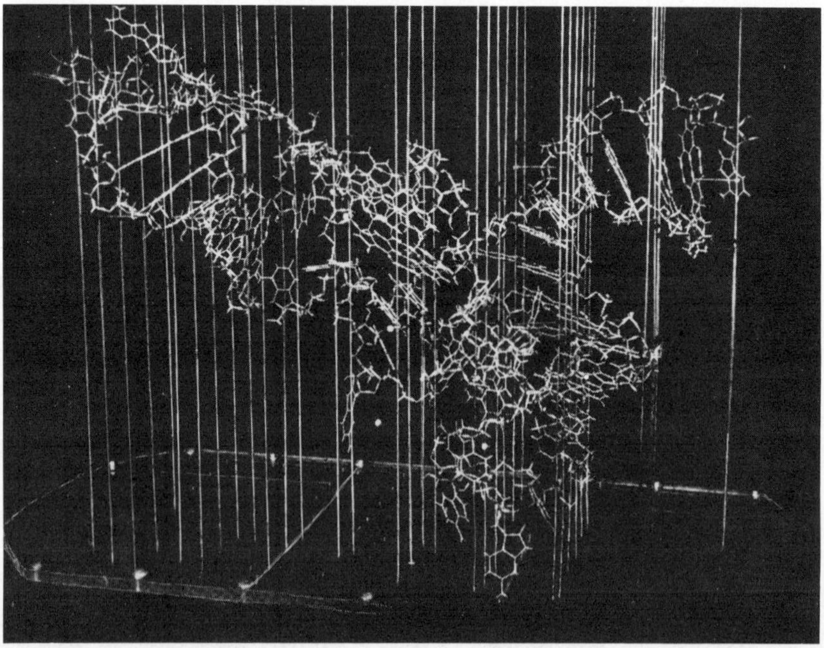

(b)

FIGURE 22-16. The nucleotide sequence of yeast phenylalanine tRNA in the conventional cloverleaf diagram. The nucleotides are numbered starting from the 5′ end. (*b*) A photograph of a Kendrew wire model of yeast phenylalanine tRNA based on the 3 Å resolution electron density map. The model is built at a scale of 2 cm/Å. The CCA stem is at the upper right. The Tψ*C* and hU loops are at the bottom and the anticodon loop is at the upper left. The L shape of the molecule is evident. [From F. L. Suddath, G. J. Quigley, A. McPherson, D. Sneden, J. J. Kim, S. H. Kim, and A. Rich, *Nature 248*, 20 (1974).]

One of the most interesting recent X-ray structures is that of transfer-ribonucleic acid (t-RNA), the molecule from which the polypeptide chain of a protein is assembled when t-RNA interacts with messenger-RNA (the reader is referred to any modern biochemistry text for details of this process). The nucleotide sequence of the molecule is shown in Fig. 22-16*a* for the particular t-RNA that codes for phenylalanine in yeast; the various nucleotides are identified by their usual abbreviations (note the abundance of modified nucleotides in addition to the usual A, G, C, and U bases). It is found that there are "stem" regions in which there is a double-helix structure holding different parts of the chain together, to give an overall L-shaped configuration for the t-RNA molecule. It is satisfying to note that the anticodon loop (the functional part of the molecule) is one in which the ribose and phosphate groups are on the *inside* of the molecule while the bases are on the *outside*, in accord with the proposed function of t-RNA. A wire model of the three-dimensional structure of yeast phenylalanine t-RNA is shown in Fig. 22-16*b*. Calculation of the structure was based on data from 4902 unique reflections and represents a spatial resolution of 3Å in the electron density map.

22.B. NEUTRON DIFFRACTION: LOCATION OF HYDROGEN ATOMS; HYDROGEN BONDING

The degree of spatial resolution provided by a scattering experiment (light scattering, X-ray diffraction, electron microscopy) is determined by the effective wavelength of the radiation employed. We have already noted that visible light can provide spatial image detail only on a rather large scale of the order of 5000 Å, while X-ray and accelerated electron wavelengths are of the order of 1 Å or less. The principal problem with all these forms of radiation is that the nature of the scattering process is a driving of the motion of bound electrons in the specimen by the oscillating electric field of the radiation, so that in comparing different specimen atoms, the scattering increases rapidly with atomic number (i.e., with the number of electrons on the specimen atom). This feature is the basis of electron microscopic staining and of X-ray heavy-atom crystallographic methods, as already observed. However, this same feature makes detection of hydrogen atoms in large molecules very difficult, since those atoms contribute so weakly to the observed scattering. On the other hand, neutrons in motion also have a characteristic wavelength (see Table 22-2) of the order of 1 Å, and so may be expected to provide atomic resolution in an experiment in which neutrons are scattered (diffracted) by a macromolecular crystal, but since the neutrons interact with the *nuclei* rather than with the *electrons* of atoms, the intensity of the scattering is essentially independent of atomic number (Table 22-3), and the hydrogen atoms (particularly if deuterium is substituted for hydrogen) are now much more prominent in the scattering pattern and can thus be resolved in the final atomic map obtained by Fourier transform of the diffraction pattern.

Table 22-2 Comparison of X-ray and Neutron Diffractometry

Property	X rays	Neutrons
1. Wavelength	0.71 Å (Mo K source) 1.54 Å (Cu K source)	0.5 to 5 Å ("Thermal" neutrons = neutrons whose kinetic energy is that of gas molecules at about room temperature.
2. Radiation source	50 keV electrons bombard a Cu or Mo target, ejecting a core electron; X rays are emitted when other electrons drop down to fill the vacant orbital.	High-energy neutrons from a particle accelerator pass through a "moderator" substance (say, D_2O at room temp.), which slows them down to speeds corresponding to wavelength given above.
3. Incoherent background	Low compared to neutron case.	High because of incoherent scattering from hydrogen atoms (can be reduced greatly by deuterium substitution).
4. Phase problem	Solvable based on greatly increased scattering by heavy atoms.	Severe problem because scattering from almost all nuclei is about the same; must use direct (trial and error) approach to determination of phases of structure factors from experimental diffraction pattern.
5. Energy of radiation (determines radiation damage to sample)	High (8 keV), 5 orders of magnitude larger than for thermal neutrons.	Radiation damage to sample is negligible, so much larger sample can be used.
6. Radiation flux from source	10^{10} quanta/cm^2/sec.	10^6 neutrons/cm^2/sec, so that much larger crystals are needed due to small intensity of radiation source.
7. Phase of radiation scattered by atom	180° out-of-phase with driving radiation (see Chapter 13); all atoms appear as positive contours in electron density map.	Usually 180° out-of-phase with driver, except for hydrogen, which is exactly in-phase with driver and thus appears as negative contours in atomic density map.

Table 22-3 Comparison of Atomic Scattering Factors for Neutron and X-ray Scattering (Both at Zero Scattering Angle)*

Atom	X-ray (Electrons)	Neutron (Fermis)[a]
H	1	−0.37
D	1	+0.67
C	6	+0.66
N	7	+0.94
O	8	+0.58
P	15	+0.51
S	16	+0.28

**Based on data from D. M. Engleman and P. B. Moore, in* Ann. Rev. Biophys. & Bioengineering 4, *219 (1975).*
[a] Fermi = 10^{-12} cm. For a neutron beam with a flux of N neutrons per unit area per unit time, the number of neutrons scattered per steradian (unit angular area) per unit time will be given by the product of N and the square of the entry in the neutron column of the table.

There are some significant differences between the limitations of the neutron versus the X-ray diffraction methods, although the basic experiment and data reduction are similar (Table 22-2). Because neutron scattering intensity is so similar in magnitude for almost all atoms, there is no real analog to the heavy-atom tricks used in X-ray diffraction analysis. (See, however, the deuteration example.) The experiment is also made more difficult by the need for a high-energy particle accelerator as radiation source (!), more background scattering (making it more difficult to locate and measure the intensity of diffraction spots), and (most important) the need for very large crystals because of the low neutron flux (smaller by a factor of about 10^5 than the corresponding X-ray source) from the source. Nevertheless, since the relative contribution of a given atom to the total scattered intensity varies as

$$\text{Fractional intensity} = n_i f_i^2 / \sum_i n_i f_i^2 \qquad (22\text{-}4)$$

in which n_i is the number of atoms of the same atomic number in the crystal, and the f_i atomic scattering factors are given in Table 22-3, it is clear that hydrogen atoms will contribute much more in neutron than in X-ray diffraction. For example, for the amino acid alanine ($C_3H_7NO_2$ chemical formula), the fractional scattered intensity for hydrogens is only 0.02 for X-ray diffraction, but is 0.25 for neutron diffraction.

Because of the severe phase problem in neutron diffraction, most molecules whose structures have been determined are relatively small (single amino acids, for example); however, attempts are currently under way to locate the hydrogen atoms in macromolecules in cases where X-ray diffraction studies have already located the other atoms. The primary interest in the neutron scattering results is in locating hydrogens in order to determine the nature and extent of hydrogen-bonding in macromolecular structures. Extensive studies for individual amino acids have shown that many of the N—H · · · 0 hydrogen bonds are *bent* [i.e., the bond angle can be significantly different (30° or more) than 180°]. It is interesting that such bent bonds, which would normally be considered unfavorable based on bond energy considerations, appear to contribute significantly to the stability of certain molecular conformations.

EXAMPLE *Hydrogen Bonding Between Complementary Nucleotides: 9-methyladenine and 1-methylthymine*

The crystal structure for this modified adenine-thymine pair has recently been worked out using neutron diffraction methods. The nuclear scattering density map (Fig. 27-17) clearly shows the positions of the purine and pyrimidine ring carbons, as well as the remaining C, H, N, and O nuclei. The interesting feature of this diagram is that while the two complementary bases are held together by one N—H · · · 0 and one N—H · · · N hydrogen bond, the conformation is different from that postulated by Watson and Crick in that the present pairing scheme involves N1 of adenine rather than N7. The Watson-Crick pairing scheme has however been verified from the X-ray high-resolution electron density map of the dinucleoside phosphate, ApU, using the normal bases.

The high interest in nucleic acid hydrogen bonding is in part connected to a suggestion by Löwdin that two protons involved in base-pairing might "jump" from one base to the other simultaneously. If such a process were to occur

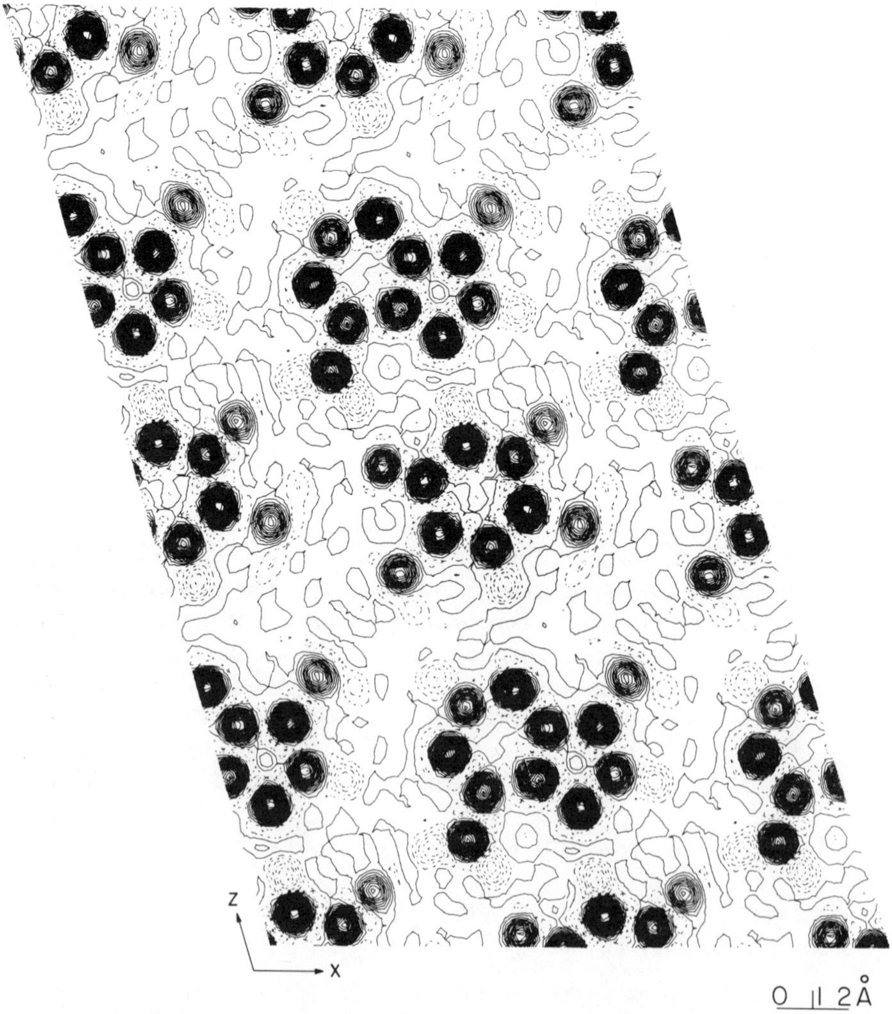

FIGURE 22-17. A section of the neutron scattering density map for 9-methyladenine: 1-methylthymine. Negative density (hydrogen atoms) is denoted by dashed contours and positive density (C,N,O atoms) by solid contours. Note the clear resolution of atomic positions. (From T. F. Koetzle, in *Spectroscopy in Biology and Chemistry: Neutron, X-Ray, Laser,* Academic Press, New York, 1974, p. 177.)

during DNA replication, the complementarity of the bases would be destroyed and a different sequence of bases (a mutation) might result. In this context, it is interesting to note that the neutron diffraction results for the base pair illustrated in Fig. 22-17 show that the protons are firmly fixed in position and exhibit no such tendency to "tunnel" through the intervening energy barrier to reach the complementary base.

EXAMPLE *Deuteration and Contrast-matching*

As shown in Table 22-4, there is a very great difference between the scattering produced by solvent H_2O and solvent D_2O. In fact, since contrast in any image (review Chapter 15.C for electron scattering analogs) is based on a difference in observed intensity from different parts of the sample, we can make almost any (normal) hydrogenated lipid, protein, or nucleic acid aggregate "disappear" when viewed by neutron scattering, simply by adjusting the ratio of H_2O/D_2O of the solvent to give the same scattered intensity ("contrast-matching") as the hydrogenated molecule of interest. Then, when a deuterated lipid (or protein or nucleic acid) is added to the mixture, it will "stand out" from the otherwise "gray" background of solvent and protonated macromolecules. (Refer to Table 22-4 for the difference in neutron scattering produced by deuterating a typical lipid, protein, or nucleic acid.)

One of the most exciting applications of contrast-matching has been in the "mapping" of the relative locations of different constituent proteins in ribo-

Table 22-4 Average Neutron Scattering (in Fermi) per Unit Volume of Some Common Biological (Protonated or Deuterated) Constituents (DDAO is Dodecadimethylamine Oxide)

Chemical Constituent	Neutron Scattering Length (Fermi)
H_2O	−0.6
D_2O	6.3
DDAO detergent, hydrogenated	−0.2
DDAO detergent, deuterated	8.0
Hydrocarbon, hydrogenated	−0.3
Hydrocarbon, deuterated	7.0
Lipid head group, hydrogenated	1.7
Lipid head group, deuterated	2.6
Proteins, hydrogenated	4.0
Proteins, 20% deuterated	5.0
Proteins, 100% deuterated	9.0
Nucleic acid, hydrogenated	4.2
Nucleic acid, 100% deuterated	7.2

[*From B. P. Schoenborn, "Neutron Scattering and Biological Structures,"* Chem. and Eng. News (*Jan. 24, 1977*), *pp. 31–41.*]

Table 22-5 Separations of Various Pairs of Ribosomal Proteins, Obtained by Neutron-Scattering from Contrast-matched Ribosomes Reconstituted from Normal and Deuterated Proteins (see text)

Protein Pair	Separation of the Proteins
S2-S5	105 A
S3-S4	61 A
S3-S5	57 A
S3-S7	115 A
S3-S8	78 A
S4-S7	95 A
S4-S8	63 A
S5-S6	111 A
S5-S8	35 A

[*From B. P. Schoenborn, "Neutron Scattering and Biological Structures,"* Chem. and Eng. News (*Jan. 24, 1977*), *pp. 31–41.*]

somes. The ribosome is an aggregate of some 54 proteins and ribonucleic acid. It is possible to disaggregate this assembly, and deuterate particular proteins, and then reconstitute the full ribosome afterward. Since the labeled proteins will appear prominently in an otherwise contrast-matched image, it is possible to determine the distance between the labeled proteins. By repeating this procedure for many protein pairs in turn (see Table 22-5), it should be possible to determine the complete quaternary structure of the ribosomal protein assembly by simple triangulation of the various protein-protein distances. This project is well under way and may succeed within the next few years.

22.C. IMAGE RECONSTRUCTION FROM PROJECTIONS

22.C.1. X-ray Tomography

In Chapter 14.G, we mentioned a technique called "ultrasonic tomography," in which an image of an object was reconstructed from the one-dimensional echo projections from incident ultrasonic sound waves directed at the object from different directions. The basic scheme for image reconstruction is the same, whether based on absorption or scattering of waves (ultrasonic—Chapter 14.G; X-ray—Chapter 22.C.1; radiofrequency magnetic—Chapter 22.C.2), and is illustrated in the simplest method for X-ray images in Fig. 22-18. In the present discussion, we are interested in images of *macroscopic*

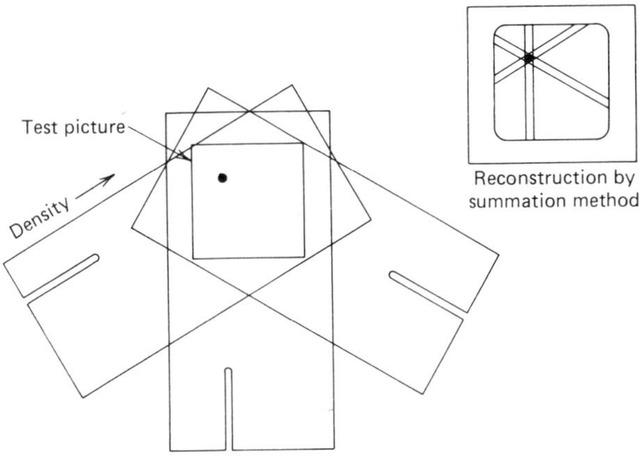

FIGURE 22-18. Summation method is a rough technique for reconstructing images from a series of projections. Here three projections are made of a simple two-dimensional test picture containing a single point. Each projection is a one-dimensional distribution of the density, or darkness, across the test picture as it is seen from a specific angle. In the case of this test picture the projection looks the same from all directions. The picture can be reconstructed from the projections: the density of each point on the reconstructed picture is estimated by adding up the densities of all the rays going through that point. The reconstruction of the single point is a "star," or spokelike image. The star is the "point-spread function" of the reconstruction technique. It approximately demonstrates the nature of summation method. [From "Image Reconstruction from Projections" by Richard Gordon, Gabor T. Herman and Steven A. Johnson. Copyright © October 1975 by Scientific American, Inc. All rights reserved.]

objects—objects whose dimensions are very much larger than the wavelength of the probing waves.

The simple "summation" method shown in Fig. 22-18 (and also used for the zeugmatography images of Chapter 22.C.2) consists of reconstructing an image by drawing lines of appropriate density across a grid at the angle at which that density was observed in viewing the original object. The obvious difficulty with this scheme is that contrast is poor, since the density in the image, while greatest at the position of the object, is still finite (rather than zero, as desired) at nearby positions. One of the several means for overcoming this difficulty is shown in Fig. 22-19, in which a digital computer constructs a hypothetical object that will produce the desired density (absorption, scattering) when viewed from a particular direction, and then repeat the process for all the viewing directions, changing the shape of the object systematically until a best fit between observed and calculated projection intensities is observed. In commercial X-ray devices, an image of a cross section of a part of the body (head, chest, abdomen usually) is obtained by collecting data at 160 equally spaced positions at a given viewing angle (i.e., 160 points on the abscissa in the projections shown in Fig. 22-20 for 180 equally spaced viewing angles (one per degree in a semicircle) around

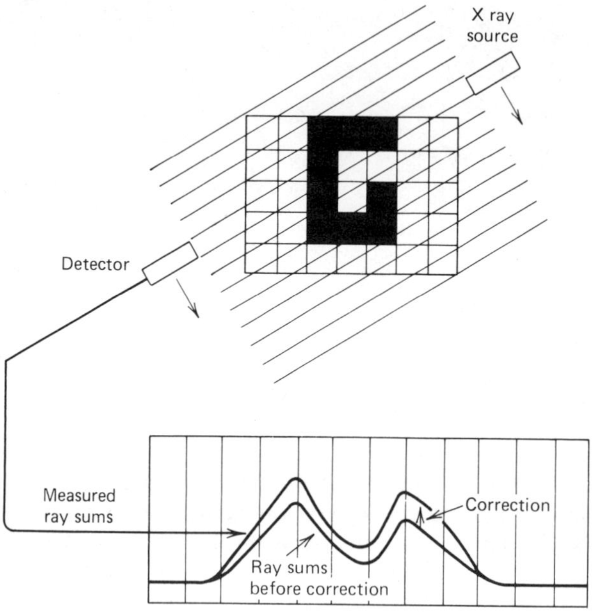

FIGURE 22-19. Algebraic reconstruction technique ("ART") has been devised to overcome the inaccuracy of the summation method. ART is executed on a digital computer, in which a picture is stored as a two-dimensional array of numbers, each number representing the X-ray density of one pixel, or small picture element (*squares*). A one-dimensional projection of the picture is stored as a list of numbers, each representing the ray sum, or total X-ray density along a ray: a narrow strip of the picture at a certain angle (*band*). ART is an iterative method that assigns an initial set of X-ray densities to the two-dimensional picture it is to reconstruct, calculates the ray sum of each point along a one-dimensional projection of the estimated picture, compares that ray sum with the ray sum of the real object stored in the computer, calculates the difference and divides it among all the pixels intersected by the ray. Modified ray sum then matches the original. Operation is repeated for all rays from all projections until a representational picture is reconstructed. [From "Image Reconstruction from Projections" by Richard Gordon, Gabor T. Herman and Steven A. Johnson. Copyright © October 1975 by Scientific American, Inc. All rights reserved.]

the body. The 28,800 (160×180) ray intensities are recorded and processed by a digital computer according to the scheme shown in Fig. 22-19 to give an image such as shown in Fig. 22-21, whose contents are explained in the corresponding caption.

The *existing* value of this technique is evident from the fact that more than 300 of these devices (at costs ranging from \$200,000 to \$700,000) have already been sold to major hospitals. Tantalizing *future* applications include study of a beating heart. Preliminary experiments (in which an isolated dog's heart was removed and rotated while beating to obtain the necessary projections) are illustrated in Fig. 22-22, in which the expansion and con-

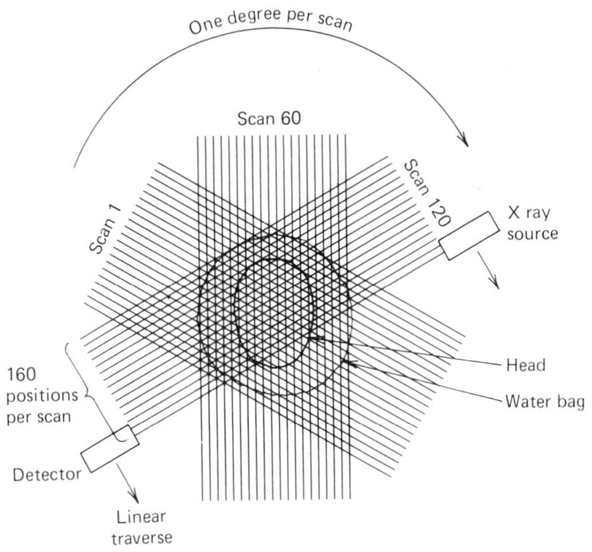

FIGURE 22-20. Scanner samples ray runs at 160 points along each projection; in five and a half minutes 180 projections are taken at one-degree intervals around the patient's head. [From "Image Reconstruction from Projections" by Richard Gordon, Gabor T. Herman and Steven A. Johnson. Copyright © October 1975 by Scientific American, Inc. All rights reserved.]

traction are clearly visible. In order to gather the required number of projections for a heart in a living human subject, it would be necessary to employ several independent X-ray sources and detectors (since it is not feasible to rotate either the subject or the X-ray source-detector system rapidly compared to the duration of a heartbeat), suggesting that this technique, if ever implemented, will be very expensive (each X-ray source-detector system costs several tens of thousands of current dollars).

Among other physical approaches to producing three-dimensional information about biological structures, holographic methods using either optical or acoustic waves (or both) appear to be attractive, and the former is beginning to look feasible. [For possibilities with acoustic waves, the interested reader is referred to articles by M. E. Cox, "Holographic Microscopy," *American Laboratory,* April, 1975, p. 17, and by P. Greguss, "Acoustical Holography," *Physics Today,* October, 1974, p. 42.] We conclude this chapter with a brief look (Chapter 22.C.2) at a promising nondestructive imaging method based on magnetic resonance power absorption.

22.C.2. Magnetic Resonance Zeugmatography: Medical Imaging

The principal constituent of biological tissue is water, so the principal proton magnetic resonance signal from tissue will be the single absorption from H_2O. (Since both protons in water are identical, the separate absorption signals from each proton effectively lie on top of each other to give a

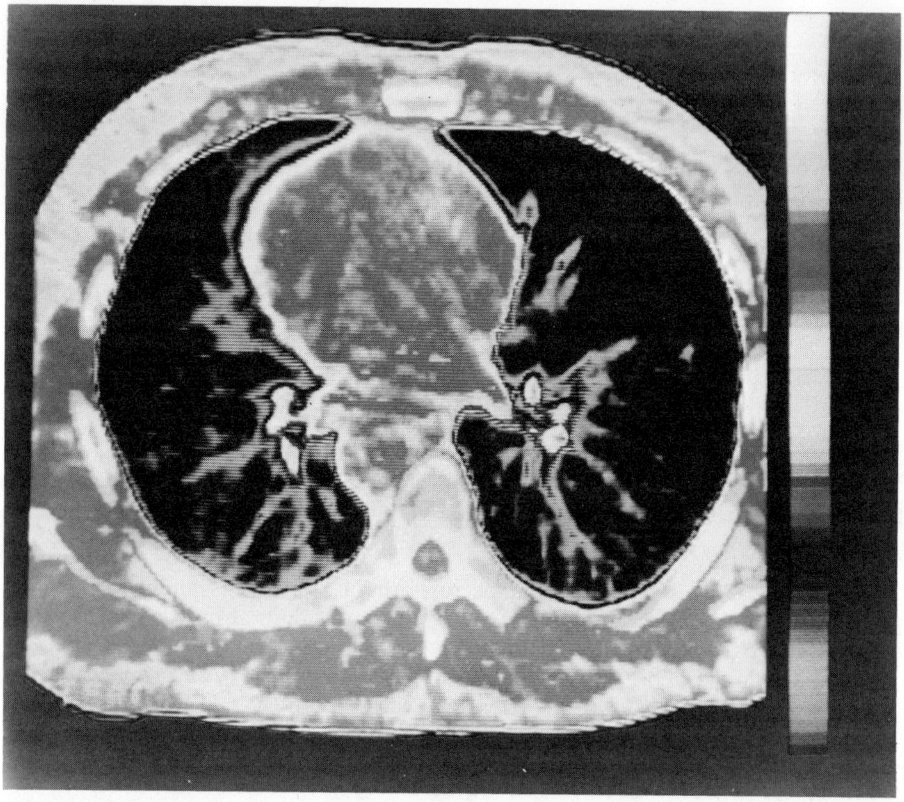

FIGURE 22-21. Cross section through the chest of a living human subject, made by the technique of reconstruction of one-dimensional X-ray projections (Figs. 22-19 and 22-20). Regions of different opaqueness to X rays are seen as different shadings. In the picture, the chest is seen in cross section as if viewed from above the subject's head. The dark spaces to the left and right are the lungs. The large gray area in the middle is the heart. The white areas are bone; below the center is the spinal column and around the lungs are sections through the ribs. The outermost regions (light gray) are fatty tissue, and the remaining gray areas are muscle. The branched areas in the lungs are blood vessels and bronchi [From the cover plate of *Scientific American 233,* October, 1975.]

single signal.) Furthermore, since nuclear magnetic resonance frequency is proportional to applied magnetic field strength, application of a magnetic field *gradient* across the tissue will result in an NMR spectrum whose signal intensity at a given frequency simply represents the relative amount of water located in the tissue segment corresponding to that particular magnetic field.

Consider first the simple case shown in Fig. 22-23, consisting of a sample with two water-filled capillaries separated by a small distance. When the applied static magnetic field magnitude varies linearly along a line connecting the two capillaries (middle bottom trace), then the proton

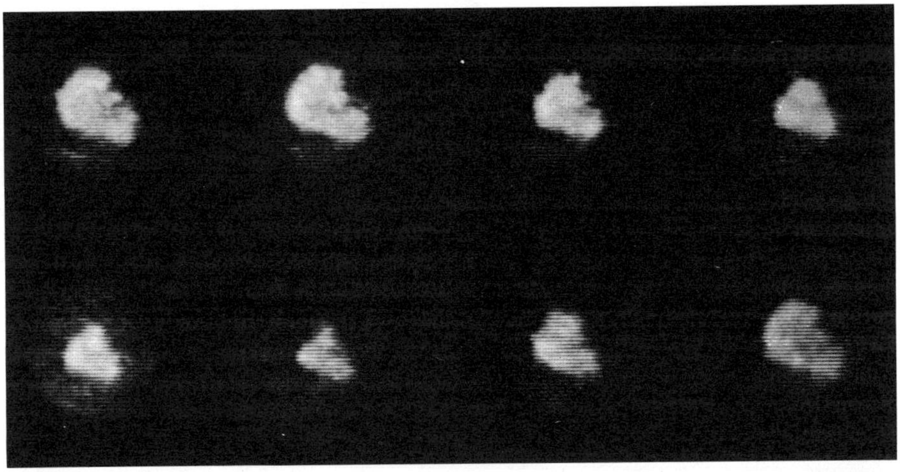

FIGURE 22-22. Isolated beating left ventricle is shown in cross section at intervals of 1/15th second during a single cardiac cycle. (The heart rate was 120 beats per minute.) All the cross sections represent the same anatomic section midway between the base and the apex of the ventricle. Images show contraction of the left ventricular chamber (*bright central area*) followed by subsequent dilation of the chamber as the heart relaxed. The reconstructed images were made by R. A. Robb and E. L. Ritman of the Mayo Clinic. [Courtesy R. A. Robb and E. L. Ritman of the Mayo Clinic, from "Image Reconstruction from Projections" by Richard Gordon, Gabor T. Herman and Steven A. Johnson, *Scientific American,* October 1975.]

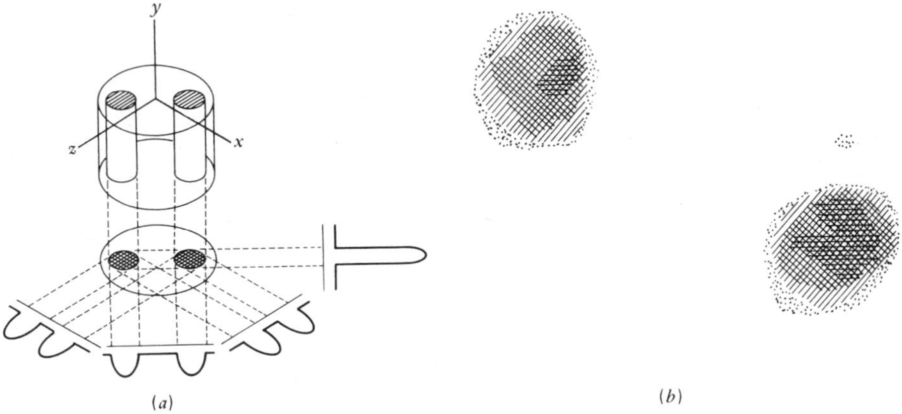

FIGURE 22-23. (*a*) Relationship between a three-dimensional object consisting of two water-filled capillaries (top), its two-dimensional cross section through the *y* axis (middle), and four one-dimensional projections in the *xz* plane (outer diagrams). The one-dimensional projections correspond to the proton NMR spectra (NMR absorption versus irradiating frequency) for the four magnetic field gradient directions indicated by lines in the figure. Figure 22-23*b* represents the image of the original object cross section, reconstructed from the four projections in (*a*). Images obtained in this way have been called "zeugmatograms" from the Greek "zeugma," meaning "that which is used for joining" (the magnetic and irradiation radiofrequency fields). [From P. C. Lauterbur, *Nature 242,* 190 (1973).]

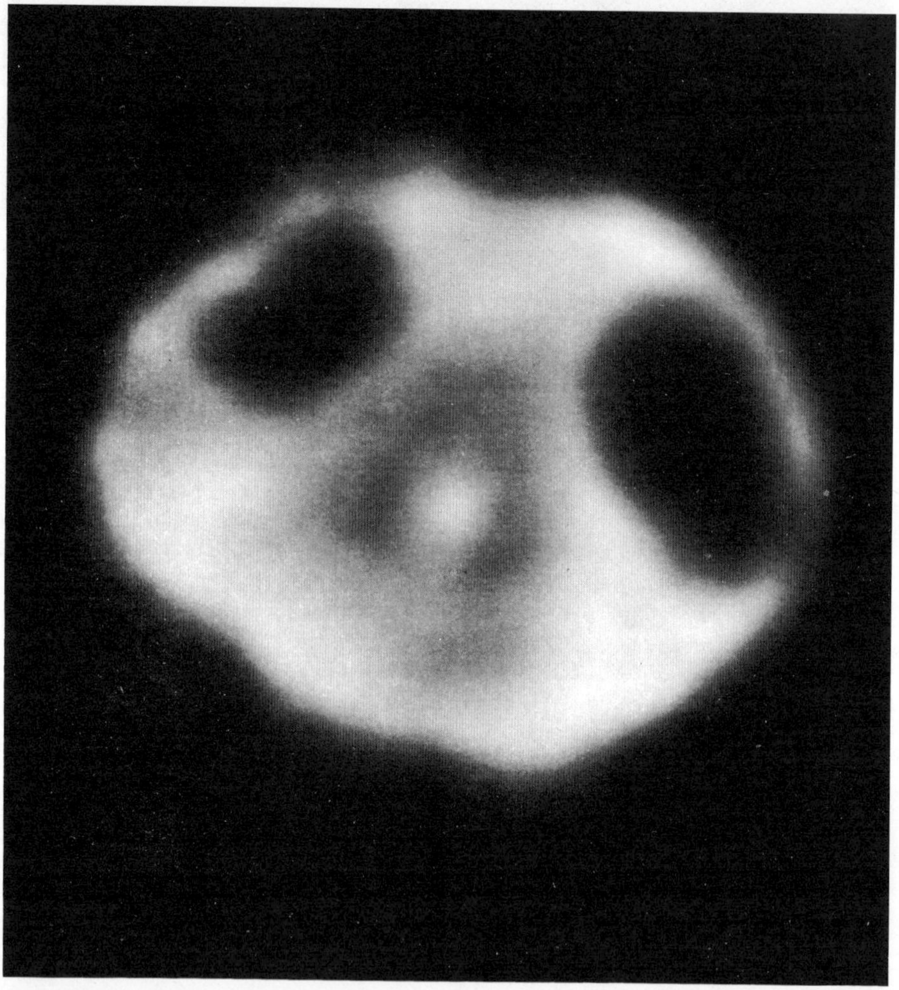

FIGURE 22-24. Zeugmatogram showing view of a live mouse along its head-to-tail axis (dorsal at top of figure). The dark areas correspond to the lungs, and the body outline is clearly visible. This is one of the earliest images obtained by this technique, and much higher resolution seems readily attainable in the near future (see Fig. 22-25 for another example). [P. C. Lauterbur, *Pure Appl. Chem. 40,* 149–157 (1974).]

NMR frequency for water in the left-hand capillary will be smaller than in the right-hand capillary, and the proton NMR spectrum for this sample will have the appearance shown in the middle bottom trace of the figure. On the other hand, if the magnetic field gradient is applied along a direction perpendicular to the line connecting the two capillaries, then the magnetic field at both capillaries will be the same and the proton NMR signals from both capillaries will fall at the same frequency (right-hand trace of Fig.

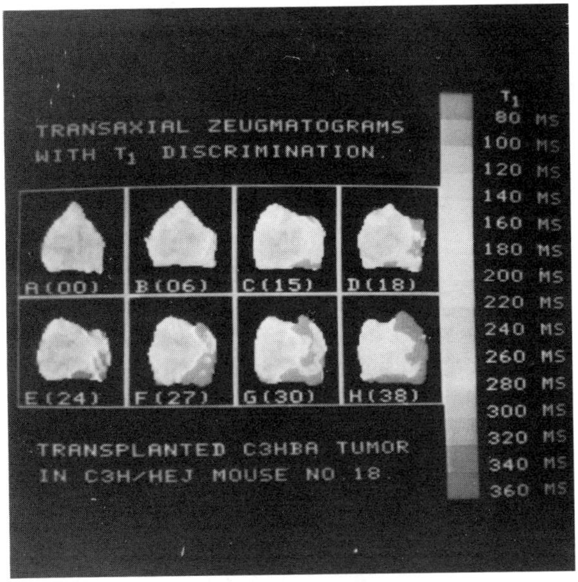

FIGURE 22-25. Series of transaxial zeugmatograms of a live C3H/HEJ mouse, into which a C3HBA tumor had been transplanted on day zero (upper left diagram). Going from left to right in the top row and ending at bottom right, the images were obtained after 6, 15, 18, 24, 27, 30, and 38 days. Differences in shading in the diagram are based on differences in proton magnetic relaxation time, T_1, as shown by the key at far right: since T_1 is much longer for the tumor cells, they are easily visible as a distinct region in the zeugmatogram images. (Photograph courtesy of Prof. Paul Lauterbur, Chemistry Department, State University of New York at Stony Brook.)

22-23). In other words, we have obtained different projection views of the objects, by using magnetic field gradients applied in different directions. By combining the projections obtained for the four field gradient directions shown in the figure, it is possible to reconstruct the original capillary cross-sections as shown in Fig. 22-23*b*.

Zeugmatography thus provides a means for resolving different parts of a macroscopic object based on differences in the amplitude of the nuclear magnetic resonance signal in different parts of the sample, as shown in Fig. 22-24. The figure presents a reconstructed image of a live mouse viewed along its head-to-tail axis. The lung cavities are clearly visible as is the body outline. More interesting and potentially diagnostically valuable information may be drawn from zeugmatograms of tissue containing a suspected tumor. Since the water proton NMR saturation parameter, T_1, is usually larger for a tumor than for normal tissue, the proton NMR signal from the tumor is more easily saturated than that for normal tissue. Thus, by irradiating the sample at a power level that saturates the tumor water NMR signal more than the surrounding tissue water NMR signal, there will be a different proton NMR signal intensity for the tumor compared to normal tissue, and when the two signals are spread out in frequency by applying a magnetic field gradient (see Fig. 22-23), it is possible to locate the tumor, as shown in Fig. 22-25. Figure 22-25 shows a zeugmatogram of a mouse into which a tumor had been implanted – the tumor outline is clearly visible in the reconstructed image. This NMR imaging technique is very new, and should be useful in much the same way that X-ray or ultrasonic "tomographs" are used to locate different types of tissue, with the added advantages that radiofrequencies contribute no known radiation damage (contrast to X rays) and that the radiofrequency signals pass easily through bone and air (contrast to ultrasonic waves, which are reflected by bone) so that images can readily be obtained for organs within the rib cage or skull regions relatively inaccessible to ultrasonic imaging.

PROBLEM

1. Using the rules given in the caption for Fig. 22-9, construct the expected Patterson function two-dimensional map for the planar molecule shown below. (You may take the carbon to be sp^2 hybridized – i.e., 120° bond angles at carbon.) Be sure to list the relative intensities of each of the spots.

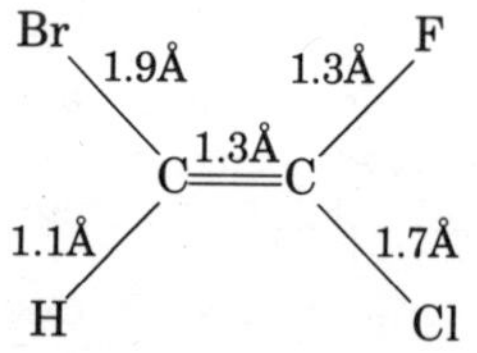

REFERENCES

X-Ray Diffraction and Crystallography

J. Waser, *J. Chem. Educ. 45,* 446 (1968). Fourier synthesis.

M. J. Buerger, *Contemporary Crystallography,* McGraw-Hill, New York (1970), Chapter 12. Good exposition of phase problem and of optical analogs for X-ray diffraction.

H. R. Wilson, *Diffraction of X-Rays by Proteins, Nucleic Acids, and Viruses,* Edward Arnold Publishers, London (1966). Treats diffraction from helix and offers clearest brief mathematical treatment of actual data reduction.

B. W. Matthews, "X-Ray Crystallographic Studies of Proteins," *Ann. Rev. Phys. Chem. 27,* 493–523 (1976). Most recent review of protein crystal structures.

Neutron Diffraction

B. P. Schoenborn, "Neutron Scattering and Biological Structures," *Chem. and Eng. News* (Jan. 24, 1977), pp. 31–41. Modern, nontechnical review of biological applications.

S.-H. Chen and S. Yip, eds., *Spectroscopy in Biology and Chemistry: Neutron, X-Ray, Laser,* Academic Press, New York (1974). Collection of papers—see especially paper #5, p. 177, by T. F. Koetzle.

Image Reconstruction from Projections: X-Ray Tomography

R. Gordon, G. T. Herman, and S. A. Johnson, *Scientific American 233,* 56–68 (1975). Excellent nontechnical monograph.

Image Reconstruction from Projections: Zeugmatography

P. C. Lauterbur, "Magnetic Resonance Zeugmatography," *Pure Appl. Chem. 40,* 149–157 (1974).

APPENDIX

LOGARITHMS

Definition: If $y = a^x$, then $x = \log_a y$ (A-1)

$\log_e x = \ln x$ (A-2)

Properties: $\log(a \cdot b) = \log a + \log b$ (A-3)

$\log(a/b) = \log a - \log b$ (A-4)

$\log(x^n) = n \log x$ (A-5)

Conversion from one base to another:

Let $z = a^x = b^y$, so that $x = \log_a z$ and $y = \log_b z$

Now take $\log_a$ of a^x and b^y:

$$\log_a a^x = \log_a b^y$$

Next, apply property (A-5):

$$x \log_a a = x = y \log_a b$$

or

$$\log_a z = (\log_a b) \cdot (\log_b z) \tag{A-6}$$

For example, $\log_{10} x = \log_{10} e \log_e x = 0.434 \log_e x$

SERIES EXPANSIONS AND APPROXIMATIONS

Taylor expansion (approximates $f(x)$ in the vicinity of $x = a$):

$$f(x) = f(a) + f'(a)(x-a) + \frac{1}{2!} f''(a)(x-a)^2 + \cdots + \frac{1}{n!} f^{(n)}(a)(x-a)^n \tag{B-1}$$

Factorials: $N! = N(N-1)(N-2)\cdots 3 \cdot 2 \cdot 1$ (B-2)

$0! = 1$ (B-3)

$\ln(N!) = N \ln N - N$ for large $N_{\text{(Stirling Approximation)}}$ (B-4)

Power series representations of simple functions:

$$e^{\pm x} = 1 \pm x + \frac{x^2}{2!} \pm \frac{x^3}{3!} + \frac{x^4}{4!} \pm \cdots + \frac{(\pm x)^n}{n!} \tag{B-5}$$

$$\cos x = 1 - \frac{x^2}{2!} + \frac{x^4}{4!} - \frac{x^6}{6!} \pm \cdots \tag{B-6}$$

$$\sin x = x - \frac{x^3}{3!} + \frac{x^5}{5!} - \frac{x^7}{7!} \pm \cdots \tag{B-7}$$

$$e^{ix} = \cos x + i \sin x = 1 + ix - \frac{x^2}{2!} - \frac{ix^3}{3!} + \frac{x^4}{4!} \pm \cdots \tag{B-8}$$

$$\ln(1 \pm x) = \pm x - \frac{x^2}{2} \pm \frac{x^3}{3} - \frac{x^4}{4} \pm \cdots, \text{ where } -1 < x < +1 \qquad \text{(B-9)}$$

Binomial expansion:

$$(a+b)^N = a^N + Na^{(N-1)}b + \cdots + \frac{N!}{m!(N-m)!}a^{(N-m)}b^m + \cdots + Nab^{(N-1)} + b^N \qquad \text{(B-10)}$$

TRIGONOMETRIC FUNCTIONS

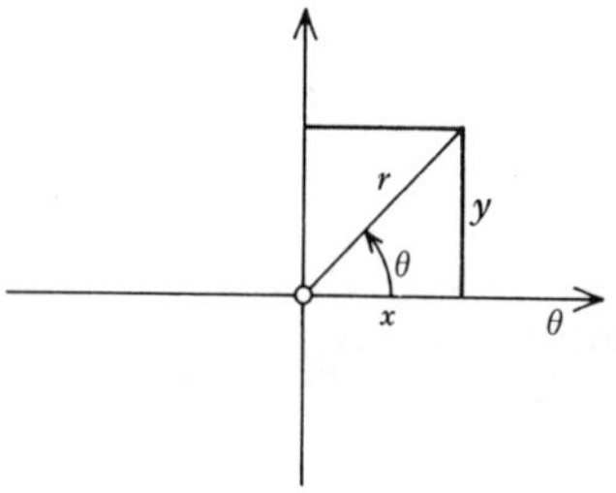

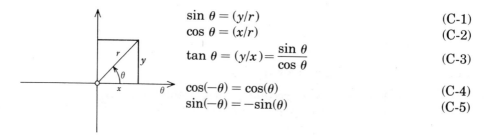

$$\sin\theta = (y/r) \qquad \text{(C-1)}$$
$$\cos\theta = (x/r) \qquad \text{(C-2)}$$
$$\tan\theta = (y/x) = \frac{\sin\theta}{\cos\theta} \qquad \text{(C-3)}$$
$$\cos(-\theta) = \cos(\theta) \qquad \text{(C-4)}$$
$$\sin(-\theta) = -\sin(\theta) \qquad \text{(C-5)}$$

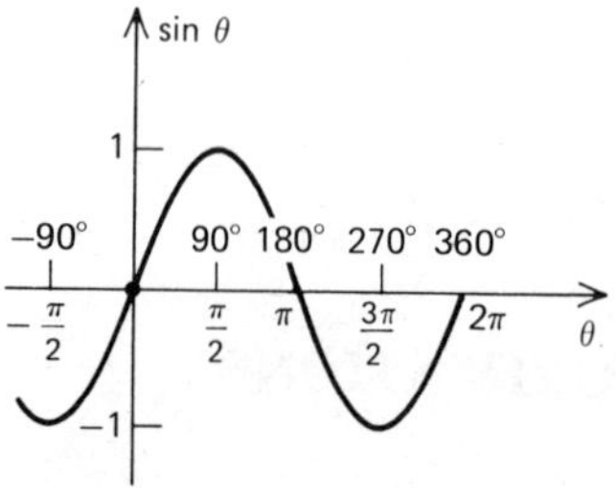

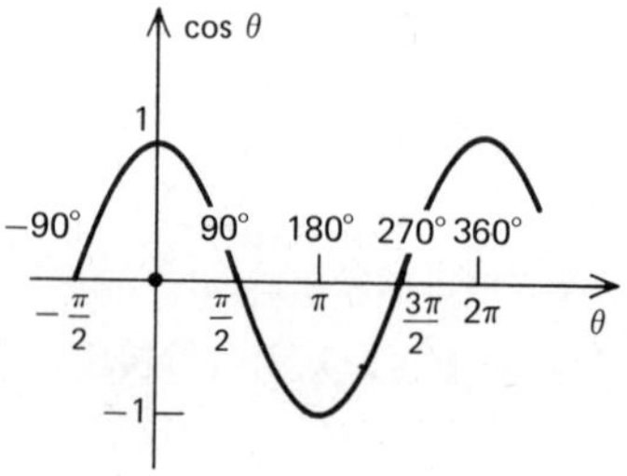

$$\text{degrees}\cdot(2\pi/360) = \text{radians} \qquad \text{(C-6)}$$

Identities:

$$\sin(A \pm B) = \sin A \cos B \pm \cos A \sin B \qquad \text{(C-7)}$$
$$\cos(A \pm B) = \cos A \cos B \mp \sin A \sin B \qquad \text{(C-8)}$$
$$\cos^2 A + \sin^2 A = 1 \qquad \text{(C-9)}$$
$$\cos\left(\frac{\pi}{2} - A\right) = \sin A \qquad \text{(C-10)}$$
$$\sin\left(\frac{\pi}{2} - A\right) = \cos A \qquad \text{(C-11)}$$

For change from cartesian to polar coordinates, see Fig. 21-6:

$$x = r\sin\theta\cos\phi \qquad \text{(C-12)}$$
$$y = r\sin\theta\sin\phi \qquad \text{(C-13)}$$
$$z = r\cos\theta \qquad \text{(C-14)}$$

CALCULUS: DERIVATIVES

Definition: $\frac{d}{dx} f(x) = f'(x)$; $\frac{d^2}{dx^2} f(x) = f''(x)$; and so on (D-1)

Derivatives of simple functions (a is a constant in Equations D-2 to D-6):

$$\frac{d}{dx} x^a = a\, x^{(a-1)} \tag{D-2}$$

$$\frac{d}{dx} e^{ax} = a\, e^{ax} \tag{D-3}$$

$$\frac{d}{dx} \ln ax = \frac{1}{x} \tag{D-4}$$

$$\frac{d}{dx} \sin ax = a \cos ax \tag{D-5}$$

$$\frac{d}{dx} \cos ax = -a \sin ax \tag{D-6}$$

Derivative of sum: $$\frac{d}{dx}\,[u(x) + v(x)] = \frac{d}{dx} u(x) + \frac{d}{dx} v(x) \tag{D-7}$$

of product: $$\frac{d}{dx}\,[u \cdot v] = v \cdot \frac{du}{dx} + u \cdot \frac{dv}{dx} \tag{D-8}$$

of quotient: $$\frac{d}{dx}\left[\frac{u}{v}\right] = \frac{v \cdot (du/dx) - u \cdot (dv/dx)}{v^2} \tag{D-9}$$

Basic rules:

$$\frac{d}{dx}\,[f(u(x))] = (df/du) \cdot (du/dx) \tag{D-10}$$

$$\frac{d}{dx}\,[f(u(x), v(x), w(x), \cdots)] = (\partial f/\partial u)(du/dx) + (\partial f/\partial v)(dv/dx) + (\partial f/\partial w)(dw/dx) + \cdots \tag{D-11}$$

CALCULUS: INTEGRALS

Basic properties:

$$\int_a^b f(x)\, dx = -\int_b^a f(x)\, dx \tag{E-1}$$

$$\int_a^b [u(x) + v(x)]\, dx = \int_a^b u(x)\, dx + \int_a^b v(x)\, dx \tag{E-2}$$

$$\int_a^b c\, f(x)\, dx = c \int_a^b f(x)\, dx, \text{ where } c \text{ is a constant} \tag{E-3}$$

$$\int_a^b f(x)\,dx = \int_a^c f(x)\,dx + \int_c^b f(x)\,dx \tag{E-4}$$

Even and odd functions:

$f(x)$ is even if $f(-x) = f(x)$ (E-5)
(Examples: x^2, $\cos x$, e^{-x^2})
$f(x)$ is odd if $f(-x) = -f(x)$ (E-6)
(Examples: x, $\sin x$)

Let E be an even function and O be an odd function. Then

$$E \cdot E = \text{even} \tag{E-7a}$$
$$E \cdot O = \text{odd} \tag{E-7b}$$
$$O \cdot O = \text{even} \tag{E-7c}$$

and integrals of even and odd functions have the properties

$$\int_{-a}^{a} E(x)\,dx = 2\int_0^a E(x)\,dx \tag{E-8}$$

$$\int_{-a}^{a} O(x)\,dx = O \tag{E-9}$$

Change of variables:

Suppose we are given the integral, $\int_{x=a}^{x=b} f(x)\,dx$, and we are asked to change variables to $u = u(x)$ and integrate over u from $u = u(a)$ to $u = u(b)$. We first express x as a function of u, $x = x(u)$, and then use the relation

$$\int_{x=a}^{x=b} f(x)\,dx = \int_{u=u(a)}^{u=u(b)} f[x(u)]\,\frac{dx}{du}\,du \tag{E-10}$$

Integration by parts:

$$\int_a^b u(x)\,v'(x)\,dx = u(x)\,v(x)\Big|_a^b - \int_a^b v(x)\,u'(x)\,dx \tag{E-11}$$

Indefinite integrals of selected functions:

$$\int x^n\,dx = \frac{x^{(n+1)}}{n+1}, \quad n \neq -1 \tag{E-12}$$

$$\int \frac{dx}{x} = \ln|x|, \quad x \neq 0 \tag{E-13}$$

$$\int \sin ax\,dx = -\frac{\cos ax}{a} \tag{E-14}$$

$$\int \cos ax\,dx = \frac{\sin ax}{a} \tag{E-15}$$

$$\int e^{ax}\,dx = \frac{e^{ax}}{a} \tag{E-16}$$

$$\int \sin ax \cos bx \, dx = -\frac{\cos (a+b)x}{2(a+b)} - \frac{\cos (a-b)x}{2(a-b)} \tag{E-17}$$

where $(a^2 \neq b^2)$

[For integrals of sin *ax* sin *bx* or cos *ax* cos *bx*, use the trig identities

$$\cos A \cos B = \tfrac{1}{2}[\cos (A+B) + \cos (A-B)]$$

and

$$\sin A \sin B = \tfrac{1}{2}[\cos (A-B) - \cos (A+B)],$$

and then use integrals E-15 and E-14.]

$$\int e^{ax} \sin bx \sin cx \, dx = \frac{e^{ax}[(b-c) \sin (b-c)x + a \cos (b-c)x]}{2[a^2 + (b-c)^2]} - \frac{e^{ax}[(b+c) \sin (b+c)x + a \cos (b+c)x]}{2[a^2 + (b+c)^2]} \tag{E-18}$$

$$\int e^{ax} \cos bx \cos cx \, dx = \frac{e^{ax}[(b-c) \sin (b-c)x + a \cos (b-c)x]}{2[a^2 + (b-c)^2]} + \frac{e^{ax}[(b+c) \sin (b+c)x + a \cos (b+c)x]}{2[a^2 + (b+c)^2]} \tag{E-19}$$

$$\int e^{ax} \sin bx \cos cx \, dx = \frac{e^{ax}[a \sin (b-c)x - (b-c) \cos (b-c)x]}{2[a^2 + (b-c)^2]} + \frac{e^{ax}[a \sin (b+c)x - (b+c) \cos (b+c)x]}{2[a^2 + (b+c)^2]} \tag{E-20}$$

$$\int \ln ax \, dx = x \ln ax - x \tag{E-21}$$

$$\int x \, e^{ax} \, dx = \frac{e^{ax}}{a^2} (ax - 1) \tag{E-22}$$

Definite integrals of selected functions:

$$\int_0^\infty e^{-ax} \, dx = \frac{1}{a} \tag{E-23}$$

$$\int_0^\pi \sin^2 mx \, dx = \int_0^\pi \cos^2 mx \, dx = \frac{\pi}{2} \tag{E-24}$$

$$\int_0^\infty e^{-a^2x^2} \, dx = \frac{\sqrt{\pi}}{2a} \tag{E-25}$$

$$\int_0^\infty x e^{-x^2} \, dx = \frac{1}{2} \tag{E-26}$$

$$\int_0^\infty x^2\, e^{-x^2}\, dx = \frac{\sqrt{\pi}}{4} \tag{E-27}$$

$$\int_0^\infty e^{-ax} \cos bx = \frac{a}{a^2 + b^2} \tag{E-28}$$

$$\int_0^\infty e^{-ax} \sin bx = \frac{b}{a^2 + b^2} \tag{E-29}$$

Index

INDEX

Selected Physical Constants[a]

Constant	*Symbol*	*Value*	*SI units*[b]	*cgs units*
Speed of light in vacuum	c	2.997925	$\times 10^{8}$ m·s^{-1}	$\times 10^{10}$ cm·s^{-1}
Elementary charge	e	1.602189	$\times 10^{-19}$ C	$\times 10^{-20}$ emu
		4.803250		$\times 10^{-10}$ esu
Avogadro's number	N_0	6.022045	$\times 10^{23}$ mol^{-1}	$\times 10^{23}$ mol^{-1}
Atomic mass unit (Dalton)	amu	1.660566	$\times 10^{-27}$ kg	$\times 10^{-24}$ g
Electron rest mass	m_e	9.109534	$\times 10^{-31}$ kg	$\times 10^{-28}$ g
		5.48580	$\times 10^{-4}$ amu	$\times 10^{-4}$ amu
Proton rest mass	m_p	1.672648	$\times 10^{-27}$ kg	$\times 10^{-24}$ g
		1.007276	amu	amu
Neutron rest mass	m_n	1.674954	$\times 10^{-27}$ kg	$\times 10^{-24}$ g
		1.008665	amu	amu
Faraday constant, N_0e	F	9.648456	$\times 10^{4}$ C mol^{-1}	$\times 10^{3}$ emu mol^{-1}
		2.892539		$\times 10^{14}$ esu mol^{-1}
Proton magnetogyric ratio	γ_p	2.6751301	$\times 10^{8}$ s^{-1}·T^{-1}	
Proton magnetic moment	μ_p	1.4106171	$\times 10^{-26}$ J·T^{-1}	
Molar gas constant	R	8.31441	J·mol^{-1}·K^{-1}	$\times 10^{7}$ erg·mol^{-1}·K^{-1}
		0.082057	$\times 10^{-3}$ m^{3}·atm·mol^{-1}·K^{-1}	l·atm·mol^{-1}·K^{-1}
		1.987		cal·K^{-1}·mol^{-1}
Boltzmann constant, $\frac{R}{N_0}$	k	1.380662	$\times 10^{-23}$ J·K^{-1}	$\times 10^{-16}$ erg·K^{-1}
Planck constant	h	6.626176	$\times 10^{-34}$ J·s	$\times 10^{-27}$ erg·s
$h/2\pi$	$\hbar$	1.054589	$\times 10^{-34}$ J·s	$\times 10^{-27}$ erg·s

[a] *For a more complete list, see E. R. Cohen and B. N. Taylor, J. Phys. Chem. Ref. Data Vol. 2, No. 4 (1973), p. 663 and U. S. Dept. of Commerce National Bureau of Standards Special Publication 398 (August, 1974).*

[b] *For discussion of S.I. units, see "Policy for NBS Usage of S.I. units," J. Chem. Educ.* 48, *569 (1971).*